Developments in Mathematics

VOLUME 28

For further volumes:
http://www.springer.com/series/5834

Hershel M. Farkas • Robert C. Gunning
Marvin I. Knopp • B.A. Taylor
Editors

From Fourier Analysis and Number Theory to Radon Transforms and Geometry

In Memory of Leon Ehrenpreis

Springer

Editors
Hershel M. Farkas
Einstein Institute of Mathematics
The Hebrew University of Jerusalem
Givat Ram, Jerusalem
Israel

Robert C. Gunning
Department of Mathematics
Princeton University
Princeton, NJ
USA

Marvin I. Knopp†
Department of Mathematics
Temple University
Philadelphia, PA
USA

B.A. Taylor
Department of Mathematics
University of Michigan
Ann Arbor, MI
USA

ISSN 1389-2177
ISBN 978-1-4614-4074-1 ISBN 978-1-4614-4075-8 (eBook)
DOI 10.1007/978-1-4614-4075-8
Springer New York Heidelberg Dordrecht London

Library of Congress Control Number: 2012944371

Printed on acid-free paper

Springer is part of Springer Science+Business Media (www.springer.com)

Leon Ehrenpreis (1930–2010)

Preface

This is a volume of papers dedicated to the memory of Leon Ehrenpreis. Although Leon was primarily an analyst, whose best known results deal with partial differential equations, he was also very interested in and made significant contributions to the fields of Riemann surfaces (both the algebraic and geometric theories), number theory (both analytic and combinatorial), and geometry in general.

The contributors to this volume are mathematicians who appreciated Leon's unique view of mathematics; most knew him well and admired his work, character, and unbounded energy. For the most part the papers are original contributions to areas of mathematics in which Leon worked; so this volume may convey a sense of the breadth of his interests.

The papers cover topics in number theory and modular forms, combinatorial number theory, representation theory, pure analysis, and topics in applied mathematics such as population biology and parallel refractors. Almost any mathematician will find articles of professional interest here.

Leon had interests that extended far beyond just mathematics. He was a student of Jewish Law and Talmud, a handball player, a pianist, a marathon runner, and above all a scholar and a gentleman. Since we would like the readers of this volume to have a better picture of the person to whom it is dedicated, we have included a biographical sketch of Leon Ehrenpreis, written by his daughter, a professional scientific journalist. We hope that all readers will find this chapter fascinating and inspirational.

Jerusalem, Israel	H.M. Farkas
Princeton, NJ	R.C. Gunning
Philadelphia, PA	M. Knopp*
Ann Arbor, MI	B.A. Taylor

*Marvin Knopp (of blessed memory) passed away on December 24, 2011, after almost the entire volume was edited by the four of us. Without him, this volume would not have appeared.

Contents

A Biography of Leon Ehrenpreis

By: Yael Nachama (Ehrenpreis) Meyer

Dr. Leon Ehrenpreis (b. May 22, 1930; d. August 16, 2010), a leading mathematician of the twentieth century, proved the Fundamental Principle that became known as the Malgrange–Ehrenpreis theorem, a foundation of the modern theory of differential equations that became the basis for many subsequent theoretical and technological developments.

He was a native New Yorker who taught and lectured throughout the USA, as well as in academic institutions in France, Israel, and Japan. Ehrenpreis made significant and novel contributions to a number of other areas of modern mathematics including differential equations, Fourier analysis, Radon transforms, integral geometry, and number theory. He was known in the mathematical community for his commitment to religious principles and to his large family, as well as for his contributions to the essence of modern mathematics.

Leon Ehrenpreis published two major works: *Fourier Analysis in Several Complex Variables* (1970) and *The Universality of the Radon Transform* (2003), authored many papers, and mentored 12 Ph.D. students in New York, Yeshiva, and Temple Universities over the course of a mathematical career that spanned over half a century. What follows is his story.

Leon Ehrenpreis was born on May 22, 1930. His mother, Ethel, née Balk, was born in Lithuania; his father, William, a native of Austria, had changed his last name from that of his own father (Kalb) to that of his mother, in order to escape the Russian draft. And so "Ehrenpreis," the German word for "prize of honor," became the family surname.

Leon, whose parents also gave him the Hebrew name "Eliezer," was born just at the close of the era during which millions of Eastern European Jews had left behind the homes where their families had lived for generations and survived eras of persecution, in order to reach the land that promised to take in all of "your tired, your poor, your huddled masses yearning to breathe free..." and give their children the opportunity to become Americans. Coming ashore in New York City, many of these new immigrants settled in Manhattan's Lower East Side, in Brooklyn, and the

Bronx. Leon's family was no exception; over the course of his childhood, he lived in all three of these boroughs. Initially, Ethel and William, their baby Leon, and his older brother Seymour, settled in a home in the Marine Park neighborhood of Brooklyn.

When Leon was 10, the family moved to the Lower East Side, a neighborhood with a large Jewish community. Leon's home was one in which the kitchen was kosher, and the Sabbath recognized, and their Jewishness the defining personal, family, and communal identity, though without knowledge or emphasis on the subtle details of religious observance. So it was only there that Leon came into contact with boys of his own age whose families were strictly observant, an introduction to religious life that started Leon on his trajectory towards full-scale observance. He also attended a Jewish studies after-school program to prepare for his bar mitzvah, his entry into Jewish adulthood. Soon after his bar mitzvah, Leon stopped attending his after-school studies, though he continued to attend Sabbath services at the local synagogue as a result of his friends' influence.

The majority of New York's Jews at that time were aiming to raise their children to be successful, high-achieving Americans, with academic success and intellectual pursuits an important priority for many, including the Ehrenpreis family. So it was that soon after his bar mitzvah, Leon followed his brother into the prestigious Stuyvesant High School in Manhattan. Leon had skipped two grades in elementary school and then skipped his initial year of high school, beginning Stuyvesant in the tenth grade.

When Leon was 16, the family moved to the Bronx. Now more interested in learning about his Jewish heritage, Leon attended the Young Israel of Clay Avenue and joined Hashomer Hadati, a youth group that would be the forerunner of the religious Zionist Bnei Akiva movement. He now traveled downtown each day to Stuyvesant, where he continued to excel in his studies, though not in his class conduct! He recalled having the highest grades in French, but failing to be awarded the French medal because of his poor behavior. He also scored the highest on the chemistry medal qualifying exam (though a teacher's error meant that he never actually received it), and he was also awarded the mathematics medal—though most of these were won by his new best friend Donald Newman, whom Leon credited with influencing him to become a mathematician. His mother, he recalled, considered the choice of mathematics a "cop out" to avoid having to do the serious lab work that a physics major would require.

Leon initially met Donald Newman on his first day at Stuyvesant, where his classmate was seated just on the other side of the aisle in their first class of the day. Almost immediately, Donald handed a clipboard to Leon with the order to "solve this problem." The board read "Sierpnerhe"—Ehrenpreis backwards. Already in ninth grade, Leon said, Donald's reputation foreshadowed his greatness. The same was said about Leon from the tenth grade onward. "He was the great man of Stuyvesant—we already knew he would be a mathematical star." The two created lifelong nicknames for each other, and "Flotzo-Flip" (Donald) and "Glockenshpiel" (Leon) formed a friendship that would last forever. "I felt like a real mathematician when Flotz and I discussed mathematics together," Leon recalled. The two friends, who were considered the best mathematicians in the class, would ultimately follow much the same path throughout their mathematical lives, remaining close personal friends throughout.

At the age of 20, Leon joined the National Guard, which involved training for 2 hours each week and 2 weeks in the summer. His youthful military duty provided him with a lifelong repertoire of "war stories." He was fond of recalling for his children how, to maintain a kosher diet, he subsisted on thrice-a-day meals of ice cream, and how his commanding officer, who initially refused his request for time off on Saturdays, finally told him to "disappear on Friday night—and don't come back until Sunday." Leon also liked to describe how his superiors eventually worked out what his strong points were—and weren't—and so assigned him to calculate the trajectory of the shots being fired instead of actually firing them. His speedy calculation ability made him popular among his fellow reservists as well, as he would finish all the work assigned to his group within the first hour of the morning—and then the entire troop would go to sleep for the rest of the day.

By this time, Leon had nearly completed his university studies, having been in college since the age of sixteen-and-a-half. Leon was enrolled at City College, the "Jewish Harvard," as it was known during those years when the Ivy League still maintained a quota of Jewish students, leaving many of the best and brightest to attend New York City's public university. In addition to his old friend Donald Newman, the class included Robert "Johnny" Aumann, Lee Rubel, Jack Schwartz, Allen Shields, Leo Flatto, Martin Davis, and David Finkelstein, a group of individuals who would go on to change the face of mathematics, computer science, and the sciences for decades to come. This high-powered group of students formed a math club and had their own table in the cafeteria—the "mathematics table"—where, Aumann recalled, the group would sit together, eating ice cream, discussing the topology of bagels, and enjoying "a lot of chess playing, a lot of math talk… that was a very intense experience."

In addition to his university studies (where handball and weightlifting competed with his mathematics major for his attention; he was the handball champion of New York City during his early college years), and his military activities, Leon expanded his Jewish education by enrolling in an evening Jewish studies program, where Bible and Talmud, as well as Hebrew language and literature, were taught entirely in Hebrew. This represented Leon's first formal attendance in an academic Jewish studies program. "It was the first time I ever studied a page of Talmud!" Leon recalled.

While attending City, Leon audited a series of lectures on probability theory given by Professor Harold Shapiro at NYU's Courant Institute. He identified an event that occurred during the course of these lectures as a "turning point" in his development into a "true" mathematician. Professor Shapiro wrote a statement on the board that he thought was obvious. Then he began writing out the proof—until he came to a step of the proof that he couldn't carry out. "You've learned more from my not knowing how to do it than by my presenting a proof," Shapiro told them. So Leon became determined to correct the proof himself. "I ate, breathed and slept correcting that step and…

Sunday: nothing.

Monday, late afternoon: Eureka! I can't fix that step in the proof because the theorem itself is wrong! So I corrected the theorem itself. Then I returned to Shapiro to inform him—it's wrong! Erdös and Chung had stated the theorem incorrectly. Although I was only eighteen, I was convinced that I was right. I showed Shapiro a counterexample to demonstrate without question that I had created the correct proof. I beat Erdös and Chung! I'm a mathematician!! No doubt anymore—I am the real thing."

That same year, Leon registered in joint mathematics–physics graduate programs at both Columbia and NYU simultaneously. He actually had not yet completed his Bachelor's degree at CUNY, during the course of which he had also "illegally" taken several advanced classes before completing the relevant prerequisites, and so for years to come would have "nightmares" that the university powers-that-be would suddenly discover his crimes and come to take away his B.S.—and his Ph.D.

Between 1952 and 1953, he worked on his doctorate with Claude Chevalley (whom Leon termed "the best in the world") as his thesis advisor. He completed his thesis, entitled, "Theory of Distributions in Locally Compact Spaces," in 1953, earning a Ph.D. from Columbia University at the age of 23.

Nearly 20 years later, Alan Taylor would ask Leon how, as a student of Chevalley, he had come to work on problems that led to what would ultimately be called the "fundamental principle." Leon explained to him that Chevalley had suggested that Leon write to Laurent Schwartz for thesis-problem suggestions. Schwartz, in turn, had responded with a list of questions about partial differential operators, along with the details he knew about them at the time, including the fundamental questions. The answers given by Leon and others, in the 1950s, would form the basis of the modern theory of linear constant-coefficient partial differential operators.

After Leon earned his doctorate, Chevalley arranged a first teaching position at Johns Hopkins University in Baltimore, Maryland, for him. It was there that Leon met Shlomo Sternberg, later a mathematics professor at Harvard, then a Johns Hopkins student, who reminisced:

> "Thinking back through the years, I can't recall a single time, no matter how trying the circumstances may have been, whether casual or serious, that his voice, his eyes, his whole demeanor conveyed less than deep warmth, profound generosity, an optimism, a hopefulness that was pure Leon. When we were young, 'pure Leon' might include a dash of madcap charm, a directness, a boyish whimsy, a ruefulness, that belied his distinguished mathematical achievements. His style was not professorial. He was not into style or image—then or ever. Leon retained and presented an honesty, a disarming forthrightness, a genuineness, a profound generosity and sheer vitality that he carried with him all of his life..."

Soon afterwards, Leon went on to the first of what would be four sabbaticals at the Institute for Advanced Study, where he remained for 3 years (1954–1957) as an assistant to Arne Beurling, a permanent professor at the Institute and the man who took over Einstein's office. Leon also renewed a friendship from his City College days with Robert Aumann, who was also doing a postdoctorate at Princeton at that time.

It was during his first 2 years at Princeton, from 1954 to 1955, that Leon proved the fundamental theorem that would forever bear his name and later that of Malgrange as well, after French mathematician Bernard Malgrange independently proved the same theorem in 1955-1956. The Malgrange–Ehrenpreis theorem, which states that every nonzero differential operator with constant coefficients has a Green's function, was a foundation of the modern theory of differential equations that would serve as the basis for a range of theoretical and technological advances in the years to come.

Leon's presence in Princeton during these years proved to be crucial for the career of a younger friend, Hillel Furstenberg, who was a graduate student then, and some years later, took a position at the Hebrew University. At that time the graduate math department was a bulwark for the prevailing mathematical currents, with a clear inclination for the fashionable. Someone not entirely attuned to this would be less than comfortable pursuing his own line of research. Furstenberg describes his experience: "I was then experimenting with certain ideas which were later to prove fundamental for my work, but these deviated from the main thrust of activity in Fine Hall. Like every other mathematician, I needed someone to bounce ideas off, and Leon turned out to be the ideal partner—someone open to everything, willing to think deeply about just about anything, and having the ability to contribute with intelligence and insight to other people's problems. I think of Leon as my mathematical 'big brother.'"

In 1957, Leon went on to a 2-year teaching stint at Brandeis, followed in 1959 by his joining the teaching staff at Yeshiva University for 2 years. Then it was back to the Institute for another year (1961–1962), followed by his appointment in 1962 to full professor of mathematics to the Courant Institute at New York University. During his tenure at Courant, Leon lived on the NYU campus in Washington Square Village.

His NYU colleague, Sylvain Cappell, a raconteur of "Leon stories," recalled one particular moment during Leon's time at Courant Institute, when Institute administrator Jay Blaire, who had heard about the brilliance of this member of the mathematics faculty, headed over to meet him. He knocked on Leon's office and when a voice said, "please come in," Jay opened the door to behold a nearly empty office in which all the furniture was piled on itself in a corner. He later learned that this was because Leon had converted his new office into a handball court—driving Professor Donsker in the next office nuts with the ping! In the midst of this otherwise empty office was Leon standing on his head, a position he maintained during their entire meeting. At its end, Leon extended an upside-down arm to shake hands and asked Jay to kindly let himself out of the office and please close the door behind himself.

The year 1970 saw the publication of what Leon considered his "best work," his first major volume, *Fourier Analysis in Several Complex Variables*, in which he developed comparison theorems to establish the fundamentals of Fourier analysis and to illustrate their applications to partial differential equations. Leon began the volume by establishing the quotient structure theorem or fundamental principle of

Fourier analysis, then focused on applications to partial differential equations, and in the final section, explored functions and their role in Fourier representation.

Alan Taylor in his memorial essay, "Remembrances of Leon Ehrenpreis," recalled following Leon's suggestion to attend Courant for a postdoctoral year, which he did in 1968. That year, which Taylor described as "the most interesting and fun year of my professional life," Leon's student Carlos Berenstein was completing his doctoral thesis at Courant while helping Leon with the final editing of his Fourier Analysis volume. Meanwhile, Leon had moved to Yeshiva University, where he was giving a course on the book, so each Thursday,

> "Carlos and I would take the A train uptown to spend the day with Leon, attending his class and talking about mathematics. I really saw Leon's style of doing mathematics in that class. He was always interested in the fundamental reasons that theorems were true and in illustrative examples, but less interested in the details. It seemed to me that he could look at almost any problem in analysis from the point of view of Fourier analysis. Indeed, his book on Fourier analysis, in addition to presenting the proof of his most important contribution, the fundamental principle, contains chapters on general boundary problems, lacunary series, and quasianalytic functions... Leon was doing mathematics 100% of the time I spent around him and I think it was true always, especially when riding the train and in his jogging...."

While his appointment at Courant had been intended as a lifetime position, Leon received a "summons" from Dr. Belkin, president of Yeshiva University, to educate the "next generation" of Jewish academics. So, in 1968, 6 years after joining the NYU faculty, Leon returned to YU, where he would remain a member of the Belfer Graduate School faculty for 18 years—riding his bicycle through the dignified halls of academia, reuniting with his old friend "Flotz," after Leon encouraged Newman to join him on the YU mathematics faculty, and impacting upon hundreds of students—until the doors of the university's graduate school of arts & sciences were shut in 1984.

As a Jewish institution, Yeshiva also provided a fertile environment for Leon's synthesis of Talmudic and mathematical concepts. He taught a class entitled, "Modern Scientific and Mathematical Concepts in the Babylonian Talmud," and also introduced his calculus class with a page from the Talmud discussing the area of a circle as it relates to the size of a *sukkah,* a temporary booth built annually for the Jewish holiday of *Sukkot.* One of the students for his "Mathematics and the Talmud" class was undergraduate student Hershel Farkas. Hershel and his wife Sara, who would host both Leon and Ahava's first date and their wedding, would become among their dearest friends, the "family" waiting to welcome them home when their oldest child was born, to celebrate their greatest joys and share their major life moments. Indeed, over 40 years later, it was Hershel, just off the plane from Israel, whom Leon would plan to meet on August 15, 2010—a final mathematical conversation that never took place.

Yeshiva's new faculty member was also known for his rather laid-back attitude to the course schedule: One of Leon's students recalled his professor informing his class on the first day that while the course was scheduled for Tuesdays and Thursdays, he couldn't make it on Thursdays—and actually Tuesdays didn't work for him either. They settled on Sunday afternoons for their weekly study of differential equations, complex analysis, and number theory.

There was the time that Leon informed his class that he would be running the New York City marathon that coming Sunday, so he might be a little late for class. True to his word, he completed the race, took a taxi uptown, showered in his nephew's dorm room and came to lecture. Leon also used to tell the story of his stint as a teacher of an undergraduate math class at Stern College, YU's women's college. This "favorite Leon story," which Peter Kuchment, of Texas A&M University, likes to relate often to his students, describes him teaching a calculus class to this new group of students. "As any good teacher would do," Kuchment tells, "he tried to lead his students, whenever possible, to the discovery of new things. So, he once said: 'Let us think, how could we try to define the slope of a curve?' 'What is there to think about?' was the reply from one smart student, 'it says on page 52 of our textbook that this is the derivative.' 'Well,' replied Leon, 'I haven't read till page 52 yet.' The result was that the class complained to the administration that they were given an unqualified teacher. So much for inspiring teaching; it can backfire!"

Meanwhile, in 1954, Leon's brother Seymour had gotten married, Leon himself had headed back to the Institute, and their parents had moved again, this time to the Brighton Beach section of Brooklyn. Leon described himself as "always in search of new vistas of knowledge," so now, at the age of 24, he took advantage of the opportunities in his family's new neighborhood to expand his Jewish textual knowledge. He bought himself a copy of the English translation of the Talmud. Leon used to read the English side of the page—and viewed himself as the very personification of a Torah scholar because he could quote from the Talmud—in English!—with ease. But he was still searching for a more intensive learning experience.

It was his mother who found the way. She asked the local kosher butcher who could teach her son and received the response that if he wanted to "study seriously" he should go to Brighton resident Rabbi Yehudah Davis. Leon headed off to Rabbi Davis, and upon seeing the long-bearded rabbi, assumed he would speak only Yiddish. But in fact, the American-born rabbi spoke perfect English, and upon hearing Leon's background, addressed him with a simple question: "Why does a negative times a negative equal a positive?" "Here I was," Leon would later tell Dr. Yitzchak Levine, a member of the Department of Mathematical Sciences at New Jersey's Stevens Institute of Technology, "a mathematician at the Institute for Advanced Study at Princeton and I could not answer his question. I still do not know why conceptually a minus times a minus is a plus—and this was not the only question about mathematics he asked which I could not answer!"

Teacher and student, renaissance men both, began to study together regularly, taking long walks on the boardwalk to discuss Jewish philosophy and the lessons to be learned from the lives of the great men of Jewish history. Three years later, when Leon took a position in Brandeis, Rabbi Davis had just been appointed as dean of a yeshiva in Boston. Leon lived in the yeshiva, and the two continued to study together. Then, when Leon returned to the Institute for the 1961–1962 academic year, he invited his Jewish studies mentor for a visit to the Faculty Tea Room, introducing him to some of the greatest mathematicians and scientists of the day, including André Weil, with whom the rabbi conversed at great length. Later, Leon hosted a group of the yeshiva's students at the Institute as well.

Leon would later credit Rabbi Davis for having had "a great influence on me and my life," establishing the foundation to his approach to Torah learning. Certain concepts in Rabbi Davis's philosophy of analysis became well-known facets of Leon's own way of viewing Biblical texts, including the idea that no two Biblical terms are synonymous; rather, each apparently similar term actually carries with it an entirely unique connotation.

During the 1960s, after Leon had begun teaching at Courant, a friend suggested that for Jewish studies on the highest intellectual level, he should attend a class by Rabbi Moshe Feinstein. Rabbi Feinstein, considered the leading rabbinic authority of the twentieth century, had established his yeshiva, Mesivta Tiferes Jerusalem (MTJ), where the highest level of intellectual study took place in the least pretentious of environments, in the nearby Lower East Side neighborhood of Manhattan. It was the perfect study environment for Leon, who was described by many as having infinite patience for academic achievement but zero patience for bureaucratic convention.

Leon attended these classes with supreme dedication, even driving to New York from Princeton when he returned for additional semesters at the Institute. Within a few years—legend has it as a mere 5 years later—Leon had received his rabbinic ordination from Rabbi Feinstein, and remained his de facto advisor on scientific and technology issues until the famed authority on Jewish law passed away in 1986.

During his first marriage, to Ruth née Bers, daughter of the renowned mathematician and human rights activist Lipman Bers, Leon became the father of Ann (b. 1962) and Naomi (b. 1965, in Boston, during her father's sabbatical at Harvard; Naomi was the only one of Leon's children not born in New York City). Leon and Ruth had been introduced by Bers at a Jerusalem mathematics conference in 1960. They were married in June 1961, spending their first year of marriage at the Institute for Advanced Study. At Princeton, Leon developed close friendships with colleagues Bernie Dwork and Eli Stein, during a period Ruth later described as one in which "we all spoke freely about our families, laughed at ourselves and shared our concerns about the conditions of the world." They subsequently returned to NYU, with an intervening 1964–1965 sabbatical at Harvard, a year, Ruth recalled, that "was exciting. Leon was delighted to be surrounded by the mathematicians at Harvard and MIT whose families welcomed us warmly and shared their love of music and good food." The marriage ended in divorce in 1968.

In January 1972 Leon met Ahava Sperka, a native of Detroit, Michigan, the daughter of the Polish-born Rabbi Joshua and Canadian-born Yetta Sperka. Both Leon and Ahava were fond of recalling the immediate "kinship" of their first meeting: Leon picked up his date, and the two headed out to the event to which they had been invited, a "math party" at the Brooklyn home of Hershel and Sara Farkas. Leon commented, "you know, I've never actually been on time to any party before," to which Ahava replied, "neither have I." So they got out of the car to pass the time drinking tea until they could head to the event, comfortably late.

Thus began a "mathematical courtship," one that consisted primarily of evenings at "math parties," at which, Ahava would often reminisce, Leon would "wander off to 'talk math,' leaving me to fend for myself." It was a good preparation for a marriage in which "vacation" would come to mean "trip to another city, state, or country, where Leon would head off to his seminar, lecture, or conference, and leave me to entertain our growing family in yet another new place." Happily, Leon had found his soulmate, a kindred spirit who shared his dedication to principle, his love of adventure, and his yearning to explore new horizons.

As their mathematical social life continued, Ahava came to know many of the academics who played a role in Leon's life, including Lewis Coburn, graduate mathematics departmental chairman at YU, where Leon had begun teaching, and his wife, Charlene—who then discovered that their wedding had been officiated by Ahava's father. Then one day, as a change from the math party scene, Leon invited Ahava to New York's Metropolitan Museum of Art—and did she mind if they were joined by his coauthor on a new paper who had come in from Paris to work with him? Ahava's new "date" turned out to be, as she recalled, "a charming gentleman by the name of Paul Malliavin, a fifth-generation French aristocrat."

Leon, having done his Ph.D. at Columbia with French mathematician Chevalley, continued to be highly involved with the French school of mathematics. He spoke a fluent French and for many years spent several weeks each spring lecturing at the University of Paris, as well as at the Institut des Hautes Études Scientifiques (IHÉS) in Bures-sur-Yvette—invariably with a visit to the Malliavin home on the exclusive Isle de Paris, where one locked cabinet, nicknamed "Leon's kosher kitchen," would be opened upon their arrival.

It was Paul Malliavin as well who accompanied Ahava on a date with Leon in November of 1972, to the third-ever New York City marathon. This competition had begun as Central Park's "Earth Day Marathon" in 1970, a small race around Central Park that attracted few participants and even less media attention. After that first marathon, Leon recalled, "I said to a fellow runner 'I'll never do this again!' I had a mathematics conference at Princeton University the next day, and I was in such excruciating pain that I had to crawl out of bed to soak in the bathtub before I could get down the steps..."

The marathon grew substantially each year to become a global phenomenon that now attracts over 35,000 runners and two million spectators and turns all of the city's five boroughs into parts of the race track. Meanwhile, the group that gathered to watch Leon run would expand to include ever more of his growing family, as his wife and children—and in later years, his sons- and daughters-in-law, and grandchildren—would stand behind the barricade at their designated stop near the end of the marathon in Manhattan's Central Park, cheering wildly for "Aba" (in the spirit of the day, nearby spectators would eagerly join them in the call), as they peered at the thousands of runners passing by, eyes seeking that one familiar figure... who would suddenly appear, wave, pause long enough to be photographed, and then continue on to the finish line.

He would train throughout the year—"I find that I can train without wasting time," Leon once explained, "because I think about mathematics while I'm running"—and in 37 years, he never missed a single marathon, despite a broken arm one year, a baby's due date another (he ran with a beeper that year, promising his wife that if summoned he would 'meet her at NYU's emergency room—after all it's right on the marathon route!'), and the commitment—which he kept—to officiate at the wedding of a fellow mathematician that same evening, completing his final 26-mile, 385-yard run at the age of 77.

A year after they met, Leon offered a romantic proposal to the woman he hoped would become his wife: "I'd really like to marry you," he explained, "but I just don't want the fuss and bother of preparing for a wedding." "So then let's just get married," his now-fiancée replied. And 10 days later, on January 25, 1973, in the Farkas home that had been the venue for their first date, they did.

The romantic times continued, as they headed off a few months later for a several-month-long honeymoon in the city of Kyoto, a distant setting in which the new couple, while eschewing the nonkosher Japanese cuisine (they subsisted there on bananas, rice, and peanut butter), thoroughly enjoyed Japanese cultural, botanical, and mathematical offerings. It was also the city in which a local physician informed Leon, in his best English, that "Mrs. Ehrenpreis would 'not, not' be giving birth in February of the following year" to the couple's first child.

It would be a number of years before they would return. Indeed it was only when that eldest child turned 15 that Leon and Ahava would take four of their children, five suitcases of clothing, and six boxes of matzah, granola, tuna fish, pasta, and other staples sufficient to feed six kosher-only individuals for 4 weeks, and fly across the horizon to the Land of the Rising Sun.

Takahiro Kawai, professor emeritus at Kyoto University, later described that first meeting between him and the man whose "fundamental principal" had been "a guiding principle for many young analysts, including me... When I first met Leon... I got the impression that he was a kind man of sincerity. The impression has continued until now.... I cannot forget the warm atmosphere full of intellectual curiousity, which led to our [joint] paper. Another incident... is that I once happened to notice that he had not taken anything [to eat] for two days and that the reason was that he was dubious about the date of the [Jewish] fast day in Kyoto due to the effect of the International Date Line...."

One of the hallmarks of Leon's uniqueness was the fact that while he remained dedicated to every detail of his religious observance, he never saw that as an obstacle to being open to all; indeed his friends, colleagues, even those who met him only briefly, would reflect on the broad spectrum of his interests—from classical music to the great works of Western literature to Aramaic grammar—and of his openness to new ideas, new people, and new experiences within the consistent framework of his steadfast principles. As his close friend Hershel Farkas would later write about him, "Ehrenpreis's diversity extended way beyond mathematics. He was a pianist, a marathon runner, a talmudic scholar, and above all a fine and gentle soul."

Over the first two decades of their marriage, Leon, the man who had told Alan Taylor that he wished to have "as many children as a baseball team," would, in partnership with his wife, Ahava, raise their three girls and three boys: Nachama Yael (b. 1974), Raphael David (b. 1975), Akiva Shammai (b. 1977), Bracha Yehosheva

(Beth, b. 1983), Saadya (b. 1984), Yocheved Yetta (b. 1986). Their gang of eight—these six children along with their two older sisters—believed that summer and semester itineraries were built around sabbaticals, university schedules, and AMS conferences in Berkeley, Bowdoin, Jerusalem, and Japan, and were bewildered to discover that their elementary-school peers did not categorize their playdate options as "commutative," count to a "google" (that was before the Internet!), or dismiss errors as "trivial" in casual conversation. Each one went through the third-grade experience of informing his or her science teacher that her explanation of Copernican heliocentrism failed to take into account new perspectives on the solar system achieved by Einstein's Theory of Relativity.

The closing of Yeshiva University's graduate arts and science school in 1984 left Leon in a quandary: What would become his new mathematical "home"? His old friend Donald Newman rose to the occasion, encouraging Leon to join him at Temple University in Philadelphia, whose faculty Newman had joined in 1976. Leon formally accepted the offer of a position at Temple, where he remained for what would be his longest period in a single university: 26 years, until his death in 2010. So Flotz & Glock were back together again, each with his own style in their shared commitment to mathematics. Jane Friedman later recalled that

> "Dr. Newman paid more attention to the little details, so Dr. Ehrenpreis might be lecturing and say something like "the answer is this, or maybe this plus or minus one" and Dr. Newman would be in the back of the room yelling at him to get it right. You don't often think of professors yelling at each other over the heads of the students, but those two did it, with affection."

The two friends, who had been born just two months apart in 1930, would ultimately travel along the same path through high school, university, and their academic positions, their lives remaining intertwined until Newman's death on March 28, 2007, a loss greatly mourned by "Glockenspiel," who eulogized his lifelong friend at Newman's memorial service.

One of the Leon's Ph.D. students at Temple, Tong Banh, recalled the details of his mentor's years at Temple, depicting him best as "a person who preferred 'soft' solutions to human problems... I remember one day when we were approached by a beggar in the street. Leon immediately drew out a handful of quarters and handed them to him..." On the other hand, Banh emphasized, Leon was not at all "soft" when it came to mathematics, reviewing papers for potential publication and writing student letters of recommendation with a characteristic intellectual integrity and perfectionism that demanded the highest standards of academic achievement.

"But at Temple University," Banh described, "people mostly saw only the 'soft' part of his personality. He was extremely flexible in trying to accommodate everybody who ever needed anything from him."

Sylvain Cappell once asked Leon whether by usually taking the local train from NY to Philadelphia he didn't risk arriving late. Leon replied that, "in all the years I've been teaching at Temple, I've never arrived late." Sylvain couldn't help but wonder how it was that Leon, not known for his impeccable promptness, had achieved such a stellar punctuality record. Replied Leon: "Because class starts when I arrive."

In 1987, Leon and Bob Gunning of Princeton University directed the American Mathematical Society Theta Functions conference, which was held at Maine's Bowdoin College. Gunning later recalled their work together in a eulogy he wrote for Leon:

> The opportunity I had to work most closely with him was in organizing and managing the Theta Functions conference at Bowdoin College in the summer of 1987. I had experienced Leon's energy and enthusiasm before, and was not too surprised, although a bit overwhelmed, by the intensity with which he threw himself into organizing the conference schedules and the participants, as well as the AMS and NSF and who knows what foreign organizations for the participants coming from abroad; but it was an exhausting effort even to keep track of what we were doing. What did surprise me, although really it should not have, was the remarkable breadth of Leon's interests, and the depth with which he really understood what was going on in so many areas that the conference covered. I could not have found a better colleague to join in running a conference on that topic; and I am sure that I learned much more from Leon about so many aspects of theta functions than he did from me. Like so many other friends and colleagues, I shall miss his wild, but surprisingly often successful, ideas about how to approach problems, and his eagerness to talk about, and think about, a wide range of mathematics.

Two years later, in June 1989, Leon's student, Carlos Berenstein, and "grand-student," Daniele Struppa, organized a 60th birthday conference for him in the southern Italian coastal town of Cetraro. At the conference, entitled, "Geometrical and Algebraical Aspects in Several Complex Variables," Leon gave the keynote speech and a beautiful presentation on extension of solutions of partial differential equations, a topic that he had investigated for many years, and to which he made lasting contributions.

Struppa, who is today chancellor of California's Chapman College, recalled how Leon used to dine in his room, since the picturesque Calabria region did not have an available source for kosher food. So, the night that the participants wished to surprise the "birthday boy" with a formal dedication of the conference to him, they had to lure him down on a pretext to the dining room where the celebration awaited. He also remembered the conference as the time Leon asked Struppa's mother for help in having a uniquely designed candelabra, with ten branches (one for each member of the family), crafted as a gift for Ahava. The result, a one-of-a-kind—immensely heavy—Italian silver showpiece, was carried by Leon from Milan back to Brooklyn, to take a place of pride as his wife's Sabbath candelabra.

That same year, Leon attended the integral geometry and tomography conference in Arcata, California, where for the first time he met Peter Kuchment, who later recalled:

> It was my very first trip outside the former USSR, and it felt like being in a dream... Another shock during my first visit and my emigration soon afterwards, was that names like Leon Ehrenpreis... which obviously existed only on book covers, or at least referred to semi-gods somewhere well above this Earth, corresponded to mere mortals... Meeting Leon in Arcata was my first experience of this kind... Just like everyone else, I loved Leon from the first encounter. His unfailing cheerful disposition and his abundant eagerness to discuss any kind of mathematics at any time made every occasion we met feel like a holiday... Leon always liked to crack or to hear a good joke. He was smiling most of the time that I saw him. It was a joy to discuss with him not only mathematics, but also religion,

> music, or anything else. What made this even more enjoyable, was that in my experience he never imposed his opinions, beliefs or personal problems (and he unfortunately had quite a few) on others. It was relaxing to talk to him. He must have been a wonderful *Rebbe* [teacher/spiritual guide]...

He was indeed "a wonderful Rebbe," asserted Temple colleague and dear friend Marvin Knopp, recalling how, "when Leon arrived at our department, he walked into each person's office and asked what work he or she was doing. If he didn't find it interesting, he never returned—but if he did, he kept coming back over and over again." Throughout his years at the university, Leon—always with a mug of tea in his hand—could constantly be found encouraging, inspiring, talking, and teaching, playing a formative role in the development of his own department and the mathematical community of his time. "He was our mentor," Knopp described, "giving us projects to do and problems to solve, spreading enthusiasm and ideas every day, and inspiring our research. That's the kind of effect of he had—and not too many people have that kind of impact.

"Leon had a quality of walking in halfway through a lecture—and rapidly understanding the material far better than the lecturer himself. This happened to me once: he came into my talk after I had already covered the board with figures, saying something about the train being late—and within two minutes he was asking me questions I couldn't answer!"

Jane Friedman, one of the Leon's Ph.D. students at Temple, later eulogized her advisor, writing:

> "I have the career and the life that I do, only because of his help, his kindness and his support. And I am truly grateful to him. As we all know, he was a brilliant mathematician. I feel tremendously privileged to have studied with him and to have had the benefit of his deep insights. Dr. Ehrenpreis was not only an inspiration to me as a mathematician, he was inspirational as a person."

Later on, Jane described Leon as someone who

> "had an amazing gift for seeing the big picture, how concepts fit together in a deep way. He was able to understand mathematics in a way which could be transformative. This was a gift he gave his students—a vision of what it was to understand deeply, to see the forest and not the trees. I was inspired by him to always try to understand deeply, not superficially, and to get beyond the details. I was also inspired by him as a person, by his evident love for his wife and children, by his commitment to his community and by his joy in his family... Nonmathematicians and beginning students have a superficial view of mathematics; they have mostly experienced math as computation and symbol manipulation. Professor Ehrenpreis helped me grow beyond this beginner's view of math. I will never understand as much and as deeply as he does, but because of him I understand more and more deeply than I would otherwise."

Jane told his daughter that "your father got all the important things right and many of the nonimportant things wrong. He always knew which was which."

In March of 1992, Leon officiated at the wedding of his eldest daughter, the first of three daughters at whose weddings he would officiate. Immediately afterwards, he was confronted with what would become the long-term illness of his son Akiva, a medical situation that would represent a major challenge to Leon and his family for years to come—although Leon, with his consummate optimism, never gave up hoping for his son's full recovery.

Peter Kuchment recalled Leon's frequent visits:

> It is well known that he was an avid runner and had run the NY marathon every year since its inception in 1970 till 2007. He also liked to run during his visits, so when he visited me in Wichita, Kansas, I would sometimes pick a room for him in a hotel seven miles away from the campus, with a sufficiently attractive route to run between the two. So, after his lecture, or just a working day, he would give me his things to take back to the hotel, while he would run. Every time I would meet him after the run, he would have some new ideas (and he had so many great ideas!) about the problem we were working on at the time. Once, when he came back and I was waiting for him in the hotel's lobby, the receptionist at the front desk asked him: "Did you really run all the way from the campus?" Leon's reply was: "What else could I do? He refused to give me a ride"—and he pointed at me. I think I lost all the receptionist's respect at that time...

On April 6 and 7, 1998, "Analysis, Geometry and Number Theory: A Conference Celebrating the Mathematics of Leon Ehrenpreis" took place in Philadelphia, under the auspices of Temple University and the National Science Foundation. The 2-day event culminated in an honorary banquet with Leon's entire family in attendance. The proceedings of the conference were published by the American Mathematical Society 2 years later.

During the decade from 1993 until its publication in 2003, Leon devoted himself to the writing of his second major work, *The Universality of the Radon Transform*. The title, his choice after deep consideration, was one he felt reflected his profound belief that "mathematics is poetry," as were the words he composed to his wife for the book's dedication:

Many are the
Inspirations of the heart
But that borne by love
Surpasses all the rest

In this volume, he expanded upon the concept of the Radon transform, an area with wide-ranging applications to X-ray technology, partial differential equations, nuclear magnetic resonance scanning, and tomography. In covering such a range of topics, Leon focused on recent research to highlight the strong relationship between the pure mathematical elements and their applications to such fields as medical imaging.

Eric Todd Quinto, a friend and collaborator, referred to the book, to which he and Peter Kuchment wrote an appendix, as reflecting Leon's "emphasis on unifying principles." Quinto explained that in the book, Leon "developed several overarching ideas and used them to understand properties of the transforms, such as range theorems and inversion methods... The book draws connections between several fields, including complex variables, PDE, harmonic analysis, number theory, and distribution—all of which benefitted from his contributions over the years."

Leon was diagnosed with prostate cancer in 2003. However, he chose to reveal this information to no one outside his immediate family, because, he stated firmly, "I don't want people to view me as a sick person." Indeed, over the next few years, he maintained his regular routine: He continued to commute on the train to Temple, 3 hours each way; he traveled to conferences and seminars; he enjoyed the births of his grandchildren. He continued running the marathon until 2007, completing this 26-mile, 385-yard race for the last time at the age of 77.

In the summer of 2008, he was invited to attend the conference in honor of Jan Bowman's birthday in Stockholm. At the age of 78 he was an honored guest who was surrounded throughout the week by young scientists eager to hear his ideas. Two months later, he took what would be his last overseas trip, to Israel, where he found opportunities for mathematical tête-à-têtes while celebrating the birth of a granddaughter. So it was that Leon continued to live life to the fullest, reflecting, as Shlomo Sternberg would later describe, that "vitality that perhaps for us best describes Leon. The years passed; life transpired with its joys and sorrows. For Leon and his family, the sorrows were of such immensity that would otherwise crush anyone. But Leon bore his with unimaginable courage and responsibility. Courage that, we dare say, none of us could have possibly comprehended, let alone mustered. But despite it all, and no matter what transpired, Leon retained every bit of the vitality of our earlier years. His mathematical work continued. His, along with Ahava's, loving care and unstinting dedication to his family continued. His kindness and loyalty to us, his friends, continued. It was who he was."

Leon spent the Fall 2008 semester on sabbatical at Rutgers University, where he had, in the words of faculty member Steve Miller, "a big fan base," with students and faculty alike affording him a deep respect. Both before and after his sabbatical term, he spent quite a bit of time at Rutgers, working primarily with both Miller and Abbas Bahri. He was active in the nonlinear analysis and PDE seminars as well as in the number theory seminar that Miller ran.

Two years later, on Tuesday, April 20, 2010, at 1:40 pm, Leon was presenting what would ultimately be his final lecture, in Rutgers mathematics department room 705, giving a continuation of earlier talks in that seminar on analytically continuing complex functions in a strip in the complex plane. A few minutes into the talk, Leon collapsed: he had suffered a stroke. Bahri and Miller rushed him to the hospital, where, as Steve Miller recalled, "many of us waited hours even without a chance to see him, just to be near this great man."

Subsequently, with his usual optimism and force of character, Leon devoted himself to restoring his health, all with his characteristic good humor. Even then he continued to "talk math" and to challenge the idea of giving up teaching, determined to "never retire." Indeed he had not yet formally retired from his position as professor of mathematics at Temple University when, on August 16, 2010, having suffered heart failure, he passed away, at Sloan-Kettering Memorial Hospital in New York City.

Six years before Leon's death, his son, Akiva, whose lifetime of ill-health, beginning with the discovery of a brainstem tumor at the age of 14, remained a relentless challenge which Leon consistently faced with the greatest optimism, had suffered a catastrophic choking episode that left him in a long-term coma. Leon, along with the entire family, had remained devoted to Akiva throughout this painful period. One year and two months after Leon's death, on October 23, 2011, Akiva too passed away.

One month after Leon's passing, Abbas Bahri of Rutgers knocked on the door of the home in Brooklyn that had been his primary residence for 30 years. In his hand was a copy of the newest issue of the journal *Advanced Nonlinear Studies* with the entry for an article, entitled "Microglobal Analysis," by Dr. Leon Ehrenpreis. To his very last day he had continued to think, to create, to develop new ideas, and to write and transmit those ideas for future generations; he would truly be, as Hershel Farkas later wrote, "sorely missed by the mathematical community as both a scholar and a gentleman."

Several months later, paying tribute to Leon at the Memorial Conference at Temple University held during the year after Leon's death, Bahri wrote:

> "There are several good mathematicians, as well as there are several important mathematicians. But the fundamental ones are few. Leon is one of them... Leon has passed away; but the influence of his mathematical work is just at its beginning. Leon, I felt, was different because he clearly has longed to be a deeper person, a person with a soul and with a quest for another world, for a better and different world.... As Leon Ehrenpreis starts to find his final place in history, these are the two fundamental facts that make him stand out among us: the importance and depth of his work in mathematics and, beyond this work, the constant search for another, a better and more moral world."

"Leon Ehrenpreis: A Mathematical Conference in Memoriam" took place at Temple University on November 15 and 16, 2010. The panel of speakers throughout the 2-day event included Charles Epstein, University of Pennsylvania; Erik Fornaess; University of Michigan; Rutgers faculty Xiaojun Huang, Henryk Iwaniec, and Francois Treves; Joseph Kohn and Eli Stein of Princeton; Temple professors Igor Rivin and Cristian Gutierrez, and Peter Sarnak of the Institute for Advanced Study. Perhaps the most powerful testament to all that he had been, as mathematician and as mentor, was expressed by one Ph.D. student: "The joy of solving a problem is gone," Tong Banh mourned, "because I cannot share the solution with Professor Ehrenpreis."

There is much more that could said about Dr. Leon Ehrenpreis, more elements to portray, more anecdotes to relate, more tales to tell. This man, who touched so many lives and shaped so much of modern mathematics, lived a personal and professional life that continues to impact, to inform, and to inspire. He truly was—and remains—the "stuff of stories," for the reason that, as Sylvain Cappell described:

> Part of what makes "Leon Stories" so memorable – and why mathematicians delight in them—is that Leon juggled two quite opposite approaches to rule and structures. To the common, nuisance strictures and structures of quotidian life, Leon paid singularly little attention. But he accorded unbounded respect and love for the structures of mathematics and Judaism, and combined these with unbounded human insight and responsiveness. We will treasure our "Leon Stories" and tell them to our students, but they can hardly convey the unbounded joy he'd shared with us.

Differences of Partition Functions: The Anti-telescoping Method

George E. Andrews

Dedicated to the memory of the great Leon Ehrenpreis.

Abstract The late Leon Ehrenpreis originally posed the problem of showing that the difference of the two Rogers–Ramanujan products had positive coefficients without invoking the Rogers–Ramanujan identities. We first solve the problem generalized to the partial products and subsequently solve several related problems. The object is to introduce the anti-telescoping method which is capable of wide generalization.

1 Introduction

At the 1987 A.M.S. Institute on Theta Functions, Leon Ehrenpreis asked if one could prove that

$$\prod_{j=1}^{\infty} \frac{1}{(1-q^{5j-4})(1-q^{5j-1})} - \prod_{j=1}^{\infty} \frac{1}{(1-q^{5j-3})(1-q^{5j-2})}$$

has nonnegative coefficients in its power series expansion without resorting to the Rogers–Ramanujan identities.

In [4], Rodney Baxter and I answered this question "sort of." Actually, the point of our paper was to show that if one begins trying to solve Ehrenpreis's problem, then there is a natural path to the solution which has the Rogers–Ramanujan identities as a corollary. Indeed, as we say there [4, p. 408]: "It may well be objected that we presented a somewhat stilted motivation. Indeed if [the Rogers–Ramanujan

G.E. Andrews (✉)
Department of Mathematics, The Pennsylvania State University, University Park, PA 16802, USA
e-mail: geal@psu.edu

H.M. Farkas et al. (eds.), *From Fourier Analysis and Number Theory to Radon Transforms and Geometry*, Developments in Mathematics 28, DOI 10.1007/978-1-4614-4075-8_1,

identities] were not in the back of our minds, we would never have thought to construct [the path to the solution of Ehrenpreis's problem]." Subsequently in 1999, Kadell [9] constructed an injection of the partitions of n whose parts are $\equiv \pm 2 \pmod 5$ into partitions of n whose parts are $\equiv \pm 1 \pmod 5$. Finally in 2005, Berkovich and Garvan [6, Sect. 5] improved upon Kadell's work by providing ingenious, injective proofs for an infinite family of partition function inequalities related to finite products (including Theorem 1 below).

In this chapter, we introduce a new method which mixes analytic and injective arguments. We illustrate the method on the most famous problem, Theorem 1. We note that Theorem 2 is also a direct corollary of [6, Sect. 5].

Theorem 1 (The Finite Ehrenpreis Problem, cf. [6]). *For $n \geq 1$, the power series expansion of*

$$\prod_{j=1}^{n} \frac{1}{(1-q^{5j-4})(1-q^{5j-1})} - \prod_{j=1}^{n} \frac{1}{(1-q^{5j-3})(1-q^{5j-2})}$$

has nonnegative coefficients.

We should note that the original question can be answered trivially if one invokes the Rogers–Ramanujan identities [5, p. 82] because

$$\begin{aligned}
&\prod_{n=1}^{\infty} \frac{1}{(1-q^{5n-4})(1-q^{5n-1})} - \prod_{n=1}^{\infty} \frac{1}{(1-q^{5n-3})(1-q^{5n-2})} \\
&= \left(1 + \sum_{n=1}^{\infty} \frac{q^{n^2}}{(1-q)(1-q^2)\cdots(1-q^n)}\right) \\
&\quad \times \left(1 + \sum_{n=1}^{\infty} \frac{q^{n^2+n}}{(1-q)(1-q^2)\cdots(1-q^n)}\right) \\
&= q + \sum_{n=2}^{\infty} \frac{q^{n^2}}{(1-q)(1-q^2)\cdots(1-q^{n-1})},
\end{aligned} \tag{1.1}$$

which clearly has nonnegative coefficients.

However, there is no possibility of proving Theorem 1 in this manner because there are no known refinements of the Rogers–Ramanujan identities fitting these finite products. A new method is required.

Our method of proof might be called "anti-telescoping." Namely, we want to write the first line of (1.1) as

$$\sum_{j=1}^{n} (P_j - P_{j-1}) \tag{1.2}$$

where each P_i is a finite product with

$$P_0 = \prod_{j=1}^{n} \frac{1}{(1-q^{5j-3})(1-q^{5j-2})}$$

and

$$P_n = \prod_{j=1}^{n} \frac{1}{(1-q^{5j-4})(1-q^{5j-1})}.$$

We construct the P_i so that they gradually change from P_1 to P_n. The proof then follows from an intricate, term-by-term analysis of (1.2).

In Sect. 2, we construct (1.2) and provide some analysis of the terms. In Sect. 3, we provide an injective map of partitions to show that each term of the constructed (1.2) has at most one negative coefficient. From there the proof of Theorem 1 is given quickly in Sect. 4.

We wish to emphasize that anti-telescoping is applicable to many problems of this nature. To make this point, we provide three further examples.

Theorem 2 (Finite Göllnitz-Gordon). *For $n \geq 1$, the power series expansion of*

$$\prod_{j=1}^{n} \frac{1}{(1-q^{8j-7})(1-q^{8j-4})(1-q^{8j-1})} - \prod_{j=1}^{n} \frac{1}{(1-q^{8j-5})(1-q^{8j-4})(1-q^{8j-3})}$$

has nonnegative coefficients.

This theorem falls to the anti-telescoping method much more easily than the finite Ehrenpreis problem (Theorem 1).

Theorem 3 (Finite little Göllnitz). *For $n \geq 1$, the power series expansion of*

$$\prod_{j=1}^{n} \frac{1}{(1-q^{8j-7})(1-q^{8j-3})(1-q^{8j-2})} - \prod_{j=1}^{n} \frac{1}{(1-q^{8j-6})(1-q^{8j-5})(1-q^{8j-1})}$$

has nonnegative coefficients.

This theorem requires a rather intricate application of anti-telescoping. We have chosen it to illustrate the breadth of this method.

We note that the partial products in Theorem 2 are from the Göllnitz-Gordon identities [2, (1.7) and (1.8) pp. 945–946] and the partial products in Theorem 3 are from identities termed by Alladi, The Little Göllnitz identities, [7, Sätze 2.3 and 2.4, pp. 166–167] (cf. [3, pp. 449–452]).

We conclude our applications of anti-telescoping by proving a finite version of differences between partition functions from the Rogers–Ramanujan–Gordon theorem ([8], cf. [1]). Again, the proof goes without difficulty; however, a few cases must be excluded including the result in Theorem 1.

Theorem 4 (Finite Rogers–Ramanujan-Gordon). *For $\frac{k}{2} > s > r \geq 1$ and $n \geq 1$, the power series expansion of*

$$\prod_{\substack{j=1 \\ j \not\equiv 0, \pm s \pmod{k}}}^{kn} \frac{1}{1-q^j} - \prod_{\substack{j=1 \\ j \not\equiv 0, \pm r \pmod{k}}}^{kn} \frac{1}{1-q^j}$$

has nonnegative coefficients except possibly in the case s prime and $s = r + 1$ and $k = 3r + 2$.

The final section of this chapter provides a number of open problems.

2 Anti-telescoping

In this short section, we construct the telescoping sum (1.2). Namely,

$$P_j = \frac{1}{(q, q^4; q^5)_j (q^{5j+2}, q^{5j+3}; q^5)_{n-j}} \tag{2.1}$$

where

$$(a; q)_s = (1-a)(1-aq)\cdots(1-aq^{s-1}),$$

and

$$(a_1, a_2, \ldots, a_r; q)_s = \prod_{i=1}^{r} (a_i; q)_s.$$

Clearly,

$$P_n = \frac{1}{(q, q^4; q^5)_n}$$

and

$$P_0 = \frac{1}{(q^2, q^3; q^5)_n}.$$

So,

$$\frac{1}{(q, q^4; q^5)_n} - \frac{1}{(q^2, q^3; q^5)_n} = \sum_{j=1}^{n} (P_j - P_{j-1}). \tag{2.2}$$

We let

$$T(n, j) := P_j - P_{j-1}.$$

So for $1 \leq j \leq n$,

$$T(n,j) = \frac{(1-q^{5j-2})(1-q^{5j-3}) - (1-q^{5j-4})(1-q^{5j-1})}{(q,q^4;q^5)_j (q^{5j-3},q^{5j-2};q^5)_{n+1-j}}$$

$$= \frac{q^{5j-4}(1-q)(1-q^2)}{(q,q^4;q^5)_j (q^{5j-3},q^{5j-2};q^5)_{n+1-j}}$$

$$= \frac{q^{5j-4}(1-q^2)}{(q^6;q^5)_{j-1}(q^4;q^5)_j (q^{5j-3},q^{5j-2};q^5)_{n+1-j}} \tag{2.3}$$

and for $2 \leq j \leq n$

$$T(n,j) = q^{5j-8}\left(\frac{q^4}{1-q^4} - \frac{q^6}{1-q^6}\right) \times \frac{1}{(q^{11};q^5)_{j-2}(q^9;q^5)_{j-1}(q^{5j-3},q^{5j-2};q^5)_{n+1-j}}. \tag{2.4}$$

So for $n \geq 1$

$$T(n,1) = \frac{q}{(1-q^3)(1-q^4)(q^7,q^8;q^5)_{n-1}}, \tag{2.5}$$

for $n \geq 2$

$$T(n,1) + T(n,2) = \frac{q+q^4+q^5+q^6+q^9}{(1-q^6)(1-q^7)(1-q^8)(1-q^9)(q^{12},q^{13};q^5)_{n-2}}, \tag{2.6}$$

for $n \geq 3$,

$$T(n,1)+T(n,2)+T(n,3) = \frac{q+q^{11}+q^{21}}{(1-q^8)(1-q^9)(1-q^{11})(1-q^{14})(q^{12},q^{13};q^5)_{n-2}} + \frac{q^4+q^{11}}{(1-q)(1-q^9)(1-q^{11})(1-q^{14})(q^{12},q^{13};q^5)_{n-2}}, \tag{2.7}$$

and for $n \geq 4$

$$T(n,1)+T(n,2)+T(n,3)+T(n,4) = \frac{q+q^{12}}{(1-q^3)(q^{12};q)_3(q^{16};q)_4(q^{22},q^{23};q^5)_{n-4}} + \frac{(2q^{11}+q^{21})(1+q^3+q^6+q^9+q^{12}+q^{15}+q^{18})}{(q^{11};q)_4(q^{16};q)_4(q^{22},q^{23};q^5)_{n-4}}$$

$$+\frac{q^5+q^{13}}{(1-q^3)(q^{11};q)_4(1-q^{16})(q^{18};q)_2(q^{22},q^{23};q^5)_{n-4}}$$
$$+\frac{q^6+q^9+q^{10}+q^{15}+q^{16}+q^{19}+q^{20}}{(1-q^3)(q^{11};q)_4(q^{16};q)_2(1-q^{19})(q^{22},q^{23};q^5)_n}. \quad (2.8)$$

Lemma 5. *For $n \geq 4$, the first terms of the power series expansion are given by*

$$T(n,1)+T(n,2)+T(n,3)+T(n,4)=q+q^4+q^5+q^6+q^7+q^8+2q^9+\cdots$$

and the remaining coefficients are all ≥ 2.

Proof. By direct computation, we may establish that the assertion of Lemma 5 is valid through the first seventeen terms.

Next, we note that the coefficients in question must all be at least as large as those of

$$\frac{q+q^{12}}{(1-q^3)}+\frac{q^5+q^6+q^9+q^{10}+q^{15}+q^{19}+q^{20}}{(1-q^3)}$$
$$=q+q^4+q^5+q^6+q^7+q^8+2q^9+2q^{10}+q^{11}+3q^{12}+2q^{13}+q^{14}$$
$$+4q^{15}+2q^{16}+q^{17}+\frac{2q^{18}}{1-q}+\frac{q^{18}(2+q)}{1-q^3},$$

and the coefficients in this last expression are all ≥ 2 beyond q^{17} owing to $\frac{2q^{18}}{(1-q)}$.

3 The Injection

Our first goal is to interpret $T(n,j)$ as given in (2.4) as the difference between two partition generating functions.

First, we define a set of integers for $n \geq j \geq 5$

$$S(n,j) := \{9,11,14,16,19,\ldots,5j-4,5j-1\}$$
$$\cup\{5j-3,5j-2,5j+2,5j+3,\ldots,5n-3,5n-2\}$$

We say that 4-*partitions* are partitions whose parts lie in $\{4\}\cup S(n,j)$ with the condition that at least one 4 is a part.

We say 6-*partitions* are partitions whose parts lie in $\{6\}\cup S(n,j)$ with the condition that at least one 6 is a part.

We let $p4_{n,j}(m)$ (resp. $p6_{n,j}(m)$) denote the number of 4-partitions (resp. 6-partitions) of m.

Thus, by (2.4) and the standard construction of product generating functions [2, p. 45], we see that for $n \geq j \geq 4$

$$T(n,j) = q^{5j-8} \sum_{m \geq 0} \left(p4_{n,j}(m) - p6_{n,j}(m)\right) q^m \tag{3.1}$$

Lemma 6. *For* $m \geq 0$, $n \geq j \geq 5$,

$$p4_{n,j}(m) - p6_{n,j}(m) = \begin{cases} -1, & \textit{if } m = 6 \\ \geq 0 & \textit{if } m \neq 6. \end{cases}$$

Proof. Clearly for $m \leq 6$, $p4_{n,j}(m) = 0$ except for $m = 4$ when it is 1, and $p6_{n,j}(m) = 0$ except for $m = 6$ when it is 1. Hence, Lemma 6 is proved for $m \leq 6$. From here on, we assume $m > 6$.

We now construct an injection of the 6-partitions of m into the 4-partitions of m to conclude our proof.

Case 1. There are $2k$ 6's in a given 6-partition. Replace these by $3k$ 4s.

Case 2. There are $(2k+1)$ 6s in a given 6-partition (with $k > 0$). Replace these by $(3k-2)$ 4s and one 14.

Case 3. The given 6-partition has exactly *one* 6. Since $m > 6$, there must be a smallest summand coming from $S(n,j)$. Call this summand the *second summand*. We must replace the unique 6, the second summand, and perhaps one or two other summands by some fours and some elements of $S(n,j)$ that are (except in the instances indicated with $(*)$) no larger than the second summand with the added proviso that either

1. The number of 4s is $\equiv 2 \pmod 3$ or
2. The number of 4s is $\equiv 1 \pmod 3$ and no 14 occurs in the image

The table below provides the replacement required in each case. The first column describes the pre-image partition; the single 6 and the second summand are always given explicitly as the first two summands. After the few summands that are to be altered are listed, there is a parenthesis such as ($\geq$11, no 14s) which means that the remaining summands are taken from $S(n,j)$, all are $\geq$11 and there are no 14s. The second column describes the image partition. The parts indicated parenthetically are unaltered in the mapping.

	pre-image partition		image partition
$(*)$	$6+9+(\geq 11$, no 14's)	$\longrightarrow$	$4+11+(\geq 11$, no 14's)
$(*)$	$6+9+9+(\geq 11$, no 14's)	$\longrightarrow$	$4+9+11+(\geq 11$, no 14's)
	$6+9+9+9+11+(\geq 9)$	$\longrightarrow$	eleven 4's $+(\geq 11)$
	$6+9+9+9+16+$ (9's or ≥ 16)	$\longrightarrow$	ten 4's $+9+$ (9's or ≥ 16)
	$6+9+9+9+$ (9's or ≥ 16)	$\longrightarrow$	$4+9+9+11+$ (9's or ≥ 16)
	$6+9+14+(\geq 9)$	$\longrightarrow$	$4+4+4+4+4+9+(\geq 9)$
	$6+11+(\geq 11)$	$\longrightarrow$	$4+4+9+(\geq 11)$

$$
\begin{array}{lll}
(*)\ 6+14+(\geq 16) & \longrightarrow & 4+16+(\geq 16)\\
\ 6+14+14+(\geq 14) & \longrightarrow & 4+4+4+4+4+14+(\geq 14)\\
\ 6+16+(\geq 16) & \longrightarrow & 4+4+14+(\geq 16)\\
\ 6+19+(\geq 19) & \longrightarrow & 4+4+4+4+9+(\geq 19)\\
\ 6+21+(\geq 21) & \longrightarrow & 4+4+19+(\geq 21)\\
\ 6+24+(\geq 24) & \longrightarrow & 4+4+11+11+(\geq 24)\\
\ 6+26+(\geq 26) & \longrightarrow & \text{eight 4's } +(\geq 26)\\
\ 6+29+(\geq 29) & \longrightarrow & 4+4+4+4+19+(\geq 29)
\end{array}
$$

Now for $j \geq i > 6$

$$
\begin{array}{lll}
6+(5i-4)+(\geq 5i-4) & \longrightarrow & 4+4+(5i-6)+(\geq 5i-4)\\
6+(5i-1)+(\geq 5i-1) & \longrightarrow & 4+4+11+(5i-14)\\
 & & \quad +(\geq 5i-1)
\end{array}
$$

(remember that $j \geq 5$)

$$
\begin{array}{lll}
(*)\ 6+(5j-3)+(\geq 5j-3) & \longrightarrow & 4+(5j-1)+(\geq 5j-3)\\
\ 6+(5j-2)+(\geq 5j-2) & \longrightarrow & 4+4+(5j-4)+(\geq 5j-2)\\
\ 6+(5j+2)+(\geq 5j+2) & & \\
\qquad\qquad\qquad\qquad \text{if } j=5 & \longrightarrow & 4+4+11+14+(\geq 5j+2)\\
\qquad\qquad\qquad\qquad \text{if } j=6 & \longrightarrow & 4+9+9+16+(\geq 5j+2)\\
\qquad\qquad\qquad\qquad \text{if } j>6 & \longrightarrow & 4+4+16+(5j-16)\\
 & & \quad +(\geq 5j+2)\\
\ 6+(5j+7)+(\geq 5j+7) & \longrightarrow & 4+4+4+4+(5j-3)\\
 & & \quad +(\geq 5j+2)
\end{array}
$$

and for $i \geq j+3$

$$
\begin{array}{lll}
6+(5i-3)+(\geq 5i-3) & \longrightarrow & \text{five 4's } +(5i-17)+(\geq 5i-3)\\
6+(5j+3)+(\geq 5j+3) & & \\
\qquad\qquad\qquad\qquad \text{if } j=5 & \longrightarrow & 4+4+4+4+9+9\\
 & & \quad +(\geq 5j+3)\\
\qquad\qquad\qquad\qquad \text{if } j\geq 6 & \longrightarrow & \text{five 4's } +(5j-11)+(\geq 5j+3)\\
6+(5i+8)+(\geq 5j+8) & \longrightarrow & 4+4+4+4+(5j-2)\\
 & & \quad +(\geq 5j+8)\\
6+(5j+13)+(\geq 5j+13) & \longrightarrow & \text{five 4's } +(5j-1)+(\geq 5j+13)
\end{array}
$$

and for $i \geq j+4$

$$
\begin{array}{lll}
6+(5i-2)+(\geq 5i-2) & \longrightarrow & 4+4+9+9+(5i-22)\\
 & & \quad +(\geq 5i-2)
\end{array}
$$

The important points to keep in mind in checking for the injection are (A) every possible pre-image is accounted for and (B) there is no overlap among the images.

Point (A) follows from direct inspection of the construction of the first column where each line accounts for every possible second summand.

Point (B) requires serious scrutiny. We note that if two partitions have a different number of 4's, then they cannot be the same partition. There are single lines where the image has 11, 10, 8, 7 fours, so these are unique. There are five lines with five 4s, and inspection of these reveals they are all different. There are five lines with four 4s, and they all are clearly different in the explicitly given parts. There are ten lines with two 4s, and inspection of these reveals only two lines of possible concern, namely,

$$6 + (5j - 2) + (\geq 5j - 2) \quad \longrightarrow \quad 4 + 4 + (5j - 4) + (\geq 5i - 2)$$

and at $j > 6$

$$6 + (5j + 2) + (\geq 5j + 2) \quad \longrightarrow \quad 4 + 4 + 16 + (5j - 16) + (\geq 5j + 2)$$

Here, the upper line if j were 4 would be $4 + 4 + 16 + (\geq 18)$ while the bottom line is $4+4+16+(5j-16)+(\geq 5j-2)$, and so we would have a possible identity of images if j were 4. Fortunately, j is specified to be ≥ 5. There are seven lines with a single 4. These seven can be displayed with their smallest parts in evidence

$$4 + 11 + \cdots$$
$$4 + 9 + 11 + \cdots$$
$$4 + 16 + \cdots$$
$$4 + (5j - 1) + \cdots$$
$$4 + 9 + 9 + 16 + \cdots$$
$$4 + 9 + 16 + \cdots;$$

so clearly, all of these lines are distinct. Thus, we have constructed the required injection.

4 Proof of Theorem 1

For $n = 1$,

$$\frac{1}{(1-q)(1-q^4)} - \frac{1}{(1-q^2)(1-q^3)} = \frac{q}{(1-q^3)(1-q^4)}. \tag{4.1}$$

For $n = 2$,

$$\frac{1}{(1-q)(1-q^4)(1-q^6)(1-q^9)} - \frac{1}{(1-q^2)(1-q^3)(1-q^7)(1-q^8)}$$
$$= \frac{q + q^4 + q^5 + q^6 + q^9}{(1-q^6)(1-q^7)(1-q^8)(1-q^9)}. \tag{4.2}$$

For $n = 3$,

$$\frac{1}{(q,q^4;q^5)_3} - \frac{1}{(q^2,q^3;q^5)_3} = T(3,1) + T(3,2) + T(3,3). \tag{4.3}$$

For $n = 4$,

$$\frac{1}{(q,q^4;q^5)_4} - \frac{1}{(q^2,q^3;q^5)_4} = T(4,1) + T(4,2) + T(4,3) + T(4,4). \tag{4.4}$$

The nonnegativity of the power series coefficients in (4.1) and (4.2) is obvious by inspection. The nonnegativity for (4.3) follows directly from (2.7) and that in (4.4) follows directly from (2.8).

So for the remainder of the proof we can assume $n \geq 5$. Hence,

$$\begin{aligned}
&\frac{1}{(q,q^4;q^5)_n} - \frac{1}{(q^2,q^3;q^5)_n} \\
&\quad = T(n,1) + T(n,2) + T(n,3) + T(n,4) + \sum_{j=5}^{n} T(n,j) \\
&\quad = T(n,1) + T(n,2) + T(n,3) + T(n,4) \\
&\qquad + \sum_{j=5}^{n} q^{5j-8} \sum_{m=0}^{\infty} \left(p4_{n,j}(m) - p6_{n,j}(m)\right) q^m.
\end{aligned}$$

With the $T(n,1) + T(n,2) + T(n,3) + T(n,4)$ term, we know by Lemma 5 that all coefficients are ≥ 2 from q^9 onward. The j-th term in the sum has exactly one negative coefficient which is -1 and occurs as the coefficient of q^{5j-2} $(5 \leq j \leq n)$, and these single subtractions of 1 occur against terms in $T(n,1) + T(n,2) + T(n,3) + T(n,4)$ where the corresponding coefficient is ≥ 2. Hence, all terms have nonnegative coefficients.

Corollary 7. *In the power series expansion of*

$$\frac{1}{(q,q^4;q^5)_n} - \frac{1}{(q^2,q^3;q^5)_n}$$

the coefficient of q^m is positive except for the cases $n = 1$ with $m = 0$, 2, 3, and 6 and $n \geq 2$ with $m = 0, 2, 3$.

Proof. For $n = 1$

$$\frac{1}{(1-q)(1-q^4)} - \frac{1}{(1-q^2)(1-q^3)} = q + q^4 + q^5 + \frac{q^7}{1-q} + \frac{q^{13}}{(1-q^3)(1-q^4)},$$

and clearly the only zero coefficients occur for q^0, q^2, q^3 and q^6.

For $n = 2$

$$\frac{1}{(1-q)(1-q^4)(1-q^6)(1-q^9)} - \frac{1}{(1-q^2)(1-q^3)(1-q^7)(1-q^8)}$$
$$= q + \frac{q^4}{1-q} + \frac{q^9+q^{10}+q^{13}+q^{14}+q^{15}}{(1-q^6)(1-q^7)(1-q^8)(1-q^9)} + \frac{q^{11}}{(1-q)(1-q^8)(1-q^9)}$$
$$\times \frac{q^{12}(1+q^5+q^7+q^{12})}{(1-q^6)(1-q^8)(1-q^9)} + \frac{q^{13}}{(1-q)(1-q^8)(1-q^9)},$$

and now the only zero coefficients occur for q^0, q^2, and q^3.

For $n = 3$, the assertion follows from (2.7).

For $n \geq 4$ we see that the proof of Theorem 1 shows that all the coefficients are positive for q^m with $m \geq 9$, and Lemma 5 together with the proof of Theorem 1 proves the result for $m < 9$.

5 Proof of Theorem 2

We define

$$g_n = (q, q^4, q^7; q^8)_n$$

and

$$h_n = (q^3, q^4, q^5; q^8)_n.$$

Then Theorem 2 is the assertion that

$$\frac{1}{g_n} - \frac{1}{h_n}$$

has nonnegative coefficients. So

$$\frac{1}{g_n} - \frac{1}{h_n} = \frac{1}{h_n}\left(\frac{h_n}{g_n} - 1\right)$$
$$= \frac{1}{h_n}\sum_{j=1}^{n}\left(\frac{h_j}{g_j} - \frac{h_{j-1}}{g_{j-1}}\right)$$
$$:= \sum_{j=1}^{n} U(n, j)$$

where

$$U(n,j) = \frac{h_{j-1}}{g_j h_n}\Big((1-q^{8j-5})(1-q^{8j-4})(1-q^{8j-3})$$

$$(1-q^{8j-7})(1-q^{8j-4})(1-q^{8j-1}\Big)$$

$$= \frac{q^{8j-7}(1+q)}{(q^9,q^{12};q^8)_{j-1}(q^7;q^8)_j(q^{8j-5},q^{8j-3};q^8)_{n+1-j}(q^{8j+4};q^8)_{n-j}}$$

and $U(n,j)$ clearly has nonnegative coefficients.

6 Proof of Theorem 3

We have chosen this third theorem to illustrate some of the problems that can arise using the anti-telescoping method and to show how to surmount arising difficulties.

If we were to follow exactly the steps in the proof of Theorem 2, we would replace g_n with $(q,q^5q^6;q^8)_n$ and h_n with $(q^2,q^3,q^7;q^8)_n$. The resulting $U(n,j)$ is fraught with difficulties. $U(n,1)$ has no negative coefficients, but for $j>1$ $U(n,j)$ has scads of negative coefficients, many of which are not just -1 or -2. Thus, the smooth ride of Sect. 5 or the "6's⟶ 4's" injection of Sect. 3 seems to become a nightmare.

The secret is to adjust the anti-telescoping. Namely, we let

$$G_n = (q^6,q^9,q^{13};q^8)_n, \tag{6.1}$$

and

$$H_n = (q^7,q^{10},q^{11};q^8)_n, \tag{6.2}$$

with

$$W(n,j) = \begin{cases} \frac{1}{H_{n-1}(1-q)(1-q^5)(1-q^{8n-2})}\left(\frac{H_j}{G_j} - \frac{H_{j-1}}{G_{j-1}}\right), & 1 \le j < n \\ \frac{1}{H_{n-1}}\left(\frac{1}{(1-q)(1-q^5)(1-q^{8n-2})} - \frac{1}{(1-q^2)(1-q^3)(1-q^{8n-1})}\right) & \text{if } j = 0. \end{cases} \tag{6.3}$$

Then,

$$\sum_{j=0}^{n-1} W(n,j) = \frac{1}{(1-q)(1-q^5)(1-q^{8n-2})G_{n-1}}$$

$$-\frac{1}{H_{n-1}(1-q)(1-q^3)(1-q^{8n-1})}$$

$$+\frac{1}{H_{n-1}}\left(\frac{1}{(1-q)(1-q^5)(1-q^{8n-2})}\right.$$
$$\left.-\frac{1}{(1-q^2)(1-q^3)(1-q^{8n-1})}\right)$$
$$=\frac{1}{(q,q^5,q^6;q^8)_n}-\frac{1}{(q^2,q^3,q^7;q^8)_n} \tag{6.4}$$

The advantage of this altered anti-telescoping is that the denominator factors $(1-q)$ and $(1-q^5)$ help reduce the terms with negative coefficients to at most one for the $W(n,j)$.

Indeed,

$$W(1,0)=\frac{q(1+q^4)}{(1-q^5)(1-q^6)(1-q^7)}, \tag{6.5}$$

$$W(2,0)=\frac{1}{H_1}\left(\frac{q(1+q^4)}{(1-q^{14})(1-q^{15})}+\frac{q^7(1+q^4)}{(1-q^3)(1-q^{14})(1-q^{15})}\right), \tag{6.6}$$

and for $n\geq 3$,

$$W(n,0)=\frac{1}{H_{n-1}}\left(\frac{q(1+q^4)(1+q^{8n-3})+q^7(1+q^3)}{(1-q^{10})(1-q^{8n-2})(1-q^{8n-1})}\right.$$
$$\left.+\frac{q^{13}(1-q^{8n-14})}{(1-q^3)(1-q^{10})(1-q^{8n-1})(1-q^{8n-2})}\right) \tag{6.7}$$

and the numerator factor $(1-q^{8n-14})$ cancels with the same factor in H_{n-1}. Hence, the nonnegativity of the coefficients of $W(n,0)$ is clear upon inspection.

Next for $n\geq 2$,

$$W(n,1)=\frac{q^6(1+q^3-q^5-q^6-q^9-q^{10}+q^{12}+q^{15}}{H_{n-1}(1-q^5)(1-q^6)(1-q^9)(1-q^{13})(1-q^{8n-2})} \tag{6.8}$$

and

$$1+q^3-q^5-q^6-q^9-q^{10}+q^{12}+q^{15}$$
$$=\frac{(1-q^5)(1-q^6)(1-q^9)+q^3(1-q^7)(1-q^{13})+q^{12}(1-q^5)(1-q^9)}{1+q^{11}}. \tag{6.9}$$

So

$$W(n,1)=\frac{q^6\left\{(1-q^5)(1-q^6)(1-q^9)+q^3(1-q^7)(1-q^{13})+q^{12}(1-q^5)(1-q^9)\right\}}{H_{n-1}(1-q^5)(1-q^6)(1-q^9)(1-q^{13})(1-q^{8n-2})(1+q^{11})} \tag{6.10}$$

and in light of the facts that (1) each factor $1-q^i$ in the numerator also appears either explicitly in the denominator or in H_{n-1} and (2) the factor of $1-q^{11}$ from H_{n-1} combines with $1+q^{11}$ to leave $1-q^{22}$ in the denominator, we see that $W(n,1)$ has nonnegative coefficients for $n \geq 2$.

Now for $n > j \geq 2$,

$$
\begin{aligned}
W(n,j) &= \frac{H_{j-1}q^{8j-2}\{(1-q^5)(1-q^{8j+2})+q^3(1-q^3)(1-q^{8j-2})\}}{H_{n-1}(1-q^5)G_j(1-q^{8n-2})} \\
&= \frac{q^{8j-2}}{(q^{8j-1},q^{8j+3};q^8)_{n-j}(q^{8j+10};q^8)_{n-j-1}G_j(1-q^{8n-2})} \\
&\quad + q^{8j-5}\left(\frac{q^6}{1-q^6}-\frac{q^9}{1-q^9}\right)\frac{1}{(q^{8j-1},q^{8j+2},q^{8j+3};q^8)_{n-j}} \\
&\quad \times \frac{1}{(1-q^5)(q^{14};q^8)_{j-2}(q^{17};q^8)_{j-1}(q^{13};q^8)_j} \\
&:= W_1(n,j)+W_2(n,j)
\end{aligned}
\tag{6.11}
$$

Now it is immediate that $W_1(n,j)$ has nonnegative coefficients. Also because $1-q^6$ is in the denominator, we see that the coefficient of q^{8j+4} is ≥ 1. In addition, because the only factors in the denominator with exponents ≤ 11 are $(1-q^6)$ and $(1-q^9)$, we see that the coefficient of q^{8j+9} in $W_1(n,j)$ is zero.

We are now in a position to show via an injection involving $W_1(n,j)$ that $W(n,j)$ has only one negative coefficient which is -1 and occurs for q^{8j+9}. This requires an analysis analogous to that in Sect. 3.

We define for $n > j \geq 2$

$$
\begin{aligned}
\Sigma(n,j) := \{&5,13,14,17,21,\ldots,8j-10,8j-7,8j-3,8j-1,8j+1,8j+2,\\
&8j+3,8j+5,\ldots,8n-9,8n-6,8n-5,8n-2\}
\end{aligned}
$$

In other words, the elements of $\Sigma(n,j)$ are the numbers that appear as exponents in the factors $1-q^x$ making up the denominator of $W_1(n,j)$ (excluding 6 and 9).

We shall say that *6-partitions* (*a new definition from that in Sect. 3*) are partitions whose parts lie in $\{6\} \cup \Sigma(n,j)$ with the condition that at least one 6 is a part.

We shall say that *9-partitions* are partitions whose parts lie in $\{9\} \cup \Sigma(n,j)$ with the condition that at least one 9 is a part.

We let $P6_{n,j}(m)$ (resp. $P9_{n,j}(m)$) denote the number of 6-partitions (resp. 9-partitions) of m. We use capital "P" so that this $P6$ will not be confused with the $p6$ of Sect. 3. Thus, by (6.11) and the standard construction of product generating functions [2, p. 45], we see that for $n > j \geq 2$

$$
W_2(n,j) = q^{8j-5}\sum_{m\geq 0}\left(P6_{n,j}(m) - P9_{n,j}(m)\right)q^m. \tag{6.12}
$$

Lemma 8. *For $m \geq 0$, $n > j \geq 2$,*

$$P6_{n,j}(m) - P9_{n,j}(m) = \begin{cases} -1 & \text{if } m = 9 \text{ or } 14 \\ \geq 0 & \text{if } m \neq 9 \text{ or } 14 \end{cases}$$

Proof. Clearly for $m \geq 14$, $P6_{n,j}(m) = 0$ except at $m = 6$ and $m = 11$ $(= 6 + 5)$ when it is 1, and $P9_{n,j}(m) = 0$ except for 9 and 14 $(= 9 + 5)$. Thus, Lemma 8 is proved for $m \leq 14$. From here on, we assume $m > 14$.

We now construct an injection of the 9-partitions into the 6-partitoins of m to conclude our proof.

Case 1. There are $2k$ 9s in a given 9-partition, replace these by $3k$ 6s.

Case 2. There are $2k + 1$ 9s in a given 9-partition (with $k > 0$). Replace these with $(3k - 2)$ 6's and one 21. (Note that there is 21 present in $\Sigma(n, j)$ because $j \geq 2$).

Case 3. The given 9-partition has exactly *one* 9. Since $m > 14$, there must either be at least two 5s in the partition or else a *second summand* (i.e. the least summand other than the one 9) coming from $\Sigma(n, j)$.

As in Sect. 3, we must replace the unique 9, the second summand (or the 5s), and perhaps one or two other summands by some 6s and some elements of $\Sigma(n, j)$ that are (except in the instances indicated with $(*)$) no larger than the original second summand with the added proviso that either

1. The number of 6's is $\equiv 2 \pmod 3$
   or
2. The number of 6's is $\equiv 1 \pmod 3$ and no 21 occurs in the image.

The table below provides the replacement in each case. As in Sect. 3, the first column describes the pre-image partition; the single 9 and the second summand are given explicitly as the first two summands. After the few summands that are to be altered are listed, there is a parenthesis such as ($\geq$14, no 21) which means that the remaining summands are taken from $\Sigma(n, j)$, all are $\geq$14 and 21 does not appear. The second column describes the image partition. The parts indicated parenthetically are unaltered by the mapping.

	pre-image partition		image partition
	$9 + 5 + 5 + (\geq 5, \text{no } 21)$	$\longrightarrow$	$6 + 13 + (\geq 5, \text{no } 21)$
	$9 + 5 + 5 + 21 + (\geq 5)$	$\longrightarrow$	$6 + 6 + 6 + 6 + 6 + 5 + 5 + (\geq 5)$
$(*)$	$9 + 14 + (\geq 14, \text{no } 21)$	$\longrightarrow$	$6 + 17 + (\geq 14, \text{no } 21)$
	$9 + 14 + 21 + (\geq 14)$	$\longrightarrow$	$6 + 6 + 6 + 6 + 6 + 14 + (\geq 14)$
	$9 + 17 + (> 21)$	$\longrightarrow$	$6 + 5 + 5 + 5 + 5 + (> 21)$
	$9 + 17 + 21 + (\geq 21)$	$\longrightarrow$	$6 + 6 + 6 + 6 + 6 + 17 + (\geq 21)$
	$9 + 21 + (\geq 21)$	$\longrightarrow$	$6 + 6 + 6 + 6 + 6 + (\geq 21)$

now for $i \geq 4$

$$9 + (8i - 7) + (\geq 8i - 7) \longrightarrow 6 + 6 + (8i - 10) + (\geq 8i - 7)$$

$$\begin{array}{lll} & 9+(8i-3)+(\geq 8i-3) & \longrightarrow\ 6+6+5+(8i-11)+(\geq 8i-3)\\ (*) & 9+(8i-2)+(\geq 8i-2) & \longrightarrow\ 6+(8i+1)+(\geq 8i-2)\\ & 9+(8j-1)+(\geq 8j-1) & \longrightarrow\ 6+5+(8j-3)+(\geq 8j-1)\\ & 9+(8j+2)+(\geq 8j+2) & \longrightarrow\ 6+6+(8j-1)+(\geq 8j-2)\\ & 9+(8j+3)+(\geq 8j+3) & \longrightarrow\ 6+5+(8j+1)+(\geq 8j+3)\end{array}$$

now for $i > j$

$$\begin{array}{ll} 9+(8i-1)+(\geq 8i-1) & \longrightarrow\ 6+6+5+(8i-9)+(\geq 8i-1)\\ 9+(8i+2)+(\geq 8i+2) & \longrightarrow\ 6+5+5+(8i-5)+(\geq 8i+2)\\ 9+(8i+3)+(\geq 8i+3) & \longrightarrow\ 6+6+5+(8i-5)+(\geq 8i+3)\end{array}$$

finally

$$9+(8n-2)+(\text{more } 8n-2\text{'s}) \longrightarrow 6+6+(8n-5)+(\text{more } 8n-2\text{'s})$$

The comments that followed the table in Sect. 3 are again relevant here. However, the task here is simpler. The subtle aspect treated in Sect. 3 was the concern with overlapping images. The two lines marked $(*)$ clearly do not coincide with each other nor with the other five lines that have a unique 6 in the image. This concludes the proof of Lemma 8.

We are now positioned to conclude the proof of Theorem 3.

The case $n = 1$ follows directly from (6.4) and (6.5). The case $n = 2$ follows from (6.4), (6.6) and (6.10).

Now suppose $n > 2$. Then by (6.4),

$$\begin{aligned}&\frac{1}{(q,q^5,q^6;q^8)_n}-\frac{1}{(q^2,q^3,q^7;q^8)_n}\\ &=\sum_{j=0}^{n-1}W(n,j)=W(n,0)+W(n,1)+\sum_{j=2}^{n-1}W(n,j)\\ &=W(n,0)+W(n,1)+\sum_{j=2}^{n-1}(W_1(n,j)+W_2(n,j)).\end{aligned}\tag{6.13}$$

By examining (6.7), we see that the coefficients of $W(n,0)$ (for $n > 2$) are at least as large as those of

$$\begin{aligned}\frac{q^{13}}{(1-q^3)(1-q^{10})}&=q^{13}+q^{16}+q^{19}+q^{22}+q^{23}+q^{25}+q^{26}+q^{28}+q^{29}\\ &\quad+\frac{q^{31}}{1-q}+\frac{q^{43}}{(1-q^3)(1-q^{10})}.\end{aligned}\tag{6.14}$$

In particular, this means that the coefficient of q^{25} in $W(n,0)$ is positive and all coefficients of q^{31} and higher powers are positive.

In addition, we know that the coefficients of $W(n,1)$ are nonnegative. We have also established that the coefficient of q^{8j+4} in $W_1(n,j)$ is at least 1. Lemma 6 establishes $W_2(n,j)$ has its only negative coefficients at q^{8j+4} and q^{8j+9} and that these negative coefficients are both -1. Thus, $W(n,j)$ $(= W_1(n,j)+W_2(n,j))$ has at most one negative coefficient which occurs at q^{8j+9} and is, at worst, -1. These occur for $j \geq 2$, i.e. the sum in (6.13) has possibly -1's as a coefficient of q^{25}, q^{33}, q^{41},.... However, the comments following (6.14) show that these -1s are all cancelled out by positive terms in $W(n,0)$.

Hence, there are no negative coefficients on the right-hand side of (6.13). Therefore, Theorem 3 is proved.

7 Proof of Theorem 4

We proceed as in Sect. 5 where injections were unnecessary. We define

$$J_n := \prod_{\substack{j=1 \\ j\not\equiv 0,\pm s \pmod{k}}}^{kn} \frac{1}{1-q^j} = \frac{(q^s, q^{k-s}, q^k; q^k)_n}{(q;q)_{kn}}$$

an

$$K_n := \prod_{\substack{j=1 \\ j\not\equiv 0,\pm r \pmod{k}}}^{kn} \frac{1}{1-q^j} = \frac{(q^r, q^{k-r}, q^k; q^k)_n}{(q;q)_{kn}},$$

where $\frac{k}{2} > s > r \geq 1$, *and* we exclude the case where s is prime and $s = r+1$ and $k = 3r+2$ hold.

The object is to prove that $J_n - K_n$ has nonnegative power series coefficients. Thus,

$$\begin{aligned}
J_n - K_n &= K_n \left(\frac{J_n}{K_n} - 1\right) \\
&= K_n \sum_{j=1}^{n} \left(\frac{J_j}{K_j} - \frac{J_{j-1}}{K_{j-1}}\right) \\
&= K_n \sum_{j=1}^{n} \frac{J_{j-1}\left(1-q^{jk}\right)}{K_j} \left(\left(1-q^{kj-k+5}\right)\left(1-q^{kj-5}\right)\right. \\
&\qquad\qquad \left. - \left(1-q^{kj-k+5}\right)\left(1-q^{kj-5}\right)\right) \\
&= K_n \sum_{j=1}^{n} \frac{J_{j-1}\left(1-q^{jk}\right)}{K_j} q^{jk-k+r}\left(1-q^{s-r}\right)\left(1-q^{k-s-r}\right)
\end{aligned}$$

$$= \frac{1}{(q)_{kn}} \sum_{j=1}^{n} \left(q^s, q^{k-s}; q^k\right)_{j-1} \left(q^{jk+r}, q^{jk+k-r}, q^k\right)_{n-j} q^{jk-k+r}$$
$$\times (1 - q^{s-r}) \left(1 - q^{k-s-r}\right).$$

There are $2n$ binary factors of the form $1 - q^i$ in the numerator where $1 \leq i < kn$. If all of the numerator factors are distinct for each j, they will cancel with the corresponding terms in the denominator and the nonnegativity of the coefficients will follow.

For every j, we see that all the factors of

$$(q^s, q^{k-s}; q^k)_{j-1}(q^{jk+r}, q^{jk+k-r}; q^k)_{n-j}$$

are distinct. So our only worry is whether $(1 - q^{s-r})$ and $(1 - q^{k-s-r})$ can overlap with other terms.

We note that $s - r \neq k - s - r$ because $\frac{k}{2} > s$.

If $j = 1$, then $(1 - q^{s-r})$ and $(1 - q^{k-s-r})$ are the only terms with exponents $<k$ and so all coefficients in the $j = 1$ term are positive.

Next, we note that

$$s - r < s < k - s,$$

and for $j > 1$ the terms with exponents less than k are

$$(1 - q^{s-r}),\ (1 - q^s),\ (1 - q^{k-s}),\ \text{and}\ (1 - q^{k-s-r}).$$

The only possible equality here occurs when $s = k - s - r$. So if $k \neq 2s + r$, then we have distinct factors in the numerator, and the jth term has nonnegative coefficients.

Suppose that $k = 2s + r$ so that there are now two factors $(1 - q^s)$ in the numerator. Noting

$$\frac{(1 - q^s)^2}{(1 - q)(1 - q^s)} = 1 + q + \cdots + q^{s-1},$$

we see that if the $(1 - q)$ has not been cancelled from the numerator, then the coefficients are again nonnegative.

Thus, the only way that we are in danger of having negative coefficients in any term is if $k = 2s + r$ and $(1 - q)$ is cancelled from the denominator by $(1 - q^{s-r})$, i.e. the cases that cannot be handled occur when both $k = 2s + r$ and $s - r = 1$, or $k = 3r + 2$ and $s = r + 1$.

This latter case can be handled if s is composite. Because then there is a $t \mid s$ with $1 < t < s$ so that

$$\frac{1 - q^s}{1 - q^t} = 1 + q^t + \cdots + q^{s-t}.$$

Thus, cancellation can still be managed in the $s = r + 1$, $k = 2s + r$ case if s is composite. Hence, the only situation not accounted for is where s is prime, and $s = r + 1$ and $k = 2s + r = 3r + 2$.

8 Conclusion

The method of anti-telescoping should be applicable in a variety of further problems. The obvious first extension would be the Gordon generalization of Rogers–Ramanujan [1, 8]:

Conjecture 9. For each $n \geq 1$ and $1 \leq j < i < \frac{k}{2}$,

$$\frac{(q^i, q^{k-i}, q^k; q^k)_n}{(q;q)_{kn}} - \frac{(q^j, q^{k-j}, q^k; q^k)_n}{(q;q)_{kn}}$$

has nonnegative power series coefficients.

The case $k = 5, i = 2, j = 1$ is Theorem 1. Theorem 4 takes care of most cases. The only open cases are for i prime and $i = j + 1$ with $k = 3j + 2$.

To make the method more easily applicable to results like Theorems 1 and 3, it would be of value to explore the following question:

Suppose that S is a set of positive integers and i and j are not in S. Let

$$T_1 = \{i\} \cup S$$
$$T_2 = \{j\} \cup S$$

with $i < j$. Let $p(S,n)$ denote the number of partitions of n whose parts are in S. Under what conditions can we assert that $p(T_1, n) \geq p(T_2, n)$ except for an explicitly given finite set of values for n?

Acknowledgements Partially supported by National Science Foundation Grant DMS-0801184

References

1. G. E. Andrews, An analytic proof of the Rogers-Ramanujan-Gordon identities, Amer. J. Math., 88 (1966), 844–846.
2. G. E. Andrews A generalization of the Göllnitz-Gordon partition theorems, Proc. Amer. Math. Soc., 18 (1967), 945–952.
3. G. E. Andrews, Applications of basic hypergeometric functinos, SIAM Review, 16 (1974), 441–484.
4. G. E. Andrews and R. J. Baxter, A motivated proof of the Rogers-Ramanujan identities, Amer. Math. Monthly, 96 (1989), 401–409.

5. G. E. Andrews and K. Eriksson, Integer Partitions, Cambridge University Press, Cambridge, 2004.
6. A. Berkovich and F. Garvan, Dissecting the Stanley partition function, J. Comb. Th. (A), 112 (2005), 277–291.
7. H. Göllnitz, Partitionen mit Differenzenbedingungen, J. reine u. angew. Math., 225 (1965), 154–190.
8. B. Gordon, A combinatorial generalization of the Rogers-Ramanujan identities, Amer. J. Math., 83 (1961), 393–399.
9. K. W. J. Kadell, An injection for the Ehrenpreis Rogers-Ramanujan problem, J. Comb. Th. (A), 86 (1999), 390–394.

The Extremal Plurisubharmonic Function for Linear Growth

David Bainbridge

Dedicated to the memory of Leon Ehrenpreis

Abstract The purpose of this chapter is to study the properties of the linear extremal function, $\Lambda_E(z)$, which is the upper envelope of plurisubharmonic functions in $\mathbb{C}^n$ that grow like $|z| + o(|z|)$ and are bounded by 0 on $E \subset \mathbb{R}^n$. The function $\Lambda_E(z)$ is an analogue of the well-known extremal plurisubharmonic function of logarithmic growth obtained when $|z|$ is replaced by $\log z$ in the definition. It arises in the study of Phragmén–Lindelöf conditions on algebraic varieties, and the interest is how the growth of $\Lambda_E(z)$ depends on the geometry of the set E. We prove that the linear extremal function can have a linear bound ($\Lambda_E(z) \leq A\,|z| + B$), or a nonlinear bound (e.g., $\Lambda_E(z) \leq A\,|z|^{3/2} + B$), or it can be unbounded ($\Lambda_E(z) \equiv +\infty$). Examples of all three cases are provided. When E is a two-sided cone in $\mathbb{R}^n$, an exact formula for $\Lambda_E(z)$ is given.

1 Introduction

A subset E of $\mathbb{R}^n$ is said to satisfy the *linear bound property* if there exist positive constants A and B such that each plurisubharmonic function u on $\mathbb{C}^n$ that satisfies the upper bounds,

$$u(z) \leq |z| + o(|z|) \text{ as } |z| \to \infty, \quad \text{and} \quad u(z) \leq 0, \ z \in E \tag{1}$$

also satisfies

$$u(z) \leq A|z| + B.$$

D. Bainbridge (✉)
Raytheon Integrated Defense Systems, 528 Boston Post Rd., Sudbury, MA 01776, USA
e-mail: David_R_Bainbridge@raytheon.com

H.M. Farkas et al. (eds.), *From Fourier Analysis and Number Theory to Radon Transforms and Geometry*, Developments in Mathematics 28, DOI 10.1007/978-1-4614-4075-8_2,

Equivalently, introduce the *extremal function for linear growth*, $\Lambda_E(z)$, defined by

$$\Lambda_E(z) = \sup\{u(z) : u \in PSH(\mathbb{C}^n) \text{ and satisfies } (1)\}^* \tag{2}$$

where the superscript * means the upper semicontinuous regularization of the function defined by the supremum. Λ_E is plurisubharmonic on any open set where it is locally bounded. The linear bound condition is the growth condition $\Lambda_E(z) \leq A|z| + B$ for all $z \in \mathbb{C}^n$.

The definition of Λ_E^* is analogous to the Lelong-Siciak-Zaharajuta extremal function, $L_E(z)$, that is defined in the same way except that the upper envelope is over the class of all plurisubharmonic functions on $\mathbb{C}^n$ of logarithmic growth, that is, plurisubharmonic functions that satisfy instead of (1), the conditions

$$u(z) \leq \log(1 + |z|) + O(1) \text{ as } |z| \to \infty, \quad \text{and} \quad u(z) \leq 0, \quad z \in E.$$

The analogy suggests investigating which properties of L_E^* are also valid for Λ_E^*. For example, L_E^* is either identically $+\infty$ or else is again of logarithmic growth, $L_E^*(z) \leq \log(1 + |z|) + O(1)$. Could something similar be true for Λ_E^*? Unfortunately, this is not the case. Even when $\Lambda_E^*(z)$ is finite for all z, it need not have linear growth (see Sect. 5). And, when it does have linear growth $\Lambda_E(z) \leq A|z| + B$, it is rare that the constant A will be equal to 1. Many of the important properties of L_E^* seem to be connected to the fact that $\log|z|$ is the minimal growth rate for plurisubharmonic functions, whereas $|z|$ seems to be one of many different growth functions one could consider, e.g., other powers of $|z|$.

Our interest in Λ_E^* comes from its connection to the study of Phragmén–Lindelöf conditions on varieties in $\mathbb{C}^d$, in particular the condition SRPL introduced in [2]. It is clear that the same definition $\Lambda_E(z) := \Lambda_E(z, V)$ can be made for subsets E of an algebraic variety V in $\mathbb{C}^d$; the upper envelope is over all the functions that are plurisubharmonic on V, ≤ 0 on E and bounded by $|z| + o(|z|)$. In this context, the variety V is said to satisfy the condition SRPL if and only the real points of V satisfy the linear bound condition. That is, $\Lambda_{\mathbb{R}^d \cap V}(z, V) \leq A|z| + B$. The classification of algebraic varieties in $\mathbb{C}^d$ that satisfy SRPL is an unsolved problem. The connection with Λ_E on $\mathbb{C}^n$ comes by considering coordinate projections of V. If V has pure dimension n, then there are coordinates so that the coordinate projection $\pi(z, w) = z$ mapping $C^d = \mathbb{C}^n \times \mathbb{C}^k$ to $\mathbb{C}^n$ is a proper analytic covering of $\mathbb{C}^n$ such that $(z, w) \in V$ implies $|w| \leq C(1 + |z|)$. If

$$E_{\mathsf{hyp}} := \{x \in \mathbb{R}^n : (x, w) \in V \implies w \in \mathbb{R}^k\}$$

denotes the projection of the "hyperbolic points" in V, and if E has the linear bound property, then it easily follows from considering the "max over the fiber function" $\tilde{u}(z) = \max\{u(z, w) : (z, w) \in V\}$ that V satisfies the condition SRPL. For example, if V denotes the variety in $\mathbb{C}^3$ defined by the equation $y(y^2 - x^2) = z^2$ and is projected onto the (x, y) variables, then

$$E_{\mathsf{hyp}} = \{(x, y) : y(x^2 - y^2) \geq 0\}$$

is the union of three cones with vertex at the origin, two of opening 45° and one of opening 90°. It is a consequence of our main Theorem 4.3 that this set has the linear bound property, so this variety V does have the SRPL property. On the other hand, if

$$E_{\text{re}} := \{x \in \mathbb{R}^n \,:\, \text{there exists } (x, w) \in V \text{ with } w \in \mathbb{R}^k\}$$

and if E_{re} fails the linear bound property, then V will fail SRPL. While the linear bound property for the set E_{hyp} is sufficient for SRPL to hold, it is not necessary. And while SRPL is sufficient in order that E_{re} have the linear bound property, we do not know if it is necessary. Therefore, our results here provide only some partial results for the SRPL characterization problem.

Our goal has been to investigate which subsets of $\mathbb{R}^n$ have the linear bound property. Our main result, Theorem 4.3, gives a characterization of this property for a very special class of subsets of $\mathbb{R}^n$, namely, those of the form

$$E = \{x \in \mathbb{R}^n \,:\, P(x) \geq 0\}$$

where P is a homogeneous polynomial in n variables. We show that E has the linear bound property if an only if E is not contained in a half-space. We also investigate several natural questions, such as

(a) If $\Lambda_E(z)$ is finite for all $z \in \mathbb{C}^n$, does it necessarily have a linear bound?
(b) If E fails the linear bound condition, does there necessarily exists a plurisubharmonic function that is ≤ 0 on E and satisfies $u(z) = o(|z|)$?
(c) Can $\Lambda_E(z)$ be finite only on a proper subset of $\mathbb{C}^n$ and infinite at other points of $\mathbb{C}^n$?

There are also very interesting particular sets E for which we do not know whether or not E has the linear bound property. We will discuss some such examples at the end of Sect. 4.

In Sect. 2, we discuss some easy and/or known properties of the extremal function. In particular, it is pretty clear that no bounded set or half-space can have a finite extremal function. We also discuss the known fact that a 2-sided cone in $\mathbb{R}^n$ has the linear bound property. In fact, we will give an explicit formula for Λ_E when E is a 2-sided cone. In Sect. 3, we will show that question (b) has a negative answer by proving that sets that are "slightly larger than a half-space," e.g. $\{(x_1, x_2) \in \mathbb{R}^2 \,:\, x_1 \geq -|x_2|^{2/3}\}$ have $\Lambda_E(z)$ infinite at many points; however, there are no nonconstant plurisubharmonic functions on $\mathbb{C}^2$ that are $o(|z|)$ and ≤ 0 in this set. In Sect. 4, we prove our main result, that the set $\{P(x) \geq 0\}$ where P is a homogeneous polynomial has the upper bound property except in the trivial case when it is contained in a half-space. Finally, in Sect. 5, we will show that the answer to question (a) is also no by giving an example of a set E for which Λ_E is continuous and bounded but has superlinear growth.

2 First Properties of Λ_E.

Definition 2.1. A subset E of $\mathbb{C}^n$ has the:

(i) *Upper bound property* if $\Lambda_E(z) < +\infty$ for all $z \in \mathbb{C}^n$;
(ii) *Linear bound property* if there exist constants A, B such that $\Lambda_E(z) \leq A|z|+B$.

For a set $E \subset \mathbb{R}^n$ to have either of these properties, it needs to be fairly large. For example, it is clear that no bounded set E can have the upper bound property since if $E \subset \{|z| \leq R\}$, then all the functions $c \log^+ \frac{|z|}{R} := c \max\{\log \frac{|z|}{R}, 0\}$ satisfy (1) for any $c > 0$, so by letting $c \to \infty$, one sees that $\Lambda_E(z) = +\infty$ if $|z| > R$. Similarly, if E is a half-space in $\mathbb{R}^n$, for example, $E = \{x = (x_1, \ldots, x_n) \in \mathbb{R}^n : x_1 \geq 0\}$, then the family of functions $c|\operatorname{Im}\sqrt{z_1}| = O(|z|^{1/2}) = o(|z|)$ also has an infinite upper bound at any point z except those with $z_1 = x_1 \geq 0$. In fact, because of the upper semicontinuous regularization of the envelope, $\Lambda_E(z) \equiv +\infty$ in this case.

On the other hand, that $E = \mathbb{R}^n$ has the linear bound property is a well-known result.

Theorem 2.2 (Phragmén–Lindelöf). *For any choice of norm* $|\cdot|$ *on* $\mathbb{C}^n$, $\Lambda_{R^n}(z) \leq |\operatorname{Im} z|$.

Proof. This follows directly from the classical Phragmén–Lindelöf theorem in one variable. When $n = 1$, this is the classical Phragmén–Lindelöf theorem: *A subharmonic function u on the complex plane that satisfies $u(z) \leq |z| + o(|z|)$ that is bounded above by 0 on the real axis satisfies $u(z) \leq |\operatorname{Im} z|$.* And, the case $n > 1$ follows from this by the following argument. Choose a point $z = x + iy \in \mathbb{C}^n$, and consider the subharmonic function of one variable $\varphi(\zeta) := u(x + \zeta y)$ where $u \in PSH(\mathbb{C}^n)$ satisfies (1). This function clearly is ≤ 0 for real ζ and

$$\frac{\varphi(\zeta)}{|y|} \leq |\zeta| + o(|\zeta|).$$

Therefore, the one-variable result shows that $\varphi(\zeta) \leq |y||\operatorname{Im}\zeta| = |\operatorname{Im}\zeta y|$. Apply this estimate for $\zeta = i$ to obtain $u(x + iy) \leq |y| = |\operatorname{Im} z|$. □

This estimate also illustrates the fact that Λ_E really does depend on the choice of norm on $\mathbb{C}^n$. We will always use the usual Euclidean norm.

There are much smaller sets than all of $\mathbb{R}^n$ that have the upper bound property. For example, the cones

$$E = C_\delta := \{x \in \mathbb{R}^n : x_1^2 \geq \delta^2(x_2^2 + \cdots + x_n^2)\}.$$

This is a direct consequence of the Sibony–Wong Theorem [6] that shows that any $u \in PSH(\mathbb{C}^n)$ that satisfies the uniform upper bound $u(z) \leq |z|$ for all z in a nonpluripolar set of complex lines in $\mathbb{C}^n$ satisfies a uniform bound $u(z) \leq A|z|$ where the constant A depends only on the set of lines, not the choice of u. To see that our assertion is a consequence of this theorem, associate to each $x \in C_\delta$, the complex line $L_x := \{\zeta x : \zeta \in \mathbb{C}\} \subset \mathbb{C}^n$. If $u \in PSH(\mathbb{C}^n)$ satisfies (1), then on each of these lines, the subharmonic function of one complex variable

$$\varphi_x(\zeta) := u(\zeta x) \leq |\zeta||x| + o(|\zeta|), \text{ and } u(\zeta x) \leq 0, \ \zeta \in \mathbb{R}.$$

so we get the upper bound

$$u(z) \leq |\operatorname{Im} \zeta||x| = |\zeta x| = |z|, \quad z \in L_x.$$

Since the collection of lines $\{L_x : x \in C_\delta, |x| \leq 1\}$ is a nonpluripolar set of lines in the projective space of all lines in $\mathbb{C}^n$, the Sibony–Wong theorem shows that $\Lambda_E(z) \leq A|z|$.

In fact, it is possible to give the exact formula for the extremal function of these cones.

Theorem 2.3. *Let $\delta \geq 0$ and let $E = \{x \in \mathbb{R}^n : x_1^2 - \delta^2\langle x', x'\rangle \geq 0\}$, where $x' = (x_2, \ldots, x_n)$. Then*

$$\Lambda_E(z) = \sqrt{\left|\operatorname{Im}\sqrt{z_1^2 - \delta^2\langle z', \overline{z'}\rangle}\right|^2 + (1+\delta^2)\,|\operatorname{Im} z'|^2}. \tag{3}$$

Proof. Consider the multivalued mappings of $\mathbb{C}^n$ to $\mathbb{C}^n$ given by

$$F_\pm(z_1, z') = \left(\pm\sqrt{z_1^2 - \delta^2(z_2^2 + \cdots + z_n^2)}, \sqrt{1+\delta^2}z'\right) \tag{4}$$

and

$$G_\pm(w_1, w') = \left(\pm\sqrt{w_1^2 + \frac{\delta^2}{1+\delta^2}(w_2^2 + \cdots + w_n^2)}, \frac{1}{\sqrt{1+\delta^2}}w'\right). \tag{5}$$

Since these functions are multiple-valued, we will use the "±" subscript to denote the two values. We will drop the subscript when both values give the same result, as in (ii) and (iii).

The functions $F_\pm$ and $G_\pm$ have the following properties:

$$\begin{aligned}
&(i)\ \ F_\pm(E) = \mathbb{R}^n, G_\pm(\mathbb{R}^n) = E\\
&(ii)\ \ F_\pm(G(w)) = (\pm w_1, w'), G_\pm(F(z)) = (\pm z_1, z')\\
&(iii)\ \ \frac{1}{\sqrt{1+2\delta^2}}|w| \leq |G(w)| \leq |w| \text{ for all } w \in \mathbb{C}^n\\
&(iv)\ \ \text{For every } a \in \mathbb{R}^n, |G(w+a)| = |G(w)| + O(1)\\
&(v)\ \ G(ix) = iG_\pm(x) \text{ for all } x \in \mathbb{R}^n\\
&(vi)\ \ |G(x)| = |x| \text{ for all } x \in \mathbb{R}^n
\end{aligned} \tag{6}$$

Let $u(z)$ be a competitor for the linear extremal function. Then $u(z)$ is plurisubharmonic, $u(z) \leq |z| + o(|z|)$, and $u(z) \leq 0$ for all $z \in E$. Define a new plurisubharmonic function $v(w)$ by

$$v(w) = \max\{u(G_+(w)), u(G_-(w))\}.$$

Then, $v(w)$ is plurisubharmonic on $\mathbb{C}^n$. Property (iii) of 6 shows that $v(w) \leq |w| + o(|w|)$, and property (i) implies that $v(w) \leq 0$ for all $w \in \mathbb{R}^n$. The Phragmén–Lindelöf theorem (Theorem 2.2) shows that $v(w) \leq |\operatorname{Im} w|$. Property (ii) of 6 shows that $u(z) \leq |\operatorname{Im} F(z)|$ and therefore $\Lambda_E(z) \leq |\operatorname{Im} F(z)|$.

Now, let $\hat{z} \in \mathbb{C}^n$ and define a plurisubharmonic function by

$$u(z) = \max_{\pm}\{|\operatorname{Im} G\,(F_{\pm}(z) - \operatorname{Re} F_+(\hat{z}))|\}. \tag{7}$$

Even though the value of $F_+(\hat{z})$ depends arbitrarily on which branch of the square root is used, this will not affect the proof. We will show that $u(\hat{z}) = |\operatorname{Im} F(\hat{z})|$ regardless of which value of $F_+(\hat{z})$ is used. Property (i) of 6 shows that $u(z) = 0$ for all $z \in E$. Since $\operatorname{Re} F_+(\hat{z})$ is constant, properties (ii) and (iv) imply that $u(z) \leq |z| + O(1)$ as $|z| \to +\infty$. Properties (v) and (vi), however, show that

$$\begin{aligned} u(\hat{z}) &\geq |\operatorname{Im} G\,(F_+(\hat{z}) - \operatorname{Re} F_+(\hat{z}))| \\ &= |\operatorname{Im} G\,(i \operatorname{Im} F_+(\hat{z}))| \\ &= |\operatorname{Im}\,[i G(\operatorname{Im} F_+(\hat{z}))]| \\ &= |G(\operatorname{Im} F_+(\hat{z}))| \\ &= |\operatorname{Im} F_+(\hat{z})|\,. \end{aligned}$$

Therefore, we have $\Lambda_E(z) = |\operatorname{Im} F(z)|$. By definition of F, this is the same formula as in (3). □

3 The "No Small Functions" Condition

We saw in the previous section that a half-space in $\mathbb{R}^n$ does not satisfy the upper bound property; in fact, its extremal function is $\equiv +\infty$. The proof used the fact that there was a plurisubharmonic function $u(z)$ that was bounded above by 0 on the half-space and satisfied $u(z) = o(|z|)$. When there are no such "small" plurisubharmonic functions that are bounded above on E, we say that it satisfies the no small functions condition.

Definition 3.1. $E \subset \mathbb{R}^n$ satisfies *the no small functions condition* if there are no nonconstant plurisubharmonic functions on $\mathbb{C}^n$ that are bounded above by 0 on E and satisfy $u(z) = o(|z|),\ \ |z| \to \infty$.

Since, by Liouville's theorem, bounded plurisubharmonic functions on $\mathbb{C}^n$ are constant, the no small functions condition for a set E is equivalent to

$$\sup\{u(z) \in PSH(\mathbb{C}^n) : u(z) = o(|z|), u(z) = 0 \text{ for all } z \in E\} = 0.$$

The proof that a half-space fails the upper bound property can be generalized to the following stronger result:

Proposition 3.2. *If E fails the no small functions conditions, then E fails the upper bound property. In particular, there exists $z_0 \in \mathbb{C}^n$ such that $\Lambda_E(z_0) = +\infty$.*

Proof. If E fails the no small functions condition, then there exists $u(z)$ plurisubharmonic with $u(z) = o(|z|) \leq |z| + o(|z|)$, $u(z) = 0$ on E, and $u(z) > 0$ for some $z_0 \in \mathbb{C}^n$. Thus, for all $C > 0$, the function $Cu(z)$ is a competitor for the linear extremal function. For any integer k, however, C can be chosen large enough so that $Cu(z_0) > k$. Therefore, $\Lambda_E(z_0) = +\infty$. □

Remark. Clearly, if $E \subset \mathbb{R}^n$ fails the no small functions condition, then $\Lambda_E(z) = +\infty$ somewhere. We do not know if this implies that $\Lambda_E(z) \equiv +\infty$ everywhere.

The next proposition shows that if E is asymptotically a half-space, then Λ_E is unbounded.

Proposition 3.3. *If $E \subset \mathbb{R}^n$ is such that there exist $a \in \mathbb{R}^n$ and an increasing, unbounded function $f : \mathbb{R}^+ \to \mathbb{R}^+$ such that $f(r) = o(r)$ and $E \subset \{x \in \mathbb{R}^n : a \cdot x + f(|x|) \geq 0\}$, then $\Lambda_E(z) \equiv +\infty$.*

Proof. Without loss of generality, we may assume that $E \subset \{x_1 + f(|x|) \geq 0\}$. Since f is increasing, we have $E \cap \{|z| \leq R\} \subset \{x_1 + f(R) \geq 0\}$. Therefore, the function $\left|\operatorname{Im}\sqrt{z_1 + f(R)}\right|$ is plurisubharmonic and identically 0 on $E \cap \{|z| \leq R\}$.

Notice that for any $R > 0$,

$$\sup_{|z| \leq R} \left|\operatorname{Im}\sqrt{z_1 + f(R)}\right| = \sqrt{R - f(R)}.$$

The maximum is where $z_1 = -R$. Hence, for $|z| \leq R$,

$$\frac{R}{2\sqrt{R - f(R)}} \left|\operatorname{Im}\sqrt{z_1 + f(R)}\right| \leq \frac{R}{2}.$$

For every $R > 0$, define a plurisubharmonic function $u_R(z)$ on $\mathbb{C}^n$ by

$$u_R(z) = \begin{cases} \max\{|\operatorname{Im} z|, \frac{R}{2\sqrt{R-f(R)}} \left|\operatorname{Im}\sqrt{z_1 + f(R)}\right| + RH\left(\frac{z}{R}\right)\} & : |z| < R \\ |\operatorname{Im} z| & : |z| \geq R \end{cases}, \tag{8}$$

where

$$H(z) = \frac{1}{2}\sum_{j=1}^{n}(\operatorname{Im} z_j)^2 - (\operatorname{Re} z_j)^2. \tag{9}$$

This pluriharmonic function was used in [4], Lemma 2.9 and it has the following properties:

$$\begin{aligned}
&(i)\ \ H(z) \le |\operatorname{Im} z| \ \text{ for } |z| \le 1\\
&(ii)\ \ H(z) \le |\operatorname{Im} z| - \frac{1}{2} \text{ for } |z| = 1\\
&(iii)\ \ H(z) = O(|z|^2) \text{ as } |z| \to 0
\end{aligned} \tag{10}$$

Property (ii) implies that u_R is plurisubharmonic on $\mathbb{C}^n$, and property (i) shows that $u_R(z) = 0$ for all $z \in E$. Since $u_R(z) = |\operatorname{Im} z| + O(1) \le |z| + o(|z|)$, each u_R is a competitor for the linear extremal function, Λ_E.

To complete the proof, we need to show that the sequence $\{u_R(z)\}$ is unbounded for almost all z. Let z be an element of $\mathbb{C}^n$ with $\operatorname{Im} z_1 \ne 0$. In order to estimate $u_R(z)$ as R goes to infinity, we need the following formula:

$$\left|\operatorname{Im}\sqrt{\zeta}\right| = \sqrt{\frac{|\zeta| - \operatorname{Re}\zeta}{2}}. \tag{11}$$

This can be easily shown using the half-angle formula for the sine. Using (11) and (8), we get a lower bound for $u_R(z)$:

$$u_R(z) \ge \frac{R}{2\sqrt{R - f(R)}}\sqrt{\frac{|z_1 + f(R)| - \operatorname{Re} z_1 - f(R)}{2}} + RH\left(\frac{z}{R}\right). \tag{12}$$

For fixed z with $|z| << f(R)$, we can estimate $|z_1 + f(R)|$ with a power series:

$$\begin{aligned}
|z_1 + f(R)| &= \sqrt{(x_1 + f(R))^2 + y_1^2}\\
&= f(R)\sqrt{1 + \frac{2x_1}{f(R)} + \frac{x_1^2 + y_1^2}{f^2(R)}}\\
&= f(R)\left[1 + \frac{1}{2}\left(\frac{2x_1}{f(R)} + \frac{x_1^2 + y_1^2}{f^2(R)}\right) - \frac{1}{8}\left(\frac{2x_1}{f(R)} + \frac{x_1^2 + y_1^2}{f^2(R)}\right)^2\right.\\
&\qquad \left. + O(1/f^3(R))\right]\\
&= f(R) + x_1 + \frac{y_1^2}{2f(R)} + O(1/f^2(R)).
\end{aligned}$$

Therefore,

$$\sqrt{\frac{|z_1 + f(R)| - \operatorname{Re} z_1 - f(R)}{2}} = \sqrt{\frac{y_1^2}{4f(R)} + O(1/f^2(R))} \tag{13}$$

Now, we can use (13) and (12) to estimate $u_R(z)$:

$$\begin{aligned} u_R(z) &\geq \frac{R}{2\sqrt{R - f(R)}} \sqrt{\frac{y_1^2}{4f(R)} + O(1/f^2(R))} + RH\left(\frac{z}{R}\right) \\ &= \frac{\sqrt{R}|\operatorname{Im} z_1|}{4\sqrt{f(R)}} \sqrt{\frac{1 + O(1/f(R))}{1 - f(R)/R}} + O(1/R) \\ &= \sqrt{\frac{R}{f(R)}} \left(\frac{|\operatorname{Im} z_1|}{4} + o(1)\right). \end{aligned} \tag{14}$$

Since $f(R) = o(R)$ as $R \to +\infty$, the right hand side of 14 is unbounded as $R \to +\infty$. Therefore, $\Lambda_E(z) = +\infty$ for all $z \in \mathbb{C}^n$ with $\operatorname{Im} z_1 \neq 0$. Upper semi-regularization forces $\Lambda_E(z) = +\infty$ for all z. □

In the proof of Proposition 3.3, we used the fact that $f(R)$ is unbounded in order to expand $|z_1 + f(R)|$ as a power series. If $f(R)$ is bounded, however, E is contained in a half-space so we still have $\Lambda_E(z) \equiv +\infty$.

Proposition 3.3 shows that, for example, the set $E = \{(x_1, x_2) \in \mathbb{R}^2 : x_1^3 + x_2^2 \geq 0\} = \{x_1 + x_2^{\frac{2}{3}} \geq 0\}$ fails the upper bound property.

Let us also note that these asymptotic half-spaces satisfy the no small functions condition.

Proposition 3.4. *Let $f : \mathbb{R}^+ \to \mathbb{R}^+$ denote an increasing, unbounded function and*

$$E = \{x = (x_1, x') \in \mathbb{R}^n : |x_1| \geq -f(|x'|)\}.$$

Then, E satisfies the no small functions condition and in fact $\Lambda_E(z) \equiv +\infty$.

Proof. For the proof, we use the fact that if $\varphi(\zeta)$ is a subharmonic function on the complex plane $\mathbb{C}$ that satisfies $\varphi(\zeta) \leq |\zeta| + o(|\zeta|)$ and $\varphi(x) \leq 0$ for all $x \in \mathbb{R}$ with $|x| \geq a$, then $\varphi(\zeta) \leq |\operatorname{Im} \sqrt{\zeta^2 - a^2}|$. That is, in one complex variable, the extremal function for the set $(\infty, -a] \cup [a, +\infty)$ is $|\operatorname{Im} \sqrt{\zeta^2 - a^2}|$. This result is proven in Sect. 5 as a Corollary of Theorem 5.1.

In the proof of the proposition, it is no loss of generality to assume that $E = \{x = (x_1, x') \in \mathbb{R}^n : x_1 \geq -f(x')\}$. If $u(z) = o(|z|)$, then for every constant $M > 0$ and each fixed value of $x' \in \mathbb{R}^{n-1}$, we have $z_1 \to Mu(z_1, x') = o(|z_1|)$ and $u(x_1, x') \leq 0$ if $|x_1| \geq f(x')$. The preceding paragraph's formula for the one-dimensional extremal function for the real line with a hole implies $Mu(z_1, x') \leq |\operatorname{Im} \sqrt{z_1^2 - f(x')^2}|$. Since this holds for every value of $M > 0$, we conclude that

$u(z_1, x') \leq 0$ so that $z_1 \to u(z_1, x')$ must be a constant plurisubharmonic function, $u(z_1, x') = u(x')$. But then, by the Phragmén–Lindelöf estimate, Theorem 2.2, we conclude that $M u(z_1, z') \leq |\operatorname{Im} z'|$. And since this holds as well for every value of M, we conclude $u(z) \leq 0$ for all $z \in \mathbb{C}^n$. Consequently, E satisfies the no small functions condition.

The last conclusion of the Proposition, that $\Lambda_E(z) \equiv +\infty$ follows from the preceding proposition. □

Corollary 3.5. *There exists a set $E \subset \mathbb{R}^n$ that satisfies the no small functions condition but fails the upper bound property; i.e., $\Lambda_E(z_0) = +\infty$ at some point of $\mathbb{C}^n$. In fact,*

$$E = \{(x_1, x_2) \in \mathbb{R}^2 : x_1 \geq -|x_2|^{2/3}\}$$

is such a set.

Proof. The set E has $\Lambda_E(z_0) = +\infty$ for some z_0 by Proposition 3.3. It satisfies the no small functions condition by Proposition 3.4. □

4 The Linear Extremal Function for the Positivity Set of a Real Homogeneous Polynomial

In this section, we will show that if P is a real homogeneous polynomial, then $E = \{x \in \mathbb{R}^n : P(x) > 0\}$ satisfies the linear bound property if and only if E is not contained in a real half-space. Since we already know that subsets of half-spaces fail the upper bound property, it then follows that the three conditions, i.e. the upper bound property, the no small functions condition, and the linear bound property, are equivalent for $\{P(x) > 0\}$, P real and homogeneous.

The following lemma shows that adding or removing a pluripolar set does not affect the linear extremal function:

Lemma 4.1. *If $E \subset \mathbb{R}^n$ and $X \subset \mathbb{C}^n$ is a pluripolar set, then $\Lambda_{E \cup X} = \Lambda_E$.*

Proof. Since $E \subset E \cup X$, we have $\Lambda_{E \cup X}(z) \leq \Lambda_E(z)$. To complete the proof, we must show that $\Lambda_{E \cup X}(z) \geq \Lambda_E(z)$.

If X is pluripolar, then Theorem 5.2.4 of [5] gives the existence of a plurisubharmonic function $v(z)$ such that $v(z) \leq \log(1 + |z|)$ and $v|_X \equiv -\infty$. Without loss of generality, we may assume that $X = \{v = -\infty\}$. Also, Lemma 2.2 of [3] shows that there exists a plurisubharmonic function $\varphi(z)$ and a constant $C > 0$ such that

$$-C \log(1 + |z|) \leq \varphi(z) \leq |\operatorname{Im} z| - \log(1 + |z|).$$

Let $u(z)$ be plurisubharmonic, $u(z) \leq |z| + o(|z|)$, and $u(z) \leq 0$ for every $z \in E$. For every $\varepsilon > 0$, define $u_\varepsilon(z)$ as

$$u_\varepsilon(z) = (1 + \varepsilon)^{-1}[u(z) + \varepsilon(\varphi(z) + v(z))]. \tag{15}$$

With this definition, we have $u_\varepsilon(z) \le |z| + o(|z|)$, $u_\varepsilon(z) \le 0$ on E, and $u_\varepsilon(z) = -\infty$ for every $z \in X$. Hence, we have $u_\varepsilon(z) \le \Lambda_{E\cup X}(z)$. Solving (15) for $u(z)$ gives

$$u(z) \le (1+\varepsilon)\Lambda_{E\cup X} - \varepsilon(\varphi(z) + v(z)).$$

In the limit as $\varepsilon \to 0$, this inequality becomes

$$u(z) \le \Lambda_{E\cup X}(z) \text{ for all } z \in \mathbb{C}^n \setminus X. \tag{16}$$

Since $\Lambda_{E\cup X}(z)$ is plurisubharmonic and upper semicontinuous and X is a set of measure 0, the bound in (16) must hold for all $z \in \mathbb{C}^n$. The fact that $u(z)$ is a competitor for Λ_E gives $\Lambda_E(z) \le \Lambda_{E\cup X}(z)$, thus completing the proof. □

Corollary 4.2. *For a real polynomial P, $\{x \in \mathbb{R}^n : P(x) \ge 0\}$ satisfies the linear bound property if and only if $\{x \in \mathbb{R}^n : P(x) > 0\}$ also satisfies it.*

Corollary 4.2 shows the linear extremal function for the set $\{xy^2 \ge 0\}$, for example, is equal to that of the set $\{xy^2 > 0\}$. The former is not contained in a half-space, while the latter is. Since the function $|\operatorname{Im}\sqrt{z_1}| = o(|z|)$ equal to 0 on the set $\{xy^2 > 0\}$, this set does not satisfy the no small functions condition. Hence, it fails the linear bound property. The set $\{xy^2 \ge 0\}$, even though it is not contained in a half-space. It is contained in a half-space plus a pluripolar set.

Theorem 4.3. *Let $E = \{x \in \mathbb{R}^n : P(x) > 0\}$, where P is a real homogeneous polynomial. E satisfies the linear bound property if and only if E is not contained in a real half-space.*

In the proof, we will use the fact that P is a homogeneous polynomial. If P is homogeneous of degree m and $P(x) > 0$, then for all $r > 0$, $P(rx) = r^m P(x) > 0$. Hence, the set $\{P(x) > 0\}$ consists of rays extending out to infinity. This property is called *outward radial* and it is essential to the proof.

Definition 4.4. A set E is called *outward radial* if for all $x \in E$ and $r \ge 1$, rx is also an element of E.

Before beginning the proof of Theorem 4.3, we need a couple of lemmas.

Lemma 4.5. *If E is outward radial and satisfies the upper bound property, then E satisfies the linear bound property.*

Proof. Let $u(z)$ be a plurisubharmonic function with $u(z) \le |z| + o(|z|)$ and $u(z) \le 0$ for all $z \in E$. Since E satisfies the upper bound property, there exists a constant $A > 0$ such that $u(z) \le A$ for all $|z| \le 1$.

Let $z_0 \in \mathbb{C}^n$ such that $|z_0| > 1$. If we let $r = |z_0|$, then the function $u_r(z) = \frac{1}{r}u(rz)$ is plurisubharmonic and satisfies $u_r(z) \le |z| + o(|z|)$. Also, the fact that E is outward radial implies $u_r(z) \le 0$ for all $z \in E$. Hence, u_r is also a competitor for the linear extremal function, $\Lambda_E(z)$ and so we must have $u_r(z) \le A$ for all $|z| \le 1$.

Since $|z_0/r| = 1$, we have $A \ge u_r(z_0/r) = \frac{1}{r}u(z_0)$ and therefore, $u(z_0) \le Ar = A|z_0|$ for all $|z| \ge 1$. Finally, the bound for $|z| \le 1$ gives $u(z) \le A|z| + A$. □

Lemma 4.6. *Let $P(x)$ be a real polynomial in n variables. If the variety $\{(z,\zeta) \in \mathbb{C}^n \times \mathbb{C} : P(z) - \zeta^2 = 0\}$ satisfies SRPL, then the set $\{x \in \mathbb{R}^n : P(x) \geq 0\}$ satisfies the upper bound property.*

Proof. Let $u(z) \leq |z| + o(|z|)$, $u(x) \leq 0$ for $x \in \{x \in \mathbb{R}^n : P(x) \geq 0\}$. Let V be the variety $\{P(z) - \zeta^2 = 0\} \subset \mathbb{C}^{n+1}$. This variety is defined so that on real points of V, we must have $P(x) \geq 0$. Therefore, the function $u(z)$ can be lifted to a function $\tilde{u}(z,\zeta) = u(z)$ with $\tilde{u}(z,\zeta) \leq |(z,\zeta)| + o(|(z,\zeta)|)$, and $\tilde{u}(z,\zeta) \leq 0$ on $V \cap \mathbb{R}^{n+1}$.

If V satisfies SRPL, there exist constants A and B independent of u such that $\tilde{u}(z,\zeta) \leq A|(z,\zeta)| + B$. Therefore, we have

$$\begin{aligned} u(z) &\leq A\sqrt{|z|^2 + |\zeta|^2} + B \\ &= \sqrt{|z|^2 + |P(z)|} + B. \end{aligned}$$

Thus, $\{P(x) \geq 0\}$ satisfies the upper bound property. □

Proof of Theorem 4.3. Since $E = \{P(x) > 0\}$ is outward radial, Lemma 4.5 shows that E satisfies the linear bound property if and only if E satisfies the upper bound property.

If E satisfies the upper bound property, then E also satisfies the no small functions condition and therefore Proposition 3.3 shows E is not contained in a half-space. To prove the other direction, we will classify the real homogeneous polynomials $P(z)$ into five categories. In each category, the set E is either contained in a half-space and therefore fails the upper bound property, or E satisfies the upper bound property and therefore is not contained in a half-space.

Case I. If $P(x) \leq 0$ for all real x, then E is empty. Hence, E fails the upper bound property and E is contained in a half-space.

Otherwise, $P(x) > 0$ for some real x. After a change of coordinates, we may assume that $x = (1, 0, \ldots, 0)$. By multiplying P by a positive constant, we may assume that $P(z) = z_1^m +$ other terms. If $Q_1, \ldots, Q_c$ are the irreducible factors of P over $\mathbb{C}$, then

$$P(z) = \prod_{j=1}^{c} (Q_j(z))^{r_j},$$

and each Q_j is unique up to multiplication by a constant. Since P is monic in z_1, we may assume that each Q_j is also monic in z_1, and therefore the Q_j's are unique.

Order the Q_j's so that $Q_1, \ldots, Q_a$ have non-real coefficients, $Q_{a+1}, \ldots, Q_b$ have real coefficients but $\dim_{\mathbb{R}}\{x \in \mathbb{R}^n : Q_j(x) = 0\} \leq n-2$ (a and b could be 0), and $Q_{b+1}, \ldots, Q_c$ have real coefficients and $\dim_{\mathbb{R}}\{x \in \mathbb{R}^n : Q_j(x) = 0\} = n-1$. Define polynomials $R_1, \ldots, R_4$ by

$$R_1(z) = \prod_{j=1}^{a} (Q_j(z))^{r_j}, \quad R_2(z) = \prod_{j=a+1}^{b} (Q_j(z))^{r_j},$$
$$R_3(z) = \prod_{j=b+1}^{c} (Q_j(z))^{\lfloor \frac{r_j}{2} \rfloor}, \quad R_4(z) = \prod_{\substack{b+1 \leq j \leq c \\ r_j \text{ odd}}} Q_j(z).$$

Then, $P = R_1 \cdot R_2 \cdot (R_3)^2 \cdot R_4$.

Obviously, $R_3^2(x) \geq 0$ for real x. Also, the factors $Q_j(z)$ for $1 \leq j \leq a$ come in conjugate pairs because P has real coefficients and each Q_j has non-real coefficients. Therefore, $R_1(x) = |\hat{R}_1(x)|^2$, for some complex polynomial $\hat{R}_1$. Thus, we also have $R_1(x) \geq 0$ for real x.

Consider the sets $\Omega_+ = \{x \in \mathbb{R}^n : R_2(x) > 0\}$ and $\Omega_- = \{x \in \mathbb{R}^n : R_2(x) < 0\}$. If neither Ω_+ nor Ω_- are empty, then $\{x \in \mathbb{R}^n : R_2(x) = 0\}$ is the boundary between them. This is impossible, however, because $\dim_{\mathbb{R}}\{x \in \mathbb{R}^n : R_2(x) = 0\} < n-1$. Hence, either Ω_- or Ω_+ must be empty. The fact that R_2 is homogeneous and monic in z_1 implies that $R_2(1,0,\ldots,0) = 1$. Therefore, $\Omega_- = \emptyset$ and so $R_2(x) \geq 0$ for all $x \in \mathbb{R}^n$.

Since R_1, R_2, and R_3^2 are all nonnegative on real points, $\{P(x) > 0\} \subset \{R_4(x) > 0\}$ and $\{R_4(x) \geq 0\} \subset \{P(x) \geq 0\}$. These inclusions and Lemma 4.2 imply that one of these four sets satisfies the upper bound property if and only if they all do. Now, we are ready to examine the other four cases:

Case II. If $\deg R_4 = 0$, then R_4 is constant. Since $R_4(1,0,\ldots,0) = 1$, we have $R_4(z) \equiv 1$. Therefore, $\{R_4(x) > 0\} = \mathbb{R}^n$, which satisfies the upper bound property. Since $\{P(x) > 0\} = \mathbb{R}^n \setminus \{P(x) = 0\}$, E is not contained in a half-space.

Case III. If $\deg R_4 = 1$, then $\{R_4(x) > 0\}$ is contained in a half-space. Hence, $\{P(x) > 0\}$ is also contained in a half-space and it does not satisfy the upper bound property.

Case IV. If $\deg R_4 = 2$, then the fact that R_4 is homogeneous and $R_4(1,0,\ldots,0) = 1$ implies that $\{x_1^2 - \delta^2(x_2^2 + \cdots + x_n^2) > 0\} \subset \{R_4(x) > 0\}$ for some $\delta > 0$. Theorem 3 and Corollary 4.2 show that $\{R_4(x) > 0\}$ satisfies the linear bound property. Hence, $\{P(x) > 0\}$ also satisfies the linear bound property and therefore it is not contained in a half-space.

Case V. If $\deg R_4 \geq 3$, then Theorem 1.1 of [2] implies that $V = \{R_4(z) - t^2 = 0\}$ satisfies SRPL. Lemma 4.6 implies that $\{P(x) > 0\}$ satisfies the upper bound property and therefore it is not contained in a half-space.

Corollary 4.7. *The linear bound property, the upper bound property, and the no small functions condition are equivalent for sets of the form* $\{P(x) > 0\}$, *where* P *is a real homogeneous polynomial.*

While Theorem 4.3 classifies the linear bound property for all sets of the form $\{P(x) \geq 0\}$, where P is real and homogeneous, a general classification for nonhomogeneous polynomials is still unknown. The next section gives examples which illustrate the differences between homogeneous and nonhomogeneous polynomials. Also, very little is known about the linear bound property if E is the set where two or more polynomials are positive. For example, the set $\{(x,y) \in \mathbb{R}^2 : y(x-y)(x+y) \geq 0, x(x-2y)(x+2y) \geq 0\}$ consists of three wedges with vertices at the origin. This set is outward radial and it is not contained in a half-space, yet it is not known if it satisfies the linear bound property. Each of the sets $\{y(x-y)(x+y) \geq 0\}$ and $\{x(x-2y)(x+2y) \geq 0\}$ satisfy the linear bound property, but it is not true in general that if E_1 and E_2 satisfy the linear bound property, then $E_1 \cap E_2$ satisfies it also. For example, the sets $E_1 = \{(x+y)(x+2y)(2x+y) \geq 0\}$ and $E_2 = \{(x-y)(x-2y)(2x-y) \geq 0\}$ each satisfy the linear bound property, yet $E_1 \cap E_2$ is contained in a half-space and so it fails the linear bound property.

5 The Linear Extremal Function for Nonhomogeneous Real Varieties

This section explores the difficulties in extending Theorem 4.3 to nonhomogeneous polynomials. We will show that, in general, the linear bound property, the upper bound property, and the no small functions condition are not equivalent. We will begin by proving a generalization of Theorem 2.3.

Theorem 5.1. *Let $c, \delta \geq 0$ and let $E_c = \{x \in \mathbb{R}^n : x_1^2 - \delta^2\langle x', x'\rangle - c^2 \geq 0\}$, where $x' = (x_2, \ldots, x_n)$. Then*

$$\Lambda_{E_c}(z) = \sqrt{\left|\operatorname{Im}\sqrt{z_1^2 - \delta^2\langle z', \overline{z'}\rangle - c^2}\right|^2 + (1+\delta^2)\,|\operatorname{Im} z'|^2}. \tag{17}$$

Proof. Theorem 2.3 gives the result for E_0. We will prove Theorem 5.1 by using a multiple-valued polynomial transformation of E_c into E_0.

Assume that $u(z)$ is a competitor for Λ_{E_c}. If we define

$$v(z) = \max_{\pm}\left\{u\left(\pm\sqrt{z_1^2 + c^2}, z_2, \ldots, z_n\right)\right\},$$

then $v(z)$ is a competitor for the linear extremal function $\Lambda_{E_0}(z)$. Hence, we have

$$\Lambda_{E_c}(z) \leq \Lambda_{E_0}\left(\pm\sqrt{z_1^2 - c^2}, z_2, \ldots, z_n\right).$$

On the other hand, if $v(z)$ is a competitor for Λ_{E_0}, then the function

$$u(z) = \max_{\pm}\left\{v\left(\pm\sqrt{z_1^2 - c^2}, z_2, \ldots, z_n\right)\right\}$$

is a competitor for the linear extremal function Λ_{E_c}. Hence, we have

$$\Lambda_{E_0}\left(\pm\sqrt{z_1^2 - c^2}, z_2, \ldots, z_n\right) \leq \Lambda_{E_c}(z).$$

Therefore, $\Lambda_{E_c}(z) = \Lambda_{E_0}(\pm\sqrt{z_1^2 - c^2}, z_2, \ldots, z_n)$. Using the formula for Λ_{E_0} in 3 completes the proof. □

Corollary 5.2. *In $\mathbb{C}^1$, let $E = \mathbb{R} \setminus (-c, c)$ for $c > 0$. Then*

$$\Lambda_E(z) = |\operatorname{Im}\sqrt{z^2 - c^2}|.$$

Corollary 5.3. *Let $P(x)$ be a real nonhomogeneous polynomial of degree m, even. If $P_m(x_0) > 0$ for some $x_0 \in \mathbb{R}^n$, then the set $\{P(x) \geq 0\}$ satisfies the linear bound property.*

Proof. With a change of coordinates, $x_0 = (1,0,\ldots,0)$. Since P_m is the highest degree part of P, $m > 0$, and $P(1,0,\ldots,0) > 0$, there exist $\delta, c > 0$ such that $\{x_1^2 - \delta^2(x_2^2 + \cdots + x_n^2) - c^2 \geq 0\} \subset \{P(x) \geq 0\}$. Theorem 5.1 then implies that $\{P(x) \geq 0\}$ satisfies the linear bound property. □

When $P(x)$ is an odd degree polynomial, it is more difficult to decide whether or not $\{P(x) \geq 0\}$ satisfies the linear bound property. Consider, for example, the set $E = \{(x, y) \in \mathbb{R}^2 : xy(x - y) - 1 \geq 0\}$. This region does not contain a subset of the form $\{x_1^2 - \delta^2(x_2^2 + \cdots + x_n^2) - c^2 \geq 0\}$, and therefore it is impossible to apply Theorem 5.1 directly. We do, however, have the following theorem:

Theorem 5.4. *Let $P(x)$ be a nonhomogeneous real polynomial of degree $m \geq 3$, odd. Let $Q_1, \ldots Q_r$ be the irreducible factors of P_m. If P_m has no repeated irreducible factors and each Q_j has the property*

$$\dim_{\mathbb{R}}\{x \in \mathbb{R}^n : Q_j(x) = 0\} = \dim_{\mathbb{C}}\{z \in \mathbb{C}^n : Q_j(z) = 0\},$$

then the set $E = \{x : P(x) \geq 0\}$ satisfies the linear bound property.

Proof. Note that $P(x) - P_m(x)$ is a polynomial of degree at most $m - 1$, which is even. Hence, there exists $C > 0$ such that $P(x) - P_m(x) \geq -C(1 + |x|^2)^{\frac{m-1}{2}}$. If we let $\tilde{P}(x) = P_m(x) - C(1 + |x|^2)^{\frac{m-1}{2}}$, then $\tilde{P}$ is a polynomial of degree m with $P(x) \geq \tilde{P}(x)$ for all $x \in \mathbb{R}^n$. Since $P \geq \tilde{P}$, $\{\tilde{P}(x) \geq 0\} \subset \{P(x) \geq 0\}$. Therefore, it is sufficient to show that $\{\tilde{P}(x) \geq 0\}$ satisfies the linear bound property.

The advantage of defining $\tilde{P}$ this way is that, while $\{P(x) \geq 0\}$ is not outward radial in general, the set $\{\tilde{P}(x) \geq 0\}$ is. If $\tilde{P}(x) \geq 0$ and $r \geq 1$, then

$$\begin{aligned}
\tilde{P}(rx) &= P_m(rx) - C(1 + |rx|^2)^{\frac{m-1}{2}} \\
&\geq r^m P_m(x) - r^{m-1}C(1 + |x|^2)^{\frac{m-1}{2}} \\
&\geq r^{m-1}(P_m(x) - C(1 + |x|^2)^{\frac{m-1}{2}} \\
&= r^{m-1}\tilde{P}(x) \geq 0.
\end{aligned}$$

Lemma 4.6 and Theorem 1.1 of [2] imply $\{\tilde{P}(x) \geq 0\}$ satisfies the upper bound property. Since this set is outward radial, Lemma 4.5 implies that it satisfies the linear bound property. Hence, $\{P(x) \geq 0\}$ satisfies the linear bound property, too. □

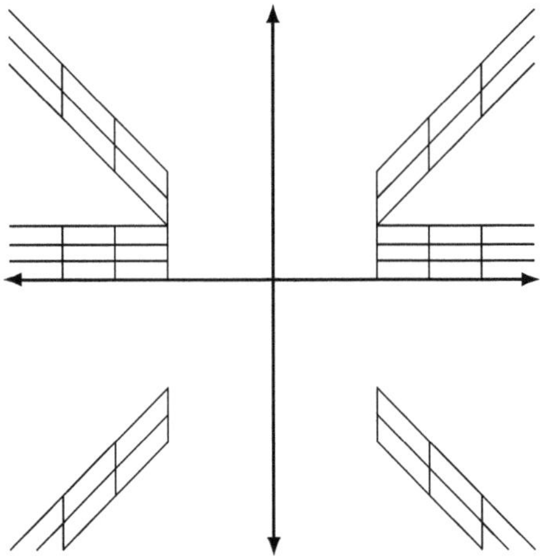

Fig. 1 The set E in Example 5.6

Theorem 5.4 shows that, for example, the set $\{xy(x-y)-1 \geq 0\}$ satisfies the linear bound property. In the case where the highest degree homogeneous part of $P(x)$ has repeated factors, e.g. $\{x^3y(x-y)-1 \geq 0\}$, Theorem 5.4 cannot be applied. Even though Theorem 4.3 implies $\{x^3y(x-y) \geq 0\}$ satisfies the linear bound property, it is not known whether or not the same holds for $\{x^3y(x-y)-1 \geq 0\}$.

One of the consequences of Theorem 4.3 is that, for homogeneous polynomials, the linear bound property, the upper bound property, and the no small functions condition are equivalent for the set $\{P(x) > 0\}$. The next pair of examples shows that this equivalence is not true in general.

Example 5.5. The set $E = \{x_1^3 + x_2^2 \geq 0\}$ fails the upper bound property but satisfies the no small functions condition.

This was proven earlier in Corollary 3.5.

Example 5.6. The set

$$E = \{(x_1, x_2) \in \mathbb{R}^2 : |x_1| \geq 2\} \cap$$
$$(\{0 \leq x_2 \leq 1\} \cup \{-x_1 - 1 \leq x_2 \leq -x_1\} \cup \{x_1 - 1 \leq x_2 \leq x_1\})$$

satisfies the upper bound property but fails the linear bound property.

The set E is not contained in a half-space nor is it outward radial. It consists of three strips with portions missing near the origin (see Fig. 1 above). The union of these three strips, including the parts with $\{|x_1| < 2\}$, also satisfies the upper bound property but fails the linear bound property. The proof is less complicated, however, if we do not have to consider the intersection of these strips. Therefore, we restrict the set E to the region outside of $\{|x_1| < 2\}$.

To prove this, we first need the following lemmas. The first is used to show that $\Lambda_E(z)$ is bounded. The second is used to show that $\Lambda_E(z)$ does not have a linear bound.

Lemma 5.7. *If* $(x_1, t) \in E_2 = \{(x_1, t) : |x_1| \geq 2$ *and* $t^2 \leq x_1^2 - 1\}$ *and* z_2 *is a root of* $z_2^3 - z_1^2 z_2 + t^2 = 0$*, then* $(x_1, z_2) \in E$.

Lemma 5.8. *If* $(x_1, x_2) \in E$*, then* $0 \leq x_1^2 x_2 - x_2^3 \leq (2|x_1| + 1)^2$.

Proof (Proof of Lemma 5.7). For $(x_1, t) \in E_2$, define $q_{(x_1,t)}(z_2) = z_2^3 - x_1^2 z_2 + t^2$. Evaluating this polynomial on the boundaries of E gives:

$$q_{(x_1,t)}(-x_1) = t^2 \; q_{(x_1,t)}(-x_1 - 1) = -2x_1^2 - 3x_1 - 1 + t^2$$
$$q_{(x_1,t)}(0) = t^2 \; q_{(x_1,t)}(1) = 1 - x_1^2 + t^2$$
$$q_{(x_1,t)}(x_1) = t^2 \; q_{(x_1,t)}(x_1 - 1) = -2x_1^2 + 3x_1 - 1 + t^2 \tag{18}$$

Since $t^2 \leq x_1^2 - 1$, we have $q_{(x_1,t)}(1) \leq 0$. We also have $q_{(x_1,t)}(0) = t^2 \geq 0$. Therefore, $q_{(x_1,t)}(z_2)$ has a real root with $0 \leq z_2 \leq 1$.

Also note that for $|x_1| \geq 2$,

$$\begin{aligned} q_{(x_1,t)}(-x_1 - 1) &= -2x_1^2 - 3x_1 - 1 + t^2 \\ &\leq -2|x_1|^2 + 3|x_1| - 1 + t^2 \\ &\leq -2|x_1|^2 + 3|x_1| - 1 + |x_1|^2 - 1 \\ &= -|x_1|^2 + 3|x_1| - 2 \\ &= -(|x_1| - 1)(|x_1| - 2) \leq 0 \end{aligned}$$

We also have $q_{(x_1,t)}(-x_1) = t^2 \geq 0$. Therefore, $q_{(x_1,t)}(z_2)$ has a real root with $-x_1 - 1 \leq z_2 \leq -x_1$. A similar calculation shows that the third root of $q_{(x_1,t)}(z_2)$ is a real value with $x_1 - 1 \leq z_2 \leq x_1$. Since $0 \leq z_2 \leq 1$ or $x_1 - 1 \leq z_2 \leq x_1$ or $-x_1 - 1 \leq z_2 \leq -x_1$, we must have $(x_1, z_2) \in E$. □

Proof of Lemma 5.8. Figure 1 shows that E is bounded by the lines $\{x_2 = 0\}$, $\{x_1 + x_2 = 0\}$, and $\{x_1 - x_2 = 0\}$. Because of this, it is easy to see that

$$E \subset \{(x_1, x_2) \in \mathbb{R}^2 : 0 \leq x_1^2 x_2 - x_2^3\}.$$

Since E consists of portions of three strips in the plane, we have three cases:

Case I. $0 \leq x_2 \leq 1$, so $x_1 - 1 \leq x_1 - x_2 \leq x_1$ and $x_1 \leq x_1 + x_2 \leq x_1 + 1$. Therefore, $x_1^2 x_2 - x_2^3 \leq |x_1 - x_2|\,|x_1 + x_2|\,|x_2| \leq (2|x_1| + 1)^2$.

Case II. $-x_1 - 1 \leq x_2 \leq -x_1$, so $2x_1 \leq x_1 - x_2 \leq 2x_1 + 1$ and $-1 \leq x_1 + x_2 \leq 0$. Therefore, $x_1^2 x_2 - x_2^3 \leq |x_1 - x_2|\,|x_1 + x_2|\,|x_2| \leq (2|x_1| + 1)^2$.

Case III. $x_1 - 1 \leq x_2 \leq x_1$, so $0 \leq x_1 - x_2 \leq 1$ and $2x_1 - 1 \leq x_1 + x_2 \leq 2x_1$. Therefore, $x_1^2 x_2 - x_2^3 \leq |x_1 - x_2|\,|x_1 + x_2|\,|x_2| \leq (2\,|x_1| + 1)^2$.

In all three cases, when $(x_1, x_2) \in E$, $0 \leq x_1^2 x_2 - x_2^3 \leq (2\,|x_1| + 1)^2$.

Proof of Example 5.6. First, we will use Lemma 5.7 to show that E_2 satisfies the upper bound property.

The roots of $z_2^3 - z_1^2 z_2 = 0$ are $z_2 = 0, \pm z_1$. Therefore, the roots of $q_{(z_1,\tau)}(z_2) = z_2^3 - z_1^2 z_2 + \tau^2 = 0$ behave like $|z_2| \leq |z_1| + o(|(z_1, \tau)|)$. Hence, there exist constants $k_1, k_2 > 0$ such that the roots of $z_2^3 - z_1^2 z_2 + \tau^2 = 0$ satisfy $|z_2| \leq k_1|(z_1, \tau)| + k_2$.

Let $u(z_1, z_2)$ be plurisubharmonic with $u(z_1, z_2) \leq |z| + o(|z|)$ and $u(z) \leq 0$ for $z \in E$. Define a plurisubharmonic function $v(z_1, \tau)$ by

$$v(z_1, \tau) = \max_{z_2}\{u(z_1, z_2) : z_2^3 - z_1^2 z_2 + \tau^2 = 0\}.$$

Then, $v(z_1, \tau) \leq k_1|(z_1, \tau)| + o(|(z_1, \tau)|)$ and Lemma 5.7 gives $v \leq 0$ on the set

$$E_2 = \{(x_1, t) \in \mathbb{R}^2 : |x_1| \geq 2, 1 - x_1^2 + t^2 \leq 0\}.$$

From Theorem 5.1, we have $\Lambda_{E_2}(z_1, \tau) \leq A|(z_1, \tau)| + B$. Hence, $v(z_1, \tau) \leq k_1 A|(z_1, \tau)| + k_1 B$, and therefore,

$$u(z) \leq k_1 A \sqrt{|z_1|^2 + |z_1^2 z_2 - z_2^3|} + k_1 B,$$

where k_1, A, and B are independent of u. Therefore, E satisfies the upper bound property.

Next, we show that Λ_E does not have a linear bound. Consider the function $u_a(z_1, z_2)$ for $a > 0$:

$$u_a(z_1, z_2) := \max_{\pm}\left\{\left|\operatorname{Im}\sqrt{z_1^2 \pm a\sqrt{z_1^2 z_2 - z_2^3} + a^2 + a}\right|\right\}.$$

Notice that $u_a(z_1, z_2) \leq |z_1| + o(|z|)$. Also, for $(x_1, x_2) \in E$, Lemma 5.8 gives $x_1^2 x_2 - x_2^3 \geq 0$, so $\pm a\sqrt{x_1^2 x_2 - x_2^3} \in \mathbb{R}$. Lemma 5.8 also gives bounds for the outer square root in the definition of $u_a(z_1, z_2)$:

$$\begin{aligned} x_1^2 \pm a\sqrt{x_1^2 x_2 - x_2^3} + a^2 + a &\geq x_1^2 - a\sqrt{(2\,|x_1| + 1)^2} + a^2 + a \\ &= |x_1|^2 - a(2\,|x_1| + 1) + a^2 + a \\ &= |x_1|^2 - 2a\,|x_1| + a^2 \\ &= (|x_1| - a)^2 \geq 0. \end{aligned}$$

Therefore, $\pm\sqrt{z_1^2 \pm a\sqrt{z_1^2 z_2 - z_2^3} + a^2 + a} \in \mathbb{R}$, so $u_a(x_1, x_2) = 0$ on the region E. Hence, $u_a(z_1, z_2)$ is a competitor for the linear extremal function, $\Lambda_E(z_1, z_2)$.

For $y > 0$, consider the functions $u_{y^{3/2}}(z_1, z_2)$ evaluated at $(0, y)$.

$$\begin{aligned} u_{y^{3/2}}(0, y) &= \max_{\pm}\left\{\left|\operatorname{Im}\sqrt{0 \pm y^{3/2}\sqrt{0 - y^3} + (y^{3/2})^2 + y^{3/2}}\right|\right\} \\ &= \max_{\pm}\left\{\left|\operatorname{Im}\sqrt{y^3 + y^{3/2} \pm i y^3}\right|\right\} \\ &= O(|y|^{3/2}) \end{aligned}$$

Therefore, $\Lambda_E(z)$ does not have a linear bound.

Examples 5.5 and 5.6 show that the linear bound property, the upper bound property, and the no small functions conditions are not equivalent in general. The sets in these examples, however, are not outward radial. Theorem 2.3 shows that these three properties are equivalent for outward radial sets of the form $E = \{P(x) > 0\}$, where P is a homogeneous real polynomial, but this might not describe all outward radial sets. This leads us to ask the following questions:

Problem 5.9. Which outward radial sets satisfy the upper bound property?

Problem 5.10. Are the upper bound property and the no small functions condition equivalent for outward radial sets?

Problem 5.11. If $E \subset \mathbb{R}^2$ is the union of three cones with vertex at the origin but not contained in a half-space, does E satisfy the linear bound property?

Acknowledgements Portions of this work were part of the author's doctoral thesis at the University of Michigan written under the direction of B. A. Taylor in 1998 [1].

References

1. Bainbridge, David, Phragmén-Lindelöf estimates for plurisubharmonic functions of linear growth, Thesis, University of Michigan, 1998
2. Braun, Rüdiger, Meise, Reinhold, Taylor, B. A., A radial Phragmén-Lindelöf Estimate for Plurisubharmonic Functions on Algebraic Varieties, Ann. Polon. Math. **72**, 1999, no. 2, 159–179
3. R. Meise, B. A. Taylor, D. Vogt: Phragmén-Lindelöf principles for algebraic varieties, J. of the Amer. Math. Soc., **11** (1998), 1–39
4. Meise, R. Taylor, B. A., Vogt, D., Extremal plurisubharmonic functions of linear growth on algebraic varieties, Math. Z., **219**, 1995. Vol 4, 515–537
5. Klimek, Maciej, Pluripotential theory, London Mathematical Society Monographs. New Series, **6**, Oxford Science Publications, The Clarendon Press Oxford University Press, New York, 1991
6. Sibony, Nessim and Wong, Pit Mann, Some results on global analytic sets, Lecture Notes in Math., Séminaire Pierre Lelong, Henri Skoda (Analyse). Années 1978/79, **822**, 221–237

Mahonian Partition Identities via Polyhedral Geometry

Matthias Beck, Benjamin Braun, and Nguyen Le

Dedicated to the memory of Leon Ehrenpreis

Abstract In a series of papers, George Andrews and various coauthors successfully revitalized seemingly forgotten, powerful machinery based on MacMahon's Ω operator to systematically compute generating functions $\sum_{\lambda \in P} z_1^{\lambda_1} \dots z_n^{\lambda_n}$ for some set P of integer partitions $\lambda = (\lambda_1, \dots, \lambda_n)$. Our goal is to geometrically prove and extend many of Andrews et al.'s theorems, by realizing a given family of partitions as the set of integer lattice points in a certain polyhedron.

Key words Composition • Integer lattice point • MacMahon's Ω operator • Partition identities • Polyhedral cone

Mathematics Subject Classification (2000): Primary 11P84; Secondary 05A15, 05A17, 11P21

1 Introduction

In a series of papers starting with [1], George Andrews and various coauthors successfully revitalized seemingly forgotten, powerful machinery based on MacMahon's Ω operator [15] to systematically compute generating functions related to various families of integer partitions. Andrews et al.'s papers concern

M. Beck (✉) • N. Le
Department of Mathematics, San Francisco State University, San Francisco, CA 94132, USA
e-mail: beck@math.sfsu.edu; nguyenle2005usa@yahoo.com

B. Braun
Department of Mathematics, University of Kentucky, Lexington, KY 40506-0027, USA
e-mail: benjamin.braun@uky.edu

H.M. Farkas et al. (eds.), *From Fourier Analysis and Number Theory to Radon Transforms and Geometry*, Developments in Mathematics 28, DOI 10.1007/978-1-4614-4075-8_3,

generating functions of the form

$$f_P(z_1,\ldots,z_n) := \sum_{\lambda\in P} z_1^{\lambda_1}\cdots z_n^{\lambda_n} \quad\text{and}\quad f_P(q) := f_P(q,\ldots,q) = \sum_{\lambda\in P} q^{\lambda_1+\cdots+\lambda_n},$$

for some set P of *partitions* $\lambda = (\lambda_1,\ldots,\lambda_n)$; i.e., we think of the integers $\lambda_n \geq \cdots \geq \lambda_1 \geq 0$ as the *parts* when some integer k is written as $k = \lambda_1 + \cdots + \lambda_n$. If we do not force an order onto the λ_j's, we call λ a *composition* of k. Below is a sample of some of these striking results.

Theorem 1 (Andrews [2]). *Let*

$$P_r := \left\{\lambda : \sum_{j=0}^{t}(-1)^j\binom{t}{j}\lambda_{k+j} \geq 0 \text{ for } k\geq 1,\ 1\leq t\leq r\right\}$$

(where we set undefined λ_j's zero). Then

$$f_{P_r}(q) = \prod_{j=1}^{\infty}\frac{1}{1-q^{\binom{j+r-1}{r}}}.$$

In words, the number of partitions of an integer k satisfying the "higher-order difference conditions" in P_r equals the number of partitions of k into parts that are r'th-order binomial coefficients.

Theorem 2 (Andrews–Paule–Riese [3]). *Let $n \geq 3$ and*

$$\tau := \{(\lambda_1,\ldots,\lambda_n)\in\mathbb{Z}^n : \lambda_n\geq\cdots\geq\lambda_1\geq 1 \text{ and } \lambda_1+\cdots+\lambda_{n-1} > \lambda_n\},$$

the set of all "n-gon partitions." Then

$$f_\tau(q) = \frac{q^n}{(1-q)(1-q^2)\cdots(1-q^n)} - \frac{q^{2n-2}}{(1-q)(1-q^2)(1-q^4)(1-q^6)\cdots(1-q^{2n-2})}.$$

More generally,

$$f_\tau(z_1,\ldots,z_n) = \frac{Z_1}{(1-Z_1)(1-Z_2)\cdots(1-Z_n)} - \frac{Z_1Z_n^{n-2}}{(1-Z_n)(1-Z_{n-1})(1-Z_{n-2}Z_n)(1-Z_{n-3}Z_n^2)\cdots(1-Z_1Z_n^{n-2})},$$

where $Z_j := z_jz_{j+1}\cdots z_n$ for $1\leq j\leq n$.

The composition analogue of Theorem 2 was inspired by a problem of Hermite [18, Ex. 31], which is essentially the case $n = 3$ of the following.

Theorem 3 (Andrews–Paule–Riese [4]). *Let*

$$H := \left\{(\lambda_1, \ldots, \lambda_n) \in \mathbb{Z}^n_{>0} : \lambda_1 + \cdots + \widehat{\lambda_j} + \cdots + \lambda_n \geq \lambda_j \text{ for all } 1 \leq j \leq n\right\}.$$

Then

$$f_H(q) = \frac{q^n}{(1-q)^n} - n\,\frac{q^{2n-1}}{(1-q)^n(1+q)^{n-1}}.$$

A natural question is whether there exist "full generating function" versions of Theorems 1 and 3, in analogy with Theorem 2; we will show that such versions (Theorems 6 and 7 below) follow effortlessly from our approach. (Xin [21, Example 6.1] previously computed a full generating function related to Theorem 3.)

Our main goal is to prove these theorems geometrically, and more, by realizing a given family of partitions as the set of integer lattice points in a certain polyhedron. This approach is not new: Pak illustrated in [16, 17] how one can obtain bijective proofs by realizing when both sides of a partition identity are generating functions of lattice points in unimodular cones (which we will define below); this included most of the identities appearing in [2], including Theorem 1. Corteel et al. [13] implicitly used the extreme-ray description of a cone (see Lemma 4 below) to derive product formulas for partition generating functions, including those appearing in [2]. Beck et al. [7] used triangulations of cones to extend results of Andrews et al. [5] on "symmetrically constrained compositions." However, we feel that each of these papers only scratched the surface of a polyhedral approach to partition identities, and we see the current paper as a further step towards a systematic study of this approach.

While the Ω-operator approach to partition identities is elegant and powerful (not to mention useful in the search for such identities), we see several reasons for pursuing a geometric interpretation of these results. As discussed in [11], partition analysis and the Ω operator are useful tools for studying partitions and compositions defined by linear constraints, which is equivalent to studying integer points in polyhedra. An explicit geometric approach to these problems often reveals interesting connections to geometric combinatorics, such as the connections and conjectures discussed in Sects. 6 and 7 below. Also, one of the great appeals of partition analysis is that it is automatic; Andrews discusses this in the context of applying the Ω operator to the four-dimensional case of lecture-hall partitions in [1]:

> The point to stress here is that we have carried off the case $j = 4$ with no effective combinatorial argument or knowledge. In other words, the entire problem is reduced by Partition Analysis to the factorization of an explicit polynomial.

As we hope to show, the geometric perspective can often provide a clear view of sometimes mysterious formulas that arise from the symbolic manipulation of the Ω operator.

2 Polyhedral Cones and Their Lattice Points

We use the standard abbreviation $\mathbf{z}^{\mathbf{m}} := z_1^{m_1} \cdots z_n^{m_n}$ for two vectors $\mathbf{z}$ and $\mathbf{m}$. Given a subset K of $\mathbb{R}^n$, the *(integer-point) generating function* of K is

$$\sigma_K(z_1, \ldots, z_n) := \sum_{\mathbf{m} \in K \cap \mathbb{Z}^n} \mathbf{z}^{\mathbf{m}} .$$

We will often encounter subsets that are cones, where a *(polyhedral) cone* C is the intersection of finitely many (open or closed) half-spaces whose bounding hyperplanes contain the origin. (Thus, the cones appearing in this chapter will not all be closed but in general partially open.) A closed cone has the alternative description (and this equivalence is nontrivial [22]) as the nonnegative span of a finite set of vectors in $\mathbb{R}^n$, the *generators* of C.

An n-dimensional cone in $\mathbb{R}^n$ is *simplicial* if we only need n half-spaces to describe it. All of our cones will be *pointed*, i.e., they do not contain lines. The following exercise in linear algebra shows how to switch between the generator and half-space descriptions of a simplicial cone.

Lemma 4. *Let* $\mathbf{A}$ *be the inverse matrix of* $\mathbf{B} \in \mathbb{R}^{n \times n}$. *Then*

$$\{\mathbf{x} \in \mathbb{R}^n : \mathbf{A}\,\mathbf{x} \geq \mathbf{0}\} = \{\mathbf{B}\,\mathbf{y} : \mathbf{y} \geq \mathbf{0}\},$$

where each inequality is understood componentwise.

The (integer-point) generating function of a simplicial cone $C \subset \mathbb{R}^n$ can be computed from first principles when C is *rational*, i.e., its generators can be chosen in $\mathbb{Z}^n$. A closed cone C is *unimodular* if its generators form a basis of $\mathbb{Z}^n$; for unimodular cones, which is all we will need in what follows, we have the following simple lemma (for much more general results, see, e.g., [8, Chap. 3]).

Lemma 5. *Suppose* $C = \sum_{j=1}^{k} \mathbb{R}_{\geq 0} \mathbf{v}_j + \sum_{i=k+1}^{n} \mathbb{R}_{>0} \mathbf{v}_i$ *is a unimodular cone in* $\mathbb{R}^n$ *generated by* $\mathbf{v}_1, \ldots, \mathbf{v}_n \in \mathbb{Z}^n$. *Then*

$$\sigma_C(z_1, \ldots, z_n) = \frac{\prod_{i=k+1}^{n} \mathbf{z}^{\mathbf{v}_i}}{\prod_{j=1}^{n} (1 - \mathbf{z}^{\mathbf{v}_j})} .$$

3 Unimodular Cones

Recall from Theorem 1 that

$$P_r = \left\{ \lambda : \sum_{j=0}^{t} (-1)^j \binom{t}{j} \lambda_{k+j} \geq 0 \text{ for } k \geq 1,\ 1 \leq t \leq r \right\}$$

(where we set undefined λ_j's zero). Let

$$P_r^n := \left\{(\lambda_1,\ldots,\lambda_n) \in \mathbb{Z}^n : \sum_{j=0}^{t}(-1)^j\binom{t}{j}\lambda_{k+j} \geq 0 \text{ for } 1 \leq k \leq n,\ 1 \leq t \leq r\right\}$$

consist of all partitions in P_r with at most n parts. As a warm-up example, we will compute the (full) generating function of P_r^n:

Theorem 6.

$$f_{P_r^n}(z_1,\ldots,z_n)$$
$$= \frac{1}{(1-z_1)\left(1-z_1^r z_2\right)\left(1-z_1^{\binom{r+1}{r-1}}z_2^r z_3\right)\left(1-z_1^{\binom{r+2}{r-1}}z_2^{\binom{r+1}{r-1}}z_3^r z_4\right)\cdots \left(1-z_1^{\binom{r+n-2}{r-1}}z_2^{\binom{r+n-3}{r-1}}\cdots z_{n-1}^r z_n\right)}.$$

Note that Theorem 1 follows upon setting $z_1 = \cdots = z_n = q$, using the identity

$$\binom{r+j-2}{r-1} + \binom{r+j-3}{r-1} + \cdots + r + 1 = \binom{r+j-1}{r},$$

and taking $n \to \infty$.

Proof. It is easy to see that the inequalities

$$\sum_{j=0}^{t}(-1)^j\binom{t}{j}\lambda_{k+j} \geq 0 \text{ for } 1 \leq k \leq n,\ 1 \leq t \leq r,$$

which define P_r^n, are implied by the inequalities for $t = r$. Thus, the cone containing P_r^n as its integer lattice points is

$$K := \left\{(x_1,\ldots,x_n) \in \mathbb{R}^n : \sum_{j=0}^{r}(-1)^j\binom{r}{j}x_{k+j} \geq 0 \text{ for } 1 \leq k \leq n\right\}$$

$$= \left\{\begin{bmatrix} 1 & r & \binom{r+1}{r-1} & \binom{r+2}{r-1} & \cdots & \binom{r+n-2}{r-1} \\ 0 & 1 & r & \binom{r+1}{r-1} & \cdots & \binom{r+n-3}{r-1} \\ 0 & 0 & 1 & r & \cdots & \binom{r+n-4}{r-1} \\ \vdots & & \ddots & \ddots & \ddots & \vdots \\ 0 & & & 0 & 1 & r \\ 0 & \cdots & & & 0 & 1 \end{bmatrix}\mathbf{y} : y_1,\ldots,y_n \geq 0\right\}$$

(whose generators we can compute, e.g., with the help of Lemma 4). Thus, K is unimodular and, by Lemma 5,

$$\sigma_K(z_1,\ldots,z_n) = \frac{1}{(1-z_1)\left(1-z_1^r z_2\right)\left(1-z_1^{\binom{r+1}{r-1}} z_2^r z_3\right)\left(1-z_1^{\binom{r+2}{r-1}} z_2^{\binom{r+1}{r-1}} z_3^r z_4\right)\cdots \left(1-z_1^{\binom{r+n-2}{r-1}} z_2^{\binom{r+n-3}{r-1}}\cdots z_{n-1}^r z_n\right)}.$$

The idea behind this approach towards Theorem 1 can be found, in disguised form, in [13, 16]. See also [10, 12] for bijective approaches to Theorem 1 and its asymptotic consequences. We included this proof here in the interest of a self-contained exposition and also because none of [2, 13, 16] contains a full generating function version of (analogues of) Theorem 1.

4 Differences of Two Cones

The key idea behind the proof of Theorem 2 is to observe that the nonsimplicial cone

$$K := \{(x_1,\ldots,x_n) \in \mathbb{R}^n : x_n \geq \cdots \geq x_1 > 0 \text{ and } x_1 + \cdots + x_{n-1} > x_n\},$$

whose integer lattice points form Andrews–Paule–Riese's set τ of n-gon partitions, can be written as a difference $K = K_1 \setminus K_2$ of two simplicial cones. Specifically, set

$$K_1 := \{(x_1,\ldots,x_n) \in \mathbb{R}^n : x_n \geq \cdots \geq x_1 > 0\}$$
$$= \left\{ \begin{bmatrix} 1 & 0 & 0 & \cdots & 0 & 0 \\ 1 & 1 & 0 & \cdots & 0 & 0 \\ 1 & 1 & 1 & \cdots & 0 & 0 \\ \vdots & \vdots & \vdots & \ddots & \vdots & \vdots \\ 1 & 1 & 1 & \cdots & 1 & 0 \\ 1 & 1 & 1 & \cdots & 1 & 1 \end{bmatrix} \mathbf{y} : \begin{array}{l} y_1 > 0, \\ y_2,\ldots,y_n \geq 0 \end{array} \right\}$$

and

$$K_2 := \{(x_1,\ldots,x_n) \in \mathbb{R}^n : x_n \geq \cdots \geq x_1 > 0 \text{ and } x_1 + \cdots + x_{n-1} \leq x_n\}$$
$$= \left\{ \begin{bmatrix} 1 & 0 & 0 & \cdots & 0 & 0 \\ 1 & 1 & 0 & \cdots & 0 & 0 \\ 1 & 1 & 1 & \cdots & 0 & 0 \\ \vdots & \vdots & \vdots & \ddots & \vdots & \vdots \\ 1 & 1 & 1 & \cdots & 1 & 0 \\ n-1 & n-2 & n-3 & \cdots & 1 & 1 \end{bmatrix} \mathbf{y} : \begin{array}{l} y_1 > 0, \\ y_2,\ldots,y_n \geq 0 \end{array} \right\}$$

(whose generators we can compute, e.g., with the help of Lemma 4). One can see immediately from the generator matrices that both K_1 and K_2 are unimodular. (In a geometric sense, this is suggested by the form of the identity in Theorem 2. A similar simplification-through-taking-differences phenomenon is described in the fifth "guideline" of Corteel et al. [11], which inspired our proof.) By Lemma 5

$$\sigma_{K_1}(z_1,\ldots,z_n) = \frac{z_1\cdots z_n}{(1-z_n)(1-z_{n-1}z_n)\cdots(1-z_1\cdots z_n)}$$

and

$$\sigma_{K_2}(z_1,\ldots,z_n)$$
$$= \frac{z_1\cdots z_{n-1}z_n^{n-1}}{\left(1-z_1\cdots z_{n-1}z_n^{n-1}\right)\left(1-z_2\cdots z_{n-1}z_n^{n-2}\right)\left(1-z_3\cdots z_{n-1}z_n^{n-3}\right)\cdots \left(1-z_{n-1}z_n\right)\left(1-z_n\right)}$$
$$= \frac{Z_1Z_n^{n-2}}{(1-Z_n)(1-Z_{n-1})(1-Z_{n-2}Z_n)(1-Z_{n-3}Z_n^2)\cdots(1-Z_1Z_n^{n-2})},$$

and the identity $\sigma_K(z_1,\ldots,z_n) = \sigma_{K_1}(z_1,\ldots,z_n) - \sigma_{K_2}(z_1,\ldots,z_n)$ completes the proof. □

5 Differences of Multiple Cones

The "cone behind" Theorem 3 is

$$K := \left\{(x_1,\ldots,x_n)\in\mathbb{R}^n_{>0} : x_j \le x_1+\cdots+\widehat{x_j}+\cdots+x_n \ \text{ for all } \ 1\le j\le n\right\};$$

Theorem 3 follows from the following result upon setting $z_1 = \cdots = z_n = q$.

Theorem 7.

$$\sigma_K(z_1,\ldots,z_n) = \frac{z_1\cdots z_n}{(1-z_1)\cdots(1-z_n)} - \sum_{k=1}^{n}\frac{z_1\cdots z_{k-1}z_k^n z_{k+1}\cdots z_n}{(1-z_k)\prod_{\substack{j=1\\ j\ne k}}^{n}(1-z_kz_j)}.$$

Proof. Let ϵ_j denote the jth unit vector in $\mathbb{R}^n$. Observe that the nonsimplicial cone K is expressible as a difference $K = O\setminus\bigcup_{k=1}^n C_k$, where $O := \sum_{j=1}^n \mathbb{R}_{>0}\,\epsilon_j$ and C_k is the cone

$$C_k := \left\{(x_1,\ldots,x_n)\in\mathbb{R}^n_{>0} : x_k > x_1+\cdots+\widehat{x_j}+\cdots+x_n\right\}$$
$$= \mathbb{R}_{>0}\,\epsilon_k + \sum_{\substack{j=1\\ j\ne k}}^{n}\mathbb{R}_{>0}\left(\epsilon_j+\epsilon_k\right)$$

Note that if $i \neq j$, then $C_i \cap C_j = \emptyset$. Thus, the closure of K is "almost" the positive orthant O, except that we have to exclude points in O that can only be written as a linear combination that requires a single $\mathbf{e}_k$ (as opposed to a linear combination of the vectors $\mathbf{e}_j + \mathbf{e}_k$). (A similar simplification-through-taking-differences phenomenon appeared in the original proof of Theorem 3.) In generating function terms, this set difference gives, by Lemma 5,

$$\begin{aligned}\sigma_K(z_1,\ldots,z_n) &= \sigma_O(z_1,\ldots,z_n) - \sum_{k=1}^{n} \sigma_{C_k}(z_1,\ldots,z_n)\\ &= \frac{z_1\cdots z_n}{(1-z_1)\cdots(1-z_n)} - \sum_{k=1}^{n} \frac{z_1\cdots z_{k-1} z_k^n z_{k+1}\cdots z_n}{(1-z_k)\prod_{\substack{j=1\\ j\neq k}}^{n}(1-z_k z_j)}. \qquad \square\end{aligned}$$

Three remarks on this theorem are in order. First, as already mentioned, Xin [21, Example 6.1] previously computed a different full generating function related to Theorem 3; Xin's generating function handles nonnegative, rather than positive, k-gon partitions. Second, the cone K is related to the second hypersimplex, a well-known object in geometric combinatorics (see Sect. 7 for more details).

Third, K is a suitable candidate for the "symmetrically constrained" approach in [7]; however, one should expect that this approach would give a different form for the generating function $\sigma_K(z_1,\ldots,z_n)$ from the one given in Theorem 7. The symmetrically constrained approach produces a triangulation of the cone K that is invariant under permutation of the standard basis vectors in $\mathbb{R}^n$ and then uses this triangulation to express $\sigma_K(z_1,\ldots,z_n)$ as a positive sum of rational generating functions for these cones (after some geometric shifting). The terms in this sum will all have $\frac{1}{1-z_1z_2\cdots z_n}$ as a factor, as each of the simplicial cones in the triangulation of K will have the all-ones vector as a ray generator; this will clearly produce a different form from that in Theorem 7.

6 Cayley Compositions

A *Cayley composition* is a composition $\lambda = (\lambda_1,\ldots,\lambda_{j-1})$ that satisfies $1 \le \lambda_1 \le 2$ and $1 \le \lambda_{i+1} \le 2\lambda_i$ for $1 \le i \le j-2$. Thus, the Cayley compositions with $j-1$ parts are precisely the integer points in

$$C_j := \left\{(\lambda_1,\ldots,\lambda_{j-1}) \in \mathbb{Z}_{>0}^{j-1} : \lambda_1 \le 2 \text{ and } \lambda_i \le 2\lambda_{i-1} \text{ for all } 2 \le i \le j-1\right\}.$$

Our apparent shift in indexing maintains continuity between our statements and [6], where Cayley compositions always begin with a $\lambda_0 = 1$ part. Let $f_{C_j}(z_1,\ldots,z_{j-1})$ be the generating function for C_j. The following theorem is quite surprising.

Theorem 8 (Andrews–Paule–Riese–Strehl [6]). *Let*

$$C_j := \left\{(\lambda_1, \ldots, \lambda_{j-1}) \in \mathbb{Z}_{>0}^{j-1} : \lambda_1 \leq 2 \text{ and } \lambda_i \leq 2\lambda_{i-1} \text{ for all } 2 \leq i \leq j-1\right\}.$$

Then for $j \geq 2$,

$$f_{C_j}(1,1,\ldots,1,q) = \sum_{h=1}^{j-2} \frac{b_{j-h-1}(-1)^{h-1} q^{2^h-1}}{(1-q)(1-q^2)(1-q^4)\cdots(1-q^{2^{h-1}})} + \frac{(-1)^j q^{2^{j-1}-1}(1-q^{2^{j-1}})}{(1-q)(1-q^2)(1-q^4)\cdots(1-q^{2^{j-2}})}$$

where b_k is the coefficient of q^{2^k-1} in the power series expansion of

$$\frac{1}{1-q} \prod_{m=0}^{\infty} \frac{1}{1-q^{2^m}}.$$

Theorem 8 is derived as a consequence of the following recurrence relation obtained via MacMahon's Ω calculus.

Theorem 9 (Andrews–Paule–Riese–Strehl [6]).

$$f_{C_j}\left(z_1,\ldots,z_{j-1}\right) = \frac{z_{j-1}}{1-z_{j-1}}\left(f_{C_{j-1}}\left(z_1,\ldots,z_{j-2}\right) - f_{C_{j-1}}\left(z_1,\ldots,z_{j-3},z_{j-2}z_{j-1}^2\right)\right).$$

Once this formula is obtained, the proof of Theorem 8 in [6] proceeds by repeatedly iterating the recurrence, specialized to $f_{C_j}(1,\ldots,1,q)$. The final step is to argue that the sum of rational functions in Theorem 8, as analytic functions, must exhibit cancellation. We remark that Corteel et al. [11, Sect. 3] gave an alternative proof of Theorem 9.

Via geometry, we can shed light on the initial recurrence relation from three perspectives. First, we recognize that the recurrence reflects expressing C_j as a difference of two subspaces of $\mathbb{R}^{j-1}$ defined by linear constraints.

Proof (First proof of Theorem 9).
As a subspace of $\mathbb{R}^{j-1}$, $C_j = K_{1,j} \setminus K_{2,j}$ where

$$K_{1,j} := \{(x_1,\ldots,x_{j-1}) \in \mathbb{R}^{j-1} : 1 \leq x_1 \leq 2, \\ 1 \leq x_{i+1} \leq 2x_i \text{ for } 1 \leq i \leq j-3, \text{ and } 1 \leq x_{j-1}\}$$

and

$$K_{2,j} := \{(x_1,\ldots,x_{j-1}) \in \mathbb{R}^{j-1} : 1 \le x_1 \le 2, \\ x_{i+1} \le 2x_i \text{ for } 1 \le i \le j-3,\ x_{j-1} > 2x_{j-2}\}.$$

If we distribute the leading multiplier in the right-hand side of the recurrence for f_{C_j}, the first term is the generating function of $K_{1,j}$, as there are no restrictions on the size of x_{j-1}. On the other hand, the integer points $\mathbf{m} \in K_{2,j}$ are precisely those in $K_{1,j}$ satisfying $x_{j-1} > 2x_{j-2}$, which is equivalent to the condition that $\mathbf{z}^{\mathbf{m}}$ be divisible by $z_{j-2}z_{j-1}^2$. The second term of the recurrence records precisely these integer points.

Our second proof amounts to a simple observation regarding the integer-point transform of C_j.

Proof (Second proof of Theorem 9). Since for any $\lambda \in C_j \cap \mathbb{Z}^{j-1}$ we have $1 \le \lambda_{j-1} \le 2\lambda_{j-2}$,

$$\begin{aligned} f_{C_j}(z_1,\ldots,z_{j-1}) &= \sum_{\lambda \in C_j \cap \mathbb{Z}^{j-1}} \mathbf{z}^\lambda \\ &= \sum_{\lambda \in C_{j-1} \cap \mathbb{Z}^{j-2}} \mathbf{z}^\lambda \left(z_{j-1} + z_{j-1}^2 + \cdots + z_{j-1}^{2\lambda_{j-2}}\right) \\ &= z_{j-1} \sum_{\lambda \in C_{j-1} \cap \mathbb{Z}^{j-2}} \mathbf{z}^\lambda \frac{1 - z_{j-1}^{2\lambda_{j-2}}}{1 - z_{j-1}} \\ &= \frac{z_{j-1}}{1 - z_{j-1}} \sum_{\lambda \in C_{j-1} \cap \mathbb{Z}^{j-2}} \mathbf{z}^\lambda - \mathbf{z}^\lambda z_{j-1}^{2\lambda_{j-2}} \\ &= \frac{z_{j-1}}{1 - z_{j-1}} \Big(f_{C_{j-1}}(z_1,\ldots,z_{j-1}) \\ &\qquad - f_{C_{j-1}}(z_1,\ldots,z_{j-3}, z_{j-2}z_{j-1}^2) \Big). \end{aligned}$$

□

Following their statement of Theorem 8, the authors of [6] make the following comment:

> It hardly needs to be pointed out that [this formula] is a surprising representation of a polynomial. Indeed, the right-hand side does not look like a polynomial at all.

Such a statement suggests that Brion's formula [9] for rational polytopes is lurking in the background; our third proof of Theorem 9 is based on this formula. Given a rational convex polytope P, we first define the *tangent cone* at a vertex $\mathbf{v}$ of P to be

$$T_P(\mathbf{v}) := \{\mathbf{v} + \alpha(\mathbf{p} - \mathbf{v}) : \alpha \in \mathbb{R}_{\ge 0},\ \mathbf{p} \in P\}.$$

Theorem 10 (Brion). *Suppose P is a rational convex polytope. Then we have the following identity of rational generating functions:*

$$\sigma_P(\mathbf{z}) = \sum_{\mathbf{v} \text{ a vertex of } P} \sigma_{T_P(\mathbf{v})}(\mathbf{z}) \,.$$

Note that the sum on the right-hand side is a sum of rational functions, while the left-hand side yields a polynomial.

Proof (Third proof of Theorem 9). To interpret the recurrence as a consequence of Brion's formula, we first assume that the $f_{C_{j-1}}$'s are expressed in the form of the right-hand side of Brion's formula, i.e., as a sum of integer-point transforms of the tangent cones at the vertices of C_{j-1}. We next rewrite the recurrence as

$$\begin{aligned} f_{C_j}\left(z_1,\ldots,z_{j-1}\right) &= \frac{z_{j-1}}{1-z_{j-1}}\, f_{C_{j-1}}\left(z_1,\ldots,z_{j-2}\right) \\ &\quad + \frac{1}{1-z_{j-1}^{-1}}\, f_{C_{j-1}}\left(z_1,\ldots,z_{j-3},z_{j-2}z_{j-1}^2\right). \end{aligned}$$

The polytope C_j is a combinatorial cube; this can be easily seen by induction on j after observing that in $C_{j-1} \times \mathbb{R}$, the hyperplanes $x_{j-1} = 1$ and $x_{j-1} = 2x_{j-2}$ do not intersect. Thus, the tangent cones for vertices of C_j can be expressed in terms of the tangent cones for vertices of C_{j-1}. Given a vertex $\mathbf{v} = \{v_1, \ldots, v_{j-2}\}$ of C_{j-1}, the two vertices of C_j obtained from $\mathbf{v}$ are $(\mathbf{v}, 1)$ and $(\mathbf{v}, 2v_{j-2})$. For the vertex $(\mathbf{v}, 1)$ in C_j, it is immediate that

$$\sigma_{T_{C_{j-1}}((\mathbf{v},1))}(\mathbf{z}) = \frac{1}{1-z_{j-1}}\, \sigma_{T_{C_{j-2}}(\mathbf{v})}(\mathbf{z}) \,.$$

Our proof will be complete after we show that for the vertex $(\mathbf{v}, 2v_{j-2})$ in C_j,

$$\sigma_{T_{C_{j-1}}((\mathbf{v},2v_{j-2}))}(\mathbf{z}) = \frac{1}{1-z_{j-1}^{-1}}\, \sigma_{T_{C_{j-2}}(\mathbf{v})}(z_1,\ldots,z_{j-3},z_{j-2}z_{j-1}^2) \,.$$

This follows from the fact that the edges in C_j emanating from $(\mathbf{v}, 2v_{j-2})$ terminate in the vertex $(\mathbf{v}, 1)$ and in the vertices $(\mathbf{w}, 2w_{j-2})$ for vertices $\mathbf{w}$ of C_{j-1} that are connected to $\mathbf{v}$ by an edge in C_{j-1}. Thus, Theorem 9 follows from Brion's formula and induction.

There is an interesting remark about Theorem 8 and Brion's formula; while one might hope that the expression in Theorem 8 is obtained by directly specializing Brion's formula to $z_1 = \cdots = z_{j-2} = 1$ and $z_{j-1} = q$, this is not the case. This specialization is not actually possible, as some of the rational functions for tangent cones in C_j have denominators that lack a z_{j-1} variable, and hence, this specialization would require evaluating rational functions at poles. The authors of [6] use the recurrence in Theorem 9 in a more subtle way, in that they first specialize the recurrence to

$$f_{C_j}(1,\ldots,1,q) = \frac{q}{1-q}\left(f_{C_{j-1}}(1,\ldots,1) - f_{C_{j-1}}\left(1,\ldots,1,q^2\right)\right)$$

and then iterate the recurrence. In doing this, they simultaneously use the interpretation of $f_{C_j}(\mathbf{z})$ as a polynomial (for the all-ones specialization) and also the interpretation of $f_{C_j}(\mathbf{z})$ as a rational function (for the specialization involving q^2). Thus, while Theorem 8 looks similar to a Brion-type result, it is obtained differently. We remark that by specializing $z_1 = \cdots = z_{j-1} = q$ in Brion's formula for C_j, one would obtain a representation of the polynomial $f_{C_j}(q,\ldots,q)$ as a sum of rational functions of q.

7 Directions for Further Investigation

7.1 Cones Over Hypersimplices

We can view the cone K of Section 5 as a cone over a "half-open" version of the second *hypersimplex*

$$\Delta(2,n) := \left\{(x_1,\ldots,x_n) \in [0,1]^n : \sum_{i=1}^{n} x_i = 2\right\},$$

in the following manner. The linear inequality $x_j \geq x_1 + \cdots + \widehat{x_j} + \cdots + x_n$ is equivalent to $\frac{\sum_{i=1}^n x_i}{2} \leq x_j$. When $\sum_{i=1}^n x_i = 1$, we are considering the "slice" of K that is constrained by $0 < x_j \leq \frac{1}{2}$ and $\sum_{i=1}^n x_i = 1$, which is $\frac{1}{2}$ of $\Delta(2,n)$ with the condition that $0 < x_j$ for all j. From this perspective, we can view the n-gon compositions of t as

$$H(t) := \left\{(\lambda_1,\ldots,\lambda_n) \in \mathbb{Z}^n_{\geq 0} : \begin{array}{l} \lambda_1 + \cdots + \lambda_n = t\,, \\ \lambda_j \leq \lambda_1 + \cdots + \widehat{\lambda_j} + \cdots + \lambda_n \quad \text{for all } 1 \leq j \leq n \end{array}\right\}$$

$$= \left\{(\lambda_1,\ldots,\lambda_n) \in \mathbb{Z}^n : \begin{array}{l} \lambda_1 + \cdots + \lambda_n = t\,, \\ 0 \leq \lambda_j \leq \frac{t}{2} \quad \text{for all } 1 \leq j \leq n \end{array}\right\}.$$

The second hypersimplex is a well-studied object; for example, in matroid theory $\Delta(2,n)$ is the matroid basis polytope for the 2-uniform matroid on n vertices, while in combinatorial commutative algebra, $\Delta(2,n)$ is the subject of [20, Chap. 9].

It would be interesting to consider analogues of Theorem 3 for the general case of the kth hypersimplex $\Delta(k,n) := \{(x_1,\ldots,x_n) \in [0,1]^n : \sum_{i=1}^{n} x_i = k\}$. The associated composition counting function has a natural interpretation: in

$$\left\{ (\lambda_1,\ldots,\lambda_n) \in \mathbb{Z}^n : \begin{array}{l} \lambda_1 + \cdots + \lambda_n = t\,, \\ 0 \le \lambda_j \le \frac{t}{k} \ \text{ for all } 1 \le j \le n \end{array} \right\}$$

are all compositions of t whose parts are at most $\frac{t}{k}$ (i.e., the parts are not allowed to be too large, where "too large" depends on k).

7.2 *Cayley Polytopes*

We refer to the polytopes C_j from Sect. 6 as *Cayley polytopes*. By taking a geometric view of Cayley compositions as integer points in C_j, we may shift our focus from combinatorial properties of the integer points to properties of C_j itself. Recall that the *normalized volume* of C_j is

$$\mathrm{Vol}(C_j) := (j-1)!\,\mathrm{vol}(C_j)\,,$$

where $\mathrm{vol}(C_j)$ is the Euclidean volume of C_j. Based on experimental data obtained using the software LattE [14] and the Online Encyclopedia of Integer Sequences [19], we make the following conjecture:

Conjecture 11. For $j \ge 2$, $\mathrm{Vol}(C_j)$ is equal to the number of labeled connected graphs on $j-1$ vertices.[1]

Acknowledgements We thank Carla Savage for pointing out several results in the literature that were relevant to our project. This research was partially supported by the NSF through grants DMS-0810105 (Beck), DMS-0758321 (Braun), and DGE-0841164 (Le).

References

1. George E. Andrews, *MacMahon's partition analysis. I. The lecture hall partition theorem*, Mathematical essays in honor of Gian-Carlo Rota (Cambridge, MA, 1996), Progr. Math., vol. 161, Birkhäuser Boston, Boston, MA, 1998, pp. 1–22.
2. ———, *MacMahon's partition analysis. II. Fundamental theorems*, Ann. Comb. **4** (2000), no. 3-4, 327–338.
3. George E. Andrews, Peter Paule, and Axel Riese, *MacMahon's partition analysis. IX. k-gon partitions*, Bull. Austral. Math. Soc. **64** (2001), no. 2, 321–329.

[1] After a preprint of the current article was made public, Matjaz Konvalinka and Igor Pak communicated to us that they resolved Conjecture 11 by a direct combinatorial argument.

4. ———, *MacMahon's partition analysis: the Omega package*, European J. Combin. **22** (2001), no. 7, 887–904.
5. ———, *MacMahon's partition analysis VII. Constrained compositions*, q-series with applications to combinatorics, number theory, and physics (Urbana, IL, 2000), Contemp. Math., vol. 291, Amer. Math. Soc., Providence, RI, 2001, pp. 11–27.
6. George E. Andrews, Peter Paule, Axel Riese, and Volker Strehl, *MacMahon's partition analysis. V. Bijections, recursions, and magic squares*, Algebraic combinatorics and applications (Gößweinstein, 1999), Springer, Berlin, 2001, pp. 1–39.
7. Matthias Beck, Ira M. Gessel, Sunyoung Lee, and Carla D. Savage, *Symmetrically constrained compositions*, Ramanujan J. **23** (2010), no. 1-3, 355–369, `arXiv:0906.5573`.
8. Matthias Beck and Sinai Robins, *Computing the continuous discretely: Integer-point enumeration in polyhedra*, Undergraduate Texts in Mathematics, Springer, New York, 2007, Electronically available at http://math.sfsu.edu/beck/ccd.html.
9. Michel Brion, *Points entiers dans les polyèdres convexes*, Ann. Sci. École Norm. Sup. (4) **21** (1988), no. 4, 653–663.
10. Rod Canfield, Sylvie Corteel, and Pawel Hitczenko, *Random partitions with non negative rth differences*, LATIN 2002: Theoretical informatics (Cancun), Lecture Notes in Comput. Sci., vol. 2286, Springer, Berlin, 2002, pp. 131–140.
11. Sylvie Corteel, Sunyoung Lee, and Carla D. Savage, *Five guidelines for partition analysis with applications to lecture hall-type theorems*, Combinatorial number theory, de Gruyter, Berlin, 2007, pp. 131–155.
12. Sylvie Corteel and Carla D. Savage, *Partitions and compositions defined by inequalities*, Ramanujan J. **8** (2004), no. 3, 357–381.
13. Sylvie Corteel, Carla D. Savage, and Herbert S. Wilf, *A note on partitions and compositions defined by inequalities*, Integers **5** (2005), no. 1, A24, 11 pp. (electronic).
14. Matthias Köppe, *A primal Barvinok algorithm based on irrational decompositions*, SIAM J. Discrete Math. **21** (2007), no. 1, 220–236 (electronic), `arXiv:math.CO/0603308`. Software `LattE macchiato` available at http://www.math.ucdavis.edu/~mkoeppe/latte/.
15. Percy A. MacMahon, *Combinatory Analysis*, Chelsea Publishing Co., New York, 1960.
16. Igor Pak, *Partition identities and geometric bijections*, Proc. Amer. Math. Soc. **132** (2004), no. 12, 3457–3462 (electronic).
17. ———, *Partition bijections, a survey*, Ramanujan J. **12** (2006), no. 1, 5–75.
18. Georg Pólya and Gábor Szegő, *Aufgaben und Lehrsätze aus der Analysis. Band I: Reihen, Integralrechnung, Funktionentheorie*, Vierte Auflage. Heidelberger Taschenbücher, Band 73, Springer-Verlag, Berlin, 1970.
19. Neil J. A. Sloane, *On-line encyclopedia of integer sequences*, http://www.research.att.com/~njas/sequences/index.html.
20. Bernd Sturmfels, *Gröbner Bases and Convex Polytopes*, University Lecture Series, vol. 8, American Mathematical Society, Providence, RI, 1996.
21. Guoce Xin, *A fast algorithm for MacMahon's partition analysis*, Electron. J. Combin. **11** (2004), no. 1, Research Paper 58, 20.
22. Günter M. Ziegler, *Lectures on polytopes*, Springer-Verlag, New York, 1995.

Second-Order Modular Forms with Characters

Thomas Blann and Nikolaos Diamantis

Dedicated to the memory of Leon Ehrenpreis

Abstract We introduce spaces of second-order modular forms for which the relevant action involves characters. We compute the dimensions of these spaces by constructing explicit bases.

Key words Second-order modular forms • Characters • Eichler cohomology

Mathematics Subject Classification (2010): 11F, 60K35

1 Introduction

The systematic study of higher-order forms was originally motivated by two main objects, Eisenstein series with modular symbols [6, 9] and certain probabilities arising in the context of percolation theory [8]. In this note, we discuss how the classification of holomorphic second-order forms given in [4] can be extended to become more relevant for the latter object. This requires the introduction of characters in second-order forms. An appropriate definition of the corresponding spaces is the focus of the next section.

The dimensions and bases of weight $k > 2$ second-order forms are given in Sects. 3 and 4. These dimensions and bases are of interest for possible applications to percolation and elsewhere. The proof relies on a setup which is more intrinsic than that of [4]. It highlights the underlying cohomology and is more consistent with the

T. Blann • N. Diamantis (✉)
School of Mathematical Sciences, University of Nottingham, Nottingham, NG7 2RD, UK
e-mail: Nikolaos.Diamantis@nottingham.ac.uk

H.M. Farkas et al. (eds.), *From Fourier Analysis and Number Theory to Radon Transforms and Geometry*, Developments in Mathematics 28, DOI 10.1007/978-1-4614-4075-8_4,

representation theoretic approach initiated by Deitmar in [2, 3]. On the other hand, the method of constructing the actual bases mainly parallels that of [4].

The higher-order objects from percolation theory investigated so far [5,8] include weight 2 forms and forms with poles at the cusps. The former is one of the subjects of work in progress by the first author [1]. There are various directions one can take in the investigation of the latter kind of object, and we intend to study it, guided by possible applications in percolation.

2 Definitions

Let $\Gamma \subset PSL_2(\mathbb{R})$ be a Fuchsian group of the first kind acting in the usual way on the upper half plane $\mathfrak{H}$ with non-compact quotient $\Gamma\backslash\mathfrak{H}$. Let $\mathfrak{F}$ be a fundamental domain. Fix representatives $\mathfrak{a}$, $\mathfrak{b}$, etc., of the inequivalent cusps in $\overline{\mathfrak{F}}$ and let $\sigma_{\mathfrak{a}}, \sigma_{\mathfrak{b}} \in \mathrm{SL}_2(\mathbb{R})$ be the corresponding scaling matrices. Specifically, $\sigma_{\mathfrak{a}}(\infty) = \mathfrak{a}$ and

$$\sigma_{\mathfrak{a}}{}^{-1}\Gamma_{\mathfrak{a}}\sigma_{\mathfrak{a}} = \Gamma_{\infty} = \left\{\pm\left(\begin{smallmatrix}1 & m\\ 0 & 1\end{smallmatrix}\right) \mid m \in \mathbb{Z}\right\}.$$

where $\Gamma_{\mathfrak{a}}$ is the set of elements of Γ fixing $\mathfrak{a}$. We let $\gamma_{\mathfrak{a}}$ denote a generator of $\Gamma_{\mathfrak{a}}$ and $T := \left(\begin{smallmatrix}1 & 1\\ 0 & 1\end{smallmatrix}\right)$. (We assume that $T \in \Gamma$)

We shall also require the generators of the group Γ. Suppose $\Gamma\backslash\mathfrak{H}$ has genus g, r elliptic fixed points and p cusps. Then there are $2g$ hyperbolic elements γ_i, r elliptic elements ϵ_i and p parabolic elements π_i generating Γ and satisfying the $r+1$ relations:

$$[\gamma_1, \gamma_{g+1}] \dots [\gamma_g, \gamma_{2g}]\epsilon_1 \dots \epsilon_r \pi_1 \dots \pi_p = 1, \qquad \epsilon_j^{e_j} = 1 \tag{1}$$

for $1 \leq j \leq r$ and integers $e_j \geq 2$. Here, $[a, b] := aba^{-1}b^{-1}$ (cf. [7] (10)).

Let χ be a (unimodular) character of Γ. Fix $k \in 2\mathbb{Z}$. The slash operator $|_{k,\chi}$ defines an action of $PSL_2(\mathbb{R})$ on functions $f : \mathfrak{H} \mapsto \mathbb{C}$ by

$$(f|_{k,\chi}\gamma)(z) = f(\gamma z)(cz + d)^{-k}\overline{\chi(\gamma)}$$

with $\gamma = \left(\begin{smallmatrix}* & *\\ c & d\end{smallmatrix}\right)$ in $PSL_2(\mathbb{R})$. Extend the action to $\mathbb{C}[PSL_2(\mathbb{R})]$ by linearity. We set $j(\gamma, z) = cz + d$ for later use. Finally, we set $|_k$ for $|_{k,\mathbf{1}}$, where $\mathbf{1}$ is the trivial character.

Let $z = x + iy$. We will say that "f is holomorphic at the cusps" if for each cusp $\mathfrak{a}$, $(f|_k\sigma_{\mathfrak{a}})(z) \ll y^c$ as $y \to \infty$ uniformly in x for some constant c. We will say that "f vanishes at the cusps" if for each cusp $\mathfrak{a}$, $(f|_k\sigma_{\mathfrak{a}})(z) \ll y^c$ as $y \to \infty$ uniformly in x for every constant c.

Definitions Let $k \in 2\mathbb{Z}$, χ, ψ be two characters of Γ and let $f : \mathfrak{H} \to \mathbb{C}$ be a holomorphic function:

1. We call f a *modular* (resp. *cusp) form of weight k with character* ψ if
   (i) $f|_{k,\psi}(\gamma - 1) = 0$ for all $\gamma \in \Gamma$.
   (ii) f is holomorphic (resp. vanishes) at the cusps.
   Their space is denoted by $M_k(\Gamma, \psi)$ (resp. $S_k(\Gamma, \psi)$).
2. We call f a *second-order modular form of weight k and type* χ, ψ if
   (i) $f|_{k,\chi}(\gamma - 1) \in M_k(\Gamma, \psi)$, for all $\gamma \in \Gamma$.
   (ii) There is a $f_0 \in M_k(\Gamma, \psi)$ such that for all parabolic $\pi \in \Gamma$, $f|_{k,\chi}(\pi - 1) = ((\psi\overline{\chi})(\pi) - 1)a_\pi f_0$ for a $a_\pi \in \mathbb{C}$.
   (iii) f is holomorphic at the cusps.
   Their space is denoted by $M_k^2(\Gamma; \chi, \psi)$.

   The meaning of condition (ii) is that $f|_{k,\chi}(\pi - 1)$ is equal to 0 whenever $\chi(\pi) = \psi(\pi)$ and, otherwise, it equals $c_\pi f_0$ for some f_0 independent of π and a $c_\pi \in \mathbb{C} \setminus \{0\}$. We formulate it in the way we do to make it more suggestive for later uses.
3. We call f a *second-order cusp form of weight k and type* χ, ψ if
   (i) $f|_{k,\chi}(\gamma - 1) \in S_k(\Gamma, \psi)$, for all $\gamma \in \Gamma$.
   (ii) There is a $f_0 \in S_k(\Gamma, \psi)$ such that for all parabolic $\pi \in \Gamma$, $f|_{k,\chi}(\pi - 1) = ((\psi\overline{\chi})(\pi) - 1)a_\pi f_0$ for some $a_\pi \in \mathbb{C}$.
   (iii) f vanishes at the cusps.
   Their space is denoted by $S_k^2(\Gamma; \chi, \psi)$.

Remark 2.1. The percolation crossing formulas $\pi_{\bar{b}}, \pi_b$ and n studied in [5] are "almost" in $M_0^2(\Gamma(2); \mathbf{1}, \chi)$, where χ is the character of $\eta(z)^4$, (η is the Dedekind eta function). They are not because they fail to be holomorphic at all cusps. This justifies the comment made in the Introduction about the need to extend the study of second-order forms to the case of poles at the cusps.

3 Cohomology Associated to $S_k^2(\Gamma; \chi, \psi)$ and $M_k^2(\Gamma; \chi, \psi)$

We recall the definition of *parabolic cohomology* as it applies to our setting. Let χ be a character of Γ. We consider the representation ρ_χ of Γ such that $\rho_\chi(\gamma)$, ($\gamma \in \Gamma$) is defined by

$$\rho_\chi(\gamma)(v) = \chi(\gamma)v \qquad \text{for all } v \in \mathbb{C}$$

Then set

$$Z^1_{\text{par}}(\Gamma, \rho_\chi) := \{f : \Gamma \to \mathbb{C};\ f(\gamma_1\gamma_2) = \rho_\chi(\gamma_1)(f(\gamma_2)) + f(\gamma_1), \forall \gamma_1, \gamma_2 \in \Gamma,$$
$$f(\pi_i) = (\rho_\chi(\pi_i) - 1)(a_i)\ (i = 1, \ldots, p) \text{ for some } a_i \in \mathbb{C}\}$$

$$B^1_{\text{par}}(\Gamma, \rho_\chi) := B^1(\Gamma, \rho_\chi)$$
$$:= \{f : \Gamma \to \mathbb{C}; \exists a \in \mathbb{C}; \forall \gamma \in \Gamma, \ f(\gamma) = (\rho_\chi(\gamma) - 1)a\}.$$

Then

$$H^1_{\text{par}}(\Gamma, \rho_\chi) := Z^1_{\text{par}}(\Gamma, \rho_\chi)/B^1_{\text{par}}(\Gamma, \rho_\chi).$$

To simplify notation, we write $H^1_{\text{par}}(\Gamma, \chi)$ instead of $H^1_{\text{par}}(\Gamma, \rho_\chi)$ and so on.

For characters χ, ψ in Γ, fix a basis of $M_k(\Gamma, \psi)$ $\{f_i\}_{i=1}^d$ where $d :=$ $\dim(M_k(\Gamma, \psi))$. Let $f \in M_k^2(\Gamma; \chi, \psi)$. Then

$$f|_{k,\chi}(\gamma - 1) = \sum_{i=1}^{d} c_i(\gamma^{-1}) f_i \tag{2}$$

for some $c_i(\gamma^{-1}) \in \mathbb{C}$. (The reason for the inversion of γ in the notation is that we want the induced cocycle to be in terms of a left action).

Since $f \in M_k^2(\Gamma; \chi, \psi)$, this implies

$$\begin{aligned} f|_{k,\chi}(\gamma - 1) &= f|_{k,\chi}(\gamma - 1)|_{k,\psi}\delta = f|_{k,\chi}((\gamma - 1)\delta)\overline{\psi(\delta)}\chi(\delta) \\ &= \left(f|_{k,\chi}(\gamma\delta - 1) - f|_{k,\chi}(\delta - 1)\right)\overline{\psi(\delta)}\chi(\delta). \end{aligned}$$

Therefore, for $i = 1, \ldots, d$, $c_i(\gamma^{-1}) = \left(c_i(\delta^{-1}\gamma^{-1}) - c_i(\delta^{-1})\right)\overline{\psi(\delta)}\chi(\delta)$, or upon replacing γ^{-1} by γ and δ^{-1} by δ,

$$c_i(\delta\gamma) = \overline{\psi(\delta)}\chi(\delta)c_i(\gamma) + c_i(\delta).$$

Further, by condition (ii) in the definition of $M_k^2(\Gamma; \chi, \psi)$, $c_i(\pi_j) \in (\rho_{\overline{\psi}\cdot\chi}(\pi_j) - 1)\mathbb{C}$ $(j = 1, \ldots, p)$ and thus c_i induces an element $[c_i]$ of $H^1_{\text{par}}(\Gamma, \overline{\psi} \cdot \chi)$.

Therefore, the map sending $f \in M_k^2(\Gamma; \chi, \psi)$ to

$$\sum_{i=1}^{d} [c_i] \otimes f_i$$

induces a linear map

$$\phi : M_k^2(\Gamma; \chi, \psi) \to H^1_{\text{par}}(\Gamma; \overline{\psi} \cdot \chi) \otimes M_k(\Gamma, \psi).$$

An analogous formula induces a map

$$\phi' : S_k^2(\Gamma; \chi, \psi) \to H^1_{\text{par}}(\Gamma; \overline{\psi} \cdot \chi) \otimes S_k(\Gamma, \psi).$$

Proposition 3.1. *The kernel of the map ϕ (resp. ϕ') is isomorphic to the image of $M_k(\Gamma,\chi)+M_k(\Gamma,\psi)$ (resp. $S_k(\Gamma,\chi)+S_k(\Gamma,\psi)$) under the natural projection into $M_k^2(\Gamma;\chi,\psi)$ (resp. $S_k^2(\Gamma;\chi,\psi)$).*

Proof. It is easily seen that $M_k(\Gamma,\chi)+M_k(\Gamma,\psi)\subset\ker(\phi)$.

In the opposite direction, suppose that $f\in\ker(\phi)$. Then we have $c_i\in B^1(\Gamma,\overline{\psi}\cdot\chi)$ or $c_i(\gamma)=a_i(\overline{\psi}(\gamma)\cdot\chi(\gamma)-1)$ for some constants $a_i\in\mathbb{C}$. Equation (2) then implies

$$f|_{k,\chi}(\gamma-1)=\left(\sum_{i=1}^{d}a_i f_i\right)(\psi(\gamma)\cdot\overline{\chi(\gamma)}-1).$$

Since $F:=\sum_{i=1}^{d}a_i f_i\in M_k(\Gamma,\psi)$, the RHS equals $\overline{\chi(\gamma)}F|_k\gamma-F=F|_{k,\chi}(\gamma-1)$. Therefore, $f-F\in M_k(\Gamma,\chi)$ which implies the assertion.

The proof of the statement for the cuspidal case is similar.

Let χ be a character in Γ. In order to estimate the dimension of $H^1_{\text{par}}(\Gamma,\chi)$, we associate to each $F=(f,\bar{g})\in S_2(\Gamma,\chi)\oplus\overline{S_2(\Gamma,\overline{\chi})}$ and $a\in\mathfrak{H}\cup\text{cusps}(\Gamma)$ a map $L_F(a,\cdot):\Gamma\to\mathbb{C}$ given by

$$L_F(a,\gamma)=\int_a^{\gamma a}f(w)\,dw+\overline{\int_a^{\gamma a}g(w)\,dw}.$$

A computation using the easy to verify identity

$$\int_{z_1}^{\gamma z_1}f(w)\,dw=\int_{z_2}^{\gamma z_2}f(w)\,dw+(\chi(\gamma)-1)\int_{z_2}^{z_1}f(w)\,dw\quad\forall z_1,z_2\in\mathfrak{H}\cup\text{cusps}(\Gamma)\tag{3}$$

shows that $L_F(a,\cdot)\in Z^1_{\text{par}}(\Gamma,\chi)$ and that it depends on a only up to coboundaries. According to a special case of the Eichler–Shimura isomorphism (cf. [11], Chap. 8), the map

$$S_2(\Gamma,\chi)\oplus\overline{S_2(\Gamma,\overline{\chi})}\to H^1_{\text{par}}(\Gamma,\chi)$$

sending F to the cohomology class $[L_F]$ of $L_F(a,\cdot)$ is an isomorphism. As a consequence of this and Proposition 3.1, we deduce that

$$\dim M_k^2(\Gamma;\chi,\psi)\le d_0\dim M_k(\Gamma,\psi)+\dim\Big(M_k(\Gamma,\chi)+M_k(\Gamma,\psi)\Big).\tag{4}$$

where $d_0:=\dim(S_2(\Gamma,\overline{\psi}\cdot\chi))+\dim(S_2(\Gamma,\overline{\chi}\cdot\psi))$. In particular, $M_k^2(\Gamma;\chi,\psi)$ is finite dimensional. Likewise,

$$\dim S_k^2(\Gamma;\chi,\psi)\le d_0\dim S_k(\Gamma,\psi)+\dim\Big(S_k(\Gamma,\chi)+S_k(\Gamma,\psi)\Big).\tag{5}$$

To fix a basis of $H^1_{\text{par}}(\Gamma,\chi)$, suppose that $\{f_i\}$ with $i=1,\dots,\dim(S_2(\Gamma,\chi))$ is a basis of $S_2(\Gamma,\chi)$ and that $\{f_{j+\dim(S_2(\Gamma,\chi))}\}$, $j=1,\dots,\dim(S_2(\Gamma,\overline{\chi}))$ is a basis of $S_2(\Gamma, \overline{\chi})$. Consider the basis of the space $S_2(\Gamma, \chi) \oplus \overline{S_2(\Gamma, \overline{\chi})}$ formed by $F_i := (f_i, 0)$ $(i = 1, \ldots, \dim(S_2(\Gamma, \chi)))$ and $F_{j+\dim(S_2(\Gamma,\chi))} := (0, \overline{f}_{j+\dim(S_2(\Gamma,\chi))})$ $(j = 1, \ldots, \dim(S_2(\Gamma, \overline{\chi})))$. Then the set

$$\{[L_i]; i = 1, \ldots, \dim(S_2(\Gamma, \chi)) + \dim(S_2(\Gamma, \overline{\chi}))\}$$

with

$$L_i := L_{F_i}(a_i, \cdot) \tag{6}$$

for a choice of $a_i \in \mathfrak{H} \cup \text{cusps}(\Gamma)$ is a basis of $H^1_{\text{par}}(\Gamma, \chi)$.

We note that it will be sometimes useful to express $L_F(a, \gamma)$ in terms of antiderivatives

$$\Lambda_h(a; z) := \int_a^z h(w)\, \mathrm{d}w \qquad \text{where } h \in S_2(\Gamma, \chi)$$

for an arbitrary $z \in \mathfrak{H} \cup \text{cusps}(\Gamma)$.

Lemma 3.2. *Let F and L_F be as above. For each $z \in \mathfrak{H} \cup \text{cusps}(\Gamma)$ and $\gamma \in \Gamma$*

$$L_F(a, \gamma) = \Lambda_f(a, \gamma z) + \overline{\Lambda_g(a, \gamma z)} - \chi(\gamma)\left(\Lambda_f(a, z) + \overline{\Lambda_g(a, z)}\right)$$

Proof. Let $z \in \mathfrak{H} \cup \text{cusps}(\Gamma)$,

$$\int_a^{\gamma a} f(w)\, \mathrm{d}w = \int_{\gamma z}^{\gamma a} f(w)\, \mathrm{d}w + \int_a^{\gamma z} f(w)\, \mathrm{d}w$$

Upon a change of variables, the first integral equals

$$\int_z^a f(\gamma w)\, \mathrm{d}(\gamma w) = -\chi(\gamma) \int_a^z f(w)\, \mathrm{d}w$$

Since we can decompose $\overline{\int_a^{\gamma a} g(w)\, \mathrm{d}w}$ $(g \in S_2(\Gamma, \bar{\chi}))$ similarly, we deduce the result.

4 Bases of $S_k^2(\Gamma; \chi, \psi)$ and $M_k^2(\Gamma; \chi, \psi)$ for $k > 2$

Let $k \geq 4$ be even, $p > 0$, $\mathfrak{a} \in \text{cusps}(\Gamma)$ and a character χ in Γ. Suppose that $\chi(\gamma_{\mathfrak{a}}) = \mathrm{e}^{2\pi \mathrm{i} y_{\mathfrak{a}}}$ for some $0 \leq y_{\mathfrak{a}} < 1$. We will call $\mathfrak{a}$ *singular* if $y_{\mathfrak{a}} = 0$ and non-singular otherwise. Let p^* denote the number of inequivalent cusps singular in χ.

For each fixed cusp $\mathfrak{a}$, the space $S_k(\Gamma, \chi)$ is spanned by the Poincaré series

$$P_{\mathfrak{a}m}(z;\chi) = \sum_{\gamma\in\Gamma_{\mathfrak{a}}\backslash\Gamma} \overline{\chi(\gamma)}\, j(\sigma_{\mathfrak{a}}{}^{-1}\gamma, z)^{-k} e((m+y_{\mathfrak{a}})\sigma_{\mathfrak{a}}{}^{-1}\gamma z) \tag{7}$$

as m ranges over the positive integers [10], Th. 5.2.4 or [7], Sect. 2. Here, $e(z) := \mathrm{e}^{2\pi i z}$.

A basis for the space $M_k(\Gamma, \chi)$ ($k \geq 4$) is comprised of the above Poincaré series together with the p^* linearly independent $P_{\mathfrak{a}0}(z, \chi)$ as $\mathfrak{a}$ varies over p^* inequivalent singular cusps.

When $m = 0$ and $\mathfrak{a}$ is non-singular in χ, the series (7) are called the holomorphic Eisenstein series. If we let $E_k(\Gamma, \chi)$ denote the space spanned by these Eisenstein series, then we have the direct sum

$$M_k(\Gamma, \chi) = E_k(\Gamma, \chi) \oplus S_k(\Gamma, \chi). \tag{8}$$

To prove that the dimensions of $S_k^2(\Gamma; \chi, \psi)$ and $M_k^2(\Gamma; \chi, \psi)$ attain the upper bounds (4) and (5), we consider

$$P_{\mathfrak{a}m}(z, L; \chi) = \sum_{\gamma\in\Gamma_{\mathfrak{a}}\backslash\Gamma} L(\mathfrak{a}, \gamma)\, j(\sigma_{\mathfrak{a}}{}^{-1}\gamma, z)^{-k} e((m+x_{\mathfrak{a}})\sigma_{\mathfrak{a}}{}^{-1}\gamma z)\overline{\chi(\gamma)} \tag{9}$$

for $m \geq 0$ and $L \in Z^1_{\text{par}}(\Gamma, \chi \cdot \overline{\psi})$ where $\psi(\gamma_{\mathfrak{a}}) = \mathrm{e}^{2\pi i x_{\mathfrak{a}}}$.

To show that these series are absolutely convergent and holomorphic for $k \geq 4$, we need to bound L. Tothis end, we prove:

Lemma 4.1. *Let χ be a character of Γ. For any f in $S_2(\Gamma, \chi)$, $z_0 \in \mathfrak{H} \cup$ cusps (Γ) all $z \in \mathfrak{H}$ and any cusp $\mathfrak{a}$,*

$$\int_{z_0}^{z} f(w)\,\mathrm{d}w \ll \mathrm{Im}(\sigma_{\mathfrak{a}}{}^{-1}z)^{\varepsilon} + \mathrm{Im}(\sigma_{\mathfrak{a}}{}^{-1}z)^{-\varepsilon} + 1$$

uniformly in x, with an implied constant depending on f, $\mathfrak{F}$, $\mathfrak{a}$, ε but independent of z.

Proof. By a change of variables

$$\int_{\infty}^{\sigma_{\mathfrak{a}} z} f(w)\mathrm{d}w = \int_{\sigma_{\mathfrak{a}}{}^{-1}\infty}^{z} (f|_2\sigma_{\mathfrak{a}})(w)\mathrm{d}w$$

However, $f|_2\sigma_{\mathfrak{a}} \in S_2(\sigma_{\mathfrak{a}}{}^{-1}\Gamma\sigma_{\mathfrak{a}}, \chi')$ for some character χ' ([10], Th. 4.3.9). Further, for every Fuchsian group of the first kind G, a character χ in G, $f \in S_2(G, \chi)$ and $z \in \mathfrak{H}$, $|yf(z)| \ll 1$. Indeed, this holds, by compactness, in the closure of a fundamental domain of $G\backslash\mathfrak{H}$. On the other hand, $|\mathrm{Im}(\gamma z) f(\gamma z)| = |yf(z)|$ for all $\gamma \in G$, and thus, the bound holds on the entire $\mathfrak{H}$. Therefore, $(f|_2\sigma_\mathfrak{a})(w) \ll \mathrm{Im}(w)^{-1}$ for all $w \in \mathfrak{H}$. This implies

$$\int_{\sigma_\mathfrak{a}^{-1}\infty}^{z} (f|_2\sigma_\mathfrak{a})(w)\mathrm{d}w = \int_{\sigma_\mathfrak{a}^{-1}\infty}^{\infty} (f|_2\sigma_\mathfrak{a})(w)\mathrm{d}w + \int_{\infty}^{z} (f|_2\sigma_\mathfrak{a})(w)\mathrm{d}w$$
$$= \int_{\sigma_\mathfrak{a}^{-1}\infty}^{\infty} (f|_2\sigma_\mathfrak{a})(w)\mathrm{d}w + \int_{\infty}^{n+x+\mathrm{i}y} (f|_2\sigma_\mathfrak{a})(w)\mathrm{d}w$$

for some $n \in \mathbb{Z}$ and $0 \leq x < 1$. The last integral equals

$$\int_{\infty}^{x+\mathrm{i}y} (f|_2\sigma_\mathfrak{a}T^n)(w)\mathrm{d}w = \mathrm{e}^{2\pi \mathrm{i}ny_\mathfrak{a}} \int_{\infty}^{x+\mathrm{i}y} (f|_2\sigma_\mathfrak{a})(w)\mathrm{d}w$$

for some $y_\mathfrak{a} \in \mathbb{R}$ since $f|_2\sigma_\mathfrak{a} \in S_2(\sigma_\mathfrak{a}^{-1}\Gamma\sigma_\mathfrak{a}, \chi')$. This implies that

$$\int_{\sigma_\mathfrak{a}^{-1}\infty}^{z} (f|_2\sigma_\mathfrak{a})(w)\mathrm{d}w = \int_{\sigma_\mathfrak{a}^{-1}\infty}^{\infty} (f|_2\sigma_\mathfrak{a})(w)\mathrm{d}w + \mathrm{e}^{2\pi \mathrm{i}nx_\mathfrak{a}} \left(\int_{\infty}^{1} (f|_2\sigma_\mathfrak{a})(x+it)\mathrm{d}t \right.$$
$$\left. + \int_{1}^{y} (f|_2\sigma_\mathfrak{a})(x+it)\mathrm{d}t \right)$$
$$\ll 1 + \int_{1}^{y} |(f|_2\sigma_\mathfrak{a})(x+it)|\mathrm{d}t$$
$$\ll 1 + \int_{1}^{y} \frac{1}{t}\mathrm{d}t = 1 + \log y$$

uniformly in x, with the implied constant depending on $\mathfrak{a}$, f and $\mathfrak{F}$. Since for all ε, $\log(y^\varepsilon) < y^\varepsilon + y^{-\varepsilon}$ for all $y > 0$, we deduce that

$$\int_{\infty}^{\sigma_\mathfrak{a} z} f(w)\mathrm{d}w \ll 1 + y^\varepsilon + y^{-\varepsilon}$$

with the implied constant further depending on ε. Upon replacing z with $\sigma_\mathfrak{a}^{-1}z$, the result follows immediately.

Proposition 4.2. *Let $4 \leq k \in 2\mathbb{Z}$ and characters χ, ψ in Γ. For each $\mathfrak{a} \in$ cusps (Γ) and $L_i(\mathfrak{a}, \cdot) \in Z^1_{\mathrm{par}}(\Gamma, \chi\cdot\overline{\psi})$ as in (6), with $i = 1, \ldots, d_0$ ($d_0 = \dim(S_2(\Gamma, \chi\cdot\overline{\psi}))+ \dim(S_2(\Gamma, \psi\cdot\overline{\chi}))$), we have*

$$P_{\mathfrak{a}0}(z, L_i(\mathfrak{a}, \cdot); \chi), \in M_k^2(\Gamma; \chi, \psi) \quad \text{if } \mathfrak{a} \text{ is singular in } \psi$$
$$P_{\mathfrak{a}m}(z, L_i(\mathfrak{a}, \cdot); \chi), \in S_k^2(\Gamma; \chi, \psi) \quad \text{if } m > 0.$$

Proof. We first show that each term of the series is independent of the representative in $\Gamma_\mathfrak{a}\backslash\Gamma$. The cocycle condition of $L_i(\mathfrak{a}, \cdot)$ implies $L_i(\mathfrak{a}, \gamma_\mathfrak{a}\gamma) = \chi(\gamma_\mathfrak{a})\overline{\psi}(\gamma_\mathfrak{a})L_i(\mathfrak{a}, \gamma)$ because clearly $L_i(\mathfrak{a}, \gamma_\mathfrak{a}) = 0$. Using the identity $\sigma_\mathfrak{a}^{-1}\gamma_\mathfrak{a} = T\sigma_\mathfrak{a}^{-1}$, we deduce

$$
\begin{aligned}
&L_i(\mathfrak{a}, \gamma_\mathfrak{a}\gamma) j(\sigma_\mathfrak{a}^{-1}\gamma_\mathfrak{a}\gamma, z)^{-k} e((m+x_\mathfrak{a})\sigma_\mathfrak{a}^{-1}\gamma_\mathfrak{a}\gamma z)\overline{\chi(\gamma_\mathfrak{a}\gamma)}\\
&\quad = L_i(\mathfrak{a}, \gamma)\chi(\gamma_\mathfrak{a})\overline{\psi(\gamma_\mathfrak{a})} j(T\sigma_\mathfrak{a}^{-1}\gamma, z)^{-k} e((m+x_\mathfrak{a})T\sigma_\mathfrak{a}^{-1}\gamma z)\overline{\chi(\gamma_\mathfrak{a}\gamma)}\\
&\quad = L_i(\mathfrak{a}, \gamma) j(\sigma_\mathfrak{a}^{-1}\gamma, z)^{-k} e((m+x_\mathfrak{a})\sigma_\mathfrak{a}^{-1}\gamma z)\overline{\chi(\gamma)}.
\end{aligned}
$$

To prove the convergence, we first note that by Lemmas 4.1 and 3.2,

$$
L_i(\mathfrak{a}, \gamma) \ll \mathrm{Im}(\sigma_\mathfrak{a}^{-1}\gamma z)^{\varepsilon} + \mathrm{Im}(\sigma_\mathfrak{a}^{-1}\gamma z)^{-\varepsilon} + \mathrm{Im}(\sigma_\mathfrak{a}^{-1} z)^{\varepsilon} + \mathrm{Im}(\sigma_\mathfrak{a}^{-1} z)^{-\varepsilon} + 1
$$

for $i = 1, \dots, d_0$. Therefore

$$
\begin{aligned}
P_{\mathfrak{a}m}(z, L_i(\mathfrak{a}, \cdot); \chi) &\ll \sum_{\gamma \in \Gamma_\mathfrak{a}\backslash\Gamma} \big(\mathrm{Im}(\sigma_\mathfrak{a}^{-1}\gamma z)^{\varepsilon} + \mathrm{Im}(\sigma_\mathfrak{a}^{-1}\gamma z)^{-\varepsilon} + \mathrm{Im}(\sigma_\mathfrak{a}^{-1} z)^{\varepsilon}\\
&\qquad + \mathrm{Im}(\sigma_\mathfrak{a}^{-1} z)^{-\varepsilon} + 1\big)|j(\sigma_\mathfrak{a}^{-1}\gamma, z)|^{-k}\\
&= y^{-k/2} \sum_{\gamma \in \Gamma_\mathfrak{a}\backslash\Gamma} (\mathrm{Im}(\sigma_\mathfrak{a}^{-1}\gamma z)^{k/2+\varepsilon} + \mathrm{Im}(\sigma_\mathfrak{a}^{-1}\gamma z)^{k/2-\varepsilon})\\
&\quad + y^{-k/2}(\mathrm{Im}(\sigma_\mathfrak{a}^{-1} z)^{\varepsilon} + \mathrm{Im}(\sigma_\mathfrak{a}^{-1} z)^{-\varepsilon} + 1)\\
&\quad \times \sum_{\gamma \in \Gamma_\mathfrak{a}\backslash\Gamma} \mathrm{Im}(\sigma_\mathfrak{a}^{-1}\gamma z)^{k/2} \qquad (10)
\end{aligned}
$$

for any $\varepsilon > 0$. (The implied constant depends on ε.) Since the non-holomorphic Eisenstein series

$$
E_\mathfrak{a}(z, s) = \sum_{\gamma \in \Gamma_\mathfrak{a}\backslash\Gamma} \mathrm{Im}(\sigma_\mathfrak{a}^{-1}\gamma z)^{s}, \qquad (11)
$$

is absolutely convergent for s with $\mathrm{Re}(s) > 1$, (10) implies the absolute and uniform (for z in compact sets in $\mathfrak{H}$) convergence of $P_{\mathfrak{a}m}(z, L_i; \chi)$ for $k/2 - \varepsilon > 1$ and hence for $k > 2$.

To determine the growth at the cusps, we recall that $E_\mathfrak{a}(z, s)$ has the Fourier expansion at the cusp $\mathfrak{b}$

$$
\begin{aligned}
E_\mathfrak{a}(\sigma_\mathfrak{b} z, s) &= \delta_{\mathfrak{a}\mathfrak{b}} y^s + \phi_{\mathfrak{a}\mathfrak{b}}(s) y^{1-s} + \sum_{m \neq 0} \phi_{\mathfrak{a}\mathfrak{b}}(m, s) W_s(mz)\\
&= \delta_{\mathfrak{a}\mathfrak{b}} y^s + \phi_{\mathfrak{a}\mathfrak{b}}(s) y^{1-s} + O(\mathrm{e}^{-2\pi y}) \qquad (12)
\end{aligned}
$$

as $y \to \infty$ with an implied constant depending only on s and Γ. Here, $W_s(z)$ is the usual Whittaker function.

This and $L_i(\mathfrak{a}, I) = 0$, for I the identity element of Γ, yields

$$
\begin{aligned}
&j(\sigma_\mathfrak{b}, z)^{-k} P_{\mathfrak{a}m}(\sigma_\mathfrak{b} z, L_i(\mathfrak{a}, \cdot); \chi)\\
&\quad = \sum_{\gamma \in \Gamma_\mathfrak{a}\backslash\Gamma} L_i(\mathfrak{a}, \gamma)\overline{\chi(\gamma)} j(\sigma_\mathfrak{a}^{-1}\gamma\sigma_\mathfrak{b}, z)^{-k} e((m+x_\mathfrak{a})\sigma_\mathfrak{a}^{-1}\gamma\sigma_\mathfrak{b} z)
\end{aligned}
$$

$$\ll y^{-k/2} \sum_{\gamma \in \Gamma_{\mathfrak{a}} \backslash \Gamma, \gamma \neq \Gamma_{\mathfrak{a}}} |L_i(\mathfrak{a}, \gamma)| \mathrm{Im}({\sigma_{\mathfrak{a}}}^{-1} \gamma \sigma_{\mathfrak{b}} z)^{k/2}$$

$$\ll y^{-k/2} \Big(\left| E_{\mathfrak{a}}(\sigma_{\mathfrak{b}} z, k/2 - \varepsilon) - \delta_{\mathfrak{a}\mathfrak{b}} y^{k/2-\varepsilon} \right| + (\mathrm{Im}({\sigma_{\mathfrak{a}}}^{-1} \sigma_{\mathfrak{b}} z)^{\varepsilon}$$
$$+ \mathrm{Im}({\sigma_{\mathfrak{a}}}^{-1} \sigma_{\mathfrak{b}} z)^{-\varepsilon} + 1) \left| E_{\mathfrak{a}}(\sigma_{\mathfrak{b}} z, k/2) - \delta_{\mathfrak{a}\mathfrak{b}} y^{k/2} \right| \Big)$$

Since $\mathrm{Im}(gz) \asymp y^{-1}$ for $g \in \mathrm{SL}_2(\mathbb{R}) \setminus \{\text{translations}\}$, this is $\ll y^{1-k+\varepsilon}$ as $y \to \infty$ uniformly in x. Therefore, $P_{\mathfrak{a}m}(z, L_i(\mathfrak{a}, \cdot); \chi)$ vanishes at the cusps for $m > 0$ as well as $m = 0$.

To verify the transformation law, we rewrite $P_{\mathfrak{a}m}(\cdot, L_i(\mathfrak{a}, \cdot); \chi)$ in the form

$$P_{\mathfrak{a}m}(\cdot, L_i; \chi) = \sum_{\gamma \in \Gamma_{\mathfrak{a}} \backslash \Gamma} \overline{\chi(\gamma)} L_i(\mathfrak{a}, \gamma) e((m + x_{\mathfrak{a}}) \cdot)|_k {\sigma_{\mathfrak{a}}}^{-1} \gamma$$

and thus

$$P_{\mathfrak{a}m}(\cdot, L_i(\mathfrak{a}, \cdot); \chi)|_{k,\chi} \delta = \sum_{\gamma \in \Gamma_{\mathfrak{a}} \backslash \Gamma} \overline{\chi(\gamma\delta)} L_i(\mathfrak{a}, \gamma) e((m + x_{\mathfrak{a}}) \cdot)|_k {\sigma_{\mathfrak{a}}}^{-1} \gamma \delta$$
$$= \sum_{\gamma \in \Gamma_{\mathfrak{a}} \backslash \Gamma} \overline{\chi(\gamma)} L_i(\mathfrak{a}, \gamma\delta^{-1}) e((m + x_{\mathfrak{a}}) \cdot)|_k {\sigma_{\mathfrak{a}}}^{-1} \gamma.$$

This and the cocycle condition of $L_i(\mathfrak{a}, \cdot)$ imply

$$P_{\mathfrak{a}m}(\cdot, L_i(\mathfrak{a}, \cdot); \chi)|_{k,\chi}(\delta - 1) = \sum_{\gamma \in \Gamma_{\mathfrak{a}} \backslash \Gamma} \overline{\chi(\gamma)} L_i(\mathfrak{a}, \delta^{-1}) \chi(\gamma) \overline{\psi(\gamma)} e((m + x_{\mathfrak{a}}) \cdot)|_k {\sigma_{\mathfrak{a}}}^{-1} \gamma$$
$$= L_i(\mathfrak{a}, \delta^{-1}) P_{\mathfrak{a}m}(\cdot, \psi). \tag{13}$$

Therefore, condition (i) of the definition of $S_k^2(\Gamma; \chi, \psi)$ (resp. $M_k^2(\Gamma; \chi, \psi)$) holds for the series $P_{\mathfrak{a}m}(z, L_i(\mathfrak{a}, \cdot); \chi)$, if $m > 0$ (resp. $P_{\mathfrak{a}0}(z, L_i(\mathfrak{a}, \cdot); \chi)$, if $\mathfrak{a}$ is singular in ψ).

Equation (13) also shows condition (ii) of the definition of second-order forms: By (3) applied with $\gamma = \pi$ parabolic, $z_1 = \mathfrak{a}$ and $z_2 =$fixed point of π, we deduce that $L_i(\mathfrak{a}, \pi) = (\chi(\pi)\overline{\psi(\pi)} - 1)a_\pi$ for some constant $a_\pi \in \mathbb{C}$. Since the cocycle condition of $L_i(\mathfrak{a}, \cdot)$ implies that $L_i(\mathfrak{a}, \pi^{-1}) = -\psi(\pi)\overline{\chi(\pi)} L_i(\mathfrak{a}, \pi)$, we deduce that $P_{\mathfrak{a}m}(\cdot, L_i(\mathfrak{a}, \cdot); \chi)|_{k,\chi}(\pi - 1)$ has the form stipulated by (ii) of the definition.

Theorem 4.3. *For $4 \le k \in 2\mathbb{Z}$ and $d_0 := \dim(S_2(\Gamma, \overline{\psi} \cdot \chi)) + \dim(S_2(\Gamma, \overline{\chi} \cdot \psi))$, we have*

$$\dim S_k^2(\Gamma; \chi, \psi) = d_0 \dim S_k(\Gamma, \psi) + \dim \Big(S_k(\Gamma, \chi) + S_k(\Gamma, \psi) \Big) \tag{14}$$

$$\dim M_k^2(\Gamma; \chi, \psi) = d_0 \dim M_k(\Gamma, \psi) + \dim \Big(M_k(\Gamma, \chi) + M_k(\Gamma, \psi) \Big) \tag{15}$$

Proof. To obtain a basis for $S_k^2(\Gamma;\chi,\psi)$, we fix a cusp $\mathfrak{a}$ and we consider the set A of series $P_{\mathfrak{a}j}(z, L_i(\mathfrak{a},\cdot);\chi)$, as $j > 0$ runs over integers yielding a basis $P_{\mathfrak{a}j}(z;\psi)$ for $S_k(\Gamma;\chi)$ and as i runs over integers in $\{1,\ldots,d_0\}$ inducing a basis $[L_i]$ of $H^1_{\text{par}}(\Gamma,\chi\cdot\overline{\psi})$. With (13), these series are all linearly independent because the linear independence of $[L_i]$ implies the linear independence of $L_i(\mathfrak{a},\cdot)$. We further consider a basis B of $S_k(\Gamma,\chi)+S_k(\Gamma,\psi)$. As such a basis, we may choose the union of bases of $S_k(\Gamma,\chi)$ and $S_k(\Gamma,\psi)$, if $\psi\not\equiv\chi$, or, otherwise, a basis of $S_k(\Gamma,\chi)$. The cardinality of the linearly independent set $A\cup B$ equals the upper bound in (5), so $A\cup B$ is a basis of $S_k^2(\Gamma;\chi,\psi)$. This proves (14).

A similar argument, using the fact that $P_{\mathfrak{a}0}(z,\psi)$ with $\mathfrak{a}$ running over the inequivalent cusps of $\Gamma\backslash\mathfrak{H}$ which are singular in terms of ψ form a basis for $E_k(\Gamma,\psi)$, yields (15).

Remark 4.2. The dimensions appearing in Theorem 4.3 can be evaluated explicitly using the formulas for the dimensions of modular forms for $k > 0$ as presented, for instance, in [7]: If χ is a character in Γ, then, with the notation used in (1), set $q = p+\sum_{j=1}^{r}(1-1/e_j)$, $\chi(\pi_i) = e(x_i)$ and $\chi(\epsilon_i) = e((k+a_j)/(2e_j))$ for some $x_i\in[0,1)$, $a_j\in[0,e_j-1]$. Then

$$\dim M_k(\Gamma,\chi) = k(g-1+q/2) - \sum_{i=1}^{p} x_i - \sum_{j=1}^{r} a_j/e_j - g + 1$$

and

$$\dim S_k(\Gamma,\chi) = k(g-1+q/2) - \sum_{i=1}^{p} x_i - \sum_{j=1}^{r} a_j/e_j - g + 1 - p^* + \delta$$

where $\delta = 0$ unless $k = 2$ and $\chi\equiv 1$.

Acknowledgements We thank Peter Kleban for several interesting comments and the referee for a very careful reading of the manuscript.

References

1. T. Blann *Bases of second-order forms with characters*, PhD Thesis, University of Nottingham (2012)
2. A. Deitmar, *Higher order group cohomology and the Eichler-Shimura map* J. reine u. angew. Math. **629** (2009), 221–235
3. A. Deitmar *Invariants, cohomology, and automorphic forms of higher order*, See. Math, (DOI 10.1007/s00029-012-0087-1)
4. N. Diamantis, C. O'Sullivan, *The dimensions of spaces of second-order automorphic forms and their cohomology*, Trans. Amer. Math. Soc., **360 (11)** (2008), 5629–5666
5. Diamantis, Nikolaos; Kleban, Peter; *New Percolation Crossing Formulas and Second-Order Modular Forms* Commun. Number Theory Phys. **3, (4)** (2009), 677–696

6. D. Goldfeld, *Zeta functions formed with modular symbols*, Proc. of the Symposia in Pure Math., Automorphic Forms, Automorphic Representations, and Arithmetic **66** (1999), 111–122
7. S. Husseini, M. Knopp, *Eichler cohomology and automorphic forms* Illinois J. Math **15** **(4)** (1971), 565–577
8. P. Kleban, D. Zagier, *Crossing probabilities and modular forms*, J. Stat. Phys., **113** (2003), 431–454
9. C. O'Sullivan, *Properties of Eisenstein series formed with modular symbols*, J. Reine Angew. Math., **518** (2000), 163–186
10. R. Rankin, *Modular forms and functions*, Cambridge University Press, Cambridge (1977)
11. G. Shimura, *Introduction to the Arithmetic Theory of Automorphic Functions*, Princeton Univ. Press, Princeton, NJ, **14** (1971), 1013–1043

Disjointness of Moebius from Horocycle Flows

J. Bourgain, P. Sarnak, and T. Ziegler

Dedicated to the memory of Leon Ehrenpreis

Abstract We formulate and prove a finite version of Vinogradov's bilinear sum inequality. We use it together with Ratner's joinings theorems to prove that the Moebius function is disjoint from discrete horocycle flows on $\Gamma\backslash SL_2(\mathbb{R})$, where $\Gamma \subset SL_2(\mathbb{R})$ is a lattice.

Key words Moebius function • Randomness principle • Vinogradov's bilinear sums • Entropy • Square-free flow • Disjointness of dynamical systems

Mathematics Subject Classification (2010): 11L20, 11N37, 37D40

1 Introduction

In this note, we establish a new case of the disjointness conjecture [Sa1] concerning the Moebius function $\mu(n)$. The conjecture asserts that for any deterministic topological dynamical system (X, T) (that is a compact metric space X with a continuous map T of zero entropy) as $N \to \infty$,

$$\sum_{n \leq N} \mu(n) f(T^n x) = o(N) \tag{1.1}$$

where $x \in X$ and $f \in C(X)$.

J. Bourgain • P. Sarnak (✉)
School of Mathematics, Institute for Advanced Study, Princeton, NJ 08540, USA
e-mail: bourgain@ias.edu; sarnak@ias.edu

T. Ziegler
Department of Mathematics, Technion - Israel Institute of Technology, Haifa 32000, Israel
e-mail: tamarzr@tx.technion.ac.il

H.M. Farkas et al. (eds.), *From Fourier Analysis and Number Theory to Radon Transforms and Geometry*, Developments in Mathematics 28, DOI 10.1007/978-1-4614-4075-8_5,

If this holds, we say that μ is disjoint from (X, T). The conjecture is known for some simple deterministic systems. For (X, T) a Kronecker flow (that is a translation in a compact abelian group), it is proven in [D] using the methods introduced in [V], while for (X, T) a translation on a compact nilmanifold, it is proved in [G-T]. It is also known for some substitution dynamics associated with the Morse sequence [M-R].[1] In all of these, the dynamics is very structured, for example, it is not mixing. Our aim is to establish the conjecture for horocycle flows for which the dynamics is much more random being mixing of all orders [M].

In more detail, let $G = SL_2(\mathbb{R})$ and $\Gamma \leq G$ a lattice, that is, a discrete subgroup of G for which $\Gamma\backslash G$ has finite volume. Let $u = \begin{bmatrix} 1 & 1 \\ 0 & 1 \end{bmatrix}$ be the standard unipotent element in G and consider the discrete horocycle flow (X, T), where $X = \Gamma\backslash G$ and T is given by

$$T(\Gamma x) = \Gamma x u. \tag{1.2}$$

Theorem 1. *Let (X, T) be a horocycle flow, then μ is disjoint from (X, T), that is given $x \in X$ and $f \in C(X)$ (if X is not compact, then f is continuous on the one-point compactification of X), as $N \to \infty$*

$$\sum_{n \leq N} \mu(n) f(T^n x) = o(N).$$

Note 1. We offer no rate in this $o(N)$ statement. For this reason, we cannot say anything about the corresponding sum over primes[2] (which the treatments in the cases of Kronecker and nilflows certainly do). The source of the lack of a rate is that we appeal to Ratner's theorem [R1] concerning joinings of horocycle flows, and her proof yields no rates.

As pointed out in [Sa1], Vinogradov's bilinear method for studying sums, over primes or correlations with $\mu(n)$, has a natural dynamical interpretation in the context of the sequences $f(T^n x)$ belonging to flows. That is, the so-called type I sums [Va] are individual Birkhoff sums for (X, T^{d_1}), and the type II sums are such Birkhoff sums for joinings of (X, T^{d_1}) with (X, T^{d_2}). The standard treatments [Va, I-K] assume that one has at least a $(\log N)^{-A}$ rate for those dynamical sums in setting up the sieving process. Our starting point is to formulate a finite version of the bilinear sums method. It applies to any multiplicative function bounded by 1.

Theorem 2. *Let $F : \mathbb{N} \to \mathbb{C}$ with $|F| \leq 1$ and let ν be a multiplicative function with $|\nu| \leq 1$. Let $\tau > 0$ be a small parameter and assume that for all primes $p_1, p_2 \leq e^{1/\tau}$, $p_1 \neq p_2$, we have that for M large enough*

---

[1] Also related to this last case is the orthogonality of μ to AC_0 functions, see [K, G, B].

[2] see [S-U] for some results on sums on primes in this case.

$$\left| \sum_{m \le M} F(p_1 m) \overline{F(p_2 m)} \right| \le \tau M. \tag{1.3}$$

Then for N large enough

$$\left| \sum_{n \le N} \nu(n) F(n) \right| \le 2\sqrt{\tau \log 1/\tau} N. \tag{1.4}$$

Note 2. There are obvious variations and extensions which allow a small set of p_1, p_2 for which (1.3) fails, but for which the conclusion (1.4) may still be drawn. We will note them as they arise.

Theorem 2 can be applied to flows (X, T) with $F(n) = f(T^n x)$ as long as we can analyze the bilinear sums $f(T^{p_1 n} x) f(T^{p_2 n} x)$. In Sects. 3 and 4, we use Ratner's theory of joinings of horocycle flows to compute the correlation limits

$$\lim_{N \to \infty} \frac{1}{N} \sum_{n \le N} f(T^{p_1 n} x) f(T^{p_2 n} x)$$

when (X, T) is a horocycle flow. This correlation is determined by a subgroup of $\mathbb{R}^*_{>0}$ denoted by $C(\Gamma, x)$ which is defined in terms of the point $x \in X$ and the commensurator, $COM(\Gamma)$ of Γ in G (see Sect. 3). After removing the mean of f (with respect to dg on X) and determining the correlation limits in (1.4), we find that for T-generic $x \in X$, namely, a point x for which the orbit $\{T^n x\}$ is equidistributed in X, (1.3) holds for τ as small as we please except for a limited number of p_1, p_2's. This leads to Theorem 1 if x is generic, while if it is not so, then thanks to Dani's theorem [Da], Theorem 1 follows from the Kronecker case.

The method used to handle $G = SL_2(\mathbb{R})$ has the potential to apply to the general Ad-unipotent flow in $\Gamma \backslash G$, with G semisimple and Γ such a lattice. For these, the correlations are still very structured by Ratner's general rigidity theorem [R2]. However, the possibilities for the correlations are more complicated, and we have not examined them in detail. There are other deterministic flows for which we can apply Theorem 2 such as various substitution flows [F] and rank one systems [Fe1]. We comment briefly on this at the end of this chapter, leaving details and work in progress for a future note.

2 A Finite Version of Vinogradov's Inequality

We prove Theorem 2. The basic idea is to decompose the set of integers in the interval $[1, N]$ into a fixed number of pieces depending on the small parameter τ. These are chosen to cover most of the interval and so that the members of the pieces have unique prime factors in suitable dyadic intervals. In this way, one can use the multiplicativity of ν, and after an application of Cauchy's inequality, one can invoke (1.3) in order to estimate the key sum in the theorem.

Let $\alpha > 0$ (small and to be chosen later to depend on the parameter τ) and set

$$j_0 = \frac{1}{\alpha}\left(\log\frac{1}{\alpha}\right)^3, \quad j_1 = j_0^2, \quad D_0 = (1+\alpha)^{j_0} \text{ and } D_1 = (1+\alpha)^{j_1}. \tag{2.1}$$

In order to decompose $[1, N]$ suitably, consider first the set S given as

$$S = \{n \in [1, N] : n \text{ has a prime factor in } (D_0, D_1)\}. \tag{2.2}$$

In what follows $N \to \infty$ with our fixed small α and $A \lesssim B$ means that asymptotically as $N \to \infty$, $A \leq B$. From the Chinese remainder theorem, it follows that (here and in what follows $[1, N]$ consists of the integers in this interval)

$$|[1, N]\backslash S| \lesssim \prod_{\substack{D_0 < \ell < D_1 \\ \ell \text{ prime}}} \left(1 - \frac{1}{\ell}\right) N. \tag{2.3}$$

We can estimate the product over primes in (2.3) using the prime number theorem and the fact that α is small and hence D_0 large,

$$\prod_{D_0<\ell<D_1} \left(1 - \frac{1}{\ell}\right) \sim \frac{\log D_0}{\log D_1} = \frac{1}{j_0}. \tag{2.4}$$

It follows that

$$|[1, N)\backslash S| \lesssim \alpha N, \tag{2.5}$$

that is, up to a fraction of α, S covers $[1, N)$.

Let P_j be the set of primes in $[(1+\alpha)^j, (1+\alpha)^{j+1}]$ for $j_0 \leq j \leq j_1$ and define S_j by

$$S_j = \left\{n \in [1, N); n \text{ has a single divisor in } P_j \text{ and no divisor in } \bigcup_{i<j} P_i\right\}. \tag{2.6}$$

The sets S_j are disjoint, and appealing again to the prime number theorem with remainder and α small, we have

$$|P_j| = \frac{(1+\alpha)^{j+1}}{(j+1)\log(1+\alpha)} - \frac{(1+\alpha)^j}{j\log(1+\alpha)} + O\big((1+\alpha)^j \mathrm{e}^{-\sqrt{\alpha j}}\big), \tag{2.7}$$

with an implied constant that is absolute.

Hence, for α small and $j_0 \leq j \leq j_1$,

$$|P_j| \leq (1+\alpha)^j \left[\frac{1}{j} + \frac{1}{\alpha j^2} + O(\mathrm{e}^{-\sqrt{\alpha j}})\right]. \tag{2.8}$$

Now from the definition of S, we have that

$$S\setminus \bigcup_{j_0\le j\le j_1} S_j \subset \bigcup_{j_0\le j\le j_1} \left\{n\in[1,N);\ \begin{matrix}\text{with } n \text{ having at least}\\ \text{two prime factors in } S_j\end{matrix}\right\}. \tag{2.9}$$

Hence,

$$\begin{aligned}|S\setminus \bigcup_{j_0\le j\le j_1} S_j| &\lesssim \sum_{j_0\le j\le j_1}\ \sum_{\ell_1,\ell_2\in P_j}\frac{N}{\ell_1\ell_2}\\ &\le N\sum_{j_0\le j\le j_1}\left(\frac{|P_j|}{(1+\alpha)^j}\right)^2.\end{aligned} \tag{2.10}$$

From (2.8), this gives for α small enough

$$\begin{aligned}|S\setminus \bigcup_{j_0\le j\le j_1} S_j| &\lesssim N\sum_{j_0\le j\le j_1}\left(\frac{1}{j}+\frac{1}{\alpha j^2}+O\left(e^{-\sqrt{\alpha j}}\right)\right)^2\\ &\lesssim N\left(\frac{1}{j_0}+\frac{1}{j_0^3\alpha^2}+O\left(\frac{1}{\alpha}(1+\sqrt{\alpha j_0})e^{-\sqrt{\alpha j_0}}\right)\right)\\ &\lesssim \alpha N.\end{aligned} \tag{2.10'}$$

So at this point, we have covered $[1,N]$ up to a fraction of α by the disjoint sets S_j, $j_0\le j\le j_1$. Finally we decompose S_j in a well-factored set and its complement. For $j_0\le j\le j_1$, let

$$Q_j=\left\{m\in\left[1,\frac{N}{(1+\alpha)^{j+1}}\right);\ m \text{ has no prime factors in } \bigcup_{i\le j} P_j\right\}. \tag{2.10''}$$

Clearly the product sets P_jQ_j satisfy

$$P_jQ_j\subset S_j \text{ for } j_0\le j\le j_1. \tag{2.11}$$

Moreover, for each such j

$$S_j\setminus(P_jQ_j)\subset P_j.\left[\frac{N}{(1+\alpha)^{j+1}},\frac{N}{(1+\alpha)^j}\right]. \tag{2.12}$$

Hence,

$$\sum_{j_0\le j\le j_1}|S_j\setminus(P_jQ_j)|\le\sum_{j_0\le j\le j_1}|P_j|.\frac{\alpha N}{(1+\alpha)^j}.$$

Applying (2.8) yields that the right-hand side above is

$$\le N\left\{\alpha\log\frac{j_1}{j_0}+\frac{1}{j_0}+O(1+\sqrt{\alpha j_0}e^{-\sqrt{\alpha j_0}})\right\}.$$

Hence, for α small enough

$$\sum_{j_0 \le j \le j_1} |S_j \backslash (P_j Q_j)| \le 2\alpha N. \tag{2.13}$$

This leads to the basic decomposition of $[1, N)$ into disjoint sets $P_j Q_j$, $j_0 \le j \le j_1$ with only a small number of points omitted. Namely, from (2.5), (2.10'), and (2.13),

$$|[1, N) \backslash \bigcup_{j_0 \le j \le j_1} P_j Q_j| \lesssim 4\alpha N. \tag{2.14}$$

Note that the map $P_j \times Q_j \to P_j Q_j$ is one-to-one and since $|F| \le 1$ and $|\nu| \le 1$, we have that

$$\Big|\sum_{n \le N} \nu(n) F(n)\Big| \lesssim \sum_{j_0 \le j \le j_1} \Big| \sum_{\substack{x \in P_j \\ y \in Q_j}} \nu(xy) F(xy)\Big| + 4\alpha N. \tag{2.15}$$

For $x \in P_j, y \in Q_j, (x, y) = 1$ so that the $\nu(xy)$ in (2.15) can be factored as $\nu(x)\nu(y)$, and hence,

$$\Big|\sum_{\nu \le N} \nu(n) F(n)\Big| \lesssim \sum_{j_0 \le j \le j_1} \sum_{y \in Q_j} \Big| \sum_{x \in P_j} \nu(x) F(xy)\Big| + 4\alpha N. \tag{2.16}$$

The inner sum may be estimated using Cauchy:

$$\begin{aligned}
&\sum_{y \in Q_j} \Big| \sum_{x \in P_j} \nu(x) F(xy)\Big| \\
&\le \Big(\sum_{y \in Q_j} 1\Big)^{1/2} \Big(\sum_{y \in Q_j} \Big|\sum_{x \in P_j} \nu(x) F(xy)\Big|^2\Big)^{1/2} \\
&\le |Q_j|^{1/2} \Big(\sum_{y \le N/(1+\alpha)^j} \Big|\sum_{x \in P_j} \nu(x) F(xy)\Big|^2\Big)^{1/2} \\
&= |Q_j|^{1/2} \Big(\sum_{y \le N/(1+\alpha)^j} \sum_{x_1, x_2 \in P_j} \nu(x_1)\overline{\nu(x_2)} F(x_1 y)\overline{F(x_2 y)}\Big)^{1/2} \\
&\le |Q_j|^{1/2} \Big(\sum_{x_1, x_2 \in P_j} \Big| \sum_{y \le N/(1+\alpha)^j} F(x_1 y)\overline{F(x_2 y)}\Big|\Big)^{1/2},
\end{aligned} \tag{2.17}$$

where we have used $|\nu| \le 1$.

Note that here

$$x_1, x_2 < (1+\alpha)^{j_1} < e^{1/\alpha^2}. \tag{2.18}$$

The diagonal contribution in (2.17), that is, $x_1 = x_2$ for each j, yields at most

$$|Q_j|^{1/2}|P_j|^{1/2}\frac{\sqrt{N}}{(1+\alpha)^{j/2}}, \tag{2.19}$$

by using that $|F| \le 1$ and the definition of Q_j. Hence, summing over j and Cauchy, it is

$$\le \sqrt{N}\left(\sum_{j_0\le j\le j_1} |P_j||Q_j|\right)^{1/2}\left(\sum_{j_0\le j\le j_1}\frac{1}{(1+\alpha)^j}\right)^{1/2}.$$

Now $|P_j Q_j| \le |S_j|$ and $\sum |S_j| \le N$; hence, the full diagonal contribution is at most

$$N\left(\sum_{j_0\le j\le j}\frac{1}{(1+\alpha)^j}\right)^{1/2} \le \alpha N. \tag{2.20}$$

For $x_1 \ne x_2$, the hypothesis in the theorem may be applied in view of (2.18), that is,

$$\left|\sum_{y\le N/(1+\alpha)^j} F(x_1 y)\overline{F(x_2 y)}\right| \le \frac{\tau N}{(1+\alpha)^j}.$$

Hence, the off-diagonal contribution is at most

$$\begin{aligned}
&\sqrt{\tau N}\sum_{j_0\le j\le j_1}|P_j||Q_j|^{1/2}(1+\alpha)^{-j/2}\\
&\quad\le \sqrt{\tau N}\left(\sum_{j_0\le j\le j_1}|P_j||Q_j|\right)^{1/2}\left(\sum_{j_0\le j\le j}|P_j|(1+\alpha)^{-j}\right)^{1/2}\\
&\quad\le \sqrt{\tau N}N^{1/2}\left(\log\frac{j_1}{j_0}+\frac{1}{j_0\alpha}+\frac{1}{\alpha}(1+\sqrt{\alpha j_0})e^{-\sqrt{j_0}}\right)^{1/2}\\
&\quad\le N\sqrt{\tau}\sqrt{\log 1/\alpha}
\end{aligned} \tag{2.21}$$

(for α small).

Putting all of these together, we have

$$\left|\sum_{n\le N}\nu(n)F(n)\right| \lesssim N\left(5\alpha+\sqrt{\tau\log 1/\alpha}\right).$$

Taking $\alpha = \sqrt{\tau}$ yields the theorem.

3 Commensurators and Correlators

As in the introduction, $G = SL_2(\mathbb{R})$ and Γ is a lattice in G. The commensurator subgroup, $COM(\Gamma)$ of Γ in G, is defined by

$$COM(\Gamma) = \{g \in G : g^{-1}\Gamma g \cap \Gamma \text{ is finite index in both } \Gamma \text{ and } g^{-1}\Gamma g\}. \quad (3.1)$$

It plays a critical role in determining the ergodic joinings of $(\Gamma\backslash G, T^a)$ with $(\Gamma\backslash G, T^b)$, where $a, b > 0$ and $T^a = \begin{bmatrix} 1 & a \\ 0 & 1 \end{bmatrix}$. Let z be a point on the projective line $\mathbb{P}^1(\mathbb{R})$ and let P_z be the stabilizer of z in G, with G acting projectively. If $z = \infty$, then

$$P_\infty = \left\{ \begin{pmatrix} \alpha & \beta \\ 0 & \delta \end{pmatrix} : \alpha\delta = 1 \right\}. \quad (3.2)$$

If $\xi \in G$ and $\xi(z) = \infty$, then

$$P_z = \xi^{-1} P_\infty \xi. \quad (3.3)$$

Define the character χ of P_∞, and hence of P_z for any z, by

$$\chi\left(\begin{pmatrix} \alpha & \beta \\ 0 & \delta \end{pmatrix}\right) = \alpha\delta^{-1} = \alpha^2. \quad (3.4)$$

χ is valued in the multiplicative group $\mathbb{R}^*_{>0}$. If Δ_z is a subgroup of P_z, we define the correlation group $C(\Delta_z)$ to be the image of Δ_z under χ_z, that is, $C(\Delta_z)$ is the subgroup of $\mathbb{R}^*$ given as $\chi_z(\Delta_z)$. We denote by $C(\Gamma, z)$ the group $C\big((COM\ \Gamma) \cap P_z\big)$, and our aim is to determine this group for Γ and z as above. Its relevance to the unipotent element u is that for $\beta \in P_\infty$,

$$\beta u \beta^{-1} = \begin{bmatrix} 1 & \chi(\beta) \\ 0 & 1 \end{bmatrix} = u^{\chi(\beta)}. \quad (3.5)$$

The explicit computation of these groups $C(\Gamma, z)$ depends on the nature of Γ, so we divide it into cases.

Case 1. In which Γ is nonarithmetic. In this case, it is known [Ma] that $COM(\Gamma)/\Gamma$ is finite, and hence, $COM(\Gamma)$ is itself a lattice in G. Hence, for $z \in \mathbb{P}^1(\mathbb{R}), COM(\Gamma) \cap P_z$ is cyclic (either trivial or infinite), and hence, what is important for us is that $C(\Gamma, z)$ is finitely generated. In particular it follows that the set of p/q with $p \neq q$ and both prime which lie in $C(\Gamma, z)$ is finite. We record this as

Lemma 1. *If Γ is nonarithmetic, then for any $z \in \mathbb{P}^1(\mathbb{R})$,*

$$\left\{ \frac{p}{q} : p, q \textit{ prime } p \neq q \right\} \bigcap C(\Gamma, z)$$

is finite (in fact consists of at most one element).

Case 2. In which Γ is arithmetic and $\Gamma\backslash G$ is compact. In this case, it is known [We] that Γ is commensurable with a unit group in a quaternion algebra A defined over a totally real number field. For simplicity, we assume that A is defined over $\mathbb{Q}$ (the general case may be analyzed similarly). Thus, $A = (\frac{a,b}{Q})$ is a four-dimensional division algebra (since $\Gamma\backslash G$ is compact) generated linearly over $\mathbb{Q}$ by $1, \omega, \Omega, \omega\Omega$. Here $\omega^2 = a, \Omega^2 = b$ with $a, b \in \mathbb{Q}$ and say $a > 0$ and a and b square-free. ω and Ω obey the usual quaternionic multiplication rules. For

$$\alpha = x_0 + x_1\omega + x_2\Omega + x_3\omega\Omega \tag{3.6}$$

with $x_j \in \mathbb{Q}$

$$\overline{\alpha} = x_0 - x_1\omega - x_2\Omega - x_3\omega\Omega, \tag{3.7}$$

$$N(\alpha) = \alpha\overline{\alpha} = x_0^2 - ax_1^2 - bx_2^2 + abx_3^2 \tag{3.8}$$

and

$$\text{trace } (\alpha) = \alpha + \overline{\alpha} = 2x_0. \tag{3.9}$$

$A/\mathbb{Q}$ being a division algebra is equivalent to the statement

$$N(\alpha) = 0 \text{ iff } \alpha = 0, \text{ for } \alpha \in A(\mathbb{Q}). \tag{3.10}$$

Let $A_1(\mathbb{Z})$ be the integral unit group:

$$A_1(\mathbb{Z}) = \{\alpha \in A(\mathbb{Z}) : N(\alpha) = 1\}. \tag{3.11}$$

We embed $A(\mathbb{Q})$ into $M_2(\mathbb{R})$ by

$$\alpha \to \phi(\alpha) = \begin{bmatrix} \overline{\xi} & \eta \\ b\overline{\eta} & \xi \end{bmatrix} \tag{3.12}$$

where $\xi = x_0 - x_1 w, \eta = x_2 + x_3\omega$, and $\omega = \sqrt{a} \in \mathbb{R}$.

Note that

$$\det \phi(\alpha) = N(\alpha) \tag{3.13}$$

$$\text{trace}\, \phi(\alpha) = \xi + \overline{\xi} = \text{ trace } (\alpha). \tag{3.14}$$

Now $\Lambda = \phi\big(A_1(\mathbb{Z})\big)$ is a cocompact lattice in G, and we are assuming that our Γ is commensurable with Λ. Hence, the commensurator of Γ (or of Λ, they are the same) consists of the $\mathbb{Q}$ points [P-R]:

$$COM(\Gamma) = \left\{ \frac{\phi(\alpha)}{(\det \alpha)^{1/2}} ; \alpha \in A^+(\mathbb{Q}) \right\}, \tag{3.15}$$

where $A^+(\mathbb{Q})$ consists of all $\alpha \in A(\mathbb{Q})$ with $N(\alpha) > 0$.

Hence, up to scalar multiples of $\begin{bmatrix} 1 & \\ & 1 \end{bmatrix}$, that is, up to the center of $GL_2(\mathbb{R})$,

$$\delta \in COM(\Gamma) \text{ iff } \delta = \begin{bmatrix} \bar{\xi} & \eta \\ b\bar{\eta} & \xi \end{bmatrix} \text{ with } \xi + \eta\Omega \in A^+(\mathbb{Q}). \tag{3.16}$$

Our interest is in $C(\Gamma, z)$, and from the description (3.16), one can check that for certain algebraic z's, this group can be an infinitely generated subgroup of K^*, where K is the corresponding algebraic extension of $\mathbb{Q}$.

What is important to us are the rationals in this group, and this is given by

Lemma 2. *For Γ as in case 2 and any $z \in \mathbb{P}^1(\mathbb{R})$*

$$C(\Gamma, z) \bigcap \mathbb{Q}^* = \{1\}.$$

Proof. Let $\hat{\delta} \in P_z \bigcap COM(\Gamma)$, then $\hat{\delta} = \phi(\delta)$, and hence, $N(\delta)^{1/2}\hat{\delta}$ in $GL_2(\mathbb{R})$ satisfies

$$\left.\begin{aligned} \text{trace } (N(\delta)^{1/2}\hat{\delta}) &= s \in \mathbb{Q} \\ \det(N(\delta)^{1/2}\hat{\delta}) &= t \in \mathbb{Q}^*_{>0} \end{aligned}\right\} \tag{3.17}$$

Also, $N(\delta)^{1/2}\hat{\delta}$ is conjugate in G to β with

$$\beta = \begin{bmatrix} \lambda & * \\ 0 & \mu \end{bmatrix},$$

where

$$\lambda\mu = t \text{ and } \lambda + \mu = 2s. \tag{3.18}$$

Now, $\chi(\hat{\delta}) = \lambda/\mu$, and if this number is in $\mathbb{Q}$, then from (3.17) and (3.18), we see that both λ and μ are in $\mathbb{Q}$. Now, $\delta \in A^+(\mathbb{Q})$, so $\delta = x_0 + x_,\omega + x_2\Omega + x_3\omega\Omega$ with $x_j \in \mathbb{Q}$, and from (3.17), we have that

$$\left(\frac{\lambda + \mu}{2}\right)^2 - ax_1^2 - bx_2^2 + abx_3^2 = \lambda\mu,$$

that is,

$$\left(\frac{\lambda - \mu}{2}\right)^2 - a_1x_1^2 - bx_2^2 + abx_3^2 = 0. \tag{3.19}$$

Now, $\lambda - \mu, x_1, x_2, x_3 \in \mathbb{Q}$, and since A is a division algebra, it follows from (3.19) that

$$\lambda - \mu = x_1 = x_2 = x_3 = 0.$$

That is, $\lambda = \mu$ and hence $\lambda/\mu = 1$ as claimed.

Case 3. In which Γ is arithmetic and $\Gamma\backslash G$ is noncompact. This time Γ is commensurable with a quaternion algebra that is split over $\mathbb{Q}$, and hence, Γ is commensurable with $SL_2(\mathbb{Z})$. Its commensurator subgroup is given by

$$COM(\Gamma) = \left\{A/(\det A)^{1/2} : A \in GL_2^+(\mathbb{Q})\right\}.$$

Now, if $z \in \mathbb{P}^1(\mathbb{Q})$, then since

$$COM(\Gamma) \cap P_\infty = \left\{\frac{1}{\sqrt{\alpha\delta}}\begin{pmatrix}\alpha & \beta\\ 0 & \delta\end{pmatrix} : \alpha, \beta, \delta \in \mathbb{Q}, \alpha\delta > 0\right\},$$

we have that

$$C(\Gamma, z) = C\big(COM(\Gamma) \cap P_z\big) = \mathbb{Q}^*. \tag{3.20}$$

So in this case, the correlator subgroup contains every rational p/q.

If $z \notin \mathbb{P}^1(\mathbb{Q})$ and z does not lie in a quadratic number field, then $(az + b)/(cz + d) = z$ has no solutions for $\begin{bmatrix}a & b\\ c & d\end{bmatrix} \in GL(2, \mathbb{Q})$ other than $\begin{bmatrix}a & b\\ c & d\end{bmatrix} = \mathbb{1}$ in PGL_2. Hence, for such a z,

$$C(\Gamma, z) = C\big(COM(\Gamma) \cap P_z\big) = \{1\}. \tag{3.21}$$

This leaves us with z quadratic in which case we have that up to scalar multiples of I:

$$COM(\Gamma) \cap P_z = \{\gamma \in GL_2^+(\mathbb{Q}) : \gamma z = z\}. \tag{3.22}$$

If z satisfies $az^2+bz+c = 0$ with a, b, c integers $(a, b, c) = 1$ and $d = b^2-4ac > 0$ and not a square, then one checks that

$$COM(\Gamma) \cap P_z = \left\{\begin{pmatrix}\frac{t+bu}{2} & cu\\ -au & \frac{t-bu}{2}\end{pmatrix} : t^2 - \mathrm{d}u^2 \in \mathbb{Q}^+\ t, u \in \mathbb{Q}\right\}.$$

Hence,

$$\chi\left(\begin{pmatrix}\frac{t+bu}{2} & cu\\ -au & \frac{t-bu}{2}\end{pmatrix}\right) = \frac{t + u\sqrt{d}}{t - u\sqrt{d}} \tag{3.23}$$

and so

$$C(\Gamma, z) = \left\{\frac{\eta}{\eta'}; \eta \in \mathbb{Q}(\sqrt{d})^*, N(\eta) > 0\right\} \tag{3.24}$$

where η' is the conjugate of η in $\mathbb{Q}(\sqrt{d})$.

While this group is infinitely generated, its intersection with $\mathbb{Q}^*$ is 1 (if $(\frac{\eta}{\eta'})' = \frac{\eta}{\eta'}$, then $\eta^2 = (\eta')^2$ or $\eta = \pm\eta'$, and since $N(\eta) > 0, \eta = \eta'$).

We summarize this with

Lemma 3. *If* Γ *is commensurable with* $SL_2(\mathbb{Z})$, *then*

$$C(\Gamma, z) \cap \mathbb{Q}^* = \begin{cases} \{1\} \text{ if } z \notin \mathbb{P}^1(\mathbb{Q}) \\ \mathbb{Q}^* \text{ if } z \in \mathbb{P}^1(\mathbb{Q}). \end{cases}$$

4 Ratner Rigidity and Moebius Disjointness

The correlator group $C(\Gamma, z)$ enters in the analysis of joinings of horocycle flows when applying Ratner's theorem [R1, R2]. According to these, we have that for $\lambda_1, \lambda_2 > 0$, and $\xi \in \Gamma\backslash G$,

$$\lim_{N\to\infty} \frac{1}{N} \sum_{n=1}^{N} f\left(\xi(u^{\lambda_1})^n\right) f\left(\xi(u^{\lambda_2})^n\right) \tag{4.1}$$

exists. Here $f \in C(\Gamma\backslash G)$ is continuous on the one-point compactification of $\Gamma\backslash G$ (if it is not compact), and $u^\lambda = \begin{bmatrix} 1 & \lambda \\ 0 & 1 \end{bmatrix}$.

The limit in (4.1) is given by

$$\int_{\Gamma\backslash G\times\Gamma\backslash G} F(\tilde{\xi}h)\mathrm{d}\nu(h), \tag{4.2}$$

where $\tilde{\xi} = (\xi, \xi) \in X \times X, F(x_1, x_2) = f(x_1)f(x_2)$, and ν is an algebraic Haar measure supported on an algebraic subgroup H of $G \times G$ and for which $(\Gamma \times \Gamma)\tilde{\xi}H$ is closed in $X \times X$. The support of ν is the closure of the orbit $\left(\xi(u^{\lambda_1})^n, \xi(u^{\lambda_2})^n\right), n = 1, 2, \ldots$.

Consider first the case that X is compact. Then any point ξ is u^λ generic (the flow (X, T^λ) is uniquely ergodic, and every point is dg equidistributed in X [Fu]). It follows that the measure ν projects onto dg on each factor X_{λ_j} of $X_{\lambda_1} \times X_{\lambda_2}$. That is, ν is a joining of X_{λ_1} with X_{λ_2}. Applying the Ratner rigidity theorems, either $\mathrm{d}\nu = \mathrm{d}g_1 \times \mathrm{d}g_2$ or it reduces to a measure on subgroups $H = \psi(G)$ where ψ is a morphism $\psi : G \to G \times G$ of the form

$$\psi(g) = \left(\psi_1(g), \psi_2(g)\right)$$

with

$$\psi_1(u) = u^{\lambda_1}, \psi_2(u) = u^{\lambda_2} \tag{4.3}$$

and $(\Gamma\xi\psi_1(g), \Gamma\xi\psi_2(g))$ is closed in $X \times X$. That is, there are $\alpha_1, \alpha_2 \in G$ such that

$$\alpha_1 u \alpha_1^{-1} = u^{\lambda_1}, \alpha_2 u \alpha_2^{-1} = u^{\lambda_2}$$

and

$$\psi(g) = \left(\alpha_1 g \alpha_1^{-1}, \alpha_2 g \alpha_2^{-1}\right). \tag{4.4}$$

In particular $\alpha_1, \alpha_2 \in P_\infty(\mathbb{R})$ and

$$\chi(\alpha_1) = \lambda_1 \text{ and } \chi(\alpha_2) = \lambda_2. \tag{4.5}$$

Now $(\xi\alpha_1 g \alpha_1^{-1}, \xi\alpha_2 g \alpha_2^{-1}); g \in G$ is closed in $\Gamma\backslash G \times \Gamma\backslash G$ iff

$$(h\alpha_1^{-1}, \xi\alpha_2\alpha_1^{-1}\xi^{-1}h\alpha_2^{-1}), h \in G \tag{4.6}$$

is closed $\Gamma\backslash G \times \Gamma\backslash G$.

The latter is equivalent to

$$\delta = \xi\alpha_2\alpha_1^{-1}\xi^{-1} \in COM(\Gamma). \tag{4.7}$$

In this case,

$$\delta \in P_{\xi(\alpha)} \cap COM(\Gamma) \text{ and } \chi(\delta) = \lambda_2/\lambda_1. \tag{4.8}$$

Thus, we have

Lemma 4. *There is a nontrivial joining (in particular ν is not* $dg_1 dg_2$*) in (4.2) iff* $\lambda_2/\lambda_1 \in C(\Gamma, \xi(\infty))$.

So if $\lambda_1 = p$ and $\lambda_2 = q$ with $p \neq q$ primes, then from Lemmas 1, 2, and 4, we have

Corollary 5. *For $\xi \in \Gamma\backslash G$ with $\Gamma\backslash G$ compact, there are most finite number of pairs of distinct primes p, q (depending only on ξ) for which the following fails:*

$$\frac{1}{N}\sum_{n=1}^{N} f(\xi u^{pn}) f(\xi u^{qn}) \to \left(\int_{\Gamma\backslash G} f(g)dg\right)^2.$$

If X is noncompact and Γ is nonarithmetic, then as long as ξ is generic for (X, T, dg), then the joinings analysis coupled with Lemma 1 leads to Corollary 5 holding for such Γ and ξ. The remaining case is that of Γ being commensurable with $SL_2(\mathbb{Z})$. In this case, by [Da], ξ is generic for (X, T, dg) iff $\xi(\infty) \notin \mathbb{P}^1(\mathbb{Q})$.

Thus, again, we can apply the joinings analysis coupled with Lemmas 3 and 4 to conclude

Corollary 6. *If X is noncompact and ξ is generic for (X, T, dg), then there are at most a finite number of pairs of distinct prime p, q, depending only on ξ, for which the following fails:*

$$\frac{1}{N}\sum_{n=1}^{N} f(\xi u^{pn}) f(\xi u^{qn}) \to \left(\int_{\Gamma\backslash G} f(g)\mathrm{d}g\right)^2.$$

We can now complete the proof of Theorem 1. If X is compact, then every ξ is generic for dg. Write

$$f(x) = f_1(x) + c \tag{4.9}$$

where $\int_{\Gamma\backslash G} f_1(x)\mathrm{d}x = 0$ and c is a constant. Then

$$\frac{1}{N}\sum_{n=1}^{N} \mu(n) f(\Gamma\xi u^n) = \frac{1}{N}\sum_{n=1}^{N} \mu(n) f_1(\Gamma\xi u^n) + o(1) \tag{4.10}$$

by the prime number theorem.

As far as the sum against f_1 is concerned, according to Corollary 5 (note $\int_{\Gamma\backslash G} f_1(x)\mathrm{d}x = 0$), the conditions of Theorem 2 are met for $F(n) = f_1(\xi u^n)$ except for finitely many pairs p, q. This causes no harm as far as concluding that the first sum in (4.10) is $o(1)$. One can certainly allow a finite number of exceptions (independent of N) in Theorem 2; in fact the proof only involves the condition for primes $p \geq D_0$ which gets large as τ gets small.

If $\Gamma\backslash G$ is not compact and ξ is generic for dg, then according to Corollary 6, everything goes through as above, and Theorem 1 follows. If ξ is not generic, then by [Da], the closure of the orbit of ξ in X is either finite or is a circle, and in the latter case, the action of u is by rotation of this circle through an angle of θ. Thus, in the first case, Theorem 1 follows from the theory of Dirichlet L-functions, while in the second case, it was proven in [D].

Note 4. The case of richest joinings of $X \times X$ of the form $(\Gamma g\alpha_1^{-1}, \Gamma\delta g\alpha_2^{-1})$, with $\Gamma = SL_2(\mathbb{Z})$ and $\delta \in COM(\Gamma)$, is not one that we had to consider directly in our analysis (since it corresponds to $\xi(\infty) \in \mathbb{P}^1(\mathbb{Q})$ so that ξ is not generic). For this joining, if $\det\delta = pq$ (taking $\delta \in GL_2^+$), the joining is

$$\frac{1}{[\Gamma:\Delta]}\int_{\Delta\backslash G} f(g\alpha_1^{-1}) f(\delta g\alpha_1^{-1})\mathrm{d}g \tag{4.11}$$

where $\Delta = \delta^{-1}\Gamma\delta \cap \Gamma$.

By the theory of correspondences (Hecke operators), if f is a Hecke eigenform (and $\int_{\Gamma\backslash G} f\mathrm{d}g = 0$) which we can assume here, the joining in (4.11) becomes

$$\frac{\lambda_f(pq)}{(p+1)(q+1)}\int_{\Gamma\backslash G} f\left(g\alpha_1^{-1}\right) f\left(g\alpha_1^{-1}\right)\mathrm{d}g \tag{4.12}$$

where $\lambda_f(n)$ is the nth Hecke eigenvalue.

So while in this case the correlation need not be zero, it is small if pq gets large. This follows from the well-known bounds for Hecke eigenvalues [Sa2]. One would expect that this would be useful in an analysis of this type, but apparently for this ineffective analysis, it is not needed.

5 Some Further Comments

The Moebius orthogonality criterion provided by Theorem 2 has applications to other systems of zero entropy. One can use it to give a "soft" proof of the qualitative Theorem 1 for Kronecker and nilflows. In what follows we will only briefly review some new consequences that are essentially immediate from a number of classical facts in ergodic theory,[3] leaving details and further research in this direction for a future paper. Some unexplained terminology below may be found in [Ka-T]. First, we mention a result due to Del Junco and Rudolph ([D-R], Cor. 6.5) asserting the disjointness of distinct powers T^m and T^n for weakly mixing transformations T with the minimal self-joinings property (MSJ). This provides another general class of systems for which Theorem 1 holds. More precisely, the disjointness statement of Theorem 1 applies to any uniquely ergodic topological model for such transformations; these exist by Jewett [J]. Next, restricting ourselves to rank-one transformations, J. King's theorem [Ki] states that mixing rank-one implies MSJ (well-known examples include the Ornstein rank-one constructions and Smorodinsky-Adams map, see [Fe1]). The condition of mixing may be weakened to "partial mixing"; see [Ki-T]. While it seems presently unknown whether any mildly mixing rank-one transformation has MSJ, this property was established in certain other cases, such as Chacon's transformation [D-R-S] (which is mildly but not partially mixing). It was shown that "typical" interval exchange transformations are never mixing [Ka], rank-one [Ve2], uniquely ergodic [Ve1, Mas], and weakly mixing [A-F]. Whether they satisfy Theorem 1 is an interesting question, especially in view of the fact that they are the immediate generalization of circle rotations. For further results on correlation of the Moebius function with rank-one systems, see this follow-up paper of Bourgain [B1].

Finally, the Moebius (and Liouville sequence) orthogonality with sequences arising from substitution dynamics has its importance from the perspective of symbolic complexity (see [Fe2] for a discussion). Our approach via Theorem 2 is applicable for sequences produced by an "admissible q-automation" (see [Quef] for definitions), provided its spectral type is of intermediate dimension. The spectral measure is indeed known to be (xq)-invariant, and disjointness of T^{p_1} and T^{p_2} for $p_1 \neq p_2$ may in this case be derived from [Ho].

---

[3]We are grateful to J-P. Thouvenot for a detailed account of the "state of the art" of various aspects of the theory of joinings.

References

[A-F] A. Avila, G. Forni, *Weak mixing for interval exchange transformations and translation flows*, Ann. Math. (2) 165, No. 2, 637–664 (2007)

[B] J. Bourgain, *On the Fourier-Walsh Spectrum on the Moebius Function*, preprint 2011, available at arxiv.org/abs/1112.1423

[B1] J. Bourgain, *On the correlation of the Moebius function with random rank-one systems*, preprint 2011, available at arxiv.org/abs/1112.1032 to appear in J. Analyse

[D] H. Davenport, *On some infinite series involving arithmetical functions. II*, Q. J. Math., Oxf. Ser. 8, 313–320 (1937)

[Da] S. Dani, *On uniformly distributed orbits of certain horocycle flows*, Ergodic Theory Dyn. Syst. 2, 139–158 (1982)

[D-R-S] A. Del Junco, A. Rahe, L. Swanson, *Chacon's automorphisms has minimal self-joinings*, J. Anal. Math. 37, 276–284 (1980)

[D-R] A. Del Junco, D.J. Rudolph, *On ergodic actions whose self-joinings are graphs*, Ergodic Theory Dyn. Syst. 7, 531–557 (1987)

[F] P. Fogg, *Substitutions in dynamics, arithmetics and combinatorics*, Lecture Notes in Math. 1794, (2002)

[Fe1] S. Ferenczi, *Systems of finite rank*, Colloq. Math. 73, No.1, 35–65 (1997)

[Fe2] S. Ferenczi, *Rank and symbolic complexity*, Ergodic Theory Dyn. Syst. 16, No.4, 663–682 (1996)

[Fu] H. Furstenberg, *The unique ergodicity of the horocycle flow*, Lecture Notes in Math 318, 95–115 (1973)

[G] B. Green, *On (not) computing the Moebius function using bounded depth circuits*, to appear in Combinatorics, Probability and Computing, preprint available at arxiv.org/abs/1103.4991

[G-T] B. Green and T. Tao, *The Moebius Function is strongly orthogonal to nilsequences*, to Ann. of Math. (7) 175, 541–566 (2012)

[Ho] M. Hochman, *Geometric rigidity of $\times m$ invariant measures*, Preprint

[I-K] H. Iwaniec and E. Kowalski, Analytic Number Theory, AMS Coll 53 (2004)

[J] R.I. Jewett, *The prevalence of uniquely ergodic systems*, J. Math. Mech. 19, 717–729 (1970)

[K] G. Kalai, *The ACO Prime Number Conjecture*, gilkalai.wordpress.com/2011/02/21/the-ac0-prime-number-conjecture/

[Ka] A. Katok, *Interval exchange transformations and some special flows are not mixing*, Israel J. Math 35, no.4, 301–310 (1980)

[Ka-T] A. Katok and J.P. Thouvenot, *Spectral properties and combinatorial constructions in ergodic theory*, Handbook of Dynamical System, Vol. 1B, 649–743 (2006)

[Ki] J. King, *Joining rank and the structure of finite rank mixing transformations*, J. Anal. Math. 51 (1988), 182–227

[Ki-T] J. King and J-P. Thouvenot, *An canonical structure theorem for finite joining-rank maps*, J. Anal. Math. 56, 211–230 (1991)

[M-R] C. Mauduit and J. Rivat, *On a problem of Gelfond: the sum of the digits of the first numbers*, Ann. Math. (2) 171, No. 3, 1591–1646 (2010)

[Ma] G. Margulis, Discrete Subgroups of Lie Groups, 1991

[M] B. Marcus, *The horocycle flow is mixing of all degrees*, Invent. Math. 46, 201–209 (1978)

[Mas] H. Masur, *Interval exchange transformations and measured foliations*, Ann. Math. (2) 115, 169–200 (1982)

[P-R] V. Platonov and A. Rapinchuk, Algebraic groups and Number Theory, A.P. 1991

[Quef] M. Queffélec, *Substitution Dynamical systems*, Spectral Analysis, Lecture Notes in Mathematics 1294

[R1] M. Ratner, *Horocycle flows, joinings and rigidity of products*, Ann. Math. (2) 118, 277–313 (1983)

[R2] M. Ratner, *On Raghunathan's measure conjecture*, Ann. Math. (2) 134, No.3, 545–607 (1991)

[Sa1] P. Sarnak, Three lectures on the Moebius Function randomness and dynamics, publications.ias.edu/sarnak/

[Sa2] P. Sarnak, *Notes on the generalized Ramanujan conjectures*, Clay Mathematics Proceedings 4, 659–685 (2005)

[S-U] P. Sarnak and A. Ubis, *Horocycle flows at prime times*, Preprint 2011, available at arxiv.org/abs/1110.0777

[Va] R. Vaughan, *Sommes trigonométriques sur les nombres premiers*, C. R. Acad. Sci., Paris, Sr. A 85, 981–983 (1977)

[Ve1] W.Veech, *Gauss measures for transformations on the space of interval exchange maps*, Ann. Math. (2) 115, 201–242 (1982)

[Ve2] W.Veech, *The metric theory of interval exchange transforms*, I, II, III, Am. J. Math. 106, 1331–1422 (1984)

[V] I.M. Vinogradov, *Some theorems concerning the theory of primes*, Rec. Math. Moscou, n. Ser. 2, 179–195 (1937)

[We] A. Weil, *Algebras with involutions and the classical groups*, J. Indian Math. Soc., n. Ser. 24, 589–623 (1961)

Duality and Differential Operators for Harmonic Maass Forms

Kathrin Bringmann, Ben Kane, and Robert C. Rhoades

In memory of Leon Ehrenpreis

Abstract Due to the graded ring nature of classical modular forms, there are many interesting relations between the coefficients of different modular forms. We discuss additional relations arising from duality, Borcherds products, and theta lifts.

Using the explicit description of a lift for weakly holomorphic forms, we realize the differential operator $D^{k-1} := (\frac{1}{2\pi i}\frac{\partial}{\partial z})^{k-1}$ acting on a harmonic Maass form for integers $k > 2$ in terms of $\xi_{2-k} := 2iy^{2-k}\overline{\frac{\partial}{\partial \overline{z}}}$ acting on a different form. Using this interpretation, we compute the image of D^{k-1}. We also answer a question arising in recent work on the p-adic properties of mock modular forms. Additionally, since such lifts are defined up to a weakly holomorphic form, we demonstrate how to construct a canonical lift from holomorphic modular forms to harmonic Maass forms.

Mathematics Subject Classification (2000): 11F37, 11F25, 11F30

1 Introduction and Statement of Results

Fourier coefficients of automorphic forms play a prominent role in mathematics (see, for instance, [26]). Kloosterman sums arise naturally in the analytic theory of such coefficients. For instance, the Kuznetsov trace formula [29] relates a certain

K. Bringmann (✉) • B. Kane
Mathematical Institute, University of Cologne, Weyertal 86-90, 50931 Cologne, Germany
e-mail: kbringma@math.uni-koeln.de; bkane@math.uni-koeln.de

R.C. Rhoades
Stanford University, Serra Mall bldg. 380, Stanford, CA 94305, USA
e-mail: rrhoades@stanford.edu

H.M. Farkas et al. (eds.), *From Fourier Analysis and Number Theory to Radon Transforms and Geometry*, Developments in Mathematics 28, DOI 10.1007/978-1-4614-4075-8_6,

infinite sum related to the Fourier coefficients of automorphic forms to an infinite sum involving Kloosterman sums. The classical Poincaré series at infinity of weight $2 < k \in \frac{1}{2}\mathbb{Z}$ on $\Gamma_0(N)$, denoted by $P(m,k,N;z)$ (see (2.2) for the definition) with $m \in \mathbb{Z}$, $N \in \mathbb{N}$, and $z \in \mathbb{H}$, play an important role in such trace formulas.

The Poincaré series $P(m,k,N;z)$ are elements of $M_k^!(N)$, the space of weakly holomorphic weight k modular forms for $\Gamma_0(N)$, i.e., those meromorphic modular forms whose poles lie only at the cusps. Furthermore, if $m \geq 0$, then $P(m,k,N;z)$ has bounded growth toward all cusps and so is in $M_k(N)$, the subspace of $M_k^!(N)$ of holomorphic modular forms. For $k > 2$, $m \in \mathbb{Z}$ with $m < 0$, and $n \in \mathbb{N}$, the nth coefficient of $P(m,k,N;z)$ equals (e.g., see [23], Chap. 3)

$$2\pi i^k \left|\frac{n}{m}\right|^{\frac{k-1}{2}} \sum_{\substack{c>0 \\ c\equiv 0 \pmod N}} \frac{K_k(m,n,c)}{c} I_{k-1}\left(\frac{4\pi\sqrt{|mn|}}{c}\right), \tag{1.1}$$

where I_{k-1} denotes the usual I-Bessel function and $K_k(m,n,c)$ denotes a certain Kloosterman sum (see (2.4) for the definition). For negative weights, certain (possibly) non-holomorphic Poincaré series $F(m,2-k,N;z)$ are natural (see (2.3) for the definition). Denote by $H_w(N)$ the space of harmonic Maass forms of weight w on $\Gamma_0(N)$ (see Sect. 2 for the definition) and let $H_w^{\infty}(N)$ be the subspace of those elements of $H_w(N)$ that are bounded at all cusps other than ∞. The nth Fourier coefficient of $F(m,2-k,N;z)$ is a sum involving Kloosterman sums $K_{2-k}(m,n,c)$ with a shape similar to (1.1). Series with Fourier expansions of this type play a prominent role in the works of Knopp, Rademacher, Zuckermann, and many others. See, for instance, [33].

Due to the obvious symmetry $|\pm mn| = |\pm nm|$ and the simple relation $\left|\pm\frac{m}{n}\right| = \left|\pm\frac{n}{m}\right|^{-1}$, (1.1) reveals that several important results about coefficients of modular forms and harmonic Maass forms manifest themselves through the symmetries of the Kloosterman sum. Firstly, whenever $k \in \mathbb{Z}$, the Kloosterman sum is symmetric in m and n. As a result, the nth Fourier coefficient of $F(m,2-k,N;z)$ equals $\left|\frac{m}{n}\right|^{k-1}$ times the mth Fourier coefficient of $F(n,2-k,N;z)$ (see [19], Theorem 3.4).

For $k \in \frac{1}{2}\mathbb{Z} \setminus \mathbb{Z}$, a slightly more complicated symmetry exists. Namely, (for a proof, see, e.g., Proposition 3.1 of [8])

$$K_k(m,n,c) = (-1)^{k+\frac{1}{2}} i K_{2-k}(n,m,c).$$

Consequentially, the nth Fourier coefficient of $F(m,2-k,N;z)$ is essentially equal to the negative of the mth Fourier coefficient of $P(n,k,N;z)$. The resulting identities among Fourier coefficients are referred to as *duality*. Duality, in this context, was studied by Zagier [39], who showed that the traces of singular moduli are Fourier coefficients of a weight $\frac{1}{2}$ weakly holomorphic modular form and then related these traces to Fourier coefficients of weight $\frac{3}{2}$ modular forms. Zagier's work gave a new perspective on a result of Borcherds [5], relating what are now known as *Borcherds products* to coefficients of weakly holomorphic modular forms. To illustrate this famous result, consider the weight 4 Eisenstein series for $\mathrm{SL}_2(\mathbb{Z})$, $(q = e^{2\pi i z})$

$$E_4(z) := 1+240\sum_{n\geq 1}\left(\sum_{d|n} d^3\right)q^n = (1-q)^{-240}\left(1-q^2\right)^{26760}\cdots = \prod_{n\geq 1}(1-q^n)^{c(n)}.$$

Borcherds related the exponents $c(n)$ to the Fourier coefficients a certain weight $\frac{1}{2}$ weakly holomorphic modular form.

The proof through Kloosterman sums of the duality shown by Zagier outlined here is due to Jenkins [24]. This was later generalized by the first author and Ono [8] to a duality in the more general setting of harmonic Maass forms.

Duality has continued to be a central theme in the literature surrounding automorphic forms. For example, Bruinier and Ono [12] have shown a natural way to map the Borcherds exponents to coefficients of a p-adic modular form through a certain differential operator. Duality was extended by Folsom and Ono, and Zwegers [20,43] to relate coefficients of different mock modular forms. Duality has also been extended by Rouse [35] to Hilbert modular forms and to Maass-Jacobi forms by the first author and Richter [10].

For every $k \in \frac{1}{2}\mathbb{Z}$, a trival change of variables (namely, $d \to -d$, see (2.4)) yields

$$K_{2-k}(m,n,c) = \overline{K_k(-m,-n,c)}, \tag{1.2}$$

from which one obtains a natural relation between the nth Fourier coefficient of $F(m, 2-k, N; z)$ and the $-n$th Fourier coefficient of $P(m,k,N;z)$. This relation plays a prominent role in the theory of harmonic Maass forms. In particular, it governs the image of $F(m, 2-k, N; z)$ under the weight $2-k$ antiholomorphic differential operator:

$$\xi_{2-k} := 2iy^{2-k}\overline{\frac{\partial}{\partial \bar{z}}}.$$

Since ξ_{2-k} is essentially the Maass weight-lowering operator (see (2.5) in Sect. 2.3), if $\mathcal{M} \in H_{2-k}^{\infty}(N)$, then $\xi_{2-k}(\mathcal{M})$ is a weight k modular form. In particular, from (1.2), we may deduce that $\xi_{2-k}(F(m, 2-k, N; z))$ equals a certain nonzero constant times $P(m,k,N;z)$ (see (2.8) for a precise statement). The surjectivity of ξ_{2-k}, first proven by Bruinier and Funke [11], follows.

Remark. We exclude the cases when the weight is $0 \leq k \leq 2$. In such cases, the convergence of the Poincaré series is delicate (see, e.g., [30] and the expository survey [15]). Moreover, the Fourier expansions of modular forms of small weight are handled by Knopp [27] and for harmonic weak Maass forms of small weight by Pribitkin [31,32].

1.1 Differential Operators via Kloosterman Sum Symmetries

We exploit another simple relation between Kloosterman sums. Whenever $k \in \mathbb{Z}$ there is an additional symmetry which occurs because the Kloosterman sum is independent of the weight $k \in \mathbb{Z}$. In particular,

$$K_k(-m,-n,c) = K_{2-k}(-m,-n,c) \tag{1.3}$$

so that (1.2) leads to a relation between the coefficients of $F(m, 2-k, N; z)$ and $F(-m, 2-k, N; z)$. We define the *flipping operator* $\mathcal{F}$ on Poincaré series by

$$F(m, 2-k, N; z) \mapsto F(-m, 2-k, N; z).$$

Since $\{F(m, 2-k, N; z) : m \in \mathbb{Z}\}$ is a basis for $H^{\infty}_{2-k}(N)$, we may extend the operator $\mathcal{F}$ to all of $H^{\infty}_{2-k}(N)$ by linearity. Moreover, when $k > 2$ and $\mathcal{M} \in H^{\infty}_{2-k}(N)$, the growth of $\mathcal{M}(z)$ as $z \to i\infty$ uniquely determines $\mathcal{M}$ as a linear combination of Poincaré series, and hence it is simple to determine the representation by this basis. Alternatively, for $f \in H^{\infty}_{2-k}(N)$, one may define $\mathcal{F}$ in terms of the weight raising operator by

$$\mathcal{F}(f) = y^{k-2}\overline{R^{k-2}_{2-k}(f)},$$

where R^{k-2}_{2-k} is the $(k-2)$-fold Maass raising operator, as defined in (2.6). We investigate this connection in Sect. 2.3:

Remarks. 1. After completion of this chapter, the authors learned that the flipping operator is independently studied from a different perspective by Fricke and will be included in his forthcoming thesis [21] advised by Zagier. Moreover, the referee pointed out that the flipping operator appears in another context in the work of Knopp [25] and Knopp–Lehner [28].

2. Denote by $M^{\infty}_{w}(N) \subseteq M^{!}_{w}(N)$ the subspace of those forms that are bounded at all cusps other than ∞. In this notation, the operator $\mathcal{F}$ gives a mapping

$$\mathcal{F} = \mathcal{F}_{k,N} : H^{\infty}_{2-k}(N) \to H^{\infty}_{2-k}(N)/M_{2-k}(N).$$

3. Although we restrict ourselves in this chapter to forms with bounded growth at cusps other than ∞, the general case would follow similarly after examining Poincaré series with growth only occurring in one of the other cusps. The cusp ∞ plays a prominent role here based on the fact that recent applications have emphasized forms with this property (see, e.g., [8, 13, 14]).

The operator D^{k-1}, where $D := \frac{1}{2\pi i}\frac{\partial}{\partial z}$, serves as a counterpart to ξ_{2-k} for $k \in 2\mathbb{N}$. The role of D^{k-1} in questions involving the algebraicity of Fourier coefficients is investigated in [14, 22]. Here, we exploit the symmetries given in (1.2) and (1.3) in order to relate the operators D^{k-1} and ξ_{2-k} through $\mathcal{F}$.

Theorem 1.1. *For $k > 2$ an integer and $\mathcal{M} \in H_{2-k}^{\infty}(N)$, we have*

$$D^{k-1}(\mathcal{M}) = (-4\pi)^{1-k}\,\Gamma(k-1)\,\xi_{2-k}(\mathcal{F}(\mathcal{M})).$$

Remark. If $\mathcal{M}(z) = \sum_{n\in\mathbb{Z}} c_n(y)e^{2\pi inx} \in H_{2-k}^{\infty}(N)$, then the operator ξ_{2-k} may (essentially) be viewed as extracting those coefficients with $n < 0$ while those with $n > 0$ are extracted by D^{k-1}.

The above discussion suggests that we could proceed by directly calculating the Fourier expansions of Poincaré series. Computing the derivatives on the Fourier expansion and using the symmetries of the Kloosterman sums then yields the theorem. Instead, we compute the derivatives on the Whittaker functions which are averaged to form the Poincaré series. This is possible because D^{k-1} and ξ_{2-k} are related to the Maass weight raising and lowering operators which commute with the action of $\Gamma_0(N)$. In fact, Bol's famous identity ([4], see also [18]) equates D^{k-1} to the $(k-1)$-fold repeated application of the weight raising operator. The technique presented here does not directly use the symmetry given in (1.3) but rather works through the raising and lowering operators.

1.2 Applications of Flipping

We revisit some existing results and some results known to experts with the fresh perspective engendered by Theorem 1.1.

In Ramanujan's last letter to Hardy (see pages 127–131 of [34]), he introduced 17 examples of functions which he called *mock theta functions*. For example, he defined

$$f(q) := 1 + \sum_{n=1}^{\infty} \frac{q^{n^2}}{\prod_{r=1}^{n}(1+q^r)^2}.$$

He noted that they satisfied properties similar to modular forms (although he referred to modular forms as "theta functions") and also stated that certain linear combinations of his mock theta functions were indeed modular forms. Although many of these properties were proven over the course of the next 80 years (e.g., see [1–3, 36, 37]), however, even a rigorous definition of Ramanujan's mock theta functions failed to present itself. Zwegers [41,42] finally placed Ramanujan's mock theta functions into a theoretical framework. In particular, if h is a mock theta function, then he constructed an associated harmonic Maass form $\mathcal{M}_h$ such that the function

$$g_h := \xi_{\frac{1}{2}}(\mathcal{M}_h) = \xi_{\frac{1}{2}}(\mathcal{M}_h - h)$$

is a unary theta function of weight $3/2$. Following Zagier [40], we call g_h the *shadow* of h.

By work of Bruinier–Funke [11], for any weakly holomorphic modular form g of weight k, there exists a "mock-like" holomorphic function h with shadow g. Following Zagier, we will call h a *mock modular form.* More precisely, there is a harmonic Maass form $\mathcal{M}_h$ naturally associated to h for which $\xi_{2-k}(\mathcal{M}_h) = \xi_{2-k}(\mathcal{M}_h - h) = g$. For each modular form g, we call the resulting harmonic Maass form a *lift* of g.

The existence of a lift or, equivalently, the surjectivity of $\xi_{2-k} : H_{2-k}^{\infty}(N) \to M_k^{\infty}(N)$, combined with Theorem 1.1, implies that the operator D^{k-1} is also surjective. Let $H_w^{\mathrm{cusp}}(N)$ be the subspace of $H_w^{\infty}(N)$ that maps to $S_{2-w}(N)$, the subspace of weight $2-w$ cusp forms, under ξ_w. This gives the following theorem which is essentially contained in Theorems 1.1 and 1.2 of [14]. In [14], Nebentypus is allowed and the restriction that growth only occurs at the cusp ∞ is not made, but the image under ξ_w is restricted to $S_{2-w}(N)$.

Theorem 1.2. *If* $2 < k \in \mathbb{Z}$ *and* $\mathcal{M} \in H_{2-k}^{\infty}(N)$, *then* $D^{k-1}(\mathcal{M}) \in M_k^{\infty}(N)$. *Moreover, in the notation of* (2.1),

$$D^{k-1}(\mathcal{M}(z)) = (-4\pi)^{1-k}(k-1)!c_{\mathcal{M}}^{-}(0) + \sum_{\substack{n \gg -\infty \\ n \neq 0}} c_{\mathcal{M}}^{+}(n)n^{k-1}q^n.$$

The image of the map

$$D^{k-1} : H_{2-k}^{\mathrm{cusp}}(N) \longrightarrow M_k^{\infty}(N)$$

consists of those $h \in M_k^{\infty}(N)$ *which are orthogonal to cusp forms (see Sect. 3 for the definition) which also have constant term 0 at all cusps of* $\Gamma_0(N)$. *Furthermore, the map*

$$D^{k-1} : H_{2-k}^{\infty}(N) \longrightarrow M_k^{\infty}(N)$$

is onto.

Implicit in the previous theorem are lifts of weakly holomorphic modular forms. Lifts of weight $3/2$ unary theta functions were given by Zwegers [42]. He gave explicit constructions in terms of Lerch sums, yielding mock modular forms of weight $1/2$. Lifts of weight $1/2$ modular forms were constructed by the first author, Folsom, and Ono [6]. The forms they construct are related to the hypergeometric series occurring in the Rogers–Fine identity. Lifts of general cusp forms in $S_k(N)$ were treated by the first author and Ono in [9], using Poincaré series. Duke et al. [17] recently constructed lifts of the weight $\frac{3}{2}$ weakly holomorphic modular forms that are Zagier's traces of singular moduli generating functions [39].

The flipping operator extends the lift in [9] to a lift for all weakly holomorphic modular forms. Given $g(z) = \sum_{n \gg -\infty} c_g(n)q^n \in M_k^{\infty}(N)$ with $k > 2$, define

$$\mathcal{P}(g)(z) := (k-1)^{-1}\overline{c_g(0)}y^{k-1} - (4\pi)^{1-k}\sum_{n \neq 0}\overline{c_g(-n)}|n|^{1-k}\Gamma(k-1; -4\pi yn)\,q^n,$$

where $\Gamma(\alpha; x) := \int_x^\infty e^{-t} t^{\alpha-1} dt$ is the *incomplete gamma function*. We note that for $g \in S_k(N)$, our definition matches that of $4^{1-k} g^*$ given by Zagier [40]. The following theorem describes the lifts of interest, which will be given in terms of Poincaré series.

Theorem 1.3. *For any $k \in \frac{1}{2}\mathbb{Z}$, $k > 2$, $N \in \mathbb{N}$, and $g \in M_k^\infty(N)$, the following are true:*

1. *There exists a harmonic Maass form $\mathcal{L}(g) \in H_{2-k}^\infty(N)$ such that*

$$\mathcal{L}(g) - \mathcal{P}(g)$$

   is a holomorphic function on $\mathbb{H}$.
2. *We have*

$$\xi_{2-k}(\mathcal{L}(g)) = \xi_{2-k}(\mathcal{P}(g)) = g.$$

Remark. The holomorphic function $\mathcal{L}(g) - \mathcal{P}(g)$ is typically not modular but mock modular. Theorem 1.3 allows us to deduce its transformation properties rather easily since the transformation properties of $\mathcal{P}(g)$ may be deduced from the transformation properties of g.

The interrelation between weakly holomorphic modular forms and their lifts have led to better understanding of arithmetic information of both modular forms and harmonic Maass forms. The forms constructed by Duke et al. [17] are related to certain cycle integrals of modular functions. Bruinier, Ono, and the third author [14] showed that the vanishing of the Hecke eigenvalues of a Hecke eigenform g implies the algebraicity of the coefficients of an appropriate lift of g. In other work, Bruinier and Ono [13] proved that the vanishing of the central value of the derivative of a weight 2 modular L-functions is related to the algebraicity of a certain Fourier coefficient of a harmonic Maass form.

Theorem 1.1 shows that for each $g \in S_k(N)$, one may find $M, M^* \in H_{2-k}^\infty(N)$ so that

$$\xi_{2-k}(M) = g \qquad \text{and} \qquad D^{k-1}(M^*) = g.$$

Recent work of Guerzhoy et al. [22] and the first two authors and Guerzhoy [7] shows that certain linear combinations of these two "lifts" are p-adic modular forms. These works lead naturally to the following question: Let M be a harmonic Maass form and set $g := \xi_{2-k}(M)$ and $h := D^{k-1}(M)$. From Theorem 1.3, we know that a harmonic Maass form M^* exists such that $h = \xi_{2-k}(M^*)$. Is $g = D^{k-1}(M^*)$?

Corollary 1.4. *Suppose that $k > 2$ is an integer, $M \in H_{2-k}^\infty(N)$, and g and h are defined as above. If $M^* \in H_{2-k}^\infty(N)$ satisfies $\xi_{2-k}(M^*) = h$, then the projection of $D^{k-1}(M^*)$ onto the space of cusp forms is g.*

Furthermore, there exists a choice of M^* such that $D^{k-1}(M^*) = g$.

Remark. In light of Theorems 1.1 and 1.2, we may write $D^{k-1}(M^*) = g + \widetilde{g}$ with $\widetilde{g} \in D^{k-1}\left(M_{2-k}^\infty(N)\right)$. The subspace $D^{k-1}\left(M_{2-k}^\infty(N)\right)$ has a number of exceptional properties. For example, the coefficients of a weakly holomorphic modular form in that space, when chosen to be algebraic, have high p-divisibility ([22], Proposition 2.1). Therefore, it is natural to factor out by $M_{2-k}^{\infty}(N)$, and the statement of Corollary 1.4 may be taken to say that $M^* \equiv \mathcal{F}(M) \pmod{M_{2-k}^{\infty}(N)}$.

1.3 Choosing a Lift

As is suggested in Corollary 1.4, lifts are not unique because the kernel of ξ_{2-k} is nontrivial. In fact, Bruinier and Funke [11] have shown that the kernel of ξ_{2-k} is $M_{2-k}^{!}(N)$. The lift described in [9] is defined on Poincaré series, and relations between the classical holomorphic Poincaré series make our lift unique up to a choice of a weakly holomorphic modular form. We present a procedure to make a choice of one such lift which is independent of the realization of $g \in M_k^{\infty}(N)$ as a linear combination of Poincaré series.

In order to describe the framework for our lift, we will need to introduce some notation. For M, a harmonic Maass form with Fourier expansion as in (2.1), there is a polynomial $G_M(z) = \sum_{n \leq 0} c_M^+(n) q^n \in \mathbb{C}\left[q^{-1}\right]$ such that $M^+(z) - G_M(z) = O\left(\mathrm{e}^{-\delta y}\right)$ as $y = \mathrm{Im}\,(z) \to \infty$ for some $\delta > 0$. Here and throughout, we denote $z = x + iy$ with $x, y \in \mathbb{R}$ $(y > 0)$. We call G_M the *principal part* of M at infinity.

Let $M_k := M_k(1)$ and define H_k, S_k, and $M_k^!$ similarly. Given a weakly holomorphic form $g \in M_k^!$, we explicitly define a harmonic Maass form $G \in H_{2-k}$ such that $\xi_{2-k}(G) - g \in S_k$. Since the principal part of g determines g modulo forms in S_k, we will obtain a lift which is explicit and well defined if for every $g \in S_k$, we construct a unique, explicit lift $\widetilde{g} \in H_{2-k}$ with $\xi_{2-k}\left(\widetilde{g}\right) - g = 0$. The difficulty in this task lies in finding a lift which commutes with the algebra of S_k so that $\widetilde{g} + \widetilde{h} = \widetilde{g + h}$ for $g, h \in S_k$. In particular, if one has two different bases for S_k, the lift must be independent of the basis representation. We call such a lift *canonical.*

Additionally, the lifts used in many applications are *good* choices of lifts (see Sect. 2 for the definition). We demonstrate a canonical lift for weakly holomorphic forms, which, in the case of normalized Hecke eigenforms, is good. To state our theorem, we introduce some notation. For $g \in S_k$, we denote the norm with respect to the usual Petersson scalar product by $||g||$. For $M \in H_{2-k}$, let

$$A(M) := \inf\{n \in \mathbb{Z} : c_M^+(n) \neq 0\}. \tag{1.4}$$

Theorem 1.5. *Let $k > 2$ and $g \in M_k^!$ be given. Choose $M \in H_{2-k}$ with $A(M)$ maximal among all $M \in H_{2-k}$ with $\xi_{2-k}(M) = g$. Then M is a canonical lift of g. Moreover, if $g \in S_k$ is a normalized Hecke eigenform, then $||g||^{-2} M$ is good for g.*

Remark. For simplicity, we have constrained ourselves to the case of level 1 forms when considering canonical lifts. We will discuss the differences in the general level case briefly at the end of Sect. 4.

This chapter is organized as follows: In Sect. 2, we recall some basic facts concerning harmonic Maass forms and Maass-Poincaré series and the relations between weight $2-k$ and weight k Poincaré series given by the operators ξ_{2-k} and D^{k-1} (when $k \in \mathbb{N}$). In Sect. 3, we prove Theorems 1.1 and 1.2 as well as Theorem 1.3 and its corollaries. In Sect. 4, we prove Theorem 1.5.

2 Harmonic Maass Forms

In this section, we recall the definition of harmonic Maass form and the properties of harmonic Maass forms which are necessary to prove our results. A good reference for much of the theory recalled in this section is [11].

2.1 Basic Notations and Definitions

As usual, it is assumed that if $k \in \frac{1}{2}\mathbb{Z} \setminus \mathbb{Z}$, then $N \equiv 0 \pmod 4$. We define the weight k *hyperbolic Laplacian* by

$$\Delta_k := -y^2\left(\frac{\partial^2}{\partial x^2} + \frac{\partial^2}{\partial y^2}\right) + iky\left(\frac{\partial}{\partial x} + i\frac{\partial}{\partial y}\right).$$

Moreover, for $\gamma = \left(\begin{smallmatrix} a & b \\ c & d \end{smallmatrix}\right) \in \mathrm{SL}_2(\mathbb{Z})$ when $k \in \mathbb{Z}$, respectively, for $\gamma \in \Gamma_0(4)$ when $k \in \frac{1}{2}\mathbb{Z} \setminus \mathbb{Z}$, and any function $g : \mathbb{H} \to \mathbb{C}$, we let

$$g \mid_k \gamma(z) := j(\gamma, z)^{-2k} g\left(\frac{az+b}{cz+d}\right),$$

where

$$j(\gamma, z) := \begin{cases} \sqrt{cz+d} & \text{if } k \in \mathbb{Z}, \gamma \in \mathrm{SL}_2(\mathbb{Z}), \\ \left(\frac{c}{d}\right) \varepsilon_d^{-1} \sqrt{cz+d} & \text{if } k \in \frac{1}{2}\mathbb{Z} \setminus \mathbb{Z}, \gamma \in \Gamma_0(4), \end{cases}$$

where for odd integers d, ε_d is defined by

$$\varepsilon_d := \begin{cases} 1 & \text{if } d \equiv 1 \pmod 4, \\ i & \text{if } d \equiv 3 \pmod 4. \end{cases}$$

Definition 2.1. A *harmonic Maass form* of weight k on $\Gamma = \Gamma_0(N)$ is a smooth function $g : \mathbb{H} \to \mathbb{C}$ satisfying:

(i) $g\mid_k \gamma(z) = g(z)$ for all $\gamma \in \Gamma$.
(ii) $\Delta_k(g) = 0$.
(iii) g has at most linear exponential growth at each cusp of Γ.

We note that $\mathcal{M} \in H_w(N)$ for $w \leq \frac{1}{2}$ has a Fourier expansion of the shape

$$\mathcal{M}(z) = c_{\mathcal{M}}^{-}(0)y^{1-w} + \sum_{\substack{n \ll +\infty \\ n\neq 0}} c_{\mathcal{M}}^{-}(n)\Gamma\left(1-w; -4\pi ny\right)q^n + \sum_{n\gg -\infty} c_{\mathcal{M}}^{+}(n)q^n. \tag{2.1}$$

We call $\mathcal{M}^+(z) := \sum_{n\gg-\infty} c_{\mathcal{M}}^{+}(n)q^n$ the *holomorphic part* of $\mathcal{M}$ and $\mathcal{M}^- := \mathcal{M} - \mathcal{M}^+$ the *non-holomorphic part* of $\mathcal{M}$.

Following [14], one says that a harmonic Maass form $f \in H_{2-k}(N)$ is *good* for a normalized Hecke eigenform $g \in S_k(N)$ if it satisfies the following properties:

1. The principal part of f at the cusp ∞ belongs to $F_g\left[q^{-1}\right]$, with F_g the number field obtained by adjoining the coefficients of g to $\mathbb{Q}$.
2. The principal parts of f at the other cusps of $\Gamma_0(N)$, defined analogously, are constant.
3. We have $\xi_{2-k}(f) = ||g||^{-2}g$.

One sees immediately by the second condition that $f \in H_{2-k}^{\infty}(N)$.

2.2 Poincaré Series

We describe two families of Poincaré series. Let m be an integer, and let $\varphi_m : \mathbb{R}^+ \to \mathbb{C}$ be a function which satisfies $\varphi_m(y) = O(y^\alpha)$, as $y \to 0$, for some $\alpha \in \mathbb{R}$. With $e(r) := \mathrm{e}^{2\pi ir}$, let

$$\varphi_m^*(z) := \varphi_m(y)e(mx).$$

Such functions are fixed by the translations, elements of $\Gamma_\infty := \left\{\pm\left(\begin{smallmatrix}1 & n\\ 0 & 1\end{smallmatrix}\right) : n \in \mathbb{Z}\right\}$. Given this data, define the generic Poincaré series

$$\mathbb{P}(m,k,\varphi_m,N;z) := \sum_{A\in\Gamma_\infty\backslash\Gamma_0(N)} \varphi_m^* \mid_k A(z).$$

We note that the Poincaré series $\mathbb{P}(m,k,\varphi_m,N;z)$ converges absolutely for $k > 2-2\alpha$, where α is the growth factor of $\varphi_m(y)$ as given above and by construction satisfies the modularity property $\mathbb{P}(m,k,\varphi_m,N;z)\mid_k \gamma(z) = \mathbb{P}(m,k,\varphi_m,N;z)$ for every $\gamma \in \Gamma_0(N)$. In this notation, the classical family of holomorphic Poincaré series (see, e.g., [23], Chap. 3) for $k \geq 2$ is given by

$$P(m,k,N;z) = \mathbb{P}(m,k,e(imy),N;z). \tag{2.2}$$

The Maass-Poincaré series (see, e.g., [19]) are defined by

$$F(m,2-k,N;z) := \mathbb{P}(-m,2-k,\varphi_{-m},N;z), \tag{2.3}$$

where

$$\varphi_{-m}(z) := \begin{cases} \mathcal{M}_{1-\frac{k}{2}}(-4\pi m y) & \text{if } k<0 \text{ and } m\neq 0,\\ |m|^{1-k}\,\mathcal{M}_{\frac{k}{2}}(-4\pi m y) & \text{if } k>2 \text{ and } m\neq 0,\\ 1 & \text{if } k<0, m=0,\\ (4\pi y)^{k-1} & \text{if } k>2, m=0. \end{cases}$$

Here, for complex s,

$$\mathcal{M}_s(y) := |y|^{\frac{k}{2}-1} M_{(1-\frac{k}{2})\mathrm{sgn}(y),\, s-\frac{1}{2}}(|y|),$$

where $M_{\nu,\mu}(z)$ is the usual M-Whittaker function.

Since φ_m^* is annihilated by the hyperbolic Laplacian and Δ_{2-k} commutes with the weight $2-k$ group action of $\Gamma_0(N)$, a consideration of the growth of φ_m^* at all of the cusps shows that $F(m,2-k,N;z) \in H_{2-k}^{\infty}(N)$. In the case $k<0$, one has

$$F(m,2-k,N;z) = P(-m,2-k,N;z).$$

In order to describe the coefficients of the Poincaré series, we define the Kloosterman sums

$$K_k(m,n,c) := \begin{cases} \sum_{d \pmod{c}^*} e\left(\frac{m\overline{d}+nd}{c}\right) & \text{if } k\in\mathbb{Z},\\ \sum_{d \pmod{c}^*} \left(\frac{c}{d}\right)^{2k} \varepsilon_d^{2k} e\left(\frac{m\overline{d}+nd}{c}\right) & \text{if } k\in\frac{1}{2}\mathbb{Z}\setminus\mathbb{Z} \text{ (and } 4\mid c), \end{cases} \tag{2.4}$$

where $\left(\frac{c}{d}\right)$ denotes the Jacobi symbol. Here, d runs through the primitive residue classes modulo c, and $\overline{d}$ is defined by the congruence $d\overline{d} \equiv 1 \pmod{c}$.

A calculation analogous to that for Theorem 3.4 of [19] yields the following result.

Lemma 2.2. *If $k>2$ and $m\in\mathbb{Z}$, then the principal part of $F(m,2-k,N;z)$ is*

$$\delta_{m>0}\Gamma(k)\,|m|^{1-k}\,q^{-m} + c(m,k,N),$$

where $\delta_{m>0}=1$ if $m>0$ and 0 otherwise, and $c(m,k,N)$ is a constant depending on k, m, and N. When $k\in 2\mathbb{Z}$, we have

$$c(m,k,N) = -(2\pi i)^k \sum_{\substack{c>0\\ c\equiv 0 \pmod{N}}} \frac{K_{2-k}(-m,0,c)}{c^k}.$$

The principal part of $P(m,k,N;z)$ is $\delta_{m\le 0} q^m$.

Remark. For $m \in \mathbb{Z}$ and $k \in 2\mathbb{Z}$, we have $c(m,k,N) = (-1)^k \overline{c(-m,k,N)}$.

Moreover, the full Fourier expansion of $F(m,k,N;z)$ is computed in Theorem 3.4 of [19]. We omit the full Fourier expansion, however, because it is not needed for our purposes.

2.3 Raising and Lowering Operators

The Maass raising and lowering operators are given by

$$R_k := 2i\frac{\partial}{\partial z} + ky^{-1} \quad \text{and} \quad L_k := -2iy^2\frac{\partial}{\partial \overline{z}}. \tag{2.5}$$

For a real analytic function f satisfying the weight k modularity property $f \mid_k \gamma(z) = f(z)$ for every $\gamma \in \Gamma_0(N)$ which is an eigenfunction under Δ_k with eigenvalue s, $R_k(f)(z)$ (respectively $L_k(f)(z)$) satisfies weight $k+2$ (resp. $k-2$) modularity and is an eigenfunction under Δ_{k+2} (resp. Δ_{k-2}) with eigenvalue $s+k$ (resp. $s-k+2$). This follows by the commutator relation

$$-\Delta_k = L_{k+2}R_k + k = R_{k-2}L_k.$$

Define for a positive integer n

$$R_k^n := R_{k+2(n-1)} \circ \cdots \circ R_{k+2} \circ R_k \tag{2.6}$$

and let R_k^0 be the identity. If $f \in H_{2-k}^{\infty}(N)$, then $f^* := y^{k-2}\overline{R_{2-k}^{k-2}(f)} \in H_{2-k}^{\infty}(N)$, as noted in Remark 7 in [14]. Furthermore, by Bol's identity ([4], see also [18]), that is

$$R_{2-k}^{k-1} = (-4\pi)^{k-1}D^{k-1}, \tag{2.7}$$

one has (for $f \in H_{2-k}^{\mathrm{cusp}}(N)$ see Remark 7 in [14]) that

$$\xi_{2-k}(f^*) = y^{-k}\overline{L_k(f^*)} = R_{2-k}^{k-1}(f) = (-4\pi)^{k-1}D^{k-1}(f).$$

So, up to a constant factor, M^* behaves as $\mathcal{F}(f)$ under ξ_{2-k}. On the other hand, one may compute the Fourier expansion of f^* and see that it is the same as that for $\mathcal{F}(f)$. In this chapter, we proceed differently and come about $\mathcal{F}$ on the level of Poincaré series.

2.4 Derivatives of Poincaré Series

The following relations, derived in the lemma below, are important for deducing the theorems of this chapter.

Lemma 2.3. *For* $m \in \mathbb{Z}$*, the action of the operators* ξ_{2-k} *and* D^{k-1} *on* $F(m, 2-k, N; z)$ *is given by*

$$\xi_{2-k}\left(F(m,2-k,N;z)\right) = (k-1)\,(4\pi)^{k-1}\,P(m,k,N;z), \tag{2.8}$$

$$D^{k-1}\left(F(m,2-k,N;z)\right) = \Gamma(k)(-1)^{k-1}P(-m,k,N;z), \tag{2.9}$$

where in (2.9) *we require k to be an integer.*

Proof. For $m > 0$, the relation (2.8) is noted (up to the constant) in Remark 3.10 of [11], while the constant is explicitly computed in Theorem 1.2 of [9]. The $m > 0$ case of (2.9) is given in (6.8) of [14].

The lemma follows from the following relations. For $k > 2$, we have

$$\xi_{2-k}\left(\varphi_m^*\right) = (k-1)\,(4\pi)^{k-1}\,q^m. \tag{2.10}$$

Additionally, whenever k is an even integer, we have

$$D^{k-1}\left(\varphi_m^*\right) = -\Gamma(k)q^{-m}. \tag{2.11}$$

The relations (2.10) and (2.11) together with the fact that ξ_{2-k} and D^{k-1} commute with the group law will immediately imply (2.8) and (2.9). Since the six calculations ($m < 0$, $m = 0$, and $m > 0$ for each) to establish (2.10) and (2.11) are all similar, we include only the case of $D^{k-1}(\varphi_{-m}(z))$ with $m < 0$. In this case, we have

$$\varphi_{-m}^*(z) = |m|^{1-k}\,\mathrm{e}^{-2\pi i m x}(4\pi|m|y)^{\frac{k}{2}-1}M_{1-\frac{k}{2},\frac{k-1}{2}}(4\pi|m|y).$$

Applying the change of variables $2\pi|m|y \to y$ and $2\pi|m|x \to x$ and relations between the W-Whittaker and M-Whittaker functions (see page 346 of [38]), we consider

$$|m|^{1-k}\,\mathrm{e}^{ix}(2y)^{\frac{k}{2}-1}\left((k-1)\exp\left(\pi i\left(1-\frac{k}{2}\right)\right)W_{\frac{k}{2}-1,\frac{k-1}{2}}(-2y)\right.$$
$$\left. -(-1)^k\Gamma(k)W_{1-\frac{k}{2},\frac{k-1}{2}}(2y)\right),$$

for which we denote the two terms as $f_1(z) + f_2(z)$. Direct computation gives

$$\frac{\partial}{\partial \bar{z}}(f_2)(z) = 0 \qquad \text{and} \qquad \frac{\partial}{\partial x}(f_2)(z) = if_2(z).$$

Hence, using $\frac{\partial}{\partial z} = \frac{\partial}{\partial x} - \frac{\partial}{\partial \overline{z}}$, we obtain

$$D^{k-1}(f_2)(2\pi z) = f_2(2\pi z).$$

Thus a change of variables and using $W_{1-\frac{k}{2},\frac{k}{2}-1}(2y) = (2y)^{1-\frac{k}{2}}e^{-y}$ yields

$$D^{k-1}(f_2)\,(2\pi|m|z) = (-1)^{k-1}\,\Gamma(k)q^{-m}.$$

It remains to show that $D^{k-1}(f_1)(z) = 0$. For this, let

$$g_r(z) := |m|^{1-k}\,e^{ix}(-2y)^{\frac{k}{2}-r}W_{\frac{k}{2}-r,\frac{k-1}{2}}(-2y).$$

From the third three-term recurrence relation

$$yW'_{k,m}(y) = \left(k - \frac{y}{2}\right)W_{k,m}(y) - \left(m^2 - \left(k - \frac{1}{2}\right)^2\right)W_{k-1,m}(y)$$

for the Whittaker function (see pages 350–352 of [38]) giving a relation for the derivative of the W-Whittaker function, we obtain

$$\frac{\partial}{\partial z}g_r(z) = -\frac{i}{2y}(k-2r)g_r(z) + i\left(r^2 - r(k-1)\right)g_{r+1}(z)$$

Hence,

$$R_{2r-k}(g_r)(2\pi z) = \left(r^2 - r(k-1)\right)g_{r+1}(2\pi z).$$

Using this for $r = k-1$ and applying Bol's identity (2.7), we have

$$D^{k-1}(f_1)(2\pi z) = (k-1)R_{2-k}^{k-1}(g_1)(2\pi z) = 0,$$

as desired.

2.5 Bol's Identity

Bol's identity (2.7) states that D^{k-1} is essentially (up to a nonzero constant multiple) equal to R_{2-k}^{k-1}. The calculations of the previous section give the action of R_{2-k}^{k-1} on the Whittaker functions which define the Poincaré series that span the spaces of forms of interest in this chapter, and then we use the commutation relation

$$\left(R_{2-k}^{k-1}(f)\right)|_k\,\gamma\,(z) = R_{2-k}^{k-1}\left(f\,|_{2-k}\,\gamma\right)(z)$$

between R_{2-k}^{k-1} and the slash operator, valid for every real analytic function f. Alternatively, we can proceed by computing the Fourier expansion of the Maass-Poincaré series, obtaining an expansion as in (2.1). Differentiating term by term yields (2.8) and (2.9). This approach does not rely on the fact that the differential operator D^{k-1} commutes with the group action (which would follow from Bol's identity). Additionally, for integral k we have

$$q^m \Gamma(k-1; -4\pi m y) = \overline{q}^m Q_{k,m}(y)$$

where $Q_{k,m}$ is a polynomial of degree at most $k-2$. Thus a direct computation of D^{k-1} avoids an application of Bol's identity.

3 Proof of Theorems 1.1–1.3 and Corollary 1.4

Having established the necessary preliminaries, we are now ready to prove Theorem 1.3.

Proof (Proof of Theorem 1.3). Since the Poincaré series $\{F(m,k,N;z)\}_{m\in\mathbb{Z}}$ span $M_k^\infty(N)$ and the series $\{F(m,2-k,N;z)\}_{m\in\mathbb{Z}}$ span $H_{2-k}^\infty(N)$, it is enough to prove the result on the level of Poincaré series. Part (1) follows from (2.8) together with (2.1) and the fact that

$$\xi_{2-k}(\mathcal{P}(P(m,k,N;z))) = P(m,k,N;z).$$

In particular,

$$\frac{(4\pi)^{1-k}}{k-1} F(m,2-k,N;z) - \mathcal{P}\left(P(m,k,N;z)\right)$$

is the desired holomorphic function associated to the modular form $P(m,k,N;z)$. Part (2) follows from (2.8).

Having established the image of the Poincaré series under the operators D^{k-1} and ξ_{2-k} in Sect. 2.4, the fact that the Poincaré series form a basis will suffice to prove Theorem 1.1.

Proof (Proof of Theorem 1.1). The proof of this result follows immediately from (2.8) and (2.9).

Borcherds [5] has defined a regularized inner product $(g,h)^{\mathrm{reg}}$ for $g,h \in M_k^\infty(N)$ from which one can define orthogonality in the more general setting of weakly holomorphic modular forms. For cusp forms g and h, the regularized inner product reduces to the classical Petersson inner product. For $M \in H_{2-k}^{\mathrm{cusp}}(N)$, we define

$$h := \Gamma(k-1)\mathcal{F}(M) \in H_{2-k}^\infty(N).$$

By Lemma 2.2 and the remark following it, the constant terms of h and M satisfy

$$c_h^+(0) = \Gamma(k-1)(-1)^k \overline{c_M^+(0)}.$$

Combining this with Theorem 4.1 of [14] immediately leads to the following lemma (with the factor $\Gamma(k-1)$ correcting a typo from the original statement of Theorem 4.1), which is the most important computation toward calculating the image of D^{k-1}.

Lemma 3.1. *If $g \in M_k(N)$ and $M \in H_{2-k}^{\mathrm{cusp}}(N)$, then*

$$(-4\pi)^{k-1}\left(g, D^{k-1}(M)\right)^{\mathrm{reg}} = \frac{(-1)^k\Gamma(k-1)}{[\mathrm{SL}_2(\mathbb{Z}) : \Gamma_0(\mathrm{N})]} \sum_{\kappa\in\Gamma_0(N)\backslash P^1(\mathbb{Q})} w_\kappa \cdot c_g(0,\kappa)\overline{c_M^+(0,\kappa)},$$

where $c_g(0,\kappa)$ (resp. $c_M^+(0,\kappa)$) denotes the constant term of the Fourier expansion of g (resp. M) at the cusp $\kappa \in P^1(\mathbb{Q})$, and w_κ is the width of the cusp κ.

Proof (Proof of Theorem 1.2). The first part of the theorem follows from Theorem 1.1 and (2.9). The surjectivity of D^{k-1} on $H_{2-k}^{\infty}(N)$ follows from the surjectivity of ξ_{2-k} (see Theorem 3.7 of [11]) and Theorem 1.1.

Additionally, if $\mathcal{M} \in H_{2-k}^{\mathrm{cusp}}(N)$, it follows from the first part of the theorem and (2.8) that there exist $\alpha_m \in \mathbb{C}$ so that

$$\mathcal{M}(z) = \sum_{m>0} \alpha_m F(m, 2-k, N; z).$$

Thus, from Lemma 3.1 and the first part of the theorem, $D^{k-1}(\mathcal{M})$ is orthogonal to cusp forms, and the constant term at each cusp of $\Gamma_0(N)$ vanishes.

Conversely, assume that $h \in M_k^{\infty}(N)$ has vanishing constant term at any cusp of $\Gamma_0(N)$ and is orthogonal to cusp forms. From (2.9), we may take

$$f(z) = \sum_{m\in\mathbb{N}} \alpha_m F(m, 2-k, N; z) \in H_{2-k}^{\mathrm{cusp}}(N)$$

such that the principal parts of $D^{k-1}(f)$ and h at the cusps agree. Consequently, $h - D^{k-1}(f) \in S_k(N)$. In view of Lemma 3.1, the hypothesis on h and (2.9), we find that $h - D^{k-1}(f)$ is orthogonal to cusp forms. Hence it vanishes identically.

We conclude with the proof of Corollary 1.4.

Proof (Proof of Corollary 1.4). Writing $M \in H_{2-k}^{\infty}(N)$ in terms of Poincaré series, we have

$$M(z) = \sum_{m\in\mathbb{Z}} \alpha_m F(m, 2-k, N; z).$$

Then Theorem 1.1 implies that $(-4\pi)^{1-k}\Gamma(k-1)\mathcal{F}(M)$ is a lift of $h = D^{k-1}(M)$ and

$$M^* - (-4\pi)^{1-k}\Gamma(k-1)\mathcal{F}(M) \in M_{2-k}^{\infty}(N),$$

where $M^* \in H_{2-k}^{\infty}(N)$ is any harmonic Maass form satisfying $\xi_{2-k}(M^*)$, as given in the statement of Corollary 1.4. Applying Theorems 1.1 and 1.2, we obtain the assertion concerning $D^{k-1}(M^*)$.

4 A Canonical Lift

When $N = 1$, we use the abbreviations $P(m,k;z) := P(m,k,1;z)$ and $F(m,k;z) := F(m,k,1;z)$. For fixed $k > 2$ integral, let $\ell := \dim S_k$ and define $f_{2-k,m} \in M_{2-k}^!$ to be the unique weakly holomorphic modular form satisfying

$$f_{2-k,m}(z) = q^{-m} + O\left(q^{-\ell}\right).$$

Such weakly holomorphic modular forms were explicitly constructed in [16] as

$$f_{2-k,m}(z) := \begin{cases} E_{k'}(z)\Delta(z)^{-\ell-1}F_m(j(z)) & \text{if } m > \ell, \\ 0 & \text{if } m \leq \ell. \end{cases} \tag{4.1}$$

Here, $k' \in \{0,4,6,8,10,14\}$ with $k' \equiv 2-k \pmod{12}$, $E_{k'}$ is the Eisenstein series of weight k', Δ is the unique normalized Hecke eigenform of weight 12, and F_m is a generalized Faber polynomial of degree $m-\ell-1$ constructed recursively in terms of $f_{2-k,m'}$ with $m' < m$ to cancel higher powers of q. Finally, for $m \in \mathbb{Z}$, define

$$G_{m,2-k}(z) := F(m,2-k;z) - \delta_{m>0}\Gamma(k)\,|m|^{1-k}\, f_{2-k,m}(z).$$

Here, $\delta_{m>0}$ is defined as in Lemma 2.2. From Lemma 2.2 and the definition of $f_{2-k,m}$, the holomorphic part $G_{m,2-k}^+(z)$ of $G_{m,2-k}(z)$ satisfies

$$G_{m,2-k}^+(z) = O\left(q^{-\ell}\right). \tag{4.2}$$

The following explicit theorem implies Theorem 1.5.

Theorem 4.1. *Suppose that $2 < k \in 2\mathbb{Z}$ and $g \in M_k^!$ and write $g(z) = \sum_{m\in I} a_m P(m,k;z)$ for some index set $I \subset \mathbb{Z}$. Then the ξ_{2-k}-preimage choice*

$$\mathcal{L}(g(z)) = \mathcal{L}_I(g(z)) := \frac{1}{k-1}\sum_{m\in I}\frac{\overline{a_m}}{(4\pi)^{k-1}}G_{m,2-k}(z)$$

defines a canonical lifting from $M_k^!$ to H_{2-k}. Moreover, when $g \in S_k$ is a normalized Hecke eigenform, the lift $\mathcal{L}(g)/||g||^2$ is good for g.

Proof. One directly obtains that $\xi_{2-k}\left(\mathcal{L}(g(z))\right) = g$ from (2.8). Consider

$$\mathcal{G}(z) := \sum_{\substack{m \in I \\ m \leq 0}} a_m P(m,k;z),$$

Then $g - \mathcal{G} \in S_k$. Set

$$\mathbb{H}(z) := \mathcal{L}\left(\mathcal{G}(z)\right) = \frac{1}{k-1} \sum_{n \leq 0} \frac{\overline{a_n}}{(4\pi)^{k-1}} F(n, 2-k; z).$$

The following lemma, which is proved after we conclude the proof of Theorem 4.1, shows that $\mathbb{H} \in H_{2-k}$ is the unique lift of $\mathcal{G}$ with $\mathbb{H}^+$ having minimal growth at the cusp ∞.

Lemma 4.2. *With g as in Theorem 4.1, the function $\mathbb{H}$ is the unique $h \in H_{2-k}$ whose holomorphic part exhibits subexponential growth at the cusp ∞ and satisfies*

$$g - \xi_{2-k}(h) \in S_k. \tag{4.3}$$

Applying Lemma 4.2, we may assume that g is a cusp form in order to prove Theorem 4.1. We write $g(z) = \sum_{m \in I} a_m P(m,k;z)$ with some index set $I \subset \mathbb{N}$. From (4.2), we obtain

$$\mathcal{L}_I(g(z)) = \frac{1}{k-1} \sum_{m \in I} \frac{\overline{a_m}}{(4\pi)^{k-1}} G_{m,2-k}(z) = O\left(q^{-\ell}\right).$$

To show that the lift is independent of the choice of the index set, let $J \subset \mathbb{N}$ be given such that $g(z) = \sum_{m' \in J} a_{m'} P(m', k; z)$. Then

$$\mathcal{L}_I(g(z)) - \mathcal{L}_J(g(z)) = O\left(q^{-\ell}\right) \tag{4.4}$$

and

$$\xi_{2-k}\left(\mathcal{L}_I(g) - \mathcal{L}_J(g)\right) = g - g = 0.$$

Hence

$$\mathcal{L}_I(g) - \mathcal{L}_J(g) \in \ker\left(\xi_{2-k}\right) = M_{2-k}^!. \tag{4.5}$$

By the valence formula, we know that a weakly holomorphic modular form h that satisfies $h(z) = O(q^{-\ell})$ must be 0. Therefore, combining (4.4) and (4.5) yields

$$\mathcal{L}_I(g) = \mathcal{L}_J(g)$$

This finishes the proof of the first statement of the theorem.

To prove the second statement, assume that $g \in S_k$ is a normalized Hecke eigenform and $h \in H_{2-k}$ is a harmonic Maass form which is good for g. Thus the principal part of h is $\sum_{n \leq 0} c_n q^n$ with $c_n \in K_g$. By comparing principal parts, we have

$$h(z) = \sum_{n>0} \frac{c_{-n} n^{k-1}}{\Gamma(k)} F(n, 2-k; z)$$

since the difference has bounded principal part and maps to a cusp form under ξ_{2-k}. We have

$$\xi_{2-k}(h)(z) = \sum_{n>0} \frac{\overline{c_{-n}}}{\Gamma(k)} (k-1)\,(4\pi n)^{k-1}\; P(n, k; z) = \frac{g(z)}{||g||^2}.$$

Set $I := \{n \leq 0 : c_n \neq 0\}$. By definition,

$$\mathcal{L}\left(\frac{g}{||g||^2}\right)(z) = \mathcal{L}_I\left(\frac{g}{||g||^2}\right)(z) = \sum_{n>0} \frac{c_{-n} n^{k-1}}{\Gamma(k)} F(n, 2-k; z) - \sum_{n>0} c_{-n} f_{2-k,n}(z).$$

It follows that

$$\left(h - \mathcal{L}\left(\frac{g}{||g||^2}\right)\right) = \sum_{n>0} c_{-n} f_{2-k,n}.$$

Since $E_{k'}$, Δ^{-1}, and $F_m(j(\tau))$ all have rational (furthermore, integer) coefficients, the weakly holomorphic modular forms $f_{2-k,n}$ have rational coefficients by (4.1). It follows that $h - \mathcal{L}\left(g/||g||^2\right)$ has coefficients in K_g. Therefore, since the coefficients of the principal part of h and the principal part of $h - \mathcal{L}\left(g/||g||^2\right)$ are both in K_g, it follows that the coefficients of the principal part of $\mathcal{L}\left(g/||g||^2\right)$ are contained in K_g. Hence $\mathcal{L}\left(g/||g||^2\right)$ is also a good lift for g.

Proof (Proof of Lemma 4.2). Using (2.8) together with the fact that $P(m, k; z) \in S_k$ for $m \geq 1$ immediately implies (4.3). To show uniqueness, let $h \in H_{2-k}$ satisfy (4.3). Since the Poincaré series $P(n, k; z)$ span the space $M_k^!$, it follows that

$$\xi_{2-k}\left(h(z)\right) = \sum_{n \in \mathbb{Z}} b_n P(n, k; z)$$

for some $b_n \in \mathbb{C}$. By (4.3), we have that

$$g(z) - \sum_{n \in \mathbb{Z}} b_n P(n, k; z) \in S_k. \tag{4.6}$$

Comparing the principal parts of both summands in (4.6), one sees that $b_n = a_n$ for every $n \leq 0$. It follows that

$$h(z) - \mathbb{H}(z) = \frac{(4\pi)^{1-k}}{k-1} \sum_{n>0} \overline{b_n} F(n, 2-k; z).$$

This has principal part (up to the constant term) equal to

$$\Gamma(k-1)\sum_{n>0}(4\pi n)^{1-k}\overline{b_n}q^{-n}$$

and hence exhibits exponential growth at ∞ unless $b_n = 0$ for every $n > 0$. This establishes the uniqueness of $\mathbb{H}$.

Remark. We now briefly discuss the canonical lift for nontrivial level. For $\mathcal{G}$ such that $g - \mathcal{G} \in S_k(N)$, one merely defines $\mathcal{L}(\mathcal{G})$ by replacing $F(n, 2-k; z)$ with $F(n, 2-k, N; z)$. In order to obtain a lift for $g \in S_k(N)$, we choose $m_N > 0$ to be minimal such that there exists $j_N^* \in M_0^{\infty}(N)$ with $j_N^*(z) = q^{-m_N} + O\left(q^{-(m_N-1)}\right)$. The condition that (1.4) is maximal among all lifts M of a form $g \in S_k$ will be further refined to the condition that

$$A(M,r) := \inf\{n \in \mathbb{Z} : n \equiv r \pmod{m_N},\ c_M^+(n) \neq 0\}$$

is maximal for every $r \in \{0, \ldots, m_N - 1\}$.

Acknowledgements The research of the first author was supported by the Alfried Krupp Prize for Young University Teachers of the Krupp Foundation and also by NSF grant DMS-0757907. The third author is supported by an NSF Postdoctoral Fellowship and was supported by the Chair in Analytic Number Theory at EPFL during part of the preparation of this chapter.

References

1. G. Andrews, *On the theorems of Watson and Dragonette for Ramanujan's Mock theta functions*, Amer. J. Math. **88** (1966), 454–490.
2. G. Andrews, *The fifth and seventh order mock theta functions*, Trans. Amer. Math Soc. **293** (1986), 113–134.
3. G. Andrews and D. Hickerson, *Ramanujan's lost notebook VII. The sixth order mock theta functions*, Adv. Math. **89**, (1991), 60–105.
4. G. Bol, *Invarianten linearer Differentialgleichungen*, Abh. Mat. Sem. Univ. Hamburg **16** (1949), 1–28.
5. R. Borcherds, *Automorphic forms with singularities on Grassmannians*, Invent. Math. **132** (1998), 491–562.
6. K. Bringmann, A. Folsom, and K. Ono, *q-series and weight 3/2 Maass forms*, Compositio Math. **145** (2009), 541–552.
7. K. Bringmann, P. Guerzhoy, and B. Kane, *Mock Modular Forms as p-adic Modular Forms*, Trans. Amer. Math. Soc. **364** (2012), 2393–2410.
8. K. Bringmann and K. Ono, *Arithmetic properties of coefficients of half-integral weight Maass-Poincaré series*, Math. Ann. **337** (2007), 591–612.
9. K. Bringmann and K. Ono, *Lifting cusp forms to Maass forms with an application to partitions*, Proc. Nat. Acad. Sci., USA **104** (2007), 3725–3731.
10. K. Bringmann and O. Richter, *Zagier-type dualities and lifting maps for harmonic Maass-Jacobi forms*, Adv. Math. **225** (2010), 2298–2315.
11. J. Bruinier and J. Funke, *On two geometric theta lifts*, Duke Math. J. **125** (2004), 45–90.

12. J. Bruinier and K. Ono, *Arithmetic of Borcherds' exponents*, Math. Ann. **327** (2003), 293–303.
13. J. Bruinier and K. Ono, *Heegner divisors, L-functions, and harmonic weak Maass forms*, Ann. of Math. **172** (2010), 2135–2181.
14. J. Bruinier, K. Ono, and R. Rhoades, *Differential Operators and Vanishing of Hecke Eigenvalues*, Math. Ann. **342** (2008), 673–693.
15. Y. Choie, M. Knopp, and W. Pribitkin, *Niebur Integrals, Mock Automorphic Forms and Harmonic Maass Forms: A Retrospective on the Work of the RademacherSchool,* preprint.
16. W. Duke and P. Jenkins, *On the zeros and coefficients of certain weakly holomorphic modular forms*, Pure Appl. Math. Q. 4 (2008), 1327–1340.
17. W. Duke, Ö. Imamoḡlu, and Á. Tóth, *Cycle Integrals of the j-function and Mock Modular Forms*, Ann. of Math. **173** (2011), 947–981.
18. M. Eichler, *Eine Verallgemeinerung der abelschen Integrale*, Math. Z. **67** (1957), 267–298.
19. J. Fay, *Fourier coefficients of the resolvent for a Fuchsian group*, J. reine und angew. Math. **293–294** (1977), 143–203.
20. A. Folsom and K. Ono, *Duality involving the mock theta function $f(q)$*, J. Lond. Math. Soc. **77** (2008), 320–334.
21. K. Fricke, *Analytische und p-adische Aspekte von klassischen und Mock-Modulformen*, Ph.D. thesis, Max-Planck Insitute of Mathematics - Bonn, Germany, in preparation.
22. P. Guerzhoy, Z. Kent, and K. Ono, *p-adic coupling of mock modular forms and shadows*, Proc. Nat. Acad. Sci. USA **107** (2010), 6169–6174.
23. H. Iwaniec, *Topics in classical automorphic forms*, Graduate studies in Mathematics **53**, Amer. Math. Soc., Providence, RI, USA, 1997.
24. P. Jenkins, *Kloosterman sums and traces of singular moduli*, J. Number Theory **117** (2006), 301–314.
25. M. Knopp, *Construction of automorphic forms on H-groups and supplementary Fourier series*, Trans. Amer. Math. Soc. **103** (1962), 168–188.
26. M. Knopp, *Modular functions in analytic number theory*, Amer. Math. Soc., Chelsea Publishing, Providence, RI, 1993.
27. M. Knopp, *On the Fourier coefficients of cusp forms having small positive weight*, Theta Functions-Bowdoin 1987, Part 2 (Brunswick, ME, 1987) eds. L. Ehrenpreis et al., Proc. Sympos. Pure Math. **49**, Part 2, Amer. Math. Soc., Providence, RI, 1989, 111–127.
28. M. Knopp and J. Lehner, *On complementary automorphic forms and supplementary Fourier series*, Illinois J. Math. 6 (1962) 98–106.
29. N. Kuznetsov, *The Petersson conjecture for cusp forms of weight zero and the Linnik conjecture. Sums of Kloosterman sums*, Mat. Sb. (N.S.) **111(153)** (1980), 334–383, 479 (in Russian); English translation: Math. USSR Sbornik **39** (1981), 299–342.
30. H. Neunhöffer, *Über die analytische Fortsetzung von Poincaréreihen*, S.-B. Heidelberger Akad. Wiss. Math.-Natur Kl. **2** (1973), 33–90.
31. W. Pribitkin, *The Fourier coefficients of modular forms and Neibur modular integrals having small positive weight, I*, Acta Arith. **91** (1999), 291–309.
32. W. Pribitkin, *The Fourier Coefficients of Modular Forms and Neibur Modular Integrals Having Small Positive Weight, II*, Acta Arith. **93** (2000), 343–358.
33. H. Rademacher, *On the expansion of the partition function in a series*, Ann. of Math. (2) **44** (1943), 416–422.
34. S. Ramanujan, *The lost notebook and other unpublished papers*, Narosa, New Delhi, 1988.
35. J. Rouse, *Zagier duality for the exponents of Borcherds products for Hilbert modular forms*, J. Lond. Math. Soc. **73** (2006), 339–354.
36. G. Watson, *The final problem: an account of the mock theta functions*, J. London Math. Soc. **11** (1936), 55–80.
37. G. Watson, *The mock theta functions (2)*, Proc. London Math. Soc. **42** (1937), 274–304.
38. E. Whittaker and G. Watson, *A course in modern analysis, 4th ed.*, Cambridge University Press, Cambridge, England (1990), 620 pp.
39. D. Zagier, *Traces of Singular Moduli*, Motives, polylogarithms and Hodge theory, Part I (Irvine, CA, 1998), 211–244, Int. Press Lect. Ser., 3, I, Int. Press, Somerville, MA, 2002.

40. D. Zagier, *Ramanujan's mock theta functions and their applications [d'après Zwegers and Bringmann-Ono]*, Sém. Bourbaki, Astérisque **326** (2009), 143–164.
41. S. Zwegers, *Mock θ-functions and Real Analytic Modular Forms*. Contemp. Math. **291** (2001), 269–277.
42. S. Zwegers, *Mock theta functions*, Ph.D. Thesis, Universiteit Utrecht, 2002.
43. S. Zwegers, *The Folsom-Ono grid contains only integers*, Proc. Amer. Math. Soc. **137** (2009), 1579–1584.

Function Theory Related to the Group $PSL_2(\mathbb{R})$

R. Bruggeman, J. Lewis, and D. Zagier

Abstract We study analytic properties of the action of $PSL_2(\mathbb{R})$ on spaces of functions on the hyperbolic plane. The central role is played by principal series representations. We describe and study a number of different models of the principal series, some old and some new. Although these models are isomorphic, they arise as the spaces of global sections of completely different equivariant sheaves and thus bring out different underlying properties of the principal series.

The two standard models of the principal series are the space of eigenfunctions of the hyperbolic Laplace operator in the hyperbolic plane (upper half-plane or disk) and the space of hyperfunctions on the boundary of the hyperbolic plane. They are related by a well-known integral transformation called the Poisson transformation. We give an explicit integral formula for its inverse.

The Poisson transformation and several other properties of the principal series become extremely simple in a new model that is defined as the space of solutions of a certain two-by-two system of first-order differential equations. We call this the *canonical model* because it gives canonical representatives for the hyperfunctions defining one of the standard models.

R. Bruggeman (✉)
Mathematisch Instituut Universiteit Utrecht, 3508 TA, Utrecht, Nederland
e-mail: r.w.bruggeman@uu.nl

J. Lewis
Massachusetts Institute of Technology, Cambridge, MA 02139, USA
e-mail: jlewis@math.mit.edu

D. Zagier
Max-Planck-Institut für Mathematik, Collège de France, 53111 Bonn, Deutschland

Collège de France, 75005 Paris, France
e-mail: don.zagier@mpim-bonn.mpg.de

H.M. Farkas et al. (eds.), *From Fourier Analysis and Number Theory to Radon Transforms and Geometry*, Developments in Mathematics 28, DOI 10.1007/978-1-4614-4075-8_7,

Another model, which has proved useful for establishing the relation between Maass forms and cohomology, is in spaces of germs of eigenfunctions of the Laplace operator near the boundary of the hyperbolic plane. We describe the properties of this model, relate it by explicit integral transformations to the spaces of analytic vectors in the standard models of the principal series, and use it to give an explicit description of the space of C^∞-vectors.

Key words Principal series • Hyperbolic Laplace operator • Hyperfunctions • Poisson transformation • Green's function • Boundary germs • Transverse Poisson transformation • Boundary splitting

Mathematics Subject Classification (2010): 22E50, 22E30, 22E45, 32A45, 35J08, 43A65, 46F15, 58C40

1 Introduction

The aim of this article is to discuss some of the analytic aspects of the group $G = \mathrm{PSL}_2(\mathbb{R})$ acting on the hyperbolic plane and its boundary. Everything we do is related in some way with the (spherical) principal series representations of the group G.

These principal series representations are among the best known and most basic objects of all of representation theory. In this chapter, we will review the standard models used to realize these representations and then describe a number of new properties and new models. Some of these are surprising and interesting in their own right, while others have already proved useful in connection with the study of cohomological applications of automorphic forms [2] and may potentially have other applications in the future. The construction of new models may at first sight seem superfluous, since by definition any two models of the same representation are equivariantly isomorphic, but nevertheless gives new information because the isomorphisms between the models are not trivial and also because each model consists of the global sections of a certain G-equivariant sheaf, and these sheaves are completely different even if they have isomorphic spaces of global sections.

The principal series representations of G are indexed by a complex number s, called the *spectral parameter*, which we will always assume to have real part between 0 and 1. (The condition $\mathrm{Re}\,(s) = \frac{1}{2}$, corresponding to unitarizability, will play no role in this chapter.) There are two basic realizations. One is the space $\mathcal{V}_s$ of functions on $\mathbb{R}$ with the (right) action of G given by

$$(\varphi\,|\,g)(t) \;=\; |ct+d|^{-2s}\,\varphi\left(\frac{at+b}{ct+d}\right) \qquad \left(t\in\mathbb{R},\ g=\begin{bmatrix} a & b\\ c & d\end{bmatrix}\in G\right). \tag{1.1}$$

The other is the space $\mathcal{E}_s$ of functions u on $\mathfrak{H}$ (complex upper half-plane) satisfying

$$\Delta\, u(z) \;=\; s(1-s)\, u(z) \qquad\qquad (z \in \mathfrak{H}), \tag{1.2}$$

where $\Delta = -y^2\big(\frac{\partial^2}{\partial x^2} + \frac{\partial^2}{\partial x^2}\big)$ $(z = x + \mathrm{i}y \in \mathfrak{H})$ is the hyperbolic Laplace operator, with the action $u \mapsto u \circ g$. They are related by Helgason's *Poisson transform* (thus named because it is the analogue of the corresponding formula given by Poisson for holomorphic functions)

$$\varphi(t) \quad\mapsto\quad (\mathsf{P}_s\varphi)(z) \;=\; \frac{1}{\pi}\int_{-\infty}^{\infty} \varphi(t)\, R(t;z)^{1-s}\, \mathrm{d}t, \tag{1.3}$$

where $R(t;z) = R_t(z) = \frac{y}{(z-t)(\bar{z}-t)}$ for $z = x+\mathrm{i}y \in \mathfrak{H}$ and $t \in \mathbb{C}$. The three main themes of this chapter are the explicit inversion of the Poisson transformation, the study of germs of Laplace eigenfunctions near the boundary $\mathbb{P}^1_{\mathbb{R}} = \mathbb{R} \cup \{\infty\}$ of $\mathfrak{H}$, and the construction of a new model of the principal series representation which is a kind of hybrid of $\mathcal{V}_s$ and $\mathcal{E}_s$. We now describe each of these briefly.

- *Inverse Poisson Transform.* We would like to describe the inverse map of P_s explicitly. The right-hand side of (1.3) can be interpreted as it stands if φ is a smooth vector in $\mathcal{V}_s$ (corresponding to a function $\varphi(x)$ which is C^∞ on $\mathbb{R}$ and such that $t \mapsto |t|^{-2s}\varphi(1/t)$ is C^∞ at $t = 0$). To get an isomorphism between $\mathcal{V}_s$ and all of $\mathcal{E}_s$, one has to allow *hyperfunctions* $\varphi(t)$. The precise definition, which is somewhat subtle in the model used in (1.1), will be reviewed in Sect. 2.2; for now we recall only that a hyperfunction on $I \subset \mathbb{R}$ is represented by a holomorphic function on $U \smallsetminus I$, where U is a neighborhood U of I in $\mathbb{C}$ with $U \cap \mathbb{R} = I$ and where two holomorphic functions represent the same hyperfunction if their difference is holomorphic on all of U. We will show in Sect. 4 that for $u \in \mathcal{E}_s$, the vector $\mathsf{P}_s^{-1}u \in \mathcal{V}_s$ can be represented by the hyperfunction

$$h_{z_0}(\zeta) \;=\; \begin{cases} u(z_0) \;+\; \displaystyle\int_{z_0}^{\zeta} \big[u(z),\, \big(R_\zeta(z)/R_\zeta(z_0)\big)^s\big] & \text{if } \zeta \in U \cap \mathfrak{H} \\ \displaystyle\int_{\bar{\zeta}}^{z_0} \big[\big(R_\zeta(z)/R_\zeta(z_0)\big)^s,\, u(z)\big] & \text{if } \zeta \in U \cap \mathfrak{H}^- \end{cases} \tag{1.4}$$

for any $z_0 \in U \cap \mathfrak{H}$, where $\mathfrak{H}^- = \{z = x + \mathrm{i}y \in \mathbb{C} \;:\; y < 0\}$ denotes the lower half-plane and $[u(z), v(z)]$ for any functions u and v in $\mathfrak{H}$ is the *Green's form*

$$[u(z), v(z)] \;=\; \frac{\partial u(z)}{\partial z}\, v(z)\, \mathrm{d}z \;+\; u(z)\, \frac{\partial v(z)}{\partial \bar{z}}\, \mathrm{d}\bar{z}, \tag{1.5}$$

which is a closed 1-form if u and v both satisfy the Laplace equation (1.2). The asymmetry in (1.4) is necessary because although $R(\zeta;z)^s$ tends to zero at $z = \zeta$ and $z = \bar{\zeta}$, both its z-derivative at ζ and its $\bar{z}$-derivative at $\bar{\zeta}$ become infinite, forcing us to change the order of the arguments in the Green's form in the two components of $U \smallsetminus I$. That the two different-looking expressions in (1.4) are nevertheless formally the same follows from the fact that $[u, v] + [v, u] = \mathrm{d}(uv)$ for any functions u and v.

• *Boundary Eigenfunctions.* If one looks at known examples of solutions of the Laplace equation (1.2), then it is very striking that many of these functions decompose into two pieces of the form $y^s A(z)$ and $y^{1-s} B(z)$ as $z = x + iy$ tends to a point of $\mathbb{R} \subset \mathbb{P}^1_{\mathbb{R}} = \partial\mathfrak{H}$, where $A(z)$ and $B(z)$ are functions which extend analytically across the boundary. For instance, the eigenfunctions that occur as building blocks in the Fourier expansions of Maass wave forms for a Fuchsian group $\mathcal{G} \subset G$ are the functions

$$k_{s,2\pi n}(z) = y^{1/2} K_{s-1/2}(2\pi|n|y)\, e^{2\pi i n x} \qquad (z = x + iy \in \mathbb{R},\ n \in \mathbb{Z},\ n \neq 0), \tag{1.6}$$

where $K_{s-1/2}(t)$ is the standard K-Bessel function which decays exponentially as $t \to \infty$. The function $K_\nu(t)$ has the form $\dfrac{\pi}{\sin \pi\nu}\big(I_\nu(t) - I_{-\nu}(t)\big)$ with

$$I_\nu(t) = \sum_{n=0}^{\infty} \frac{(-1/4)^n\, t^{2n+\nu}}{n!\,\Gamma(n+\nu)},$$

so $k_{s,2\pi n}(z)$ decomposes into two pieces of the form $y^s \times$(analytic near the boundary) and $y^{1-s} \times$(analytic near the boundary). The same is true for other elements of $\mathcal{E}_s$, involving other special functions like Legendre or hypergeometric functions, that play a role in the spectral analysis of automorphic forms. A second main theme of this chapter is to understand this phenomenon. We will show that to every analytic function φ on an interval $I \subset \mathbb{R}$, there is a unique solution u of (1.2) in $U \cap \mathfrak{H}$ (where U as before is a neighborhood of I in $\mathbb{C}$ with $U \cap \mathbb{R} = I$, supposed simply connected and sufficiently small) such that $u(x+iy) = y^s\, \Phi(x+iy)$ for an analytic function Φ on U with restriction $\Phi|_I = \varphi$. In Sect. 5 we will call the (locally defined) map $\varphi \mapsto u$ the *transverse Poisson transform* of φ and will show that it can be described by both a Taylor series in y and an integral formula, the latter bearing a striking resemblance to the original (globally defined) Poisson transform (1.3):

$$(\mathsf{P}^\dagger_s \varphi)(z) = \frac{-i\,\Gamma(s+\frac{1}{2})}{\Gamma(s)\Gamma(\frac{1}{2})} \int_{\bar z}^{z} \varphi(\zeta)\, R(\zeta; z)^{1-s}\, d\zeta, \tag{1.7}$$

where the function $\varphi(\zeta)$ in the integral is the unique holomorphic extension of $\varphi(t)$ to U and the integral is along any path connecting $\bar z$ and z within U. The transverse Poisson map produces an eigenfunction u from a real-analytic function φ on an interval I in $\mathbb{P}^1_{\mathbb{R}}$. We also give an explicit integral formula representing the holomorphic function φ in U in terms of the eigenfunction $u = \mathsf{P}^\dagger_s \varphi$.

As an application, we will show in Sect. 7 that the elements of $\mathcal{E}_s$ corresponding under the Poisson transform to analytic vectors in $\mathcal{V}_s$ (which in the model (1.1) are represented by analytic functions φ on $\mathbb{R}$ for which $t \mapsto |t|^{-2s}\varphi(1/t)$ is analytic at $t = 0$) are precisely those which have a decomposition $u = \mathsf{P}^\dagger_s \varphi_1 + \mathsf{P}^\dagger_{1-s} \varphi_2$ near the boundary of $\mathfrak{H}$, where φ_1 and φ_2, which are uniquely determined by u, are analytic functions on $\mathbb{P}^1_{\mathbb{R}}$.

• *Canonical Model.* We spoke above of two realizations of the principal series, as $\mathcal{V}_s$ (functions on $\partial\mathfrak{H} = \mathbb{P}^1_{\mathbb{R}}$) and as $\mathcal{E}_s$ (eigenfunctions of the Laplace operator in $\mathfrak{H}$). In fact $\mathcal{V}_s$ comes in many different variants, discussed in detail in Sect. 1, each of which resolves various of the defects of the others at the expense of introducing new ones. For instance, the "line model" (1.1) which we have been using up to now has a very simple description of the group action but needs special treatment of the point $\infty \in \mathbb{P}^1_{\mathbb{R}}$, as one could already see several times in the discussion above (e.g., in the description of smooth and analytic vectors or in the definition of hyperfunctions). One can correct this by working on the projective rather than the real line, but then the description of the group action becomes very messy, while yet other models (circle model, plane model, induced representation model, ...) have other drawbacks. In Sect. 4, we will introduce a new realization $\mathcal{C}_s$ ("canonical model") that has many advantages:

- All points in hyperbolic space, and all points on its boundary, are treated in an equal way.
- The formula for the group action is very simple.
- Its objects are actual functions, not equivalence classes of functions.
- The Poisson transformation is given by an extremely simple formula.
- The canonical model $\mathcal{C}_s$ coincides with the image of a canonical inversion formula for the Poisson transformation.
- The elements of $\mathcal{C}_s$ satisfy differential equations, discussed below, which lead to a sheaf $\mathcal{D}_s$ that is interesting in itself.
- It uses *two* variables, one in $\mathfrak{H}$ and one in $\mathbb{P}^1_{\mathbb{C}} \smallsetminus \mathbb{P}^1_{\mathbb{R}}$, and therefore gives a natural bridge between the models of the principal series representations as eigenfunctions in $\mathfrak{H}$ or as hyperfunctions in a deleted neighborhood of $\mathbb{P}^1_{\mathbb{R}}$ in $\mathbb{P}^1_{\mathbb{C}}$.

The elements of the space $\mathcal{C}_s$ are precisely the functions $(z, z_0) \mapsto h_{z_0}(z)$ arising as in (1.4) for some eigenfunction $u \in \mathcal{E}_s$, but also have several intrinsic descriptions, of which perhaps the most surprising is a characterization by a system of two linear differential equations:

$$\frac{\partial h}{\partial z} = -s\,\frac{\zeta - \bar{z}}{z - \bar{z}}\,h^*, \qquad \frac{\partial h^*}{\partial \bar{z}} = \frac{s}{(\zeta - \bar{z})(z - \bar{z})}\,h, \tag{1.8}$$

where $h(\zeta, z)$ is a function on $\left(\mathbb{P}^1_{\mathbb{C}} \smallsetminus \mathbb{P}^1_{\mathbb{R}}\right) \times \mathfrak{H}$ which is holomorphic in the first variable and where $h^*(\zeta, z) := (h(\zeta, z) - h(z, z))/(\zeta - z)$. The "Poisson transform" in this model is very simple: it simply assigns to $h(\zeta, z)$ the function $u(z) = h(z, z)$, which turns out to be an eigenfunction of the Laplace operator. The name "canonical model" refers to the fact that $\mathcal{V}_s$ consists of hyperfunctions and that in $\mathcal{C}_s$ we have chosen a family of canonical representatives of these hyperfunctions, indexed in a G-equivariant way by a parameter in the upper half-plane: $h(\,\cdot\,, z)$ for each $z \in \mathfrak{H}$ is the unique representative of the hyperfunction $\varphi(t) R(t; z)^{-s}$ on $\mathbb{P}^1_{\mathbb{R}}$ which is holomorphic in all of $\mathbb{P}^1_{\mathbb{C}} \smallsetminus \mathbb{P}^1_{\mathbb{R}}$ and vanishes at $\bar{z}$.

• *Further Remarks.* The known or potential applications of the ideas in this chapter are to automorphic forms in the upper half-plane. When dealing with such forms, one needs to work with functions of general weight, not just weight 0 as considered here. We expect that many of our results can be modified to the context of general weights, where the group $G = \mathrm{PSL}_2(\mathbb{R})$ has to be replaced by $\mathrm{SL}_2(\mathbb{R})$ or its universal covering group.

Some parts of what we do in this chapter are available in the literature, but often in a different form or with another emphasis. In Sect. 4 of the introduction of [6], Helgason gives an overview of analysis on the upper half-plane. One finds there the Poisson transformation; the injectivity is proved by a polar decomposition. As far as we know, our approach in Theorem 4.2 with the Green's form is new, and in [2], it is an essential tool to build cocycles. Helgason gives also the asymptotic expansion near the boundary of eigenfunctions of the Laplace operator, from which the results in Sect. 7 may also be derived. For these asymptotic expansions, one may also consult the work of Van den Ban and Schlichtkrull, [1]. A more detailed and deeper discussion can be found in [7], where Section 0 discusses the inverse Poisson transformation in the context of the upper half-plane. Our presentation stresses the transverse Poisson transformation, which also seems not to have been treated in the earlier literature and which we use in [2] to recover Maass wave forms from their associated cocycles. Finally, the hybrid models in Sect. 4 and the related sheaf $\mathcal{D}_s$ are, as far as we know, new.

This chapter ends with an appendix giving a number of explicit formulas, including descriptions of various eigenfunctions of the Laplace operator and tables of Poisson transforms and transverse Poisson transforms.

Acknowledgements The two first-named authors would like to thank the Max Planck Institute in Bonn and the Collège de France in Paris for their repeated hospitality and for the excellent working conditions they provided. We thank YoungJu Choie for her comments on an earlier version.

Conventions and Notations. We work with the Lie group

$$G = \mathrm{PSL}_2(\mathbb{R}) = \mathrm{SL}_2(\mathbb{R})/\{\pm \mathrm{Id}\}.$$

We denote the element $\pm\begin{pmatrix} a & b \\ c & d \end{pmatrix}$ of G by $\begin{bmatrix} a & b \\ c & d \end{bmatrix}$. A maximal compact subgroup is $K = \mathrm{PSO}(2) = \{k(\theta) \ : \theta \in \mathbb{R}/\pi\mathbb{Z}\}$, with

$$k(\theta) = \begin{bmatrix} \cos\theta & \sin\theta \\ -\sin\theta & \cos\theta \end{bmatrix}. \tag{1.9a}$$

We also use the Borel subgroup NA, with the unipotent subgroup $N = \{n(x) \ : x \in \mathbb{R}\}$ and the torus $A = \{a(y) \ : \ y > 0\}$, with

$$a(y) = \begin{bmatrix} \sqrt{y} & 0 \\ 0 & 1/\sqrt{y} \end{bmatrix}, \qquad n(x) = \begin{bmatrix} 1 & x \\ 0 & 1 \end{bmatrix}. \tag{1.9b}$$

We use $\mathbb{H}$ as a generic letter to denote the hyperbolic plane. We use two concrete models: the unit disk $\mathbb{D} = \{w \in \mathbb{C} \ : \ |w| < 1\}$ and the upper half-plane

$\mathfrak{H} = \{z \in \mathbb{C} \;:\; \operatorname{Im} z > 0\}$. We will denote by x and y the real and imaginary parts of $z \in \mathfrak{H}$, respectively. The boundary $\partial\mathbb{H}$ of the hyperbolic plane is in these models: $\partial\mathfrak{H} = \mathbb{P}^1_{\mathbb{R}} = \mathbb{R} \cup \{\infty\}$, the real projective line, and $\partial\mathbb{D} = \mathbb{S}^1$, the unit circle. Both models of $\mathbb{H} \cup \partial\mathbb{H}$ are contained in $\mathbb{P}^1_{\mathbb{C}}$, on which G acts in the upper half-plane model by $\left[\begin{smallmatrix} a & b \\ c & d \end{smallmatrix}\right] : z \mapsto \frac{az+b}{cz+d}$ and in the disk model by $\left[\begin{smallmatrix} a & b \\ c & d \end{smallmatrix}\right] : w \mapsto \frac{Aw+B}{\bar{B}w+\bar{A}}$, with $\left[\begin{smallmatrix} A & B \\ \bar{B} & \bar{A} \end{smallmatrix}\right] = \left[\begin{smallmatrix} 1 & -i \\ 1 & i \end{smallmatrix}\right]\left[\begin{smallmatrix} a & b \\ c & d \end{smallmatrix}\right]\left[\begin{smallmatrix} 1 & -i \\ 1 & i \end{smallmatrix}\right]^{-1}$.

All the representations that we discuss in the first five sections depend on $s \in \mathbb{C}$, the *spectral parameter*; it determines the eigenvalue $\lambda_s = s - s^2$ of the Laplace operator Δ, which is given in the upper half-plane model by $-y^2\partial_x^2 - y^2\partial_y^2$ and in the disk model by $-(1-|w|^2)^2\,\partial_w\,\partial_{\bar{w}}$. We will always assume $s \notin \mathbb{Z}$ and usually restrict to $0 < \operatorname{Re}(s) < 1$. We work with right representations of G, denoted by $v \mapsto v|_{2s}\,g$ or $v \mapsto v\,|\,g$.

2 The Principal Series Representation $\mathcal{V}_s$

This section serves to discuss general facts concerning the principal series representation. Much of this is standard, but quite a lot of it is not, and the material presented here will be used extensively in the rest of the chapter. We will therefore give a self-contained and fairly detailed presentation.

The principal series representations can be realized in various ways. One of the aims of this chapter is to gain insight by combining several of these models. Section 2.1 gives six standard models for the continuous vectors in the principal series representation. Section 2.2 presents the larger space of hyperfunction vectors in some of these models, and in Sect. 2.3, we discuss the isomorphism (for $0 < \operatorname{Re} s < 1$) between the principal series representations with the values s and $1-s$ of the spectral parameter.

2.1 *Six Models of the Principal Series Representation*

In this subsection, we look at six models to realize the principal series representation $\mathcal{V}_s$, each of which is the most convenient in certain contexts. Three of these models are realized on the boundary $\partial\mathbb{H}$ of the hyperbolic plane. Five of the six models have easy algebraic isomorphisms between them. The sixth has a more subtle isomorphism with the others but gives explicit matrix coefficients. In later sections we will describe more models of $\mathcal{V}_s$ with a more complicated relation to the models here. We also describe the duality between $\mathcal{V}_s$ and $\mathcal{V}_{1-s}$ in the various models. (*Note*: We will use the letter $\mathcal{V}_s$ somewhat loosely to denote "the" principal series representation in a generic way or when the particular space of functions under consideration plays no role. The spaces $\mathcal{V}_s^{\infty}$ and $\mathcal{V}_s^{\omega}$ of smooth and analytic vectors, and the spaces $\mathcal{V}_s^{-\infty}$ and $\mathcal{V}_s^{-\omega}$ of distributions and hyperfunctions introduced in

Sect. 2.2, will be identified by the appropriate superscript. Other superscripts such as $\mathbb{P}$ and $\mathbb{S}$ will be used to distinguish vectors in the different models when needed.)

- *Line Model.* This well-known model of the principal series consists of complex-valued functions on $\mathbb{R}$ with the action of G given by

$$\varphi\big|_{2s}\begin{bmatrix} a & b \\ c & d \end{bmatrix}(t) = |ct+d|^{-2s}\,\varphi\Big(\frac{at+b}{ct+d}\Big). \tag{2.1}$$

Since G acts on $\mathbb{P}^1_{\mathbb{R}} = \mathbb{R} \cup \{\infty\}$, and not on $\mathbb{R}$, the point at infinity plays a special role in this model, and a more correct description requires the use of a pair $(\varphi, \varphi_\infty)$ of functions $\mathbb{R} \to \mathbb{C}$ related by $\varphi(t) = |t|^{-2s}\varphi_\infty(-1/t)$ for $t \neq 0$, and with the right-hand side in (2.1) replaced by $|at+b|^{-2s}\varphi_\infty\big(-\frac{ct+d}{at+b}\big)$ if $ct+d$ vanishes, together with the obvious corresponding formula for φ_∞. However, we will usually work with φ alone and leave the required verification at ∞ to the reader.

The space $\mathcal{V}_s^\infty$ of *smooth vectors* in this model consists of the functions $\varphi \in C^\infty(\mathbb{R})$ with an asymptotic expansion

$$\varphi(t) \sim |t|^{-2s}\sum_{n=0}^{\infty} c_n\, t^{-n} \tag{2.2}$$

as $|t| \to \infty$. Similarly, we define the space $\mathcal{V}_s^\omega$ of *analytic vectors* as the space of $\varphi \in C^\omega(\mathbb{R})$ (real-analytic functions on $\mathbb{R}$) for which the series appearing on the right-hand side of (2.2) converges to $\varphi(t)$ for $|t| \geq t_0$ for some t_0. Replacing $C^\infty(\mathbb{R})$ or $C^\omega(\mathbb{R})$ by $C^p(\mathbb{R})$ and the expansion (2.2) with a Taylor expansion of order p, we define the space $\mathcal{V}_s^p$ for $p \in \mathbb{N}$.

- *Plane Model.* The line model has the advantage that the action (2.1) of G is very simple and corresponds to the standard formula for its action on the complex upper half-plane $\mathfrak{H}$, but the disadvantage that we have to either cover the boundary $\mathbb{R} \cup \{\infty\}$ of $\mathfrak{H}$ by two charts and work with pairs of functions or else give a special treatment to the point at infinity, thus breaking the inherent G-symmetry. Each of the next five models eliminates this problem at the expense of introducing complexities elsewhere. The first of these is the plane model, consisting of even functions $\Phi : \mathbb{R}^2 \smallsetminus \{0\} \to \mathbb{C}$ satisfying $\Phi(tx, ty) = |t|^{-2s}\Phi(x, y)$ for $t \neq 0$, with the action

$$\Phi\Big|\begin{bmatrix} a & b \\ c & d \end{bmatrix}(x,y) = \Phi(ax+by, cx+dy). \tag{2.3}$$

The relation with the line model is

$$\varphi(t) = \Phi(t,1), \qquad \varphi_\infty(t) = \Phi(-1,t),$$
$$\Phi(x,y) = \begin{cases} |y|^{-2s}\,\varphi(x/y) & \text{if } y \neq 0, \\ |x|^{-2s}\,\varphi_\infty(-y/x) & \text{if } x \neq 0, \end{cases} \tag{2.4}$$

and of course the elements in $\mathcal{V}_s^p$, for $p = 0, 1, \dots, \infty, \omega$, are now just given by $\Phi \in C^p(\mathbb{R}^2 \smallsetminus \{0\})$. This model has the advantage of being completely G-symmetric, but requires functions of two variables rather than just one.

- *Projective Model.* If $(\varphi, \varphi_\infty)$ represents an element of the line model, we put

$$\varphi^{\mathbb{P}}(t) = \begin{cases} (1+t^2)^s\,\varphi(t) & \text{if } t \in \mathbb{P}^1_{\mathbb{R}} \smallsetminus \{\infty\} = \mathbb{R}, \\ (1+t^{-2})^s\,\varphi_\infty(-1/t) & \text{if } t \in \mathbb{P}^1_{\mathbb{R}} \smallsetminus \{0\} = \mathbb{R}^* \cup \{\infty\}. \end{cases} \tag{2.5}$$

The functions $\varphi^{\mathbb{P}}$ form the projective model of $\mathcal{V}_s$, consisting of functions f on the real projective line $\mathbb{P}^1_{\mathbb{R}}$ with the action

$$f\big|^{\mathbb{P}}_{2s}\begin{bmatrix} a & b \\ c & d \end{bmatrix}(t) = \left(\frac{t^2+1}{(at+b)^2+(ct+d)^2}\right)^s f\left(\frac{at+b}{ct+d}\right). \tag{2.6}$$

Note that the factor $\left(\frac{t^2+1}{(at+b)^2+(ct+d)^2}\right)^s$ is real-analytic on the whole of $\mathbb{P}^1_{\mathbb{R}}$ since the factor in parentheses is analytic and strictly positive on $\mathbb{P}^1_{\mathbb{R}}$. This model has the advantage that all points of $\mathbb{P}^1_{\mathbb{R}}$ get equal treatment but the disadvantage that the formula for the action is complicated and unnatural.

- *Circle Model.* The transformation $\xi = \frac{t-\mathrm{i}}{t+\mathrm{i}}$ in $\mathbb{P}^1_{\mathbb{C}}$, with inverse $t = \mathrm{i}\frac{1+\xi}{1-\xi}$, maps $\mathbb{P}^1_{\mathbb{R}}$ isomorphically to the unit circle $\mathbb{S}^1 = \{\xi \in \mathbb{C} : |\xi| = 1\}$ in $\mathbb{C}$ and leads to the circle model of $\mathcal{V}_s$, related to the three previous models by

$$\varphi^{\mathbb{S}}(\mathrm{e}^{-2\mathrm{i}\theta}) = \varphi^{\mathbb{P}}(\cot\theta) = \Phi(\cos\theta, \sin\theta) = |\sin\theta|^{-2s}\,\varphi(\cot\theta). \tag{2.7}$$

The action of $g = \left[\begin{smallmatrix} a & b \\ c & d \end{smallmatrix}\right] \in \mathrm{PSL}_2(\mathbb{R})$ is described by $\tilde{g} = \left[\begin{smallmatrix} 1 & -\mathrm{i} \\ 1 & \mathrm{i} \end{smallmatrix}\right] g \left[\begin{smallmatrix} 1 & -\mathrm{i} \\ 1 & \mathrm{i} \end{smallmatrix}\right]^{-1} = \left[\begin{smallmatrix} A & B \\ \bar{B} & \bar{A} \end{smallmatrix}\right]$ in $\mathrm{PSU}(1,1) \subset \mathrm{PSL}_2(\mathbb{C})$, with $A = \frac{1}{2}(a + \mathrm{i}b - \mathrm{i}c + d)$, $B = \frac{1}{2}(a - \mathrm{i}b - \mathrm{i}c - d)$:

$$f\big|^{\mathbb{S}}_{2s}\, g\,(\xi) = |A\xi + B|^{-2s} f\left(\frac{A\xi + B}{\bar{B}\xi + \bar{A}}\right) \qquad (|\xi| = 1). \tag{2.8}$$

Since $|A|^2 - |B|^2 = 1$, the factor $|A\xi + B|$ is nonzero on the unit circle.

Note that in both the projective and circle models, the elements in $\mathcal{V}_s^p$ are simply the elements of $C^p(\mathbb{P}^1_{\mathbb{R}})$ or $C^p(\mathbb{S}^1)$, so that as vector spaces these models are independent of s.

- *Induced Representation Model.* The principal series is frequently defined as the induced representation from the Borel group NA to G of the character $n(x)a(y) \mapsto y^{-s}$, in the notation in (1.9b). (See for instance Chap. VII in [8].) This is the space of functions F on G transforming on the right according to this character of AN, with G acting by left translation. Identifying G/N with $\mathbb{R}^2 \smallsetminus \{(0,0)\}$ leads to the plane model, via $F\left[\begin{smallmatrix} a & b \\ c & d \end{smallmatrix}\right] = \Phi(a,c)$. On the other hand, the functions in the induced representation model are determined by their values on K, leading to the relation $\varphi^{\mathbb{S}}(\mathrm{e}^{2\mathrm{i}\theta}) = F(k(\theta))$ with the circle model, with $k(\theta)$ as in (1.9a).

We should warn the reader that in defining the induced representation, one often considers functions whose restrictions to K are square integrable, obtaining a Hilbert space isomorphic to $L^2(K)$. The action of G in this space is a bounded representation, unitary if $\mathrm{Re}\, s = \frac{1}{2}$. Since not all square integrable functions are continuous, this Hilbert space is larger than $\mathcal{V}_s^0$. For $p \in \mathbb{N}$, the space of p times differentiable vectors in this Hilbert space is larger than our $\mathcal{V}_s^p$. (It is between $\mathcal{V}_s^{p-1}$ and $\mathcal{V}_s^p$.) However, $\mathcal{V}_s^\infty$ and $\mathcal{V}_s^\omega$ coincide with the spaces of infinitely-often differentiable, respectively analytic, vectors in this Hilbert space.

- *Sequence Model.* We define elements $\mathbf{e}_{s,n} \in \mathcal{V}_s^\omega$, $n \in \mathbb{Z}$, represented in our five models as follows:

$$\mathbf{e}_{s,n}(t) = (t^2+1)^{-s}\left(\frac{t-\mathrm{i}}{t+\mathrm{i}}\right)^n, \tag{2.9a}$$

$$\mathbf{e}^{\mathbb{R}^2}_{s,n}(x,y) = (x^2+y^2)^{-s}\left(\frac{x-\mathrm{i}y}{x+\mathrm{i}y}\right)^n, \tag{2.9b}$$

$$\mathbf{e}^{\mathbb{P}}_{s,n}(t) = \left(\frac{t-\mathrm{i}}{t+\mathrm{i}}\right)^n, \tag{2.9c}$$

$$\mathbf{e}^{\mathbb{S}}_{s,n}(\xi) = \xi^n, \tag{2.9d}$$

$$\mathbf{e}^{\text{ind repr}}_{s,n}\left(\begin{bmatrix} a & b \\ c & d \end{bmatrix}\right) = \left(a^2+c^2\right)^{-s}\left(\frac{a-\mathrm{i}c}{a+\mathrm{i}c}\right)^n. \tag{2.9e}$$

Fourier expansion gives a convergent representation $\varphi^{\mathbb{S}}(\xi) = \sum_n c_n \mathbf{e}_{s,n}(\xi)$ for each element of $\mathcal{V}_s^0$. This gives the sequence model, consisting of the sequences of coefficients $\mathbf{c} = (c_n)_{n\in\mathbb{Z}}$. The action of G is described by $\mathbf{c} \mapsto \mathbf{c}'$ with $\mathbf{c}'_m = \sum_n A_{m,n}(g) c_n$, where the matrix coefficients $A_{m,n}(g)$ are given (by the binomial theorem) in terms of $\tilde{g} = \left[\begin{smallmatrix} A & B \\ \bar{B} & \bar{A} \end{smallmatrix}\right]$ as

$$A_{m,n}(g) = \frac{(A/B)^m\,(A/\bar{B})^n}{|A|^{2s}} \sum_{l \ge \max(m,n)} \binom{n-s}{l-m}\binom{-n-s}{l-n}\left|\frac{B}{A}\right|^{2l}, \tag{2.10}$$

which can be written in closed form in terms of hypergeometric functions as

$$A_{m,n}(g) = \begin{cases} \frac{A^{n+m}\bar{B}^{m-n}}{|A|^{2s+2m}}\binom{-s-n}{m-n} F\left(s-n, s+m; m-n+1; \left|\frac{B}{A}\right|^2\right) & \text{if } m \ge n, \\ \frac{A^{n+m}B^{n-m}}{|A|^{2s+2n}}\binom{-s+n}{n-m} F\left(s+n, s-m; n-m+1; \left|\frac{B}{A}\right|^2\right) & \text{if } n \ge m. \end{cases} \tag{2.11}$$

The description of the smooth and analytic vectors is easy in the sequence model:

$$\mathcal{V}_s^\omega = \left\{(c_n) \,:\, c_n = \mathrm{O}\left(\mathrm{e}^{-a|n|}\right) \text{ for some } a > 0\right\},$$

$$\mathcal{V}_s^\infty = \{(c_n) \,:\, c_n = \mathrm{O}((1+|n|)^{-a}) \text{ for all } a \in \mathbb{R}\}. \tag{2.12}$$

The precise description of $\mathcal{V}_s^p$ for finite $p \in \mathbb{N}$ is less obvious in this model, but at least we have $(c_n) \in \mathcal{V}_s^p \Rightarrow c_n = \mathrm{o}(|n|^{-p})$ as $|n| \to \infty$, and, conversely, $c_n = \mathrm{O}(|n|^{-\rho})$ with $\rho > p+1$ implies $(c_n) \in \mathcal{V}_s^p$.

• *Duality.* There is a duality between $\mathcal{V}_s^0$ and $\mathcal{V}_{1-s}^0$, given in the six models by the formulas

$$\langle \varphi, \psi \rangle = \frac{1}{\pi} \int_{\mathbb{R}} \varphi(t)\, \psi(t)\, \mathrm{d}t, \tag{2.13a}$$

$$\langle \Phi, \Psi \rangle = \frac{1}{2\pi} \int_0^{2\pi} \Phi(\cos\theta, \sin\theta)\, \Psi(\cos\theta, \sin\theta)\, \mathrm{d}\theta, \tag{2.13b}$$

$$\langle \varphi^{\mathbb{P}}, \psi^{\mathbb{P}} \rangle = \frac{1}{\pi} \int_{\mathbb{P}^1_{\mathbb{R}}} \varphi^{\mathbb{P}}(t)\, \psi^{\mathbb{P}}(t)\, \frac{\mathrm{d}t}{1+t^2}, \tag{2.13c}$$

$$\langle \varphi^{\mathbb{S}}, \psi^{\mathbb{S}} \rangle = \frac{1}{2\pi \mathrm{i}} \int_{\mathbb{S}^1} \varphi^{\mathbb{S}}(\xi)\, \psi^{\mathbb{S}}(\xi)\, \frac{\mathrm{d}\xi}{\xi}, \tag{2.13d}$$

$$\langle F, F_1 \rangle = \int_0^{\pi} F\big(k(\theta)\big)\, F_1\big(k(\theta)\big)\, \frac{\mathrm{d}\theta}{\pi}, \tag{2.13e}$$

$$\big\langle \mathbf{c}, \mathbf{d} \big\rangle = \sum_n c_n d_{-n}. \tag{2.13f}$$

This bilinear form on $\mathcal{V}_s^0 \times \mathcal{V}_{1-s}^0$ is G-invariant:

$$\langle \varphi|_{2s}\, g,\ \psi|_{2-2s}\, g \rangle = \langle \varphi, \psi \rangle \qquad (g \in G). \tag{2.14}$$

Furthermore we have for $n, m \in \mathbb{Z}$:

$$\big\langle \mathbf{e}_{1-s,n}, \mathbf{e}_{s,m} \big\rangle = \delta_{n,-m}. \tag{2.15}$$

• *Topology.* The natural topology of $\mathcal{V}_s^p$ with $p \in \mathbb{N} \cup \{\infty\}$ is given by seminorms which we define with use of the action $\varphi \mapsto \varphi|\mathbf{W} = \frac{\mathrm{d}}{\mathrm{d}t} \varphi|\mathrm{e}^{t\mathbf{W}}\big|_{t=0}$ where $\mathbf{W} = \left(\begin{smallmatrix} 0 & -1 \\ 1 & 0 \end{smallmatrix}\right)$ is in the Lie algebra. The differential operator $\mathbf{W}$ is given by $2\mathrm{i}\xi\, \partial_\xi$ in the circle model, by $(1+t^2)\, \partial_t$ in the projective model, and by $(1+x^2)\, \partial_x + 2sx$ in the line model. For $p \in \mathbb{N}$, the space $\mathcal{V}_s^p$ is a Banach space with norm equal to the sum over $j = 0, \ldots, p$ of the seminorms

$$\|\varphi\|_j = \sup_{x \in \partial\mathbb{H}} \big|\varphi|\mathbf{W}^j(x)\big|. \tag{2.16}$$

The collection of all seminorms $\|\cdot\|_j$, $j \in \mathbb{N}$, gives the natural topology of $\mathcal{V}_s^\infty = \bigcap_{p \in \mathbb{N}} \mathcal{V}_s^p$. In Sect. 2.2 we shall discuss the natural topology on $\mathcal{V}_s^\omega$.

Although we have strict inclusions $\mathcal{V}_s^\infty \subset \cdots \subset \mathcal{V}_s^1 \subset \mathcal{V}_s^0$, all these representation spaces of G are irreducible as topological G-representations due to our standing assumptions $0 < \mathrm{Re}\, s < 1$, which implies $s \notin \mathbb{Z}$.

• *Sheaf Aspects.* In the line model, the projective model, and the circle model, we can extend the definition of the G-equivariant spaces $\mathcal{V}_s^p$ for $p = 0, 1, \ldots, \infty, \omega$ of functions on $\partial\mathbb{H}$ to G-equivariant sheaves on $\partial\mathbb{H}$. For instance, in the circle model, we can define $\mathcal{V}_s^\omega(I)$ for any open subset $I \subset \mathbb{S}^1$ as the space of real-analytic functions on I. The action of G induces linear maps $f \mapsto f \mid g$, from $\mathcal{V}_s^\omega(I)$ to $\mathcal{V}_s^\omega(g^{-1}I)$, so that $I \mapsto \mathcal{V}_s^\omega(I)$ is a G-equivariant sheaf on the G-space $\mathbb{S}^1$ whose space of global sections is the representation $\mathcal{V}_s^\omega$ of G. For the line model and the projective model, we proceed similarly.

2.2 Hyperfunctions

So far we have considered $\mathcal{V}_s$ as a space of functions. We now want to include generalized functions: distributions and hyperfunctions. We shall be most interested in hyperfunctions on $\partial\mathbb{H}$, in the projective model and the circle model.

• $\mathcal{V}_s^\omega$ *and Holomorphic Functions.* Before we discuss hyperfunctions, let us first consider $\mathcal{V}_s^\omega$. In the *circle model*, it is the space $C^\omega(\mathbb{S}^1)$ of real-analytic functions on $\mathbb{S}^1$, with the action (2.8). Since the restriction of a holomorphic function on a neighborhood of $\mathbb{S}^1$ in $\mathbb{C}$ to $\mathbb{S}^1$ is real-analytic, and since every real-analytic function on $\mathbb{S}^1$ is such a restriction, $C^\omega(\mathbb{S}^1)$ can be identified with the space $\varinjlim \mathcal{O}(U)$, where U in the inductive limit runs over all open neighborhoods of $\mathbb{S}^1$ and $\mathcal{O}(U)$ denotes the space of holomorphic functions on U.

• *Hyperfunctions.* We can also consider the space $\mathbf{H}(\mathbb{S}^1) = \varinjlim \mathcal{O}(U \smallsetminus \mathbb{S}^1)$ (with U running over the same sets as before) of germs of holomorphic functions in deleted neighborhoods of $\mathbb{S}^1$ in $\mathbb{C}$. The space $C^{-\omega}(\mathbb{S}^1)$ of *hyperfunctions* on $\mathbb{S}^1$ is the quotient in the exact sequence

$$0 \longrightarrow C^\omega(\mathbb{S}^1) \longrightarrow \mathbf{H}(\mathbb{S}^1) \longrightarrow C^{-\omega}(\mathbb{S}^1) \longrightarrow 0; \tag{2.17}$$

see, e.g., Sect. 1.1 of [11]. So $C^{-\omega}(\mathbb{S}^1) = \varinjlim_U \mathcal{O}(U \setminus \mathbb{S}^1)/\mathcal{O}(U)$ where U is as above and where restriction gives an injective map $\mathcal{O}(U) \to \mathcal{O}(U \smallsetminus \mathbb{S}^1)$. Actually, the quotient $\mathcal{O}(U \smallsetminus \mathbb{S}^1)/\mathcal{O}(U)$ does not depend on the choice of U, so it gives a model for $C^{-\omega}(\mathbb{S}^1)$ for any choice of U. Intuitively, a hyperfunction is the jump across $\mathbb{S}^1$ of a holomorphic function on $U \smallsetminus \mathbb{S}^1$.

• *Embedding.* The image of $C^\omega(\mathbb{S}^1)$ in $C^{-\omega}(\mathbb{S}^1)$ in (2.17) is of course zero. There is an embedding $C^\omega(\mathbb{S}^1) \to C^{-\omega}(\mathbb{S}^1)$ induced by

$$(\varphi \in \mathcal{O}(U)) \mapsto (\varphi_1 \in \mathbf{H}_s(\mathbb{S}^1)), \quad \varphi_1(w) = \begin{cases} \varphi(w) & \text{if } w \in U,\ |w| < 1, \\ 0 & \text{if } w \in U,\ |w| > 1. \end{cases} \tag{2.18}$$

• *Pairing.* We next define a pairing between hyperfunctions and analytic functions on $\mathbb{S}^1$. We begin with a pairing on $\mathbf{H}(\mathbb{S}^1) \times \mathbf{H}(\mathbb{S}^1)$. Let $\varphi,\ \psi \in \mathbf{H}(\mathbb{S}^1)$ be represented by $f,\ h \in \mathcal{O}(U \setminus \mathbb{S}^1)$ for some U. Let C_+ and C_- be closed curves in $U \setminus \mathbb{S}^1$ which are small deformations of $\mathbb{S}^1$ to the inside and outside, respectively, traversed in the positive direction, e.g., $C_\pm = \{|w| = e^{\mp\varepsilon}\}$ with ε sufficiently small. Then the integral

$$\langle \varphi, \psi \rangle \;=\; \frac{1}{2\pi i}\left(\int_{C_+} - \int_{C_-}\right) f(w)\, h(w)\, \frac{dw}{w} \tag{2.19}$$

is independent of the choice of the contours $C_\pm$ and of the neighborhood U. Moreover, if f and h are both in $\mathcal{O}(U)$, then Cauchy's theorem gives $\langle \varphi, \psi \rangle = 0$. Hence, if $\psi \in C^\omega(\mathbb{S}^1)$, then the right-hand side of (2.19) depends only on the image (also denoted φ) of φ in $C^{-\omega}(\mathbb{S}^1)$ and we get an induced pairing $C^{-\omega}(\mathbb{S}^1) \times C^\omega(\mathbb{S}^1) \to \mathbb{C}$, which we also denote by $\langle \cdot, \cdot \rangle$. Similarly, $\langle \cdot, \cdot \rangle$ gives a pairing $C^\omega(\mathbb{S}^1) \times C^{-\omega}(\mathbb{S}^1) \to \mathbb{C}$. Finally, if φ belongs to the space $C^\omega(\mathbb{S}^1)$, embedded into $C^{-\omega}(\mathbb{S}^1)$ as explained in the preceding paragraph, then it is easily seen that $\langle \varphi, \psi \rangle$ is the same as the value of the pairing $C^\omega(\mathbb{S}^1) \times C^\omega(\mathbb{S}^1) \to \mathbb{C}$ already defined in (2.13d).

• *Group Action.* We now define the action of G. We had identified $\mathcal{V}_s^\omega$ in the circle model with $C^\omega(\mathbb{S}^1)$ together with the action (2.8) of $G = PSL_2(\mathbb{R}) \cong PSU(1,1)$. For $\tilde{g} = \left[\begin{smallmatrix} A & B \\ \bar{B} & \bar{A} \end{smallmatrix}\right]$ and $\xi \in \mathbb{S}^1$, we have $|A\xi + B|^2 = (A + B\xi^{-1})(\bar{A} + \bar{B}\xi)$, which is holomorphic and takes values near the positive real axis for ξ close to $\mathbb{S}^1$ (because $|A| > |B|$). So if we rewrite the automorphy factor in (2.8) as $\left[(\bar{A} + \bar{B}\xi)(A + B/\xi)\right]^{-s}$, then we see that it extends to a single-valued and holomorphic function on a neighborhood of $\mathbb{S}^1$ (in fact, outside a path from 0 to $-B/A$ and a path from ∞ to $-\bar{A}/\bar{B}$). In other words, in the description of $\mathcal{V}_s^\omega$ as $\varinjlim \mathcal{O}(U)$, the G-action becomes

$$\varphi|_{2s}g\,(w) \;=\; \left[(\bar{A} + \bar{B}w)(A + B/w)\right]^{-s} \varphi(\tilde{g}w). \tag{2.20}$$

This description makes sense on $\mathcal{O}(U \setminus \mathbb{S}^1)$ and hence also on $\mathbf{H}(\mathbb{S}^1)$ and $C^{-\omega}(\mathbb{S}^1)$. We define $\mathcal{V}_s^{-\omega}$ as $C^{-\omega}(\mathbb{S}^1)$ together with this G-action. It is then easy to check that the embedding $\mathcal{V}_s^\omega \subset \mathcal{V}_s^{-\omega}$ induced by the embedding $C^\infty(\mathbb{S}^1) \subset C^{-\omega}(\mathbb{S}^1)$ described above is G-equivariant and also that the pairing (2.19) satisfies (2.14) and hence defines an equivariant pairing $\mathcal{V}_s^{-\omega} \times \mathcal{V}_{1-s}^\omega \to \mathbb{C}$ extending the pairing (2.13d) on $\mathcal{V}_s^\omega \times \mathcal{V}_{1-s}^\omega$.

Note also that if we denote by $\mathbf{H}_s$ the space $\mathbf{H}(\mathbb{S}^1)$ equipped with the action (2.20), then (2.17) becomes a short exact sequence

$$0 \longrightarrow \mathcal{V}_s^\omega \longrightarrow \mathbf{H}_s \xrightarrow{\;\pi\;} \mathcal{V}_s^{-\omega} \longrightarrow 0 \tag{2.21}$$

of G-modules and (2.19) defines an equivariant pairing $\mathbf{H}_s \times \mathbf{H}_{1-s} \to \mathbb{C}$.

The equivariant duality identifies $\mathcal{V}_s^{-\omega}$ with a space of linear forms on $\mathcal{V}_{1-s}^\omega$, namely (in the circle model), the space of all linear forms that are continuous for the inductive limit topology on $C^{-\omega}(\mathbb{S}^1)$ induced by the topologies on the spaces $\mathcal{O}(U)$ given by supremum norms on annuli $1-\varepsilon < |w| < 1+\varepsilon$. Similarly, the space $\mathcal{V}_s^{-\infty}$ of distributional vectors in $\mathcal{V}_s$ can be defined in the circle model as the space of linear forms on $\mathcal{V}_s^p$ that are continuous for the topology with supremum norms of all derivatives as its set of seminorms. We thus have an increasing sequence of spaces:

$$\begin{aligned} &\mathcal{V}_s^{\omega} \text{ (analytic functions) } \subset \mathcal{V}_s^{\infty} \text{ (smooth functions) } \subset \cdots \\ &\subset \mathcal{V}_s^{-\infty} \text{ (distributions) } \subset \mathcal{V}_s^{-\omega} \text{ (hyperfunctions)}, \end{aligned} \tag{2.22}$$

where all of the inclusions commute with the action of G.

- *Hyperfunctions in Other Models.* The descriptions of the spaces $\mathcal{V}_s^{-\omega}$ and $\mathcal{V}_s^{-\infty}$ in the *projective model* are similar. The space of hyperfunctions $C^{-\omega}(\mathbb{P}^1_{\mathbb{R}})$ is defined similarly to (2.17), where we now let U run through neighborhoods of $\mathbb{P}^1_{\mathbb{R}}$ in $\mathbb{P}^1_{\mathbb{C}}$. The formula (2.6) describing the action of G on functions on $\mathbb{P}^1_{\mathbb{R}}$ makes sense on a neighborhood of $\mathbb{P}^1_{\mathbb{R}}$ in $\mathbb{P}^1_{\mathbb{C}}$ and can be rewritten

$$f|_{2s}^{\mathbb{P}}\begin{bmatrix} a & b \\ c & d \end{bmatrix}(\tau) = \left(a^2+c^2\right)^{-s}\left(\frac{\tau-\mathrm{i}}{\tau-g^{-1}(i)}\right)^{s}\left(\frac{\tau+\mathrm{i}}{\tau-g^{-1}(-i)}\right)^{s} f\left(\frac{a\tau+b}{c\tau+d}\right). \tag{2.23}$$

where the automorphy factor now makes sense and is holomorphic and single-valued outside a path from i to $g^{-1}(\mathrm{i})$ and a path from $-\mathrm{i}$ to $g^{-1}(-\mathrm{i})$. The duality in this model is given by

$$\langle \varphi, \psi \rangle = \frac{1}{\pi}\left(\int_{C_+} - \int_{C_-}\right)\varphi(\tau)\,\psi(\tau)\,\frac{\mathrm{d}\tau}{1+\tau^2} \tag{2.24}$$

where the contour C_+ runs in the upper half-plane $\mathfrak{H}$, slightly above the real axis in the positive direction, and returns along a wide half circle in the positive direction and the contour C_- is defined similarly, but in the lower half-plane $\mathfrak{H}^-$, going clockwise. Everything else goes through exactly as before.

The kernel function

$$k(\zeta,\tau) = \frac{(\zeta+\mathrm{i})(\tau-\mathrm{i})}{2\mathrm{i}(\tau-\zeta)} \tag{2.25}$$

can be used to obtain a representative in $\mathbf{H}_s$ (in the projective model) for any $\alpha \in \mathcal{V}_s^{-\omega}$: if we think of α as a linear form on $\mathcal{V}_{1-s}^{\omega}$, then

$$g(\zeta) = \langle k(\zeta,\cdot),\alpha \rangle \tag{2.26}$$

is a holomorphic function on $\mathbb{P}^1_{\mathbb{C}} \smallsetminus \mathbb{P}^1_{\mathbb{R}}$ such that $\pi(g) = \alpha$. Cauchy's theorem implies that g and any representative $\psi \in \mathbf{H}_s$ of α differ by a holomorphic function on a neighborhood of $\mathbb{P}^1_{\mathbb{R}}$. The particular representative g has the nice properties of being holomorphic on $\mathfrak{H} \cup \mathfrak{H}^-$ and being normalized by $g(-\mathrm{i}) = 0$.

If one wants to handle hyperfunctions in the line model, one has to use both hyperfunctions φ and φ_∞ on $\mathbb{R}$, glued by $\varphi(\tau) = (\tau^2)^{-s}\varphi_\infty(-1/\tau)$ on neighborhoods of $(0,\infty)$ and $(-\infty,0)$. For instance, for $\mathrm{Re}\, s < \frac{1}{2}$, the linear form $\varphi \mapsto \frac{1}{\pi}\int_{-\infty}^{\infty}\varphi(t)\,dt$ on $\mathcal{V}_{1-s}^{0}$ defines a distribution $\mathbf{1}_s \in \mathcal{V}_s^{-\infty}$. In Sect. A.2 we use (2.26) to describe $\mathbf{1}_s \in \mathcal{V}_s^{-\omega}$ in the line model. The plane model seems not to be convenient for working with hyperfunctions.

Finally, in the *sequence model*, there is the advantage that one can describe all four of the spaces in (2.22) very easily since the descriptions in (2.12) applied to $\mathcal{V}_{1-s}^{\omega}$ and $\mathcal{V}_{1-s}^{\infty}$ lead immediately to the descriptions

$$\mathcal{V}_s^{-\omega} = \Big\{(c_n)\ :\ c_n = \mathrm{O}\big(\mathrm{e}^{a|n|}\big) \text{ for all } a > 0\Big\},$$
$$\mathcal{V}_s^{-\infty} = \{(c_n)\ :\ c_n = \mathrm{O}\left((1+|n|)^a\right) \text{ for some } a \in \mathbb{R}\} \tag{2.27}$$

of their dual spaces, where a sequence $\mathbf{c}$ corresponds to the hyperfunction represented by the function which is $\sum_{n\ge 0} c_n w^n$ for $1-\varepsilon < |w| < 1$ and $-\sum_{n<0} c_n w^n$ for $1 < |w| < 1+\varepsilon$; the action of G still makes sense here because the matrix coefficients as given in (2.11) decay exponentially (like $(|B|/|A|)^{|n|}$) as $|n| \to \infty$ for any $g \in G$. Thus in the sequence model, the four spaces in (2.22) correspond to sequences $\{c_n\}$ of complex numbers having exponential decay, superpolynomial decay, polynomial growth, or subexponential growth, respectively. (See (2.12) and (2.27).)

2.3 The Intertwining Map $\mathcal{V}_s^{-\omega} \to \mathcal{V}_{1-s}^{-\omega}$

The representations $\mathcal{V}_s^{-\omega}$ and $\mathcal{V}_{1-s}^{-\omega}$, with the same eigenvalue $s(1-s)$ for the Casimir operator, are not only dual to one another but are also isomorphic (for $s \notin \mathbb{Z}$). Suppose first that $F \in C^p(G)$ is in the induced representation model of $\mathcal{V}_s^p$ with $\mathrm{Re}\, s > \frac{1}{2}$ and $p = 0, 1, \ldots, \infty$. With $n(x) = \left[\begin{smallmatrix}1 & x\\ 0 & 1\end{smallmatrix}\right]$ as in (1.9b) and $w = \left[\begin{smallmatrix}0 & -1\\ 1 & 0\end{smallmatrix}\right]$, we define

$$I_s F(g) = \frac{1}{b(s-\frac{1}{2})}\int_{-\infty}^{\infty} F(gn(x)w)\,dx, \qquad b(s) = \mathrm{B}\Big(s, \frac{1}{2}\Big) = \frac{\Gamma(s)\Gamma(\frac{1}{2})}{\Gamma(s+\frac{1}{2})}, \tag{2.28}$$

where the gamma factor $b(s-\frac{1}{2})$ is a normalization, the reason of which will become clear later. The shift over $\frac{1}{2}$ is chosen since we will meet the same gamma factor unshifted in Sect. 5. From $n(x)w \in k(-\mathrm{arccot}\, x)a\big(\sqrt{1+x^2}\big)N$, we find

$$I_s F(g) = b\left(s - \frac{1}{2}\right)^{-1}\int_{-\infty}^{\infty} F\left(gk(-\mathrm{arccot}\, x)\right)(1+x^2)^{-s}\,dx,$$

which shows that the integral converges absolutely for $\mathrm{Re}\, s > \frac{1}{2}$. By differentiating under the integral in (2.28), we see that $I_s F \in C^p(G)$. From $a(y)n(x)w = n(yx)wa(y)^{-1}$, it follows that $I_s F(ga(y)n(x')) = y^{s-1} F(g)$. The action of G in the induced representation model is by left translation; hence, I_s is an intertwining operator $\mathcal{V}_s^p \to \mathcal{V}_{1-s}^p$:

$$(I_s F)\,|_{1-s}\, g_1 = I_s(F\,|_s\, g_1) \qquad \text{for } g_1 \in G. \tag{2.29}$$

To describe I_s in the plane model, we choose for a given $(\xi,\eta) \in \mathbb{R}^2 \smallsetminus \{0\}$ the element $g_{\xi,\eta} = \left[\begin{smallmatrix} \xi & -\eta/(\xi^2+\eta^2) \\ \eta & \xi/(\xi^2+\eta^2) \end{smallmatrix}\right] \in G$ to obtain

$$\begin{aligned} I_s \Phi(\xi,\eta) &= b\left(s-\frac{1}{2}\right)^{-1} \int_{-\infty}^{\infty} \Phi\left(g_{\xi,\eta}\begin{bmatrix} x & -1 \\ 1 & 0 \end{bmatrix}\right) \mathrm{d}x \\ &= b\left(s-\frac{1}{2}\right)^{-1} \int_{-\infty}^{\infty} \Phi\left(x\,(\xi,\eta) + \frac{1}{\xi^2+\eta^2}(-\eta,\xi)\right) \mathrm{d}x. \end{aligned} \tag{2.30a}$$

By relatively straightforward computations, we find that the formulas for I_s in the other models (still for $\mathrm{Re}\, s > \frac{1}{2}$) are given by

$$I_s\varphi(t) = b\left(s-\frac{1}{2}\right)^{-1} \int_{-\infty}^{\infty} |t-x|^{2s-2}\,\varphi(x)\,\mathrm{d}x, \tag{2.30b}$$

$$I_s\varphi^{\mathbb{P}}(t) = b\left(s-\frac{1}{2}\right)^{-1} \int_{\mathbb{P}^1_{\mathbb{R}}} \left(\frac{(t-x)^2}{(1+t^2)(1+x^2)}\right)^{s-1} \varphi^{\mathbb{P}}(x)\,\frac{\mathrm{d}x}{x^2+1}, \tag{2.30c}$$

$$I_s\varphi^{\mathbb{S}}(\xi) = \frac{2^{1-2s}}{\mathrm{i}}\, b(s-\tfrac{1}{2})^{-1} \int_{\mathbb{S}^1} (1-\xi/\eta)^{s-1}(1-\eta/\xi)^{s-1}\varphi^{\mathbb{S}}(\eta)\,\frac{\mathrm{d}\eta}{\eta}, \tag{2.30d}$$

$$(I_s\mathbf{c})_n = \frac{\Gamma(s)}{\Gamma(1-s)}\,\frac{\Gamma(1-s+n)}{\Gamma(s+n)}\,c_n = \frac{(1-s)_{|n|}}{(s)_{|n|}}\,c_n, \tag{2.30e}$$

with in the last line the Pochhammer symbol given by $(a)_k = \prod_{j=0}^{k-1}(a+j)$ for $k \geq 1$ and $(a)_0 = 1$. The factor $(1-s)_{|n|}/(s)_{|n|}$ is holomorphic on $0 < \mathrm{Re}\, s < 1$. Hence, $I_s\mathbf{e}_{s,n}$ is well defined for these values of the spectral parameter. The polynomial growth of the factor shows that I_s extends to a map $I_s : \mathcal{V}_s^p \to \mathcal{V}_{1-s}^p$ for $0 < \mathrm{Re}\, s < 1$ for $p = \omega, \infty, -\infty, -\omega$, but for finite p, we have only $I_s\mathcal{V}_s^p \subset \mathcal{V}_{1-s}^{p-1}$ if $0 < \mathrm{Re}\, s < 1$. See the characterizations (2.12) and (2.27). The intertwining property (2.29) extends holomorphically. The choice of the normalization factor in (2.28) implies that $I_{1-s} \circ I_s = \mathrm{Id}$, as is more easily seen from formula (2.30e). From this formula we also see that $\langle I_{1-s}\varphi, I_s\alpha\rangle = \langle \varphi, \alpha\rangle$ for $\varphi \in \mathcal{V}_{1-s}^{\omega}$, $\alpha \in \mathcal{V}_s^{-\omega}$ and that $I_{1/2} = \mathrm{Id}$.

For $\varphi \in \mathcal{V}_s^p$, $p \geq 1$, we have in the line model $\varphi'(x) = O(|x|^{-2s-1})$ as $|x| \to \infty$. For $\operatorname{Re} s > \frac{1}{2}$, integration by parts gives

$$I_s\varphi(t) = \frac{-\Gamma(s)}{2\sqrt{\pi}\,\Gamma(s+\frac{1}{2})} \int_{-\infty}^{\infty} \operatorname{sign}(t-x)\,|t-x|^{2s-1}\,\varphi'(x)\,\mathrm{d}x, \tag{2.31}$$

and this now defines $I_s\varphi$ for $\operatorname{Re} s > 0$ and shows that $I_s\mathcal{V}_s^1 \subset \mathcal{V}_{1-s}^0$.

We can describe the operator $I_s : \mathcal{V}_s^{-\omega} \to \mathcal{V}_{1-s}^{-\omega}$ on the representatives of hyperfunctions in $\mathbf{H}_s$ by sending a Laurent series $\sum_{n\in\mathbb{Z}} b_n w^n$ on an annulus $\alpha < |w| < \beta$ in $\mathbb{C}^*$ to $\sum_{n\in\mathbb{Z}} \frac{(1-s)_{|n|}}{(s)_{|n|}} b_n w^n$ converging on the same annulus. One can check that this gives an intertwining operator $I_s : \mathbf{H}_s \to \mathbf{H}_{1-s}$. (Since G is connected, it suffices to check this for generators of the Lie algebra, for which the action on the $\mathbf{e}_{s,n}$ is relatively simple. See Sect. A.5.)

3 Laplace Eigenfunctions and the Poisson Transformation

The principal series representations can also be realized as the space of eigenfunctions of the Laplace operator Δ in the hyperbolic plane $\mathbb{H}$. This model has several advantages: the action of G involves no automorphy factor at all; the model does not give a preferential treatment to any point; all vectors correspond to actual functions, with no need to work with distributions or hyperfunctions; and the values s and $1-s$ of the spectral parameter give the same space. The isomorphism from the models on the boundary used so far to the hyperbolic plane model is given by a simple integral transform (Poisson map). Before discussing this transformation in Sect. 3.3, we consider in Sect. 3.1 eigenfunctions of the Laplace operator on hyperbolic space and discuss in Sect. 3.2 the Green's form already used in [10].

Finally, in Sect. 3.4, we consider second-order eigenfunctions, i.e., functions on $\mathbb{H}$ that are annihilated by $\big(\Delta - s(1-s)\big)^2$.

3.1 *The Space $\mathcal{E}_s$ and Some of Its Elements*

We use $\mathbb{H}$ as general notation for the hyperbolic 2-space. For computations, it is convenient to work in a realization of $\mathbb{H}$. In this chapter, we use the realization as the complex upper half-plane and a realization as the complex unit disk.

The upper half-plane model of $\mathbb{H}$ is $\mathfrak{H} = \{z = x + iy : y > 0\}$, with boundary $\mathbb{P}^1_{\mathbb{R}}$. Lengths of curves in $\mathfrak{H}$ are determined by integration of $y^{-1}\sqrt{(\mathrm{d}x)^2 + (\mathrm{d}y)^2}$. To this metric are associated the Laplace operator $\Delta = -y^2(\partial_x^2 + \partial_y^2) = (z-\bar{z})^2\partial_z\partial_{\bar{z}}$ and the volume element $\mathrm{d}\mu = \frac{\mathrm{d}x\,\mathrm{d}y}{y^2}$. The hyperbolic distance $\mathrm{d}(z,z')$ between two points $z, z' \in \mathbb{H}$ is given in the upper half-plane model by

$$\cosh \mathrm{d}^{\mathfrak{H}}(z,z') = \rho^{\mathfrak{H}}(z,z') = 1 + \frac{|z-z'|^2}{2yy'} \qquad (z,z' \in \mathfrak{H}). \tag{3.1}$$

The isometry group of $\mathfrak{H}$ is the group $G = \mathrm{PSL}_2(\mathbb{R})$, acting as usual by fractional linear transformations $z \mapsto \frac{az+b}{cz+d}$. The subgroup leaving fixed i is $K = \mathrm{PSO}(2)$. So $G/K \cong \mathfrak{H}$. The action of G leaves invariant the metric and the volume element and commutes with Δ.

We use also the disk model $\mathbb{D} = \{w \in \mathbb{C} : |w| < 1\}$ of $\mathbb{H}$, with boundary $\mathbb{S}^1$. It is related to the upper half-plane model by $w = \frac{z-\mathrm{i}}{z+\mathrm{i}}$, $z = \mathrm{i}\frac{1+w}{1-w}$. The corresponding metric is $\frac{2\sqrt{(\mathrm{d}\,\mathrm{Re}\,w)^2+(\mathrm{d}\,\mathrm{Im}\,w)^2}}{1-|w|^2}$, and the Laplace operator $\Delta = -\left(1-|w|^2\right)^2 \partial_w \partial_{\bar{w}}$. The formula for hyperbolic distance becomes

$$\cosh \mathrm{d}^{\mathbb{D}}(w,w') = \rho^{\mathbb{D}}(w,w') = 1 + \frac{2|w-w'|^2}{\left(1-|w|^2\right)\left(1-|w'|^2\right)}. \tag{3.2}$$

Here the group of isometries, still denoted G, is the group $\mathrm{PSU}(1,1)$ of matrices $\left[\begin{smallmatrix} A & B \\ \bar{B} & \bar{A} \end{smallmatrix}\right]$ ($A, B \in \mathbb{C}$, $|A|^2 - |B|^2 = 1$), again acting via fractional linear transformations.

By $\mathcal{E}_s$ we denote the space of solutions of $\Delta u = \lambda_s u$ in $\mathbb{H}$, where $\lambda_s = s(1-s)$. Since Δ is an elliptic differential operator with real-analytic coefficients, all elements of $\mathcal{E}_s$ are real-analytic functions. The group G acts by $(u|g)(z) = u(gz)$. (We will use z to denote the coordinate in both $\mathfrak{H}$ and $\mathbb{D}$ when we make statements applying to both models of $\mathbb{H}$.) Obviously, $\mathcal{E}_s = \mathcal{E}_{1-s}$. If U is an open subset of $\mathbb{H}$, we denote by $\mathcal{E}_s(U)$ the space of solutions of $\Delta u = \lambda_s u$ on U.

There are a number of special elements of $\mathcal{E}_s$ which we will use in the sequel. Each of these elements is invariant or transforms with some character under the action of a one-parameter subgroup $H \subset G$. The simplest are $z = x + \mathrm{i}y \mapsto y^s$ and $z \mapsto y^{1-s}$, which are invariant under $N = \left\{\left[\begin{smallmatrix} 1 & x \\ 0 & 1 \end{smallmatrix}\right] : x \in \mathbb{R}\right\}$, and transform according to a character of $A = \left\{\left[\begin{smallmatrix} \sqrt{y} & 0 \\ 0 & 1/\sqrt{y} \end{smallmatrix}\right] : y > 0\right\}$. More generally, the functions in $\mathcal{E}_s$ transforming according to nontrivial characters of N are written in terms of Bessel functions. These are important in describing Maass forms with respect to a discrete subgroup of G that contains $\left[\begin{smallmatrix} 1 & 1 \\ 0 & 1 \end{smallmatrix}\right]$. The functions transforming according to a character of A are described in terms of hypergeometric functions. (The details, and properties of all special functions used, are given in Sect. A.1.)

If we choose the subgroup H to be $K = \mathrm{PSO}(2)$, we are led to the functions $P_{s,n}$ described in the disk model with polar coordinates $w = r\mathrm{e}^{\mathrm{i}\theta}$ by

$$P_{s,n}(r\mathrm{e}^{\mathrm{i}\theta}) := P^n_{s-1}\left(\frac{1+r^2}{1-r^2}\right)\mathrm{e}^{in\theta} \qquad (n \in \mathbb{Z}), \tag{3.3}$$

where P^n_{s-1} denotes the Legendre function of the first kind. Note the shift of the spectral parameter in P^n_{s-1} and $P_{s,n}$. If $n = 0$, one usually writes P_{s-1} instead of P^0_{s-1}, but to avoid confusion, we will not omit the 0 in $P_{s,0}$.

Every function in $\mathcal{E}_s$ can be described in terms of the $P_{s,n}$: if we write the Fourier expansion of $u \in \mathcal{E}_s$ as $u(r\mathrm{e}^{\mathrm{i}\theta}) = \sum_{n\in\mathbb{Z}} A_n(r)\mathrm{e}^{in\theta}$, then $A_n(r)$ has the form $a_n\, P^n_{s-1}\left(\frac{1+r^2}{1-r^2}\right)$ for some $a_n \in \mathbb{C}$, so we have an expansion

$$u(w) \;=\; \sum_{n\in\mathbb{Z}} a_n\, P_{s,n}(w), \qquad a_n \in \mathbb{C}. \tag{3.4}$$

Sometimes it will be convenient to consider also subgroups of G conjugate to K. For a given $z' = x' + \mathrm{i}y' \in \mathfrak{H}$, we choose $g_{z'} = \left[\begin{smallmatrix} \sqrt{y'} & x'/\sqrt{y'} \\ 0 & 1/\sqrt{y'} \end{smallmatrix}\right] \in NA \subset G$ to obtain an automorphism of $\mathfrak{H}$ sending i to z'. If we combine this with our standard identification of $\mathfrak{H}$ and $\mathbb{D}$, we get a new identification sending the chosen point z' to $0 \in \mathbb{D}$, and the function $P_{s,n}$ on $\mathbb{D}$ becomes the following function on $\mathfrak{H} \times \mathfrak{H}$:

$$p_{s,n}(z,z') \;:=\; P_{s,n}\left(\frac{z-z'}{z-\overline{z'}}\right). \tag{3.5}$$

This definition of $P_{s,n}$ depends in general on the choice of $g_{z'}$ in the coset $g_{z'}K$. In the case $n = 0$, the choice has no influence, and we obtain the very important *point-pair invariant* $p_s(z,z')$, defined, in either the disk or the upper half-plane, by the formula

$$p_s(z,z') \;:=\; p_{s,0}(z,z') \;=\; P_{s-1}\big(\rho^{\mathbb{H}}(z,z')\big) \qquad (z, z' \text{ in } \mathbb{H}), \tag{3.6}$$

with the argument $\rho(z,z') = \cosh \mathrm{d}(z,z')$ of the Legendre function $P_{s-1} = P^0_{s-1}$ being given algebraically in terms of the coordinates of z and z' by formulas (3.1) or (3.2), respectively. This function is defined on the product $\mathbb{H} \times \mathbb{H}$, is invariant with respect to the diagonal action of G on this product, and satisfies the Laplace equation with respect to each variable separately.

The Legendre function Q^n_{s-1} in (A.8) in the appendix provides elements of $\mathcal{E}_s(\mathbb{D} \smallsetminus \{0\})$:

$$Q_{s,n}\big(r\mathrm{e}^{\mathrm{i}\theta}\big) \;=\; Q^n_{s-1}\left(\frac{1+r^2}{1-r^2}\right)\mathrm{e}^{\mathrm{i}n\theta} \qquad (n \in \mathbb{Z}). \tag{3.7}$$

The corresponding point-pair invariant with $Q^0_{s-1} = Q_{s-1}$

$$q_s(z,z') \;=\; Q_{s-1}\big(\rho^{\mathbb{H}}(z,z')\big) \qquad (z, z' \text{ in } \mathbb{H}) \tag{3.8}$$

is the well-known Green's function for Δ (integral kernel function of $(\Delta - \lambda_s)^{-1}$), has a logarithmic singularity as $z \to z'$, and grows like the sth power of the Euclidean distance (in the disk model) from z to the boundary as $z \to \partial\mathbb{H}$ with z' fixed. This latter property will be crucial in Sect. 5, where we will study a space $\mathcal{W}^\omega_s$ of germs of eigenfunctions near $\partial\mathbb{H}$ having precisely this boundary behavior.

The eigenfunction $R(t;\cdot)^s$, given in the $\mathfrak{H}$-model by

$$R(t;z)^s \;=\; \frac{y^s}{|t-z|^{2s}} \qquad (t \in \mathbb{R},\ z = x + \mathrm{i}y \in \mathfrak{H}), \tag{3.9}$$

is the image under the action of $\left[\begin{smallmatrix} 0 & 1 \\ -1 & t \end{smallmatrix}\right] \in G$ on the eigenfunction $z \mapsto y^s$. This function was already used extensively in [10] (Sects. 2 and 5 of Chap. II). For fixed

$t \in \mathbb{R}$, the functions $R(t;\,\cdot)^s$ and $R(t;\,\cdot)^{1-s}$ are both in $\mathcal{E}_s$. For fixed $z \in \mathfrak{H}$, we have $R(\cdot\,; z)^s$ in the line model of $\mathcal{V}_s^\omega$. The basic invariance property

$$|ct+d|^{-2s} R(gt; gz)^s \;=\; R(t,z)^s \qquad \left(g = \begin{bmatrix} a & b \\ c & d \end{bmatrix} \in G\right) \tag{3.10}$$

may be viewed as the statement that $(t,z) \mapsto R(t;z)^s$ belongs to $(\mathcal{V}_s^\omega \otimes \mathcal{E}_s)^G$. The function $R(\cdot\,;\,\cdot)^{1-s}$ is the kernel function of the Poisson transform in Sect. 3.3.

We may allow t to move off $\mathbb{R}$ in such a way that $R(t,z)^s$ becomes holomorphic in this variable:

$$R(\zeta;z)^s \;=\; \left(\frac{y}{(\zeta-z)(\zeta-\bar{z})}\right)^s \qquad (\zeta \in \mathbb{C},\ z = x+\mathrm{i}y \in \mathfrak{H}). \tag{3.11}$$

However, this not only has singularities at $z=\zeta$ or $z=\bar{\zeta}$ but is also many-valued. To make a well-defined function, we have to choose a path C from ζ to $\bar{\zeta}$, in which case $R(\zeta;\,\cdot)^s$ becomes single-valued on $U = \mathfrak{H} \smallsetminus C$ and lies in $\mathcal{E}_s(U)$. (Cf. [10], Chap. II, Sect. 1.) Sometimes it is convenient to write R_ζ^s instead of $R(\zeta;\,\cdot)^s$.

Occasionally, we will choose other branches of the multivalued function $R(\cdot\,;\,\cdot)^s$. We have

$$\partial_z R(\zeta;z)^s \;=\; \frac{s}{z-\bar{z}}\,\frac{\zeta-\bar{z}}{\zeta-z}\,R(\zeta;z)^s, \qquad \partial_{\bar{z}} R(\zeta;z)^s \;=\; -\frac{s}{z-\bar{z}}\,\frac{\zeta-z}{\zeta-\bar{z}}\,R(\zeta;z)^s, \tag{3.12}$$

provided we use the same branch on the left and the right.

3.2 The Green's Form and a Cauchy Formula for $\mathcal{E}_s$

Next we recall the bracket operation from [10], which associates to a pair of eigenfunctions of Δ with the same eigenvalue[1] a closed 1-form (Green's form). It comes in two versions, differing by an exact form:

$$[u,v] \;=\; u_z\,v\,\mathrm{d}z \,+\, u\,v_{\bar{z}}\,\mathrm{d}\bar{z}, \qquad \{u,v\} \;=\; 2\mathrm{i}\,[u,v] \,-\, i\,\mathrm{d}(uv). \tag{3.13}$$

Because $[u|g, v|g] = [u,v]\circ g$ for any locally defined holomorphic map g (cf. [10], lemma in Sect. 2 of Chap. II), these formulas make sense and define the same

[1]More generally, for any differentiable functions u and v on $\mathbb{H}$ we have

$$\mathrm{d}\{u,v\} \;=\; 2\mathrm{i}\,\mathrm{d}[u,v] \;=\; \Big((\Delta u)\,v - u\,(\Delta v)\Big)\,\mathrm{d}\mu,$$

where $\mathrm{d}\mu$ ($= y^{-2}\,\mathrm{d}x\,\mathrm{d}y$ in the upper half-plane model) is the invariant measure in $\mathbb{H}$.

1-form whether we use the $\mathfrak{H}$- or $\mathbb{D}$-model of $\mathbb{H}$, and define G-equivariant maps $\mathcal{E}_s \times \mathcal{E}_s \to \Omega^1(\mathbb{H})$ (or $\mathcal{E}_s(U) \times \mathcal{E}_s(U) \to \Omega^1(U)$ for any open subset U of $\mathbb{H}$). The $\{u, v\}$-version of the bracket, which is antisymmetric, is given in (x, y)-coordinates $z = x + \mathrm{i}y \in \mathfrak{H}$ by

$$\{u,\, v\} \;=\; \begin{vmatrix} u & u_x & u_y \\ v & v_x & v_y \\ 0 & \mathrm{d}x & \mathrm{d}y \end{vmatrix} \tag{3.14}$$

and in (r, θ)-coordinates $w = r\mathrm{e}^{\mathrm{i}\theta} \in \mathbb{D}$ by

$$\{u,\, v\} \;=\; \begin{vmatrix} u & r\,u_r & u_\theta \\ v & r\,v_r & v_\theta \\ 0 & \mathrm{d}r/r & \mathrm{d}\theta \end{vmatrix}. \tag{3.15}$$

We can apply the Green's form in particular to any two of the special functions discussed above, and in some cases, the resulting closed form can be written as the total differential of an explicit function. A trivial example is $2\mathrm{i}\,[y^s, y^{1-s}] = s\,\mathrm{d}z - (1-s)\,\mathrm{d}\bar{z}$, $\{y^s, y^{1-s}\} = (2s-1)\,\mathrm{d}x$. A less obvious example is

$$[R_a^s,\, R_b^{1-s}](z) \;=\; \frac{1}{b-a}\,\mathrm{d}\left(\frac{(\bar{z}-a)(z-b)}{z-\bar{z}}\,R_a^s(z)\,R_b^{1-s}(z)\right), \tag{3.16}$$

where a and b are either distinct real numbers or distinct complex numbers and $z \notin \{a, b, \bar{a}, \bar{b}\}$. On both sides we take the same branches of R_a^s and R_b^{1-s}. This formula, which can be verified by direct computation, can be used to prove the Poisson inversion formula discussed below (cf. Remark 1, Sect. 4.2). Some other examples are given in Sect. A.4.

We can also consider the brackets of any function $u \in \mathcal{E}_s$ with the point-pair invariants $p_s(z, z')$ or $q_s(z, z')$. The latter is especially useful since it gives us the following $\mathcal{E}_s$-analogue of Cauchy's formula:

Theorem 3.1. *Let C be a piecewise smooth simple closed curve in $\mathbb{H}$ and u an element of $\mathcal{E}_s(U)$, where $U \subset \mathbb{H}$ is some open set containing C and its interior. Then for $w \in \mathbb{H} \smallsetminus C$, we have*

$$\frac{1}{\pi \mathrm{i}} \int_C [u, q_s(\,\cdot\,, w)] \;=\; \begin{cases} u(w) & \text{if } w \text{ is inside } C, \\ 0 & \text{if } w \text{ is outside } C, \end{cases} \tag{3.17}$$

where the curve C is traversed is the positive direction.

Proof. Since $[u, q_s(\,\cdot\,, w)]$ is a closed form, the value of the integral in (3.17) does not change if we deform the path C, so long as we avoid the point w where the form becomes singular. The vanishing of the integral when w is outside of C is therefore clear, since we can simply contract C to a point. If w is inside C, then we can deform C to a small hyperbolic circle around w. We can use the G-equivariance to put $w = 0$, so that this hyperbolic circle is also a Euclidean one, say $z = \varepsilon e^{i\theta}$. We can also replace $[u, q_s(\cdot, 0)]$ by $\{u, q_s(\cdot, 0)\}/2i$, since their difference is exact. From (3.15) and the asymptotic result (A.11), we find that the closed form $-\frac{i}{2}\{u, q_s(\cdot, 0)\}$ equals $\left(\frac{i}{2}u(0) + O(\varepsilon \log \varepsilon)\right) d\theta$ on the circle. The result follows. ■

The method of the proof just given can also be used to check that for a contour C in $\mathbb{D}$ encircling 0 once in positive direction, we have for all $n \in \mathbb{Z}$

$$\int_C [P_{s,n}, Q_{s,m}] = \pi i (-1)^n \delta_{n,-m}. \tag{3.18}$$

Combining this formula with the expansion (3.4), we arrive at the following generalization of the standard formula for the Taylor expansion of holomorphic functions:

Proposition 3.2. *For each $u \in \mathcal{E}_s$:*

$$u(w) = \sum_{n \in \mathbb{Z}} \frac{(-1)^n}{\pi i} P_{s,n}(w) \int_C [u, Q_{s,-n}]. \tag{3.19}$$

If $u \in \mathcal{E}_s(A)$, where A is some annulus of the form $r_1 < |w| < r_2$ in $\mathbb{D}$, there is a more complicated expansion of the form

$$u(w) = \sum_{n \in \mathbb{Z}} \bigl(a_n P_{s,n}(w) + b_n Q_{s,n}(w)\bigr). \tag{3.20}$$

For fixed $w' \in \mathbb{D}$, the function $w \mapsto q_s(w, w')$ has only one singularity, at $w = w'$. So both on the disk $|w| < |w'|$ and on the annulus $|w| > |w'|$ the function $q_s(\cdot, w')$ has a polar Fourier expansion, which can be given explicitly:

Proposition 3.3. *For $w, w' \in \mathbb{D}$ with $|w| \neq |w'|$:*

$$q_s(w, w') = \begin{cases} \sum_{n \in \mathbb{Z}} (-1)^n P_{s,-n}(w') Q_{s,n}(w) & \text{if } |w| > |w'|, \\ \sum_{n \in \mathbb{Z}} (-1)^n P_{s,-n}(w) Q_{s,n}(w') & \text{if } |w| < |w'|. \end{cases} \tag{3.21}$$

Proof. Apply (3.20) to $q_s(\cdot, w')$ on the annulus $A = \{w \in \mathbb{D} : |w'| < |w|\}$. Since $q_s(\cdot, w')$ represents an element of $\mathcal{W}_s^\omega$, the expansion becomes

$$q_s(w, w') = \sum_{n \in \mathbb{Z}} b_n(w') Q_{s,n}(w) \qquad (|w| > |w'|).$$

From $q_s(e^{i\theta}w, e^{i\theta}w') = q_s(w, w')$, it follows that $b_n(e^{i\theta}w') = e^{-in\theta} b_n(w)$. For $w \in \mathbb{D} \setminus \{0\}$, we have $q_s(w, \cdot) \in \mathcal{E}_s(B)$ with $B = \{w' \in \mathbb{D} : |w'| < |w|\}$. Then the coefficients b_n are also in $\mathcal{E}_s(B)$. Since $Q_{s,-n}$ has a singularity at $0 \in \mathbb{D}$, the coefficients have the form $b_n(w') = c_n\, P_{s,-n}(w')$. Now we apply (3.17) and (3.18) to obtain with a path C inside the region A:

$$\begin{aligned} P_{s,m}(w') &= \frac{1}{\pi \mathrm{i}} \int_C \left[P_{s,m}, q_s(\cdot\,, w')\right] = \frac{1}{\pi \mathrm{i}} \sum_{n\in\mathbb{Z}} c_n\, P_{s,-n}(w') \int_C \left[P_{s,m}, Q_{s,n}\right] \\ &= c_{-m}\, P_{s,m}(w')\,(-1)^m. \end{aligned}$$

Hence, $c_m = (-1)^m$, and the proposition follows, with the symmetry of q_s. ■

3.3 The Poisson Transformation

There is a well-known isomorphism P_s from $\mathcal{V}_s^{-\omega}$ to $\mathcal{E}_s$. This enables us to view $\mathcal{E}_s$ as a model of the principal series. We first describe P_s abstractly and then more explicitly in various models of $\mathcal{V}_s^{-\omega}$. In Sect. 4.2 we will describe the inverse isomorphism from $\mathcal{E}_s$ to $\mathcal{V}_s^{-\omega}$.

For $\alpha \in \mathcal{V}_s^{-\omega}$ and $g \in G$

$$(\mathsf{P}_s\alpha)(g) = \langle \alpha|_{2s}\, g,\, \mathbf{e}_{1-s,0}\rangle = \langle \alpha,\, \mathbf{e}_{1-s,0}|_{2-2s}\, g^{-1}\rangle \tag{3.22}$$

describes a function on G that is K-invariant on the right. Hence, it is a function on $G/K \cong \mathbb{H}$. The center of the enveloping algebra is generated by the Casimir operator. It gives rise to a differential operator on G that gives, suitably normalized, the Laplace operator Δ on the right-K-invariant functions. Since the Casimir operator acts on $\mathcal{V}_{1-s}^{-\omega}$ as multiplication by $\lambda_s = (1-s)s$, the function $\mathsf{P}_s\alpha$ defines an element of $\mathcal{E}_s$. We write in the upper half-plane model

$$\mathsf{P}_s\alpha(z) = \mathsf{P}_s\alpha\big(n(x)a(y)\big), \tag{3.23}$$

with the notation in (1.9b). The definition in (3.22) implies that the Poisson transformation is G-equivariant:

$$\mathsf{P}_s(\alpha|_{2s}g)(z) = \mathsf{P}_s\alpha(gz). \tag{3.24}$$

The fact that the intertwining operator $I_s : \mathcal{V}_s^{-\omega} \to \mathcal{V}_{1-s}^{-\omega}$ preserves the duality implies that the following diagram commutes:

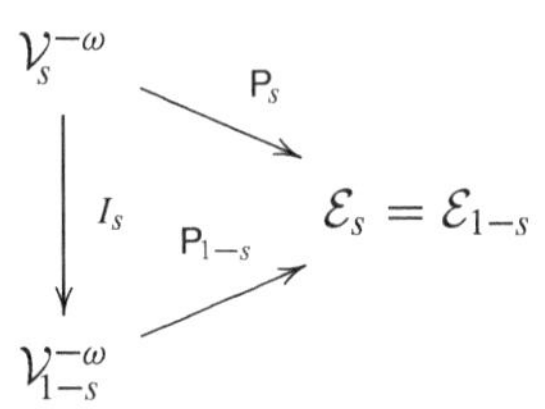

(3.25)

If $\alpha \in \mathcal{V}_s^0$, we can describe $\mathsf{P}_s\alpha$ by a simple integral formula. In the line model, this takes the form

$$\begin{aligned} \mathsf{P}_s\alpha(z) &= \left\langle \mathbf{e}_{s,0}, \alpha|_{2s} n(x)a(y) \right\rangle = \frac{1}{\pi}\int_{-\infty}^{\infty} \mathbf{e}_{s,0}|_{2-2s} \left(n(x)a(y)\right)^{-1}(t)\,\alpha(t)\,\mathrm{d}t \\ &= \frac{1}{\pi}\int_{-\infty}^{\infty} y^{-1+s}\left(\left(\frac{t-x}{y}\right)^2+1\right)^{s-1} \alpha(t)\,\mathrm{d}t = \frac{1}{\pi}\int_{-\infty}^{\infty} R(t;z)^{1-s}\alpha(t)\,\mathrm{d}t, \end{aligned} \tag{3.26}$$

so that R^{1-s} is the kernel of the Poisson transformation in the line model. If α is a hyperfunction, the pairing in (3.22) has to be interpreted as discussed in Sect. 2.2 as the difference of two integrals over contours close to and on opposite sides of $\partial\mathbb{H}$ ((2.19) in the circle model), with $R(\cdot\,;z)^{1-s}$ extended analytically to a neighborhood of $\partial\mathbb{H}$.

In the projective model and the circle model, we find

$$\mathsf{P}_s\alpha(z) = \langle R(\cdot\,;z)^{1-s}, \alpha\rangle, \tag{3.27a}$$

with $R(\cdot\,;z)^{1-s}$ in the various models given by

$$R^{\mathbb{P}}(\zeta;z)^{1-s} = y^{s-1}\left(\frac{\zeta-\mathrm{i}}{\zeta-z}\right)^{1-s}\left(\frac{\zeta+\mathrm{i}}{\zeta-\bar{z}}\right)^{1-s} = \left(\frac{R(\zeta;z)}{R(\zeta;i)}\right)^{1-s}, \tag{3.27b}$$

$$R^{\mathbb{S}}(\xi;w)^{1-s} = \left(\frac{1-|w|^2}{(1-w/\xi)(1-\bar{w}\xi)}\right)^{1-s}. \tag{3.27c}$$

By $R(\cdot\,,\cdot)^{1-s}$, without superscript on the R, we denote the Poisson kernel in the line model (as in (3.11)). We take the branch for which $\arg R(\zeta;z) = 0$ for ζ on $\mathbb{R}$.

In the circle model, we have for each $\alpha \in \mathcal{V}_s^{-\omega}$:

$$\mathsf{P}_s\alpha(w) = \frac{(1-|w|^2)^{1-s}}{2\pi\mathrm{i}}\left(\int_{C_+} - \int_{C_-}\right) g(\xi)\left((1-w/\xi)(1-\bar{w}\xi)\right)^{s-1}\frac{\mathrm{d}\xi}{\xi}, \tag{3.28}$$

with C_+ and C_- as in (2.19), adapted to the domain of the representative $g \in \mathbf{H}_s$ of the hyperfunction α.

For the values of s we are interested in, Helgason has shown that the Poisson transformation is an isomorphism:

Theorem 3.4. (Theorem 4.3 in [5]). *The Poisson map $P_s : \mathcal{V}_s^{-\omega} \to \mathcal{E}_s$ is a bijection for $0 < \operatorname{Re} s < 1$.*

The usual proof of this uses the K-Fourier expansion, where K ($\cong \mathrm{PSO}(2)$) is the standard maximal compact subgroup of G. One first checks by explicit integration the formula

$$\mathsf{P}_s\mathbf{e}_{s,m} = (-1)^m\frac{\Gamma(s)}{\Gamma(s+m)}P_{s,m} \qquad (n \in \mathbb{Z}), \tag{3.29}$$

with $\mathbf{e}_{s,m}$ and $P_{s,m}$ as defined in (2.9d) and (3.3) respectively. (Indeed, with (3.27b) and (3.28), we obtain the Poisson integral

$$\mathsf{P}_s\mathbf{e}_{s,m}(w) \;=\; \frac{(1-|w|^2)^{1-s}}{2\pi\mathrm{i}} \int_{|\xi|=1} \xi^m \left((1-w/\xi)(1-\bar{w}\xi)\right)^{s-1} \frac{\mathrm{d}\xi}{\xi}.$$

Since $|w/\xi| < 1$ and $|\bar{w}\xi| < 1$, this leads to the expansion

$$(1-|w|^2)^{1-s} \sum_{n_1,n_2\ge 0,\, -n_1+n_2=m} \frac{(1-s)_{n_1}\,(1-s)_{n_2}}{n_1!\,n_2!}\, w^{n_1}\,\bar{w}^{n_2}$$

$$= (1-|w|^2)^{1-s}\,\frac{(1-s)_{|m|}}{|m|!} \sum_{n\ge 0} \frac{(1-s)_n\,(1-s+|m|)_n}{(1+|m|)_n\,n!}\,|w|^{2n} \cdot \begin{cases} w^m & \text{if } m \ge 0, \\ \bar{w}^{-m} & \text{if } m \le 0. \end{cases}$$

This is $(-1)^m\Gamma(s)/\Gamma(s+m)$ times $P_{s,m}$ as defined in (A.8) and (A.9).) Then one uses the fact that the elements of $\mathcal{V}_s^{-\omega}$ are given by sums $\sum c_n\,\mathbf{e}_{s,n}$ with coefficients c_n of subexponential growth ((2.27)) and shows that the coefficients in the expansion (3.19) also have subexponential growth for each $u \in \mathcal{E}_s$. This is the analogue of the fact that a holomorphic function in the unit disk has Taylor coefficients at 0 of subexponential growth and can be proved the same way. An alternative proof of Theorem 3.4 will follow from the results of Sect. 4.2, where we shall give an explicit inverse map for P_s.

Thus, $\mathcal{E}_s$ is a model of the principal series representation $\mathcal{V}_s^{-\omega}$, and also of $\mathcal{V}_{1-s}^{-\omega}$, that does not change under the transformation $s \mapsto 1-s$ of the spectral parameter. It is completely G-equivariant. The action of G is simply given by $u \,|\, g = u \circ g$.

As discussed in Sect. 1, the space $\mathcal{V}_s^{-\omega}$ (hyperfunctions on $\partial\mathbb{H}$) contains three canonical subspaces $\mathcal{V}_s^{-\infty}$ (distributions), $\mathcal{V}_s^{\infty}$ (smooth functions), and $\mathcal{V}_s^{\omega}$ (analytic functions on $\partial\mathbb{H}$), and we can ask whether there is an intrinsic characterization of the corresponding subspaces $\mathcal{E}_s^{-\infty}$, $\mathcal{E}_s^{\infty}$, and $\mathcal{E}_s^{\omega}$ of $\mathcal{E}_s$. For $\mathcal{E}_s^{-\infty}$, the answer is simple and depends only on the asymptotic properties of the eigenfunctions near the boundary, namely,

Theorem 3.5. ([9], Theorems 4.1 and 5.3) *Let* $0 < \operatorname{Re} s < 1$. *The space* $\mathcal{E}_s^{-\infty} = \mathsf{P}_s\big(\mathcal{V}_s^{-\infty}\big)$ *consists of the functions in* $\mathcal{E}_s$ *having at most polynomial growth near the boundary.*
("At most polynomial growth near the boundary" means $\ll \big(1-|w|^2\big)^{-C}$ for some C in the disk model and $\ll \big((|z+\mathrm{i}|^2)/y\big)^C$ in the upper half-plane model.)

The corresponding theorems for the spaces $\mathcal{E}_s^{\infty}$ and $\mathcal{E}_s^{\omega}$, which do not only involve estimates of the speed of growth of functions near $\partial\mathbb{H}$, are considerably more complicated. We will return to the description of these spaces in Sect. 7.

- *Explicit Examples.* One example is given in (3.29). Another example is

$$\mathsf{P}_s\delta_{s,\infty}(z) \;=\; R^{\mathbb{P}}(\infty;z)^{1-s} \;=\; y^{1-s}, \tag{3.30}$$

where $\delta_{s,\infty} \in \mathcal{V}_s^{-\infty}$ is the distribution associating to $\varphi \in \mathcal{V}_{1-s}^{\infty}$, in the projective model, its value at ∞. As a third example, we consider the element $R(\cdot\,;z_0)^s \in \mathcal{V}_s^{\omega}$. For convenience we use the circle model. Then $a(w,w') = \big(\mathsf{P}_s R^{\mathbb{S}}(\cdot\,;w')^s\big)(w)$ satisfies the relation $a(gw',gw) = a(w',w)$ for all $g \in G$, by equivariance of the Poisson transform and of the function R^s. So a is a point-pair invariant. Since $a(w,\cdot) \in \mathcal{E}_s$, it has to be a multiple of p_s. We compute the factor by taking $w' = w = 0 \in \mathbb{D}$:

$$\begin{aligned}\mathsf{P}_s R^{\mathbb{S}}(\cdot\,;0)^s(0) &= \frac{1}{2\pi \mathrm{i}}\int_{\mathbb{S}^1} R^{\mathbb{S}}(\xi;0)^s\, R^{\mathbb{S}}(\xi;0)^{1-s}\,\frac{d\xi}{\xi}\\ &= \frac{1}{2\pi \mathrm{i}}\int_{\mathbb{S}^1} 1\cdot 1\,\frac{d\xi}{\xi} = 1.\end{aligned}$$

Thus we have

$$\mathsf{P}_s R(\cdot\,;w')^s(w) = p_s(w',w). \tag{3.31}$$

With (3.25) and the fact that $P_{1-s,0} = P_{s,0}$, this implies

$$I_s\, R(\cdot\,;w')^s = R(\cdot\,;w')^{1-s}. \tag{3.32}$$

3.4 Second-Order Eigenfunctions

The Poisson transformation allows us to prove results concerning the space

$$\mathcal{E}_s' := \operatorname{Ker}\big((\Delta-\lambda_s)^2 : C^{\infty}(\mathbb{H}) \longrightarrow C^{\infty}(\mathbb{H})\big). \tag{3.33}$$

Proposition 3.6. *The following sequence is exact:*

$$0 \longrightarrow \mathcal{E}_s \longrightarrow \mathcal{E}_s' \overset{\Delta-\lambda_s}{\longrightarrow} \mathcal{E}_s \longrightarrow 0. \tag{3.34}$$

Proof. Only the surjectivity of $\mathcal{E}_s' \to \mathcal{E}_s$ is not immediately clear.

Let $0 < \operatorname{Re} s_0 < 1$. Suppose we have a family $s \mapsto f_s$ on a neighborhood of s_0 such that $f_s \in \mathcal{E}_s$ for all s near s_0, and suppose that this family is C^{∞} in (s,z) and holomorphic in s. Then

$$(\Delta-\lambda_{s_0})\big(\partial_s f_s|_{s=s_0}\big) - (1-2s_0)f_{s_0} = 0.$$

For $s_0 \neq \frac{1}{2}$, this gives an element of $\mathcal{E}_{s_0}'$ that is mapped to f_{s_0} by $\Delta-\lambda_{s_0}$. If $s_0 = \frac{1}{2}$, we replace f_s by $\frac{1}{2}(f_s + f_{1-s})$ and differentiate twice.

To produce such a family, we use the Poisson transformation. By Theorem 3.4, there is a unique $\alpha \in \mathcal{V}_{s_0}^{-\omega}$ such that $f = \mathsf{P}_{s_0}\alpha$. We fix a representative $g \in \mathcal{O}(U \smallsetminus \mathbb{S}^1)$

of α in the circle model, which represents a hyperfunction α_s for all $s \in \mathbb{C}$. (The projective model works as well.) We put

$$f_s(w) = \mathsf{P}_s\alpha_s(w) = \frac{1}{2\pi\mathrm{i}}\left(\int_{C_+} - \int_{C_-}\right) R^{\mathbb{S}}(\zeta;w)^{1-s} g(\zeta)\,\frac{\mathrm{d}\zeta}{\zeta}.$$

The contours C_+ and C_- have to be adapted to w but can stay the same when w varies through a compact subset of $\mathfrak{H}$. Differentiating this family provides us with a lift of f in $\mathcal{E}'_{s_0}$. ■

This proof gives an explicit element

$$\tilde{f}(w) = \frac{-1}{2\pi\mathrm{i}}\left(\int_{C_+} - \int_{C_-}\right) R^{\mathbb{S}}(\zeta;w)^{1-s_0} \left(\log R^{\mathbb{S}}(\zeta;w)\right) g(\zeta)\,\frac{\mathrm{d}\zeta}{\zeta} \tag{3.35}$$

of $\mathcal{E}'_s$ with $(\Delta - \lambda_s)\tilde{f} = (1-2s)f$. Note that for $s = \frac{1}{2}$, the function $\tilde{f}$ belongs to $\mathcal{E}_{1/2}$, giving an interesting map $\mathcal{E}_{1/2} \to \mathcal{E}_{1/2}$. As an example, if $f(z) = y^{1/2}$, then we can take $g(\zeta) = \frac{\zeta}{2\mathrm{i}}$ as the representative of the hyperfunction $\alpha = \delta_{1/2,\infty}$ with $\mathsf{P}_{1/2}\alpha = h$, and by deforming the contours C_+ and C_- into one circle $|\zeta| = R$ with R large, we obtain (in the projective model)

$$\begin{aligned}\tilde{f}(z) &= \frac{-1}{\pi}\int_{|\zeta|=R} R^{\mathbb{P}}(\zeta;z)^{1/2}\left(\log y + \log\frac{\zeta^2+1}{(\zeta-z)(\zeta-\bar{z})}\right)\frac{\zeta}{2\mathrm{i}}\,\frac{\mathrm{d}\zeta}{\zeta^2+1}\\ &= -y^{1/2}\log y.\end{aligned} \tag{3.36}$$

In part C of Table A.1 in Sect. A.2, we describe the distribution in $\mathcal{V}_{1/2}^{-\infty}$ corresponding to this element of $\mathcal{E}_{1/2}$.

Theorem 3.5 shows that the subspace $\mathcal{E}_s^{-\infty}$ corresponding to $\mathcal{V}_s^{-\infty}$ under the Poisson transformation is the space of elements of $\mathcal{E}_s$ with polynomial growth. We define $(\mathcal{E}'_s)^{-\infty}$ as the subspace of $\mathcal{E}'_s$ of elements with polynomial growth. The following proposition, including the somewhat technical second statement, is needed in Chap. V of [2].

Proposition 3.7. *The sequence*

$$0 \longrightarrow \mathcal{E}_s^{-\infty} \longrightarrow (\mathcal{E}'_s)^{-\infty} \xrightarrow{\Delta-\lambda_s} \mathcal{E}_s^{-\infty} \longrightarrow 0 \tag{3.37}$$

is exact. All derivatives $\partial_w^l \partial_{\bar{w}}^m f(w)$, $l,m \geq 0$, of $f \in (\mathcal{E}'_s)^{-\infty}$, in the disk model, have polynomial growth.

Proof. We use the construction in the proof of Proposition 3.6. We use

$$f_s(w) = \langle R^{\mathbb{S}}(\cdot\,;w)^{1-s}, \alpha\rangle,$$

with $\alpha \in \mathcal{V}_{s_0}^{-\infty}$. For $\zeta \in \mathbb{S}^1$ we obtain by differentiating the expression for $R^{\mathbb{S}}$ in (3.27c)

$$(\zeta\partial\zeta)^n \partial_w^l \partial_{\bar{w}}^m \, R^{\mathbb{S}}(\zeta; w)^{1-s} \ll_{n,l,m} (1-|w|^2)^{s-l-m-n}.$$

With the seminorm $\|\cdot\|_n$ in (2.16), we can reformulate this as

$$\|\partial_w^l \partial_{\bar{w}}^m R^{\mathbb{S}}(\cdot\,; w)^{1-s}\|_n \ll_{n,l,m} (1-|w|^2)^{s-l-m-n}. \tag{3.38}$$

Since α determines a continuous linear form on $\mathcal{V}_s^p$ for some $p \in \mathbb{N}$, this gives an estimate

$$\partial_w^l \partial_{\bar{w}}^m f(w) \ll_{\alpha,l,m} (1-|w|^2)^{\mathrm{Re}\, s-1-l-m-p}$$

for $f \in \mathcal{E}_{s_0}^{-\infty}$.

Differentiating $R^{\mathbb{S}}(\cdot\,; w)^{1-s}$ once or twice with respect to s multiplies the estimate in (3.38) with at most a factor $\left|\log(1-|w|^2)\right|^2$. The lift $\tilde{f} \in \mathcal{E}'_{s_0}$ of f_{s_0} in the proof of Proposition 3.6 satisfies

$$\partial_w^l \partial_{\bar{w}}^m \tilde{f}(w) \ll_{\alpha,l,m,\varepsilon} (1-|w|^2)^{\mathrm{Re}\, s-1-l-m-p-\varepsilon}$$

for each $\varepsilon > 0$. ∎

4 Hybrid Models for the Principal Series Representation

In this section we introduce the canonical model of the principal series, discussed in the introduction. In Sect. 4.1 we define first two other models of $\mathcal{V}_s$ in functions or hyperfunctions on $\partial\mathbb{H} \times \mathbb{H}$, which we call *hybrid models*, since they mix the properties of the model of $\mathcal{V}_s$ in eigenfunctions, as discussed in Sect. 3, with the models discussed in Sect. 2. The second of these, called the flabby hybrid model, contains the *canonical model* as a special subspace. The advantage of the canonical model becomes very clear in Sect. 4.2, where we give an explicit inverse for the Poisson transformation whose image coincides exactly with the canonical model.

In Sect. 4.3 we will characterize the canonical model as a space of functions on $(\mathbb{P}^1_{\mathbb{C}} \smallsetminus \mathbb{P}^1_{\mathbb{R}}) \times \mathfrak{H}$ satisfying a certain system of differential equations. We use these differential equations to define a sheaf $\mathcal{D}_s$ on $\mathbb{P}^1_{\mathbb{C}} \times \mathfrak{H}$, the *sheaf of mixed eigenfunctions*. The properties of this sheaf and of its sections over other natural subsets of $\mathbb{P}^1_{\mathbb{C}} \times \mathfrak{H}$ are studied in the remainder of the subsection and in more detail in Sect. 6.

4.1 The Hybrid Models and the Canonical Model

The line model of principal series representations is based on giving $\infty \in \partial\mathfrak{H}$ a special role. The projective model eliminated the special role of the point at infinity in the line model at the expense of a more complicated description of the action of $G = PSL_2(\mathbb{R})$, but it also broke the G-symmetry in a different way by singling out the point $i \in \mathfrak{H}$. The corresponding point $0 \in \mathbb{D}$ plays a special role in the circle model. The sequence model is based on the characters of the specific maximal compact subgroup $K = PSO(2) \subset G$ and not of its conjugates, again breaking the G-symmetry. The induced representation model depends on the choice of the Borel group NA. Thus none of the one-variable models for $\mathcal{V}_s$ discussed in Sect. 2 reflects fully the intrinsic symmetry under the action of G.

To remedy these defects, we will replace our previous functions φ on $\partial\mathbb{H}$ by functions $\widetilde{\varphi}$ on $\partial\mathbb{H} \times \mathbb{H}$, where the second variable plays the role of a base point, with $\widetilde{\varphi}(\cdot, i)$ being equal to the function $\varphi^{\mathbb{P}}$ of the projective model. This has the disadvantage of replacing functions of one variable by functions of two, but gives a very simple formula for the G-action, is completely symmetric, and will also turn out to be very convenient for the Poisson transform. Explicitly, given $(\varphi, \varphi_\infty)$ in the line model, we define $\widetilde{\varphi} : \mathbb{P}^1_{\mathbb{R}} \times \mathfrak{H} \to \mathbb{C}$ by

$$\widetilde{\varphi}(t,z) = \begin{cases} \left(\dfrac{|z-t|^2}{y}\right)^s \varphi(t) & \text{if } t \in \mathbb{P}^1_{\mathbb{R}} \smallsetminus \{\infty\}, \\ \left(\dfrac{|1+z/t|^2}{y}\right)^s \varphi_\infty\left(-\dfrac{1}{t}\right) & \text{if } t \in \mathbb{P}^1_{\mathbb{R}} \smallsetminus \{0\} \end{cases} \tag{4.1}$$

(here $y = \mathrm{Im}\,(z)$ as usual), generalizing (2.5) for $z = \mathrm{i}$. The function $\widetilde{\varphi}$ then satisfies

$$\widetilde{\varphi}(t,z_1) = \left(\frac{|z_1-t|^2/y_1}{|z_2-t|^2/y_2}\right)^s \widetilde{\varphi}(t,z_2) = \left(\frac{R(t;z_2)}{R(t;z_1)}\right)^s \widetilde{\varphi}(t,z_2) \tag{4.2}$$

for $t \in \mathbb{P}^1_{\mathbb{R}}$ and $z_1, z_2 \in \mathfrak{H}$. A short calculation, with use of (3.10), shows that the action of G becomes simply

$$\widetilde{\varphi|g}(t,z) = \widetilde{\varphi}(gt, gz) \qquad (t \in \mathbb{P}^1_{\mathbb{R}},\ z \in \mathfrak{H},\ g \in G) \tag{4.3}$$

in this model. From (4.1), (2.5), and (4.2), we find

$$\varphi^{\mathbb{P}}(t) = \varphi(t,\mathrm{i}), \qquad \widetilde{\varphi}(t,z) = \left(\frac{(t-z)(t-\bar{z})}{(t^2+1)\,y}\right)^s \varphi^{\mathbb{P}}(t), \tag{4.4}$$

giving the relation between the new model and the projective model. And we see that only the complicated factor relating $\widetilde{\varphi}$ to $\varphi^{\mathbb{P}}$ is responsible for the complicated action of G in the projective model.

We define the *rigid hybrid model* to be the space of functions $h : \mathbb{P}^1_{\mathbb{R}} \times \mathfrak{H} \to \mathbb{C}$ satisfying (4.2) with $\widetilde{\varphi}$ replaced by h. The G-action is given by $F \mapsto F \circ g$, where G acts diagonally on $\mathbb{P}^1_{\mathbb{R}} \times \mathfrak{H}$. The smooth (resp. analytic) vectors are those for which $F(\cdot, z)$ is smooth (resp. analytic) on $\mathbb{P}^1_{\mathbb{R}}$ for any $z \in \mathfrak{H}$; this is independent of the choice of z because the expression in parentheses in (4.2) is analytic and strictly positive on $\mathbb{P}^1_{\mathbb{R}}$. These spaces are models for $\mathcal{V}_s^\infty$ and $\mathcal{V}_s^\omega$, respectively, but when needed will be denoted $\mathcal{V}_s^{\infty,\mathrm{rig}}$ and $\mathcal{V}_s^{\omega,\mathrm{rig}}$ to avoid confusion. We may view the elements of the rigid hybrid model as a family of functions $t \mapsto \widetilde{\varphi}(t,z)$ in projective models with a varying special point $z \in \mathfrak{H}$. The isomorphism relating the rigid hybrid model and the line (respectively projective) model is then given by (4.1) (respectively (4.4)).

In the case of $\mathcal{V}_s^{\omega,\mathrm{rig}}$, we can replace t in (4.2) or (4.4) by a variable ζ on a neighborhood of $\mathbb{P}^1_{\mathbb{R}}$. We observe that, although $R(\zeta;z)^s$ is multivalued in ζ, the quotient $\left(\frac{R(\zeta;z_1)}{R(\zeta;z)}\right)^s$ in (4.2) is holomorphic in ζ on a neighborhood (depending on z and z_1) of $\mathbb{P}^1_{\mathbb{R}}$ in $\mathbb{P}^1_{\mathbb{C}}$. In the rigid hybrid model, the space $\mathbf{H}_s^{\mathrm{rig}}$ consists of germs of functions h on a deleted neighborhood $U \smallsetminus (\mathbb{P}^1_{\mathbb{R}} \times \mathfrak{H})$ which are holomorphic in the first variable and satisfy

$$h(\zeta,z_1) \;=\; \left(\frac{R(\zeta;z_2)}{R(\zeta;z_1)}\right)^s h(\zeta,z_2) \qquad (z_1,z_2 \in \mathfrak{H},\ \zeta \text{ near } \mathbb{P}^1_{\mathbb{R}}), \tag{4.5}$$

where "near $\mathbb{P}^1_{\mathbb{R}}$" means that ζ is sufficiently far in the hyperbolic metric from the geodesic joining z_1 and z_2. This condition ensures that (ζ,z_1) and (ζ,z_2) belong to U and the multiplicative factor in (4.5) is a power of a complex number not in $(-\infty,0]$ and is therefore well defined. The action of G on $\mathbf{H}_s^{\mathrm{rig}}$ is given by $h(\zeta,z) \mapsto h(g\zeta,gz)$. In this model, $\mathcal{V}_s^{-\omega}$ is represented as $\mathbf{H}_s^{\mathrm{rig}}/\mathcal{V}_s^{\omega,\mathrm{rig}}$. The pairing between hyperfunctions and test functions in this model is given by

$$\langle h,\, \widetilde{\psi}\rangle \;=\; \frac{1}{\pi}\left(\int_{C_+} - \int_{C_-}\right) h(\zeta,z)\; \widetilde{\psi}(\zeta,z)\; R(\zeta;z)\, \mathrm{d}\zeta \tag{4.6}$$

with the contours C_+ and C_- as in (2.24). Provided we adapt the contours to z, we can use any $z \in \mathfrak{H}$ in this formula for the pairing.

The rigid hybrid model, as described above, solves all of the problems of the various models of $\mathcal{V}_s$ as function spaces on $\partial\mathbb{H}$, but it is in some sense artificial, since the elements h depend in a fixed way on the second variable, and the use of this variable is therefore in principle superfluous. We address the remaining artificiality by replacing the rigid hybrid model by another model. The intuition is to replace functions satisfying (4.5) by hyperfunctions satisfying this relation.

Specifically, we define the *flabby hybrid model* as

$$\mathcal{M}_s^{-\omega} \;:=\; \mathcal{H}_s/\mathcal{M}_s^{\omega},$$

where $\mathcal{H}_s$ is the space of functions[2] $h(\zeta,z)$ that are defined on $U \smallsetminus \left(\mathbb{P}^1_{\mathbb{R}} \times \mathfrak{H}\right)$ for some neighborhood U of $\mathbb{P}^1_{\mathbb{R}} \times \mathfrak{H}$ in $\mathbb{P}^1_{\mathbb{C}} \times \mathfrak{H}$, are holomorphic in ζ, and satisfy

[2]Here one has the choice to impose any desired regularity conditions (C^0, C^∞, C^ω, ...) in the second variable or in both variables jointly. We do not fix any such choice since none of our considerations depend on which choice is made and since in any case the most interesting elements of this space, like the canonical representative introduced below, are analytic in both variables.

$$\zeta \mapsto h(\zeta, z_1) - \left(\frac{R(\zeta; z_2)}{R(\zeta; z_1)}\right)^s h(\zeta, z_2) \in \mathcal{O}(U_{z_1, z_2}) \quad \text{for all } z_1, z_2 \in \mathfrak{H}, \qquad (4.7)$$

where $U_{z_1,z_2} = \{\zeta \in \mathbb{P}^1_{\mathbb{C}} : (\zeta, z_1), (\zeta, z_2) \in U\}$, while $\mathcal{M}_s^{\omega}$ consists of functions defined on a neighborhood U of $\mathbb{P}^1_{\mathbb{R}} \times \mathfrak{H}$ in $\mathbb{P}^1_{\mathbb{C}} \times \mathfrak{H}$ and holomorphic in the first variable. The action of G in $\mathcal{H}_s$ is by $h \mid g\,(\zeta, z) = h(g\zeta, gz)$. The pairing between hyperfunctions and analytic functions is given by the same formula (4.6) as in the rigid hybrid model.

An element $h \in \mathcal{H}_s$ can thus be viewed as a family $\{h(\cdot\,, z)\}_{z \in \mathfrak{H}}$ of representatives of hyperfunctions parametrized by $\mathfrak{H}$. Adding an element of $\mathcal{M}_s^{\omega}$ does not change this family of hyperfunctions. The requirement (4.7) on h means that the family of hyperfunctions satisfies (4.5) in hyperfunction sense.

Finally, we describe a subspace $\mathcal{C}_s \subset \mathcal{H}_s$ which maps isomorphically to $\mathcal{V}_s^{-\omega}$ under the projection $\mathcal{H}_s \twoheadrightarrow \mathcal{V}_s^{-\omega}$ and hence gives a canonical choice of representatives of the hyperfunctions in $\mathcal{M}_s^{-\omega}$. We will call $\mathcal{C}_s$ the *canonical hybrid model*, or simply the *canonical model*, for the principal series representation $\mathcal{V}_s^{-\omega}$. To define $\mathcal{C}_s$, we recall that any hyperfunction on $\mathbb{P}^1_{\mathbb{R}}$ can be represented by a holomorphic function on $\mathbb{P}^1_{\mathbb{C}} \smallsetminus \mathbb{P}^1_{\mathbb{R}}$ with the freedom only of an additive constant. One usually fixes the constant by requiring $h(\mathrm{i}) = 0$ or $h(\mathrm{i}) + h(-\mathrm{i}) = 0$, which is of course not G-equivariant. Here we can exploit the fact that we have two variables to make the normalization in a G-equivariant way by requiring that

$$h(\bar{z}, z) = 0. \qquad (4.8)$$

We thus define $\mathcal{C}_s$ as the space of functions on $\left(\mathbb{P}^1_{\mathbb{C}} \smallsetminus \mathbb{P}^1_{\mathbb{R}}\right) \times \mathfrak{H}$ that are holomorphic in the first variable and satisfy (4.7) and (4.8). We will see below (Theorem 4.2) that the Poisson transform $\mathsf{P}_s : \mathcal{V}_s^{-\omega} \to \mathcal{E}_s$ becomes extremely simple when restricted to $\mathcal{C}_s$ and also that $\mathcal{C}_s$ coincides with the image of a canonical lifting of the inverse Poisson map $\mathsf{P}_s^{-1} : \mathcal{E}_s \to \mathcal{V}_s^{-\omega}$ from the space of hyperfunctions to the space of hyperfunction representatives.

Remark. We will also occasionally use the slightly larger space $\mathcal{C}_s^{+}$ (no longer mapped injectively to $\mathcal{E}_s$ by P_s) consisting of functions in $\mathcal{H}_s$ that are defined on all of $\left(\mathbb{P}^1_{\mathbb{C}} \smallsetminus \mathbb{P}^1_{\mathbb{R}}\right) \times \mathfrak{H}$, without the requirement (4.8). Functions in this space will be called *semicanonical* representatives of the hyperfunctions they represent. The decomposition $h(\zeta, z) = \left(h(\zeta, z) - h(\bar{z}, z)\right) + h(\bar{z}, z)$ gives a canonical and G-equivariant splitting of $\mathcal{C}_s^{+}$ as the direct sum of $\mathcal{C}_s$ and the space of functions on $\mathfrak{H}$, so that there is no new content here, but specific hyperfunctions sometimes have a particularly simple semicanonical representative (an example is given below), and it is not always natural to require (4.8).

- *Summary.* We have introduced a "rigid", a "flabby", and a "canonical" hybrid model, related by

$$\mathcal{V}_s^{-\omega} \cong \mathbf{H}_s^{\mathrm{rig}} / \mathcal{V}_s^{\omega,\mathrm{rig}} \cong \mathcal{M}_s^{-\omega} = \mathcal{H}_s / \mathcal{M}_s^{\omega} \cong \mathcal{C}_s \subset \mathcal{H}_s. \qquad (4.9)$$

In the flabby hybrid model, the space $\mathcal{H}_s$ consists of functions on a deleted neighborhood $U \smallsetminus (\mathbb{P}^1_{\mathbb{R}} \times \mathfrak{H})$ that may depend on the function, holomorphic in the first variable, and satisfying (4.7). The subspace $\mathcal{M}_s^\omega$ consists of the functions on the whole of some neighborhood U of $\mathbb{P}^1_{\mathbb{R}} \times \mathfrak{H}$, holomorphic in the first variable.

For the elements of $\mathcal{H}_s$ and $\mathcal{M}_s^\omega \subset \mathcal{H}_s$, we do not require any regularity in the second variable. In the rigid hybrid model, the spaces $\mathbf{H}_s^{\mathrm{rig}} \subset \mathcal{H}_s$ and $\mathcal{V}_s^{\omega,\mathrm{rig}} \subset \mathcal{M}_s^\omega$ are characterized by the condition in (4.5), which forces a strong regularity in the second variable. The canonical hybrid model $\mathcal{C}_s$ consists of a specific element from each class of $\mathcal{H}_s/\mathcal{M}_s^\omega$ that is defined on $(\mathbb{P}^1_{\mathbb{C}} \times \mathfrak{H}) \smallsetminus (\mathbb{P}^1_{\mathbb{R}} \times \mathfrak{H})$ and is normalized by (4.8). In Sect. 4.3 we will see that this implies analyticity in both variables jointly.

- *Examples.* As an example we represent the distribution $\delta_{s,\infty}$ in all three hybrid models. This distribution, which was defined by $\varphi^{\mathbb{P}} \mapsto \varphi^{\mathbb{P}}(\infty)$ in the projective model (cf. (3.30)), is represented in the projective model by $h^{\mathbb{P}}(\zeta) = \frac{1}{2i}\zeta$, and hence, by (4.4), by

$$\widetilde{h}(\zeta, z) = \frac{\zeta}{2i}\, y^{-s} \left(\frac{(\zeta - z)(\zeta - \bar{z})}{(\zeta - i)(\zeta + i)} \right)^s \tag{4.10}$$

in the rigid hybrid model. Since the difference $\frac{\zeta}{2i y^s}\left(\left(\frac{1-z/\zeta}{1-i/z}\right)^s \left(\frac{1-\bar{z}/\zeta}{1+i\zeta}\right)^s - 1 \right)$ is holomorphic in ζ on a neighborhood of $\mathbb{P}^1_{\mathbb{R}}$ in $\mathbb{P}^1_{\mathbb{C}}$ for each z, we obtain the much simpler semicanonical representative

$$h_{\mathrm{s}}(\zeta, z) = \frac{\zeta}{2i\, y^s}, \tag{4.11}$$

of $\delta_{s,\infty}$ in the flabby hybrid model. Finally, subtracting $h_{\mathrm{s}}(\bar{z}, z)$, we obtain the (unique) representative of $\delta_{s,\infty}$ in the canonical hybrid model:

$$h_{\mathrm{c}}(\zeta, z) = \frac{\zeta - \bar{z}}{2i}\, y^{-s}. \tag{4.12}$$

We obtain other elements of $\mathcal{C}_s$ by the action of G. For $g \in G$ with $g\infty = a \in \mathbb{R}$ we get

$$h_{\mathrm{c}}|g^{-1}(\zeta, z) = \frac{\zeta - \bar{z}}{z - \bar{z}}\, \frac{z - a}{\zeta - a}\, R(a; z)^{1-s}. \tag{4.13}$$

Here property (4.8) is obvious, and (4.7) holds because the only singularity of (4.13) on $\mathbb{P}^1_{\mathbb{R}}$ is a simple pole of residue $(i/2)\, R(a; z)^{-s}$ at $\zeta = a$.

- *Duality and Poisson Transform.* From (2.24) we find that if $h \in \mathcal{H}_s$ and $f \in \mathcal{M}_{1-s}^\omega$ are defined on $U \smallsetminus (\mathbb{P}^1_{\mathbb{R}} \times \mathfrak{H})$, respectively U, for the same neighborhood U of $\mathbb{P}^1_{\mathbb{R}} \times \mathfrak{H}$, then

$$\langle f, h \rangle = \frac{1}{\pi} \left(\int_{C_+} - \int_{C_-} \right) f(\zeta, z)\, h(\zeta, z)\, R(\zeta; z)\, d\zeta, \tag{4.14}$$

where C_+ and C_- are contours encircling z in $\mathfrak{H}$ and $\bar{z}$ in $\mathfrak{H}^-$, respectively, such that $C_+ \times \{z\}$ and $C_- \times \{z\}$ are contained in U. The result $\langle f, h\rangle$ does not change if we replace h by another element of $h + \mathcal{M}_s^\omega \subset \mathcal{H}_s$.

We apply this to the Poisson kernel $f_z^{\mathbb{P}}(\zeta) = R^{\mathbb{P}}(\zeta; z)^{1-s} = \left(\frac{R(\zeta;z)}{R(\zeta;i)}\right)^{1-s}$, for $z \in \mathfrak{H}$. The corresponding element in the rigid pair model is

$$\tilde{f}_z(\zeta, z_1) = \left(\frac{R(\zeta; z)}{R(\zeta; i)}\right)^{1-s} \left(\frac{R(\zeta; i)}{R(\zeta; z_1)}\right)^{1-s} = \left(\frac{R(\zeta; z)}{R(\zeta, z_1)}\right)^{1-s}.$$

Applying (4.14), we find for $z, z_1 \in \mathfrak{H}$:

$$\begin{aligned} \mathsf{P}_s h(z) &= \frac{1}{\pi}\left(\int_{C_+} - \int_{C_-}\right)\left(\frac{R(\zeta; z)}{R(\zeta; z_1)}\right)^{1-s} h(\zeta, z_1) R(\zeta; z_1)\, d\zeta \\ &= \frac{1}{\pi}\left(\int_{C_+} - \int_{C_-}\right)\left(\frac{R(\zeta; z_1)}{R(\zeta; z)}\right)^{s} h(\zeta, z_1)\, R(\zeta; z)\, d\zeta, \end{aligned} \tag{4.15}$$

where C_+ encircles z and z_1 and C_- encircles $\bar{z}$ and $\overline{z_1}$. Since this does not depend on z_1, we can choose $z_1 = z$ to get

$$\begin{aligned} \mathsf{P}_s h(z) &= \frac{1}{\pi}\left(\int_{C_+} - \int_{C_-}\right) h(\zeta, z)\ R(\zeta; z)\, d\zeta \\ &= \frac{1}{2\pi i}\left(\int_{C_+} - \int_{C_-}\right) h(\zeta, z)\, \frac{(z - \bar{z})}{(\zeta - z)(\zeta - \bar{z})}\, d\zeta. \end{aligned} \tag{4.16}$$

The representation of the Poisson transformation given by formula (4.16) has a very simple form. The dependence on the spectral parameter s is provided by the model, not by the Poisson kernel. But a really amazing simplification occurs if we assume that the function $h \in \mathcal{H}_s$ belongs to the subspace $\mathcal{C}_s$ of canonical hyperfunction representatives. In that case, $h(\zeta, z)$ is holomorphic in ζ in all of $\mathbb{C} \smallsetminus \mathbb{R}$, so we can evaluate the integral by Cauchy's theorem. In the lower half-plane, there is no pole since $h(\bar{z}, z)$ vanishes, so the integral over C_- vanishes. In the upper half-plane, there is a simple pole of residue $h(z, z)$ at $\zeta = z$. Hence, we obtain

Proposition 4.1. *The Poisson transform of a function $h \in \mathcal{C}_s$ is the function*

$$\mathsf{P}_s h(z) = h(z, z), \tag{4.17}$$

defined by restriction to the diagonal.

As examples of the proposition, we set $\zeta = z$ in (4.12) and (4.13) to get

$$\begin{aligned} u(z) = y^{1-s} \quad &\Rightarrow \quad (\mathsf{P}_s^{-1} u)_{\mathrm{can}}(\zeta, z) = \frac{\zeta - \bar{z}}{2i}\, y^{-s} \\ u(z) = R(a; z)^{1-s} \quad &\Rightarrow \quad (\mathsf{P}_s^{-1} u)_{\mathrm{can}}(\zeta, z) = \frac{\zeta - \bar{z}}{z - \bar{z}}\, \frac{z - a}{\zeta - a}\, u(z). \end{aligned} \tag{4.18}$$

Finally, we remark that on the larger space $\mathcal{C}_s^+$ introduced in the Remark above, we have *two* restriction maps

$$\mathsf{R}_s^+ h(z) = h(z,z), \qquad \mathsf{R}_s^- h(z) = h(\bar{z},z) \tag{4.19}$$

to the space of functions on $\mathfrak{H}$. The analogue of the proposition just given is then that the restriction of P_s to $\mathcal{C}_s^+$ equals the difference $\mathsf{R}_s = \mathsf{R}_s^+ - \mathsf{R}_s^-$.

4.2 Poisson Inversion and the Canonical Model

The canonical model is particularly suitable to give an integral formula for the inverse Poisson transformation, as we see in the main result of this subsection, Theorem 4.2. In Proposition 4.4 we give an integral formula for the canonical representative of a hyperfunction in terms of an arbitrary representative in $\mathcal{H}_s$. Proposition 4.6 relates, for $u \in \mathcal{E}_s$, the Taylor expansions in the upper and lower half-plane of the canonical representative of $\mathsf{P}_s^{-1}u$ to the polar expansion of u with the functions $p_{s,n}$.

To determine the image $\mathsf{P}_s^{-1}u$ under the inverse Poisson transform for a given $u \in \mathcal{E}_s$, we have to construct a hyperfunction on $\partial\mathbb{H}$ which maps under P_s to u. A first attempt, based on [10], Chap. II, Sect. 2, would be (in the line model) to integrate the Green's form $\{u, R(\zeta;\cdot)^s\}$ from some base point to ζ. This does not make sense at ∞ since $R(\zeta;\cdot)$ has a singularity there and one cannot take a well-defined sth power of it, so we should renormalize by dividing $R(\zeta;\cdot)^s$ by $R(\zeta;i)^s$, or better, to avoid destroying the G-equivariance of the construction, by $R(\zeta;z)^s$ with a variable point $z \in \mathfrak{H}$. This suggests the formula

$$h(\zeta,z) = \begin{cases} \displaystyle\int_{z_0}^{\zeta} \{u, (R_\zeta(\cdot)/R_\zeta(z))^s\} & \text{if } \zeta \in \mathfrak{H}, \\ \displaystyle\int_{z_0}^{\bar{\zeta}} \{u, (R_\zeta(\cdot)/R_\zeta(z))^s\} & \text{if } \zeta \in \mathfrak{H}^-, \end{cases} \tag{4.20}$$

in the hybrid model, where $z_0 \in \mathfrak{H}$ is a base point, as a second attempt. This almost works: the fact that the Green's form is closed implies that the integrals are independent of the path of integration, and changing the base-point z_0 changes $h(\cdot,z)$ by a function holomorphic near $\mathbb{P}^1_{\mathbb{R}}$ and hence does not change the hyperfunction it represents. The problem is that both integrals in (4.20) diverge because $R(\zeta;z')^s$ has a singularity like $(\zeta - z')^{-s}$ near ζ and like $(\zeta - \overline{z'})^{-s}$ near $\bar{\zeta}$ and the differentiation implicit in the bracket $\{\cdot,\cdot\}$ turns these into singularities like $(\zeta - z')^{-s-1}$ and $(\zeta - \overline{z'})^{-s-1}$ which are no longer integrable at $z' = \zeta$ or $z' = \bar{\zeta}$, respectively. To remedy this in the upper half plane, we replace $\{u, (R_\zeta(\cdot)/R_\zeta(z))^s\}$ by $[u, (R_\zeta(\cdot)/R_\zeta(z))^s]$, which differs from it by a harmless exact 1-form but is now integrable at ζ. (The same trick was already used in Sect. 2, Chap. II of [10], where z_0 was ∞.) In the lower half-plane,

$[u, (R_\zeta(\cdot)/R_\zeta(z)^s]$ is not small near $z' = \bar{\zeta}$, so here we must replace the differential form $\{u, (R_\zeta(\cdot)/R_\zeta(z))^s\} = -\{(R_\zeta(\cdot)/R_\zeta(z))^s, u\}$ by $-[(R_\zeta(\cdot)/R_\zeta(z))^s, u]$ instead. (We recall that $\{\cdot, \cdot\}$ is antisymmetric but $[\cdot, \cdot]$ is not.) However, since the differential forms $[u, (R_\zeta(\cdot)/R_\zeta(z))^s]$ and $-[(R_\zeta(\cdot)/R_\zeta(z))^s, u]$ differ by the exact form $\mathrm{d}(u\,(R_\zeta(\cdot)/R_\zeta(z))^s)$, this change requires correcting the formula in one of the half-planes. (We choose the upper half-plane.) This gives the formula

$$h(\zeta; z) = \begin{cases} u(z_0)\left(\dfrac{R(\zeta; z_0)}{R(\zeta; z)}\right)^s + \displaystyle\int_{z_0}^{\zeta}\left[u, \left(\dfrac{R(\zeta; \cdot)}{R(\zeta; z)}\right)^s\right] & \text{if } \zeta \in \mathfrak{H}, \\ \displaystyle\int_{\bar{\zeta}}^{z_0}\left[\left(\dfrac{R(\zeta; \cdot)}{R(\zeta; z)}\right)^s, u\right] & \text{if } \zeta \in \mathfrak{H}^-. \end{cases} \tag{4.21}$$

We note that in this formula, $h(\overline{z_0}, z) = 0$. So we can satisfy (4.8) by choosing $z_0 = z$, at the same time restoring the G-symmetry which was broken by the choice of a base point z_0. We can then choose the continuous branch of $\left(R_\zeta/R_\zeta(z)\right)^s$ that equals 1 at the end point z of the path of integration. Thus we have arrived at the following Poisson inversion formula, already given in the Introduction (1.4):

Theorem 4.2. *Let $u \in \mathcal{E}_s$. Then the function $\mathsf{B}_s u \in \mathcal{H}_s$ defined by*

$$(\mathsf{B}_s u)(\zeta, z) = \begin{cases} u(z) + \displaystyle\int_z^{\zeta}\left[u, \left(R_\zeta/R_\zeta(z)\right)^s\right] & \text{if } \zeta \in \mathfrak{H}, \\ \displaystyle\int_{\bar{\zeta}}^{z}\left[\left(R_\zeta/R_\zeta(z)\right)^s, u\right] & \text{if } \zeta \in \mathfrak{H}^- \end{cases} \tag{4.22}$$

along any piecewise C^1-path of integration in $\mathfrak{H} \smallsetminus \{\zeta\}$, respectively $\mathfrak{H} \smallsetminus \{\bar{\zeta}\}$, with the branch of $\left(R_\zeta/R_\zeta(z)\right)^s$ chosen to be 1 at the end point z, belongs to $\mathcal{C}_s$ and is a representative of the hyperfunction $\mathsf{P}_s^{-1}u \in \mathcal{M}_s^{-\omega} = \mathcal{H}_s/\mathcal{M}_s^{\omega}$.

Corollary 4.3. *The maps $\mathsf{B}_s : \mathcal{E}_s \to \mathcal{C}_s$ and $\mathsf{R}_s : \mathcal{C}_s \to \mathcal{E}_s$ defined by* (4.22) *and* (4.17) *are inverse isomorphisms, and we have a commutative diagram*

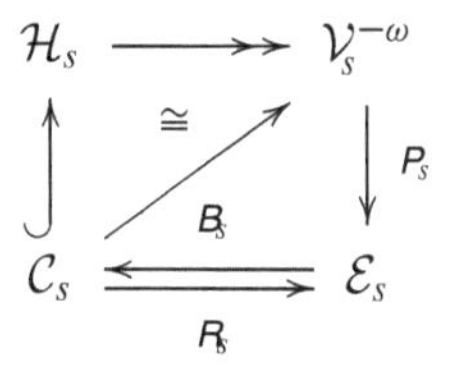

(4.23)

Proof. Let $u \in \mathcal{E}_s$. First we check that $h = \mathsf{B}_s u$ is well defined and determines an element of $\mathcal{C}_s$. The convergence of the integrals in (4.22) requires an estimate of the integrand at the boundaries. For $\zeta \in \mathfrak{H} \smallsetminus \{z\}$, we use

$$\left[u(z'), \left(R_\zeta(z')/R_\zeta(z)\right)^s\right]_{z'} = \left(\frac{R(\zeta;z')}{R(\zeta;z)}\right)^s \left(u_z(z')\,\mathrm{d}z' + \frac{\mathrm{i}s}{2y'}u(z')\,\frac{\zeta - z'}{\zeta - \overline{z}'}\,\mathrm{d}\overline{z}'\right). \tag{4.24}$$

The factor in front is 1 for $z' = z$ and $\mathrm{O}\left((\zeta - z')^{-s}\right)$ for z' near ζ. The other contributions stay finite, so the integral for $\zeta \in \mathfrak{H} \smallsetminus \{z\}$ converges. (Recall that $\mathrm{Re}\,(s)$ is always supposed to be < 1.) For $\zeta \in \mathfrak{H}^- \smallsetminus \{\bar{z}\}$, we use in a similar way

$$\left[\left(R_\zeta(z')/R_\zeta(z)\right)^s, u(z')\right]_{z'} = \left(\frac{R(\zeta;z')}{R(\zeta;z)}\right)^s \left(\frac{\mathrm{i}s}{2y'}u(z')\frac{\zeta - \overline{z}'}{\zeta - z'}\,\mathrm{d}z' + u_{\bar{z}}(z')\,\mathrm{d}\overline{z}'\right).$$

We have normalized the branch of $\left(R(\zeta;z')/R(\zeta;z)\right)^s$ by prescribing the value 1 at $z' = z$. This choice fixes $\left(R(\zeta;z')/R(\zeta;z)\right)^s$ as a continuous function on the paths of integration. The result of the integration does not depend on the path, since the differential form is closed and since we have convergence at the other end point ζ or $\bar{\zeta}$. Any continuous deformation of the path within $\mathfrak{H} \smallsetminus \{\zeta\}$ or $\mathfrak{H} \smallsetminus \{\bar{\zeta}\}$ is allowed, even if the path intersects itself with different values of $\left(R(\zeta;z')/R(\zeta;z)\right)^s$ at the intersection point.

If we choose the geodesic path from z to ζ, and if ζ is very near the real line, then the branch of $\left(R_\zeta(z')/R_\zeta(z)\right)^s$ near $z' = \zeta$ is the principal one (argument between $-\pi$ and π).

The holomorphy in ζ follows from a reasoning already present in [10], Chap. II, Sect. 2, and hence given here in a condensed form. Since the form (4.24) is holomorphic in ζ, a contribution to $\partial_{\bar{\zeta}} h$ could only come from the upper limit of integration, but in fact vanishes since $\mathrm{O}\left((\zeta - z')^{-s}\right)(\zeta - z') = \mathrm{o}(1)$ as $\zeta \to z'$. Hence, $h(\cdot, z)$ is holomorphic on $\mathfrak{H} \smallsetminus \{z\}$. For ζ near z, we integrate a quantity $\mathrm{O}\left((\zeta - z')^{-s}\right)$ from z to ζ, which results in an integral estimated by $\mathrm{O}\left((\zeta - z)^{1-s}\right)$. So $(\mathsf{B}_s u)(\zeta, z)$ is bounded for ζ near z. Hence, $h(\zeta, z) = (\mathsf{B}_s u)(\zeta, z)$ is holomorphic at $\zeta = z$ as well. For the holomorphy on $\mathfrak{H}^-$, we proceed similarly. This also shows that $h(\bar{z}, z) = 0$, which is condition (4.8) in the definition of $\mathcal{C}_s$.

For condition (4.7), we note that

$$h(\zeta;z) - \left(\frac{R(\zeta;z_1)}{R(\zeta;z)}\right)^s h(\zeta;z_1) = u(z) - u(z_1)\left(\frac{R(\zeta;z_1)}{R(\zeta;z)}\right)^s + \int_z^{z_1}\left[u, \left(\frac{R(\zeta;\cdot)}{R(\zeta;z)}\right)^s\right]$$

if $\zeta \in \mathfrak{H} \smallsetminus \{z, z_1\}$ and

$$h(\zeta; z) - \left(\frac{R(\zeta; z_1)}{R(\zeta; z)}\right)^s h(\zeta; z_1) = \int_{z_1}^{z} \left[\left(\frac{R(\zeta; \cdot)}{R(\zeta; z)}\right)^s, u\right]$$

if $\zeta \in \mathfrak{H}^- \smallsetminus \{\bar{z}, \bar{z}_1\}$. The right-hand sides both have holomorphic extensions in ζ to a neighborhood of $\mathbb{P}^1_{\mathbb{R}}$, and the difference of these two extensions is seen, using (3.13) and the antisymmetry of $\{\cdot, \cdot\}$, to be equal to

$$u(z) - u(z_1)\left(\frac{R(\zeta; z_1)}{R(\zeta; z)}\right)^s + \int_z^{z_1} \mathrm{d}\left(u(z')\left(\frac{R(\zeta; z')}{R(\zeta; z)}\right)^s\right) = 0.$$

In summary, the function $\mathsf{B}_s u$ belongs to $\mathcal{H}_s$, is defined in all of $\left(\mathbb{P}^1_{\mathbb{C}} \smallsetminus \mathbb{P}^1_{\mathbb{R}}\right) \times \mathfrak{H}$, and vanishes on the antidiagonal, so $\mathsf{B}_s u \in \mathcal{C}_s$, which is the first statement of the theorem. The second follows immediately from Proposition 4.1, since it is obvious from (4.22) that $\mathsf{R}_s \mathsf{B}_s u = u$ and the proposition says that R_s is the restriction of P_s to $\mathcal{C}_s$. ■

The corollary follows immediately from the theorem if we use Helgason's result (Theorem 3.4) that the Poisson transformation is an isomorphism. However, given that we have now constructed an explicit inverse map for the Poisson transformation, we should be able to give a more direct proof of this result, not based on polar expansions, and indeed this is the case. Since $\mathsf{P}_s \mathsf{B}_s u = \mathsf{R}_s \mathsf{B}_s u = u$, it suffices to show that R_s is injective. To see this, assume that $h \in \mathcal{C}_s$ satisfies $h(z, z) = 0$ for all $z \in \mathfrak{H}$. For fixed $z_1, z_2 \in \mathfrak{H}$, let $c(\zeta)$ denote the difference in (4.7). This function is holomorphic near $\mathbb{P}^1_{\mathbb{R}}$ and extends to $\mathbb{P}^1_{\mathbb{C}}$ in a multivalued way with branch points of mild growth, $(\zeta - \zeta_0)^{\pm s}$ with $0 < \operatorname{Re} s < 1$, at z_1, $\overline{z_1}$, z_2, and $\overline{z_2}$. Moreover, $c(\zeta)$ tends to 0 as ζ tends to z_1 or $\overline{z_1}$ (because $\operatorname{Re} s > 0$) and also as ζ tends to z_2 or $\overline{z_2}$ (because $h(z_2, z_2) = h(\overline{z_2}, z_2) = 0$ and $\operatorname{Re} s < 1$). Suppose that c is not identically zero. The differential form $\mathrm{d} \log c(\zeta)$ is meromorphic on all of $\mathbb{P}^1_{\mathbb{C}}$ and its residues at $\zeta_0 \in \{z_z, z_2, \overline{z_1}, \overline{z_2}\}$ have positive real part. Since c is finite elsewhere on $\mathbb{P}^1_{\mathbb{C}}$, any other residue is nonnegative. This contradicts the fact that the sum of all residues of a meromorphic differential on $\mathbb{P}^1_{\mathbb{C}}$ is zero. Hence, we conclude that $c = 0$. Then the local behavior of $h(\zeta, z_1) = h(\zeta, z_2)\left(R_\zeta(z_2)/R_\zeta(z_1)\right)^s$ at the branch points shows that both $h(\cdot, z_1)$ and $h(\cdot, z_2)$ vanish identically.

Remarks.

1. It is also possible to prove that $\mathsf{B}_s \mathsf{P}_s \varphi = \varphi$ and $\mathsf{P}_s \mathsf{B}_s u = u$ by using complex contour integration and (3.16), and our original proof that $\mathsf{B}_s = \mathsf{P}_s^{-1}$ went this way, but the above proof using the canonical space $\mathcal{C}_s$ is much simpler.
2. Taking $z = \mathrm{i}$ in formula (4.22) gives a representative for $\mathsf{P}_s^{-1} u$ in the projective model, and using the various isomorphisms discussed in Sect. 2, we can also adapt it to the other $\partial\mathbb{H}$ models of the principal representation.

We know that each element of $\mathcal{V}_s^{-\omega}$ has a unique canonical representative lying in $\mathcal{C}_s$. The following proposition, in which $k(\tau, \zeta; z)$ denotes the kernel function

$$k(\tau, \zeta; z) = \frac{1}{2\mathrm{i}\,(\tau - \zeta)} \frac{\zeta - \bar{z}}{\tau - \bar{z}}, \tag{4.25}$$

tells us how to determine it starting from an arbitrary representative.

Proposition 4.4. *Suppose that $g \in \mathcal{H}_s$ represents $\alpha \in \mathcal{V}_s^{-\omega}$. The canonical representative $g_c \in \mathcal{C}_s$ of α is given, for each $z_0 \in \mathfrak{H}$ by*

$$g_c(\zeta, z) = \frac{1}{\pi}\left(\int_{C_+} - \int_{C_-}\right) g(\tau, z_0)\left(\frac{R(\tau; z_0)}{R(\tau; z)}\right)^s k(\tau, \zeta; z)\,\mathrm{d}\tau, \tag{4.26}$$

with contours C_+ and C_- homotopic to $\mathbb{P}^1_{\mathbb{R}}$ inside the domain of g, encircling z and z_0, respectively $\bar{z}$ and $\overline{z_0}$, with C_+ positively oriented in $\mathfrak{H}$ and C_- negatively oriented in $\mathfrak{H}^-$, and ζ inside C_+ or inside C_-.

Note that this can be applied when a representative g_0 of α in the projective model is given: simply apply the proposition to the corresponding representative in the rigid hybrid model as given by (4.4).

Proof. Consider $k(\cdot, \zeta; z)$ as an element of $\mathcal{V}_s^{\omega}$ in the projective model. Then $g_c(\zeta, z) = \langle \alpha, k(\cdot, \zeta; z)\rangle$. Adapting the contours, we see that $g_c(\cdot, z)$ is holomorphic on $\mathfrak{H} \cup \mathfrak{H}^-$.

For a fixed $\zeta \in \operatorname{dom} g$, we deform the contours such that ζ is between the new contours. This gives a term $g(\zeta, z)$ plus the same integral, but now representing a holomorphic function in ζ on the region between C_+ and C_-, which is a neighborhood of $\mathbb{P}^1_{\mathbb{R}}$. So g and g_c represent the same hyperfunction. Condition (4.8) follows from $k(\tau, \bar{z}; z) = 0$. ■

Choosing $z_0 = z$ in (4.26) gives the simpler formula

$$g_c(\zeta, z) = \frac{1}{\pi}\left(\int_{C_+} - \int_{C_-}\right) g(\tau, z)\, k(\tau, \zeta; z)\,\mathrm{d}\tau, \tag{4.27}$$

(which is, of course, identical to (4.26) if g belongs to $\mathcal{V}_s^{\omega,\mathrm{rig}}$). In terms of $\alpha \in \mathcal{V}_s^{-\omega}$ as a linear form on $\mathcal{V}_{1-s}^{\omega}$, we can write this as

$$g_c(\zeta, z) = \langle f_\zeta, \alpha\rangle \quad \text{with} \quad f_\zeta(\tau, z) = \frac{(\zeta - \bar{z})(\tau - z)}{(z - \bar{z})(\tau - \zeta)}. \tag{4.28}$$

The integral representation (4.26) has the following consequence:

Corollary 4.5. *All elements of $\mathcal{C}_s$ are real-analytic on $(\mathbb{P}^1_{\mathbb{C}} \smallsetminus \mathbb{P}^1_{\mathbb{R}}) \times \mathfrak{H}$.*

• *Expansions in the Canonical Model.* For $u \in \mathcal{E}_s$, the polar expansion (3.4) can be generalized, with the shifted functions $p_{s,n}$ in (3.5), to an arbitrary central point:

$$u(z) = \sum_{n\in\mathbb{Z}} a_n(u,z')\, p_{s,n}(z,z') \qquad (z' \in \mathfrak{H} \text{ arbitrary}). \tag{4.29}$$

Let $h = \mathsf{B}_s u \in \mathcal{C}_s$ be the canonical representative of $\mathsf{P}_s^{-1} u \in \mathcal{V}_s^{-\omega}$. For $z' \in \mathfrak{H}$ fixed, $h(\zeta,z')$ is a holomorphic function of $\zeta \in \mathbb{C} \smallsetminus \mathbb{R}$ and has Taylor expansions in $\frac{\zeta-z'}{\zeta-\overline{z'}}$ on $\mathfrak{H}$ and in $\frac{\zeta-\overline{z'}}{\zeta-z'}$ on $\mathfrak{H}^-$. Since $h(\bar z, z) = 0$, the constant term in the expansion on $\mathfrak{H}^-$ vanishes. Thus there are $A_n(h,z') \in \mathbb{C}$ such that

$$h(\zeta,z') = \begin{cases} \displaystyle\sum_{n\ge 0} A_n(h,z') \left(\frac{\zeta - z'}{\zeta - \overline{z'}}\right)^n & \text{for } \zeta \in \mathfrak{H}, \\ \displaystyle -\sum_{n<0} A_n(h,z') \left(\frac{\zeta - z'}{\zeta - \overline{z'}}\right)^n & \text{for } \zeta \in \mathfrak{H}^-. \end{cases} \tag{4.30}$$

(We use a minus sign in the expansion on $\mathfrak{H}^-$ because then

$$(\zeta,z') \mapsto \sum_{n\ge n_0} A_n(h,z') \left(\frac{\zeta-z'}{\zeta-\overline{z'}}\right)^n \text{ for } \zeta \in \mathfrak{H}, \quad -\sum_{n<n_0} A_n(h,z') \left(\frac{\zeta-z'}{\zeta-\overline{z'}}\right)^n \text{ for } \zeta \in \mathfrak{H}^-,$$

represents the same hyperfunction $\mathsf{P}_s^{-1} u$ for any choice of $n_0 \in \mathbb{Z}$.) From $\frac{g\zeta-gz'}{g\zeta-g\overline{z'}} = \frac{c\overline{z'}+d}{cz'+d}\,\frac{\zeta-z'}{\zeta-\overline{z'}}$ for $g = \left[\begin{smallmatrix} a & b \\ c & d \end{smallmatrix}\right] \in G$, it follows that

$$A_n(h \,|\, g, z') = \left(\frac{c\overline{z'}+d}{cz'+d}\right)^n A_n(h, gz'). \tag{4.31}$$

Similarly, we have from (3.5) and (3.3):

$$a_n(u \,|\, g, z') = \left(\frac{c\overline{z'}+d}{cz'+d}\right)^n a_n(u, gz'). \tag{4.32}$$

In fact, the coefficients $A_n(\,)$ and $a_n(\,)$ are proportional:

Proposition 4.6. *For $u \in \mathcal{E}_s$ and $h = \mathsf{B}_s u \in \mathcal{C}_s$, the coefficients in the expansions* (4.29) *and* (4.30) *are related by*

$$a_n(u,z') = (-1)^n \frac{\Gamma(s)}{\Gamma(s+n)} A_n(h,z'). \tag{4.33}$$

Proof. The expansion (4.30) for $z' = i$ shows that the hyperfunction $\mathsf{P}_s^{-1} u$ has the expansion $\sum_n A_n(h,i)\, \mathbf{e}_{s,n}$ in the basis functions in (2.9). Then (3.28) gives

$$u(z) = \sum_{n\in\mathbb{Z}} (-1)^n \frac{\Gamma(s)}{\Gamma(s+n)} A_n(h,i)\, P_{s,n}(z).$$

This gives the relation in the proposition if $z' = i$, and the general case follows from the transformation rules (4.31) and (4.32). ∎

The transition $s \leftrightarrow 1-s$ does not change $\mathcal{E}_s = \mathcal{E}_{1-s}$ or $p_{s,n} = p_{1-s,n}$. So the coefficients $a_n(u,z')$ stay the same under $s \mapsto 1-s$. With the commutative diagram (3.25), we get

Corollary 4.7. *The operator $\mathcal{C}_s \to \mathcal{C}_{1-s}$ corresponding to $I_s : \mathcal{V}_s^{-\omega} \to \mathcal{V}_{1-s}^{-\omega}$ acts on the coefficients in* (4.30) *by*

$$A_n(I_s h, z') = \frac{(1-s)_{|n|}}{(s)_{|n|}} A_n(h, z').$$

We remark that Proposition 4.6 can also be used to give an alternative proof of Corollary 4.5, using (3.19) with u replaced by $u \circ g_{z'}$ to obtain the analyticity of $a_n(u,z')$ in z' and then (4.33) to control the speed of convergence in (4.30).

4.3 Differential Equations for the Canonical Model and the Sheaf of Mixed Eigenfunctions

The canonical model provides us with an isomorphic copy $\mathcal{C}_s$ of $\mathcal{V}_s^{-\omega} \cong \mathcal{E}_s$ inside the flabby hybrid model $\mathcal{H}_s$. We now show that the elements of the canonical model are real-analytic in both variables jointly and satisfy first-order differential equations in the variable $z \in \mathfrak{H}$ with ζ as a parameter.

The same differential equations can be used to define a sheaf $\mathcal{D}_s$ on $\mathbb{P}^1_{\mathbb{C}} \times \mathfrak{H}$. In Proposition 4.10 and Theorem 4.13, we describe the local structure of this sheaf. It turns out that we can identify the space $\mathcal{V}_s^{\omega,\mathrm{rig}}$ of the rigid hybrid model with a space of sections of this sheaf of a special kind. There is a sheaf morphism that relates $\mathcal{D}_s$ to the sheaf $\mathcal{E}_s : U \mapsto \mathcal{E}_s(U)$ of λ_s-eigenfunctions on $\mathfrak{H}$. For elements of the full space $\mathcal{E}_s = \mathcal{E}_s(\mathfrak{H})$, the canonical model gives sections of $\mathcal{D}_s$ over $(\mathbb{P}^1_{\mathbb{C}} \smallsetminus \mathbb{P}^1_{\mathbb{R}}) \times \mathfrak{H}$.

Theorem 4.8. *Each $h \in \mathcal{C}_s$ and its corresponding eigenfunction $u = P_s h = R_s h \in \mathcal{E}_s$ satisfy, for $\zeta \in \mathbb{P}^1_{\mathbb{C}} \smallsetminus \mathbb{P}^1_{\mathbb{R}}$, $z \in \mathfrak{H}$, $\zeta \notin \{z, \bar z\}$, the differential equations*

$$(z-\bar z)\,\partial_z h(\zeta,z) + s\,\frac{\zeta-\bar z}{\zeta - z}\big(h(\zeta,z) - u(z)\big) = 0, \qquad (4.34a)$$

$$(z-\bar z)\,\partial_{\bar z}\big(h(\zeta,z) - u(z)\big) - s\,\frac{\zeta - z}{\zeta-\bar z}\,h(\zeta,z) = 0. \qquad (4.34b)$$

Conversely, any continuous function h on $(\mathbb{P}^1_{\mathbb{C}} \smallsetminus \mathbb{P}^1_{\mathbb{R}}) \times \mathfrak{H}$ that is holomorphic in the first variable, continuously differentiable in the second variable, and satisfies the differential equations (4.34) *for some $u \in C^1(\mathfrak{H})$ belongs to $\mathcal{C}_s$, and u is $P_s h$.*

The differential equations (4.34) look complicated but in fact are just the dz- and $d\bar{z}$-components of the identity

$$\big[R(\zeta;\cdot)^s,\ u(\zeta;\cdot)\big] = \mathrm{d}\big(R(\zeta;\cdot)^s h(\zeta;\cdot)\big) \tag{4.35}$$

between 1-forms, as one checks easily. (The function $R(\zeta;z)^s$ is multivalued, but if we take the same branch on both sides of the equality, then it makes sense locally.)

Proof of Theorem 4.8. The remark just made shows almost immediately that the function $h = \mathsf{B}_s u \in \mathcal{C}_s$ defined by (4.22) satisfies the differential equations (4.34): differentiating (4.22) in z gives

$$\mathrm{d}_z\big(h(\zeta,z)R(\zeta;s)^s\big) = \begin{cases} \mathrm{d}\big(u(z)R(\zeta;z)^s\big) - [u(z), R(\zeta;z)^s]_z & \text{if } z \in \mathfrak{H}, \\ \qquad\qquad\qquad\quad + [R(\zeta;z)^s, u(z)]_z & \text{if } z \in \mathfrak{H}^-, \end{cases}$$

and the right-hand side equals $[R(\zeta;z)^s, u(z)]$ in both cases by virtue of (3.13).

An alternative approach, not using the explicit Poisson inversion formula (4.22), is to differentiate (4.7) with respect to z_1 (resp. $\overline{z_1}$) and then set $z_1 = z_2 = z$ to see that the expression on the left-hand side of (4.34a) (resp. (4.34b)) is holomorphic in ζ near $\mathbb{P}^1_{\mathbb{R}}$. (Here that we use the result proved above that elements of the canonical model are analytical in both variables jointly.) The equations $h(z,z) = u(z)$, $h(\bar{z},z) = 0$ then show that the expressions in (4.34), for z fixed, are holomorphic in ζ on all of $\mathbb{P}^1_{\mathbb{C}}$ and hence constant. To see that both constants vanish, we set $\zeta = \bar{z}$ in (4.34a) (resp. $\zeta = z$ in (4.34b)) and use

$$\partial_z\big(h(\zeta,z)\big)|_{\zeta=\bar{z}} = \partial_z\big(h(\bar{z},z)\big) = \partial_z(0) = 0, \quad \partial_{\bar{z}}\big(h(\zeta,z)\big)|_{\zeta=z} = \partial_{\bar{z}}\big(h(z,z)\big) = \partial_{\bar{z}}u(z).$$

This proves the forward statement of Theorem 4.8. Instead of proving the converse immediately, we first observe that the property of satisfying the differential equations in the theorem is a purely local one and therefore defines a sheaf of functions.

We now give a formal definition of this sheaf and then prove some general statements about its local sections that include the second part of Theorem 4.8.

We note that the differential equations (4.34) make sense, not only on $(\mathbb{P}^1_{\mathbb{C}} \setminus \mathbb{P}^1_{\mathbb{R}}) \times \mathfrak{H}$ but on all of $\mathbb{P}^1_{\mathbb{C}} \times \mathfrak{H}$, with singularities on the "diagonal" and "antidiagonal" defined by

$$\Delta^+ = \{(z,z) : z \in \mathfrak{H}\}, \qquad \Delta^- = \{(\bar{z},z) : z \in \mathfrak{H}\}. \tag{4.36}$$

We, therefore, define our sheaf on open subsets of this larger space.

Definition 4.9. For every open subset $U \subset \mathbb{P}^1_{\mathbb{C}} \times \mathfrak{H}$, we define $\mathcal{D}_s(U)$ as the space of pairs (h,u) of functions on U such that:

(a) h and u are continuous on U.
(b) h is holomorphic in its first variable.

(c) Locally u is independent of the first variable.
(d) h and u are continuously differentiable in the second variable and satisfy the differential equations (4.34) on $U \smallsetminus (\Delta^+ \cup \Delta^-)$, with $u(z)$ replaced by $u(\zeta, z)$.

This defines $\mathcal{D}_s$ as a sheaf of pairs of functions on $\mathbb{P}^1_{\mathbb{C}} \times \mathfrak{H}$, the sheaf of *mixed eigenfunctions*. In this language, the content of Theorem 4.8 is that $\mathcal{C}_s$ can be identified via $h \mapsto (h, \mathsf{R}_s h)$ with the space of global sections of $\mathcal{D}_s\big((\mathbb{P}^1_{\mathbb{C}} \smallsetminus \mathbb{P}^1_{\mathbb{R}}) \times \mathfrak{H}\big)$. The following proposition gives a number of properties of the local sections.

Proposition 4.10. *Let $(h,u) \in \mathcal{D}_s(U)$ for some open set $U \subset \mathbb{P}^1_{\mathbb{C}} \times \mathfrak{H}$. Then*

(i) *The functions h and u are real-analytic on U. The function u is determined by h and satisfies $\Delta u = s(1-s)u$.*
(ii) *If U intersects $\Delta^+ \cup \Delta^-$, then we have $h = u$ on $U \cap \Delta^+$ and $h = 0$ on $U \cap \Delta^-$.*
(iii) *If $u = 0$, then the function h locally has the form $h(\zeta, z) = \varphi(\zeta)\, R(\zeta; z)^{-s}$ for some branch of $R(\zeta; z)^{-s}$, with φ holomorphic.*
(iv) *The function h is determined by u on each connected component of U that intersects $\Delta^+ \cup \Delta^-$.*

Proof. The continuity of h and u allows us to consider them and their derivatives as distributions. We obtain from (4.34) the following equalities of distributions on $U \smallsetminus (\Delta^+ \cup \Delta^-)$:

$$\partial_z \partial_{\bar z} h = \partial_z\left(\partial_{\bar z} u + \frac{s}{z - \bar z}\,\frac{\zeta - z}{\zeta - \bar z}\, h\right) = \partial_z \partial_{\bar z} u - \frac{s}{(z-\bar z)^2}\, h - \frac{s^2}{(z-\bar z)^2}\,(h-u),$$

$$\partial_{\bar z} \partial_z h = \partial_{\bar z}\left(\frac{-s}{z - \bar z}\,\frac{\zeta - \bar z}{\zeta - z}\,(h-u)\right) = \frac{-s}{(z-\bar z)^2}\,(h-u) - \frac{s^2}{(z-\bar z)^2}\, h.$$

The differential operators ∂_z and $\partial_{\bar z}$ on distributions commute. In terms of the hyperbolic Laplace operator $\Delta = (z - \bar z)^2\, \partial_z \partial_{\bar z}$, we have in distribution sense

$$\big(\Delta - \lambda_{s+1}\big)\, h = (\Delta + s^2) u = su. \tag{4.37}$$

Since u is an eigenfunction of the elliptic differential operator $\Delta - \lambda_s$ with real-analytic coefficients, u and also h are real-analytic functions in the second variable. To conclude that h is real-analytic in both variables jointly, we note that it is also a solution of the following elliptic differential equation with analytic coefficients

$$\big(-\partial_\zeta \partial_{\bar\zeta} + \Delta - \lambda_{s+1}\big) h = su.$$

Near $\infty \in \mathbb{P}^1_{\mathbb{C}}$, we replace ζ by $\upsilon = 1/\zeta$ in the last step.

Since u is locally independent of ζ, we conclude that u is real-analytic on the whole of U and satisfies $\Delta u = s(1-s)u$ on U. Then (4.37) gives the analyticity of h on U. Now we use (4.34a) to obtain

$$u(\zeta, z) = h(\zeta, z) + \frac{z - \bar z}{s}\,\frac{\zeta - z}{\zeta - \bar z}\,\partial_z h(\zeta, z).$$

So h determines u on $U \smallsetminus \Delta^-$ and then by continuity on the whole of U. Furthermore, $u = h$ on Δ^+. Similarly, (4.34b) implies $h = 0$ on $U \cap \Delta^-$. This proves parts (4.10) and (4.10) of the proposition.

Under the assumption $u = 0$ in part (4.10), the differential equations (4.34) become homogeneous in h. For fixed ζ, the solutions are multiples of $z \mapsto R(\zeta; z)^{-s}$, as is clear from (4.35). Hence, h locally has the form $h(\zeta, z) = \varphi(\zeta)\, R(\zeta; z)^{-s}$, where φ is holomorphic by condition b) in the definition of $\mathcal{D}_s$. It also follows that h vanishes on any connected component of U on which $R(\zeta; z)^{-s}$ is multivalued and, in particular, on any component that intersects $\Delta^+ \cup \Delta^-$. Part 4.10) now follows by linearity. ■

Proof of Theorem 4.8, converse direction. Functions h and u with the properties assumed in the second part of the theorem determine a section $(h, u) \in \mathcal{D}_s\big((\mathbb{P}^1_{\mathbb{C}} \smallsetminus \mathbb{P}^1_{\mathbb{R}}) \times \mathfrak{H}\big)$. Proposition 4.10 shows that $u \in \mathcal{E}_s$. By the first part of the theorem, we have $(\mathsf{B}_s u, u) \in \mathcal{D}_s\big((\mathbb{P}^1_{\mathbb{C}} \smallsetminus \mathbb{P}^1_{\mathbb{R}}) \times \mathfrak{H}\big)$. Since this has the same second component as (h, u), part (4.10) of the proposition shows that $h = \mathsf{B}_s u \in \mathcal{C}_s$, and then part (4.10) gives $u = \mathsf{R}_s h = \mathsf{P}_s h$. ■

- *Local Description of h Near the Diagonal.* Part (4.10) of Proposition 4.10 says that the first component of a section (h, u) of $\mathcal{D}_s$ near the diagonal or antidiagonal is completely determined by the second component, but does not tell us explicitly how. We would like to make this explicit. We can do this in two ways, in terms of Taylor expansions or by an integral formula. We will use this in Sect. 6.

We first consider an arbitrary real-analytic function u in a neighborhood of a point $z_0 \in \mathfrak{H}$ and a real-analytic solution h of (4.34a) near (z_0, z_0) which is holomorphic in the first variable. Then h has a power series expansion $h(\zeta, z) = \sum_{n=0}^{\infty} h_n(z)(\zeta - z)^n$ in a neighborhood of (z_0, z_0), and (4.34a) is equivalent to the recursive formulas

$$h_n(z) = \begin{cases} u(z) & \text{if } n = 0, \\ \dfrac{1}{1-s}\, \dfrac{\partial h_0(z)}{\partial z} & \text{if } n = 1, \\ \dfrac{1}{n-s} \left(\dfrac{\partial h_{n-1}(z)}{\partial z} + \dfrac{s}{z - \bar{z}}\, h_{n-1}(z) \right) & \text{if } n \geq 2, \end{cases}$$

which we can solve to get the expansion

$$h(\zeta, z) = u(z) + y^{-s} \sum_{n=1}^{\infty} \frac{\partial^{n-1}}{\partial z^{n-1}} \left(y^s \frac{\partial u}{\partial z} \right) \frac{(\zeta - z)^n}{(1-s)_n}, \tag{4.38}$$

where $(1-s)_n = (1-s)(2-s)\cdots(n-s)$ as usual is the Pochhammer symbol. Conversely, for any real-analytic function $u(z)$ in a neighborhood of z_0, the series in (4.38) converges and defines a solution of (4.34a) near (z_0, z_0). Thus there is a bijection between germs of real-analytic functions u near z_0 and germs of real-analytic solutions of (4.34a), holomorphic in ζ, near (z_0, z_0). If u further satisfies $\Delta u = \lambda_s u$, then a short calculation shows that the function defined by (4.38) satisfies (4.34b), so we get a bijection between germs of λ_s-eigenfunctions u near z_0 and the stalk of $\mathcal{D}_s$ at (z_0, z_0). An exactly similar argument gives, for any λ_s-eigenfunction u near z_0, a unique solution

$$h(\zeta, z) \;=\; -y^{-s} \sum_{n=1}^{\infty} \frac{\partial^{n-1}}{\partial \bar{z}^{n-1}} \left(y^s \, \frac{\partial u}{\partial \bar{z}} \right) \frac{(\zeta - \bar{z})^n}{(1-s)_n}, \tag{4.39}$$

of (4.34a) and (4.34b) near the point $(\bar{z}_0, z_0) \in \Delta^-$. This proves:

Proposition 4.11. *Let $u \in \mathcal{E}_s(U)$ for some open set $U \subset \mathfrak{H}$. Then there is a unique section (h, u) of $\mathcal{D}_s$ in a neighborhood of $\{\, (z, z) \mid z \in U\} \cup \{\, (\bar{z}, z) \mid z \in U\}$, given by* (4.38) *and* (4.39).

The second way of writing h in terms of u near the diagonal or antidiagonal is based on (4.22). This equation was used to lift a global section $u \in \mathcal{E}_s$ to a section $(\mathsf{B}_s u, u)$ of $\mathcal{D}_s$ over all of $\left(\mathbb{P}^1_{\mathbb{C}} \smallsetminus \mathbb{P}^1_{\mathbb{R}}\right) \times \mathfrak{H}$, but its right-hand side can also be used for functions $u \in \mathcal{E}_s(U)$ for open subsets $U \subset \mathfrak{H}$ to define h near points (z, z) or $(\bar{z}, z)$ with $z \in U$. This gives a new proof of the first statement in Proposition 4.11, with the advantage that we now also get some information off the diagonal and antidiagonal:

Proposition 4.12. *If U is connected and simply connected, then the section (h, u) given in Proposition 4.11 extends analytically to $\left(U \cup \bar{U}\right) \times U$.*

- *Formulation with Sheaves.* Proposition 4.10 shows that the component h of a local section (h, u) of $\mathcal{D}_s$ determines the component u, which is locally independent of the second variable and satisfies the Laplace equation. So there is a map from sections of $\mathcal{D}_s$ to sections of $\mathcal{E}_s$. To formulate this as a sheaf morphism, we need to have sheaves on the same space. We denote the projections from $\mathbb{P}^1_{\mathbb{C}} \times \mathfrak{H}$ on $\mathbb{P}^1_{\mathbb{C}}$, respectively $\mathfrak{H}$, by p_1. We use the inverse image sheaf $p_2^{-1}\mathcal{E}_s$ on $\mathbb{P}^1_{\mathbb{C}} \times \mathfrak{H}$, associated to the presheaf $U \mapsto \mathcal{E}_s(p_2 U)$. (See, e.g., Sect. 1, Chap. II, in [4].) The map p_2 is open, so we do not need a limit over open $V \supset p_2 U$ in the description of the presheaf. Note that the functions in $\mathcal{E}_s(p_2 U)$ depend only on z, but that the sheafification of the presheaf adds sections to $p_2^{-1}\mathcal{E}_s$ that may depend on the first variable. In this way, $(h, u) \mapsto u$ corresponds to a sheaf morphism $C : \mathcal{D}_s \to p_2^{-1}\mathcal{E}_s$. We call the kernel $\mathcal{K}_s$.

We denote the sheaf of holomorphic functions on $\mathbb{P}^1_{\mathbb{C}}$ by $\mathcal{O}$. Then $p_1^{-1}\mathcal{O}$ is also a sheaf on $\mathbb{P}^1_{\mathbb{C}} \times \mathfrak{H}$. The following theorem describes $\mathcal{K}_s$ in terms of $p_1^{-1}\mathcal{O}$ and shows that the morphism C is surjective.

Theorem 4.13. *The sequence of sheaves on $\mathbb{P}^1_{\mathbb{C}} \times \mathfrak{H}$*

$$0 \longrightarrow \mathcal{K}_s \longrightarrow \mathcal{D}_s \overset{C}{\longrightarrow} p_2^{-1}\mathcal{E}_s \longrightarrow 0 \tag{4.40}$$

is exact. If a connected open set $U \subset \mathbb{P}^1_{\mathbb{C}} \times \mathfrak{H}$ satisfies $U \cap (\Delta^+ \cup \Delta^-) \neq \emptyset$, then $\mathcal{K}_s(U) = \{0\}$. The restriction of $\mathcal{K}_s$ to $\left(\mathbb{P}^1_{\mathbb{C}} \times \mathfrak{H}\right) \smallsetminus \left(\Delta^+ \cup \Delta^-\right)$ is locally isomorphic to $p_1^{-1}\mathcal{O}$ where holomorphic functions φ correspond to $(\zeta, z) \mapsto \left(\varphi(\zeta)\, R(\zeta; z)^{-s}, 0\right)$.

The inductive limit of $\mathcal{K}_s(U)$ over all neighborhoods U of $\mathbb{P}^1_{\mathbb{R}} \times \mathfrak{H}$ in $\mathbb{P}^1_{\mathbb{C}} \times \mathfrak{H}$ is canonically isomorphic to the space $\mathcal{V}_s^{\omega,\mathrm{rig}}$.

Proof. For the exactness, we only have to check the surjectivity of $C : \mathcal{D}_s \to p_2^{-1}\mathcal{E}_s$. For this we have to verify that for any point $P_0 = (\zeta_0, z_0) \in \mathbb{P}^1_{\mathbb{C}} \times \mathfrak{H}$, any solution of $\Delta u = \lambda_s u$ lifts to a section $(h, u) \in \mathcal{D}_s(U)$ for some sufficiently small neighborhood U of P_0. If $P_0 \in \Delta^+ \cup \Delta^-$, then this is precisely the content of the first statement of Proposition 4.11. If $P_0 \notin \Delta^+ \cup \Delta^-$, then we define h near P_0 by the formula

$$h(\zeta, z) = \int_{z_0}^{z} \left[\left(R_\zeta(\cdot)/R_\zeta(z) \right)^s, u \right] \tag{4.41}$$

instead, again with $\left(R_\zeta(z_1)/R_\zeta(z)\right)^s = 1$ at $z_1 = z$. The next two assertions of the theorem follow from Proposition 4.10. The relation with the rigid hybrid model is based on (4.5). ■

We end this section by making several remarks about the equations (4.34) and their solution spaces $\mathcal{C}_s$ and $D_s(U)$.

The first is that there are apparently very few solutions of these equations that can be given in "closed form." One example is given by the pair $h(\zeta, z) = \frac{\zeta - \bar{z}}{2i} y^{-s}$, $u(z) = y^{1-s}$ (cf. (4.12)). Of course one also has the translations of this by the action of G, and in Example 2 after Theorem 5.6, we will give further generalizations where h is still a polynomial times y^{-s}. One also has the local solutions of the form $(\varphi(\zeta) R(\zeta; z)^{-s}, 0)$ for arbitrary holomorphic functions $\varphi(\zeta)$, as described in Theorem 4.13.

The second observation is that the description of $\mathcal{C}_s$ in terms of differential equations can be generalized in a very simple way to the space $\mathcal{C}_s^+$ of semicanonical hyperfunction representatives introduced in the Remark in Sect. 4.1: these are simply the functions h on $\left(\mathbb{P}^1_{\mathbb{C}} \smallsetminus \mathbb{P}^1_{\mathbb{R}}\right) \times \mathfrak{H}$ that satisfy the system of differential equations:

$$\begin{aligned} \partial_z\big(h(\zeta, z) - u_-(z)\big) &= -s \, \frac{\zeta - \bar{z}}{(z - \bar{z})(\zeta - z)} \big(h(\zeta, z) - u_+(z)\big), \\ \partial_{\bar{z}}\big(h(\zeta, z) - u_+(z)\big) &= s \, \frac{\zeta - z}{(z - \bar{z})(\zeta - \bar{z})} \big(h(\zeta, z) - u_-(z)\big) \end{aligned} \tag{4.35}$$

for some function u_+ and u_- of z alone. This defines a sheaf $\mathcal{D}_s^+$ which projects to $\mathcal{D}_s$ by $(h, u_+, u_-) \mapsto (h, u_+ - u_-)$, and we have a map from $\mathcal{C}_s^+$ to the space of global sections of $\mathcal{D}_s^+$ defined by $h \mapsto (h, \mathsf{R}_s^+ h, \mathsf{R}_s^- h)$ with $\mathsf{R}^\pm$ defined as in (4.19). In some ways, $\mathcal{C}_s^+$ is a more natural space than $\mathcal{C}_s$, but we have chosen to normalize once and for all by $u_-(z) = 0$ in order to have something canonical.

The third remark concerns the surjectivity of $C : \mathcal{D}_s \to p_2^{-1}\mathcal{E}_s$. We know from Theorem 4.13 that any solution u of the Laplace equation can be completed locally to a solution (h, u) of the differential equations (4.34). We now show that such a lift does not necessarily exist for a u defined on a non-simply connected subset of $\mathfrak{H}$.

Specifically, we will show that there is no section of $\mathcal{D}_s$ of the form $(h, q_{1-s}(z,i))$ on any open set $U \subset \mathbb{P}^1_{\mathbb{C}} \times \mathfrak{H}$ whose image under p_2 contains a hyperbolic annulus with center i.

Now the disk model is more appropriate. We work with coordinates $\xi = \mathrm{i}\frac{1+\zeta}{1-\zeta} \in \mathbb{C}^*$ and $w = \mathrm{i}\frac{z-\mathrm{i}}{z+\mathrm{i}} \in \mathbb{D}$. The differential equations (4.34a) and (4.34b) take the form

$$(1-r^2)\,\partial_w h + s\,\frac{1-\bar{w}\xi}{\xi - w}\,(h-u) \;=\; 0, \tag{4.36a}$$

$$(1-r^2)\,\partial_{\bar{w}}(h-u) + s\,\frac{\xi - w}{1-\bar{w}\xi}\,h \;=\; 0, \tag{4.36b}$$

with $r = |w|$, and (4.35) becomes

$$\left[R^{\mathbb{S}}(\xi;\,\cdot)^s, u(\xi,\,\cdot)\right] \;=\; \mathrm{d}\big(R^{\mathbb{S}}(\xi;\,\cdot)\,h(\xi,\,\cdot)\big), \tag{4.36c}$$

with the Poisson kernel $R^{\mathbb{S}}$ in the circle model, as in (3.27c).

Proposition 4.14. *Let $A \subset \mathbb{D}$ be an annulus of the form $r_1 < |w| < r_2$ with $0 \le r_1 < r_2 \le 1$, and let $V \subset \mathbb{C}^*$ be a connected open set that intersects the region $r_1 < |\xi| < r_1^{-1}$ in $\mathbb{C}^*$. Then $\mathcal{D}_s(V \times A)$ does not contain sections of the form $(h, Q_{1-s,n})$ for any $n \in \mathbb{Z}$.*

Proof. Suppose that such a section $(h, Q_{1-s,n})$ exists. Take $\rho \in (r_1, r_2)$ such that V intersects the annulus $A_\rho = \{\rho < |\xi| < \rho^{-1}\}$. Let C be the contour $|w| = \rho$. Then the function f given by

$$f(\xi) \;=\; \int_C \left[R^{\mathbb{S}}(\xi;\,\cdot)^s, Q_{1-s,n}\right]$$

is defined and holomorphic on A_ρ. For $\xi \in V \cap A_\rho$ we know from (4.36c) that the closed differential form $\left[R^{\mathbb{S}}(\xi;\,\cdot)^s, Q_{1-s,n}\right]$ on A has a potential. Hence, $f(\xi) = 0$ for $\xi \in V \cap A_\rho$, and then $f = 0$ in A_ρ. In particular, $f(\xi) = 0$ for $\xi \in \mathbb{S}^1$. In view of (3.19), this implies that the expansion $R^{\mathbb{S}}(\xi;\,\cdot)^s = \sum_{m\in\mathbb{Z}} a_m(\xi)\,P_{s,m}$ (with $\xi \in \mathbb{S}^1$) satisfies $a_{-n}(\xi) = 0$. The function $R^{\mathbb{S}}(\xi;\,\cdot)^s$ is the Poisson transform $\mathsf{P}_{1-s}\,\delta_{1-s,\xi}$ of the distribution $\delta_{1-s,\xi} : \varphi^{\mathbb{S}} \mapsto \varphi^{\mathbb{S}}(\xi)$ on $\mathcal{V}^{\omega}_{1-s}$. This delta distribution has the expansion $\delta_{1-s,\xi} = \sum_{m\in\mathbb{Z}} \xi^{-m}\,\mathbf{e}_{1-s,m}$. Hence, $R^{\mathbb{S}}(\xi;\,\cdot)^s = \sum_{m\in\mathbb{Z}} \xi^{-m}\,\frac{(-1)^m\,\Gamma(1-s)}{\Gamma(1-s+m)}\,P_{1-s,m}$, in which all coefficients are nonzero. Since $P_{1-s,m} = P_{s,m}$, this contradicts the earlier conclusion. ∎

This nonexistence result is a monodromy effect. In a small neighborhood of a point $(\xi_0, w_0) \in \mathbb{S}^1 \times A$, we can construct a section $(h, Q_{1-s,n})$ of $\mathcal{D}_s$ as in (4.41):

$$h(\xi, w) \;=\; \int_{w_0}^{w} \left[\big(R^{\mathbb{S}}(\xi;w')/R^{\mathbb{S}}(\xi;w)\big)^s, Q_{1-s,n}(w')\right]_{w'}. \tag{4.37}$$

If we now let the second variable go around the annulus A, then $h(\xi, w)$ is changed to $h(\xi, w) + h_0(\xi, w)$, where h_0 is defined by the same integral as h but with the path of integration being the circle $|w'| = |w_0|$. Using

$$\left[\left(R^{\mathbb{S}}(\xi; w')/R^{\mathbb{S}}(\xi; w)\right)^s, Q_{1-s,n}(w')\right]_{w'} = R^{\mathbb{S}}(\xi; w)^{-s}\left[R^{\mathbb{S}}(\xi; w')^s, Q_{1-s,n}(w')\right]_{w'}$$

and the absolutely convergent expansion $R^{\mathbb{S}}(\xi; w')^s = \sum_{m\in\mathbb{Z}} \xi^{-m} \frac{(-1)^m \Gamma(1-s)}{\Gamma(1-s+m)} P_{s,m}(w')$ from the proof above, we find from the explicit potentials in Table A.3 in Sect. A.4 that only the term $m = -n$ contributes and that h_0 is given by

$$h_0(\xi, w) = \pi \mathrm{i} \frac{(-1)^n \Gamma(1-s)}{\Gamma(1-s-n)} R^{\mathbb{S}}(\xi; w)^{-s}. \tag{4.38}$$

(Here we have also used (3.13) to replace [,] by { , }.)

5 Eigenfunctions Near $\partial\mathbb{H}$ and the Transverse Poisson Transform

The space $\mathcal{E}_s$ of λ_s-eigenfunctions of the Laplace operator embeds canonically into the larger space $\mathcal{F}_s$ of germs of eigenfunctions near the boundary of $\mathbb{H}$. In Sect. 5.1 we introduce the subspace $\mathcal{W}_s^\omega$ of $\mathcal{F}_s$ consisting of eigenfunction germs that have the behavior $y^s \times$ (analytic across $\mathbb{R}$) near $\mathbb{R}$, together with the corresponding property near $\infty \in \mathbb{P}^1_{\mathbb{R}}$, and show that $\mathcal{F}_s$ splits canonically as the direct sum of $\mathcal{E}_s$ and $\mathcal{W}_s^\omega$. In Sect. 5.2 the space $\mathcal{W}_s^\omega$ is shown to be isomorphic to $\mathcal{V}_s^\omega$ by integral transformations, one of which is called the *transverse Poisson transformation* because it is given by the same integral as the usual Poisson transformation $\mathcal{V}_s^\omega \to \mathcal{E}_s$, but with the integral taken across rather than along $\mathbb{P}^1_{\mathbb{R}}$. This transformation gives another model $\mathcal{W}_s^\omega$ of the principal series representation $\mathcal{V}_s^\omega$, which has proved to be extremely useful in the cohomological study of Maass forms in [2]. In Sect. 5.3 we describe the duality of $\mathcal{V}_s^\omega$ and $\mathcal{V}_{1-s}^{-\omega}$ in (2.19) in terms of a pairing of the isomorphic spaces $\mathcal{W}_s^\omega$ and $\mathcal{E}_{1-s}$. In Sect. 5.4 we construct a smooth version $\mathcal{W}_s^\infty$ of $\mathcal{W}_s^\omega$ isomorphic to $\mathcal{V}_s^\infty$ by using jets of λ_s-eigenfunctions of the Laplace operator. This space is also used in [2].

5.1 *Spaces of Eigenfunction Germs*

Let $\mathcal{F}_s$ be the space of germs of eigenfunctions of Δ, with eigenvalue $\lambda_s = s(1-s)$, near the boundary of $\mathbb{H}$, i.e.,

$$\mathcal{F}_s = \varinjlim_U \mathcal{E}_s(U \cap \mathfrak{H}), \tag{5.1}$$

where the direct limit is taken over open neighborhoods U of $\mathbb{H}$ in $\mathbb{P}^1_{\mathbb{C}}$ (for either of the realizations $\mathfrak{H} \subset \mathbb{P}^1_{\mathbb{C}}$ or $\mathbb{D} \subset \mathbb{P}^1_{\mathbb{C}}$). This space canonically contains $\mathcal{E}_s$ because an eigenfunction in $\mathfrak{H}$ is determined by its values near the boundary (principle of analytic continuation). The action of G in $\mathcal{F}_s$ is by $f \mid g(z) = f(gz)$. The functions $Q_{s,n}$ and $Q_{1-s,n}$ in (3.7) represent elements of $\mathcal{F}_s$ not lying in $\mathcal{E}_s$. Clearly we have $\mathcal{F}_{1-s} = \mathcal{F}_s$.

Consider $u, v \in \mathcal{F}_s$, represented by elements of $\mathcal{E}_s(U \cap \mathfrak{H})$ for some neighborhood U of $\partial\mathbb{H}$ in $\mathbb{P}^1_{\mathbb{C}}$. Then the Green's form $[u, v]$ is defined and closed in U, and for a positively oriented closed path C in U which is homotopic to $\partial\mathbb{H}$ in $U \cap (\mathbb{H} \cup \partial\mathbb{H})$, the integral

$$\beta(u,v) = \frac{1}{\pi \mathrm{i}} \int_C [u,v] = \frac{2}{\pi} \int_C \{u,v\} \tag{5.2}$$

is independent of the choice of C or of the set U on which the representatives of u and v are defined. This defines a G-equivariant antisymmetric bilinear pairing

$$\beta : \mathcal{F}_s \times \mathcal{F}_s \longrightarrow \mathbb{C}. \tag{5.3}$$

If both u and v are elements of $\mathcal{E}_s$, we can contract C to a point, thus arriving at $\beta(u,v) = 0$. Hence, β also induces a bilinear pairing $\mathcal{E}_s \times (\mathcal{F}_s/\mathcal{E}_s) \to \mathbb{C}$.

For each $z \in \mathbb{H}$, the element $q_s(\cdot, z)$ of $\mathcal{F}_s$ is not in $\mathcal{E}_s$. By $(\Pi_s u)(z) = \beta\big(u, q_s(\cdot, z)\big)$ we define a G-equivariant linear map $\Pi_s : \mathcal{F}_s \to \mathcal{E}_s$. Explicitly, $u_{\mathrm{in}}(z) := \Pi_s u(z)$ is given by an integral $\frac{1}{\pi \mathrm{i}} \int_C [u(z'), q_s(z', z)]_{z'}$, where z is inside the path of integration C. By deforming C, we, thus, obtain $u_{\mathrm{in}}(z)$ for all $z \in \mathfrak{H}$, so $u_{\mathrm{in}} \in \mathcal{E}_s$. We can also define $u_{\mathrm{out}}(z) := \frac{-1}{\pi \mathrm{i}} \int_C [u(z'), q_s(z', z)]_{z'}$ where now z is between the boundary of $\mathbb{H}$ and the path of integration. For $u \in \mathcal{E}_s$ we see that $u_{\mathrm{out}} = 0$. More generally, Theorem 3.1 shows that

$$u = u_{\mathrm{out}} + u_{\mathrm{in}} \qquad (\forall\, u \in \mathcal{F}_s). \tag{5.4}$$

The G-equivariance of $[\cdot, \cdot]$ implies that the maps Π_s and $1 - \Pi_s$ are G-equivariant. This gives the following result.

Proposition 5.1. *The G-equivariant maps $\Pi_s : u \mapsto u_{\mathrm{in}}$ and $1 - \Pi_s : u \mapsto u_{\mathrm{out}}$ split the exact sequence of G-modules*

$$0 \longrightarrow \mathcal{E}_s \underset{\Pi_s}{\overset{}{\rightleftarrows}} \mathcal{F}_s \underset{1-\Pi_s}{\rightleftarrows} \mathcal{F}_s/\mathcal{E}_s \longrightarrow 0 \tag{5.5}$$

We now define the subspace $\mathcal{W}_s^\omega$ of $\mathcal{F}_s$. It is somewhat easier in the disk model:

Definition 5.2. The space $\mathcal{W}_s^\omega$ consists of those boundary germs $u \in \mathcal{F}_s$ that are of the form

$$u(w) = 2^{-2s}(1 - |w|^2)^s A^{\mathbb{S}}(w)$$

where $A^{\mathbb{S}}$ is a real-analytic function on a two-sided neighborhood of $\mathbb{S}^1$ in $\mathbb{P}^1_{\mathbb{C}}$.

In other words, representatives of elements of $\mathcal{W}_s^\omega$, divided by the factor $(1-|w|)^s$, extend analytically across the boundary $\mathbb{S}^1$. (The factor 2^{-2s} is included for compatibility with other models.)

The next proposition shows that $\mathcal{W}_s^\omega$ is the canonical direct complement of $\mathcal{E}_s$ in $\mathcal{F}_s$.

Proposition 5.3. *The kernel of $\Pi_s : \mathcal{F}_s \to \mathcal{E}s$ is equal to the space $\mathcal{W}_s^\omega$, and we have the direct sum decomposition of G-modules*

$$\mathcal{F}_s = \mathcal{E}_s \oplus \mathcal{W}_s^\omega, \tag{5.6}$$

given by $u \leftrightarrow (u_{\mathrm{in}}, u_{\mathrm{out}})$.

Notice that all the spaces in the exact sequence (5.5) are the same for s and $1-s$, but that Π_s and Π_{1-s} give different splittings and that $\mathcal{W}_{1-s}^\omega \neq \mathcal{W}_s^\omega$ (for $s \neq \frac{1}{2}$).

Proof. In view of Proposition 5.1, it remains to show that $\mathcal{W}_s^\omega$ is equal to the image of $u \mapsto u_{\mathrm{out}}$.

The asymptotic behavior of Q_{s-1} in (A.13) gives for w' on the path of integration C and w outside C in the definition of $u_{\mathrm{out}}(w)$

$$q_s(w,w') = \left(\frac{2}{\rho^{\mathbb{D}}(w',w)+1}\right)^s f_s\left(\frac{2}{\rho^{\mathbb{D}}(w',w)+1}\right),$$

where f_s is analytic at 0. With (3.2),

$$\frac{2}{\rho(w',w)+1} = \frac{(1-|w'|)^2}{|w-w'|^2+(1-|w'|^2)(1-|w|^2)}\,(1-|w|^2).$$

We conclude that if w' stays in the compact set C, and w tends to $\mathbb{S}^1$, we have

$$u_{\mathrm{out}}(w) = (1-|w|^2)^s \left(\text{analytic function of } 1-|w|^2\right).$$

So $u_{\mathrm{out}} \in \mathcal{W}_s^\omega$.

For the converse inclusion, it suffices to show that $\mathcal{E}_s \cap \mathcal{W}_s^\omega = \{0\}$. This follows from the next lemma, which is slightly stronger than needed here. ■

Lemma 5.4. *Let u be a solution of $\Delta u = \lambda_s u$ on some annulus $1-\delta \leq |w|^2 < 1$ with $\delta > 0$. Suppose that u is of the form*

$$u(w) = (1-|w|^2)^s A(w) + \mathrm{O}\big((1-|w|^2)^{s+1}\big), \tag{5.7}$$

with a continuous function A on the closed annulus $1-\delta \leq |w|^2 \leq 1$. Then $u = 0$.

Proof. On the annulus the function u is given by its polar Fourier series, with terms

$$u_n(w) = \int_0^{2\pi} \mathrm{e}^{-2in\theta} f(\mathrm{e}^{i\theta}w)\,\frac{d\theta}{2\pi}.$$

Each u_n satisfies the estimate (5.7), with A replaced by its Fourier term A_n. Moreover, the G-equivariance of Δ implies that u_n is a λ_s-eigenfunction of Δ. It is the term of order n in the expansion (3.19). In particular, u_n is a multiple of $P_{s,n}$. In Sect. A.1.2 we see that $P_{s,n}$ has a term $(1-|w|^2)^{1-s}$ in its asymptotic behavior near the boundary, or a term $(1-|w|^2)^{1/2}\log(1-|w|^2)$ if $s=\frac{1}{2}$. So u_n can satisfy (5.7) only if it is zero. ■

Remark. The proof of the lemma gives the stronger assertion: If $u \in \mathcal{F}_s$ satisfies (5.7), then $u \in \mathcal{W}_s^\omega$ and $\Pi_s u = 0$.

Returning to the definition of $\mathcal{W}_s^\omega$, we note that the action $g : w \mapsto \frac{Aw+B}{\bar{B}w+\bar{A}}$ in $\mathbb{D}$ gives for the function $A^{\mathbb{S}}$

$$A^{\mathbb{S}}|g(w) = \left|\bar{B}w+\bar{A}\right|^{-2s} A^{\mathbb{S}}\left(\frac{Aw+B}{\bar{B}w+\bar{A}}\right), \tag{5.8}$$

first for $w \in \mathbb{D}$ near the boundary and by real-analytic continuation on a neighborhood of $\mathbb{S}^1$ in $\mathbb{P}^1_{\mathbb{C}}$. On the boundary, where $|w|=1$, this coincides with the action of G in the circle model, as given in (2.8). In (2.20) the action in the circle model of $\mathcal{V}_s^\omega$ is extended to holomorphic functions on sets in $\mathbb{P}^1_{\mathbb{C}}$. That action and the action in (5.8) coincide only on $\mathbb{S}^1$, but are different elsewhere. This reflects that $A^{\mathbb{S}}$ is real-analytic, but not holomorphic.

The restriction of $A^{\mathbb{S}}$ to $\mathbb{S}^1$ induces the *restriction map*

$$\rho_s : \mathcal{W}_s^\omega \longrightarrow \mathcal{V}_s^\omega, \tag{5.9}$$

which is G-equivariant.

Examples of elements of $\mathcal{W}_s^\omega$ are the functions $Q_{s,n}$, represented by elements of $\mathcal{E}_s(\mathbb{D} \smallsetminus \{0\})$, whereas the functions $Q_{1-s,n}$ belong to $\mathcal{F}_s$ but not to $\mathcal{W}_s^\omega$.

We note that the factor $2^{-2s}(1-|w|^2)^s$ corresponds to $\left(\frac{y}{|z+i|^2}\right)^s$ on the upper half plane. So in the upper half-plane model, the elements of $\mathcal{W}_s^\omega$ are represented by functions of the form $u(z) = \left(\frac{y}{|z+i|^2}\right)^s A^{\mathbb{P}}(z)$ with $A^{\mathbb{P}}$ real-analytic on a neighborhood of $\mathbb{P}^1_{\mathbb{R}}$ in $\mathbb{P}^1_{\mathbb{C}}$. The transformation behavior for $A^{\mathbb{P}}$ turns out to coincide on $\mathbb{P}^1_{\mathbb{R}}$ with the action of G in the projective model of $\mathcal{V}_s^\omega$ in (2.6). Outside $\mathbb{P}^1_{\mathbb{R}}$ it differs from the action in (2.23) on holomorphic functions. The restriction map $\rho_s : \mathcal{W}_s^\omega \to \mathcal{V}_s^\omega$ is obtained by $u \mapsto A^{\mathbb{P}}|_{\mathbb{P}^1_{\mathbb{R}}}$.

In the line model, we have $u(z) = y^s A(z)$ near $\mathbb{R}$ and $u(z) = (y/|z|^2)^s A^\infty(-1/z)$ near ∞, with A and A^∞ real-analytic on a neighborhood of $\mathbb{R}$ in $\mathbb{C}$. The action on A is given by

$$A \left| \begin{bmatrix} a & b \\ c & d \end{bmatrix} \right.(z) = |cz+d|^{-2s} A\left(\frac{az+b}{cz+d}\right), \tag{5.10}$$

coinciding on $\mathbb{R}$ with the action in the line model. Restriction of A to $\mathbb{R}$ induces the description of ρ_s in the line model. The factors $2^{-2s}(1-|w|^2)^s$, $\left(y/|z+i|^2\right)^s$, y^s, and $\left(y/|z|\right)^s$ have been chosen in such a way that $A^{\mathbb{S}}$, $A^{\mathbb{P}}$, A, and A^∞ restrict to elements of the circle, projective, and line models, respectively, of $\mathcal{V}_s^\omega$, related by (2.8) and (2.5).

The space $\mathcal{W}_s^\omega$ is the space of global sections of a sheaf, also denoted $\mathcal{W}_s^\omega$, on $\partial\mathbb{H}$, where in the disk model $\mathcal{W}_s^\omega(I)$ for an open set I in $\mathbb{S}^1$ corresponds to the real-analytic functions $A^{\mathbb{S}}$ on a neighborhood of I in $\mathbb{P}^1_{\mathbb{C}}$ such that $(1-|w|^2)^s A^{\mathbb{S}}(w)$ is annihilated by $\Delta - \lambda_s$. Restriction gives $\rho_s : \mathcal{W}_s^\omega(I) \to \mathcal{V}_s^\omega(I)$ for each $I \subset \mathbb{S}^1$. In the line model, $\mathcal{W}_s^\omega(I)$ for $I \subset \mathbb{R}$ can be identified with the space of real-analytic functions A on a neighborhood U of I with $y^s A(z) \in \mathrm{Ker}\,(\Delta - \lambda_s)$ on $U \cap \mathfrak{H}$; for $I \smallsetminus \mathbb{P}^1_{\mathbb{R}} \smallsetminus \{0\}$, we use $(y/|z|^2)^s A^\infty(-1/z)$. The function $z \mapsto y^s$ is an element of $\mathcal{W}_s^\omega(\mathbb{R})$, but not of $\mathcal{W}_s^\omega = \mathcal{W}_s^\omega(\mathbb{P}^1_{\mathbb{R}})$.

Another example is the function $z \mapsto y^{1-s}$, which represents an element of $\mathcal{W}_{1-s}^\omega(\mathbb{R})$, but not of $\mathcal{W}_{1-s}^\omega = \mathcal{W}_{1-s}^\omega(\mathbb{P}^1_{\mathbb{R}})$. It is the Poisson transform of the distribution $\delta_{s,\infty}$, which has support $\{\infty\}$.

The *support* $\mathrm{Supp}\,(\alpha)$ of a hyperfunction $\alpha \in \mathcal{V}_s^{-\omega}$ is the smallest closed subset X of $\partial\mathbb{H}$ such that each $g \in \mathbf{H}_s$ representing α extends holomorphically to a neighborhood of $\partial\mathbb{H} \smallsetminus X$.

Proposition 5.5. *The Poisson transform of a hyperfunction $\alpha \in \mathcal{V}_s^{-\omega}$ represents an element of $\mathcal{W}_{1-s}^\omega\big(\partial\mathbb{H} \smallsetminus \mathrm{Supp}\,(\alpha)\big)$.*

This statement is meaningful only if $\mathrm{Supp}\,(\alpha)$ is not the whole of $\partial\mathbb{H}$. In Theorem 6.4, we will continue the discussion of the relation between support of a hyperfunction and the boundary behavior of its Poisson transform.

Proof. Let $g \in \mathbf{H}_s$ be a representative of $\alpha \in \mathcal{V}_s^{-\omega}$. In the Poisson integral in (3.28), we can replace the integral over C_+ and C_- by the integral

$$\mathsf{P}_s\alpha(w) = \frac{(1-|w|^2)^{1-s}}{2\pi i} \int_C g(w)\big((1-w/\xi)(1-\bar{w}\xi)\big)^{s-1}\, \frac{d\xi}{\xi}, \tag{5.11}$$

where C is a path inside the domain of g encircling $\mathrm{Supp}\,(\alpha)$. For w outside C, the integral defines a real-analytic function on a neighborhood of $\partial\mathbb{D}$, so there the boundary behavior is $(1-|w|^2)^{1-s} \times$ (analytic). Adapting C, we can arrange that any point of $\partial\mathbb{D} \smallsetminus \mathrm{Supp}\,(\alpha)$ is inside this neighborhood. ■

• *Decomposition of Eigenfunctions.* We close this subsection by generalizing the decomposition (5.4) from $\mathcal{F}_s$ to $\mathcal{E}_s(R)$, where R is any annulus $0 \le r_1 < |w| < r_2 \le 1$ in $\mathbb{D}$. For $u \in \mathcal{E}_s(R)$ we define

$$u_{\mathrm{in}} \in \mathcal{E}_s\big(\{|w| < r_2\}\big) \qquad \text{and } u_{\mathrm{out}} \in \mathcal{E}_s\big(\{|w| > r_1\}\big),$$

by $u_{\mathrm{in}}(z) = \frac{1}{\pi i}\int_C [u, q_s(\cdot\,, z)]$ and $u_{\mathrm{out}}(z) = \frac{-1}{\pi i}\int_C [u, q_s(\cdot\,, z)]$, where $C \subset R$ is a circle containing the argument of u_{in} in its interior, respectively the argument of u_{out} in its exterior. Then (5.4) holds in the annulus R. Explicitly, any $u \in \mathcal{E}_s(R)$ has an expansion of the form

$$u = \sum_{n\in\mathbb{Z}} (a_n Q_{s,n} + b_n P_{s,n}) \qquad \text{on } r_1 < |w| < r_2, \tag{5.12}$$

and u_{in} and u_{out} are then given by

$$u_{\text{in}} = \sum_{n\in\mathbb{Z}} b_n P_{s,n}, \qquad u_{\text{out}} = \sum_{n\in\mathbb{Z}} a_n Q_{n,s}. \tag{5.13}$$

5.2 The Transverse Poisson Map

In the last subsection, we defined restriction maps $\rho_s : \mathcal{W}_s^\omega \to \mathcal{V}_s^\omega$, and more generally $\mathcal{W}_s^\omega(I) \to \mathcal{V}_s^\omega(I)$. We now show that these restriction maps are isomorphisms and construct the explicit inverse maps. We in fact give two descriptions of ρ_s^{-1}, one in terms of power series and one defined by an integral transform (transverse Poisson map); the former is simpler and also applies in the C^∞ setting (treated in Sect. 5.4), while the latter (which is motivated by the power series formula) gives a much stronger statement in the context of analytic functions.

- *Power Series Version.* Let $u \in \mathcal{W}_s^\omega(I)$, where we work in the line model and can assume that $I \subset \mathbb{R}$ by locality. Write z as $x + iy$ and for $x \in I$ expand the real-analytic function A such that $u(z) = y^s A(z)$ as a power series $\sum_{n=0}^\infty a_n(x) y^n$ in y, convergent in some neighborhood of I in $\mathbb{C}$. By definition, the constant term $a_0(x)$ in this expansion is the image $\varphi = \rho_s(u)$ of u under the restriction map. The differential equation $\Delta u = \lambda_s u$ of u translates into the differential equation

$$y\left(A_{xx} + A_{yy}\right) + 2s\,A_y = 0. \tag{5.14}$$

Applying this to the power series expansion of A, we find that

$$a_{n-2}''(x) + n(n+2s-1)\,a_n(x) = 0$$

for $n \geq 2$ and that $a_1 \equiv 0$. Together with the initial condition $a_0 = \varphi$, this gives

$$a_n(x) = \begin{cases} \dfrac{(-1/4)^k\,\Gamma(s+\frac{1}{2})}{k!\,\Gamma(k+s+\frac{1}{2})}\,\varphi^{(2k)}(x) & \text{if } n = 2k, \\ 0 & \text{if } 2 \nmid n, \end{cases} \tag{5.15}$$

and hence a complete description of A in terms of φ. Conversely, if φ is any analytic function in a neighborhood of $x \in \mathbb{R}$, then its Taylor expansion at x has a positive radius of convergence r_x and we have $\varphi^{(n)}(x) = \mathrm{O}(n!\,c^n)$ for any $c > r_x^{-1}$. From Stirling's formula or the Legendre duplication formula, we see that $4^{-k}/k!\,\Gamma(k+s+\frac{1}{2}) = \mathrm{O}\big(k^{-\mathrm{Re}(s)}/(2k)!\big)$, so the power series $\sum_{n\geq 0} a_n(x) y^n$ with $a_n(x)$ defined by (5.15) converges for $|y| < r_x$. By a straightforward uniform convergence argument, the function $A(x+iy)$ defined by this power series is real-analytic in a neighborhood of I, and of course it satisfies the differential equation

$A_{xx} + A_{yy} + 2sy^{-1}A_y = 0$, so the function $u(z) = y^s A(z)$ is an eigenfunction of Δ with eigenvalue λ_s. This proves:

Theorem 5.6. *Let I be an open subset of $\mathbb{R}$. Define a map from analytic functions on I to the germs of functions on a neighborhood of I in $\mathbb{C}$ by*

$$(\mathsf{P}_s^\dagger\varphi)(x+\mathrm{i}y) \;=\; y^s \sum_{k=0}^{\infty} \frac{\varphi^{(2k)}(x)}{k!\,\left(s+\frac{1}{2}\right)_k}(-y^2/4)^k, \tag{5.16}$$

with the Pochhammer symbol $\left(\frac{1}{2}+s\right)_k = \prod_{j=0}^{k-1}\left(\frac{1}{2}+s+j\right)$. Then $\mathsf{P}_s^\dagger$ is an isomorphism from $\mathcal{V}_s^\omega(I)$ to $\mathcal{W}_s^\omega(I)$ with inverse ρ_s.

Of course, we can now use the G-equivariance to deduce that the local restriction map $\rho_s : \mathcal{W}_s^\omega(I) \to \mathcal{V}_s^\omega(I)$ is an isomorphism for every open subset $I \subset \mathbb{P}^1_{\mathbb{R}}$ and that the global restriction map $\rho_s : \mathcal{W}_s^\omega \to \mathcal{V}_s^\omega$ is an equivariant isomorphism. The inverse maps, which we still denote $\mathsf{P}_s^\dagger$, can be given explicitly in a neighborhood of infinity using the functions φ_∞ and A^∞ as usual for the line model or by the corresponding formulas in the circle model. The details are left to the reader.

Example 1. Take $\varphi(x) = 1$. Then (5.16) gives $\mathsf{P}_s^\dagger\varphi(z) = y^s$ in $\mathcal{W}_s^\omega(\mathbb{R})$. More generally, if $\varphi(x) = \mathrm{e}^{\mathrm{i}\alpha x}$ with $\alpha \in \mathbb{R}$, then $\mathsf{P}_s^\dagger\varphi$ is the function $i_{s,\alpha}$ defined in (A.3b).

Example 2. We can generalize Example 1 from $\varphi = 1$ to arbitrary polynomials:

$$\varphi(x) \;=\; \binom{-2s}{m}\,x^m \quad\Rightarrow\quad \mathsf{P}_s^\dagger\varphi(z) \;=\; y^s \sum_{k+\ell=m} \binom{-s}{k}\binom{-s}{\ell}\,z^k\,\bar{z}^\ell. \tag{5.17}$$

This can be checked either from formula (5.16) or, using the final statement of Theorem 5.6, by verifying that the expression on the right belongs to $\mathcal{E}_s$ and that its quotient by y^s is analytic near $\mathbb{R}$ and restricts to $\binom{-2s}{m}x^m$ when $y = 0$.

Example 3. Let $a \in \mathbb{C} \smallsetminus I$. Then (5.16) and the binomial theorem give

$$\varphi(x) = (x-a)^{-2s} \quad\Rightarrow\quad \mathsf{P}_s^\dagger\varphi(z) = y^s \sum_{k=0}^{\infty} \binom{-s}{k} \frac{y^{2k}}{(x-a)^{2s+2k}} \;=\; R(a;z)^s, \tag{5.18}$$

(Here the branches in $(x-a)^{-2s}$ and $R(a;z)^s$ have to be taken consistently.) Again, we could skip this calculation and simply observe that $R(a;\cdot)^s \in \mathcal{W}_s^\omega(I)$ and that $\varphi(x)$ is the restriction $y^{-s}R(a;x+iy)^s\big|_{y=0}$. If $|a| > |x|$, then expanding the two sides of (5.18) by the binomial theorem gives another proof of (5.17) and makes clear where the binomial coefficients in that formula come from.

Example 4. Our fourth example is

$$\varphi(x) = R(x;z_0)^s \quad\Rightarrow\quad \mathsf{P}_s^\dagger\varphi(z) \;=\; b(s)^{-1}\,q_s(z,z_0) \qquad (z_0 \in \mathfrak{H}), \tag{5.19}$$

where the constant $b(s)$ is given in terms of beta or gamma functions by

$$b(s) \;=\; \mathsf{B}\big(s,\frac{1}{2}\big) \;=\; \frac{\Gamma(s)\,\Gamma(\frac{1}{2})}{\Gamma(s+\frac{1}{2})}. \tag{5.20}$$

Formula (5.19) is proved by remarking that the function on the right belongs to $\mathcal{W}_s^{\omega}$ and that its image under ρ_s is the function on the left (as one sees easily from the asymptotic behavior of the Legendre function $Q_{s-1}(t)$ as $t \to \infty$). Obtaining it from the power series in (5.16) would probably be difficult, but we will see at the end of the section how to get it from the integral formula for $\mathsf{P}_s^{\dagger}$ given below.

Remark. Equation (5.15) shows that the function $y^{-s} \cdot \mathsf{P}_s^{\dagger}\varphi(x+iy)$ is even in y (as is visible in Examples 1 and 2 above). In the projective model, $A^{\mathbb{P}}(z) = \big(\frac{y}{|z+\mathrm{i}|^2}\big)^{-s} u(z) = |z+\mathrm{i}|^{2s} A(z)$ is not even in y. In the circle model, related to the projective model by $w = \frac{z-\mathrm{i}}{z+\mathrm{i}}$, the reflection $z \mapsto \bar z$ corresponds to $w \mapsto 1/\bar w$ (or $r \mapsto r^{-1}$ in polar coordinates $w = re^{i\theta}$), and the function $A^{\mathbb{S}}(w) = 2^{2s}(1-|w|^2)^{-s}u(w)$ satisfies $A^{\mathbb{S}}(1/\bar w) = |w|^{2s} A^{\mathbb{S}}(w)$. For example, (A.8) and (A.9) say that the function in $\mathcal{W}_s^{\omega}\big(\mathbb{D} \smallsetminus \{0\}\big)$ whose image under ρ_s is w^n ($n \in \mathbb{Z}$) corresponds to $A^{\mathbb{S}}(w) = \bar w^{-n} F\big(s-n, s; 2s; 1-|w|^2\big)$, and this equals $w^n |w|^{-2s} F\big(s-n, s; 2s; 1-|w|^{-2}\big)$ by a Kummer relation. Note that if we had used the factor $\big(\frac{1-|w|}{1+|w|}\big)^s$ instead of $2^{-2s}(1-|w|^2)^s$ in Definition 5.2, we would have obtained functions $A^{\mathbb{S}}$ that are invariant under $w \mapsto 1/\bar w$.

- *Integral Version.* If φ is a real-analytic function on an interval $I \subset \mathbb{R}$, then we can associate to it two extensions, both real-analytic on a sufficiently small complex neighborhood of I: the holomorphic extension, which we will denote by the same letter, and the solution A of the differential equation (5.14) given in Theorem 5.6. The following result shows how to pass explicitly from φ to A, and from A to φ, and show that their domains coincide.

Theorem 5.7. *Let $\varphi \in \mathcal{V}_s^{\omega}(I)$ for some open interval $I \subset \mathbb{R}$, and write $\mathsf{P}_s^{\dagger}\varphi(z) = y^s A(z)$ with a real-analytic function A defined in some neighborhood of I. Let $U = \bar U \subset \mathbb{C}$ be a connected and simply connected subset of $\mathbb{C}$, with $I = U \cap \mathbb{R}$. Then the following two statements are equivalent:*

(i) *φ extends holomorphically to all of U.*
(ii) *A extends real-analytically to all of U.*

Moreover, the two functions define one another in the following way.

(a) *Suppose that φ is holomorphic in U. Then the function $u = \mathsf{P}_s^{\dagger}\varphi$ is given for $z \in U \cap \mathfrak{H}$ by*

$$u(z) \;=\; \frac{1}{\mathrm{i}\,b(s)} \int_{\bar z}^{z} R(\zeta; z)^{1-s}\, \varphi(\zeta)\, d\zeta, \tag{5.21}$$

where $b(s)$ is given by (5.20) and the integral is taken along any piecewise C^1-path in U from $\bar z$ to z intersecting I only once, with the branch of $R(\zeta; z)^{1-s}$

continuous on the path and equal to its standard value at the intersection point with I.

(b) *Suppose that $u(z) = \mathsf{P}_s^\dagger\varphi(z) = y^s A(z)$ with A real-analytic in U. Then the holomorphic extension of φ to U is given by*

$$\varphi(\zeta) = \begin{cases} \dfrac{2\,b(s)\sin\pi s}{\pi} \displaystyle\int_{\xi_0}^{\zeta} \big[u(\cdot), \big(-R(\zeta;\,\cdot)\big)^s\big] & \textit{if } \zeta \in U \cap \mathfrak{H}, \\ A(\zeta) & \textit{if } \zeta \in I, \\ \dfrac{2\,b(s)\sin\pi s}{\pi} \displaystyle\int_{\xi_0}^{\bar\zeta} \big[\big(-R(\zeta;\,\cdot)\big)^s, u(\cdot)\big] & \textit{if } \zeta \in U \cap \mathfrak{H}^-, \end{cases} \tag{5.22}$$

where the integrals are along piecewise C^1-paths in $U \cap \mathfrak{H}$ from any $\xi_0 \in I$ to ζ, respectively $\bar\zeta$, with the branch of $\big(-R(\zeta;z)\big)^s$ fixed by $|\arg\big(-R(\zeta;z)\big)| < \pi$ for z near ξ_0.

We note the formal similarity between the formula (5.21) for $\mathsf{P}_s^\dagger\varphi$ and the formula (3.26) for the Poisson map: the integrand is exactly the same, but in the case of P_s the integration is over $\mathbb{P}^1_{\mathbb{R}}$ (or $\mathbb{S}^1$), while in the formula for $\mathsf{P}_s^\dagger$ it is over a path which crosses $\mathbb{P}^1_{\mathbb{R}}$. We therefore call $\mathsf{P}_s^\dagger$ the *transverse Poisson map.*

We have stated the theorem only for neighborhoods of intervals in $\mathbb{R}$, but because everything is G-equivariant, they can easily be transferred to any interval in $\mathbb{P}^1_{\mathbb{R}}$. (Details are left to the reader.) Alternatively, one can work in the projective or the circle model. This will be discussed after we have given the proof.

Proof of Theorem 5.7. First we show that (5.21) gives A on $U \cap \mathfrak{H}$ starting from a holomorphic φ on U. Define $\mathsf{P}_s^\dagger\varphi$ locally by (5.16). For $x \in I$ we denote by r_x the radius of the largest open disk with center x contained in U. Using the identity

$$\frac{(2k)!}{4^k\,k!\,\Gamma(k+s+\frac12)} = \frac{\Gamma(k+\frac12)}{\Gamma(\frac12)\,\Gamma(k+s+\frac12)} = \frac{1}{\Gamma(s)\,\Gamma(\frac12)} \int_0^1 (1-t)^{s-1}\,t^{k-\frac12}\,dt$$

(duplication formula and beta function), we find for $x \in I$ and $0 < y < r_x$ the formula

$$\begin{aligned} b(s)\,(\mathsf{P}_s^\dagger\varphi)(x+iy) &= y^s \int_0^1 (1-t)^{s-1}\,t^{-1/2} \left(\sum_{k=0}^{\infty} \frac{\varphi^{(2k)}(x)}{(2k)!}(-ty^2)^k\right) dt \\ &= \frac12\,y^s \int_0^1 (1-t)^{s-1}\,t^{-1/2} \Big(\varphi(x+iy\sqrt{t}) + \varphi(x-iy\sqrt{t})\Big)\,dt \\ &= y^s \int_{-1}^1 (1-t^2)^{s-1}\varphi(x+iyt)\,dt \qquad (t=t^2) \\ &= y^s \int_{-y}^{y} \left(\frac{y^2}{y^2-\eta^2}\right)^{1-s} \varphi(x+i\eta)\,y^{-1}\,d\eta \qquad (t=\eta/y) \end{aligned}$$

$$= -i \int_{x-\mathrm{i}y}^{x+\mathrm{i}y} \left(\frac{y}{(\zeta - z)(\zeta - \bar{z})} \right)^{1-s} \varphi(\zeta)\, \mathrm{d}\zeta, \tag{5.23}$$

where the path of integration is the vertical line from $x - \mathrm{i}y$ to $x + \mathrm{i}y$. The integral converges at the end points. The value of the factor $\big(y/(\zeta - z)(\zeta - \bar{z})\big)^{1-s}$ is based on the positive value y^{-1} of $y/(\zeta - z)(\zeta - \bar{z})$ at $\zeta = x$. Continuous deformation of the path does not change the integral, as long as we anchor the branch of the factor $\big(y/(\zeta - z)(\zeta - \bar{z})\big)^{1-s}$ at the intersection point with I. (This holds even though that factor is multivalued on $U \smallsetminus \{z, \bar{z}\}$. We could also allow multiple crossings of I, but then would have to prescribe the crossing point at which the choice of the branch of the Poisson kernel is anchored.) This proves (5.21) for points $z \in U \cap \mathfrak{H}$ sufficiently near to I, and the extension to all of $U \cap \mathfrak{H}$ is then automatic since the integral makes sense in the whole of that domain and is real-analytic in z.

To show that (5.22) gives φ on U if we start from a given A, we also consider first the case that $\zeta = X + iY \in U \cap \mathfrak{H}$ and that the vertical segment from X to ζ is contained in U. Since we want to integrate up to $z = \zeta$, we will use the green's form $\omega_\zeta(z) = \big[u(z), (-R(\zeta;z))^s\big]$ rather than $\big[(-R(\zeta;z))^s, u(z)\big]$ or $\big\{u(z), (-R(\zeta;z))^s\big\}$, which would have nonintegrable singularities at this end point. (The minus sign is included because $R(\zeta;z)$ is negative on the segment.) Explicitly, this Green's form is given for $z = x + \mathrm{i}y \in U \cap \mathfrak{H}$ by

$$\begin{aligned}
\omega_\zeta(z) &= \big(-R(\zeta;z)\big)^s \left(\frac{\partial u}{\partial z}\, \mathrm{d}z + \frac{\mathrm{i}s}{2y} \frac{z-\zeta}{\bar{z}-\zeta}\, u\, \mathrm{d}\bar{z} \right) \\
&= \big(-y\, R(\zeta;z)^s\big) \left(\frac{\partial A}{\partial z}\, \mathrm{d}z - \frac{\mathrm{i}s}{2y} A\, \mathrm{d}z + \frac{\mathrm{i}s}{2y} \frac{z-\zeta}{\bar{z}-\zeta} A\, \mathrm{d}\bar{z} \right) \\
&= \big(-y\, R(\zeta;z)\big)^s \left[\left(\frac{\partial A}{\partial z} - \frac{s}{\bar{z}-\zeta} A \right) \mathrm{d}x + \left(\mathrm{i}\frac{\partial A}{\partial z} + \frac{s}{y} \frac{x-\zeta}{\bar{z}-\zeta} A \right) \mathrm{d}y \right].
\end{aligned} \tag{5.24}$$

If we restrict this to the vertical line $z = X + \mathrm{i}tY$ $(0 < t < 1)$ joining X and ζ, it becomes

$$\begin{aligned}
\omega_\zeta(X + \mathrm{i}tY) &= \left(\frac{\mathrm{i}Y}{2} A_x(X + \mathrm{i}tY) + \frac{Y}{2} A_y(X + \mathrm{i}tY) + \frac{s}{t(1+t)} A(X + \mathrm{i}tY) \right) \\
&\qquad \frac{t^{2s}\, \mathrm{d}t}{(1-t^2)^s} \\
&= \sum_{k=0}^{\infty} \frac{\Gamma(s+\frac{1}{2})}{k!\, \Gamma(k+s+\frac{1}{2})} \Big[\varphi^{(2k+1)}(X)\, \big(iY/2\big)^{2k+1} \\
&\qquad + \left(\frac{k}{t} + \frac{s}{t(1+t)} \right) \varphi^{(2k)}(X)\, \big(iY/2\big)^{2k} \Big] t^{2k} \frac{t^{2s}}{(1-t^2)^s}\, \mathrm{d}t,
\end{aligned}$$

where φ is the holomorphic function near I with $\varphi = A$ on I.

Now we use the beta integrals

$$\int_0^1 t^{2k}\,\frac{t^{2s}}{(1-t^2)^s}\,\mathrm{d}t = \frac{1}{2}\int_0^1 t^{k+s-\frac{1}{2}}(1-t)^{-s}\mathrm{d}t = \frac{\Gamma(k+s+\frac{1}{2})\,\Gamma(1-s)}{2\,\Gamma(k+\frac{3}{2})}$$

$$= \frac{\Gamma(s+\frac{1}{2})\,\Gamma(1-s)}{2\,\Gamma(\frac{1}{2})}\cdot\frac{k!\,\Gamma(k+s+\frac{1}{2})}{\Gamma(s+\frac{1}{2})}\cdot\frac{2^{2k+1}}{(2k+1)!},$$

$$\int_0^1\Big(\frac{k}{t}+\frac{s}{t(1+t)}\Big)t^{2k}\,\frac{t^{2s}}{(1-t^2)^s}\,\mathrm{d}t = \int_0^1\Big(k\,\frac{t^{2k+2s-1}}{(1-t^2)^s}+s\,\frac{t^{2k+2s-1}-t^{2k+2s}}{(1-t^2)^{s+1}}\Big)\,\mathrm{d}t$$

$$= \frac{k}{2}\int_0^1 t^{k+s-1}(1-t)^{-s}\mathrm{d}t+\frac{s}{2}\int_0^1\big(t^{k+s-1}-t^{k+s-\frac{1}{2}}\big)\,(1-t)^{-s-1}\,\mathrm{d}t$$

$$= \frac{k}{2}\,\frac{\Gamma(k+s)\,\Gamma(1-s)}{\Gamma(k+1)}+\frac{s}{2}\Big(\frac{\Gamma(k+s)\,\Gamma(-s)}{\Gamma(k)}-\frac{\Gamma(k+s+\frac{1}{2})\,\Gamma(-s)}{\Gamma(k+\frac{1}{2})}\Big)$$

$$= \frac{\Gamma(s+\frac{1}{2})\,\Gamma(1-s)}{2\,\Gamma(\frac{1}{2})}\cdot\frac{k!\,\Gamma(k+s+\frac{1}{2})}{\Gamma(s+\frac{1}{2})}\cdot\frac{2^{2k}}{(2k)!}$$

(the second calculation is valid initially for $\mathrm{Re}\,(s) < 0$, $\mathrm{Re}\,(k+s) > 0$, but then by analytic continuation for $\mathrm{Re}\,(s) < 1$, $\mathrm{Re}\,(k+s) > 0$, where the left-hand side converges) to get

$$\int_X^\zeta \omega_\zeta = \frac{\Gamma(s+\frac{1}{2})\,\Gamma(1-s)}{2\,\Gamma(\frac{1}{2})}\sum_{n=0}^\infty \varphi^{(n)}(X)\frac{(\mathrm{i}Y)^n}{n!} = \frac{\pi}{2\,b(s)\,\sin\pi s}\,\varphi(\zeta). \quad (5.25)$$

Furthermore, we see from (5.24) that the $\mathrm{d}x$-component of the 1-form $\omega_\zeta(x+\mathrm{i}y)$ extends continuously to $U\cap\overline{\mathfrak{H}}$ and vanishes on I, so $\int_{x_0}^X \omega_\zeta$ vanishes for any $\xi_0 \in I$ and we can replace the right-hand side of (5.25) by $\int_{\xi_0}^\zeta \omega_\zeta$. On the other hand, the fact that the 1-form is closed means that we can integrate along any path from ξ_0 to ζ inside $U\cap\mathfrak{H}$, not just along the piecewise linear path just described, and hence also that we can move ζ anywhere within $U\cap\mathfrak{H}$, thus obtaining the analytic continuation of φ to this domain as stated in (5.22).

If $\zeta = X-\mathrm{i}Y$ $(Y>0)$ belongs to $\mathfrak{H}^-\cap U$, then the calculation is similar. We suppose that the segment from X to $\bar\zeta$ is in U, and parametrize it by $z = X+\mathrm{i}tY$. The differential form is

$$[(-R(\zeta;z))^s, u]$$

$$= (-y\,R(\zeta;z))^s\Big(\Big(\frac{\partial A}{\partial\bar z}(z)+\frac{s}{\zeta-z}\,A(z)\Big)\,\mathrm{d}x+\Big(-\mathrm{i}\frac{\partial A}{\partial\bar z}(z)+\frac{s}{y}\,\frac{\zeta-x}{\zeta-z}\,A(z)\Big)\mathrm{d}y\Big),$$

which leads to the integral

$$\int_X^{\bar\xi} \omega_\zeta = \int_0^1 \frac{t^{2s}}{(1-t^2)^s} \biggl(-\frac{\mathrm{i}Y}{2} A_x(X+\mathrm{i}tY) + \frac{Y}{2} A_y(X+\mathrm{i}tY) + \frac{s}{t}\frac{1}{1+t} A(X+\mathrm{i}tY) \biggr)\, \mathrm{d}t$$

$$= \sum_{k\ge 0} \frac{\Gamma(s+\frac12)}{k!\,\Gamma(s+\frac12+k)} \biggl(\varphi^{(2k+1)}(X)\,(-\mathrm{i}Y/2)^{2k+1} + \Bigl(\frac{k}{t} + \frac{s}{t(1+t)}\Bigr) \varphi^{(2k)}(X)\,(-\mathrm{i}Y/2)^{2k} \biggr) \frac{t^{2s+2k}}{(1-t^2)^s}\, \mathrm{d}t,$$

which is the expression that we obtained in the previous case with Y replaced by $-Y$. We replace Y by $-Y$ in (5.25) and obtain the statement in (5.22) on $U \cap \mathfrak{H}^-$ as well. ■

It is not easy to find examples that illustrate the integral transformation (5.22) explicitly, i.e., examples of functions in $\mathcal{W}_s^\omega$ for which the Green's form $[u(\cdot), R(\zeta;\cdot)^s]$ can be written explicitly as $\mathrm{d}F$ for some potential function $F(\cdot)$. One case which works, though not without some effort, is $u(z) = y^s = \mathsf{P}_s^\dagger(1)$ (Example 1). Here the needed potential function is given by Entry 6 in Table A.3 in Sect. A.4, and a somewhat lengthy calculation, requiring careful consideration of the branches and of the behavior at the end points of the integral, lets us deduce from (5.22) that the inverse transverse Poisson transform of the function $y^s \in \mathcal{W}_s^\omega(\mathbb{R})$ is indeed the constant function 1.

- *Other Models.* The two integral formulas above were formulated in the line model. To go to the projective model, we consider first $U \subset \mathbb{C}$ as in the theorems not intersecting the half-line $i\,[1,\infty)$. In that case we find by (2.5) and (3.27b)

$$\mathsf{P}_s^\dagger \varphi^{\mathbb{P}}(z) = \frac{1}{\mathrm{i}\,b(s)} \int_{\bar z}^{z} R^{\mathbb{P}}(\zeta;z)^{1-s}\, \varphi^{\mathbb{P}}(\zeta)\, \frac{\mathrm{d}\zeta}{1+\zeta^2} \qquad (z \in U \cap \mathfrak{H}), \tag{5.26a}$$

$$\varphi^{\mathbb{P}}(\zeta) = \begin{cases} \dfrac{2\,b(s)\sin\pi s}{\pi} \displaystyle\int_{\xi_0}^{\zeta} \bigl[u(\cdot), \bigl(-R^{\mathbb{P}}(\zeta;\cdot)\bigr)^s\bigr] & \text{if } \zeta \in U \cap \mathfrak{H}, \\ A^{\mathbb{P}}(\zeta) & \text{if } \zeta \in U \cap \mathbb{R} = I, \\ \dfrac{2\,b(s)\sin\pi s}{\pi} \displaystyle\int_{\xi_0}^{\bar\zeta} \bigl[\bigl(-R^{\mathbb{P}}(\zeta;\cdot)\bigr)^s, u(\cdot)\bigr] & \text{if } \zeta \in U \cap \mathfrak{H}^-, \end{cases} \tag{5.26b}$$

with $u(z) = \bigl(\frac{y}{|z+\mathrm{i}|^2}\bigr)^s A^{\mathbb{P}}(z)$, where the paths of integration and the choices of branches in the Poisson kernels are as in the theorems, suitably adapted.

These formulas then extend by G-equivariance to any connected and simply connected open set $U = \bar{U} \subset \mathbb{P}^1_{\mathbb{C}} \smallsetminus \{\mathrm{i}, -\mathrm{i}\}$ and any $\xi_0 \in U \cap \mathbb{P}^1_{\mathbb{R}}$, giving a local description of the isomorphism $\mathcal{V}^\omega_s \cong \mathcal{W}^\omega_s$ on all of $\mathbb{P}^1_{\mathbb{R}}$. Note that the integrals in (5.26) make sense if we take for U an annulus $1-\varepsilon < \left|\frac{z-\mathrm{i}}{z+\mathrm{i}}\right| < 1+\varepsilon$ in $\mathbb{P}^1_{\mathbb{C}}$, which is not simply connected, but the theorem then has to be modified. We will explain this in a moment.

In the circle model, we have

$$\mathsf{P}^\dagger_s \varphi^{\mathbb{S}}(w) = \frac{1}{2\,b(s)} \int_w^{1/\bar{w}} R^{\mathbb{S}}(\eta; w)^{1-s}\, \varphi^{\mathbb{S}}(\eta)\, \frac{d\eta}{\eta} \qquad (w \in U \cap \mathbb{D}), \tag{5.27a}$$

$$\varphi^{\mathbb{S}}(\eta) = \begin{cases} \dfrac{2\,b(s)\sin \pi s}{\pi} \displaystyle\int_{\xi_0}^{\eta} \left[u(\cdot), \left(-R^{\mathbb{S}}(\eta;\cdot)\right)^s\right] & \text{if } \eta \in U,\ |\eta| < 1, \\ A^{\mathbb{S}}(\zeta) & \text{if } \eta \in U \cap \mathbb{S}^1, \\ \dfrac{2\,b(s)\sin \pi s}{\pi} \displaystyle\int_{\xi_0}^{1/\bar{\eta}} \left[\left(-R^{\mathbb{S}}(\eta;\cdot)\right)^s, u(\cdot)\right] & \text{if } \eta \in U,\ |\eta| > 1, \end{cases} \tag{5.27b}$$

with $u(w) = 2^{-2s}(1-|w|)^s A^{\mathbb{S}}(w)$, for $w \in U \cap \mathbb{D}$, with U open in $\mathbb{C} \smallsetminus \{0\}$, connected, simply connected, and invariant under $w \mapsto 1/\bar{w}$, and with $\xi_0 \in U \cap \mathbb{S}^1$, with the paths of integration and the choice of branches of the Poisson kernel again suitably adapted from the versions in the line model.

If U is an annulus of the form $\varepsilon < |w| < \varepsilon^{-1}$ with $\varepsilon \in (0,1)$, we still can apply the relations in (5.27), provided we take in (5.27a) the path from w to $1/\bar{w}$ homotopic to the shortest path. If we change to a path that goes around a number of times, the result differs from $\mathsf{P}^\dagger_s\varphi(w)$ by an integral multiple of $\frac{\pi \mathrm{i}}{b(s)}\,\mathsf{P}_s\varphi^{\mathbb{S}}(w)$. In (5.27b) we can freely move the point ξ_0 in $\partial\mathbb{D}$, without changing the outcome of the integral.

Let us use (5.27a) to verify the formula for $\mathsf{P}^\dagger_s\left(R(\cdot\,; z_0)^s\right)$ given in Example 3. By G-equivariance, we can suppose that $z_0 = \mathrm{i}$. Now changing to circle model coordinates, we find with the help of (3.27c) that the function $\varphi(x) = R(x;\mathrm{i})^s$ corresponds to $\varphi^{\mathbb{S}}(\xi) = 1$ and that the content of formula (5.19) is equivalent to the formula

$$\int_r^{1/r} \left(\frac{(1-r/\eta)(1-r\eta)}{1-r^2}\right)^{s-1} \frac{d\eta}{\eta} = (1-r^2)^s \int_0^1 \frac{(1-t)^{s-1}\, t^{s-1}\, dt}{\left(1-t(1-r^2)\right)^s}$$
$$= (1-r^2)^s\, \frac{\Gamma(s)^2}{\Gamma(2s)}\, F(s,s;2s;1-r^2) = 2\,Q_{s-1}\!\left(\frac{1+r^2}{1-r^2}\right),$$

where in the first line we have made the substitution $\eta = (1-t)r^{-1} + tr$.

5.3 Duality

We return to the bilinear form β on $\mathcal{F}_s$ defined in (5.2). We have seen that β is zero on $\mathcal{E}_s \times \mathcal{E}_s$. The next result describes β on other combinations of elements of $\mathcal{F}_s$ in terms of the duality map $\langle\ ,\ \rangle : \mathcal{V}_s^{\omega} \times \mathcal{V}_{1-s}^{-\omega} \to \mathbb{C}$ defined in (2.19).

Proposition 5.8. *Let $u, v \in \mathcal{F}_s$.*

(a) *If $u \in \mathcal{E}_s$ and $v \in \mathcal{W}_s^{\omega}$, then*

$$\beta(u,v) = b(s)^{-1} \langle \varphi, \alpha \rangle \tag{5.28}$$

with $b(s)$ as in (5.20), where $u = \mathsf{P}_{1-s}\alpha$ with $\alpha \in \mathcal{V}_{1-s}^{-\omega}$ and $v = \mathsf{P}_s^{\dagger}\varphi$ with $\varphi \in \mathcal{V}_s^{\omega}$.

(b) *If $u, v \in \mathcal{W}_s^{\omega}$, then $\beta(u,v) = 0$.*

(c) *If $u \in \mathcal{W}_{1-s}^{\omega}$ and $v \in \mathcal{W}_s^{\omega}$, then*

$$\beta(u,v) = \left(s - \frac{1}{2}\right) \langle \varphi, \psi \rangle, \tag{5.29}$$

with $\varphi \in \mathcal{V}_{1-s}^{\omega}$, $\psi \in \mathcal{V}_s^{\omega}$ such that $u = \mathsf{P}_{1-s}^{\dagger}\varphi$ and $v = \mathsf{P}_s^{\dagger}\psi$.

Proof. The bijectivity of the maps $\mathsf{P}_{1-s} : \mathcal{V}_s^{-\omega} \to \mathcal{E}_{1-s} = \mathcal{E}_s$ and $\mathsf{P}_s^{\dagger} : \mathcal{V}_s^{\omega} \to \mathcal{W}_s^{\omega}$ implies that we always have φ and α as indicated in (5.8). All transformations involved are continuous for the topologies of $\mathcal{V}_{1-s}^{-\omega}$ and $\mathcal{V}_s^{\omega}$, so it suffices to check the relation for $\varphi = \mathbf{e}_{s,m}$ and $\alpha = \mathbf{e}_{1-s,m}$. Now we use (3.29), the result for $\mathsf{P}_s^{\dagger} e_{s,n}$ in Sect. A.3, and (3.18) and (2.15) to get the factor in (5.28).

For part (5.8) we write $u = (1-|w|^2)^s A(w)$ and $v = (1-|w|^2)^s B(w)$ with A and B extending in a real-analytic way across $\partial\mathbb{D}$. If we take for C a circle $|w| = r$ with r close to 1, then

$$[u,v] = \frac{1}{2\mathrm{i}} r(1-r^2)^{2s} \left(AB_r - BA_r\right) \mathrm{d}\theta.$$

It follows that the integral is $\mathrm{O}\big((1-r^2)^s\big)$ as $r \uparrow 1$ and hence vanishes.

In view of b), we can restrict ourselves for c) to the case $s \neq \frac{1}{2}$. As in part 5.8), it suffices to consider the relation for basis vectors. We derive the relation from (A.14):

$$\beta(Q_{1-s,m}, Q_{s,n}) = \beta\left(\pi \cot \pi s\ P_{s,m} + Q_{s,m}, Q_{s,n}\right) = \pi \cot \pi s\, (-1)^n \delta_{n,-m}. \qquad \blacksquare$$

5.4 Transverse Poisson Map in the Differentiable Case

The G-module $\mathcal{W}_s^{\omega}$, which is isomorphic to $\mathcal{V}_s^{\omega}$, turns out to be very useful for the study of cohomology with coefficients in $\mathcal{V}_s^{\omega}$, discussed in detail in [2]. There we also study cohomology with coefficients in $\mathcal{V}_s^{p}$, with $p = 2, 3, \ldots, \infty$, and for this we need an analogue $\mathcal{W}_s^p$ of $\mathcal{W}_s^\omega$ related to p times differentiable functions. In this subsection, we define such a space and show that there is an equivariant isomorphism $\mathsf{P}_s^\dagger : \mathcal{V}_s^p \to \mathcal{W}_s^p$. To generalize the restriction $\rho_s : \mathcal{W}_s^\omega \to \mathcal{V}_s^\omega$, we will define $\mathcal{W}_s^p$ not as a space of boundary germs but as a quotient of G-modules. In fact, we give a uniform discussion, covering also the case $p = \omega$ treated in the previous subsections.

Definition 5.9. For $p = 2, 3, \ldots, \infty, \omega$ we define $\mathcal{G}_s^p$ and $\mathcal{N}_s^p$ as spaces of functions $f \in C^2(\mathbb{D})$ for which there is a neighborhood U of $\partial\mathbb{D} = \mathbb{S}$ in $\mathbb{C}$ such that the function $\tilde{f}(w) = (1-|w|)^{-s} f(w)$ extends as an element of $C^p(U)$ and satisfies on U the conditions

p	For $\mathcal{G}_s^p$	For $\mathcal{N}_s^p$
$\in \mathbb{Z}_{\geq 2}$	$\tilde{\Delta}_s \tilde{f}(w) = o\big((1-\vert w\vert^2)^p\big)$	$\tilde{f}(w) = o\big((1-\vert w\vert^2)^p\big)$
∞	The above condition for all $p \in \mathbb{N}$	The above condition for all $p \in \mathbb{N}$
ω	$\tilde{\Delta}_s \tilde{f}(w) = 0$	$\tilde{f}(w) = 0$

where $\tilde{\Delta}_s$ is the differential operator corresponding to $\Delta - \lambda_s$ under the transformation $f \mapsto \tilde{f}$.

In the analytic case $p = \omega$, the space $\mathcal{G}_s^\omega$ consists of C^2-representatives of germs in $\mathcal{W}_s^\omega$, and $\mathcal{N}_s^\omega$ consists of C^2-representatives of the zero germ in $\mathcal{W}_s^\omega$, i.e., $\mathcal{N}_s^\omega = C_c^2(\mathbb{D})$. Any representative of a germ can be made into a C^2-germ by multiplying it by a suitable cutoff function. Thus $\mathcal{W}_s^\omega$ as as in Definition 5.2 is isomorphic to $\mathcal{G}_s^\omega/\mathcal{N}_s^\omega$. We take C^2-representatives to be able to apply Δ without the need to use a distribution interpretation.

In the upper half-plane model, there is an equivalent statement with $f^{\mathbb{S}}$ replaced by $f^{\mathbb{P}}$, and $2^{-2}(1-|w|^2)$ by $\frac{y}{|z+i|^2}$. The group G acts on $\mathcal{G}_s^p$ and $\mathcal{N}_s^p$ for $p = 2, \ldots, \infty, \omega$, by $f \mid g(z) = f(gz)$, and the operator corresponding to $\Delta - \lambda_s$ is $\tilde{\Delta}_s = -y^2(\partial_y^2 + \partial_x^2(- 2sy\, \partial_y$ (cf. (5.14)).

The definition works locally: $\mathcal{G}_s^p(I)$ and $\mathcal{N}_s^p(I)$, with $I \subset \partial\mathbb{H}$ open, are defined in the same way, with Ω now a neighborhood of I in $\mathbb{P}^1_{\mathbb{C}}$. In the case that $I \subset \mathbb{R}$ in the upper half-plane model, we have $f(z) = y^s \tilde{f}(z)$ on $\Omega \cap \mathfrak{H}$ with $\tilde{f} \in C^p(\Omega)$. On can check that $\mathcal{G}_s^p$ and $\mathcal{N}_s^p$ are sheaves on $\partial\mathbb{H}$.

- *Examples.* The function $z \mapsto y^s$ is in $\mathcal{G}_s^\omega(\mathbb{R})$. The function $Q_{s,n}$ in (3.7) has the right boundary behavior, but is not defined at $w = 0 \in \mathbb{D}$. We can multiply it by $re^{i\theta} \mapsto \chi(r)$ with a smooth function χ that vanishes near zero and is equal to one on a neighborhood on 1. In this way we obtain an element of $\mathcal{G}_s^\omega$.
- *Restriction to the Boundary.* For $f \in \mathcal{G}_s^p$ the corresponding function $f^{\mathbb{S}}$ on Ω has a restriction to $\mathbb{S}^1$ that we denote by $\rho_s f$. It is an element of $\mathcal{V}_s^p$. In this way, restriction to the boundary gives a linear map

$$\rho_s : \mathcal{G}_s^p \longrightarrow \mathcal{V}_s^p \tag{5.30}$$

that turns out to intertwine the actions of G and that behaves compatibly with respect to the inclusions $\mathcal{G}_s^p \to \mathcal{G}_s^q$ and $\mathcal{V}_s^p \to \mathcal{V}_s^q$ if $q < p$. This restriction map can be localized to give $\rho_s : \mathcal{G}_s^p(I) \to \mathcal{V}_s^p(I)$ for open intervals $I \subset \partial\mathbb{H}$.

Lemma 5.10. *Let $I \subset \partial\mathbb{H}$ be open. For $p = 2, \dots, \infty, \omega$ the space $\mathcal{N}_s^p(I)$ is a subspace of $\mathcal{G}_s^p(I)$. It is equal to the kernel of $\rho_s : \mathcal{G}_s^p(I) \to \mathcal{V}_s^p(I)$.*

Proof. The sheaf properties imply that we can work with $I \neq \partial\mathbb{H}$. The action of G can be used to arrange $I \subset \mathbb{R}$ in the upper half-plane model.

Let first $p \in \mathbb{N}$, $p \geq 2$. Suppose that $f(z) = y^s \tilde{f}(z)$ on $\Omega \cap \mathfrak{H}$ for some $\tilde{f} \in C^p(\Omega)$, with Ω a neighborhood of I in $\mathbb{C}$. The Taylor expansion at $x \in I$ gives for $i, j \geq 0, i + j \leq p$

$$\partial_x^i \partial_y^j \tilde{f}(x + iy) = \sum_{n=i+j}^{p} \frac{(n-i)!}{(n-i-j)!} a_{n-i}^{(i)}(x)\, y^{n-i-j} + \mathrm{o}\big(y^{p-i-j}\big) \tag{5.31}$$

on Ω, with

$$a_n(x) = \frac{1}{n!} \partial_y^n \tilde{f}(x).$$

The differential operator $\Delta - \lambda_s$ applied to f corresponds to the operator $\tilde{\Delta}_s = -y^2\partial_x^2 - y^2\partial_y^2 - 2sy\partial_y$ applied to $\tilde{f}$ on the region $\Omega \cap \mathfrak{H}$. Thus we find

$$\tilde{\Delta}_s \tilde{f}(x + iy) = -2sa_1(x) - \sum_{n=2}^{p} \Big(a_{n-2}''(x) + n(n + 2s - 1)a_n(x)\Big) y^n + \mathrm{o}(y^p). \tag{5.32}$$

If $f \in \mathcal{N}_s^p(I)$, then $a_n = 0$ for $0 \leq n \leq p$, and $\tilde{\Delta}_s \tilde{f}(z) = \mathrm{o}(y^p)$. So $f \in \mathcal{G}_s^p(I)$, and $\rho_s f(x) = \tilde{f}(x) = a_0(x)$. Hence, $\mathcal{N}_s^p(I) \subset \operatorname{Ker} \rho_s$.

Suppose that $f \in \mathcal{G}_s^p(I)$ is in the kernel of ρ_s. Then $a_0 = 0$. From (5.32) we have $a_1 = 0$ and $a_{n-2}'' = n(1 - 2s - n)a_n$ for $2 \leq n \leq p$. Hence, $a_n = 0$ for all $n \leq p$, and $f \in \mathcal{N}_s^p(I)$.

The case $p = \infty$ follows directly from the result for $p \in \mathbb{N}$.

In the *analytic case*, $p = \omega$, the inclusions $\mathcal{N}_s^\omega(I) \subset \mathcal{G}_s^\omega(I)$ and $\mathcal{N}_s^p(I) \subset \operatorname{Ker} \rho_s$ are clear. If $f \in \mathcal{G}_s^\omega(I) \cap \operatorname{Ker} \rho_s$, then $\tilde{f}$ is real-analytic on Ω, and instead of the Taylor expansion (5.31), we have a power series expansion with the same structure. Since $(\operatorname{Ker} \rho_s) \cap \mathcal{G}_s^\omega(I) \subset \mathcal{N}_s^\infty(I)$, we have $a_n = 0$ for all n; hence, the analytic function $\tilde{f}$ vanishes on the connected component of Ω containing I. Thus, $f \in \mathcal{N}_s^\omega$. ■

Relation (5.32) in this proof also shows that any $f \in \mathcal{G}_s^p(I)$ with $I \subset \mathbb{R}$ has the expansion

$$f(x + iy) = \sum_{0 \leq k \leq p/2} \frac{(-1/4)^k\, \Gamma(s + \frac{1}{2})}{k!\, \Gamma(s + k + \frac{1}{2})} \varphi^{(2k)}(x)\, y^{s+2k} + \mathrm{o}(y^{s+p}) \quad (y \downarrow 0,\ x \in I), \tag{5.33}$$

with $\varphi = \rho_s f \in \mathcal{V}_s^p(I)$.

- *Boundary Jets.* For $p = 2, \ldots, \infty$ we define $\mathcal{W}_s^p$ as the quotient in the exact sequence of sheaves on $\partial\mathbb{H}$

$$0 \longrightarrow \mathcal{N}_s^p \longrightarrow \mathcal{G}_s^p \longrightarrow \mathcal{W}_s^p \longrightarrow 0. \tag{5.34}$$

In the analytic case, $p = \omega$, we have already seen that $\mathcal{W}_s^\omega$ is the quotient of $\mathcal{G}_s^\omega / \mathcal{N}_s^\omega$.

In the differentiable case $p = 2, \ldots, \infty$, an element of $\mathcal{W}_s^p(I)$ is given on a covering $I = \bigcup_j I_j$ by open intervals I_j by a collection of $f_j \in \mathcal{G}_s^p(I_j)$ such that $f_j \equiv f_j' \bmod \mathcal{N}_s^p(I_j \cap I_{j'})$ in $\mathcal{G}_s^p(I_j \cap I_{j'})$ if $I_j \cap I_{j'} \neq \emptyset$. To each j is associated a neighborhood Ω_j of I_j in $\mathbb{P}^1_{\mathbb{C}}$ on which $f_j^{\mathbb{S}}$ is p times differentiable. Add an open set $\hat{\Omega} \subset \mathbb{H}$ such that $\mathbb{H} \subset \hat{\Omega} \cup \bigcup_j \Omega_j$. With a partition of unity subordinate to the collection $\{\hat{\Omega}\} \cup \{\Omega_j : j\}$, we build one function f on $\mathbb{H}$ such that $f^{\mathbb{S}} = (1 - |w|^2)^{-s} f(w)$ differs from $f_j^{\mathbb{S}}$ on Ω_j by an element of $\mathcal{N}_s^p(I_j)$. In this way we obtain $\mathcal{W}_s^p(I) = \mathcal{G}_s^p(I)/\mathcal{N}_s^p(I)$ in the differentiable case as well.

We have also

$$0 \longrightarrow \mathcal{N}_s^p(\partial\mathbb{H}) \longrightarrow \mathcal{G}_s^p(\partial\mathbb{H}) \longrightarrow \mathcal{W}_s^p(\partial\mathbb{H}) \longrightarrow 0$$

as an exact sequence of G modules. We call elements of $\mathcal{W}_s^p$ *boundary jets* if $p = 2, \ldots, \infty$. The G-morphism ρ_s induces a G-morphism $\rho_s : \mathcal{W}_s^p(I) \to \mathcal{V}_s^p(I)$ for $p = 2, \ldots, \infty, \omega$. The morphism is injective by Lemma 5.10. In fact it is also surjective:

Theorem 5.11. *The restriction map $\rho_s : \mathcal{W}_s^p(I) \to \mathcal{V}_s^p(I)$ is an isomorphism for every open set $I \subset \partial\mathbb{H}$, for $p = 2, \ldots, \infty, \omega$.*

The case $p = \omega$ was the subject of Sect. 5.2. Theorem 5.7 described the inverse $\mathsf{P}_s^\dagger$ explicitly with a transverse Poisson integral, and Theorem 5.6 works with a power series expansion. It is the latter approach that suggests how to proceed in the differentiable and smooth cases. We denote the inverse by $\mathsf{P}_s^\dagger$ or by $\mathsf{P}_{s,p}^\dagger$ if it is desirable to specify p.

Proof. In the *differentiable case* $p \in \mathbb{N} \cup \{\infty\}$, it suffices to consider $\varphi \in C_c^p(I)$ where I is an interval in $\mathbb{R}$. The obvious choice would be to define near I

$$f(z) = \sum_{0 \le k \le p/2} \frac{(-1)^k}{4^k\, k!\, \left(s + \frac{1}{2}\right)_k}\, \varphi^{(2k)}(x)\, y^{s+2k}. \tag{5.35}$$

However, this is in general not in $C^p(\mathfrak{H})$ because each term $\varphi^{(2k)}(x)\, y^{s+2k}$ is only in C^{p-2k}. Instead we set

$$f(z) = y^s \int_{-\infty}^{\infty} \omega(t)\, \varphi(x + yt)\, \mathrm{d}t = y^{s-1} \int_{-\infty}^{\infty} \omega\Big(\frac{t-x}{y}\Big)\, \varphi(t)\, \mathrm{d}t, \tag{5.36}$$

where φ has been extended by zero outside its support and where ω is an even real-analytic function on $\mathbb{R}$ with quick decay that has prescribed moments

$$M_{2k} := \int_{-\infty}^{\infty} t^{2k}\,\omega(t)\,\mathrm{d}t = \frac{(-1)^k(\frac{1}{2})_k}{(s+\frac{1}{2})_k} \qquad \text{for even } k \geq 0. \tag{5.37}$$

(For instance, we could take ω to be the Fourier transform of the product of the function $u \mapsto \Gamma(s+\frac{1}{2})\left(\frac{|u|}{2}\right)^{\frac{1}{2}-s} I_{s-1/2}(|u|)$ and an even function in $C_c^\infty(\mathbb{R})$ that is equal to 1 on a neighborhood of 0 in $\mathbb{R}$. This choice is even real-analytic.) Replacing φ in (5.36) by its Taylor expansion up to order p, we see that this formally matches the expansion (5.35), but it now makes sense and is C^∞ in all of $\mathfrak{H}$, as we see from the second integral. The first integral shows that

$$\tilde{f}(z) = y^{-s} f(z) = \int_{-\infty}^{\infty} \omega(t)\varphi(x+yt)\,\mathrm{d}t \tag{5.38}$$

extends as a function in $C^\omega(\mathbb{C})$.

Inserting the power series expansion of order p of φ at $x \in I$ in (5.38), we arrive at $\tilde{\Delta}_s \tilde{f}(z) = \mathrm{O}(y^p)$. This finishes the proof in the differentiable and smooth cases. ∎

In the proof of Theorem 5.11, we have chosen a real-analytic Schwartz function ω with prescribed moments. In the case $p = 2, 3, \ldots$ we may use the explicit choice in the following lemma, which will be used in the next chapter:

Lemma 5.12. *For any $s \notin \frac{1}{2}\mathbb{Z}$ and any integer $N \geq 0$, there is a unique decomposition*

$$(t^2+1)^{s-1} = \frac{\mathrm{d}^N\alpha(t)}{\mathrm{d}t^N} + \beta(t) \tag{5.39}$$

where $\alpha(t) = \alpha_{N,s}(t)$ is $(t^2+1)^{s-1}$ times a polynomial of degree N in t and $\beta(t) = \beta_{N,s}(t)$ is $\mathrm{O}(t^{2s-N-3})$ as $|t| \to \infty$.

We omit the easy proof. The first few examples are

$$(t^2+1)^{s-1} = \frac{\mathrm{d}}{\mathrm{d}t}\left[\frac{t(t^2+1)^{s-1}}{2s-1}\right] + \frac{2s-2}{2s-1}(t^2+1)^{s-2},$$

$$(t^2+1)^{s-1} = \frac{\mathrm{d}^2}{\mathrm{d}t^2}\left[\frac{(t^2+1)^{s-1}}{(2s-1)(2s-3)} + \frac{(t^2+1)^s}{2s(2s-1)}\right] + \frac{4(s-1)(s-2)}{(2s-1)(2s-3)}(t^2+1)^{s-3},$$

$$(t^2+1)^{s-1} = \frac{\mathrm{d}^3}{\mathrm{d}t^3}\left[\frac{2t(t^2+1)^{s-1}}{(2s+1)(2s-1)(2s-3)} + \frac{t(t^2+1)^s}{2s(2s+1)(2s-1)}\right]$$
$$+ \frac{4(s-1)(s-2)}{(2s+1)(2s-1)}\left(\frac{2s+3}{2s-3} + 3t^2\right)(t^2+1)^{s-4}.$$

In general we have

$$\alpha_{N,s}(t) = \frac{1}{2^N}\sum_{j=0}^{N/2}\binom{N-j-1}{N/2-1}\frac{(t^2+1)^{s-1+j}}{(s)_j\,(s-\frac{N+1}{2}+j)_{N-j}}$$

if $N \geq 2$ is even, where $(s)_j = s(s+1)\cdots(s+j-1)$ is the ascending Pochhammer symbol, and a similar formula if N is odd, as can be verified using the formula

$$\frac{1}{n!}\frac{\mathrm{d}^n}{\mathrm{d}t^n}(t^2+1)^{s-1} = \sum_{0\leq j\leq n/2}\binom{n-j}{j}\binom{s-1}{n-j}(2t)^{n-2j}\,(t^2+1)^{s-n+j}.$$

Let us compute the moments of $\beta = \beta_{N,s}$ as in (5.39). For $0 \leq n < N$, we have

$$\int_{-\infty}^{\infty} t^n\,\beta(t)\,\mathrm{d}t = \int_{-\infty}^{\infty}\left((t^2+1)^{s-1} - \frac{\mathrm{d}^N\alpha(t)}{\mathrm{d}t^N}\right)t^n\,\mathrm{d}t.$$

This is a holomorphic function of s on $\mathrm{Re}\, s < 1$. We compute it by considering $\mathrm{Re}\, s < -\frac{n}{2}$:

$$\int_{-\infty}^{\infty} t^n(t^2+1)^{s-1}\,\mathrm{d}t = \begin{cases} \int_0^1 x^{\frac{n-1}{2}}(1-x)^{-s-\frac{n+1}{2}}\,\mathrm{d}x & \text{if } n \text{ is even,}\\ 0 & \text{if } n \text{ is odd;}\end{cases}$$

$$= \sqrt{\pi}\tan\pi s\,\frac{\Gamma(s)}{\Gamma(s+\frac{1}{2})}\cdot\begin{cases}(-1)^k\frac{(\frac{1}{2})_k}{(s+\frac{1}{2})_k} & \text{if } n = 2k \text{ is even,}\\ 0 & \text{if } n \text{ is odd.}\end{cases} \tag{5.40}$$

So a multiple of $\beta_{N,s}$ has the moments that we need in the proof of Theorem 5.11.

6 Boundary Behavior of Mixed Eigenfunctions

In this section we combine ideas from Sects. 4 and 5. Representatives u of elements of $\mathcal{W}_{1-s}^{\omega}$ have the special property that $(1-|w|^2)^{s-1}\,u(w)$ (in the circle model) or $y^{s-1}u(z)$ (in the line model) extends analytically across the boundary $\partial\mathbb{H}$. If such an eigenfunction occurs in a section (h,u) of the sheaf $\mathcal{D}_s$ of mixed eigenfunctions, we may ask whether a suitable multiple of h also extends across the boundary. In Sect. 6.3 we will show that this is true locally (Theorem 6.2), but not globally (Proposition 6.5).

In Sect. 6.1 we use the differential equations satisfied by $y^{-s}u$ for representatives u of elements of $\mathcal{W}_s^{\omega}$ to define an extension $\mathcal{A}_s$ of the sheaf $\mathcal{E}_s$ from $\mathfrak{H}$ to $\mathbb{P}^1_{\mathbb{C}}$. We also extend the sheaf $\mathcal{D}_s$ on $\mathbb{P}^1_{\mathbb{C}}\times\mathfrak{H}$ to a sheaf $\mathcal{D}_s^*$ on $\mathbb{P}^1_{\mathbb{C}}\times\mathbb{P}^1_{\mathbb{C}}$ that has the same relation to $\mathcal{A}_{1-s}$ as the relation of $\mathcal{D}_s$ to $\mathcal{E}_s = \mathcal{E}_{1-s}$. In Sect. 6.2 we show that the power series expansion of sections of $\mathcal{A}_s$ leads in a natural way to sections of $\mathcal{D}^*_{1-s}$, a result which is needed for the proofs in Sect. 6.3, and in Sect. 6.4 we give the generalization of Theorem 4.13 to the sheaf $\mathcal{D}^*_s$. Finally, in Sect. 6.5 we consider the sections of $\mathcal{D}_s$ near $\mathbb{P}^1_{\mathbb{R}} \times \mathfrak{H}$.

6.1 Interpolation Between Sheaves on $\mathfrak{H}$ and Its Boundary

In this subsection we formulate results from Sect. 5.2 in terms of a sheaf on $\mathbb{P}^1_{\mathbb{C}}$ that is an extension of the sheaf $\mathcal{E}_s$. This will be used in the rest of this section to study the behavior of mixed eigenfunctions near the boundary $\mathbb{P}^1_{\mathbb{C}} \times \mathbb{P}^1_{\mathbb{R}}$ of $\mathbb{P}^1_{\mathbb{C}} \times \mathfrak{H}$ and to extend them across this boundary. We also define an extension $\mathcal{D}^*_s$ of the sheaf $\mathcal{D}_s$ of mixed eigenfunctions.

For an open set $U \subset \mathbb{C}$, let $\mathcal{A}_s(U)$ be the space of real-analytic solutions $A(z)$ of (5.14) in U. For $U \subset \mathbb{P}^1_{\mathbb{C}}$ containing ∞, the definition is the same except that the solutions have the form $A(z) = |z|^{-2s} A^{\infty}(-1/z)$ for some real-analytic function A^{∞} near 0 (which then automatically satisfies the same equation). The action (5.10) makes $\mathcal{A}_s$ into a G-equivariant sheaf: $A \mapsto A|g$ defines an isomorphism $\mathcal{A}_s(U) \cong \mathcal{A}_s(g^{-1}U)$ for any open $U \subset \mathbb{P}^1_{\mathbb{C}}$ and $g \in G$.

For any $U \subset \mathbb{P}^1_{\mathbb{C}}$, the space $\mathcal{A}_s(U)$ can be identified via $A(z) \mapsto u(z) = |y|^s A(z)$ with a subspace of the space $\mathcal{E}_s(U \smallsetminus \mathbb{P}^1_{\mathbb{R}})$ of λ_s-eigenfunctions of the Laplace operator $\Delta = -y^2(\partial_x^2 + \partial_y^2)$ in $\mathbb{P}^1_{\mathbb{C}} \smallsetminus \mathbb{P}^1_{\mathbb{R}}$ (up to now we have considered the operator Δ and the sheaf $\mathcal{E}_s$ only on $\mathfrak{H}$), namely, the subspace consisting of functions which are locally of the form $|y|^s \times$ (analytic) near $\mathbb{R}$ and of the form $|y/z^2|^s \times$ (analytic) near ∞.

If $U \subset \mathbb{P}^1_{\mathbb{C}} \smallsetminus \mathbb{P}^1_{\mathbb{R}}$, then the map $A \mapsto u$ is an isomorphism between $\mathcal{A}_s(U)$ and $\mathcal{E}_s(U)$. (In this case, the condition "real-analytic" in the definition of $\mathcal{A}_s(U)$ can be dropped, since C^2 or even distributional solutions of the differential equation are automatically real-analytic.) At the opposite extreme, if U meets $\mathbb{P}^1_{\mathbb{R}}$ in a nonempty set I, then any section of $\mathcal{A}_s$ over U restricts to a section of $\mathcal{V}^{\omega}_s$ over I, and for any $I \subset \mathbb{P}^1_{\mathbb{R}}$, we obtain from Theorem 5.6 an identification between $\mathcal{V}^{\omega}_s(I)$ and the inductive limit of $\mathcal{A}_s(U)$ over all neighborhoods $U \supset I$. The sheaf $\mathcal{A}_s$, thus, "interpolates" between the sheaf $\mathcal{E}_s$ on $\mathbb{P}^1_{\mathbb{C}} \smallsetminus \mathbb{P}^1_{\mathbb{R}}$ and the sheaf $\mathcal{V}^{\omega}_s$ on $\mathbb{P}^1_{\mathbb{R}}$. At points outside $\mathbb{P}^1_{\mathbb{R}}$, the stalks of $\mathcal{A}_s$ are the same as those of $\mathcal{E}_s$, while at points in $\mathbb{P}^1_{\mathbb{R}}$, the stalks of the sheaves $\mathcal{A}_s$, $\mathcal{V}^{\omega}_s$, and $\mathcal{W}^{\omega}_s$ are all canonically isomorphic. At the level of open sets rather than stalks, Theorem 5.7 says that the space $\mathcal{A}_s(U)$ for suitable U intersecting $\mathbb{P}^1_{\mathbb{R}}$ is isomorphic to $\mathcal{O}(U)$ by a unique isomorphism compatible with restriction to $U \cap \mathbb{R}$, the isomorphisms in both directions being given by explicit integral transforms. Finally, from (5.15) we see that if U is connected and invariant under conjugation, then any $A \in \mathcal{A}_s(U)$ is invariant under $z \mapsto \bar{z}$. In the language of sheaves, this says that if we denote by $c : \mathbb{P}^1_{\mathbb{C}} \to \mathbb{P}^1_{\mathbb{C}}$ the complex conjugation, the induced isomorphism $c : c^{-1}\mathcal{A}_s \to \mathcal{A}_s$ is the identity when restricted to $\mathbb{P}^1_{\mathbb{R}}$.

We now do the same construction for the sheaf $\mathcal{D}_s$ of mixed eigenfunctions, defining a sheaf $\mathcal{D}_s^*$ on $\mathbb{P}^1_{\mathbb{C}} \times \mathbb{P}^1_{\mathbb{C}}$ which bears the same relation to $\mathcal{D}_s$ as $\mathcal{A}_s$ has to $\mathcal{E}_s$. (We could therefore have used the notation $\mathcal{E}_s^*$ instead of $\mathcal{A}_s$, but since $\mathcal{A}_s$ interpolates between two very different subsheaves $\mathcal{E}_s$ and $\mathcal{V}_s^\omega$, we preferred a neutral notation which does not favor one of these aspects over the other. Also, $\mathcal{E}_s = \mathcal{E}_{1-s}$, but $\mathcal{A}_s \neq \mathcal{A}_{1-s}$.)

Let (h,u) be a section of the sheaf $\mathcal{D}_s$ in $U \cap (\mathbb{C}\times\mathfrak{H})$, where U is a neighborhood in $\mathbb{C}\times\mathbb{C}$ of a point (x_0,x_0) with $x_0 \in \mathbb{R}$. The function $u(z)$ is a λ_s-eigenfunction of Δ, and we can ask whether it ever has the form $y^s\,A(z)$ or $y^{1-s}\,A(z)$ with $A(z)$ (real-)analytic near x_0. It turns out that the former does not happen, but the latter does, and moreover that in this case, the function $h(\zeta,z)$ has the form $y^{-s}\,B(\zeta,z)$ where $B(\zeta,z)$ is also analytic in a neighborhood of $(x_0,x_0) \in \mathbb{C}\times\mathbb{C}$. To see this, we make the substitution

$$u(z) \;=\; y^{1-s}\,A(z), \qquad h(\zeta,z) \;=\; y^{-s}\,B(\zeta,z) \tag{6.1}$$

in the differential equations (4.34) to obtain that these translate into the differential equations

$$(\zeta - z)\,\partial_z B \;=\; -s\,B - \frac{is}{2}\,(\zeta-\bar z)\,A, \tag{6.2a}$$

$$(\zeta-\bar z)\,\partial_{\bar z}\big(B - yA\big) \;=\; -s\,B - \frac{is}{2}\,(\zeta-\bar z)\,A, \tag{6.2b}$$

for A and B, in which there is no singularity at $y=0$. (This would not work if we had used $u = y^s A$, $h = y^* B$ instead.)

As long as z is in the upper half plane, the equations (6.1) define a bijection between pairs (h,u) and pairs (B,A), and it makes no difference whether we study the original differential equations (4.34) or the new ones (6.2). The advantage of the new system is that it makes sense for all $z \in \mathbb{C}$ and defines a sheaf $\mathcal{D}_s^*$ on $\mathbb{C}\times\mathbb{C}$ whose sections over $U \subset \mathbb{C}\times\mathbb{C}$ are real-analytic solutions (B,A) of (6.2) in U with B holomorphic in the first variable and A locally constant in the first variable. This sheaf is G-equivariant with respect to the action $(B,A)|g = (B|g, A|g)$ given for $g = \left[\begin{smallmatrix} a & b \\ c & d\end{smallmatrix}\right]$ by

$$B|g(\zeta,z) \;=\; |cz+d|^{2s}\,B(g\zeta,gz), \qquad A|g(z) \;=\; |cz+d|^{2s-2}A(gz), \tag{6.3}$$

so it extends to a sheaf on all of $\mathbb{P}^1_{\mathbb{C}}\times\mathbb{P}^1_{\mathbb{C}}$ by setting $\mathcal{D}_s^*(U) = \mathcal{D}_s^*(g^{-1}U)|g$ if U is a small neighborhood of a point (ζ_0,∞) or (∞,z_0) and g is chosen with $g^{-1}U \subset \mathbb{C}\times\mathbb{C}$.

In (4.38) and (4.39), we give a formula for h in terms of u near the diagonal and the antidiagonal where (h,u) is a section of $\mathcal{D}_s$. In terms of $A = y^{s-1}u$ and $B = y^s h$, this formula becomes

$$B(\zeta,z) \;=\; \begin{cases} -\dfrac{i}{2}(\zeta-\bar z)\displaystyle\sum_{n\ge 0}\frac{\partial^n A}{\partial z^n}(z)\,\frac{(\zeta-z)^n}{(1-s)_n} & \text{for } \zeta \text{ near } z, \\[2ex] -\dfrac{i}{2}(\zeta-z)\displaystyle\sum_{n\ge 1}\frac{\partial^n A}{\partial \bar z^n}(z)\,\frac{(\zeta-\bar z)^n}{(1-s)_n} - \dfrac{i}{2}(\zeta-\bar z)A(z) & \text{for } \zeta \text{ near } \bar z. \end{cases} \tag{6.4}$$

Now inspection shows that the right-hand side of (6.4) satisfies the differential equations (6.2), whether $z \in \mathfrak{H}$ or not, so (B, A) with B as in (6.4) gives a section of $\mathcal{D}_s^*$ on neighborhoods of points (z, z) and $(\bar{z}, z)$ for all $z \in \mathbb{C}$. From (6.4) it is not clear that for $z \in \mathbb{R}$ both expressions define the same function on a neighborhood of z. In the next subsection, we will see that they do.

6.2 Power Series Expansion

Sections of $\mathcal{A}_s$ are real-analytic functions of one complex variable and hence can be seen as power series in two variables. In this subsection, we show that the coefficients in these expansions have interesting properties. They will be used in Sect. 6.3 to study the structure of sections of $\mathcal{D}_s$ and $\mathcal{D}_s^*$ near the diagonal of $\mathbb{P}^1_{\mathbb{R}} \times \mathbb{P}^1_{\mathbb{R}}$.

Let $U \subset \mathbb{C}$ be open, and let $z_0 \in U$. We write the expansion of a section A of $\mathcal{A}_s$ at a point z_0 in the strange form (the reason for which will become apparent in a moment)

$$A(z) = \sum_{m,n \geq 0} \binom{m+s-1}{m} \binom{n+s-1}{n} c_{m,n}(z_0)\,(z-z_0)^m\,(\overline{z-z_0})^n. \tag{6.5}$$

Then we have the following result.

Theorem 6.1. *Let $U \subset \mathbb{C}$, $A \in \mathcal{A}_s(U)$, and for $z_0 \in U$ define the coefficients $c_{m,n}(z_0)$ for $m, n \geq 0$ by* (6.5). *Let $r : U \to \mathbb{R}_+$ be continuous. Then the series* (6.5) *converges in $|z - z_0| < r(z_0)$ if and only if the series*

$$\Phi_A(z_0; v, w) := \sum_{m,n \geq 0} c_{m,n}(z_0)\, v^m w^n \tag{6.6}$$

converges for $|v|, |w| < r(z_0)$. The function defined by (6.6) *has the form*

$$\Phi_A(z_0; v, w) = \frac{B(z_0 + v, z_0) - B(\bar{z}_0 + w, z_0)}{y_0 + (v-w)/2i} \tag{6.7}$$

for a unique analytic function B on

$$U' = \{(\zeta, z) \in \mathbb{C} \times U : |\zeta - z| < r(z)\} \cup \{(\zeta, z) \in \mathbb{C} \times U : |\zeta - \bar{z}| < r(z)\}$$

satisfying $B(\zeta, z) = y\,A(z)$ and $B(\bar{z}, z) = 0$ for $z \in U$, and the pair (B, A) is a section of $\mathcal{D}_{1-s}^*$ over U'.

Proof. The fact that $\binom{m+s-1}{m} = m^{O(1)}$ as $m \to \infty$ implies the relation between the convergence of (6.5) and (6.6). (We use here that a power series $\sum c_{mn,n} v^n w^m$ in two variable converges for $|v|, |w| < r$ if and only if its restriction to $w = \bar{v}$

converges for $|v| < r$.) The differential equation (5.14) is equivalent to the very simple recursion

$$2\mathrm{i}y_0\, c_{m,n}(z_0) \;=\; c_{m,n-1}(z_0) - c_{m-1,n}(z_0) \qquad (m,n \geq 1) \tag{6.8}$$

for the coefficients $c_{m,n}(z_0)$. (This was the reason for the choice of the normalization in (6.5).) This translates into the fact that $(2\mathrm{i}y_0 + v - w)\, \Phi_A(z_0; v, w)$ is the sum of a function of v alone and a function of w alone, i.e., we have

$$\Phi_A(z_0; v, w) \;=\; \frac{L_A(z_0; v) - R_A(z_0; w)}{y_0 + (v-w)/2\mathrm{i}}, \tag{6.9}$$

where, if we use the freedom of an additive constant to normalize $R_A(z_0; 0) = 0$, the functions L_A and R_A are given explicitly in terms of the boundary coefficients $\{c_{j,0}(z_0)\}_{j\geq 0}$ and $\{c_{0,j}(z_0)\}_{j\geq 1}$ by

$$2\mathrm{i}\, L_A(z_0; v) = (v + 2\mathrm{i}y_0) \sum_{m\geq 0} c_{m,0}(z_0)\, v^m,$$

$$2\mathrm{i}\, R_A(z_0; w) = c_{0,0}(z_0)\, w + (w - 2\mathrm{i}y_0) \sum_{n\geq 1} c_{0,n}(z_0)\, w^n. \tag{6.10}$$

(Multiplied out, this says that coefficients $c_{m,n}(z_0)$ satisfying (6.8) are determined by their boundary values by

$$c_{m,n}(z_0) \;=\; \sum_{j=1}^{m} \frac{(-1)^n}{(2iy_0)^{m+n-j}} \binom{m-j+n-1}{m-j} c_{j,0}(z_0)$$

$$+ \sum_{j=1}^{n} \frac{(-1)^{n-j}}{(2iy_0)^{m+n-j}} \binom{n-j+m-1}{n-j} c_{0,j}(z_0), \tag{6.11}$$

which of course can be checked directly.)

We define B on U' (now writing z instead of z_0) by

$$B(\zeta, z) \;=\; \begin{cases} L_A(z, \zeta - z) & \text{if } |\zeta - z| < r(z), \\ R_A(z, \zeta - \bar{z}) & \text{if } |\zeta - \bar{z}| < r(z). \end{cases} \tag{6.12}$$

These two definitions are compatible if the disks in question overlap (which happens if $r(z) > |y_0|$) because the convergence of (6.6) for $|v|$, $|w| < r(z)$ implies that the fraction in (6.9) is holomorphic in this region and hence that its numerator vanishes if $z_0 + v = \bar{z}_0 + w$.

Surprisingly, the function B thus defined constitutes, together with the given section A of $\mathcal{A}_s$, a section (B, A) of $\mathcal{D}^*_{1-s}$ for ζ near z or $\bar{z}$. To see this, we apply the formulas (6.4), with s replaced by $1-s$, and express the derivatives of A in the coefficients $c_{m,n}(z)$ with help of (6.5). We find that the first expression in (6.4) is equal to $L_A(z; \zeta - z)$, and the second one to $R_A(z; \zeta - \bar{z})$. ∎

Example 1. Let $A(z) \in \mathcal{A}_s(\mathbb{C} \smallsetminus \{0\})$ be the function $|z|^{-2s}$. For any $z_0 \neq 0$ the binomial theorem gives $c_{m,n}(z) = (-1)^{m+n}\,|z_0|^{-2s}\,z_0^{-m}\,\bar{z}_0^{-n}$ and

$$\Phi_A(z_0; v, w) \;=\; \frac{|z_0|^{2-2s}}{(z_0+v)(\bar{z}_0+w)} \;=\; \frac{1}{2iy_0+v-w}\left(\frac{|z_0|^{2-2s}}{\bar{z}_0+w} - \frac{|z_0|^{2-2s}}{z_0+v}\right), \tag{6.13}$$

in accordance with (6.7) with the solution $B(\zeta, z) = \frac{\mathrm{i}}{2}\,|z|^{2-2s}\,\left(\zeta^{-1} - \bar{z}^{-1}\right)$, defined on $z \neq 0$, $\zeta \neq 0$. (The regions $|\zeta - z| < |z|$ and $|\zeta - \bar{z}| < z$ do not overlap.)

Example 2. If $z_0 \in \mathbb{R}$, then (6.8) says that $c_{m,n}(z_0)$ depends only on $n+m$, so the generating function Φ_A has an expansion of the form

$$\Phi_A(z_0; v, w) \;=\; \sum_{N=0}^{\infty} C_N(z_0)\,\frac{v^{N+1} - w^{N+1}}{v-w}.$$

Hence, in this case, we have $A(z) \;=\; \sum_{N\geq 0} C_N(z_0) P_N(z - z_0)$ where P_N is the section of $\mathcal{A}_s$ defined by

$$P_N(z) \;:=\; (-1)^N \sum_{m,n\geq 0,\, m+n=N} \binom{-s}{m}\binom{-s}{n}\, z^m\,\bar{z}^n, \tag{6.14}$$

a polynomial that already occurred in (5.17).

Example 3. Let $A(z) = y^{-s} p_{s,k}(z, \mathrm{i})$, defined in (3.5), with $z_0 = \mathrm{i}$ and $k \geq 0$. We describe $A(z) = \frac{\Gamma(s+k)}{k!\,\Gamma(s-k)}\,\tilde{A}(w)$ first in the coordinate $w = \frac{z-\mathrm{i}}{z+\mathrm{i}}$ of the disk model. Taking into account (A.8) and (A.9), we obtain

$$\tilde{A}(w) \;=\; w^k\left(\frac{1-w\bar{w}}{|1-w|^2}\right)^{-s}(1-w\bar{w})^s\;F\big(s, s+k; 1+k, w\bar{w}\big).$$

Set $p = (z-\mathrm{i})/2\mathrm{i}$, so that $w = p/(p+1)$. Then

$$\begin{aligned}
\tilde{A}(w) &= (1-w)^s\,(1-\bar{w})^s \sum_{l\geq 0} \frac{(s)_\ell\,(s+k)_\ell}{(1+k)_\ell\,\ell!}\,w^{k+\ell}\bar{w}^\ell \\
&= \sum_{\ell\geq 0}\binom{-s}{\ell}\binom{-s-k}{\ell}\binom{\ell+k}{\ell}^{-1} p^{k+\ell}\,\bar{p}^\ell\,(1+p)^{-s-k-\ell}\,(1+\bar{p})^{-s-\ell} \\
&= \sum_{\ell,i,j\geq 0}\binom{-s}{\ell}\binom{-s-k}{\ell}\binom{\ell+k}{\ell}^{-1}\binom{-s-k-\ell}{i}\binom{-s-\ell}{j} p^{k+\ell+i}\,\bar{p}^{\ell+j} \\
&= \sum_{m\geq k,\, n\geq 0} p^m\,\bar{p}^n\,\binom{-s-k}{m-k}\binom{-s}{n}\binom{n+k}{k}^{-1}\sum_{l=0}^{n}\binom{m-k}{\ell}\binom{n+k}{n-\ell}
\end{aligned}$$

$$= \sum_{m\ge k,\, n\ge 0} \left(\frac{z-\mathrm{i}}{2\mathrm{i}}\right)^m \left(\frac{\bar z+\mathrm{i}}{-2\mathrm{i}}\right)^n \binom{-s-k}{m-k}\binom{-s}{n}\binom{n+k}{k}^{-1}\binom{m+n}{n}.$$

Hence, A has an expansion as in (6.5) with

$$c_{m,n}^{[k]}(\mathrm{i}) := c_{m,n}(\mathrm{i}) = (-1)^m\,(2\mathrm{i})^{-m-n}\,(1-s)_k \binom{m+n}{n+k} \tag{6.15a}$$

($= 0$ if $m < k$), which satisfies the recursion (6.8).

The analogous computation for $k < 0$ gives

$$c_{m,n}^{[k]}(\mathrm{i}) = (-1)^m(2\mathrm{i})^{-m-n}\,(1-s)_k \binom{m+n}{m-k} \tag{6.15b}$$

($= 0$ if $n < -k$).[3]

In this example we can describe the form of the function B up to a factor without computation by equivariance: since $z \mapsto p_{s,n}(z,\mathrm{i})$ transforms according to the character $\left[\begin{smallmatrix}\cos\theta & \sin\theta\\ -\sin\theta & \cos\theta\end{smallmatrix}\right] \mapsto \mathrm{e}^{2ik\theta}$, the function $h = y^{s-1}B$ should do the same near points of the diagonal or the antidiagonal. Thus, for $k \ge 0$, we know that $B(\zeta,\mathrm{i})$ is a multiple of $\left(\frac{\zeta-\mathrm{i}}{\zeta+\mathrm{i}}\right)^k$ near $\zeta = \mathrm{i}$ and vanishes near $\zeta = -\mathrm{i}$, while for $k < 0$, we have $B(\zeta,\mathrm{i}) = 0$ for ζ near i, and $B(\zeta,\mathrm{i})$ is a multiple of $\left(\frac{\zeta-\mathrm{i}}{\zeta+\mathrm{i}}\right)^k$ for ζ near $-\mathrm{i}$. The explicit computation using (6.12), (6.10), and (6.15) confirms these predictions, giving $B(\zeta,\mathrm{i}) = (-1)^k\,(1-s)_k\left(\frac{\zeta-\mathrm{i}}{\zeta+\mathrm{i}}\right)^k$ if $k \ge 0$ and ζ is near i, and $B(\zeta,\mathrm{i}) = -(-1)^k\,\frac{\Gamma(k+1-s)}{\Gamma(1-s)}\left(\frac{\zeta-\mathrm{i}}{\zeta+\mathrm{i}}\right)^k$ if $k < 0$ and ζ is near $-\mathrm{i}$.

Note that since any holomorphic function of ζ near i (resp. $-\mathrm{i}$) can be written as a power series in $\frac{\zeta-\mathrm{i}}{\zeta+\mathrm{i}}$ (resp. $\frac{\zeta+\mathrm{i}}{\zeta-\mathrm{i}}$), we see that this example is generic for the expansions of A and B for any section (B,A) of $\mathcal{D}_s$ near $(\zeta,z) = (\pm\mathrm{i},\mathrm{i})$, and hence by G-equivariance for z near any $z_0 \in \mathfrak{H}$ and ζ near z_0 or $\bar z_0$.

Remark. We wrote formula (6.5) as the expansion of a fixed section $A \in \mathcal{A}_s(U)$ around a variable point $z_0 \in U$. If we simply define a function $A(z)$ by (6.5), where z_0 (say in $\mathfrak{H}$) is fixed, then we still find that the differential equation $(\Delta - \lambda_s)\big(y^s A(\cdot, z_0)\big) = 0$ is equivalent to the recursion (6.8) and to the splitting (6.9) of the generating function Φ_A defined by (6.6). In this way, we have constructed a very large family of (locally defined) λ_s-eigenfunctions of Δ: for any $z_0 \in \mathfrak{H}$ and any holomorphic functions $L(v)$ and $R(w)$ defined on disks of radius $r \le y_0$ around 0, we define coefficients $c_{m,n}$ either by (6.9) and (6.6) or by (6.10) and (6.11); then the function $u(z) = y^s A(z)$ with A given by (6.5) is a λ_s-eigenfunction of Δ in the disk of radius r around z_0.

---

[3]The Pochhammer symbol $(x)_k$ is defined for $k < 0$ as $(x-1)^{-1}\cdots(x-|k|)^{-1}$, so that $(x)_k = \Gamma(x+k)/\Gamma(x)$ in all cases.

6.3 Mixed Eigenfunctions Near the Diagonal of $\mathbb{P}^1_{\mathbb{R}} \times \mathbb{P}^1_{\mathbb{R}}$

Parts (4.10) and (4.10) of Proposition 4.10 show that if (h, u) is a section of $\mathcal{D}_s$ near a point $(z, z) \in \mathfrak{H} \times \mathfrak{H}$ of the diagonal or a point $(\bar{z}, z) \in \mathfrak{H}^- \times \mathfrak{H}$ of the antidiagonal, then the function h and u determine each other. Diagonal points $(\xi, \xi) \in \mathbb{P}^1_{\mathbb{R}} \times \mathbb{P}^1_{\mathbb{R}}$ are not contained in the set $\mathbb{P}^1_{\mathbb{C}} \times \mathfrak{H}$ on which the sheaf $\mathcal{D}_s$ is defined. Nevertheless, there is a relation between the analytic extendability of h and u near such points, which we now study.

Theorem 6.2. *Let $\xi \in \mathbb{P}^1_{\mathbb{R}}$. Suppose that (h, u) is a section of $\mathcal{D}_s$ over $U \cap (\mathfrak{H} \times \mathfrak{H})$ for some neighborhood U of (ξ, ξ) in $\mathbb{P}^1_{\mathbb{C}} \times \mathbb{P}^1_{\mathbb{C}}$. Then the following statements are equivalent:*

(a) *The function $y^{s-1}u$ extends real-analytically to a neighborhood of ξ in $\mathbb{P}^1_{\mathbb{C}}$.*
(b) *The function $y^s h$ extends real-analytically to a neighborhood of (ξ, ξ) in $\mathbb{P}^1_{\mathbb{C}} \times \mathbb{P}^1_{\mathbb{C}}$.*
(c) *The function $y^s h$ extends real-analytically to $U' \cap (\mathbb{P}^1_{\mathbb{C}} \times \mathfrak{H})$ for some neighborhood U' of (ξ, ξ) in $\mathbb{P}^1_{\mathbb{C}} \times \mathbb{P}^1_{\mathbb{C}}$.*

The theorem can be formulated partly in terms of stalks of sheaves. In particular, the functions u in a) represent elements of the stalk $(\mathcal{W}^{\omega}_{1-s})_\xi$, and the pairs $(y^s h, y^{s-1}u)$ with $y^{s-1}u$ as in a) and $y^s h$ as in b) represent germs in the stalk $(\mathcal{D}^*_s)_{(\xi,\xi)}$. The theorem has the following consequence:

Corollary 6.3. *For each $\xi \in \mathbb{P}^1_{\mathbb{R}}$ the morphism $C : \mathcal{D}_s \to p_2^{-1}\mathcal{E}_s$ in Theorem 4.13 induces a bijection*

$$\varinjlim_{U} \mathcal{D}_s\big(U \cap (\mathbb{P}^1_{\mathbb{R}} \times \mathfrak{H})\big) \;\cong\; (\mathcal{W}^{\omega}_{1-s})_\xi,$$

where U runs over the open neighborhoods of (ξ, ξ) in $\mathbb{P}^1_{\mathbb{C}} \times \mathbb{P}^1_{\mathbb{C}}$, and

$$(\mathcal{W}^{\omega}_{1-s})_\xi \;\cong\; (\mathcal{D}^*_s)_{(\xi,\xi)}.$$

Proof of Theorem 6.2. We observe that since $U \cap (\mathfrak{H} \times \mathfrak{H})$ intersects the diagonal, the functions h and u in the theorem determine each other near (ξ, ξ) by virtue of parts 4.10) and 4.10) of Proposition 4.10. Hence, the theorem makes sense.

Clearly (b) $\Rightarrow$ (c). We will prove (a) $\Rightarrow$ (b) and (c) $\Rightarrow$ (a). By G-equivariance we can assume that $\xi = 0$.

For (a) $\Rightarrow$ (b) we write $u = y^{1-s}A$ with A real-analytic on a neighborhood of 0 in $\mathbb{C}$. We apply Theorem 6.1. The power series (6.5) converges for $|z_0| \le R$, $|z - z_0| < r$ for some $r, R > 0$. (Choose r to be the minimum of $r(z_0)$ in $|z_0| \le R$ for R small.) The theorem gives us an analytic function B on the region $W = \{(\zeta, z) \in \mathbb{C} \times \mathbb{C} \,:\, |z| < R, |\zeta - z| < r\}$ such that $(B, A) \in \mathcal{D}^*_s(W)$. By the uniqueness clause of Proposition 4.11, the restriction of B to $W \cap (\mathbb{C} \times \mathfrak{H})$ is $y^s h$. Since $(0, 0) \in W$ this gives (b).

For c) $\Rightarrow$ a) we start with a section $(y^s B, y^{1-s} A)$ of $\mathcal{D}_s(U_R \times U_R^+)$ for some $R > 0$, where $U_R \equiv \{z \in \mathbb{C} \; : \; |z| < R\}$ and $U_R^+ = U_R \cap \mathfrak{H}$. Then $A \in \mathcal{A}_{1-s}(U_R^+)$. We apply Theorem 6.1 again, with $z_0 \in U_R^+$. By the uniqueness clause in Proposition (4.11), the function B appearing in (6.7) is the same as the given B in a neighborhood of $\{(z, z) \; : \; z \in U_R^+\} \cup \{(\bar{z}, z) \; : \; z \in U_R^+\}$. Since $B(\cdot, z_0)$ is holomorphic in U_R for each $z_0 \in U_R^+$, the right-hand side of (6.7) is holomorphic for all v, w with $|z_0 + v|, |\bar{z}_0 + w| < R$. (The denominator does not produce any poles since the numerator vanishes whenever the denominator does.) Hence, the first statement of Theorem 6.1 shows that the series (6.5) represents $A(z)$ on the open disk $|z - z_0| < R - |z_0|$. For $|z_0| < \frac{1}{2}R$, this disk contains 0, so A is real-analytic at 0. ■

In Proposition 5.5 we showed that the Poisson transform of a hyperfunction represents an element of $\mathcal{W}_{1-s}^{\omega}$ outside the support of the hyperfunction. With Theorem 6.2 we arrive at the following more complete result.

Theorem 6.4. *Let $I \subset \partial\mathbb{H}$ be open, and let $\alpha \in \mathcal{V}_s^{-\omega}$. Then $\mathsf{P}_s\alpha$ represents an element of $\mathcal{W}_{1-s}^{\omega}(I)$ if and only if $I \cap \mathrm{Supp}(\alpha) = \emptyset$.*

Proof. Proposition 5.5 gives the implication $\Leftarrow$. For the other implication, suppose that $\mathsf{P}_s\alpha$ represents an element of $\mathcal{W}_{1-s}^{\omega}(I)$. Let g_{can} be the canonical representative of α, defined in Sect. 4.1. Then $(g_{\mathrm{can}}, \mathsf{P}_s\alpha) \in \mathcal{D}_s((\mathbb{P}_{\mathbb{C}}^1 \smallsetminus \mathbb{P}_{\mathbb{R}}^1) \times \mathfrak{H})$ by Theorem 4.8 and Definition 4.9. The implication (a) $\Rightarrow$ (b) in Theorem 6.2 gives the analyticity of $y^s g_{\mathrm{can}}$ on a neighborhood of (ξ, ξ) in $\mathbb{P}_{\mathbb{C}}^1 \times \mathbb{P}_{\mathbb{C}}^1$ for each $\xi \in I$. It follows that for $z_0 \in \mathfrak{H}$ sufficiently close to ξ, the function $g_{\mathrm{can}}(\cdot, z_0)$ is holomorphic at ξ. It then follows from the definition of the mixed hybrid model in Sect. 4.1 that $g_{\mathrm{can}}(\cdot, z)$ is holomorphic at ξ for all $z \in \mathfrak{H}$. Thus ξ cannot be in $\mathrm{Supp}(\alpha)$. ■

Theorem 6.2 is a local statement. We end this subsection with a generalization of Proposition 4.14, which shows that the results of Theorem 6.2 have no global counterpart. For convenience we use the disk model.

Proposition 6.5. *Let $A \subset \mathbb{D}$ be an annulus of the form $r_1 < |w| < 1$ with $0 \le r_1 < 1$, and let $V \subset \mathbb{P}_{\mathbb{C}}^1$ be a connected open set that intersects the region $r_1 < |w| < r_1^{-1}$. Then $\mathcal{D}_s(V \times A)$ does not contain nonzero sections of the form (h, u) where $u \in \mathcal{E}_s(A)$ represents an element of $\mathcal{W}_{1-s}^{\omega}$.*

Proof. The proof is similar to that of Proposition 4.14. Suppose that $(h, u) \in \mathcal{D}_s(V \times A)$ where u represents an element of $\mathcal{W}_{1-s}^{\omega}$. By (4.36c) the holomorphic function $\xi \mapsto \int_C [R^{\mathbb{S}}(\xi; \cdot)^s, u]$ is identically zero on some neighborhood $\rho < |\xi| < \rho^{-1}$ of the unit circle. We have the absolutely convergent representation $u = \sum_n b_n Q_{1-s,n}$ on A for a sequence (b_n) of complex numbers. Combining this with the expansion $R^{\mathbb{S}}(\xi; \cdot)^s = \sum_m \frac{(-\xi)^{-m}}{(1-s)_m} P_{1-s,m}$ and (3.18), we obtain

$$\sum_n b_n \frac{(-\xi)^n}{(1-s)_m} = 0$$

for all $\xi \in \mathbb{S}^1$. Hence, all b_n vanish, so u and hence also h are zero. ■

Corollary 6.6. *If V is a neighborhood of $\partial\mathbb{D}$ in $\mathbb{P}^1_{\mathbb{C}}$, then $\mathcal{D}_s\big(V \times (V \cap \mathfrak{H})\big) = \{0\}$.*

Proof. Let $(h,u) \in \mathcal{D}_s\big(V \times (V \cap \mathfrak{H})\big)$. Corollary 6.3 implies that $u \in \mathcal{E}_s(B \cap \mathbb{D})$ represents an element of $\mathcal{W}^{1-s}_\omega$. The neighborhood V contains an annulus of the form $r_1 < |w| < r_1^{-1}$, and Proposition 6.5 shows that $(h,u) = (0,0)$. ■

6.4 The Extended Sheaf of Mixed Eigenfunctions

In Sect. 6.1 we defined an extension $\mathcal{D}^*_s$ of the sheaf of mixed eigenfunctions from $\mathbb{P}^1_{\mathbb{C}} \times \mathfrak{H}$ to $\mathbb{P}^1_{\mathbb{C}} \times \mathbb{P}^1_{\mathbb{C}}$. We now prove an analogue of Theorem 4.13, the main result on the sheaf $\mathcal{D}_s$, for $\mathcal{D}^*_s$.

We denote by $\mathcal{O}$ the sheaf of holomorphic functions on $\mathbb{P}^1_{\mathbb{C}}$, by $p_j : \mathbb{P}^1_{\mathbb{C}} \times \mathbb{P}^1_{\mathbb{C}} \to \mathbb{P}^1_{\mathbb{C}}$ the projection of the jth factor ($j = 1,2$), and put $\Delta^{\pm} = \big\{(z,z)\big\}_{z \in \mathbb{P}^1_{\mathbb{C}}} \cup \big\{(z,\bar z)\big\}_{z \in \mathbb{C}}$. We define $\mathcal{K}^*_s$ to be the subsheaf of $\mathcal{D}^*_s$ whose sections have the form $(B,0)$.

Theorem 6.7. *The sheaf $\mathcal{K}^*_s$ is the kernel of the surjective sheaf morphism $C : \mathcal{D}^*_s \to p_2^{-1}\mathcal{A}_s$ that sends $(B,A) \in \mathcal{D}^*_s(U)$ ($U \subset \mathbb{P}^1_{\mathbb{C}} \times \mathbb{P}^1_{\mathbb{C}}$ open) to A. The restriction of $\mathcal{K}^*_s$ to $\Delta^{\pm}$ vanishes, and its restriction to $\big(\mathbb{P}^1_{\mathbb{C}} \times \mathbb{P}^1_{\mathbb{C}}\big) \smallsetminus \Delta^{\pm}$ is locally isomorphic to $p_1^{-1}\mathcal{O}$.*

This theorem gives us the exact sequence

$$0 \longrightarrow \mathcal{K}^*_s \longrightarrow \mathcal{D}^*_s \xrightarrow{\ C\ } p_2^{-1}\mathcal{A}_{1-s} \longrightarrow 0$$

generalizing the exact sequence in Theorem 4.13.

Proof. By G-equivariance we can work on open $U \subset \mathbb{C} \times \mathbb{C}$. The differential equations (6.2) imply that sections $(B,0)$ of $\mathcal{D}^*_s$ on U have the form $B(\zeta,z) = \varphi(\zeta)$ $(\zeta - z)^s$ $(\zeta - \bar z)^s$ for some function φ. The analyticity of B implies that $\varphi = 0$ near points of $\Delta^{\pm}$, and the holomorphy of B in its first variable implies that φ is holomorphic. Thus $\mathcal{K}^*_s$ is locally isomorphic to $\partial_1^{-1}\mathcal{O}$ outside Δ^* and its stalks at points of Δ^* vanish.

Let (h,u) be a section of $\mathcal{D}^*_s$ over some open $U \subset \mathbb{C} \times \mathbb{C}$. Denote by $D_{(a)}$ and $D_{(b)}$ the expressions in the left-hand sides of (6.2). A computation shows that

$$\big((\zeta - \bar z)\partial_{\bar z} + s\big)\, D_{(a)} - \big((\zeta - z)\partial_z + s\big)\, D_{(b)}$$

is $\frac{1}{2i}\,(\zeta - z)\,(\zeta - \bar z)$ times $(z - \bar z)A_{z\bar z} - (1-s)\ A_z + (1-s)\ A_{\bar z}$. The vanishing of the latter is the differential equation defining $\mathcal{A}_{1-s}$. So A is a section of $\mathcal{A}_{1-s}$ on $p_2 U \smallsetminus \Delta^*$. By analyticity it is in $\mathcal{A}_{1-s}(p_2 U)$. Hence, $C : (B,A) \mapsto A$ determines a sheaf morphism between the restrictions of $\mathcal{D}^*_s$ and $p_2^{-1}\mathcal{A}_{1-s}$ on $\mathbb{C} \times \mathbb{C}$, and by G-equivariance on $\mathbb{P}^1_{\mathbb{C}} \times \mathbb{P}^1_{\mathbb{C}}$.

To prove the surjectivity of C, we constructed for each $(\zeta_0, z_0) \in \mathbb{C}\times\mathbb{C}$ and each $A \in \mathcal{A}_s(U)$ for some neighborhood U of z_0 a section (B, A) of $\mathcal{D}_s^*$ on a possibly smaller neighborhood of (ζ_0, z_0). This suffices by G-equivariance.

For $(\zeta_0, z_0) \in \Delta^*$, this construction is carried out in (6.4). Let $(\zeta_0, z_0) \notin \Delta^*$. The integral in (4.41) suggests that we should consider the differential form

$$\begin{aligned}\omega &= y^s\,\left[\left(R(\zeta;z_1)/R(\zeta;z)\right)^s, y_1^{1-s}\,A(z_1)\right]_{z_1}\\ &= \left(\frac{(\zeta-z)(\zeta-\bar z)}{(\zeta-z_1)(\zeta-\bar z_1)}\right)^s\left(\frac{s(\zeta-\bar z_1)}{2i(\zeta-z_1)}\,A(z_1)\,dz_1+\left(\frac{i}{2}(1-s)A(z_1)+y_1\,A_{\bar z}(z_1)\right)d\bar z_1\right).\end{aligned}$$

Choosing continuous branches of $\left(\frac{(\zeta-z)(\zeta-\bar z)}{(\zeta-z_1)(\zeta-\bar z_1)}\right)^s$ near (ζ_0, z_0), we obtain $B(\zeta, z) = \int_{z_0}^{z}\omega$ such that (B, A) satisfies (6.2) near (ζ_0, z_0), which can be checked by a direct computation, and follows from the proof of Theorem 4.13 if $z_0 \in \mathfrak{H}$. ∎

Remark. We defined $\mathcal{D}_s^*$ in such a way that the restriction of $\mathcal{D}_s^*$ to $\mathbb{P}^1_{\mathbb{C}} \times \mathfrak{H}$ is isomorphic to $\mathcal{D}_s$. Let $c : (\zeta, z) \mapsto (\zeta, \bar z)$. An isomorphism $\mathcal{D}_s^* \to c^{-1}\mathcal{D}_s^*$ is obtained by $\tilde B(\zeta, z) = B(\zeta, \bar z) + yA(\bar z)$, $\tilde A(z) = A(\bar z)$. So the restriction of $\mathcal{D}_s^*$ to $\mathbb{P}^1_{\mathbb{C}} \times \mathfrak{H}^-$ is isomorphic to $c^{-1}\mathcal{D}_s$. New in the theorem is the description of $\mathcal{D}_s^*$ along $\mathbb{P}^1_{\mathbb{C}}\times\mathbb{P}^1_{\mathbb{R}}$. In points (ξ, ξ) with $\xi \in \mathbb{P}^1_{\mathbb{R}}$, the surjectivity of C is the step (a) $\Rightarrow$ (b) in Theorem 6.2.

6.5 Boundary Germs for the Sheaf $\mathcal{D}_s$

In Sect. 6.3 we considered sections of $\mathcal{D}_s$ that extend across $\partial\mathbb{H}$ and established a local relation between these sections and the sheaf $\mathcal{W}_{1-s}^{\omega}$. In this subsection we look instead at the sections of $\mathcal{D}_s$ along the inverse image $p_1^{-1}\mathbb{P}^1_{\mathbb{R}}$, where $p_1 : \mathbb{P}^1_{\mathbb{C}} \times \mathfrak{H} \to \mathbb{P}^1_{\mathbb{C}}$ is the projection on the first component. The proofs will be omitted or sketched briefly.

A first natural thought would be to consider the inductive limit $\varinjlim \mathcal{D}_s\big(U \cap (\mathbb{P}^1_{\mathbb{C}} \times \mathfrak{H})\big)$, where U runs through the collection of all neighborhoods of $\mathbb{P}^1_{\mathbb{R}} \times \mathbb{P}^1_{\mathbb{R}}$ in $\mathbb{P}^1_{\mathbb{C}} \times \mathbb{P}^1_{\mathbb{C}}$, but Corollary 6.6 shows that this space is zero. Instead, we define

$$\mathsf{d}_s = \varinjlim \mathcal{D}_s\big(U \smallsetminus (\mathbb{P}^1_{\mathbb{R}} \times \mathfrak{H})\big), \qquad \mathsf{h}_s = \varinjlim \mathcal{D}_s\big(U\big), \tag{6.16}$$

where the open sets U run over either:

(a) The collection of open neighborhoods of $\mathbb{P}^1_{\mathbb{R}} \times \mathfrak{H}$ in $\mathbb{P}^1_{\mathbb{C}} \times \mathfrak{H}$, or
(b) The larger collection of open neighborhoods of $\mathbb{P}^1_{\mathbb{R}} \times \mathfrak{H}'$ in $\mathbb{P}^1_{\mathbb{C}} \times \mathfrak{H}'$ with $\mathfrak{H}'$ the complement of some compact subset of $\mathfrak{H}$

It turns out that the direct limits in (6.16) are the same for both choices. Clearly d_s contains h_s and the group G acts on both spaces. The canonical model $\mathcal{C}_s$ is a subspace of the space d_s.

In Theorem 4.13 we considered the sheaf morphism $C : \mathcal{D}_s \to p_2^{-1}\mathcal{E}_s$ that sends a pair (h, u) to its second coordinate $u = u(\zeta, z)$, which is locally constant in ζ and a λ_s-eigenfunction in z. This morphism induces a surjective map $C : \mathsf{h}_s \to \mathcal{E}_s$ whose kernel is the space $\mathcal{V}_s^{\omega,\mathrm{rig}}$ introduced in Sect. 4.1. It also induces a map (still called C) from the larger space d_s to $\mathcal{E}_s \oplus \mathcal{E}_s$ by sending (h, u) to the pair (u_+, u_-), where $u_\pm(\cdot) = u(\zeta_\pm, \cdot)$ for any $\zeta_\pm \in \mathfrak{H}^\pm$. This map is again surjective and its kernel is the space $\mathbf{H}_s^{\mathrm{rig}}$ studied in Sect. 4.1. Moreover, the results of that subsection show that the kernels of these two maps C are related by the exact sequence

$$0 \longrightarrow \mathcal{V}_s^{\omega,\mathrm{rig}} \longrightarrow \mathbf{H}_s^{\mathrm{rig}} \xrightarrow{\mathsf{P}_s} \mathcal{E}_s \longrightarrow 0,$$

where the Poisson map P_s is given explicitly by

$$\mathsf{P}_s h(z, z_1) = \frac{1}{\pi}\left(\int_{C_+} - \int_{C_-}\right)\left(\frac{R(\zeta; z_1)}{R(\zeta; z)}\right)^s h(\zeta, z_1)\, R(\zeta; z)\, \mathrm{d}\zeta \qquad (z,\ z_1 \in \mathfrak{H}),$$

(4.15). Here C_+ (resp. C_-) is a closed path in $\mathfrak{H}$ (resp. $\mathfrak{H}^-$) encircling z and z_1 (resp. $\bar{z}$ and $\bar{z}_1$), and the right-hand side is independent of z_1. Now consider an element of h_s represented by the pair $(h, u) \in \mathcal{D}_s\big(U \smallsetminus (\mathbb{P}^1_{\mathbb{R}} \times \mathfrak{H})\big)$ for some open neighborhood U of $\mathbb{P}^1_{\mathbb{R}} \times \mathfrak{H}$ in $\mathbb{P}^1_{\mathbb{C}} \times \mathfrak{H}$ and define $\mathsf{P}_s h(z, z_1)$ by the same formula, where C_+ and C_- we now required to lie in the neighborhood $\{\zeta \in \mathbb{P}^1_{\mathbb{C}} \mid (\zeta, z_1) \in U\}$ of $\mathbb{P}^1_{\mathbb{R}}$ and to be homotopic to $\mathbb{P}^1_{\mathbb{R}}$ in this neighborhood. The right-hand side is still independent of the choice of contours $C_\pm$ and is also independent of the choice of representative (h, u) of $[(h, u)] \in \mathsf{d}_s$, but it is no longer independent of z_1. Instead, we have that the function $\mathsf{P}_s h(\cdot, z_1)$ belongs to $\mathcal{E}_s$ for each fixed $z_1 \in \mathfrak{H}$ and that its dependence on z_1 is governed by

$$\mathrm{d}_{z_1}\big(\mathsf{P}_s(h, u)(z, z_1)\big) = \big[p_s(\cdot, z),\ u_+ - u_-\big] \tag{6.17}$$

with the Green's form as in (3.13) and the point-pair invariant $p_s(\cdot, \cdot)$ as in (3.6).

We therefore define a space $\mathcal{E}_s^+$ consisting of pairs (f, v) where v belongs to $\mathcal{E}_s$ and $f : \mathfrak{H} \times \mathfrak{H} \to \mathbb{C}$ satisfies

$$f(\cdot, z_1) \in \mathcal{E}_s \text{ for each } z_1 \in \mathfrak{H}, \tag{6.18a}$$

$$\mathrm{d}\big(f(z, \cdot)\big) = [p_s(\cdot, z), v] \text{ on } \mathfrak{H} \text{ for each } z \in \mathfrak{H}. \tag{6.18b}$$

The group G acts on this space by composition (diagonally in the case of f). By the discussion above, we can define an equivariant and surjective map $\mathsf{P}_s^+ : \mathsf{h}_s \to \mathcal{E}_s^+$ with kernel h_s by $[(h, u)] \mapsto \big(\mathsf{P}_s h, u_+ - u_-\big)$. Finally, the space $\mathcal{E}_s^+$ is mapped to $\mathcal{E}_s$ by $(f, v) \mapsto v$ with kernel $\mathcal{E}_s$ (because $f(\cdot, z_1)$ is constant if $v = 0$). (In fact, the space $\mathcal{E}_s^+$ is isomorphic to $\mathcal{E}_s \times \mathcal{E}_s$ as a vector space, though not as a G-module, by the map sending (f, v) to $\big(f(\cdot, i), v\big)$.) Putting all these maps together, we can summarize the interaction of the morphisms C and P_s by the following commutative diagram with exact rows and columns:

$$
\begin{array}{ccccccccc}
 & & 0 & & 0 & & 0 & & \\
 & & \downarrow & & \downarrow & & \downarrow & & \\
0 & \longrightarrow & \mathcal{V}_s^{\omega,\mathrm{rig}} & \longrightarrow & \mathsf{h}_s & \xrightarrow{C} & \mathcal{E}_s & \longrightarrow & 0 \\
 & & \downarrow & & \downarrow & & \downarrow{\scriptstyle \mathrm{diag}} & & \\
0 & \longrightarrow & \mathbf{H}_s^{\mathrm{rig}} & \longrightarrow & \mathsf{d}_s & \xrightarrow{C} & \mathcal{E}_s \oplus \mathcal{E}_s & \longrightarrow & 0 \\
 & & \downarrow{\scriptstyle \mathsf{P}_s} & & \downarrow{\scriptstyle \mathsf{P}_s^+} & & \downarrow{\scriptstyle (1,-1)} & & \\
0 & \longrightarrow & \mathcal{E}_s & \longrightarrow & \mathcal{E}_s^+ & \xrightarrow{\mathrm{pr}_2} & \mathcal{E}_s & \longrightarrow & 0 \\
 & & \downarrow & & \downarrow & & \downarrow & & \\
 & & 0 & & 0 & & 0 & &
\end{array}
$$

7 Boundary Splitting of Eigenfunctions

In the introduction we mentioned that eigenfunctions often have the local form $y^s \times (\text{analytic}) + y^{1-s} \times (\text{analytic})$ near points of $\mathbb{R}$. Here we consider this phenomenon more systematically in both the analytic context (Sect. 7.1) and the differentiable context (Sect. 7.2). This will lead in particular to a description of both $\mathcal{E}_s^\omega = \mathsf{P}_s(\mathcal{V}_s^\omega)$ and $\mathcal{E}_s^\infty = \mathsf{P}_s(\mathcal{V}_s^\infty)$ in terms of boundary behavior.

As stated in the introduction, results concerning the boundary behavior of elements of $\mathcal{E}_s$ are known (also for more general groups; see, eg., [1, 7]). However, our approach is more elementary and also includes several formulas that do not seem to be in the literature and that are useful for certain applications (such as those in [2]).

7.1 Analytic Case

In Proposition 5.3 we showed that the space $\mathcal{F}_s$ of boundary germs is the direct sum of $\mathcal{E}_s$ (the functions that extend to the interior) and $\mathcal{W}_s^\omega$ (the functions that extend across the boundary). We now look at the relation of these spaces with $\mathcal{E}_s^\omega$, the image in $\mathcal{E}_s$ of $\mathcal{V}_s^\omega$ under the Poisson transformation.

If $s \neq \frac{1}{2}$, we denote by $\mathcal{F}_s^\omega$ the direct sum of $\mathcal{W}_s^\omega$ and $\mathcal{W}_{1-s}^\omega$. (That this sum is direct is obvious since for $s \neq \frac{1}{2}$, an eigenfunction u cannot have the behavior $y^s \times (\text{analytic})$ and at the same time $y^{1-s} \times (\text{analytic})$ near points of $\mathbb{R}$.) For $s = \frac{1}{2}$, we will define $\mathcal{F}_{1/2}^\omega$ as a suitable limit of these spaces in the following sense. If $s \neq \frac{1}{2}$, an element of $\mathcal{F}_s^\omega$ is locally (near $x_0 \in \mathbb{R}$) represented by a linear combination of

y^s and y^{1-s} with coefficients that are analytic in a neighborhood of x_0. Replacing y^s and y^{1-s} by $\frac{1}{2}\left(y^s + y^{1-s}\right)$ and $\frac{1}{2s-1}\left(y^s - y^{1-s}\right)$, we see that an element of $\mathcal{F}_{1/2}^{\omega}$ should (locally) have the form $A(z)\, y^{1/2} \log y + B(z)\, y^{1/2}$ with A and B analytic at x_0. We, therefore, define $\mathcal{F}_{1/2}^{\omega}$ (now using disk model coordinates to avoid special explanations at ∞) as the space of germs in $\mathcal{F}_{1/2}$ represented by

$$f(w) \;=\; (1-|w|^2)^{1/2}\left(A(w)\log(1-|w|^2) + B(w)\right), \tag{7.1}$$

with A and B real-analytic on a neighborhood of $\mathbb{S}^1$ in $\mathbb{C}$. We have a G-equivariant exact sequence

$$0 \longrightarrow \mathcal{W}_{1/2}^{\omega} \longrightarrow \mathcal{F}_{1/2}^{\omega} \stackrel{\tau}{\longrightarrow} \mathcal{W}_{1/2}^{\omega} \longrightarrow 0 \tag{7.2}$$

where τ sends f in (7.1) to A. The surjectivity of τ is a consequence of the following proposition, which we will prove below. This proposition shows that for all s with $0 < \operatorname{Re} s < 1$, the space $\mathcal{F}_s^{\omega}$ is isomorphic as a G-module to the sum of two copies of $\mathcal{V}_s^{\omega}$.

Proposition 7.1. *The exact sequence* (7.2) *splits G-equivariantly.*

The splittings $\mathcal{F}_s = \mathcal{E}_s \oplus \mathcal{W}_s^{\omega} = \mathcal{E}_s \oplus \mathcal{W}_{1-s}^{\omega}$ show that nonzero elements of $\mathcal{E}_s$ cannot belong to $\mathcal{W}_s^{\omega}$ or $\mathcal{W}_{1-s}^{\omega}$. The following theorem shows that they *can* be in $\mathcal{F}_s^{\omega}$, and that this happens if and only if they belong to $\mathcal{E}_s^{\omega}$.

Theorem 7.2. *Let* $0 < \operatorname{Re} s < 1$. *Then*

$$\mathcal{E}_s^{\omega} = \mathcal{E}_s \cap \mathcal{F}_s^{\omega},$$

and $\mathcal{F}_s^{\omega} \;=\; \mathcal{E}_s^{\omega} \oplus \mathcal{W}_s^{\omega} = \mathcal{E}_s^{\omega} \oplus \mathcal{W}_{1-s}^{\omega}$.

So for $s \neq \frac{1}{2}$, the space $\mathcal{F}_s^{\omega}$ is the direct sum of each two of the three isomorphic subspaces $\mathcal{E}_s^{\omega}$, $\mathcal{W}_s^{\omega}$, and $\mathcal{W}_{1-s}^{\omega}$. For $s = \frac{1}{2}$, two of these subspaces coincide.

We discuss the cases $s \neq \frac{1}{2}$ and $s = \frac{1}{2}$ separately.

Proposition 7.3. *Let* $s \neq \frac{1}{2}$. *For each* $\varphi \in \mathcal{V}_s^{\omega}$, *we have*

$$P_s\varphi \;=\; c(s)\, P_s^{\dagger}\varphi + c(1-s)\, P_{1-s}^{\dagger} I_s\varphi, \tag{7.3}$$

where, with $b(s)$ *as in* (5.20), *the factor* $c(s)$ *is given by*

$$c(s) \;=\; \frac{\tan \pi s}{\pi}\, b(s) \;=\; \frac{1}{\sqrt{\pi}}\, \frac{\Gamma(\frac{1}{2}-s)}{\Gamma(1-s)}. \tag{7.4}$$

Proof. Since φ is given by a Fourier expansion which converges absolutely uniformly on the paths of integration in the transformation occurring in (7.3), it is sufficient to prove this relation in the spacial case $\mathbf{e}_{s,n}$ ($n \in \mathbb{Z}$). We have

$\mathsf{P}_s\mathbf{e}_{s,n} = (-1)^n \frac{\Gamma(s)}{\Gamma(s+n)} P_{s,n}$ and $\mathsf{P}_s^{\dagger}\mathbf{e}_{s,n} = (-1)^n \frac{\Gamma(s+\frac{1}{2})}{\sqrt{\pi}\,\Gamma(s+n)} Q_{s,n}$. See Sect. A.2. The relations (A.14) and (2.30e) give the lemma for $\varphi = \mathbf{e}_{s,n}$ for all $n \in \mathbb{Z}$. ■

Remark. One can also give a direct (but more complicated) proof of (7.3) for arbitrary $\varphi \in \mathcal{V}_s^{\omega}$, without using the basis $\{\mathbf{e}_{s,n}\}$, by writing all integral transforms explicitly and moving the contours suitably.

The proof of Theorem 7.2 (for $s \neq \frac{1}{2}$) follows from Proposition 7.3. The inclusion $\mathcal{E}_s^{\omega} \subset \mathcal{F}_s^{\omega}$ is a consequence of the more precise formula (7.3). For the reverse inclusion we write an arbitrary $u \in \mathcal{F}_s^{\omega}$ in the form $c(s)\,\mathsf{P}_s^{\dagger}\varphi + v$ with $v \in \mathcal{W}_{1-s}^{\omega}$ and $\varphi \in \mathcal{V}_s^{\omega}$. If $u \in \mathcal{E}_s$, then $u - \mathsf{P}_s\varphi = v - c(1-s)\mathsf{P}_{s-1}^{\dagger} I_s\varphi \in \mathcal{E}_s \cap \mathcal{W}_{1-s}^{\omega} = \{0\}$, so $u = \mathsf{P}_s\varphi \in \mathcal{E}_s^{\omega}$. This completes the proof.

We can summarize this discussion and its relation with the Poisson transformation in the following commutative diagram of G-modules and canonical G-equivariant morphisms

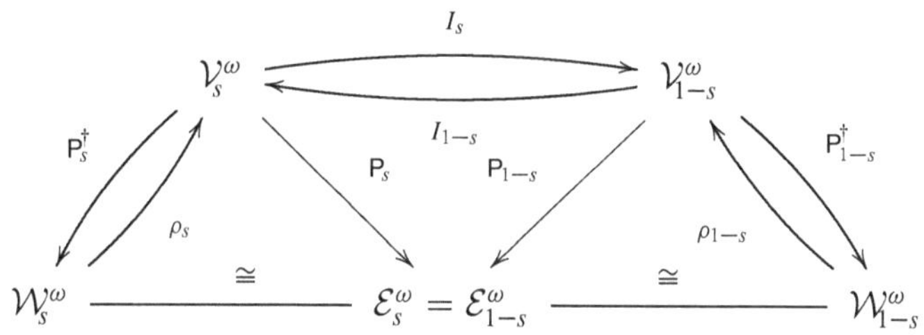

together with the fundamental examples (and essential ingredient in the proof):

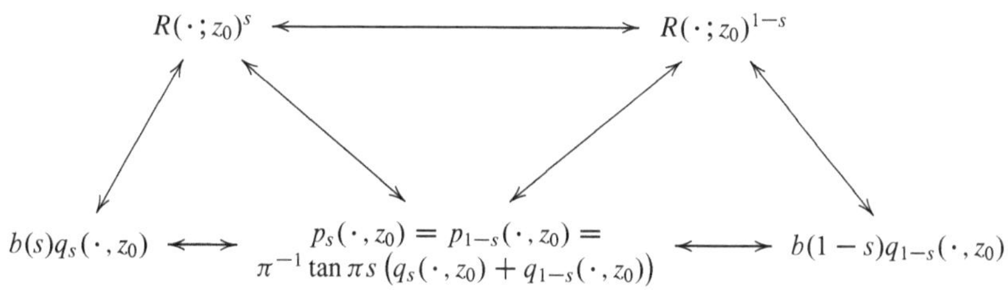

We now turn to the case $s = \frac{1}{2}$. We have to prove Proposition 7.1 and Theorem 7.2 in this case.

To construct a splitting $\sigma : \mathcal{W}_{1/2}^{\omega} \to \mathcal{F}_{1/2}^{\omega}$ of the exact sequence (7.2), we put $\sigma Q_{1/2,n} = -\frac{\pi^2}{2} P_{1/2,n} \in \mathcal{E}_s$ for $n \in \mathbb{Z}$. Since $P_{1/2,n} \in \mathsf{P}_{1/2}\mathcal{V}_{1/2}^{\omega}$, we have $\sigma Q_{n,1/2} \in \mathcal{E}_s^{\omega}$. Further, $\tau\sigma Q_{1/2,n} = Q_{1/2,n}$ by (A.13) and (A.15). The $Q_{1/2,n} \in \mathcal{W}_{1/2}^{\omega}$ with $n \in \mathbb{Z}$ generate a dense linear subspace of $\mathcal{W}_{1/2}^{\omega}$ for the topology of $\mathcal{V}_{1/2}^{\omega}$ transported to $\mathcal{W}_{1/2}^{\omega}$ by $\mathsf{P}_{1/2}^{\dagger} : \mathcal{V}_{1/2}^{\omega} \to \mathcal{W}_{1/2}^{\omega}$. Hence, there is a continuous linear extension

$\sigma : \mathcal{W}^\omega_{1/2} \to \mathcal{E}^\omega_{1/2}$. The generators $\mathbf{E}^+$ and $\mathbf{E}^-$ of the Lie algebra of G act in the same way on the system $\left(Q_{s,n}\right)_n$ as on the system $\left(P_{s,n}\right)_n$. (See Sect. A.5 and use case **G** in Table A.1 of Sect. A.2 and case **a** in Table A.2 of Sect. A.3.) So σ is an infinitesimal G-morphism and since G is connected, a G-morphism. The splitting $\sigma : \mathcal{W}^\omega_{1/2} \to \mathcal{F}^\omega_{1/2}$ also gives the surjectivity of τ and hence the exactness of the sequence (7.2).

Since $\sigma Q_{1/2,n}$ belongs to $\mathcal{E}^\omega_{1/2}$, we have $\sigma(\mathcal{W}^\omega_{1/2}) \subset \mathcal{E}^\omega_{1/2}$. Since $\mathcal{E}^\omega_{1/2}$ is an irreducible G-module, this inclusion is an equality. This gives $\mathcal{F}^\omega_{1/2} = \mathcal{E}^\omega_{1/2} \oplus \mathcal{W}^\omega_{1/2}$ and $\mathcal{E}^\omega_{1/2} \subset \mathcal{E}_{1/2} \cap \mathcal{F}^\omega_{1/2}$. The reverse inclusion then follows by the same argument as for $s \neq \frac{1}{2}$. ■

Remark. The case $s = \frac{1}{2}$ could also have been done with explicit elements. For each s with $0 < \operatorname{Re} s < 1$ and each $n \in \mathbb{Z}$, the subspace $\mathcal{F}^\omega_{s,n}$ of $\mathcal{F}^\omega_s$ in which the elements $\left[\begin{smallmatrix} \cos\theta & \sin\theta \\ -\sin\theta & \cos\theta \end{smallmatrix}\right]$ act as multiplication by $e^{2in\theta}$ has dimension 2. In the family $s \mapsto \mathcal{F}^\omega_{s,n}$, there are three families of nonzero eigenfunctions: $s \mapsto P_{s,n} \in \mathcal{E}^\omega_s$, $s \mapsto Q_{s,n} \in \mathcal{W}^\omega_s$, and $s \mapsto Q_{1-s,n} \in \mathcal{W}^\omega_{1-s}$. For $s \neq \frac{1}{2}$, each of these functions can be expressed as a linear combination of the other two, as given by (A.14), which is at the basis of our proof of Proposition 7.3. At $s = \frac{1}{2}$, the elements $Q_{s,n}$ and $Q_{1-s,n}$ coincide. This is reflected in the singularities at $s = \frac{1}{2}$ in the relation (A.14). The families $s \mapsto P_{s,n}$ and $s \mapsto Q_{s,n}$ provide a basis of $\mathcal{F}^\omega_s$ for all s, corresponding to the decomposition $\mathcal{F}^\omega_s = \mathcal{E}^\omega_s \oplus \mathcal{W}^\omega_s$. Relation (A.14) implies

$$P_{1/2,n} = \frac{-2}{\pi^2} \frac{d}{ds} Q_{s,n}\Big|_{s=1/2},$$

which explains the logarithmic behavior at the boundary.

7.2 Differentiable Case

In the previous subsection, we described the boundary behavior of elements of $\mathcal{E}^\omega_s = \mathsf{P}_s \mathcal{V}^\omega_s$ in terms of convergent expansions. In the differentiable case, the spaces $\mathcal{W}^p_s$ consist of boundary jets, not of boundary germs. So a statement like that in Theorem 7.2 seems impossible. Nevertheless, we have the following generalization of Proposition 7.3:

Proposition 7.4. (i) *Let $p \in \mathbb{N}$, $p \geq 2$, and $s \neq \frac{1}{2}$. For each $\varphi \in \mathcal{V}^p_s$ there are $b \in \mathcal{G}^p_s$ representing $c(s)\, \mathsf{P}^\dagger_s \varphi \in \mathcal{W}^p_s$ and $a \in \mathcal{G}^{p-1}_{1-s}$ representing $c(1-s)\, \mathsf{P}^\dagger_{1-s} I_s \varphi \in \mathcal{W}^{p-1}_{1-s}$ such that*

$$\mathsf{P}_s\varphi(w) = b(w) + a(w) + \mathrm{O}\big((1-|w|^2)^{p-s}\big) \qquad (|w| \uparrow 1). \qquad (7.5)$$

(ii) *Let $s \neq \frac{1}{2}$. For each $\varphi \in \mathcal{V}^\infty_s$ there are $b \in \mathcal{G}^\infty_s$ and $a \in \mathcal{G}^\infty_{1-s}$ representing $c(s)\, \mathsf{P}^\dagger_s \varphi$ and $c(1-s)\, \mathsf{P}^\dagger_{1-s} \varphi$, respectively, such that for each $N \in \mathbb{N}$*

$$\mathsf{P}_s\varphi(w) \sim b(w) + a(w) + \mathrm{o}\big((1-|w|^2)^N\big) \qquad (|w| \uparrow 1). \qquad (7.6)$$

Proof. The proof of Proposition 7.3 used the fact that the $\mathbf{e}_{s,n}$ generate a dense subspace of $\mathcal{V}_s^\omega$ and that the values of Poisson transforms and transverse Poisson transforms are continuous with respect to this topology. That reasoning seems hard to generalize when we work with boundary jets. Instead, we use the explicit Lemma 5.12.

A given $\varphi \in \mathcal{V}_s^p$ can be written as a sum of elements in $\mathcal{V}_s^p$ each with support in a small interval in $\partial\mathbb{H}$. With the G-action, this reduces the situation to be considered to $\varphi \in C_c^p(I)$ where I is a finite interval in $\mathbb{R}$. Proposition 5.5 shows that $\mathsf{P}_s\varphi$ represents an element of $\mathcal{W}_{1-s}^\omega(\mathbb{P}^1_{\mathbb{R}} \smallsetminus I)$. So we can restrict our attention to $\mathsf{P}_s\varphi(z)$ with z near I and work in the line model.

We take α and β as in Lemma 5.12 with $N = p$. Then

$$\begin{aligned}\mathsf{P}_s\varphi(z) &= \pi^{-1}\, y^{1-s} \int_{-\infty}^{\infty} (t^2+y^2)^{s-1}\, \varphi(t+x)\, dt \\ &= \frac{1}{\pi}\, y^s \int_{-\infty}^{\infty} (t^2+1)^{s-1}\, \varphi(x+yt)\, dt \;=\; y^s A(z) + y^{1-s} B(z),\end{aligned}$$

with

$$B(z) = \pi^{-1} \int_{-\infty}^{\infty} \beta(t)\, \varphi(x+yt)\, dt, \qquad A(z) = \pi^{-1} y^{2s-1} \int_{-\infty}^{\infty} \alpha^{(p)}(t)\, \varphi(x+yt)\, dt.$$

We consider $B(z)$ and $A(z)$ for $x \in I$ and $0 < y \le 1$. The decay of β implies that

$$B(z) = \frac{1}{\pi} \sum_{n=0}^{p} \frac{\varphi^{(n)}(x)}{n!} y^n \int_{-\infty}^{\infty} t^n\, \beta(t)\, dt + \mathrm{o}(y^p).$$

In (5.40) we have computed the integrals. We arrive at

$$B(z) = c(s) \sum_{k=0}^{[p/2]} \frac{(-1/4)^k\, \Gamma(s+\frac{1}{2})}{k!\, \Gamma(k+s+\frac{1}{2})}\, \varphi^{(2k)}(x)\, y^{2k} + \mathrm{o}(y^p). \tag{7.7}$$

A comparison with (5.33) shows that $y^s\, B(z)$ has the asymptotic behavior near I of representatives of $c(s)\, \mathsf{P}_s^\dagger \varphi$.

In the second term, we apply p-fold integration by parts:

$$A(z) = (-1)^p \pi^{-1} y^{2s-1+p} \int_{-\infty}^{\infty} \alpha(t)\, \varphi^{(p)}(x+yt)\, dt.$$

For fixed φ, this expression is a holomorphic function of s on the region $\mathrm{Re}\, s > 0$. In the computation we shall work with $\mathrm{Re}\, s$ large.

The function $h \mapsto (1+h)^{s-1}$ has a Taylor expansion at $h = 0$ of any order R, with a remainder term $\mathrm{O}(h^{R+1})$ that is uniform for $h \ge 0$. This implies that $\alpha(t)$

has an expansion of the form $\alpha(t) = \sum_{n=0}^{R} b_n |t|^{p+2s-2-2n} + \mathrm{O}(|t|^{p+2s-2R-4})$, uniformly for $t \in \mathbb{R} \smallsetminus \{0\}$. We take $R = [p/2]$ and use the relation $\partial_t^p \alpha(t) = (1+t^2)^{s-1} - \beta(t)$ and the decay of $\beta(t)$ to conclude that

$$\alpha(t) = \sum_{0 \le n \le p/2} \binom{s-1}{n} \frac{(\operatorname{sign} t)^p |t|^{2s-2n+p-2}}{(2s-2n-1)_p} + \mathrm{O}\big(|t|^{2s-3}\big). \tag{7.8}$$

We compute this with $\operatorname{Re} s > 1$. The error term contributes to $A(z)$:

$$y^{2s-1+p} \int_{t=-\infty}^{\infty} \mathrm{O}\big(|t|^{2s-3}\big)\, \varphi^{(p)}(x+yt)\, \mathrm{d}t = \mathrm{O}\big(y^{p+1}\big). \tag{7.9}$$

(We have replaced t by t/y in the integral.) The term of order n contributes

$$\begin{aligned} &\frac{(-1)^p y^{2n}}{\pi} \binom{s-1}{n} \frac{y^{-2s-p+2n+1}}{(2s-2n-1)_p} \int_{-\infty}^{\infty} (\operatorname{sign} t)^p |t|^{2s+p-2n-2} \varphi^{(p)}(x+t)\, \mathrm{d}t \\ &= \frac{(-1)^p y^{2n}\, \Gamma(s)\, \Gamma(2s-2n-1)}{\pi\, n!\, \Gamma(s-n)\, \Gamma(2s-2n-1+p)} (-1)^{p-2n} (2s-1)_{p-2n} \\ &\quad \cdot \int_{-\infty}^{\infty} |t|^{2s-2} \varphi^{(2n)}(x+t)\, \mathrm{d}t \qquad \text{(partial integration } p-2n \text{ times).} \end{aligned}$$

In (2.30b) we see that the holomorphic function $\int_{-\infty}^{\infty} |t|^{2s-2} \varphi^{(2n)}(x+t)\, \mathrm{d}t$ continued to the original value of s gives us $b(s-\frac{1}{2})\, (I_s\varphi)^{(2n)}(x)$, provided $2n < p$. We have $I_s \mathcal{V}_s^p \subset \mathcal{V}_{1-s}^{p-1}$, but not necessarily $I_s\varphi \in \mathcal{V}_s^p$. For even p, we move the contribution $\mathrm{O}(y^p)$ to the error term. The terms of order $n < p/2$ give

$$\begin{aligned} &\frac{y^{2n}\, \Gamma(s)\, \Gamma(2s-2n-1)}{\pi\, n! \Gamma(s-n)\, \Gamma(2s-1)} \frac{\sqrt{\pi}\, \Gamma(s-\frac{1}{2})}{\Gamma(s)} (I_s\varphi)^{(2n)}(x) \\ &= \frac{\tan \pi(1-s)}{\sqrt{\pi}} \frac{\Gamma(1-s)}{\Gamma(\frac{3}{2}-s)} \frac{(-1/4)^k\, \Gamma(\frac{3}{2}-s)}{n!\, \Gamma(\frac{3}{2}-s+n)} (I_s\varphi)^{(2n)}(x)\, y^{2k}. \end{aligned}$$

Thus we arrive at

$$A(z) = c(1-s) \sum_{0 \le n < p/2} \frac{(-1/4)^n\, \Gamma(\frac{3}{2}-s)}{n!\, \Gamma(\frac{3}{2}-s+n)} (I_s\varphi)^{(2n)}(x) y^{2k} + \mathrm{O}\big(y^{2[\frac{p+1}{2}]}\big). \tag{7.10}$$

Again we have arrived at the expansion a representative of $\mathcal{W}_{1-s}^{p-1}$ should have according to (5.33). This completes the proof of part (4.10).

In view of Definition 5.9, the estimate (7.5) holds for all representatives $b \in \mathcal{G}_s^p$ and $a \in \mathcal{G}_{1-s}^p$ of $c(s)\, \mathsf{P}_s^\dagger \varphi$, respectively $c(1-s)\, \mathsf{P}_{1-s}^\dagger \varphi$. In particular, for $\varphi \in \mathcal{V}_s^\infty$, this estimate holds for each $p \in \mathbb{N}$, $p \ge 2$, for representatives $b_\infty \in \mathcal{G}_s^\infty$ of $c(s)\, \mathsf{P}_s^\dagger \varphi$ and $a_\infty \in \mathcal{G}_{1-s}^\infty$ of $c(1-s)\, \mathsf{P}_{1-s}^\dagger \varphi$. This implies part (4.10) of the proposition. ∎

Appendix: Examples and Explicit Formulas

We end by giving a collection of definitions and formulas that were needed in the main body of this chapter or that illustrate its results. In particular, we describe a number of examples of eigenfunctions of the Laplace operator (in A.1), of Poisson transforms (in A.2), of transverse Poisson transforms (in A.3), and of explicit potentials of the Green's form $\{u, v\}$ for various special choices of u and v (in A.4), as well as some formulas for the action of the Lie algebra of G (in A.5).

A.1 Special Functions and Equivariant Elements of $\mathcal{E}_s$

Let $H \subset G$ be one of the subgroups $N = \{n(x) \,:\, x \in \mathbb{R}\}$, $A = \{a(y) \,:\, y > 0\}$ or $K = \{k(\theta) \,:\, \theta \in \mathbb{R}/\mathbb{Z}\}$ with

$$n(x) = \begin{bmatrix} 1 & x \\ 0 & 1 \end{bmatrix}, \qquad a(y) = \begin{bmatrix} \sqrt{y} & 0 \\ 0 & 1/\sqrt{y} \end{bmatrix}, \qquad k(\theta) = \begin{bmatrix} \cos\theta & \sin\theta \\ -\sin\theta & \cos\theta \end{bmatrix}. \tag{A.1}$$

For each character χ of H, we determine the at most two-dimensional subspace $\mathcal{E}_{s,\chi}^{H}$ of $\mathcal{E}_s$ transforming according to this character.

A.1.1 Equivariant Eigenfunctions for the Unipotent Group N

The characters of N are $\chi_\alpha \,:\, n(x) \mapsto \mathrm{e}^{\mathrm{i}\alpha x}$ with $\alpha \in \mathbb{R}$. If $u \in \mathcal{E}_{s,\alpha}^{N}$ (we write $\mathcal{E}_{s,\alpha}^{N}$ instead of $\mathcal{E}_{s,\chi_\alpha}^{N}$), then $u(z) = \mathrm{e}^{\mathrm{i}\alpha x} f(y)$, where f satisfies the differential equation

$$y^2 f''(y) = (s^2 - s + \alpha^2 y^2) f(y). \tag{A.2}$$

This can also be applied to $\mathcal{E}_{s,\alpha}^{N}(U)$ for any connected N-invariant subset U of $\mathfrak{H}$. For the trivial character, i.e., $\alpha = 0$, this leads to the basis $z \mapsto y^s$, $z \mapsto y^{1-s}$ of $\mathcal{E}_{s,0}^{N}$ if $s \neq \frac{1}{2}$, and $z \mapsto y^{1/2}$, $z \mapsto y^{1/2} \log y$ if $s = \frac{1}{2}$. For nonzero α, we have

$$\begin{aligned} k_{s,\alpha}(z) &= \sqrt{y}\, K_{s-1/2}(|\alpha| y)\, \mathrm{e}^{\mathrm{i}\alpha x} \\ &= \frac{2^{s-\frac{3}{2}}\, \Gamma(s)}{\sqrt{\pi}\, |\alpha|^{s-\frac{1}{2}}}\, \mathrm{e}^{\mathrm{i}\alpha x} \int_{-\infty}^{\infty} \mathrm{e}^{\mathrm{i}\alpha t} \frac{y^s\, \mathrm{d}t}{(y^2 + t^2)^s}, \end{aligned} \tag{A.3a}$$

$$i_{s,\alpha}(z) = \frac{\Gamma\left(s + \frac{1}{2}\right)}{|\alpha/2|^{s-\frac{1}{2}}} \sqrt{y}\, I_{s-1/2}(|\alpha| y)\, \mathrm{e}^{\mathrm{i}\alpha x}, \tag{A.3b}$$

with the modified Bessel functions

$$I_u(t) = \sum_{n=0}^{\infty} \frac{(t/2)^{u+2n}}{n!\,\Gamma(u+1+n)}, \qquad K_u(t) = \frac{\pi}{2}\,\frac{I_{-u}(t) - I_u(t)}{\sin \pi u}. \tag{A.4}$$

The element $i_{\alpha,s}$ represents a boundary germ in $\mathcal{W}_s^{\omega}(\mathbb{R})$. The normalization of $i_{s,\alpha}$ is such that the restriction $\rho_s i_{\alpha,s}(x) = \mathrm{e}^{\mathrm{i}\alpha x}$ in the line model.

The elements $k_{s,\alpha}$ and $i_{s,\alpha}$ form a basis of $\mathcal{E}_{s,\alpha}^N$ for all s with $0 < \mathrm{Re}\, s < 1$. For $s \neq \frac{1}{2}$ another basis is $i_{s,\alpha}$ and $i_{1-s,\alpha}$. The element $k_{s,\alpha}$ is invariant under $s \mapsto 1-s$, and

$$k_{s,\alpha} = \frac{\Gamma\left(\frac{1}{2}-s\right)}{|\alpha|^{1/2-s}2^{s+1/2}}\, i_{s,\alpha} + \frac{\Gamma\left(s-\frac{1}{2}\right)}{|\alpha|^{s-1/2}2^{3/2-s}}\, i_{1-s,\alpha} \tag{A.5}$$

gives (for $s \neq \frac{1}{2}$) a local boundary splitting as an element of $\mathcal{W}_s^{\omega}(\mathbb{R}) \oplus \mathcal{W}_{1-s}^{\omega}(\mathbb{R})$.

For the trivial character, $k_{s,\alpha}$ may be replaced by

$$\ell_s(z) = \frac{y^s - y^{1-s}}{2s-1} \quad \text{for } s \neq \frac{1}{2}, \qquad \ell_{1/2}(z) = y^{1/2}\log y. \tag{A.6}$$

A.1.2 Equivariant Eigenfunctions for the Compact Group K

The characters of K are $k(\theta) \mapsto \mathrm{e}^{\mathrm{i}n\theta}$ with $n \in \mathbb{Z}$ and $k(\theta)$ as in (A.1). If $u(r\mathrm{e}^{\mathrm{i}\theta}) = f(r)\mathrm{e}^{\mathrm{i}n\theta}$ is in $\mathcal{E}_{s,n}^K(U)$, with a K-invariant subset $U \subset \mathfrak{H}$, then f satisfies the differential equation

$$-\frac{1}{4}\left(1-r^2\right)^2\left(f''(r) + r^{-1}f'(r) - n^2r^{-2}f(r)\right) = s(1-s)f(r). \tag{A.7}$$

For general annuli in $\mathbb{H}$, the solution space has dimension 2, with basis

$$\begin{aligned} P_{s,n}(r\mathrm{e}^{\mathrm{i}\theta}) = P_{1-s,n}(r\mathrm{e}^{\mathrm{i}\theta}) &= P_{s-1}^n\left(\frac{1+r^2}{1-r^2}\right)\mathrm{e}^{\mathrm{i}n\theta}, \\ Q_{s,n}(r\mathrm{e}^{\mathrm{i}\theta}) = Q_{s-1}^n\left(\frac{1+r^2}{1-r^2}\right)\mathrm{e}^{\mathrm{i}n\theta}, \end{aligned} \tag{A.8}$$

with the Legendre functions

$$\begin{aligned} P_{s-1}^m\left(\frac{1+r^2}{1-r^2}\right) &= \frac{\Gamma(s+m)}{|m|!\,\Gamma(s-|m|)}\, r^{|m|} F\left(1-s, s; 1+|m|; \frac{r^2}{r^2-1}\right) \\ &= \frac{\Gamma(s+m)}{|m|!\,\Gamma(s-|m|)}\, r^{|m|}\left(1-r^2\right)^s F\left(s, s+|m|; 1+|m|; r^2\right) \\ &= \frac{\Gamma(s+m)}{|m|!\,\Gamma(s-|m|)}\, r^{|m|}\left(1-r^2\right)^{1-s} F\left(1-s, 1-s+|m|; 1+|m|; r^2\right), \end{aligned}$$

$$Q^m_{s-1}\left(\frac{1+r^2}{1-r^2}\right) = \frac{(-1)^m}{2}\,\frac{\Gamma(s)\Gamma(s+m)}{\Gamma(2s)}\,\frac{(1-r^2)^s}{r^m}\,F\left(s-m,s;2s;1-r^2\right)$$
$$= \frac{(-1)^m}{2}\,\frac{\Gamma(s)\Gamma(s+m)}{\Gamma(2s)}\,\frac{(1-r^2)^s}{r^{2s-m}}\,F\left(s-m,s;2s;1-r^{-2}\right), \tag{A.9}$$

and the hypergeometric function $F = {}_2F_1$ given for $|z| < 1$ by

$$F\left(a,b;c;z\right) = \sum_{n\geq 0} \frac{(a)_n(b)_n}{(c)_n}\,\frac{z^n}{n!}, \qquad \text{with} \quad (a)_n = \prod_{j=0}^{n-1}(a+j). \tag{A.10}$$

(See [3], 2.1, 3.2 (3), 3.3.1 (7), 3.3.1 (1), 3.2 (36), and 2.9 (2) and (3).) The space $\mathcal{E}^K_{s,n}(\mathbb{H})$ is spanned by $P_{s,n}$ alone, since $Q_{s,n}(r)$ has a singularity as $r \downarrow 0$:

$$Q_{s,n}(r) = \begin{cases} -\log r\,\left(1 + r\cdot(\text{analytic in } r)\right) + (\text{analytic in } r) & \text{if } n = 0, \\ \frac{1}{2}\,(|n|-1)!\,\frac{\Gamma(s+n)}{\Gamma(s+|n|)} r^{-|n|}(1 + r\cdot(\text{analytic in } r)) & \\ \qquad + \log r \cdot (\text{analytic in } r) & \text{otherwise.} \end{cases} \tag{A.11}$$

See [3], 3.9.2 (5)–(7) for the leading terms, and 2.3.1 for more information. Directly from (A.9), we find for $r \downarrow 0$

$$P_{s,n}(r) = \frac{\Gamma(s+n)}{|n|!\;\Gamma(s-|n|)}\,r^{|n|}\left(1 + r\cdot(\text{analytic in } r)\right). \tag{A.12}$$

The solution $Q_{s,n}$ is special near the boundary $\mathbb{S}^1$ of $\mathbb{D}$. As $r \uparrow 1$:

$$Q_{s,n}(r) = (-1)^n\,\frac{\sqrt{\pi}\,\Gamma(s+n)}{\Gamma(s+\frac{1}{2})}\,2^{-2s}(1-r^2)^s\left(1 + (1-r)\cdot(\text{analytic in } 1-r)\right). \tag{A.13}$$

Thus, $Q_{s,n} \in \mathcal{W}^\omega_s$, and $\rho_s Q_{s,n}(\xi) = (-1)^n\,\frac{\sqrt{\pi}\,\Gamma(s+n)}{\Gamma(s+\frac{1}{2})}\,\xi^n$ on $\mathbb{S}^1$.
For $s \neq \frac{1}{2}$, we have

$$P_{s,n} = \frac{1}{\pi}\tan \pi s\,\left(Q_{s,n} - Q_{1-s,n}\right). \tag{A.14}$$

(The formula in [3], 3.3.1, (3) gives this relation with a minus sign in front of $\frac{1}{\pi}$.) This relation confirms that $P_{1-s,n} = P_{s,n}$ and forms the basis of the boundary splitting in (7.3). It shows that in the asymptotic expansion of $P_{s,n}(r)$ as $r \uparrow 0$, there are nonzero terms with $(1-r)^s$ and with $(1-r)^{1-s}$. At $s = \frac{1}{2}$, we have as $r \uparrow 1$

$$P_{1/2,n}\left(re^{i\theta}\right) = -\frac{(-1)^n\,\Gamma(\frac{1}{2}+n)}{\pi^{3/2}}\,(1-r^2)^{1/2}\log(1-r^2) + O(1). \qquad \text{(A.15)}$$

So $P_{s,n}$ is not in $\mathcal{W}_s^\omega(I)$ for any interval $I \subset \mathbb{S}^1$.

A.1.3 Equivariant Eigenfunctions for the Torus A

The characters of A are of the form $a(t) \mapsto t^{i\alpha}$ with $\alpha \in \mathbb{R}$. We use the coordinates $z = \rho e^{i\phi}$ on $\mathfrak{H}$, for which $a(t)$ acts as $(\rho,\phi) \mapsto (t\rho,\phi)$. If $u(\rho e^{i\phi}) = \rho^{id} f(\cos\phi)$ is in $\mathcal{E}_{s,\alpha}^A$, then f satisfies on $(-1,1)$ the differential equation

$$-\left(1-t^2\right)^2 f''(t) + t\left(1-t^2\right) f'(t) + \left(\alpha^2(1-t^2) - s(1-s)\right) f(t) = 0. \quad \text{(A.16)}$$

This leads to the following basis of the space $\mathcal{E}_{s,\alpha}^A$:

$$f_{s,\alpha}^+(\rho e^{i\phi}) = \rho^{i\alpha}(\sin\phi)^s F\left(\frac{s+i\alpha}{2}, \frac{s-i\alpha}{2}; \frac{1}{2}; \cos^2\phi\right),$$

$$f_{s,\alpha}^-(\rho e^{i\phi}) = \rho^{i\alpha}\cos\phi\,(\sin\phi)^s F\left(\frac{s+i\alpha+1}{2}, \frac{s-i\alpha+1}{2}; \frac{3}{2}; \cos^2\phi\right). \quad \text{(A.17)}$$

The $+$ or $-$ indicates the parity under $z \mapsto -\bar{z}$. In particular

$$f_{s,\alpha}^+(i) = 1, \quad \frac{\partial f_{s,ff}^+}{\partial\phi}(i) = 0, \quad f_{s,ff}^-(i) = 0, \quad \frac{\partial f_{s,ff}^-}{\partial\phi}(i) = -1. \qquad \text{(A.18)}$$

Relation (2), Sect. 2.9 in [3] shows that $f_{1-s,\alpha}^+ = f_{s,\alpha}^+$ and $f_{1-s,\alpha}^- = f_{s,\alpha}^-$.

For the boundary behavior, it is better to apply the Kummer relation (33) in Sect. 2.9 of [3] to the following function in $\mathcal{E}_s\,(\mathfrak{H} \setminus i\mathbb{R}_+)$

$$\rho^{i\alpha}(\sin\phi)^s F\left(\frac{s+i\alpha}{2}, \frac{s-i\alpha}{2}; s+\frac{1}{2}; \sin^2\phi\right). \qquad \text{(A.19)}$$

One has to choose $\sqrt{\cos^2\phi}$. Denote by $f_{s,\alpha}^R$ the restriction to $0 < \phi < \frac{\pi}{2}$ and by $f_{s,\alpha}^L$ the restriction to $\frac{\pi}{2} < \phi < \pi$. The Kummer relation implies the following equalities:

$$\begin{aligned}
f^R_{s,\alpha} &= \frac{\sqrt{\pi}\,\Gamma\left(s+\frac{1}{2}\right)}{\Gamma\left(\frac{s+\mathrm{i}\alpha+1}{2}\right)\Gamma\left(\frac{s-\mathrm{i}\alpha+1}{2}\right)}\,f^+_{\alpha,s} - \frac{2\sqrt{\pi}\,\Gamma\left(s+\frac{1}{2}\right)}{\Gamma\left(\frac{s+\mathrm{i}\alpha}{2}\right)\Gamma\left(\frac{s-\mathrm{i}\alpha}{2}\right)}\,f^-_{\alpha,s},\\
f^L_{s,\alpha} &= \frac{\sqrt{\pi}\,\Gamma\left(s+\frac{1}{2}\right)}{\Gamma\left(\frac{s+\mathrm{i}\alpha+1}{2}\right)\Gamma\left(\frac{s-\mathrm{i}\alpha+1}{2}\right)}\,f^+_{\alpha,s} + \frac{2\sqrt{\pi}\,\Gamma\left(s+\frac{1}{2}\right)}{\Gamma\left(\frac{s+\mathrm{i}\alpha}{2}\right)\Gamma\left(\frac{s-\mathrm{i}\alpha}{2}\right)}\,f^-_{\alpha,s}.
\end{aligned} \tag{A.20}$$

Thus, we see that $f^R_{s,\alpha}$ and $f^L_{s,\alpha}$ extend as elements of $\mathcal{E}_s$; that $f^R_{s,\alpha}$ represents an element of $\mathcal{W}^\omega_s(\mathbb{R}_+)$ with, in the line model, $\rho_s f^R_{s,\alpha}(x) = x^{\mathrm{i}\alpha}$; and that $f^L_{s,\alpha}$ represents an element of $\mathcal{W}^\omega_s(\mathbb{R}_-)$ with $\rho_s f^L_{s,\alpha}(x) = (-x)^{\mathrm{i}\alpha}$. Inverting the relation in (A.20) one finds, for $s \neq \frac{1}{2}$, the following expressions for $f^+_{s,\alpha}$ and $f^-_{s,\alpha}$ as a linear combination of $f^R_{s,\alpha}$ and $f^R_{1-s,\alpha}$,

$$\begin{aligned}
f^+_{s,\alpha} &= \frac{\sqrt{\pi}\,\Gamma\left(\frac{1}{2}-s\right)}{\Gamma\left(\frac{1-s+\mathrm{i}\alpha}{2}\right)\Gamma\left(\frac{1-s-\mathrm{i}\alpha}{2}\right)}\,f^R_{s,\alpha} + \frac{\sqrt{\pi}\,\Gamma\left(s-\frac{1}{2}\right)}{\Gamma\left(\frac{s+\mathrm{i}\alpha}{2}\right)\Gamma\left(\frac{s-\mathrm{i}\alpha}{2}\right)}\,f^R_{1-s,\alpha},\\
f^-_{s,\alpha} &= \frac{\sqrt{\pi}\,\Gamma\left(\frac{1}{2}-s\right)}{2\Gamma\left(1-\frac{s+\mathrm{i}\alpha}{2}\right)\Gamma\left(1-\frac{s-\mathrm{i}\alpha}{2}\right)}\,f^R_{s,\alpha} + \frac{\sqrt{\pi}\,\Gamma\left(s-\frac{1}{2}\right)}{2\Gamma\left(\frac{s+\mathrm{i}\alpha+1}{2}\right)\Gamma\left(\frac{s-\mathrm{i}\alpha+1}{s}\right)}\,f^R_{1-s,\alpha},
\end{aligned} \tag{A.21}$$

and similarly of $f^L_{s,\alpha}$ and $f^L_{1-s,\alpha}$, showing that each of these elements belongs to the direct sums $\mathcal{W}^\omega_s(\mathbb{R}_+) \oplus \mathcal{W}^\omega_{1-s}(\mathbb{R}_+)$ and $\mathcal{W}^\omega_s(\mathbb{R}_-) \oplus \mathcal{W}^\omega_{1-s}(\mathbb{R}_-)$, but not to $\mathcal{W}^\omega_s(I) \oplus \mathcal{W}^\omega_{1-s}(I)$ for any neighborhood I of 0 or ∞ in $\mathbb{P}^1_{\mathbb{R}}$; in other words, just as for the Bessel functions $i_{s,\alpha}$ and $k_{s,\alpha}$, we have a local but not a global boundary splitting.

A.2 *Poisson Transforms*

Almost all of the special elements in Sect. A.1 belong to $\mathcal{E}_s$ and hence are the Poisson transform of some hyperfunction by Helgason's Theorem 3.4. Actually in all cases except one, the function has polynomial growth and hence is the Poisson transform of a distribution (Theorem 3.5). In Table A.1 and the discussion below, we give explicit representations of these eigenfunctions as Poisson transforms of distributions and/or hyperfunctions.

A. In (3.30) we have shown that y^{1-s} is the Poisson transform of the distribution $\delta_{s,\infty}$. See (3.30) for an explicit description of $\delta_{s,\infty}$ as a hyperfunction.

B. The description of y^s as a Poisson transform takes more work. For $\mathrm{Re}\, s < \frac{1}{2}$ the linear form $\mathbf{1}_s : \varphi \mapsto \frac{1}{\pi}\int_{-\infty}^{\infty}\varphi(t)\,dt$ is continuous on $\mathcal{V}^0_{1-s}$, in the line model. Note that the constant function 1 is not in $\mathcal{V}^\omega_s$ since it does not satisfy the asymptotic behavior (2.2) at ∞. Application of (3.26) gives the Poisson transform $\mathsf{P}_s\mathbf{1}_s$ indicated in the table.

Table A.1 Poisson representation of elements of $\mathcal{E}_s$

	$u \in \mathcal{E}_s$	$\mathsf{P}_s^{-1} u \in \mathcal{V}_s^{-\omega}$	Model
A	y^{1-s}	$\delta_{s,\infty} : \varphi^{\mathbb{P}} \mapsto \varphi^{\mathbb{P}}(\infty)$	Proj.
B	$\frac{\Gamma(\frac{1}{2}-s)}{\sqrt{\pi}\,\Gamma(1-s)}\, y^s$	$\mathbf{1}_s \;=\;$ integration against 1 for $\operatorname{Re} s < \frac{1}{2}$, with meromorphic continuation	Line
C	$-2\frac{\Gamma(\frac{3}{2}-s)}{\sqrt{\pi}\,\Gamma(1-s)}\ell_s(z)$ ℓ_s as in (A.6)	$\varphi \mapsto \frac{-1}{2}\int_{-\infty}^{\infty}(\varphi(t) - \frac{\varphi_\infty}{\sqrt{1+t^2}})\,dt$ with $\varphi_\infty = \lim_{t\to\infty} \lvert t\rvert^{2s-2}\varphi(t)$	Line
D	$R(t;z)^{1-s} \quad (t \in \mathbb{R})$	$\delta_{s,t} : \varphi \mapsto \varphi(t)$	Line
E	$\frac{2^{s+\frac{1}{2}}\lvert\alpha\rvert^{\frac{1}{2}-s}}{\sqrt{\pi}\,\Gamma(1-s)} k_{s,\alpha}(z)$ $(\alpha \in \mathbb{R} \smallsetminus \{0\})$	Integration against $e^{i\alpha t}$ for $\operatorname{Re} s < \frac{1}{2}$, or integration of $-\varphi'$ against $\frac{e^{i\alpha x}}{i\alpha}$ for $0 < \operatorname{Re} s < 1$	Line
F	$i_{1-s,\alpha}(z)$ $(\alpha \in \mathbb{R} \smallsetminus \{0\})$	Support $\{\infty\}$; representative near ∞: $-\frac{1}{2}\tau\,(1+\tau^{-2})^s F\big(1; 2-2s; i\alpha\tau\big)$	Proj.
G	$\frac{(-1)^n\Gamma(s)}{\Gamma(s+n)}\, p_{s,n}$	$\mathbf{e}_{s,n}$	Circle
H	$p_s(w', \cdot)$	$R^{\mathbb{S}}(\cdot\,; w')^s$	Circle
I	$\frac{\Gamma(1+i\alpha-s)\Gamma(1-i\alpha-s)}{\pi\,\Gamma(2-2s)}\, f^L_{1-s,\alpha}$	Integration against $x^{i\alpha-s}$ on $\mathbb{R}_+$	Line
J	$\frac{\Gamma(1+i\alpha-s)\Gamma(1-i\alpha-s)}{\pi\,\Gamma(2-2s)}\, f^R_{1-s,\alpha}$	Integration against $(-x)^{i\alpha-s}$ on $\mathbb{R}_-$	Line

To describe $\mathbf{1}_s$ as a hyperfunction in the line model (and also to continue it in s), we want to give representatives $g_{\mathbb{R}}$ and g_∞ of $\mathbf{1}_s$ on $\mathbb{R}$ and $\mathbb{P}^1_{\mathbb{R}} \smallsetminus \{0\}$, related by $g_\infty(\zeta) = \zeta^{-2s} g_{\mathbb{R}}(-1/\zeta)$ up to a holomorphic function on a neighborhood of $\mathbb{R} \smallsetminus \{0\}$.

Formula (2.26) gives a representative $g^{\mathbb{P}}$ in the projective model:

$$g^{\mathbb{P}}(\zeta) \;=\; \frac{1}{2\pi i}\int_{-\infty}^{\infty} \frac{\zeta + i}{t - \zeta}\,(t-i)^s(t+i)^{s-1}\,dt \qquad (\zeta \in \mathbb{P}^1_{\mathbb{C}} \smallsetminus \mathbb{P}^1_{\mathbb{R}}).$$

(The factor $(t^2+1)^{s-1}$ comes from passage between models.) This function extends both from $\mathfrak{H}$ and from $\mathfrak{H}^-$ across the real axis. An application of Cauchy's formula shows that the difference of both extension is given by $(\zeta^2+1)^s$, corresponding to the function 1 in the line model. See (2.5).

To get a representative near ∞, we write

$$\big(1 + e^{2\pi i s}\big)\, g^{\mathbb{P}}(\zeta) \;=\; \frac{1}{2\pi i}\int_C \frac{\zeta + i}{(z-\zeta)(z+i)}\,(z^2+1)^s\,dz, \tag{A.22}$$

where C is the contour shown below. The factor $(z^2+1)^s$ is multivalued on the contour and is fixed by choosing $\arg(z^2+1) \in [0, 2\pi)$. On the part of the contour just above $(0,\infty)$, the argument of z^2+1 is approximately zero, and just below $(0,\infty)$, the argument is approximately 2π. Near $(-\infty, 0)$, the argument is approximately 2π just above the real line and approximately 0 below the real line. We take the contour so large that $\zeta \in \mathfrak{H} \cup \mathfrak{H}^-$ is inside one of the loops of C. If we let the contour grow, the arcs in the upper and lower half planes give a contribution $o(1)$. In the limit, for $\mathrm{Re}\, s < \frac{1}{2}$, we are left with twice the integral along $(0,\infty)$ and along $(-\infty,0)$, both once with the standard value and and once with $e^{2\pi i s}$ times the standard value. This gives the equality (A.22) and the continuation of $g^{\mathbb{P}}$ as a meromorphic function of s.

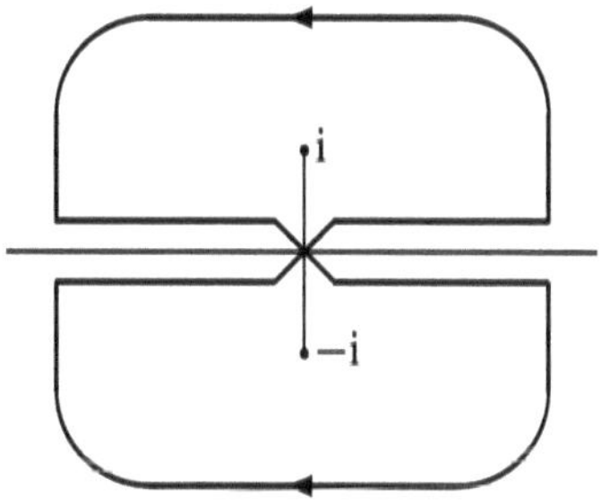

Now consider $\zeta \in \mathfrak{H}^{\pm}$ with $|\zeta| > 1$. Moving the path of integration across ζ, we obtain with Cauchy's theorem that $\left(1 + e^{2\pi i s}\right) g^{\mathbb{P}}(\zeta)$ is equal to $\pm(\zeta^2+1)^s$ plus a holomorphic function of ζ on a neighborhood of ∞. The term $\pm(\zeta^2+1)^s$ obeys the choice of the argument discussed above. To bring it back to the standard choice of arguments in $(-\pi, \pi]$, we write it as $\zeta^{2s}\left(1+\zeta^{-2}\right)^s$ for $\zeta \in \mathfrak{H}$ and as $-(-\zeta)^{2s}\left(1+\zeta^{-2}\right)^s$ for $\zeta \in \mathfrak{H}^-$. The factor $\left(1+\zeta^{-2}\right)^s$ is what we need to go back to the line model with (2.5). Thus we arrive at the following representatives in the line model.

$$g_{\mathbb{R}}(\zeta) = \begin{cases} 1 & \text{on } \mathfrak{H}, \\ 0 & \text{on } \mathfrak{H}^-; \end{cases} \qquad g_{\infty}(\zeta) = \pm\zeta^{-2s}\left(1 + e^{\mp 2\pi i s}\right)^{-1} \text{ on } \mathfrak{H}^{\pm}. \tag{A.23}$$

Finally one checks that $g_{\mathbb{R}}(\zeta) - (\zeta^2)^{-s} g_{\infty}(-1/\zeta)$ extends holomorphically across both $\mathbb{R}_+$ and $\mathbb{R}_-$, thus showing that the pair $(g_{\mathbb{R}}, g_{\infty})$ determines the hyperfunction $\mathbf{1}_s$. These representatives also show that $\mathbf{1}_s$ extends meromorphically in s, giving $\mathbf{1}_s \in \mathcal{V}_s^{-\omega}$ for all $s \neq \frac{1}{2}$ with $0 < \mathrm{Re}\, s < 1$.

For the relation between the cases **A** and **B**, we use (3.25) to get

$$\mathsf{P}_s I_{1-s} \delta_{1-s,\infty}(z) = \mathsf{P}_{1-s} \delta_{1-s,\infty}(z) = y^s.$$

The fact that the Poisson transformation is an isomorphism $\mathcal{V}_s^{-\omega} \to \mathcal{E}^s$ implies

$$\mathbf{1}_s = \frac{\Gamma(\frac{1}{2}-s)}{\sqrt{\pi}\,\Gamma(1-s)} I_s \delta_{s,\infty}. \tag{A.24}$$

C. For $\operatorname{Re} s < \frac{1}{2}$ we have

$$\langle \varphi, \mathbf{1}_s \rangle = \frac{1}{\pi} \int_{-\infty}^{\infty} \left(\varphi(t) - \varphi^{\mathbb{P}}(\infty)(1+t^2)^{s-1} \right) \mathrm{d}x + \frac{\Gamma(\frac{1}{2}-s)}{\sqrt{\pi}\,\Gamma(1-s)} \delta_{s,\infty}(\varphi^{\mathbb{P}}).$$

So the distribution L_s given by

$$L_s : \varphi \mapsto \frac{1}{\pi} \int_{-\infty}^{\infty} \left(\varphi(t) - \varphi^{\mathbb{P}}(\infty)(1+t^2)^{s-1} \right) \mathrm{d}t,$$

which is well defined for $\operatorname{Re} s < 1$, is equal to $\mathbf{1}_s - \frac{\Gamma(\frac{1}{2}-s)}{\sqrt{\pi}\,\Gamma(1-s)} \delta_{s,\infty}$ for $\operatorname{Re} s < \frac{1}{2}$. The results of the cases **A** and **B** give the expression of $\mathsf{P}_s L_s$ as a multiple of ℓ_s defined in (A.6). Going over to the line model, we obtain the statement in the table.

D. This is simply the definition of the Poisson transformation in (3.22) and (3.23) applied to the delta distribution at t. It also follows from Case **A**, using the G-equivariance.

The latter method involves a transition between the models. We explain some of the steps to be taken. In the projective model, $\delta_{s,t}^{\mathbb{P}} : \varphi^{\mathbb{P}} \mapsto (1+t^2)^{s-1} \varphi^{\mathbb{P}}(t)$. We have

$$\left\langle \delta_{s,t}^{\mathbb{P}} \big|_{2s} \begin{bmatrix} t & -1 \\ 1 & 0 \end{bmatrix}, \varphi^{\mathbb{P}} \right\rangle = \left\langle \delta_{s,t}^{\mathbb{P}}, \varphi^{\mathbb{P}} \big|_{2-2s} \begin{bmatrix} 0 & 1 \\ -1 & t \end{bmatrix} \right\rangle = \cdots = \delta_{s,\infty}^{\mathbb{P}}(\varphi^{\mathbb{P}}).$$

Hence,

$$\mathsf{P}_s(\delta_{s,t})(z) = \mathsf{P}_s \left(\delta_{s,\infty} \Big| \begin{bmatrix} 0 & 1 \\ -1 & t \end{bmatrix} \right)(z) = (\mathsf{P}_s \delta_{s,\infty})(1/(t-z)) = \left(\frac{y}{|t-z|^2} \right)^{1-s}.$$

E. For $\alpha \neq 0$, we need no complicated contour integration. When $\operatorname{Re} s < \frac{1}{2}$, the distribution $\varphi \mapsto \frac{1}{\pi} \int_{-\infty}^{\infty} \varphi(t) e^{i\alpha t}\, \mathrm{d}t$ in the line model is equal to $\varphi \mapsto \frac{-1}{\pi i \alpha} \int_{-\infty}^{\infty} \varphi'(t) e^{i\alpha t}\, \mathrm{d}t$. The latter integral converges absolutely for $\operatorname{Re} s < 1$.

F. Since $i_{s,\alpha}$ has exponential growth, we really need a hyperfunction. The representative in the table does not behave well near 0. However it is holomorphic on a deleted neighborhood of ∞, and represents a hyperfunction on $\mathbb{P}^1_{\mathbb{R}} \setminus \{0\}$ in the projective model. We extend it by zero to obtain a hyperfunction on $\mathbb{P}^1_{\mathbb{R}}$.

The path of integration $\int_{C_+} - \int_{C_-}$ can be deformed into a large circle $|\tau| = R$, such that we can replace τ by $\tau - x$ in the integration. We obtain

$$\begin{aligned} &\frac{1}{\pi} \int_{|\tau|=R} \frac{i\tau}{-2} (1+\tau^{-2})^s F\left(1; 2-2s; i\alpha\tau\right) \left(\frac{y(1+\tau^2)}{(\tau-z)(\tau-\bar{z})} \right)^{s-1} \frac{\mathrm{d}\tau}{1+\tau^2} \\ &= \frac{1}{2\pi i} y^{1-s} \int_{|\tau|=R} \left(1 + \frac{x}{\tau}\right)^{1-2s} \left(1 + \frac{y^2}{\tau^2}\right)^{s-1} F\left(1; 2-2s; i\alpha(\tau + x)\right) \frac{\mathrm{d}\tau}{\tau}. \end{aligned}$$

Expand the factors $\left(1+\frac{x}{\tau}\right)^{1-2s}$ and $\left(1+\frac{y^2}{\tau^2}\right)^{s-1}$ and the hypergeometric function into power series and carry out the integration term by term. In the resulting sum, we recognize the power series of $e^{i\alpha x}$ and, after some standard manipulations with gamma factors, also the expansion of the modified Bessel function $I_{1/2-s}(|\alpha|y)$.

G. See the discussion after Theorem 3.4.

H. See (3.31).

I and **J**. Integration against $x \mapsto x^{i\alpha-s}$ on $(0,\infty)$ and against $x \mapsto (-x)^{i\alpha-s}$, in the line model, defines distributions. For $\rho e^{i\phi} \in \mathfrak{H}$, the Poisson integral leads to

$$\frac{\rho^{i\alpha}(\sin\phi)^{1-s}}{\pi}\int_0^\infty t^{i\alpha-s}\left(t^2+1+2Ct\right)^{s-1}\,dt,$$

with $C=\mp\cos\phi$. Let us consider this for small values of C, i.e., for points near $i\mathbb{R}_+$ in $\mathfrak{H}$. Expanding the integrand in powers of C gives a series in which one may separate the even and odd terms and arrive at

$$\frac{\rho^{i\alpha}\sqrt{1-C^2}^{\,1-s}}{2\pi\Gamma(1-s)}\left(\Gamma\left(\frac{1-i\alpha-s}{2}\right)\Gamma\left(\frac{1+i\alpha-s}{2}\right)F\left(\frac{1-i\alpha-s}{2},\frac{1+i\alpha-s}{2};\frac{1}{2};C^2\right)\right.$$
$$\left.-2C\,\Gamma\left(1-\frac{i\alpha+s}{2}\right)\Gamma\left(1+\frac{i\alpha-s}{2}\right)F\left(1-\frac{i\alpha+s}{2},1+\frac{i\alpha-s}{2};\frac{3}{2};C^2\right)\right).$$

Now take $C=-\cos\phi$, respectively $C=\cos\phi$, and conclude that we have a multiple of $f^L_{1-s,\alpha}$, respectively $f^R_{1-s,\alpha}$.

A.3 *Transverse Poisson Transforms*

In Table A.2 we give examples of pairs $u=\mathsf{P}^\dagger_s\varphi$, $\varphi=\rho_s u$, where $\varphi\in\mathcal{V}^\omega_s(I)$ for some $I\subset\partial\mathbb{H}$.

In Cases **c**, **d**, **f**, **g**, and **h** in the table, the eigenfunction u is in $\mathcal{E}_s=\mathcal{E}_s(\mathbb{H})$; hence, it is also a Poisson transform. If we write $u=\mathsf{P}_{1-s}\alpha$, then entries **A**, **D**, **F**, **J**, and **I**, respectively, in Table A.1 (with the s replaced by $1-s$ in most cases) show that the support of α is the complement of the set I in $\partial\mathbb{H}$ for each of these cases, illustrating Theorem 6.4.

A.4 *Potentials for Green's Forms*

If $u,v\in\mathcal{E}_s(U)$ for some $U\subset\mathbb{H}$, then the Green's forms $\{u,v\}$ and $[u,v]$ are closed. So if U is simply connected, there are well-defined potentials of $[u,v]$ and $\{u,v\}$ in $C^\omega(U)$, related according to (3.13). We list some examples of potentials F of $\{u,v\}$ in Table A.3. Then $\frac{1}{2i}F+\frac{1}{2}uv$ is a potential of the other Green's form $[u,v]$.

Table A.2 Transverse Poisson representations of boundary germs

	$u = \mathsf{P}_s^\dagger \varphi \in \mathcal{W}_s^\omega(I)$	$\varphi = \rho_s u \in \mathcal{V}_s^\omega(I)$	I	Model
a	$Q_{s,n}$	$(-1)^n \frac{\sqrt{\pi}\,\Gamma(s+n)}{\Gamma(s+\frac{1}{2})}\, \mathbf{e}_{s,n}$	$\mathbb{S}^1$	Circle
b	$q_s(\cdot, w')$	$\frac{\sqrt{\pi}\,\Gamma(s)}{2^{2s}\,\Gamma(s+1/2)} \frac{(1-\lvert w'\rvert^2)^s}{\lvert\xi - w'\rvert^{2s}}$	$\mathbb{S}^1$	Circle
c	y^s	1	$\mathbb{R}$	Line
d	$R(t;z)^s \quad (t \in \mathbb{R})$	$\lvert t-x\rvert^{-2s}$	$\mathbb{R} \smallsetminus \{t\}$	Line
e	$R(\zeta;z)^s \quad (\zeta \in \mathbb{C} \smallsetminus \mathbb{R})$	$(\zeta - x)^{2s}$ (multivalued)	$\mathbb{R}$	Line
f	$i_{s,\alpha}$	$e^{i\alpha x}$	$\mathbb{R}$	Line
g	$f_{s,\alpha}^R$	$x^{i\alpha - s}$	$(0, \infty)$	Line
h	$f_{s,\alpha}^L$	$(-x)^{i\alpha-s}$	$(-\infty, 0)$	Line

We found most of these potentials by writing down $\{u, v\}$, guessing F, and checking our guess.

Case **3** is essentially (3.16). In Case **6** we needed the following function:

$$F_s(r) = 2s \int_r^\infty (1+q^2)^{-s-1} \tag{A.25}$$

have used that $(\operatorname{Im} gz)^s = R(t;z)^s$ and $R(0;gz)^s = |p - R(t;z)^s$ and $R(0;gz)^s = |p-t|^{2s} R(p;z)^s$ with $g = \left[\begin{smallmatrix} \frac{-1}{p-t} & \frac{p}{p-t} \\ -1 & t \end{smallmatrix}\right]$ with $t, p \in \mathbb{R}$. So **6** leads to the potential in **7** if $p \neq t$ are real. We write $\left((p-t)^2\right)^{-s}$ and not $|p-t|^{-2s}$ to allow holomorphic continuation in p and t. For Case **8** we use that if $u(z) = e^{i\alpha x} f(y)$ and $v(z) = e^{i\alpha x} g(y)$, then

$$\{u, v\} = e^{2i\alpha x} (f'g - fg')\, dx,$$

and that the Wronskian $fg' - f'g$ is constant if $u, v \in \mathcal{E}_s$. Cases **9–12** are obtained in a similar way. In **9** and **11** the potentials are multivalued if U is not simply connected.

Cases **3–5** are valid on $\mathfrak{H}$ if t and p are real. Otherwise $\{u, v\}$ and F are multivalued with branch points at t and at p in **3**. We have to chose the same branch in $\{u, v\}$ and F. Also in **7**, the branches have to be chosen consistently. In **4** there are singularities at $t = z$ and $t = \bar{z}$, but $\{u, v\}$ and F are univalued.

A.5 *Action of the Lie Algebra*

The real Lie algebra of G has $\mathbf{H} = \left(\begin{smallmatrix} 1 & 0 \\ 0 & -1 \end{smallmatrix}\right)$, $\mathbf{V} = \left(\begin{smallmatrix} 0 & 1 \\ 1 & 0 \end{smallmatrix}\right)$, $\mathbf{W} = \left(\begin{smallmatrix} 0 & 1 \\ -1 & 0 \end{smallmatrix}\right)$ as a basis. Any $\mathbf{Y}$ in the Lie algebra acts on $\mathcal{V}_s^\infty$ by $f|_{2s}\mathbf{Y} = \partial_t f|_{2s} e^{t\mathbf{Y}}\big|_{t=0}$. Note that for right actions, we have $f \mid [\mathbf{Y}_1, \mathbf{Y}_2] = (f \mid \mathbf{Y}_2) \mid \mathbf{Y}_1 - (f \mid \mathbf{Y}_1) \mid \mathbf{Y}_2$.

In the projective model,

Table A.3 Potentials for Green's forms

	u	v	F such that $\mathrm{d}F = \{u, v\}$	Domain
1	y^s	y^{1-s}	$(2s-1)x$	$\mathfrak{H}$
2	$y^{1/2}$	$y^{1/2}\log y$	$-x$	$\mathfrak{H}$
3	$R(t;z)^s$ $t \in \mathbb{R}$	$R(p;z)^{1-s}$ $p \in \mathbb{R} \smallsetminus \{t\}$	$\frac{(t-x)(p-x)+y^2}{y(p-t)} R(t;z)^s R(p;z)^{1-s}$	$\mathfrak{H}$
4	$R(t;z)^s$ $t \in \mathbb{R}$	$R(t;z)^{1-s}$	$\frac{s}{t-z} + \frac{s-1}{t-\bar z} - \mathrm{i}R(t;z)$	$\mathfrak{H}$
5	y^s	$R(t;z)^{1-s}$ $t \in \mathbb{R}$	$-(\mathrm{i}y^s + (t-z)y^{s-1})\, R(t;z)^{1-s}$	$\mathfrak{H}$
6	y^s	$R(t;z)^s$	$-F_s((x-t)/y)$, F_s as in (A.25)	$\mathfrak{H}$
7	$R(t;z)^s$ $t \in \mathbb{R}$	$R(p;z)^s$ $p \in \mathbb{R} \smallsetminus \{t\}$	$\left((p-t)^2\right)^{-s} F_s\left(\frac{(p-x)(t-x)+y^2}{y(p-t)}\right)$	$\mathfrak{H}$
8	$k_{s,\alpha}$	$l_{s,\alpha}$	$\frac{\mathrm{i}\,\Gamma(s+1/2)}{2^{3/2-s}\alpha\lvert\alpha\rvert^{s-1/2}}\, \mathrm{e}^{2\mathrm{i}\alpha x}$	$\mathfrak{H}$
9	$P_{s,0}(r\mathrm{e}^{\mathrm{i}\theta})$	$Q_{s,0}(r\mathrm{e}^{\mathrm{i}\theta})$	$-\theta$	$U \subset \mathbb{D} \smallsetminus \{0\}$ simply connected
10	$P_{s,n}(r\mathrm{e}^{\mathrm{i}\theta})$ $n \in \mathbb{Z} \smallsetminus \{0\}$	$Q_{s,n}(r\mathrm{e}^{\mathrm{i}\theta})$	$-\frac{(-1)^n\,\Gamma(s+n)}{2\mathrm{i}n\,\Gamma(s-n)}\, \mathrm{e}^{2\mathrm{i}n\theta}$	$\mathbb{D} \smallsetminus \{0\}$
11	$P_{s,-n}(r\mathrm{e}^{\mathrm{i}\theta})$ $n \in \mathbb{Z} \smallsetminus \{0\}$	$Q_{s,n}(r\mathrm{e}^{\mathrm{i}\theta})$	$-2\mathrm{i}n \int P_{s,-m}(r)\, Q_{s,n}(r)\, \frac{\mathrm{d}r}{r}$ $-(-1)^n\theta$	$U \subset \mathbb{D} \smallsetminus \{0\}$ simply connected
12	$P_{s,m}(r\mathrm{e}^{\mathrm{i}\theta})$ $m \in \mathbb{Z}$	$Q_{s,n}(r\mathrm{e}^{\mathrm{i}\theta})$ $n \in \mathbb{Z} \smallsetminus \{-m\}$	$\mathrm{e}^{\mathrm{i}(m+n)\theta)} r\, \big(Q_{s,n}(r)\, \partial_r P_{s,m}(r)$ $- P_{s,m}(r)\, \partial_r Q_{s,n}(r)\big)/\mathrm{i}(m+n)$	$\mathbb{D} \smallsetminus \{0\}$

$$
\begin{aligned}
f|_{2s}\mathbf{H}(\tau) &= \left(2s\frac{1-\tau^2}{1+\tau^2} + 2\tau\partial_\tau\right) f(\tau),\\
f|_{2s}\mathbf{V}(\tau) &= \left(-4s\frac{\tau}{1+\tau^2} + (1-\tau^2)\partial_\tau\right) f(\tau),\\
f|_{2s}\mathbf{W}(\tau) &= (1+\tau^2)\partial_\tau f(\tau).
\end{aligned}
\tag{A.26}
$$

For the elements $\mathbf{E}^+ = \mathbf{H} + \mathrm{i}\mathbf{V}$ and $\mathbf{E}^- = \mathbf{H} - \mathrm{i}\mathbf{V}$ in the complexified Lie algebra, we find

$$f|_{2s}\mathbf{E}^+(\tau) = \left(-2s\frac{\tau+\mathrm{i}}{\tau-\mathrm{i}} - \mathrm{i}(\tau+\mathrm{i})^2\partial_\emptyset\right)f(\emptyset),$$
$$f|_{2s}\mathbf{E}^-(\tau) = \left(-2s\frac{\tau-\mathrm{i}}{\tau+\mathrm{i}} + \mathrm{i}(\tau-\mathrm{i})^2\partial_\tau\right)f(\tau). \tag{A.27}$$

In particular

$$\mathbf{e}_{s,n}|_{2s}\mathbf{W} = 2in\,\mathbf{e}_{s,n}, \qquad \mathbf{e}_{s,n}|_{2s}\mathbf{E}^\pm = -2(s\mp n)\,\mathbf{e}_{s,n\mp 1}. \tag{A.28}$$

By transposition, these formulas are also valid on hyperfunctions.

The Lie algebra generates the universal enveloping algebra, which also acts on $\mathcal{V}_s^\infty$. The center of this algebra is generated by the Casimir operator $\omega = -\frac{1}{4}\mathbf{E}^+\mathbf{E}^- + \frac{1}{4}\mathbf{W}^2 - \frac{\mathrm{i}}{2}\mathbf{W}$. It acts on $\mathcal{V}_s$ as multiplication by $s(1-s)$.

For the action of G by left translation on functions on $\mathfrak{H}$:

$$\begin{aligned}\mathbf{W} &= (1+z^2)\partial_z + (1+\bar{z}^2)\partial_{\bar{z}}, \qquad \mathbf{E}^\pm = \mp\mathrm{i}(z\pm\mathrm{i})^2\partial_z \mp \mathrm{i}(\bar{z}\pm\mathrm{i})^2\partial_{\bar{z}},\\ \omega &= (z-\bar{z})^2\partial_z\partial_{\bar{z}} = \Delta,\end{aligned} \tag{A.29}$$

and on $\mathbb{D}$:

$$\begin{aligned}\mathbf{W} &= 2\mathrm{i}w\partial_w - 2\mathrm{i}\bar{w}\partial_{\bar{w}}, \qquad \mathbf{E}^+ = 2\partial_w - 2\bar{w}^2\partial_{\bar{w}},\\ \mathbf{E}^- &= -2w^2\partial_w + 2\partial_{\bar{w}}, \qquad \omega = -(1-|w|^2)^2\,\partial_w\partial_{\bar{w}}.\end{aligned} \tag{A.30}$$

A counterpart of (A.28) is

$$\begin{aligned}P_{s,n}\,|\,\mathbf{W} &= 2in\,P_{s,n}, & P_{s,n}\,|\,\mathbf{E}^+ &= 2(s-n)(s+n-1)\,P_{s,n-1},\\ P_{s,n}\,|\,\mathbf{E}^- &= 2\,P_{s,n+1}, & Q_{s,n}\,|\,\mathbf{E}^+ &= 2(s-n)(s+n-1)\,Q_{s,n-1},\\ Q_{s,n}\,|\,\mathbf{E}^- &= 2\,Q_{s,n+1}, & Q_{s,n}\,|\,\mathbf{W} &= 2in\,Q_{s,n}.\end{aligned} \tag{A.31}$$

References

1. E. P. van den Ban, H. Schlichtkrull, *Asymptotic expansions and boundary values of eigenfunctions on Riemannian spaces.* J. reine angew. Math. **380** (1987), 108–165.
2. R. Bruggeman, J. Lewis and Preprint.
3. A. Erdélyi et al., *Higher transcendental functions, Vol. I.* McGraw-Hill, 1953.
4. R. Hartshorne, *Algebraic geometry.* Springer-Verlag, 1977.
5. S. Helgason, *Topics in harmonic analysis on homogeneous spaces.* Progr. in Math. **13**, Birkhäuser, 1981.
6. S. Helgason, *Groups and geometric analysis.* Math. Surv. and Monogr. **83**, AMS, 1984.
7. M. Kashiwara, A. Kowata, K. Minekura, K. Okamoto, T. Oshima, M. Tanaka, *Eigenfunctions of invariant differential operators on a symmetric space.* Ann. Math. **107** (1978), 1–39.

8. A. W. Knapp, *Representation theory of semisimple Lie groups, an overview based on examples.* Princeton University Press, Princeton, 1986.
9. J. B. Lewis, *Eigenfunctions on symmetric spaces with distribution-valued boundary forms.* J. Funct. Anal. **29** (1978), 287–307.
10. J. Lewis, D. Zagier, *Period functions for Maass wave forms. I.* Ann. Math. **153** (2001), 191–258.
11. H. Schlichtkrull, *Hyperfunctions and harmonic analysis on symmetric spaces.* Progr. in Math. **49**, Birkhäuser, 1984.

Analysis of Degenerate Diffusion Operators Arising in Population Biology

Charles L. Epstein and Rafe Mazzeo

This paper is dedicated to the memory of Leon Ehrenpreis, a giant in the field of partial differential equations

Abstract In this chapter, we describe our recent work on the analytic foundations for the study of degenerate diffusion equations which arise as the infinite population limits of standard models in population genetics. Our principal results concern existence, uniqueness, and regularity of solutions when the data belong to anisotropic Hölder spaces, adapted to the degeneracy of these operators. These results suffice to prove the existence of a strongly continuous $\mathcal{C}^0$-semigroup. The details of the definitions and complete proofs of these results can be found in [8].

Key words Wright-Fisher process • Diffusion limit • Degenerate diffusion • Hölder estimate

1 Introduction

In natural haploid population, three principal forces govern the evolution of the frequencies of different types within the population:

1. *Genetic drift:* The manifestation of the randomness in the number of offspring/generation each individual produces

C.L. Epstein (✉)
Departments of Mathematics and Radiology, University of Pennsylvania, PA, USA
e-mail: cle@math.upenn.edu

R. Mazzeo
Department of Mathematics, Stanford University, CA 94305, USA
e-mail: mazzeo@math.stanford.edu

H.M. Farkas et al. (eds.), *From Fourier Analysis and Number Theory to Radon Transforms and Geometry*, Developments in Mathematics 28, DOI 10.1007/978-1-4614-4075-8_8,

2. *Mutation:* The possibility of an individual spontaneously changing from one type to another
3. *Selection:* The fact that some types are better adapted to their environment than others and hence have more offspring

R.A. Fischer and Sewall Wright were among the first to model and quantify these effects. The simplest form of their model considers a population, of fixed size Q, with two variants (alleles) a and A at a single locus. In this model, the entire population reproduces at once, with the generations labeled by a nonnegative integer. Let X_j denote the number of individuals of type A at time j. The model is a Markov chain, with transition probabilities:

$$\text{Prob}(X_{j+1} = k | X_j = l) = p_{kl}(\mu_0, \mu_1, s), \tag{1}$$

where μ_0 is the rate of mutation from type a to type A, μ_1 the rate of mutation from type A to type a, and s the selective advantage of type A over a. If $\mu_0 = \mu_1 = s = 0$, then only the randomness of mating remains and we see that:

$$p_{kl} = \binom{Q}{k} \frac{l^k (Q-l)^{Q-k}}{Q^Q}. \tag{2}$$

This model has variants, for example, there can be multiple alleles at a single locus as well as many loci with several alleles.

As discrete models are difficult to analyze, the Markov chain models are often replaced, following Feller and Kimura by limiting, continuous in time and space, stochastic processes; see [13]. There is a precise sense in which the paths of the limiting process are limits of those of the discrete processes; see [10]. This limit is achieved by allowing the population size to tend to infinity and rescaling both the state space and the time variable. In the simple 1-site, 2-allele model described above, one may take the limit of the rescaled process $Q^{-1} X_{[\![Qt]\!]}$, as $Q \to \infty$, to get a Markov process on the unit interval $[0, 1]$. The formal generator of this process (the "forward" Kolmogorov operator) is the second-order operator:

$$L = \frac{1}{2} \partial_x^2 x(1-x) - m_0 \partial_x (1-x) + m_1 \partial_x x - \sigma \partial_x x(1-x); \tag{3}$$

here m_0, m_1, σ are scaled versions of μ_0, μ_1, and s.

If there are $N+1$ possible types, then a typical configuration space for the resulting continuous Markov process is the N-simplex

$$\mathcal{S}_N = \{(x_1, \ldots, x_N) : x_j \geq 0 \text{ and } x_1 + \cdots + x_N \leq 1\}. \tag{4}$$

It is possible to obtain different, and sometimes noncompact, domains if the limit is taken with a different scaling. For example, using a different scaling, we consider the sequence $< Q^{-\frac{1}{2}} X_{[\![\sqrt{Q}t]\!]} >$, whose limit is a process on $[0, \infty)$ used in the study of "rare" alleles.

The limiting operator of the Wright-Fisher process with $N + 1$ types, without mutation or selection, is the Kimura diffusion operator with formal generator:

$$L_{\text{Kim}} = \sum_{i,j=1}^{N} x_i(\delta_{ij} - x_j)\partial_{x_i}\partial_{x_j}. \tag{5}$$

This is the "backward" Kolmogorov operator for the limiting Markov process. This operator is elliptic in the interior of $\mathcal{S}_N$, but the coefficient of the second-order normal derivative along each codimension one boundary component vanishes simply. We can introduce local coordinates $(r, y_1, \ldots, y_{N-1})$ near the interior of a point on one of the boundary faces so that the boundary is given locally by the equation $r = 0$, and the second-order part of the operator then takes the form

$$r\partial_r^2 + \sum_{\ell=1}^{N-1} c_\ell r\partial_r\partial_{y_\ell} + \sum_{\ell,m=1}^{N-1} c_{\ell m}\partial_{y_m}\partial_{y_\ell}, \tag{6}$$

where the matrices $c_{\ell m}(r, y)$ are positive definite. The key feature here is the fact that the coefficient of ∂_r^2 vanishes to order exactly 1. This leads to a further difficulty in applications to Markov processes since the square root of the coefficient of the second-order terms is not Lipschitz continuous up to the boundary—indeed, this square root is Hölder continuous of order $\frac{1}{2}$. It is therefore impossible to apply standard methods to obtain uniqueness of solutions to either the forward Kolmogorov equation or the associated Martingale problem.

As a geometric object, the simplex is fairly complicated; its boundary is not a smooth manifold, but is instead a union of boundary hypersurfaces

$$\begin{aligned} \Sigma_{1,l} &= \{x_j = 0\} \cap \mathcal{S}_N \text{ for } l = 1, \ldots, N, \quad \text{and} \\ \Sigma_{1,0} &= \{x_1 + \cdots + x_N = 1\} \cap \mathcal{S}_N, \end{aligned} \tag{7}$$

which meet along higher codimension edges. Components of the edge of codimension l are the intersections

$$\Sigma_{1,i_1} \cap \cdots \cap \Sigma_{1,i_l}, \tag{8}$$

for any choice of integers $0 \leq i_1 < \cdots < i_l \leq N$. The simplex is an example of a *manifold with corners*, which seems to be the most natural setting for this class of operators. This singular structure of the boundary significantly complicates the analysis of differential operators on such spaces.

The basic existence theory for the operator L_{Kim} on $\mathcal{S}_N$ was initially obtained by Karlin and Kimura. Their analysis rests on the fact that L_{Kim} preserves the space of polynomials of degree less than or equal to d for each d. This is used to show the existence of a complete basis of polynomial eigenfunctions for this operator, which leads in turn to the existence of a polynomial (in space) solution to the initial value problem for $(\partial_t - L_{\text{Kim}})v = 0$ with polynomial initial data. Using the maximum principle, this suffices to prove the existence of a strongly continuous $\mathcal{C}^0$-semigroup and to establish many of its basic properties; see [14].

To include the effects of mutation and selection, one typically adds a vector field of the form:

$$V = \sum_{i=1}^{N} b_i(x)\partial_{x_i}, \tag{9}$$

where V is inward pointing along the boundary of $\mathcal{S}_N$. In the classical models, when including only the effect of mutation, the coefficients $\{b_i(x)\}$ can be taken as linear polynomials, but if selection is included, then these coefficients are at least quadratic polynomials and can be quite complicated; see [5]. Using the Trotter product formula and the fact that V is inward pointing, Ethier [9] showed that $L_{\text{Kim}} + V$ is the generator of strongly continuous semigroup on $\mathcal{C}^0$. Various extensions of these results have been obtained in the intervening years, e.g., by Sato, Cerrami and Clément, and Bass and Perkins; see [1–4], but these all place fairly restrictive assumptions on the domain and the operator. For example, Cerrai and Clément consider diffusions of this type acting on $\mathcal{C}^0([0,1]^N)$ assuming that the coefficients a_{ij} of $\partial_{x_i}\partial_{x_j}$ have the form

$$a_{ij}(x) = m(x)A_{ij}(x_i, x_j), \tag{10}$$

where $m(x)$ is strictly positive. Bass and Perkins considered a similar class of operators to those considered herein, but restricted their attention $\mathbb{R}_+^n$. Before the work reported here, very little was known about the true regularity of solutions, or the basic existence theory, outside of these special cases.

We have not yet said anything about boundary conditions. This would seem to be a serious omission since, in the absence of boundary conditions, an elliptic PDE on a manifold with boundary has an infinite dimensional null space. Somewhat remarkably, in this setting, a seemingly innocuous requirement that solutions have a certain regularity at the boundary is tantamount to imposing a boundary condition and ensures uniqueness of solutions of the parabolic problem with given initial data. We illustrate this in the simplest 1-dimensional case,

$$\partial_t v - [x(1-x)\partial_x^2 + b(x)\partial_x]v = 0 \text{ and } v(x,0) = f(x), \tag{11}$$

with $b(0) \geq 0$, $b(1) \leq 0$. If we simply *assume* that $\partial_x v(x,t)$ extends continuously to $[0,1] \times (0,\infty)$ and in addition that

$$\lim_{x\to 0+} x(1-x)\partial_x^2 v(x,t) = \lim_{x\to 1-} x(1-x)\partial_x^2 v(x,t) = 0, \tag{12}$$

then a simple argument using the maximum principle shows that (11) has a unique solution. We explain this in slightly more detail below.

2 Generalized Kimura Diffusion Operators

In his seminal work [11], Feller analyzed the most general closed extensions of one-dimensional operators of the form (11) which generate Feller semigroups. However, as already noted, despite Ethier's abstract existence theorem, until now very little has been determined about the finer analytic properties of the solution to the initial value problem for the heat equation

$$\partial_t v - (L_{\mathrm{Kim}} + V)v = 0 \text{ in } (0, \infty) \times \mathcal{S}_N \text{ with } v(0, x) = f(x) \tag{13}$$

in higher dimensions. Indeed, if one replaces L_{Kim} by a qualitatively similar second-order operator, not of one of the forms described above, then even the existence of a solution had not been established. To address these issues, we introduce in [8] a very flexible analytic framework for studying a large class of equations of this type, including all the standard models appearing in population genetics, and the SIR model for epidemics, as well as many models that arise in Mathematical Finance. This approach extends our work in [7] on the one-dimensional case.

We allow the configuration space P to be any manifold with corners, and we study a class of *generalized Kimura diffusion* operators $\partial_t - L$, where L is locally of the form given below in (15)–(18). Working in this generality is not just a convenience or an idle generalization, but is actually indispensable for the proofs of our basic estimates and existence results.

As part of our approach, we introduce nonstandard Hölder spaces naturally adapted to this class of operators. On this scale of spaces, we establish sharp existence and regularity results for the solutions to the inhomogeneous and homogeneous heat equations, as well as for the corresponding elliptic operators. The Lumer-Phillips theorem then gives the existence of a strongly continuous semigroup on $\mathcal{C}^0(P)$ with the given formal generator (backward Kolmogorov operator). As consequences of this, we conclude the uniqueness of the solution to the forward Kolmogorov equation, and this in turn establishes the uniqueness-in-law for associated SDE and the existence of a strongly continuous Markov process with paths confined to P.

An example of a manifold with corners is a subset of $\mathbb{R}^N$ defined by inequalities:

$$P = \bigcap_{k=1}^{K} \{x \in B_1(0) : \ p_k(x) \geq 0\}, \tag{14}$$

where the p_k are smooth functions, $k = 1, \ldots, K\}$, with $\{dp_{i_k} : \ 1 \leq k \leq n\}$ linearly independent at each point p where $p_k(p) = 0$, $k = 1, \ldots, n$. (Note that this last condition implies that $K \leq N$.) More generally, a manifold with corners P is a topological space for which every point has neighborhood diffeomorphic to a model orthant $\mathbb{R}^n_+ \times \mathbb{R}^m$, with $n + m = N$. The boundary hypersurfaces of P (in the example above, these are the sets $\Sigma_k = P \cap \{p_k(x) = 0\}$) are themselves manifolds with boundary or corners. Their connected components are called faces. As in (8), the codimension ℓ stratum of bP is formed from intersections of ℓ faces.

The *formal* generator of a generalized Kimura diffusion operator is a degenerate elliptic operator

$$L = \sum_{i,j=1}^{N} a_{ij}(x)\partial_{x_i}\partial_{x_j} + \sum_{j=1}^{N} b_j(x)\partial_{x_j} \tag{15}$$

satisfying certain conditions. The coefficients are all smooth; $(a_{ij}(x))$ is a symmetric matrix-valued function on P which is positive definite in the interior of P and degenerates along the hypersurface boundary components in a rather specific way. Again using the notation of the example, we require that

$$\sum_{i,j=1}^{N} a_{ij}(x)\partial_{x_i} p_k(x)\partial_{x_j} p_k(x) \propto p_k(x) \text{ as } x \text{ approaches } \Sigma_k, \tag{16}$$

while

$$\sum_{i,j=1}^{N} a_{ij}(x)v_i v_j > 0 \text{ for } x \in \text{int}\,\Sigma_k \text{ and } v \neq 0 \in T_x\Sigma_k. \tag{17}$$

The first-order part of L is an inward pointing vector field

$$Vp_k(x) = \sum_{j=1}^{N} b_j(x)\partial_{x_j} p_k(x) \geq 0 \text{ for } x \in \Sigma_k. \tag{18}$$

We call a second-order partial differential operator on P which satisfies all of these conditions a *generalized Kimura diffusion operator*.

Let P be a manifold with corners and L a generalized Kimura diffusion operator on P. Our goal is to prove the existence, uniqueness, and regularity of solutions to the equation

$$\begin{aligned} (\partial_t - L)u &= g \text{ in } P \times (0,\infty) \\ \text{with } u(p,0) &= f(p), \end{aligned} \tag{19}$$

where we specify certain boundary conditions along $bP \times [0,\infty)$ and for all data g and f satisfying appropriate regularity conditions.

3 Model Problems

The problem of proving the existence of solutions to a class of PDEs is essentially a matter of finding a good class of model problems, for which existence and regularity can be established, more or less directly, and then finding a functional analytic setting in which to do a perturbative analysis of the equations of interest. The model operators for Kimura diffusions are the differential operators, defined on $\mathbb{R}_+^n \times \mathbb{R}^m$ by

$$L_{b,m} = \sum_{j=1}^{n} [x_j \partial_{x_j}^2 + b_j \partial_{x_j}] + \sum_{k=1}^{m} \partial_{y_k}^2. \tag{20}$$

Here $b = (b_1, \ldots, b_n)$ is a vector with nonnegative entries. This class of model operators was also considered in [1].

The boundary conditions imposed along $b\mathbb{R}_+^n \times \mathbb{R}^m$ can be defined by a local differential expression on bP which is of a generalized "Neumann" type. We explain the one-dimensional case. Suppose that $b > 0$, $b \notin \mathbb{N}$; then the one-dimensional model operator $L_b = x\partial_x^2 + b\partial_x$ has two indicial roots

$$\beta_0 = 0, \beta_1 = 1 - b. \tag{21}$$

These are, by definition, the values of β determined by the equation $L_b x^\beta = 0$. A general regularity theorem states that any solution of $L_b u = 0$ has the form

$$u = u_0(x)x^{\beta_0} + u_1(x)x^{\beta_1} = u_0(x) + u_1(x)x^{1-b}, \tag{22}$$

where u_0 and u_1 are smooth down to $x = 0$. To exclude the second term on the right, we require that u satisfy the boundary condition

$$\lim_{x\to 0+} [\partial_x(x^b u(x,t)) - bx^{b-1}u(x,t)] = 0. \tag{23}$$

This condition ensures that $u_1 = 0$ and hence the solution has the maximum regularity allowed by the data: for example, if $g = 0$ and f is m-times continuously differentiable at $x = 0$, then the same is true when $0 \leq t$ for the solution u satisfying (23), and furthermore u is infinitely differentiable up to $x = 0$ when $t > 0$.

A convenient way to encode this boundary condition uses the function space $\mathcal{C}_{\mathrm{WF}}^2(\mathbb{R}_+)$. By definition, the function f belongs to $\mathcal{C}_{\mathrm{WF}}^2(\mathbb{R}_+)$, if $f \in \mathcal{C}^1(\mathbb{R}_+) \cap \mathcal{C}^2((0,\infty))$ and in addition

$$\lim_{x\to 0+} x\partial_x^2 f(x) = 0. \tag{24}$$

The boundary condition (23) is equivalent to the requirement that $u(\cdot,t) \in \mathcal{C}_{\mathrm{WF}}^2(\mathbb{R}_+)$ for $t > 0$. We call this solution, or its analogue in higher dimensions, the *regular* solution to the generalized Kimura diffusion operator.

Following [11], there is another natural boundary condition:

$$\lim_{x\to 0+} [\partial_x(xu(x,t)) - (2-b)u] = 0; \tag{25}$$

this one is associated to the adjoint operator. Solutions of the adjoint problem satisfying this boundary condition are not smooth up to the boundary, even when the data is. Because of the application to Markov processes, the adjoint L^t is naturally defined as an operator on $\mathcal{M}_1(P)$, the space of finite Borel measures on P, which explains why one is interested in the semigroup generated by L on $\mathcal{C}^0$. In any case, the study of regular solutions of L is naturally approached using the tools of PDE and is considerably simpler than the study of solutions to the adjoint problem, which is more naturally approached using techniques from probability theory; see [15]. For example, the null space of L is represented by smooth functions on P, whereas the null space of L^t is represented by nonnegative measures supported on certain components of the stratification of bP.

The solution operators for the one-dimensional model problems are given by simple explicit formulæ. If $b > 0$, then the heat kernel is

$$k_t^b(x,y)\mathrm{d}y = \left(\frac{y}{t}\right)^b \mathrm{e}^{-\frac{x+y}{t}} \psi_b\left(\frac{xy}{t^2}\right)\frac{\mathrm{d}y}{y}, \tag{26}$$

where

$$\psi_b(z) = \sum_{j=0}^{\infty} \frac{z^j}{j!\Gamma(j+b)}. \tag{27}$$

If $b = 0$, then

$$k_t^0(x,y) = \mathrm{e}^{-\frac{x}{t}}\delta_0(y) + \left(\frac{x}{t}\right)\mathrm{e}^{-\frac{x+y}{t}}\psi_2\left(\frac{xy}{t^2}\right)\frac{\mathrm{d}y}{t}. \tag{28}$$

Notably, the character of the kernel changes dramatically as $b \to 0$, but nonetheless, the regular solutions to this family of heat equations satisfy estimates which are uniform in b even as $b \to 0^+$. This is essential for the success of our approach.

For the higher dimensional model problems $(\partial_t - L_{b,m})v = 0$, the solution kernel is the product of one-dimensional kernels:

$$\prod_{l=1}^{n} k_t^{b_l}(x_l, x_l') \times \frac{1}{(4\pi t)^{\frac{m}{2}}} \mathrm{e}^{-\frac{|y-y'|^2}{4t}}. \tag{29}$$

In [8] we obtain the existence of a strongly continuous semigroup on the space $\mathcal{C}^0(\mathbb{R}_+^n \times \mathbb{R}^m)$ generated by $L_{b,m}$ (and then, more generally, to any general Kimura diffusion operator L on $\mathcal{C}^0(P)$). To study the refined mapping properties of this semigroup and its adjoint, however, we consider the problem (19), specialized to the model operator $L_{b,m}$, with f and g belonging to a certain family of anisotropic Hölder spaces associated to the singular, incomplete metric on $\mathbb{R}_+^n \times \mathbb{R}^m$

$$\mathrm{d}s_{\mathrm{WF}}^2 = \sum_{j=1}^{n} \frac{\mathrm{d}x_j^2}{x_j} + \sum_{k=1}^{m} \mathrm{d}y_m^2. \tag{30}$$

The use of function spaces associated to a certain singular metric in the study of a class of degenerate operators has been used in many other settings; see in particular [6, 12]. In the latter of these sources, Goulaouic and Shimakura obtain a priori estimates in similar Hölder spaces, and for an operator with the same type of degeneracy we are studying, but assuming that the boundary is smooth. As in these earlier works, we introduce two separate families of anisotropic Hölder spaces, $\mathcal{C}^{k,\gamma}_{\mathrm{WF}}(P)$, and $\mathcal{C}^{k,2+\gamma}_{\mathrm{WF}}(P)$, for $k \in \mathbb{N}_0$, and $0 < \gamma < 1$. It turns out that $\mathcal{C}^{k,2+\gamma}_{\mathrm{WF}}(P) \subseteq \mathcal{C}^{k+1,\gamma}_{\mathrm{WF}}(P)$, but $\mathcal{C}^{k+2,\gamma}_{\mathrm{WF}}(P) \nsubseteq \mathcal{C}^{k,2+\gamma}_{\mathrm{WF}}(P)$, which explains the need for considering both families of spaces.

In the passage from this family of model problems to the general problem, we must patch together these model problems with smoothly varying parameters b. Thus it is necessary to prove estimates for solutions of the model problems with data in these Hölder spaces, uniformly for $b \geq \mathbf{0}$, and notably, it is possible to do this. These estimates are obtained using the explicit formulæ for the fundamental solutions; the required analysis is time consuming but elementary. The solution of the homogeneous Cauchy problem

$$\begin{aligned}(\partial_t - L_{b,m})u &= 0 \text{ in } P \times (0,\infty)\\ \text{with } u(p,0) &= f(p),\end{aligned} \tag{31}$$

has an analytic extension to $\operatorname{Re} t > 0$, which satisfies many useful estimates. To obtain a gain of derivatives in a manner that can be extended beyond the model problems, one must address the inhomogeneous problem, which has somewhat simpler analytic properties. By this device, one can also estimate the Laplace transform of the heat semigroup, which is the resolvent operator:

$$(\mu - L_{b,m})^{-1} = \int_0^\infty e^{tL_{b,m}} e^{-\mu t} dt. \tag{32}$$

The estimates for the inhomogeneous problem show that for each $k \in \mathbb{N}_0$, the operator $(\mu - L_{b,m})^{-1}$ maps $\mathcal{C}^{k,\gamma}_{\mathrm{WF}}(P)$ to $\mathcal{C}^{k,2+\gamma}_{\mathrm{WF}}(P)$, gaining two derivatives in the scale of spaces above. It is analytic in $\mu \in \mathbb{C} \setminus (-\infty, 0]$, and one can resynthesize the heat operator from the resolvent operator via contour integration:

$$e^{tL_{b,m}} = \frac{1}{2\pi i} \int_C (\mu - L_{b,m})^{-1} e^{\mu t} d\mu. \tag{33}$$

Here C is of the form $|\arg \mu| = \frac{\pi}{2} + \alpha$, for an $0 < \alpha < \frac{\pi}{2}$. This shows that for t with positive real part, $e^{tL_{b,m}}$ also maps $\mathcal{C}^{k,\gamma}_{\mathrm{WF}}(P)$ to $\mathcal{C}^{k,2+\gamma}_{\mathrm{WF}}(P)$.

4 Perturbation Theory

The next step is to use the analysis of the model operators in a perturbative argument to prove existence and regularity for a generalized Kimura diffusion operator L on a manifold with corners P. There are several difficulties in doing so:

1. The principal part of L degenerates at the boundary.
2. The boundary of P is not smooth.
3. The "indicial roots" of L vary with the location of the point on bP.
4. The character of the solution operator is quite different at points where the vector field V is tangent to bP.

Let us expand on some of these further.

The boundary of a manifold with corners is a stratified space. To handle this, we use an induction on the maximal codimension of the strata of bP. It is for this reason that we need to consider domains that are not simply polyhedra in $\mathbb{R}^N$.

An additional complication when studying a generalized Kimura diffusion operator in dimensions greater than 1 is that the coefficient of the normal first derivative typically varies as one moves along the boundary. This behavior turns out to be mostly invisible in the study of L, but leads to the thorny issue of a smoothly varying indicial root when studying the adjoint operator. This places the analysis of this problem beyond what has been achieved using the detailed kernel methods familiar in geometric microlocal analysis. This means that we must carefully analyze the dependence of the model kernels on b and, in particular, must include the case where some of the b_i vanish on some portion of bP. The uniformity of the estimates in b plays a role precisely here.

The induction starts with the simplest case where bP is a manifold (and P is a manifold with smooth boundary). In this case, we can use the model operators to build a parametrix $\widehat{Q}_t^b$ for the solution operator to the heat equation in a neighborhood of the boundary. Using classical arguments and the ellipticity of L in the interior of P, there is an exact solution operator $\widehat{Q}_t^i$ defined on the complement of a neighborhood of the boundary. We then "glue these together" using a partition of unity to define a parametrix, $\widehat{Q}_t$ for the solution operator. The Laplace transform

$$\widehat{R}(\mu) = \int_0^\infty e^{\mu t} \widehat{Q}_t \mathrm{d}t \tag{34}$$

is then a right parametrix for $(\mu - L)^{-1}$. Using the estimates and analyticity for the model problems, and the properties of the interior solution operator, we can show that

$$(\mu - L)\widehat{R}(\mu) = \mathrm{Id} + E(\mu), \tag{35}$$

where $E(\mu)$ is analytic in $\mathbb{C}\setminus(-\infty, 0]$ with values in the space of bounded operators on $\mathcal{C}_{\mathrm{WF}}^{k,\gamma}$. For any $\alpha > 0$, the Neumann series for $(\mathrm{Id} + E(\mu))^{-1}$ converges in the operator norm topology for μ in sectors $|\arg \mu| \leq \pi - \alpha$, provided $|\mu|$ sufficiently large. This allows us to show that

$$(\mu - L)^{-1} = \widehat{R}(\mu)(\mathrm{Id} + E(\mu))^{-1} \tag{36}$$

is analytic and satisfies certain estimates in

$$T_{\alpha,R} = \{\mu : |\arg \mu| < \pi - \alpha, \quad |\mu| > R\}, \tag{37}$$

for any $0 < \alpha$ and R depending on α.

For t in the right half plane, we can now reconstruct the heat semigroup acting on the Hölder spaces:

$$\mathrm{e}^{tL} = \frac{1}{2\pi \mathrm{i}} \int\limits_{bT_{\alpha,R_\alpha}} (\mu - L)^{-1} \mathrm{e}^{\mu t} \mathrm{d}\mu \tag{38}$$

for any choice of $\alpha > 0$. This allows us to verify that e^{tL} has an analytic continuation to $\mathrm{Re}\, t > 0$, which satisfies the desired estimates with respect to the anisotropic Hölder spaces defined above.

The proof for the general case now proceeds by induction on the maximal codimension of the strata of bP. In all cases we use the model operators to construct a boundary parametrix $\widehat{Q}_t^b$ on a neighborhood of the union of these maximal codimensional strata. By the induction hypothesis, we also obtain an exact solution operator $\widehat{Q}_t^i$ on the complement of a neighborhood of these same maximal codimensional strata. We glue these together as before to obtain a parametrix $\widehat{Q}_t$. A crucial point in this argument is to verify that the heat operator we eventually obtain satisfies a set of hypotheses which allow the induction to be continued. The representation of e^{tL} in (38) is a critical part of this argument.

5 Main Results

We can state our main results. The sharp estimates for the operators e^{tL} and $(\mu - L)^{-1}$ are phrased in terms of the two families of Hölder spaces mentioned earlier. For $k \in \mathbb{N}_0$ and $0 < \gamma < 1$, we define the spaces $\mathcal{C}^{k,\gamma}_{\mathrm{WF}}(P)$, $\mathcal{C}^{k,2+\gamma}_{\mathrm{WF}}(P)$, and their "heat-space" analogues, $\mathcal{C}^{k,\gamma}_{\mathrm{WF}}(P \times [0,T]), \mathcal{C}^{k,2+\gamma}_{\mathrm{WF}}(P \times [0,T])$. For example, in the one-dimensional case, $f \in \mathcal{C}^{0,\gamma}_{\mathrm{WF}}([0,\infty))$ if f is continuous and

$$\sup_{x \neq y} \frac{|f(x) - f(y)|}{|\sqrt{x} - \sqrt{y}|^\gamma} < \infty. \tag{39}$$

It belongs to $\mathcal{C}^{0,2+\gamma}_{\mathrm{WF}}([0,\infty))$ if $f, \partial_x f$, and $x\partial_x^2 f$ all belong to $\mathcal{C}^{0,\gamma}_{\mathrm{WF}}([0,\infty))$, with

$$\lim_{x \to 0+} x\partial_x^2 f(x) = 0.$$

For $k \in \mathbb{N}$, we say that $f \in \mathcal{C}^{k,\gamma}_{\mathrm{WF}}([0,\infty))$, if $f \in \mathcal{C}^k([0,\infty))$, and $\partial_x^k f \in \mathcal{C}^{0,\gamma}_{\mathrm{WF}}([0,\infty))$. A function $g \in \mathcal{C}^{0,\gamma}_{\mathrm{WF}}([0,\infty)\times[0,\infty))$, if $g \in \mathcal{C}^0([0,\infty)\times[0,\infty))$, and

$$\sup_{(x,t)\neq(y,s)} \frac{|g(x,t)-g(y,s)|}{[|\sqrt{x}-\sqrt{y}|+\sqrt{|t-s|}]^{\gamma}} < \infty, \tag{40}$$

etc.

To describe the uniqueness properties for solutions to these equations, consider the geometric structure of the boundary of P. This boundary is a stratified space, with hypersurface boundary components $\{\Sigma_{1,j} : j = 1,\dots,N_1\}$. A boundary component of codimension n is a component of an intersection

$$\Sigma_{1,i_1} \cap \dots \cap \Sigma_{1,i_n}, \tag{41}$$

where $1 \leq i_1 < \dots < i_n \leq N_1$. A component of bP is *minimal* if it is an isolated point or a positive dimensional manifold without boundary. We denote the union of minimal components by bP_{min}. Fix a generalized Kimura diffusion operator L. Let $\{\rho_j : j = 1,\dots,N_1\}$ be defining functions for the hypersurface boundary components. We say that L is tangent to $\Sigma_{1,j}$ if $L\rho_j \restriction_{\Sigma_{1,j}} = 0$, and transverse if there is a $c > 0$ so that

$$L\rho_j \restriction_{\Sigma_{1,j}} > c. \tag{42}$$

If $\Sigma = \Sigma_{1,i_1} \cap \dots \cap \Sigma_{1,i_k}$, then L is transverse to Σ if there is a $c > 0$ so that

$$L\rho_{i_j} \restriction_{\Sigma_{1,i_j}} > c \text{ for } j = 1\dots,k. \tag{43}$$

We say that a stratum Σ of the boundary is *terminal* if $\Sigma \in bP_{\mathrm{min}}$ and L is tangent to Σ, or L is tangent to Σ and L_Σ, its restriction to Σ, is transverse to $b\Sigma$. We denote these components by $bP_{\mathrm{ter}}(L)$. Using a variant of the Hopf maximum principle, we can prove

Theorem 5.1. *Suppose that L is either tangent or transverse to every hypersurface boundary component of bP. The cardinality of the $bP_{\mathrm{ter}}(L)$ equals the dimension of the null space of L acting on $\mathcal{C}^2(P)$.*

Much of [8] is concerned with proving detailed estimates for the model problems with respect to these Hölder spaces. We state the results for the general case.

Theorem 5.2. *Let P be a manifold with corners, L a generalized Kimura diffusion operator on P, $k \in \mathbb{N}_0$ and $0 < \gamma < 1$. If $f \in \mathcal{C}^{k,\gamma}_{\mathrm{WF}}(P)$, then there is a unique solution*

$$v \in \mathcal{C}^{k,\gamma}_{\mathrm{WF}}(P \times [0,\infty)) \cap \mathcal{C}^\infty(P \times (0,\infty))$$

to the initial value problem

$$(\partial_t - L)v = 0 \text{ with } v(p,0) = f(p). \tag{44}$$

This solution has an analytic continuation to t with $\mathrm{Re}\, t > 0$.

Theorem 5.3. *Let P be a manifold with corners, L a generalized Kimura diffusion operator on P, $k \in \mathbb{N}_0$, $0 < \gamma < 1$, and $T > 0$. If $g \in \mathcal{C}^{k,\gamma}_{\mathrm{WF}}(P \times [0,T])$, then there is a unique solution*

$$u \in \mathcal{C}^{k,2+\gamma}_{\mathrm{WF}}(P \times [0,T])$$

to

$$(\partial_t - L)u = g \text{ with } u(p,0) = 0. \tag{45}$$

There is a $C_{k,\gamma}$ so that this solution satisfies

$$\|u\|_{\mathrm{WF},k,2+\gamma,T} \leq C_{k,\gamma}(1+T)\|g\|_{\mathrm{WF},k,\gamma,T}. \tag{46}$$

We also have a result for the resolvent of L acting on the spaces $\mathcal{C}^{k,2+\gamma}_{\mathrm{WF}}(P)$.

Theorem 5.4. *Let P be a manifold with corners, L a generalized Kimura diffusion operator on P, $k \in \mathbb{N}_0$, $0 < \gamma < 1$. The spectrum, E, of the unbounded, closed operator L, with domain*

$$\mathcal{C}^{k,2+\gamma}_{\mathrm{WF}}(P) \subset \mathcal{C}^{k,\gamma}_{\mathrm{WF}}(P),$$

is independent of k, and γ. It lies in a conic neighborhood of $(-\infty, 0]$. The eigenfunctions belong $\mathcal{C}^{\infty}(P)$.

Remark 5.3. Note that $\mathcal{C}^{k,2+\gamma}_{\mathrm{WF}}(P)$ is *not* a dense subspace of $\mathcal{C}^{k,\gamma}_{\mathrm{WF}}(P)$.

Using the Lumer-Phillips theorem, these results suffice to prove that the $\mathcal{C}^0(P)$-graph closure of L acting on $\mathcal{C}^2(P)$ is the generator of a strongly continuous contraction semigroup. This in turn suffices to prove that the solution to the adjoint problem is unique; therefore L^* is the generator of a strongly continuous semigroup, and the associated Martingale problem has a unique solution. A standard argument then shows that the paths for associated Markov process remain, almost surely, within P. From this we can deduce a wide variety of results about the forward Kolmogorov equation. The precise nature of these results depends on the behavior of the vector field V along bP.

We refer to the monograph [8] for detailed definitions, explanations, and proofs of these theorems.

Acknowledgements We would like to acknowledge the generous financial and unflagging personal support provided by Ben Mann and the DARPA FunBio project. It is certainly the case that without Ben's encouragement, we would never have undertaken this project. We would like to thank our FunBio colleagues who provided us with the motivation and knowledge base to pursue this project and Simon Levin for his leadership and inspiration. CLE would like to thank Warren Ewens, Josh Plotkin, and Ricky Der, from whom he has learned most of what he knows about population genetics. We would both like to thank Charlie Fefferman for showing us an explicit formula for $k^0_t(x,y)$, which set us off in the very fruitful direction pursued herein. Work of Charles L. Epstein research partially supported by NSF grant DMS06-03973, and DARPA grants HR00110510057 and HR00110910055. Work of Rafe Mazzeo research partially supported by NSF grant DMS08-05529, and DARPA grants HR00110510057 and HR00110910055. Any opinions, findings, and conclusions or recommendations expressed in this material are those of the authors, and do not nessarily, reflect the views of either the National Science Foundation or DARPA.

References

1. R. F. Bass and E. A. Perkins, *Degenerate stochastic differential equations with Hölder continuous coefficients and super-Markov chains*, Transactions of the American Math Society, 355 (2002), pp. 373–405.
2. S. Cerrai and P. Clément, *Schauder estimates for a class of second order elliptic operators on a cube*, Bull. Sci. Math., 127 (2003), pp. 669–688.
3. ———, *Well-posedness of the martingale problem for some degenerate diffusion processes occurring in dynamics of populations*, Bull. Sci. Math., 128 (2004), pp. 355–389.
4. ———, *Schauder estimates for a degenerate second order elliptic operator on a cube*, J. Differential Equations, 242 (2007), pp. 287–321.
5. F. A. C. C. Chalub and M. O. Souza, *The frequency-dependent Wright-Fisher model: diffusive and non-diffusive approximations*, arXiv:, 1107.1549v1 (2011).
6. P. Daskalopoulos and R. Hamilton, *Regularity of the free boundary for the porous medium equation*, Jour. of the AMS, 11 (1998), pp. 899–965.
7. C. L. Epstein and R. Mazzeo, *Wright-Fisher diffusion in one dimension*, SIAM J. Math. Anal., 42 (2010), pp. 568–608.
8. ———, *Degenerate Diffusion Operators Arising in Population Biology*, to appear in Annals of Math. Studies, Princeton U. Press, 2013.
9. S. N. Ethier, *A class of degenerate diffusion processes occurring in population genetics*, CPAM, 29 (1976), pp. 417–472.
10. S. N. Ethier and T. G. Kurtz, *Markov processes: Characterization and convergence*, Wiley Series in Probability and Mathematical Statistics: Probability and Mathematical Statistics, John Wiley & Sons Inc., New York, 1986.
11. W. Feller, *The parabolic differential equations and the associated semi-groups of transformations*, Ann. of Math., 55 (1952), pp. 468–519.
12. C. Goulaouic and N. Shimakura, *Regularité Hölderienne de certains problèmes aux limites elliptiques dégénérés*, Ann. Scuola Norm. Sup. Pisa Cl. Sci. (4), 10 (1983), pp. 79–108.
13. M. Kimura, *Diffusion models in population genetics*, Journal of Applied Probability, 1 (1964), pp. 177–232.
14. N. Shimakura, *Formulas for diffusion approximations of some gene frequency models*, J. Math. Kyoto Univ., 21 (1981), pp. 19–45.
15. D. W. Stroock, *Partial differential equations for probabilists*, vol. 112 of Cambridge Studies in Advanced Mathematics, Cambridge University Press, Cambridge, 2008.

A Matrix Related to the Theorem of Fermat and the Goldbach Conjecture

Hershel M. Farkas

Dedicated to the Memory of Leon Ehrenpreis

Abstract In this chapter, we show how converting a Lambert series to a Taylor series introduces a matrix similar to the Redheffer matrix, whose inverse is determined by the Mobius function. A variant of the Mobius function which generalizes the Littlewood function along with this matrix allows one to count the integral solutions to the equation $x^l + y^l = r$. Similar ideas hold for the Goldbach conjecture.

Key words Fermat's theorem • Goldbach conjecture • Mobius function • Littlewood function

Mathematics Subject Classification: 11A25, 11A41

1 Introduction

In this chapter, we show how a matrix which is a "variant" of the Redheffer matrix [B,F,P], [V1] is connected to the theorem of Fermat and the Goldbach conjecture. A generalization of the classical Liouville function, which itself is a "variant" of the Mobius function [H,W], appears and allows a reformulation of Fermat's theorem. We hasten to admit that we do not give a new proof of Fermat or prove Goldbach

H.M. Farkas (✉)
Institute of Mathematics, The Hebrew University of Jerusalem, Jerusalem, Israel
e-mail: farkas@math.huji.ac.il

H.M. Farkas et al. (eds.), *From Fourier Analysis and Number Theory to Radon Transforms and Geometry*, Developments in Mathematics 28, DOI 10.1007/978-1-4614-4075-8_9,

here but hope that there is some possibility of using the ideas here presented to do so. In order to motivate what we shall be doing in the sequel, we begin with a simple example.

Let us begin with a function defined by the Lambert series

$$f(z) = \sum_{k=1}^{\infty} \frac{a_k z^k}{1 - z^k}.$$

For the sake of simplicity, we assume that this series is uniformly convergent in the unit disc and therefore defines therein a holomorphic function which vanishes at the origin. This assumption is of course unnecessary since we will be treating these as formal series. This is explained in the book by Stanley [S] in Chap. 1 where generating functions are discussed. The Taylor series of this function centered at the origin is easily seen to be

$$\begin{aligned}
\sum_{k=1}^{\infty} a_k z^k (1 + z^k + z^{2k} + \cdots + z^{nk} + \cdots) &= \sum_{k=1}^{\infty} a_k (z^k + z^{2k} + \cdots) \\
&= a_1(z + z^2 + z^3 + \cdots) + a_2(z^2 + z^4 + z^6 + \cdots) + a_3(z^3 + z^6 + z^9 + \cdots) \\
&\quad + \cdots + a_l(z^l + z^{2l} + \cdots) + \cdots \\
&= \sum_{n=1}^{\infty} \left(\sum_{d|n} a_d \right) z^n = \sum_{n=1}^{\infty} c_n z^n
\end{aligned}$$

where

$$c_n = \sum_{d|n} a_d.$$

We have thus written a formula which converts the Lambert series to a Taylor series and have given the Taylor coefficients around the origin in terms of the coefficients a_k of the Lambert series.

The above calculation can be put into matrix form. We denote by A the matrix whose first row has a 1 in the first place and all other entries 0, whose second row has a 1 in the first and second place and the remaining entries 0, whose kth row has a one in every place which is a divisor of k and zeros elsewhere. This matrix has an infinite number of rows and columns and clearly, by construction, satisfies the matrix equation

$$\begin{pmatrix} c_1 \\ c_2 \\ c_3 \\ \cdots \\ c_n \\ \cdots \end{pmatrix} = A \begin{pmatrix} a_1 \\ a_2 \\ a_3 \\ \cdots \\ a_n \\ \cdots \end{pmatrix}.$$

The matrix A will be our starting point in this discussion, and we have here tried to motivate its discussion. In the final section of this note, we shall show that this matrix A is a "close relative" of the Redheffer matrix mentioned above.

2 Properties of the Matrix A

A closer look at the matrix A shows that the matrix can alternatively be defined in the following way. The first column consists of a 1 in each row. The second column has a 1 in each row which is a multiple of two and 0 elsewhere, while the kth column has a 1 in each row which is a multiple of k and 0 elsewhere. If we cut the matrix off after n rows and n columns obtaining a square matrix with n rows and n columns, since the matrix is lower triangular with ones on the major diagonal, it is clear that for each n, the determinant of the matrix is 1 and thus, that for each n, the matrix is nonsingular and has an inverse. Let us denote the finite square matrix with n rows and n columns by A_n so that for $n = 5$, we have

$$A_5 = \begin{pmatrix} 1\,0\,0\,0\,0 \\ 1\,1\,0\,0\,0 \\ 1\,0\,1\,0\,0 \\ 1\,1\,0\,1\,0 \\ 1\,0\,0\,0\,1 \end{pmatrix}.$$

The inverse is easily computed to be

$$B_5 = A_5^{-1} = \begin{pmatrix} 1 & 0 & 0\,0\,0 \\ -1 & 1 & 0\,0\,0 \\ -1 & 0 & 1\,0\,0 \\ 0 & -1 & 0\,1\,0 \\ -1 & 0 & 0\,0\,1 \end{pmatrix}.$$

In fact, it is not hard to see that the infinite matrix A has an inverse whose description is most easily given using the classical Mobius function. We recall that the classical Mobius function $\mu(n)$ is defined as follows: $\mu(1) = 1$ and if p is a prime $\mu(p) = -1$. If n is a positive integer whose prime decomposition is $p_1^{n_1}, \ldots, p_k^{n_k}$, then if each $n_i = 1$, we define $\mu(n) = \prod_{i=1}^{k} \mu(p_i)$, while if any n_i is greater than 1, we define $\mu(n) = 0$. We can thus say that if n is not square free, $\mu(n) = 0$, while if n is square free, $\mu(n) = 1$ if n has an even number of prime factors and $= -1$ if n has an odd number of prime factors. From this description, we see that the matrix B_5 above can be described as follows:

$$B_5 = \begin{pmatrix} \mu(1) & 0 & 0 & 0 & 0 \\ \mu(2) & \mu(1) & 0 & 0 & 0 \\ \mu(3) & 0 & \mu(1) & 0 & 0 \\ \mu(4) & \mu(2) & 0 & \mu(1) & 0 \\ \mu(5) & 0 & 0 & 0 & \mu(1) \end{pmatrix}.$$

From the above, it is not hard to guess that the inverse to the general matrix A should be the matrix whose first column has the value $\mu(n)$ in the nth row, whose second column has the value 0 in all odd numbered rows and the value $\mu(k)$ in the $2k$th row, and whose lth column has the value $\mu(k)$ in the lkth row and all other values 0.

If we extend the definition of the Mobius function to be zero on the positive rationals which are not integers, then the picture we have given above is very easy to describe. The element B_{mn} of the matrix B is simply $\mu(\frac{m}{n})$.

Lemma 1. *Let* $B_{mn} = \mu(\frac{m}{n})$. *Then*

$$A * B = I.$$

Proof. Clearly, the element

$$(A * B)_{ij} = \sum_k A_{ik} B_{kj} = \sum_k A_{ik} \mu\left(\frac{k}{j}\right).$$

It is clear that if $j > i$, then since $\mu(l/j) = 0$ for each $l \leq i$, the result vanishes. It is also clear that if $j = i$, the sum is just equal to 1. Hence, we need only show that for $j < i$, the sum vanishes. The reason this will be true is the well-known property of the Mobius function which asserts

$$\sum_{d|n} \mu(d) = 0$$

for all $n \geq 2$ and equals 1 when $n = 1$.

Let us denote the jth column of the matrix B by B^j. It is then clear from the above that $A * B^1 = \mathrm{e}^1$ where e^1 is as usual the vector with a 1 in the first place and zeros elsewhere. Let us note immediately that if j does not divide i, $A_{i,lj} = 0$ for all l. Since these are the only terms which appear in the sum, the sum vanishes. We can therefore assume that j does divide i and that in fact $i = mj$. Our sum now reduces to

$$A_{i,j}\mu(1) + A_{i,2j}\mu(2) + \cdots + A_{i,mj}\mu(m).$$

The expressions $A_{i,kj}$ do not vanish precisely when k divides m. Hence, the above can be rewritten as

$$\sum_{d|m} \mu(d)$$

which vanishes for all $m \geq 2$. This concludes the proof showing that indeed $B = A^{-1}$.

The matrix A has some interesting properties which are worth mentioning at this point. For all integers k, let us denote the vector $(1^k, 2^k, 3^k, \ldots, n^k, \ldots)$ by N_k. Then

$$A * N_k^t = (\sigma_k(1), \sigma_k(2), \ldots, \sigma_k(n), \ldots) = \Sigma_k$$

where $\sigma_k(n)$ is the sum of the kth powers of the divisors of n. In particular, $\sigma_0(n) = d(n)$ the number of divisors of n.

The interest in this is from the fact that we now have the following relation between Lambert series and Taylor series:

$$\sum_{n=1}^{\infty} \frac{n^k z^n}{1 - z^n} = \sum_{n=1}^{\infty} \sigma_k(n) z^n.$$

More important for what we wish to do here is the fact which follows from the above lemma that $A * B^k = \mathrm{e}^k$. From this, we can conclude that

$$z = \sum_{k=1}^{\infty} \frac{\mu(k) z^k}{1 - z^k}$$

$$z^4 = \sum_{k=1}^{\infty} \frac{\mu(\frac{k}{4}) z^k}{1 - z^k}$$

$$z^8 = \sum_{k=1}^{\infty} \frac{\mu(\frac{k}{8}) z^k}{1 - z^k}$$

and in general that for any positive integers m, l we have

$$z^{m^l} = \sum_{k=1}^{\infty} \frac{\mu(\frac{k}{m^l}) z^k}{1 - z^k}.$$

If we now fix $l \geq 1$, we have $z + z^{2^l} + z^{3^l} + \cdots + z^{n^l} + \cdots$ can be written as

$$\sum_{n=1}^{\infty} \frac{\mu(n) z^n}{1 - z^n} + \sum_{n=1}^{\infty} \frac{\mu(\frac{n}{2^l}) z^n}{1 - z^n} + \cdots + \sum_{n=1}^{\infty} \frac{\mu(\frac{n}{k^l}) z^n}{1 - z^n} + \cdots$$

$$= \sum_{n=1}^{\infty} \frac{\phi_l(n) z^n}{1 - z^n}$$

where

$$\phi_l(n) = \sum_{k=1}^{\infty} \mu\left(\frac{n}{k^l}\right).$$

It turns out that the case $l = 1$ is exceptional here, so we shall at this point assume that l is at least 2.

With the hypothesis $l \geq 2$ in force, it is clear that there is at most one nonzero term in this sum. This is the term $\mu(\frac{n}{m^l})$ where m^l is the largest lth power which divides n. Every other term will have to vanish either because the quotient is not a positive integer or because the integer is not square free.

There is another way to define this function $\phi_l(n)$ which we now describe.

Definition 1. For p a prime and l and α positive integers with l at least 2, we define

$$\phi_l(p^\alpha) = \begin{matrix} -1 & \alpha \equiv 1 \mod l \\ 0 & \alpha \equiv 2, \ldots, l-1 \mod l \\ 1 & \alpha \equiv 0 \mod l \end{matrix}$$

Extend $\phi_l(n)$ to be a multiplicative function on the positive integers with as usual $\phi_l(1) = 1$.

It is clear that for $l \geq 2$, the above definition of $\phi_l(n)$ coincides with the original definition.

If we wish to also consider the case $l = 1$, which we probably should, then the appropriate definition is $\phi_1(1) = 1$ and $\phi_1(n) = 0$ for all $n \geq 2$.

The case $l = 2$ is the function defined by Liouville and called the Liouville function. It is generally denoted by $\lambda(n)$ and, as we have already seen, satisfies

$$\sum_{n=1}^{\infty} z^{n^2} = \sum_{n=1}^{\infty} \frac{\phi_2(n) z^n}{1 - z^n} = \sum_{n=1}^{\infty} \frac{\lambda(n) z^n}{1 - z^n}.$$

In fact, the Mobius function and Liouville functions also satisfy identities related to the zeta function. The identities in question are [H,W, Chap. 17]

$$\sum_{n=1}^{\infty} \frac{\mu(n)}{n^s} = \frac{1}{\zeta(s)}, \sum_{n=1}^{\infty} \frac{\lambda(n)}{n^s} = \frac{\zeta(2s)}{\zeta(s)}.$$

While the following is not important for our discussion, here we point out the not too surprising fact that

Proposition 1. *For $l \geq 2$, we have*

$$\sum_{n=1}^{\infty} \frac{\phi_l(n)}{n^s} = \frac{\zeta(ls)}{\zeta(s)}.$$

Proof. Begin by recalling that

$$\zeta(s) = \prod_{p,\text{prime}} \frac{1}{1-p^{-s}}.$$

Hence, we have

$$\frac{\zeta(ls)}{\zeta(s)} = \prod_p \frac{1-p^{-s}}{1-p^{-ls}} = \prod_p (1-p^{-s}) \left(\sum_{n=0}^{\infty} p^{-lsn} \right)$$

$$= \prod_p \left(\sum_{n=0}^{\infty} p^{-lsn} - p^{-(nl+1)s} \right) = \prod_p \sum_{n=0}^{\infty} \phi_l(p^n) p^{-ns} = \sum_{n=1}^{\infty} \frac{\phi_l(n)}{n^s}.$$

The last equality is just the uniqueness of representations of integers as a product of primesand the multiplicativity of the function $\phi_l(n)$.

We remarked previously that the case $l = 1$ is exceptional but note that if we would take $l = 1$ here, the correct definition for $\phi_1(n)$ is as given previously. The Mobius function can be thought of also as the limit of ϕ_l as l tends to ∞.

As a consequence of the above, we observe once again that

Proposition 2. $\sum_{d,d|n} \phi_l(d) = 1$ *if n is an lth power and vanishes otherwise.*

Proof.

$$\sum_{n=1}^{\infty} \frac{\phi_l(n)}{n^s} \cdot \sum_{n=1}^{\infty} \frac{1}{n^s} = \zeta(ls) = \sum_{n=1}^{\infty} \frac{1}{n^{ls}}.$$

In this chapter, we think of the matrix A as an operator on the space of Lambert series transforming them to power series. However, the above suggests that we can also think of A as an operator on the space of Dirichlet series taking the Dirichlet series

$$\sum_{n=1}^{\infty} \frac{a_n}{n^s} \to \sum_{n=1}^{\infty} \frac{c_n}{n^s}$$

where $c_n = \sum_{d|n} a_d$. If one does this, then one immediately sees that the operator A is simply multiplication by $\zeta(s)$ and explains the above formula $\sum_{n=1}^{\infty} \frac{\mu(n)}{n^s} = \frac{1}{\zeta(s)}$. This point of view shows the following: If we recall Euler's ϕ function, $\phi(n)$, the number of positive integers less than n which are relatively prime to n, and the fact that

$$\sum_{d|n} \phi(d) = n$$

we see two things. The first from the point of view of Dirichlet series that

$$\sum_{n=1}^{\infty} \frac{\phi(n)}{n^s} \zeta(s) = \sum_{n=1}^{\infty} \frac{n}{n^s} = \zeta(s-1)$$

and the second that

$$\sum_{n=1}^{\infty} \frac{\phi(n)z^n}{1-z^n} = \sum_{n=1}^{\infty} nz^n$$

the Koebe function expressed as a Lambert series.

In general, we also observe that if we let $\sum_{n=1}^{\infty} \frac{n^k}{n^s} = \zeta(s-k)$, then $\zeta(s)\zeta(s-k) = \sum_{n=1}^{\infty} \frac{\sigma_k(n)}{n^s}$.

3 Sums of lth Powers

We now ask the following question!

Let N and l be positive integers. How many nontrivial representations does N have as a sum of two lth powers. By nontrivial we mean $N = x^l + y^l$ with x, y positive integers. For example, if $N = 4$ and $l = 2$, the representation $0^2 + 2^2$ is trivial and is not counted. The answer to the above question is remarkably easy in the sense that there is a simple algorithm for the solution.

We construct a vector in Z^{N-1} which consists of zeros and ones. We put a 1 in the first place and in every other place that is an lth power. Put zeros in the remaining places. Let us denote this vector by $\chi_l(N)$. As an example, take $l = 3$, $N = 5, 10$ so that $\chi_3(5) = (1,0,0,0)$, $\chi_3(10) = (1,0,0,0,0,0,0,1,0)$. Let D_{N-1} be the square $N-1$ by $N-1$ matrix satisfying

$$(D_{N-1})_{k,l} = \begin{matrix} 1 & k+l=N \\ 0 & \text{otherwise} \end{matrix}.$$

We now consider the quadratic form $\chi_l(N)D_{N-1}\chi_l^t(N)$.

Theorem 1. *The number of nontrivial representations of N as a sum of two lth powers is given by*

$$\chi_l(N)D_{N-1}\chi_l^t(N).$$

Proof.

$$\chi_l(N)D_{N-1}\chi_l^t(N) = (a_1, a_2, \ldots, a_{N-1})D_{N-1}(a_1, a_2, \ldots, a_{N-1})^t$$

$$= \sum_{i,j=1}^{N-1} a_i a_j (D_{N-1})_{i,j} = \sum_{k=1}^{N-1} a_k a_{N-k}.$$

The summands $a_k a_{N-k}$ are equal to 0 or 1 with the value 1 assumed only when both the indices k and $N-k$ are lth powers. If this occurs, we have $k = m^l$, $N-k = r^l$ and $m^l + r^l = N$. The above sum counts the number of times this happens.

Fermat's theorem says that when $l \geq 3$ and $N = c^l$, we have $\chi_l(N)D_{N-1}\,\chi_l^t(N) = \chi_l(c^l)D_{c^l-1}\chi_l^t(c^l) = 0$.

There does not seem to be any simple way to conclude Fermat's theorem from the above even though we have an explicit formula for the number of representations. The geometric statement is that if we write down the vector $\chi_l(N)$ as a vector in Z^{N-1} and then write it down backwards as a vector in Z^{N-1}, the inner product of these vectors must vanish when N is an lth power.

The fact that we know almost everything when $l = 2$ allows us to conclude many things. For example, if $N \equiv 3 \bmod 4$, then we clearly have

$$\chi_2(N)D_{N-1}\chi_2^t(N) = 0.$$

Since there are also exact formulas for the number of representations in terms of the number of divisors congruent to 1 and 3 mod 4 (although these formulae do not demand nontrivial representations), we can also get formulas for the above expression in those cases. Finally, we make the point that we at least understand that $l = 2$ is different than $l > 2$ in the sense that when $l = 2$, $\phi_l(k)$ never vanishes. This is not true for l larger than 2. It is of course not clear how to use this to conclude Fermat.

4 Sums of Two Primes

In the preceding section, we asked the question whether a number N is the sum of two lth powers. In this section, we ask the question whether a given number N is the sum of two primes. The famous Goldbach conjecture is that every even number larger than 2 is the sum of two primes. The same ideas of the previous section apply only now we use the vector $\chi_{\text{prime}}(N)$ which is a vector with a 0 in the first place and a 1 in every place that is a prime. All other entries are 0. Hence, the vector $\chi_{\text{prime}}(10) = (0, 1, 1, 0, 1, 0, 1, 0, 0)$. It is clear that the number of representations of N as a sum of two primes is now given by $\chi_{\text{prime}}(N)D_{N-1}\chi_{\text{prime}}^t(N)$. This formula gives the number of representations of 12 as 2 with $12 = 5 + 7 = 7 + 5$. Geometrically, we take the inner product of (0,1,1,0,1,0,1,0,0,0,1) with (1,0,0,0,1,0,1,0,1,1,0). It is clear that the inner product in general is positive only when there is at least one prime p in the kth place and also one companion prime p' in the $N - k$ place. In this case, we get $p + p' = N$. In this language, the assertion of the Goldbach conjecture is that if $N = 2m$ with $m \geq 2$, then there is at least one prime p with $2 \leq p \leq m$ such that $2m - p$ is also a prime. In fact, there is another way to say this which is maybe more suggestive. It is that given any positive integer N at least two, there is a nonnegative integer x such that $N - x$ and $N + x$ are primes. In other words, that every integer lies equidistant between two primes. If N is itself a prime, then $x = 0$. If $x = 1$, then N is an even integer which lies between a pair of twin primes.

Once again, there does not seem to be any simple way of deducing Goldbach from the above simple (in some sense) explicit formula. Namely, to conclude that when $N = 2m$ the quadratic form does not vanish.

5 Return of the Mobius Function

In Sect. 2 above, we have introduced the Mobius function $\mu(n)$ and saw that it was intimately related to the matrix $B = A^{-1}$. We have already also seen above that

$$\sum_{n=1}^{\infty} z^{n^l} = \sum_{n=1}^{\infty} \frac{\phi_l(n) z^n}{1 - z^n}$$

thus concluding that if for each N we denote the vector

$$(\phi_l(1), \phi_l(2), \ldots, \phi_l(N-1))$$

by $\Phi_l(N)$, we clearly have the relation

$$A_{N-1} * \Phi_l(N) = \chi_l(N).$$

We now return to our matrix A and use it to define a new finite square matrix with $N - 1$ rows and columns. We define a matrix

$$P_{N-1} = A_{N-1}^t D_{N-1} A_{N-1}$$

where A_{N-1} is the matrix A cut off after $N - 1$ rows and columns. It is clear that P_{N-1} is a symmetric matrix with positive integer entries.

Theorem 2. *The quadratic form*

$$\Phi_l(N) P_{N-1} \Phi_l^t(N)$$

counts the number of nontrivial representations of N as a sum of two lth powers.

Proof.

$$\Phi_l(N) P_{N-1} \Phi_l^t(N) = \Phi_l(N) A_{N-1}^t D_{N-1} A_{N-1} \Phi_l^t(N) = \chi_l(N) D_{N-1} \chi_l^t(N)$$

and this, as we have already seen above, counts the number of nontrivial representations of N as a sum of two lth powers.

In particular, we see that if we take $N = c^l$, the quadratic form vanishes by Fermat's theorem. We of course would like to show that the form vanishes and thus prove Fermat this way.

The matrix P_N is easily constructed from the matrix A_N. Since $D_N * A_N$ is simply the matrix A written upside down, i.e., with the last row of A_N being the first row of $D_N * A_N$, we will denote $D_N * A_N$ by $\tilde{A}_N$. It thus follows that we can write P as $A^t * \tilde{A}$. As an example, consider the case $N = 6$. we have

$$A_5 = \begin{pmatrix} 1\,0\,0\,0\,0 \\ 1\,1\,0\,0\,0 \\ 1\,0\,1\,0\,0 \\ 1\,1\,0\,1\,0 \\ 1\,0\,0\,0\,1 \end{pmatrix}, A_5^t = \begin{pmatrix} 1\,1\,1\,1\,1 \\ 0\,1\,0\,1\,0 \\ 0\,0\,1\,0\,0 \\ 0\,0\,0\,1\,0 \\ 0\,0\,0\,0\,1 \end{pmatrix}, \tilde{A}_5 = \begin{pmatrix} 1\,0\,0\,0\,1 \\ 1\,1\,0\,1\,0 \\ 1\,0\,1\,0\,0 \\ 1\,1\,0\,0\,0 \\ 1\,0\,0\,0\,0 \end{pmatrix}.$$

It thus follows that

$$P_5 = A_5^t * \tilde{A}_5 = \begin{pmatrix} 5\,2\,1\,1\,1 \\ 2\,2\,0\,1\,0 \\ 1\,0\,1\,0\,0 \\ 1\,1\,0\,0\,0 \\ 1\,0\,0\,0\,0 \end{pmatrix}.$$

Let us note the following about the matrix P_5.

The element $P_{11} = 5$ and is equal to the number of representations of 6 as $x \cdot 1 + y \cdot 1$ with x and y positive integers. The element $P_{12} = 2$ and is equal to the number of representations of 6 as $x \cdot 1 + y \cdot 2$ again with x and y positive integers. The element $P_{22} = 2$ and is equal to the number of representations of 6 as $x \cdot 2 + y \cdot 2$ again with x and y positive integers. In fact, the reader can easily check that the entry P_{ij} of the matrix is the number of representations of 6 as $x \cdot i + y \cdot j$ with x and y positive integers. This is a general fact about the matrix P_N.

Theorem 3. *The entry P_{ij} of the matrix P_{N-1} is the number of representations of N as $x \cdot i + y \cdot j$ with x and y positive integers. In particular, the elements $P_{k,N-k} = 1$ for all $k = 1, \ldots, N-1$ and $P_{ij} = 0$ for all pairs i, j with $i + j \geq N+1$.*

Proof. We recall that we have already seen that

$$\sum_{n=1}^{\infty} z^{n^l} = \sum_{n=1}^{\infty} \frac{\phi_l(n) z^n}{1 - z^n}.$$

It thus follows that

$$\left(\sum_{n=1}^{\infty} z^{n^l} \right)^2 = \sum_{n=1}^{\infty} r_l(n) z^n$$

where $r_l(n)$ is the number of nontrivial representations of n as a sum of two lth powers. It is easy to see that

$$\left(\sum_{n=1}^{\infty} z^{n^l} \right)^2 = \sum_{n=1}^{\infty} \chi_l(N) D_{N-1} \chi_l(N)^t z^n$$

which we have already seen is the same as

$$\sum_{n=1}^{\infty} \Phi_l(n) P_{N-1} \Phi_l(N)^t z^n.$$

We now simply compute

$$\sum_{k=1}^{\infty} \frac{\phi_l(k) z^k}{1-z^k} \cdot \sum_{m=1}^{\infty} \frac{\phi_l(m) z^m}{1-z^m} = \sum_{k,m=1}^{\infty} \frac{\phi_l(k)\phi_l(m) z^{k+m}}{(1-z^k)(1-z^m)}$$

$$= \sum_{k,m=1}^{\infty} \phi_l(k)\phi_l(m)(z^k + z^{2k} + \cdots)(z^m + z^{2m} + \cdots)$$

$$= \sum_{N=1}^{\infty} \left(\sum_{k,m=1}^{N-1} \phi_l(k)\phi_l(m)(P_{\tilde{N}-1})_{km} z^N \right)$$

with $(P_{\tilde{N}-1})_{kl}$ the number of representations of N as stated. Finally, though we see that $P_{\tilde{N}-1} = P_{N-1}$ by uniqueness of power series coefficients.

In fact, returning now to our remarks at the end of Sect. 2, we see that if we define $\Phi(N) = (\phi(1), \phi(2), \ldots, \phi(N-1))$ with ϕ Euler's function, we see that

$$\Phi(N) P_{N-1} \Phi(N)^t = (1, 2, \ldots, N-1) D_{N-1} (1, 2, \ldots, N-1)^t = \frac{N^3 - N}{6}.$$

Our final remarks in this section are that Fermat's claim was that the equation

$$x^n + y^n = z^n$$

has no nontrivial integer solutions when n is at least 3. This claim took a very long time to prove. There are easier claims though which have very simple proofs.

Proposition 3. *Let n be a positive odd integer at least 3. Let p be any odd prime. Then the equation*

$$x^n + y^n = p$$

has no positive integer solutions.

Proof. Suppose there were a solution (x_0, y_0) with both x_0, y_0 positive integers. Then we would have

$$p = x_0^n + y_0^n = (x_0 + y_0)(x_0^{n-1} - x_0^{n-2} y + x_0^{n-3} y_0^2 - \ldots - x_0 y_0^{n-2} + y_0^{n-1})$$

and thus $x_0 + y_0$ would have to divide p and therefore since p is prime would have to equal p. We are, however, given that $x_0^n + y_0^n = p$ and clearly unless $(x_0, y_0) = (1, 1)$ and $p = 2$

$$x_0 + y_0 < x_0^n + y_0^n$$

which is a contradiction.

An immediate corollary is

Corollary 1.

$$\chi_n(p)D_{p-1}\chi_n^t(p) = 0$$

$$\Phi_n(p)P_{p-1}\Phi_n^t(p) = 0$$

for every odd $n \geq 3$ and odd prime p.

Note that for every n, it is true that $1^n + 1^n = 2$.

In particular, we see that the number $p-1$ for an odd prime p can never be an lth power when l is at least 3. Of course, when $l = 2$, it can be and in fact the question of whether there are an infinite number of primes of the form $n^2 + 1$ is still open. The answer to the same question for $n^{2k+1} + 1$ with k at least 1 is much easier. There are no such primes.

There is another case of Fermat's theorem which is easy to prove.

Proposition 4. *Let n be a positive odd integer at least 3 and let p be any prime. Then there are no nontrivial solutions to*

$$x^n + y^n = p^n.$$

Proof. The proof is very similar to the proof of the previous proposition. Suppose

$$p^n = x_0^n + y_0^n = (x_0 + y_0)(x_0^{n-1} - x_0^{n-2}y + x_0^{n-3}y_0^2 - \cdots - x_0y_0^{n-2} + y_0^{n-1}).$$

Then $(x_0 + y_0)$ must divide p^n and since the divisors of p^n are $1, p, p^2, \ldots, p^n$, $x_0 + y_0 = p^k$ for some $k = 1, \ldots, n$.

If $k = n$, then $(x_0 + y_0) = p^n$, and this contradicts $x_0^n + y_0^n = p^n$ since unless $(x_0, y_0) = (1, 1)$

$$(x_0 + y_0) < x_0^n + y_0^n.$$

The reader easily sees though that (1,1) is not a solution so we have a contradiction.

If $k = 1$, then $x_0 + y_0 = p$ and therefore

$$p^n = (x_0 + y_0)^n = x_0^n + y_0^n + \epsilon(x_0, y_o)$$

with $\epsilon(x_0, y_o)$ positive and thus $x_0^n + y_0^n < p^n$ which is again a contradiction.

Finally, if $k = 2, \ldots, n-1$, we have $x_0 + y_0 = p^k$. This, however, is already a contradiction since if $x_0^n + y_o^n = p^n$, it must be the case that $x_0 < p, y_0 < p$. This implies that

$$x_0 + y_0 < 2p < p^k$$

unless $p = k = 2$. However, in this case, we have $x_0^n + y_0^n = 2^n$ which clearly has no solutions.

Thus, Fermat's theorem is simple also when the right-hand side is an nth power of a prime. This of course yields the following result:

Corollary 2.

$$\Phi_l(p^l)P_{p^l-1}\Phi_l^t(p^l) = 0.$$

The same ideas just used also give us a way of converting the quadratic form

$$\chi_{\text{prime}}(N)D_{N-1}\chi^t_{\text{prime}}(N)$$

into a form involving the matrix P_{N-1}. The only issue is what is the vector we need to use to replace the vector χ_{prime}. If we denote the components of the vector $\chi_{\text{prime}}(N)$ by $\chi(j)$, then the sought after vector is just $A_{N-1}^{-1} * \chi_{\text{prime}}(N)$. Denoting the above by the vector $(a_1, \ldots, a_{N-1})$, it is clear that

$$a_k = \sum_{j=1}^{N-1} \mu\left(\frac{k}{j}\right)\chi(j) = \sum_{p|k} \mu\left(\frac{k}{p}\right)$$

where p of course is a prime.

Let us take an example to see how this works: Recall that we have already computed

$$P_6 = \begin{pmatrix} 5\,2\,1\,1\,1 \\ 2\,2\,0\,1\,0 \\ 1\,0\,1\,0\,0 \\ 1\,1\,0\,0\,0 \\ 1\,0\,0\,0\,0 \end{pmatrix}.$$

It thus follows that $A_6^{-1} * \chi_{\text{prime}}(6) = (a_1, \ldots, a_5)^t$ with a_k defined above. We therefore have

$$a_1 = 0, a_2 = \mu(2/2) = 1, a_3 = \mu(3/3) = 1, a_4 = \mu(4/2) = \mu(2) = -1,$$
$$a_5 = \mu(5/5) = 1.$$

We therefore find that

$$(0,1,1,-1,1) * P_6 * (0,1,1,-1,1)^t$$
$$= (0,1,1,-1,1) * \begin{pmatrix} 5\,2\,1\,1\,1 \\ 2\,2\,0\,1\,0 \\ 1\,0\,1\,0\,0 \\ 1\,1\,0\,0\,0 \\ 1\,0\,0\,0\,0 \end{pmatrix} * (0,1,1,-1,1)^t$$
$$= (3,1,1,1,0) * (0,1,1,-1,1)^t = 1.$$

This gives the fact that 6 is representable as a sum of two primes in precisely one way, namely, $6 = 3 + 3$.

We also point out that in the statement of Theorem 3 above, there was no reason to stop with the computation $\sum_{k=1}^{\infty} \frac{\phi_l(k)z^k}{1-z^k} * \sum_{m=1}^{\infty} \frac{\phi_l(m)z^m}{1-z^m}$. We could just as well computed the product of r sums and obtained

$$\sum_{i_1,\dots,i_r} \phi_l(i_1),\dots,\phi_l(i_r)P_{i_1,i_2,\dots,i_r}(N)$$

where $P_{i_1,\dots,i_r}(N)$ is the number of representations of N as $x_1 * i_1 + \dots + x_r * i_r$ with all x_i positive integers as the number of nontrivial representations of N as a sum of r lth powers. In fact, even for a simple product of two sums $\sum \frac{\phi_l(n)z^n}{1-z^n} * \sum \frac{\phi_m(n)z^n}{1-z^n}$ we can obtain $\Phi_l(N)P_{n=1}\Phi_m^t(N)$ is the number of solutions to

$$x^l + y^m = N.$$

6 Concluding Remarks

In this section, we will show two things: The first a quadratic identity for the Mobius function and then why there is some possibility that one could prove either Fermat or Goldbach from these considerations. In addition, as promised in the introduction, we point out how the Redheffer matrix is also easily constructed from the matrix A. We begin with the quadratic identity.

Let us denote the vector $(\mu(1),\dots,\mu(N-1))$ by $\mu(N)$. Then it is clear that

$$\mu(N) * P_{N-1} * \mu^t(N) = 0.$$

This is so because we have already seen that $A_{N-1} * \mu^t(N) = \mathrm{e}^1$. It thus follows that $\mu(N) * P_{N-1} * \mu^t(N) = (\mathrm{e}^1)^t * D_{N-1} * \mathrm{e}^1$ which clearly vanishes.

It is also clear that for any l with $l \geq 2$, we have whenever $\mu(k) \neq 0$ that $\phi_l(k) = \mu(k)$. It thus follows that

$$\begin{aligned}\mu(N) * P_{N-1} * \mu^t(N) &= \sum_{i,k=1}^{N-1} \mu(i)\mu(k)(P_{N-1})_{ik} \\ &= \sum_{m,n=1}^{N-1} \phi_l(m)\phi_l(n)(P_{N-1})_{mn}\end{aligned}$$

where the sum is taken over all m,n with $\mu(m)$ and $\mu(n)$ both $\neq 0$. It thus also follows that

$$\sum_{p,q=1}^{N-1} \phi_l(p)\phi_l(q)(P_{N-1})_{pq}$$

where at least one of $\mu(p), \mu(q)$ equals zero counts the number of nontrivial representations of N as a sum of two lth powers. For example, when $l = 2$ and $N = 5$, we would get that the number of representation of 5 as a sum of two squares is given by

$$2\phi_2(1)\phi_2(4)(P_4)_{14} + 2\phi_2(2)\phi_2(4)(P_4)_{24} + 2\phi_2(3)\phi_2(4)(P_4)_{34} + \phi_2^2(4)(P_4)_{44}.$$

All terms but the first vanish so the result as expected is 2.

Our final remarks concern the relation of our matrix A to the Redheffer matrix. We remark that the fact that the Mobius function is related to the Riemann hypothesis and prime number theorem has been known for a long time. The connection is via estimates on $\sum_k \mu(k)$. The Redheffer matrix is a square n by n matrix whose determinant equals $\sum_{k=1}^{n} \mu(k)$. It is defined see [V2] as follows: The Redheffer matrix $B_n = (b_{ij})_{i,j=1,\ldots,n}$ is defined by $b_{ij} = 1$ when i divides j and is zero otherwise. Our final remark here is that our matrix A_N can easily be converted into a matrix with the same property.

We define a matrix $\tilde{R}$ by changing all the zeros in the first row of A_N to ones. It is then evident that the product $\tilde{R} * A_N^{-1} = C$, where C is a matrix with $C_{11} = \sum_k \mu(k)$ and $C_{k1} = 0$ for all $k \geq 2$. Striking out the first row and column of C leaves the $N-1$ by $N-1$ identity matrix so clearly $\det(C) = \sum_{k=1}^{N} \mu(k)$. Finally, since

$$\det C = \det\tilde{R} * \det A_N^{-1} = \det\tilde{R}$$

because $\det A_N^{-1} = 1$, we are done.

While we have not been able to prove either Fermat or Goldbach from the above considerations, we have at least shown that the Mobius function, or perhaps more correctly a "variant" of the Mobius function, is also enmeshed in their solution.

References

[B,F,P] Barett, W. W. Forcade, R. W. Pollington, A. D. On the Spectral Radius of a (0,1) Matrix Related to Mertns' Function. Linear Algebra Appl. 107 (1988) pp. 151–159

[H,W] Hardy, G. H. Wright,E. M. An Introduction to the Theory of Numbers, fourth edition, Oxford, Clarendon Press (1960)

[S] Stanley, R. Enumerative Combinatorics Volume 1, Wadsworth & Brooks/Cole (1986)

[V1] Vaughan, R. C. On the Eigenvalues of Redheffer's Matrix I. Lecture Notes in Pure and Applied Math. 147 Dekker, N.Y. (1993)

[V2] ——— On the Eigenvalues of Redheffer's Matrix II. J. Austral. Math. Soc. Ser A 60 (1996) pp. 260–273

Continuous Solutions of Linear Equations

Charles Fefferman and János Kollár

Abstract We provide necessary and sufficient conditions for existence of continuous solutions of a system of linear equations whose coefficients are continuous functions.

1 Introduction

Consider a system of linear equations $A \cdot \mathbf{y} = \mathbf{b}$ where the entries of

$$A = \big(a_{ij}(x_1, \dots, x_n)\big) \quad \text{and of} \quad \mathbf{b} = \big(b_i(x_1, \dots, x_n)\big)$$

are themselves continuous functions on $\mathbb{R}^n$. Our aim is to decide whether the system $A \cdot \mathbf{y} = \mathbf{b}$ has a solution $\mathbf{y} = \big(y_j(x_1, \dots, x_n)\big)$, where the $y_j(x_1, \dots, x_n)$ are also continuous functions on $\mathbb{R}^n$.

More generally, if the a_{ij} and the b_i have some regularity property, can we chose the y_j to have the same (or some weaker) regularity properties?

There are two cases when the answer is rather straightforward. If A is invertible over a dense open subset $U \subset \mathbb{R}^n$, then $\mathbf{y} = A^{-1}\mathbf{b}$ holds over U. Thus, there is a continuous solution iff $A^{-1}\mathbf{b}$ extends continuously to $\mathbb{R}^n$. The case when rank A is constant on $\mathbb{R}^n$ can also be treated by standard linear algebra.

By contrast, if the system is underdetermined and rank A varies, the problem seems quite subtle. In fact, the hardest case appears to be when there is only one equation in many unknowns. It can be restated as follows.

Question 1 *Let $f_1, \dots, f_r$ be continuous functions on $\mathbb{R}^n$. Which continuous functions ϕ can be written in the form*

C. Fefferman (✉) • J. Kollár
Princeton University, Princeton, NJ 08544-1000, USA
e-mail: cf@math.princeton.edu; kollar@math.princeton.edu

H.M. Farkas et al. (eds.), *From Fourier Analysis and Number Theory to Radon Transforms and Geometry*, Developments in Mathematics 28, DOI 10.1007/978-1-4614-4075-8_10,

$$\phi = \sum_i \phi_i f_i \tag{1.1}$$

where the ϕ_i are continuous functions? Moreover, if ϕ and the f_i have some regularity properties, can we chose the ϕ_i to have the same (or some weaker) regularity properties?

If the f_i have no common zero, then a partition of unity argument shows that every $\phi \in C^0(\mathbb{R}^n)$ can be written this way and the ϕ_i have the same regularity properties (e.g., being Hölder, Lipschitz, or C^m) as ϕ and the f_i. By Cartan's theorem B, if ϕ and the f_i are real analytic, then the ϕ_i can also be chosen real analytic.

None of these hold if the common zero set $Z := (f_1 = \cdots = f_r = 0)$ is not empty. Even if ϕ and the f_i are polynomials, the best one can say is that the ϕ_i can be chosen to be Hölder continuous; see (30.1). Thus, the interesting aspects happen near the common zero set Z.

The C^∞-version of Question 1 was studied extensively (see, e.g., [Mal67, Tou72]) and it played a rôle in the work of Ehrenpreis (see [Ehr70]). The continuous version studied here is closer in spirit to the following question for L^∞_{loc}:

Which functions can be written in the form $\sum_i \psi_i f_i$ where $\psi_i \in L^\infty_{\mathrm{loc}}$?

The answer to the latter variant turns out to be rather simple. If ϕ is such, then $\phi / \sum_i |f_i| \in L^\infty_{\mathrm{loc}}$. Conversely, if this holds, then

$$\phi = \sum_i \phi_i f_i \quad \text{where} \quad \phi_i := \frac{\phi}{\sum_j |f_j|} \cdot \frac{\bar{f_i}}{|f_i|} \in L^\infty_{\mathrm{loc}}.$$

Equivalently, the obvious formulas

$$\sum_i |f_i| = \sum_i \frac{\bar{f_i}}{|f_i|} f_i \quad \text{and} \quad \phi = \frac{\phi}{\sum_i |f_i|} \sum_i |f_i| \tag{1.2}$$

show that $L^\infty_{\mathrm{loc}}(\mathbb{R}^n) \cdot (f_1, \dots, f_r)$ is the principal ideal generated by $\sum_i |f_i|$. For many purposes, it is even better to write ϕ as

$$\phi = \sum_i \psi_i f_i \quad \text{where} \quad \psi_i := \frac{\phi \bar{f_i}}{\sum_j |f_j|^2} \in L^\infty_{\mathrm{loc}}. \tag{1.3}$$

Note that if ϕ is continuous (resp. differentiable), then the ψ_i given in (1.3) are continuous (resp. differentiable) outside the common zero set Z; again indicating the special role of Z.

The above formulas also show that the discontinuity of the ψ_i along Z can be removed for certain functions.

Lemma 2 *For a continuous function ϕ, the following are equivalent:*

(1) $\phi = \sum_i \phi_i f_i$ *where the ϕ_i are continuous functions such that* $\lim_{x\to z} \phi_i = 0$ *for every i and every $z \in Z$.*
(2) $\lim_{x\to z} \frac{\phi}{\sum_i |f_i|} = 0$ *for every $z \in Z$.* □

Similar conditions do not answer Question 1. First, if the ψ_i defined in (1.3) are continuous, then $\phi = \sum_i \psi_i f_i$ is continuous, but frequently, one can write $\phi = \sum_i \phi_i f_i$ with ϕ_i continuous yet the formula (1.3) defines discontinuous functions ψ_i. This happens already in very simple examples, like $f_1 = x, f_2 = y$. For $\phi = x$ (1.3) gives

$$x = \frac{x^2}{x^2+y^2}\cdot x + \frac{xy}{x^2+y^2}\cdot y$$

whose coefficients are discontinuous at the origin.

An even worse example is given by $f_1 = x^2, f_2 = y^2$ and $\phi = xy$. Here, ϕ cannot be written as $\phi = \phi_1 f_1 + \phi_2 f_2$, but every inequality that is satisfied by x^2 and y^2 is also satisfied by $\phi = xy$. We believe that there is no universal test or formula as above that answers Question 1. At least it is clear that $C^0(\mathbb{R}^n)\cdot(x, y)$ is not a principal ideal in $C^0(\mathbb{R}^n)$.

Nonetheless, these examples and the concept of axis closure defined by [Bre06] suggest several simple necessary conditions. These turn out to be equivalent to each other, but they do not settle Question 1.

The algebraic version of Question 1 was posed by Brenner, which led him to the notion of the *continuous closure* of ideals [Bre06]. We learned about it from a lecture of Hochster. It seems to us that the continuous version is the more basic variant. In turn, the methods of the continuous case can be used to settle several of the algebraic problems [Kol10].

3 (Pointwise Tests). *For a continuous function ϕ and for a point $p \in \mathbb{R}^n$, the following are equivalent:*

(1) *For every sequence $\{x_j\}$ converging to p, there are $\psi_{ij} \in \mathbb{C}$ such that $\lim_{j\to\infty} \psi_{ij}$ exists for every i and $\phi(x_j) = \sum_i \psi_{ij} f_i(x_j)$ for every j.*
(2) *We can write $\phi = \sum_i \psi_i^{(p)} f_i$ where the $\psi_i^{(p)}(x)$ are continuous at p.*
(3) *We can write $\phi = \phi^{(p)} + \sum_i c_i^{(p)} f_i$ where $c_i^{(p)} \in \mathbb{C}$ and $\lim_{x\to p} \frac{\phi^{(p)}}{\sum_i |f_i|} = 0$.*

If $\phi = \sum_i \phi_i f_i$ where the ϕ_i are continuous functions, then we obtain the $\psi_{ij}, \psi_i^{(p)}$ by restriction and $\phi = \left(\sum_i(\phi_i - \phi_i(p))f_i\right) + \sum_i \phi_i(p) f_i$ shows that ϕ satisfies the third test. Conversely, if ϕ satisfies (3), then $\phi_p := \phi - \sum_i c_i^{(p)} f_i$ is continuous and $\lim_{x\to p} \frac{\phi_p}{\sum_i |f_i|} = 0$. By Lemma 2, we can write

$$\phi = \sum_i \psi_i^{(p)} f_i \quad \textit{where} \quad \psi_i^{(p)} := c_i^{(p)} + \frac{\phi_p \bar{f_i}}{\sum_j |f_j|^2}$$

and the $\psi_i^{(p)}(x)$ *are continuous at* p*. Thus, (2) and (3) are equivalent. One can see their equivalence with (1) directly, but for us, it is more natural to obtain it by showing that they are all equivalent to the finite set test to be introduced in (26).*

If the common zero set $Z := (f_1 = \cdots = f_r = 0)$ *consists of a single point* p*, then the* $\psi_i^{(p)}(x)$ *constructed above are continuous everywhere. More generally, if* Z *is a finite set of points, then these tests give necessary and sufficient conditions for Question 1. However, the following example of Hochster shows that the pointwise test for every* p *does not give a sufficient condition in general.*

3.4 Example. [Hoc10] Take $\{f_1, f_2, f_3\} := \{x^2, y^2, xyz^2\}$ *and* $\phi := xyz$*.*

Pick a point $p = (a, b, c) \in \mathbb{R}^3$*. If* $c \neq 0$*, then we can write*

$$xyz = \tfrac{1}{c}xyz^2 + \tfrac{1}{c}(c-z)xyz \quad \text{and} \quad \lim_{(x,y,z)\to(a,b,c)} \frac{(c-z)xyz}{|x^2| + |y^2| + |xyz^2|} = 0,$$

thus (3.3) holds. Note that if $a = b = 0$*, then* $\frac{1}{c}xyz^2$ *is the only possible constant coefficient term that works. As* $c \to 0$*, the coefficient* $\frac{1}{c}$ *is not continuous; thus,* xyz *cannot be written as* $xyz = \phi_1 x^2 + \phi_2 y^2 + \phi_3 xyz^2$ *where the* ϕ_i *are continuous. Nonetheless, if* $c = 0$*, then*

$$\lim_{(x,y,z)\to(a,b,0)} \frac{xyz}{|x^2| + |y^2| + |xyz^2|} = 0.$$

shows that (3.3) is satisfied (with all $c_i^{(a,b,0)} = 0$*).*

One problem is that the coefficients $c_i^{(p)}$ *are not continuous functions of* p*. In general, they are not even functions of* p *since a representation as in (3.2) or (3.3) is not unique. Still, this suggests a possibility of reducing Question 1 to a similar problem on the lower dimensional set* $Z = (f_1 = \cdots = f_r = 0)$*.*

We present two methods to answer Question 1.

The first method starts with $f_1, \ldots, f_r$ and ϕ and decides if $\phi = \sum_i \phi_i f_i$ is solvable or not. The union of the graphs of all discontinuous solutions $(\phi_1, \ldots, \phi_r)$ is a subset $\mathcal{H} \subset \mathbb{R}^n \times \mathbb{R}^r$. Then we use the tests (3.1–3) repeatedly to get smaller and smaller subsets of $\mathcal{H}$. After $2r + 1$ steps, this process stabilizes. This follows [Fef06, Lem.2.2]. It was adapted from a lemma in [BMP03], which in turn was adapted from a lemma in [Gla58]. At the end, we use Michael's theorem [Mic56] to get a necessary and sufficient criterion. The dependence on ϕ is somewhat delicate.

The second method considers the case when the f_i are polynomials (or real-analytic functions). The method relies on the observation that formulas like (1.2) and (1.3) give a continuous solution to $\phi = \sum_i \phi_i f_i$, albeit not on $\mathbb{R}^n$ but on some real algebraic variety mapping to $\mathbb{R}^n$. Following this idea, we transform the original Question 1 on $\mathbb{R}^n$ to a similar problem on a real algebraic variety Y for which the solvability on any finite subset is equivalent to continuous solvability.

The algebraic method also shows that if ϕ is Hölder continuous (resp. semialgebraic and continuous) and the (1.1) has a continuous solution, then there is also a solution where the ϕ_i are Hölder continuous (resp. semialgebraic and continuous) (29). By contrast, if can happen that ϕ is a continuous rational function on $\mathbb{R}^3$, (1.1) has a continuous semialgebraic solution but has no continuous rational solutions [Kol11].

Both of the methods work for any linear system of equations $A \cdot \mathbf{y} = \mathbf{b}$.

2 The Glaeser–Michael Method

Fix positive integers n, r and let Q be a compact metric space.

4 (Singular Affine Bundles). *By a singular affine bundle (or bundle for short), we mean a family $\mathcal{H} = (H_x)_{x \in Q}$ of affine subspaces $H_x \subseteq \mathbb{R}^r$, parametrized by the points $x \in Q$. The affine subspaces H_x are the fibers of the bundle $\mathcal{H}$. (Here, we allow the empty set $\emptyset$ and the whole space $\mathbb{R}^r$ as affine subspaces of $\mathbb{R}^r$.) A section of a bundle $\mathcal{H} = (H_x)_{x \in Q}$ is a continuous map $f : Q \to \mathbb{R}^r$ such that $f(x) \in H_x$ for each $x \in Q$. We ask:*

$$\text{How can we tell whether a given bundle of } \mathcal{H} \text{ has a section?} \tag{2.1}$$

For instance, let $f_1, \ldots, f_r$ and φ be given real-valued functions on Q. For $x \in Q$, we take

$$H_x = \{(\lambda_1, \ldots, \lambda_r) \in \mathbb{R}^r : \lambda_1 f_1(x) + \cdots + \lambda_r f_r(x) = \varphi(x)\}. \tag{2.2}$$

Then a section $(\phi_1, \ldots, \phi_r)$ of the bundle (2.2) is precisely an r-tuple of continuous functions solving the equation

$$\phi_1 f_1 + \cdots + \phi_r f_r = \varphi \text{ on } Q. \tag{2.3}$$

To answer Question (2.1), we introduce the notion of "Glaeser refinement." (Compare with [Gla58], [BMP03], [Fef06].) Let $\mathcal{H} = (H_x)_{x \in Q}$ be a bundle. Then the Glaeser refinement of $\mathcal{H}$ is the bundle $\mathcal{H}' = (H'_x)_{x \in Q}$, where, for each $x \in Q$,

$$H'_x = \{\lambda \in H_x : \operatorname{dist}(\lambda, H_y) \to 0 \text{ as } y \to x \quad (y \in Q)\}. \tag{2.4}$$

One checks easily that

$$\mathcal{H}' \text{ is a subbundle of } \mathcal{H}, \text{ i.e., } H'_x \subseteq H_x \text{ for each } x \in Q \tag{2.5}$$

and

$$\text{the bundles } \mathcal{H} \text{ and } \mathcal{H}' \text{ have the same sections.} \tag{2.6}$$

Starting from a given bundle $\mathcal{H}$, and iterating the above construction, we obtain a sequence of bundles $\mathcal{H}^0, \mathcal{H}^1, \mathcal{H}^2, \ldots$, where $\mathcal{H}^0 = \mathcal{H}$ and $\mathcal{H}^{i+1}$ is the Glaeser refinement of $\mathcal{H}^i$ for each i. In particular, $\mathcal{H}^{i+1}$ is a subbundle of $\mathcal{H}^i$, and all the bundles $\mathcal{H}^i$ have the same sections.

We will prove the following results.

Lemma 5 (Stabilization Lemma) $\mathcal{H}^{2r+1} = \mathcal{H}^{2r+2} = \cdots$

Lemma 6 (Existence of Sections) *Let $\mathcal{H} = (H_x)_{x\in Q}$ be a bundle. Suppose that $\mathcal{H}$ is its own Glaeser refinement and suppose each fiber H_x is nonempty. Then $\mathcal{H}$ has a section.*

The above results allow us to answer Question (2.1). Let $\mathcal{H}$ be a bundle, let $\mathcal{H}^0, \mathcal{H}^1, \mathcal{H}^2, \ldots$ be its iterated Glaeser refinements, and let $\mathcal{H}^{2r+1} = (\widetilde{H}_x)_{x\in Q}$. Then $\mathcal{H}$ has a section if and only if each fiber $\widetilde{H}_x$ is nonempty.

The bundle (2.2) provides an interesting example. One checks that its Glaeser refinement is given by $\mathcal{H}^1 = (H_x^1)_{x\in Q}$, where

$$H_x^1 = \Big\{(\lambda_1, \ldots, \lambda_r) \in \mathbb{R}^r : \big|\textstyle\sum_1^r \lambda_i f_i(y) - \varphi(y)\big| = o\big(\sum_1^r |f_i(y)|\big) \text{ as } y \to x\Big\}.$$

Thus, the necessary condition (3) for the existence of continuous solutions of (2.3) asserts precisely that the fibers H_x^1 are all nonempty.

In Hochster's example (3.4), (2.3) has no continuous solutions, because the second Glaeser refinement $\mathcal{H}^2 = (H_x^2)_{x\in Q}$ has an empty fiber, namely, H_0^2.

We present self-contained proofs of (5) and (6), for the reader's convenience. A terse discussion would simply note that the proof of [Fef06, Lem.2.2] also yields (5) and that one can easily prove (6) using Michael's theorem [Mic56], [BL00].

7 (Proof of the Stabilization Lemma) *Let $\mathcal{H}^0, \mathcal{H}^1, \mathcal{H}^2, \cdots$ be the iterated Glaeser refinements of $\mathcal{H}$ and let $\mathcal{H}^i = (H_x^i)_{x\in Q}$ for each i.*

We must show that $H_x^\ell = H_x^{2r+1}$ for all $x \in Q$, $\ell \geq 2r+1$. If $H_x^{2r+1} = \emptyset$, then the desired result is obvious.

For nonempty H_x^{2r+1}, it follows at once from the following.

Claim 7.1$_k$. Let $x \in Q$. If $\dim H_x^{2k+1} \geq r-k$, then $H_x^\ell = H_x^{2k+1}$ for all $\ell \geq 2k+1$.

We prove (7.1$_k$) for all $k \geq 0$, by induction on k. In the case $k = 0$, (7.1$_k$) asserts that

$$\text{If } H_x^1 = \mathbb{R}^r, \text{ then } H_x^\ell = \mathbb{R}^r \text{ for all } \ell \geq 1. \tag{2.7}$$

By definition of Glaeser refinement, we have

$$\dim H_x^{\ell+1} \leq \liminf_{y\to x} \dim H_y^\ell. \tag{2.8}$$

Hence, if $H_x^1 = \mathbb{R}^r$, then $H_y^0 = \mathbb{R}^r$ for all y in a neighborhood of x. Consequently, $H_y^\ell = \mathbb{R}^r$ for all y in a neighborhood of x and for all $\ell \geq 0$. This proves (7.1$_k$)

in the base case $k = 0$. For the induction step, we fix k and assume (7.1_k) for all $x \in Q$. We will prove (7.1_{k+1}). We must show that

$$\text{If } \dim H_x^{2k+3} \geq r - k - 1, \text{ then } H_x^\ell = H_x^{2k+3} \text{ for all } \ell \geq 2k + 3. \tag{2.9}$$

If $\dim H_x^{2k+1} \geq r - k$, *then (2.9) follows at once from (7.1_k). Hence, in proving (2.9), we may assume that* $\dim H_x^{2k+1} \leq r - k - 1$. *Thus,*

$$\dim H_x^{2k+1} = \dim H_x^{2k+2} = \dim H_x^{2k+3} = r - k - 1. \tag{2.10}$$

We now show that

$$H_y^{2k+2} = H_1^{2k+1} \text{ for all } y \text{ near enough to } x. \tag{2.11}$$

If fact, suppose that (2.11) fails, i.e., suppose that

$$\dim H_y^{2k+2} \leq \dim H_y^{2k+1} - 1 \text{ for } y \text{ arbitrarily close to } x. \tag{2.12}$$

For y as in (2.12), our inductive assumption (7.1_k) shows that $\dim H_y^{2k+1} \leq r - k - 1$. *Therefore, for y arbitrarily near x, we have*

$$\dim H_y^{2k+2} \leq \dim H_y^{2k+1} - 1 \leq r - k - 2.$$

Another application of (2.8) now yields $\dim H_x^{2k+3} \leq r - k - 2$, *contradicting (2.10). Thus, (2.11) cannot fail.*

From (2.11), we see easily that $H_y^\ell = H_y^{2k+3}$ for all y near enough to x and for all $\ell \geq 2k + 3$.

This completes the inductive step (2.9) and proves the Stabilization Lemma. □

8 (Proof of Existence of Sections) *We give the standard proof of Michael's theorem in the relevant special case. We start with a few definitions. If $H \subset \mathbb{R}^r$ is an affine subspace and $v \in \mathbb{R}^r$ is a vector, then $H - v$ denotes the translate $\{w - v : w \in H\}$. If $\mathcal{H} = (H_x)_{x \in Q}$ is a bundle, and if $f : Q \to \mathbb{R}^r$ is a continuous map, then $\mathcal{H} - f$ denotes the bundle $(H_x - f(x))_{x \in Q}$. Note that if $\mathcal{H}$ is its own Glaeser refinement and has nonempty fibers, then the same is true of $\mathcal{H} - f$.*

Let $\mathcal{H} = (H_x)_{x \in Q}$ be any bundle with nonempty fibers. We define the norm $\|\mathcal{H}\| := \sup_{x \in Q} \operatorname{dist}(0, H_x)$. *Thus, $\|\mathcal{H}\|$ is a nonnegative real number or $+\infty$.*

Now suppose that $\mathcal{H} = (H_x)_{x \in Q}$ is a bundle with nonempty fibers and suppose that $\mathcal{H}$ is its own Glaeser refinement.

Proposition 9 $\|\mathcal{H}\| < +\infty$.

Proof. *Given $x \in Q$, we can pick $w_x \in H_x$ since H_x is nonempty. Also,* $\operatorname{dist}(w_x, H_y) \to 0$ *as $y \to x$ ($y \in Q$), since $\mathcal{H}$ is its own Glaeser refinement.*

Hence, there exists an open ball B_x centered at x, such that $\mathrm{dist}(w_x, H_y) \leq 1$ *for all $y \in Q \cap B_x$. It follows that* $\mathrm{dist}(0, H_y) \leq |w_x| + 1$ *for all $y \in Q \cap B_x$. We can cover the compact space Q by finitely many of the open balls B_x $(x \in Q)$; say,*

$$Q \subset B_{x_1} \cup B_{x_2} \cup \cdots \cup B_{x_N}.$$

Since $\mathrm{dist}(0, H_y) \leq |w_{x_i}| + 1$ *for all $y \in Q \cap B_{x_i}$, it follows that*

$$\mathrm{dist}(0, H_y) \leq \max\{|w_{x_i}| + 1 : i = 1, 2, \ldots, N\} \text{ for all } y \in Q.$$

Thus, $\|\mathcal{H}\| < +\infty$.

Proposition 10 *Given $\varepsilon > 0$, there exists a continuous map $g : Q \to \mathbb{R}^r$ such that*

$$\mathrm{dist}(g(y), H_y) \leq \varepsilon \text{ for all } y \in Q,$$

and

$$|g(y)| \leq \|\mathcal{H}\| + \varepsilon \text{ for all } y \in Q.$$

Proof. Given $x \in Q$, we can find $w_x \in H_x$ such that $|w_x| \leq \|\mathcal{H}\| + \varepsilon$. We know that $\mathrm{dist}(w_x, H_y) \to 0$ as $y \to x$ $(y \in Q)$, since $\mathcal{H}$ is its own Glaeser refinement. Hence, there exists an open ball $B(x, 2r_x)$ centered at x, such that

$$\mathrm{dist}(w_x, H_y) < \varepsilon \text{ for all } y \in Q \cap B(x, 2r_x).$$

The compact space Q may be covered by finitely many of the open balls $B(x, r_x)$ $(x \in Q)$; say

$$Q \subset B(x_1, r_{x_1}) \cup \cdots \cup B(x_N, r_{x_N}).$$

For each $i = 1, \ldots, N$, we introduce a nonnegative continuous function $\widetilde{\varphi}_i$ on $\mathbb{R}^n$, supported in $B(x_i, 2r_{x_i})$ and equal to one on $B(x_i, r_{x_i})$. We then define $\varphi_i(x) = \widetilde{\varphi}_i(x)/(\widetilde{\varphi}_1(x) + \cdots + \widetilde{\varphi}_N(x))$ for $i = 1, \ldots, N$ and $x \in Q$. (This makes sense, thanks for (8).)

The φ_i form a partition of unity on Q:

- Each φ_i is a nonnegative continuous function on Q, equal to zero outside $Q \cap B(x_i, 2r_{x_i})$.
- $\sum_{i=1}^{N} \varphi_i = 1$ on Q.

We define

$$g(y) = \sum_{i=1}^{N} w_{x_i} \varphi_i(y) \text{ for } y \in Q.$$

Thus, g is a continuous map from Q into $\mathbb{R}^r$. Moreover, (8) *shows that* $\operatorname{dist}(w_{x_i}, H_y) \leq \varepsilon$ *whenever* $\varphi_i(y) \neq 0$. *Therefore,*

$$\begin{aligned}\operatorname{dist}(g(y), H_y) &\leq \sum_{i=1}^{N} \operatorname{dist}(w_{x_i}, H_y)\varphi_i(y)\\ &\leq \varepsilon \sum_{i=1}^{N} \varphi_i(y) = \varepsilon \text{ for all } y \in Q.\end{aligned}$$

Also, for each $y \in Q$, we have

$$|g(y)| \leq \sum_{i=1}^{N} |w_{x_i}|\varphi_i(y) \leq \sum_{i=1}^{N} (\|\mathcal{H}\| + \varepsilon)\varphi_i(y) = \|\mathcal{H}\| + \varepsilon.$$

The proof of Proposition 10 is complete.

Corollary 11 *Let $\mathcal{H}$ be a bundle with nonempty fibers, equal to its own Glaeser refinement. Then there exists a continuous map $g : Q \to \mathbb{R}^r$, such that $\|\mathcal{H} - g\| \leq \frac{1}{2}\|\mathcal{H}\|$, and $|g(y)| \leq 2\|\mathcal{H}\|$ for all $y \in Q$.*

Proof. *If $\|\mathcal{H}\| > 0$, then we can just take $\varepsilon = \frac{1}{2}\|\mathcal{H}\|$ in Proposition 10. If instead $\|\mathcal{H}\| = 0$, then we can just take $g = 0$.*

Now we can prove the existence of sections. Let $\mathcal{H} = (H_x)_{x\in Q}$ be a bundle. Suppose the H_x are all nonempty and assume that $\mathcal{H}$ is its own Glaeser refinement. By induction on $i = 0, 1, 2, \dots$, we define continuous maps $f_i, g_i : Q \to \mathbb{R}^r$. We start with $f_0 = g_0 = 0$. Given f_i and g_i, we apply Corollary 11 to the bundle $\mathcal{H} - f_i$, to produce a continuous map $g_{i+1} : Q \to \mathbb{R}^r$, such that $\|(\mathcal{H} - f_i) - g_{i+1}\| \leq \frac{1}{2}\|\mathcal{H} - f_i\|$, and $|g_{i+1}(y)| \leq 2\|\mathcal{H} - f_i\|$ for all $y \in Q$.

We then define $f_{i+1} = f_i + g_{i+1}$. This completes our inductive definition of the f_i and g_i. Note that $f_0 = 0$, $\|\mathcal{H} - f_{i+1}\| \leq \frac{1}{2}\|\mathcal{H} - f_i\|$ for each i, and $|f_{i+1}(y) - f_i(y)| \leq 2\|\mathcal{H} - f_i\|$ for each $y \in Q$, $i \geq 0$. Therefore, $\|\mathcal{H} - f_i\| \leq 2^{-i}\|\mathcal{H}\|$ for each i, and $|f_{i+1}(y) - f_i(y)| \leq 2^{1-i}\|\mathcal{H}\|$ for each $y \in Q$, $i \geq 0$. In particular, the f_i converge uniformly on Q to a continuous map $f : Q \to \mathbb{R}^r$, and $\|\mathcal{H} - f_i\| \to 0$ as $i \to \infty$.

Now, for any $y \in Q$, we have

$$\begin{aligned}\operatorname{dist}(f(y), H_y) &= \lim_{i\to\infty} \operatorname{dist}(f_i(y), H_y)\\ &= \lim_{i\to\infty} \operatorname{dist}\big(0, H_y - f_i(y)\big) \leq \liminf_{i\to\infty} \|\mathcal{H} - f_i\| = 0.\end{aligned}$$

Thus, $f(y) \in H_y$ for each $y \in Q$. Since also $f : Q \to \mathbb{R}^r$ is a continuous map, we see that f is a section of $\mathcal{H}$. This completes the proof of existence of sections. $\square$

12 (Further Problems and Remarks) *We return to the equation*

$$\phi_1 f_1 + \cdots + \phi_r f_r = \varphi \ \text{on} \ \mathbb{R}^n, \tag{2.13}$$

where $f_1, \ldots, f_r$ are given polynomials.

Let X be a function space, such as $C^m_{\mathrm{loc}}(\mathbb{R}^n)$ or $C^\alpha_{\mathrm{loc}}(\mathbb{R}^n)$ $(0 < \alpha \leq 1)$. It would be interesting to know how to decide whether (2.13) admits a solution $\phi_1, \ldots, \phi_r \in X$. Some related examples are given in (30). If φ is real analytic, and if (2.13) admits a continuous solution, then we can take the continuous functions ϕ_i to be real analytic outside the common zeros of the f_i. To see this, we invoke the following

Theorem 13 (Approximation Theorem, see [Nar68]) *Let $\phi, \sigma : \Omega \to \mathbb{R}$ be continuous functions on an open set $\Omega \subset \mathbb{R}^n$ and suppose $\sigma > 0$ on Ω. Then there exists a real-analytic function $\tilde{\phi} : \Omega \to \mathbb{R}$ such that $|\tilde{\phi}(x) - \phi(x)| \leq \sigma(x)$ for all $x \in \Omega$.*

Once we know the Approximation Theorem, we can easily correct a continuous solution $\phi_1, \ldots, \phi_r$ of (2.13) so that the functions ϕ_i are real analytic outside the common zeros of $f_1, \ldots, f_r$. We take $\Omega = \{x \in \mathbb{R}^n : f_i(x) \neq 0$ for some $i\}$ and set $\sigma(x) = \sum_i (f_i(x))^2$ for $x \in \Omega$.

We obtain real-analytic functions $\tilde{\phi}_i$ on Ω such that $|\tilde{\phi}_i - \phi_i| \leq \sigma$ on Ω. Setting $h = \sum_i \tilde{\phi}_i f_i - \varphi = \sum_i (\tilde{\phi}_i - \phi_i) f_i$ on Ω and then defining

$$\left\{ \begin{array}{ll} \phi_i^\# = \tilde{\phi}_i - \frac{h f_i}{f_1^2 + \cdots + f_r^2} & \text{on } \Omega \\ \phi_i^\# = \phi_i \ \text{ on } \mathbb{R}^n \setminus \Omega & \end{array} \right\},$$

we see that $\sum_i \phi_i^\# f_i = \varphi$, with $\phi_i^\#$ continuous on $\mathbb{R}^n$ and real analytic on Ω.

3 Computation of the Solutions

In this section, we show how to compute a continuous solution $(\phi_1, \ldots, \phi_r)$ of the equation

$$\phi_1 f_1 + \cdots + \phi_r f_r = \phi, \tag{3.1}$$

assuming such a solution exists. We start with an example, then spend several sections explaining how to compute Glaeser refinements and sections of bundles, and finally return to (3.1) in the general case.

For our example, we pick Hochster's equation

$$\phi_1 x^2 + \phi_2 y^2 + \phi_3 xyz^2 = \phi \quad \text{on} \ Q = [-1, 1]^3, \tag{3.2}$$

where ϕ is a given, continuous, real-valued function on Q. Our goal here is to compute a continuous solution of (3.2), assuming such a solution exists.

Suppose ϕ_1, ϕ_2, ϕ_3 satisfy (3.2). Then, for every positive integer ν, we have

$$\phi_1\left(\frac{1}{\nu}, 0, z\right) \cdot \frac{1}{\nu^2} = \phi\left(\frac{1}{\nu}, 0, z\right),$$

$$\phi_2\left(0, \frac{1}{\nu}, z\right) \cdot \frac{1}{\nu^2} = \phi\left(0, \frac{1}{\nu}, z\right), \text{ and}$$

$$\phi_1\left(\frac{1}{\nu}, \frac{1}{\nu}, z\right) \cdot \frac{1}{\nu^2} + \phi_2\left(\frac{1}{\nu}, \frac{1}{\nu}, z\right) \cdot \frac{1}{\nu^2} + \phi_3\left(\frac{1}{\nu}, \frac{1}{\nu}, z\right) \cdot \frac{z^2}{\nu^2} = \phi\left(\frac{1}{\nu}, \frac{1}{\nu}, z\right)$$

for all $z \in [-1, 1]$. Hence, it is natural to define

$$\xi_1(z) = \lim_{\nu\to\infty} \nu^2 \cdot \phi\left(\frac{1}{\nu}, 0, z\right), \tag{3.3}$$

$$\xi_2(z) = \lim_{\nu\to\infty} \nu^2 \cdot \phi\left(0, \frac{1}{\nu}, z\right) \quad \text{and} \tag{3.4}$$

$$\xi_3(z) = \lim_{\nu\to\infty} \nu^2 \cdot \phi\left(\frac{1}{\nu}, \tfrac{1}{\nu}, z\right) \quad \text{for } z \in [-1, 1]. \tag{3.5}$$

If (3.2) has a continuous solution $\vec{\phi} = (\phi_1, \phi_2, \phi_3)$, then the limits (3.3) exist, and our solution $\vec{\phi}$ satisfies

$$\phi_1(0, 0, z) = \xi_1(z), \quad \phi_2(0, 0, z) = \xi_2(z), \quad \text{and} \tag{3.6}$$

$$\phi_1(0, 0, z) + \phi_2(0, 0, z) + z^2\phi_3(0, 0, z) = \xi_3(z) \tag{3.7}$$

for $z \in [-1, 1]$, so that

$$\phi_3(0, 0, z) = z^{-2} \cdot [\xi_3(z) - \xi_1(z) - \xi_2(z)] \quad \text{for } z \in [-1, 1] \setminus \{0\}. \tag{3.8}$$

To recover $\phi_3(0, 0, 0)$, we just pass to the limit in (3.8). Let us define

$$\xi = \lim_{\nu\to\infty} \nu^2 \cdot \xi_3(\tfrac{1}{\nu}) - \xi_1(\tfrac{1}{\nu}) - \xi_2(\tfrac{1}{\nu}). \tag{3.9}$$

If (3.2) has a continuous solution $\vec{\phi}$, then the limit (3.9) exists, and we have

$$\phi_3(0, 0, 0) = \xi. \tag{3.10}$$

Thus, $\vec{\phi}(0, 0, z)(z \in [-1, 1])$ can be computed from the given function ϕ. Note that $\phi_3(0, 0, 0)$ arises from ϕ by taking an iterated limit.

Since we assumed that $\vec{\phi}$ is continuous, we have in particular

$$\text{the functions } \phi_i(0, 0, z) \ (i = 1, 2, 3) \text{ are continuous on } [-1, 1]. \tag{3.11}$$

From now on, we regard $\vec{\phi}(0,0,z) = (\phi_1(0,0,z), \phi_2(0,0,z), \phi_3(0,0,z))$ as known.

Let us now define

$$\vec{\phi}^{\#}(x,y,z) = \vec{\phi}(x,y,z) - \vec{\phi}(0,0,z) = (\phi_1^{\#}(x,y,z), \phi_2^{\#}(x,y,z), \phi_3^{\#}(x,y,z)) \tag{3.12}$$

and

$$\phi^{\#}(x,y,z) = \phi(x,y,z) - [\phi_1(0,0,z)\cdot x^2 + \phi_2(0,0,z)\cdot y^2 + \phi_3(0,0,z)\cdot xyz^2] \tag{3.13}$$

on Q. Then, since $\vec{\phi}$ is a continuous solution of (3.2), we see that

$$\phi^{\#} \text{ and all the } \phi_i^{\#} \text{ are continuous functions on } Q; \tag{3.14}$$

$$\phi_i^{\#}(0,0,z) = 0 \qquad \text{for all } z \in [-1,1], \quad i = 1,2,3; \text{ and} \tag{3.15}$$

$$\phi_1^{\#}(x,y,z)\cdot x^2 + \phi_2^{\#}(x,y,z)\cdot y^2 + \phi_3^{\#}(x,y,z)\cdot xyz^2 = \phi^{\#}(x,y,z) \text{ on } Q. \tag{3.16}$$

We don't know the functions $\phi_i^{\#}(i = 1,2,3)$, but $\phi^{\#}$ may be computed from the given function ϕ in (3.2), since we have already computed $\phi_i(0,0,z)(i = 1,2,3)$. (See (3.13).)

We now define $\vec{\Phi}^{\#}(x,y,z) = (\Phi_1^{\#}(x,y,z), \Phi_2^{\#}(x,y,z), \Phi_3^{\#}(x,y,z))$ to be the shortest vector $(v_1, v_2, v_3) \in \mathbb{R}^3$ such that

$$v_1\cdot x^2 + v_2\cdot y^2 + v_3\cdot xyz^2 = \phi^{\#}(x,y,z). \tag{3.17}$$

Thus,

$$\Phi_1^{\#}(x,y,z)\cdot x^2 + \Phi_2^{\#}(x,y,z)\cdot y^2 + \Phi_3^{\#}(x,y,z)\cdot xyz^2 = \phi^{\#}(x,y,z) \text{ on } Q. \tag{3.18}$$

Unless $x = y = 0$, we have

$$\Phi_1^{\#}(x,y,z) = \frac{x^2}{x^4+y^4+x^2y^2z^4}\cdot\phi^{\#}(x,y,z),$$

$$\Phi_2^{\#}(x,y,z) = \frac{y^2}{x^4+y^4+x^2y^2z^4}\cdot\phi^{\#}(x,y,z),$$

$$\Phi_3^{\#}(x,y,z) = \frac{xyz^2}{x^4+y^4+x^2y^2z^4}\cdot\phi^{\#}(x,y,z) \tag{3.19}$$

$$\text{If } x = y = 0, \quad \text{then } \Phi_i^{\#}(x,y,z) = 0 \quad \text{for } i = 1,2,3. \tag{3.20}$$

Since $\phi^{\#}$ may be computed from ϕ, the functions $\Phi_i^{\#}$ $(i = 1,2,3)$ may also be computed from ϕ.

Recall that $\vec{\phi}^{\#} = (\phi_1^{\#}, \phi_2^{\#}, \phi_3^{\#})$ satisfies (3.16). Since $\vec{\Phi}(x,y,z)$ was defined as thc shortest vector satisfying (3.17), we learn that

$$\left|\, \vec{\Phi}^{\#}(x,y,z) \,\right| \le \left|\, \vec{\phi}^{\#}(x,y,z) \,\right| \quad \text{for all } (x,y,z) \in Q. \tag{3.21}$$

Since also $\vec{\phi}^{\#}$ satisfies (3.14) and (3.15), it follows that

$$\Phi_i^{\#}(x, y, z) \to 0 \text{ as } (x, y, z) \to (0, 0, z'), \text{ for each } i = 1, 2, 3. \tag{3.22}$$

Here, $z' \in [-1, 1]$ is arbitrary.

We will now check that

$$\Phi_1^{\#}, \Phi_2^{\#}, \Phi_3^{\#} \text{ are continuous functions on } Q. \tag{3.23}$$

Indeed, the $\Phi_i^{\#}$ are continuous at each $(x, y, z) \in Q$ such that $(x, y) \neq (0, 0)$, as we see at once from (3.14) and (3.19). On the other hand, (3.20) and (3.22) tell us that the $\Phi_i^{\#}$ are continuous at each $(x, y, z) \in Q$ such that $(x, y) = (0, 0)$. Thus, (3.23) holds.

Next, we set

$$\Phi_i(x, y, z) = \Phi_i^{\#}(x, y, z) + \phi_i(0, 0, z) \text{ for } (x, y, z) \in Q, \quad i = 1, 2, 3. \tag{3.24}$$

Since $\Phi_i^{\#}(x, y, z)$ and $\phi_i(0, 0, z)$ can be computed from ϕ, the same is true of $\Phi_i(x, y, z)$.

Also, (3.11) and (3.23) imply

$$\Phi_1, \Phi_2, \Phi_3 \text{ are continuous functions on } Q. \tag{3.25}$$

From (3.13), (3.18) and (3.24), we have

$$\Phi_1(x, y, z) \cdot x^2 + \Phi_2(x, y, z) \cdot y^2 + \Phi_3(x, y, z) \cdot xyz^2 = \phi(x, y, z) \text{ on } Q. \tag{3.26}$$

Note also that the Φ_i satisfy the estimate

$$\max_{x \epsilon Q,\ i=1,2,3} \left| \Phi_i(x) \right| \leq C \max_{x \epsilon Q,\ i=1,2,3} \left| \phi_i(x) \right| \tag{3.27}$$

for an absolute constant C, as follows from (3.13), (3.21), and (3.24).

Let us summarize the above discussion of (3.2). Given a function $\phi : Q \to \mathbb{R}$, we proceed as follows:

Step 1: We compute the limits (3.3), (3.4), (3.5) for each $z \in [-1, 1]$, to obtain the functions $\xi_i(z) \quad (i = 1, 2, 3)$.
Step 2: We compute the limit (3.9), to obtain the number ξ.
Step 3: We read off the functions $\phi_i(0, 0, z) \quad (i = 1, 2, 3)$ from (3.6), (3.7), (3.8), (3.10).
Step 4: We compute the function $\phi^{\#}(x, y, z)$ from (3.13).
Step 5: We compute the functions $\Phi_i^{\#}(x, y, z) \quad (i = 1, 2, 3)$ from (3.19) $\cdots$ (3.20).
Step 6: We read off the functions $\Phi_i(x, y, z) \quad (i = 1, 2, 3)$ from (3.24).

If, for our given ϕ, (3.2) has a continuous solution (ϕ_1, ϕ_2, ϕ_3), then the limits exist in Steps 3 and 3, and the above procedure produces continuous functions Φ_1, Φ_2, Φ_3 that solve (3.2) and satisfy estimate (3.27).
If instead (3.2) has no continuous solutions, then we cannot guarantee that the limits in Steps 3 and 3 exist. It may happen that those limits exist, but the functions Φ_1, Φ_2, Φ_3 produced by our procedure are discontinuous.

This concludes our discussion of example (3.2). We devote the next several sections to making calculations with bundles. We show how to pass from a given bundle to its iterated Glaeser refinements by means of formulas involving iterated limits. After recalling the construction of "Whitney cubes" (which will be used below), we then provide additional formulas to compute a section of a given Glaeser stable bundle with nonempty fibers. These results together allow us to compute a section of any given bundle for which a section exists. Finally, we apply our results on bundles, to provide a discussion of (3.1) in the general case, analogous to the discussion given above for example (3.2).

3.1 Computation of the Glaeser Refinement

We use the standard inner product on $\mathbb{R}^r$. We define a *homogeneous bundle* to be a family $\mathcal{H}^0 = (H_x^0)_{x\in Q}$ of vector subspaces $H_x^0 \subset \mathbb{R}^r$, indexed by the points x of a closed cube $Q \subset \mathbb{R}^n$. We allow $\{0\}$ and $\mathbb{R}^r$, but not the empty set, as vector subspaces of $\mathbb{R}^r$. Note that the fibers of a homogeneous bundle are vector subspaces of $\mathbb{R}^r$, while the fibers of a bundle are (possibly empty) affine subspaces of $\mathbb{R}^r$.

Any bundle $\mathcal{H}$ with nonempty fibers may be written uniquely in the form

$$\mathcal{H} = (H_x)_{x\in Q} = (v(x) + H_x^0)_{x\in Q}, \tag{3.28}$$

where $\mathcal{H}^0 = (H_x^0)_{x\in Q}$ is a homogeneous bundle, and $v(x) \perp H_x^0$ for each $x \in Q$.

Let $\widetilde{\mathcal{H}}$ be the Glaeser refinement of $\mathcal{H}$, and suppose $\widetilde{\mathcal{H}}$ has nonempty fibers. Just as $\mathcal{H}$ may be written in the form (3.28), we can express $\widetilde{\mathcal{H}}$ uniquely in the form

$$\widetilde{\mathcal{H}} = (\widetilde{v}(x) + \widetilde{H}_x^0)_{x\in Q}, \tag{3.29}$$

where $\widetilde{\mathcal{H}}^0 = (\widetilde{H}_x^0)_{x\in Q}$ is a homogeneous bundle, and $\widetilde{v}(x) \perp \widetilde{H}_x^0$ for each $x \in Q$.

One checks easily that $\widetilde{\mathcal{H}}^0$ is the Glaeser refinement of $\mathcal{H}^0$. The goal of this section is to understand how the vectors $\widetilde{v}(x)(x \in Q)$ depend on the vectors $v(y)(y \in Q)$ for fixed $\mathcal{H}^0$.

To do so, we introduce the sets

$$E = \{(x,\lambda) \in Q \times \mathbb{R}^r : \lambda \perp H_x^0\}, \text{ and} \tag{3.30}$$

$$\Lambda(x) = \{\widetilde{\lambda} \in \mathbb{R}^r : (x, \widetilde{\lambda}) \text{ belongs to the closure of } E\} \text{ for } x \in Q.. \tag{3.31}$$

The following is immediate from the definitions (3.30), (3.31).

Claim 14 *Given* $\widetilde{\lambda} \in \Lambda(x)$, *there exist points* $y^\nu \in Q$ *and vectors* $\lambda^\nu \in \mathbb{R}^r (\nu \geq 1)$, *such that* $y^\nu \to x$ *and* $\lambda^\nu \to \widetilde{\lambda}$ *as* $\nu \to \infty$, *and* $\lambda^\nu \perp H^0_{y^\nu}$ *for each* ν. □

Note that E and $\Lambda(x)$ depend on $\mathcal{H}^0$, but not on the vectors $v(y)$, $y \in Q$. The basic properties of $\Lambda(x)$ are given by the following result:

Lemma 15 *Let* $x \in Q$. *Then:*

(1) *Each* $\widetilde{\lambda} \in \Lambda(x)$ *is perpendicular to* $\widetilde{H}^0_x$.
(2) *Given any vector* $\widetilde{v} \in \mathbb{R}^r$ *not belonging to* $\widetilde{H}^0_x$, *there exists a vector* $\lambda \in \Lambda(x)$ *such that* $\lambda \cdot \widetilde{v} \neq 0$.
(3) *The vector space* $(\widetilde{H}^0_x)^\perp \subset \mathbb{R}^r$ *has a basis* $\widetilde{\lambda}_1(x), \ldots, \widetilde{\lambda}_s(x)$ *consisting entirely of vectors* $\widetilde{\lambda}_l(x) \in \Lambda(x)$.

Proof. To check (15), let $\widetilde{\lambda} \in \Lambda(x)$ and let $\widetilde{v} \in \widetilde{H}^0_x$. We must show that $\widetilde{\lambda} \cdot \widetilde{v} = 0$. Let $y^\nu \in Q$ and $\lambda^\nu \in \mathbb{R}^r$ $(\nu \geq 1)$ be as in (3.9). Since $\widetilde{v} \in \widetilde{H}^0_x$ and $(\widetilde{H}^0_y)_{y \in Q}$ is the Glaeser refinement of $(H^0_y)_{y \in Q}$, we know that distance $(\widetilde{v}, H^0_y) \to 0$ as $y \to x$. In particular, distance $(\widetilde{v}, H^0_{y^\nu}) \to 0$ as $\nu \to \infty$. Hence, there exist $v^\nu \in H^0_{y^\nu}$ $(\nu \geq 1)$ such that $v^\nu \to \widetilde{v}$ as $\nu \to \infty$. Since $v^\nu \in H^0_{y^\nu}$ and $\lambda^\nu \perp H^0_{y^\nu}$, we have $\lambda^\nu \cdot v^\nu = 0$ for each ν. Since $\lambda^\nu \to \widetilde{\lambda}$ and $v^\nu \to \widetilde{v}$ as $\nu \to \infty$, it follows that $\widetilde{\lambda} \cdot \widetilde{v} = 0$, proving (15).

To check (15), suppose $\widetilde{v} \in \mathbb{R}^r$ does not belong to $\widetilde{H}^0_x$. Since $(\widetilde{H}^0_y)_{y \in Q}$ is the Glaeser refinement of $(H^0_y)_{y \in Q}$, we know that distance $(\widetilde{v}, H^0_y)$ does not tend to zero as $y \in Q$ tends to x. Hence, there exist $\epsilon > 0$ and a sequence of points $y^\nu \in Q$ $(\nu \geq 1)$, such that

$$y^\nu \to x \text{ as } \nu \to \infty, \quad \text{but} \quad \mathrm{dist}(\widetilde{v}, H^0_{y^\nu}) \geq \epsilon \text{ for each } \nu. \tag{3.32}$$

Thanks to (3.14), there exist unit vectors $\lambda^\nu \in \mathbb{R}^r$ $(\nu \geq 1)$, such that

$$\lambda^\nu \perp H^0_{y^\nu} \quad \text{and} \quad \lambda^\nu \cdot \widetilde{v} \geq \epsilon \quad \text{for each } \nu. \tag{3.33}$$

Passing to a subsequence, we may assume that the vectors λ^ν tend to a limit $\widetilde{\lambda} \in \mathbb{R}^r$ as $\nu \to \infty$.

Comparing (3.33) to (3.30), we see that $(y^\nu, \lambda^\nu) \in E$ for each ν. Since $y^\nu \to x$ and $\lambda^\nu \to \widetilde{\lambda}$ as $\nu \to \infty$, the point $(x, \widetilde{\lambda})$ belongs to the closure of E; hence, $\widetilde{\lambda} \in \Lambda(x)$. Also, $\widetilde{\lambda} \cdot \widetilde{v} = \lim_{\nu \to \infty} \lambda^\nu \cdot \widetilde{v} \geq \epsilon$ by (3.16); in particular, $\widetilde{\lambda} \cdot \widetilde{v} \neq 0$. The proof of (15) is complete. Finally, to check (15), we note that

$$\bigcap_{\widetilde{\lambda} \in \Lambda(x)} (\widetilde{\lambda}^\perp) = \widetilde{H}^0_x, \quad \text{thanks to (3.10) and (3.11)}.$$

Assertion (15) now follows from linear algebra. The proof of Lemma 15 is complete.

Let $\widetilde{\lambda}_1(x), \ldots, \widetilde{\lambda}_s(x)$ be the basis for $(\widetilde{H}_x^0)^{\perp}$ given by (15) and let $\widetilde{\lambda}_{s+1}(x), \ldots,$ $\widetilde{\lambda}_r(x)$ be a basis for $\widetilde{H}_x^0$. Thus,

$$\widetilde{\lambda}_1(x), \ldots, \widetilde{\lambda}_r(x) \text{ form a basis for } \mathbb{R}^r. \tag{3.34}$$

For $1 \leq i \leq s$, the vector $\widetilde{\lambda}_i(x)$ belongs to $\Lambda(x)$. Hence, by (14), there exist vectors $\lambda_i^\nu(x) \in \mathbb{R}^r$ and points $y_i^\nu(x) \in Q$ $(\nu \geqslant 1)$, such that

$$y_i^\nu(x) \to x \text{ as } \nu \to \infty, \tag{3.35}$$

$$\lambda_i^\nu(x) \to \widetilde{\lambda}_i(x) \text{ as } \nu \to \infty, \text{ and} \tag{3.36}$$

$$\lambda_i^\nu(x) \perp H^0_{y_i^\nu(x)} \text{ for each } \nu. \tag{3.37}$$

For $s+1 \leq i \leq r$, we take $y_i^\nu(x) = x$ and $\lambda_i^\nu(x) = 0$ $(\nu \geqslant 1)$. Thus, (3.19) holds also for $s+1 \leq i \leq r$, although (3.36) holds only for $1 \leqslant i \leqslant s$.

We now return to the problem of computing $\widetilde{v}(x) (x \in Q)$ for the bundles given by (3.28) and (3.29). The answer is as follows.

Lemma 16 *Given $x \in Q$, we have $\widetilde{\lambda}_i(x) \cdot \widetilde{v}(x) = \lim_{\nu\to\infty} \lambda_i^\nu(x) \cdot v(y_i^\nu(x))$ for $i = 1, \ldots, r$. In particular, the limit in (16) exists.*

Remarks. Since $\widetilde{\lambda}_1(x), \ldots, \widetilde{\lambda}_r(x)$ form a basis for $\mathbb{R}^r$, (16) completely specifies the vector $\widetilde{v}(x)$. Note that the points $y_i^\nu(x)$ and the vectors $\widetilde{\lambda}_i(x)$, $\lambda_i^\nu(x)$ depend only on $\mathcal{H}^0$, not on the vectors $v(y)$ $(y \in Q)$.

Proof. First, suppose that $1 \leq i \leq s$. Since $\widetilde{v}(x)$ belongs to the fiber $\widetilde{v}(x) + \widetilde{H}_x^0$ of the Glaeser refinement of $(v(y)+H_y^0)_{y\in Q}$, we know that $\text{dist}(\widetilde{v}(x), v(y)+H_y^0) \to 0$ as $y \to x$ $(y \in Q)$. In particular, $\text{dist}(\widetilde{v}(x), v(y_i^\nu(x)) + H^0_{y_i^\nu(x)}) \to 0$ as $\nu \to \infty$. Hence, there exist vectors $w_i^\nu(x) \in H^0_{y_i^\nu(x)}$ such that $v(y_i^\nu(x)) + w_i^\nu(x) \to \widetilde{v}(x)$ as $\nu \to \infty$. Since also $\lambda_i^\nu(x) \to \widetilde{\lambda}_i(x)$ as $\nu \to \infty$, it follows that $\widetilde{\lambda}_i(x) \cdot \widetilde{v}(x) = \lim_{\nu\to\infty} \lambda_i^\nu(x) \cdot [v(y_i^\nu(x)) + w_i^\nu(x)]$. However, since $w_i^\nu(x) \in H^0_{y_i^\nu(x)}$ and $\lambda_i^\nu(x) \perp H^0_{y_i^\nu(x)}$, we have $\lambda_i^\nu(x) \cdot w_i^\nu(x) = 0$ for each ν.

Therefore, $\widetilde{\lambda}_i^\nu(x) \cdot \widetilde{v}(x) = \lim_{\nu\to\infty} \lambda_i^\nu(x) \cdot v(y_i^\nu(x))$, i.e., (3.20) holds for $1 \leq i \leq s$.

On the other hand, suppose $s+1 \leq i \leq r$. Then since $\widetilde{\lambda}_i(x) \in \widetilde{H}_x^0$ and $\widetilde{v}(x) \perp \widetilde{H}_x^0$, we have $\widetilde{\lambda}_i(x) \cdot \widetilde{v}(x) = 0$. Also, in this case, we defined $\lambda_i^\nu(x) = 0$. Hence, $\lambda_i^\nu(x) \cdot v(y_i^\nu(x)) = 0$ for each ν. Therefore, $\widetilde{\lambda}_i(x) \cdot \widetilde{v}(x) = 0 = \lim_{\nu\to\infty} \lambda_i^\nu(x) \cdot v(y_i^\nu(x))$, so that (16) holds also for $s+1 \leq i \leq r$. The proof of Lemma 16 is complete.

3.2 *Computation of Iterated Glaeser Refinements*

In this section, we apply the results of the preceding section to study iterated Glaeser refinements. Let $\mathcal{H} = (v(x) + H_x^0)_{x\in Q}$ be a bundle, given in the form (3.28). We assume that $\mathcal{H}$ has a section. Therefore, $\mathcal{H}$ and all its iterated Glaeser refinements have nonempty fibers. For $\ell \geq 0$, we write the ℓth iterated Glaeser refinement in the form

$$\mathcal{H}^{(\ell)} = (v^{\ell}(x) + H_x^{0,\ell})_{x \in Q}, \tag{3.38}$$

where $\mathcal{H}^{0,\ell} = (H_x^{0,\ell})_{x \in Q}$ is a homogeneous bundle, and $v^{\ell}(x) \perp H_x^{0,\ell}$ for each $x \in Q$. (Again, we use the standard inner product on $\mathbb{R}^r$.) In particular, $\mathcal{H}^{(0)} = \mathcal{H}$, and

$$\mathcal{H}^{0,0} = (H_x^0)_{x \in Q}, \text{ with } H_x^0 \text{ as in (3.1)}. \tag{3.39}$$

One checks easily that $\mathcal{H}^{0,\ell}$ is the ℓth iterated Glaeser refinement of $\mathcal{H}^{0,0}$. Our goal here is to give formulas computing $v^{\ell}(x)$ in terms of the $v(y) (y \in Q)$ in (3.1).

We proceed by induction on ℓ. For $\ell = 0$, we have

$$v^0(x) = v(x) \text{ for all } x \in Q. \tag{3.40}$$

For $\ell \geq 1$, we apply the results of the preceding section, to pass from $(v^{\ell-1}(x))_{x \in Q}$ to $(v^{\ell}(x))_{x \in Q}$.

Claim 17 *We obtain points $y_i^{\ell,\nu}(x) \in Q (\nu \geq 1,\ 1 \leq i \leq r,\ x \in Q)$; and vectors $\widetilde{\lambda}_i^{\ell}(x) \in \mathbb{R}^r\ (1 \leq i \leq r,\ x \in Q), \widetilde{\lambda}_i^{\ell,\nu}(x)\ (1 \leq i \leq r,\ \nu \geq 1,\ x \in Q)$ with the following properties:*

(1) The above points and vectors depend only on $\mathcal{H}^{0,0}$, not on the family of vectors $(v(x))_{x \in Q}$,
(2) $\widetilde{\lambda}_1^{\ell}(x), \ldots, \widetilde{\lambda}_r^{\ell}(x)$ form a basis of $\mathbb{R}^r$, for each $\ell \geq 1, x \in Q$.
(3) $y_i^{\ell,\nu}(x) \to x$ as $\nu \to \infty$ for each $\ell \geq 1,\ 1 \leq i \leq r,\ x \in Q$.
(4) $[\widetilde{\lambda}_i^{\ell}(x) \cdot v^{\ell}(x)] = \lim_{\nu \to \infty} [\widetilde{\lambda}_i^{\ell,\nu}(x) \cdot v^{\ell-1}(y_i^{\ell,\nu}(x))]$ for each $\ell \geq 1,\ 1 \leq i \leq r,$ $x \in Q$.

The last formula computes the $v^{\ell}(x)\ (x \in Q)$ in terms of the $v^{\ell-1}(y)\ (y \in Q)$ for $\ell \geq 1$, completing our induction on ℓ.

Note that we have defined the basis vectors $\widetilde{\lambda}_1^{\ell}(x), \ldots, \widetilde{\lambda}_r^{\ell}(x)$ only for $\ell \geq 1$. For $\ell = 0$, it is convenient to use the standard basis vectors for $\mathbb{R}^r$, i.e., we define

$$\widetilde{\lambda}_i^0(x) = (0, 0, \ldots, 0, 1, 0, \ldots, 0) \in \mathbb{R}^r, \text{with the 1 in the } i\text{th slot.} \tag{3.41}$$

It is convenient also to set

$$\xi_i^{\ell}(x) = \widetilde{\lambda}_i^{\ell}(x) \cdot v^{\ell}(x) \text{ for } x \in Q,\ \ell \geq 0, \quad 1 \leq i \leq r, \tag{3.42}$$

and to expand $\widetilde{\lambda}_i^{\ell,\nu}(x) \in \mathbb{R}^r$ in terms of the basis $\widetilde{\lambda}_1^{\ell-1}(y), \ldots, \widetilde{\lambda}_r^{\ell-1}(y)$ for $y = y_i^{\ell,\nu}(x)$. Thus, for suitable coefficients $\beta_{ij}^{\ell,\nu}(x) \in \mathbb{R}\ (\ell \geq 1,\ \nu \geq 1,\ 1 \leq i \leq r,$ $1 \leq j \leq r,\ x \in Q)$, we have

$$\widetilde{\lambda}_i^{\ell,\nu}(x) = \sum_{ij}^{r} \beta_{ij}^{\ell,\nu}(x) \cdot \widetilde{\lambda}_j^{\ell-1}(y_i^{\ell,\nu}(x)) \text{ for } x \in Q,\ \ell \geq 1, \nu \geq 1,\ 1 \leq i \leq r. \tag{3.43}$$

Note that the coefficients $\beta_{ij}^{\ell,\nu}(x)$ depend only on $\mathcal{H}^{0,0}$, not on the vectors $v(y)$ $(y \in Q)$.

Putting (3.42) and (3.43) into (17.4), we obtain a recurrence relation for the $\xi_i^\ell(x)$:

$$\xi_i^\ell(x) = \lim_{\nu\to\infty} \sum_{j=1}^{r} \beta_{ij}^{\ell,\nu}(x) \cdot \xi_j^{\ell-1}\left(y_i^{\ell,\nu}(x)\right) \quad \text{for} \quad \ell \geq 1,\ 1 \leq i \leq r,\ x \in Q. \tag{3.44}$$

For $\ell = 0$, (3.40)–(3.42) give

$$\xi_i^0(x) = [i\text{th component of } v(x)]. \tag{3.45}$$

Since $\beta_{ij}^{\ell,\nu}(x)$ and $y_i^{\ell,\nu}(x)$ are independent of the vectors $v(y)(y \in Q)$, our formulas (3.44), (3.18) express each $\xi_i^\ell(x)$ as an iterated limit in terms of the vectors $v(y)(y \in Q)$. In particular, the $\xi_i^\ell(x)$ depend linearly on the $v(y)$ $(y \in Q)$.

We are particularly interested in the case $\ell = 2r + 1$, since the bundle $\mathcal{H}^{2r+1}$ is Glaeser stable, as we proved in section X.

Since $\widetilde{\lambda}_1^{2r+1}(x), \ldots, \widetilde{\lambda}_r^{2r+1}(x)$ form a basis of $\mathbb{R}^r$ for each $x \in Q$, there exist vectors $w_1(x), \ldots, w_r(x) \in \mathbb{R}^r$ for each $x \in Q$, such that

$$v = \sum_{l=1}^{r} \widetilde{\lambda}_l^{2r+1}(x) \cdot v w_l(x) \text{ for any vector } v \in \mathbb{R}^r, \text{ and for any } x \in Q. \tag{3.46}$$

Note that the vectors $w_1(x), \ldots, w_r(x) \in \mathbb{R}^r$ depend only on $\mathcal{H}^{0,0}$, not on the vectors $v(y)(y \in Q)$.

Taking $v = v^{2r+1}(x)$ in (3.46), and recalling (3.42), we see that

$$v^{2r+1}(x) = \sum_{l=1}^{r} \xi_l^{2r+1}(x) w_l(x) \qquad \text{for each } x \in Q. \tag{3.47}$$

Thus, we determine the $\xi_i^\ell(x)$ by the recursion (3.44), (3.45), and then compute $v^{2r+1}(x)$ from formula (3.47). Since also $(H_x^{0,2r+1})_{x \in Q}$ is simply the $(2r+1)^{rst}$.

Glaeser refinement of $\mathcal{H}^{0,0}$, we have succeeded in computing the Glaeser stable bundle $(v^{2r+1}(x) + H_x^{0,2r+1})_{x\in Q}$ in terms of the initial bundle as in (3.28).

Our next task is to give a formula for a section of a Glaeser stable bundle. To carry this out, we will use "Whitney cubes," a standard construction which we explain below.

3.3 *Whitney Cubes*

In this section, for the reader's convenience, we review "Whitney cubes" (see [Mal67, Ste70, Whi34]). We will work with closed cubes $Q \subset \mathbb{R}^n$ whose sides are parallel to the coordinate axes. We write ctr(x) and δ_Q to denote the center and side length of Q, respectively, and we write Q^* to denote the cube with center ctr(Q) and side length 3δ.

To "bisect" Q is to write it as a union of 2^n subcubes, each with side length $\frac{1}{2}\delta_Q$, in the obvious way; we call those 2^n subcubes the "children" of Q.

Fix a cube Q^o. The "dyadic cubes" are the cube Q^o, the children of Q^o, the children of the children of Q^o, and so forth. Each dyadic Q is a subcube of Q^o. If Q is a dyadic cube other than Q^o, then Q is a child of one and only one dyadic cube, which we call Q^+. Note that $Q^+ \subset Q^*$.

Now let E_1 be a nonempty closed subset of Q^o. A dyadic cube $Q \neq Q^o$ will be called a "Whitney cube" if it satisfies

$$\text{dist}(Q^*, E_1) \geq \delta_Q, \text{ and} \tag{3.48}$$

$$\text{dist}((Q^+)^*, E_1) < \delta_{Q^+}. \tag{3.49}$$

The next result gives a few basic properties of Whitney cubes. In this section, we write c, C, C', etc. to denote constants depending only on the dimension n. These symbols need not denote the same constant in different occurrences.

Lemma 18 *For each Whitney cube Q, we have:*

(1) $\delta_Q \leq \text{dist}(Q^*, E_1) \leq C\delta_Q$.
(2) In particular, $Q^* \cap E_1 = \phi$.
(3) The union of all Whitney cubes is $Q^o \setminus E_1$.
(4) Any given $y \in Q^o \setminus E_1$ *has a neighborhood that meets* Q^* *for at most* C *distinct Whitney cubes* Q.

Proof. Estimates (1) follow at once from (1) and (2), and (4) is immediate from (3).

To check (3), we note first that each Whitney cube Q is contained in $Q^o \setminus E_1$, thanks to (2) and our earlier remark that every dyadic cube is contained in Q^o. Conversely, let $x \in Q^o \setminus E_1$ be given. Any small enough dyadic cube $\widehat{Q}$ containing x will satisfy (3.48). Fix such a $\widehat{Q}$. There are only finitely many dyadic cubes Q containing x with side length greater than or equal to $\delta_{\widehat{Q}}$. Hence, there exists a dyadic cube $Q \ni x$ satisfying (3.48), whose side length is at least as large as that of any other dyadic cube $Q' \ni x$ satisfying (3.48). We know that $Q \neq Q^o$, since (3.48) fails for Q^o. Hence, Q has a dyadic parent Q^+. We know that (3.48) fails for Q^+, since the side length of Q^+ is greater than that of Q. It follows that Q satisfies (3.49). Thus, $Q \ni x$ is a Whitney cube, completing the proof of (3).

We turn our attention to (4). Let $y \in Q^o \setminus E_1$. We set $r = 10^{-3}$ distance (y, E_1), and we prove that there are at most C distinct Whitney cubes Q for which Q^* meets the ball $B(x, r)$.

Indeed, let Q be such a Whitney cube. Then there exists $z \in B(y, r) \cap Q^*$. By (3.55), we have

$$\delta_Q \leq \text{dist}(z, E_1) \leq C\delta_Q. \tag{3.50}$$

Since $z \in B(y,r)$, we know that $|\mathrm{dist}(z, E_1) - \mathrm{dist}(y, E_1)\ \ | \le 10^{-3}\,\mathrm{dist}(y, E_1)$. Hence

$$(1 - 10^{-3})\,\mathrm{dist}(y, E_1) \le \mathrm{dist}(z, E_1) \le (1 + 10^{-3})\,\mathrm{dist}(y, E_1). \tag{3.51}$$

From (3.50), (3.51) we learn that

$$c\,\mathrm{dist}(y, E_1) \le \delta_Q \le C\,\mathrm{dist}(y, E_1). \tag{3.52}$$

Since $z \in B(y,r) \cap Q^*$, we know also that

$$\mathrm{dist}(y, Q^*) \le \mathrm{dist}(y, E_1). \tag{3.53}$$

For fixed y, there are at most C distinct dyadic cubes that satisfy (3.52), (3.53).

Thus, (3.6) holds and Lemma 18 is proven.

The next result provides a partition of unity adapted to the geometry of the Whitney cubes.

Lemma 19 *There exists a collection of real-valued functions θ_Q on Q^o, indexed by the Whitney cubes Q, satisfying the following conditions:*

(1) Each θ_Q is a nonnegative continuous function on Q^o.
(2) For each Whitney cube Q, the function θ_Q is zero on $Q^o \setminus Q^*$.
(3) $\sum_Q \theta_Q = 1$ on $Q^o \setminus E_1$.

Proof. Let $\widetilde{\theta}(x)$ be a nonnegative, continuous function on $\mathbb{R}^n$, such that $\widetilde{\theta}(x) = 1$ for $x = (x_1, \dots, x_n)$ with $\max\{|x_1|, \dots, |x_n|\} \le \frac{1}{2}$ and $\widetilde{\theta}(x) = 0$ for $x = (x_1, \dots, x_n)$ with $\max\{|x_1\ |, \dots, |x_n|\} \ge 1$.

For each Whitney cube Q, define $\widetilde{\theta}_Q(x) = \widetilde{\theta}(\frac{x - \mathrm{ctr}(Q)}{\delta_Q})$, for $x \in \mathbb{R}^n$. Thus, $\widetilde{\theta}_Q$ is a nonnegative continuous function on $\mathbb{R}^n$, equal to 1 on Q and equal to 0 outside Q^*. It follows easily, thanks to (3) and (4), that $\sum\limits_{Q'} \widetilde{\theta}_{Q'}$ is a nonnegative continuous function on $Q^o \setminus E_1$, greater than or equal to one at every point of $Q^o \setminus E_1$.

Consequently, the functions θ_Q, defined by $\theta_Q(x) = \widetilde{\theta}_Q(x) / \sum\limits_{Q'} \widetilde{\theta}_{Q'}(x)$ for $x \in Q^o \setminus E_1$, $\theta_Q(x) = 0$ for $x \in E_1$, are easily seen to satisfy (1)–(3).

Additional basic properties of Whitney cubes and sharper versions of Lemma 19 may be found in [Mal67, Ste70, Whi34].

The partition of unity $\{\theta_Q\}$ on $Q^o \setminus E_1$ is called the "Whitney partition of unity."

3.4 The Glaeser–Stable Case

In this section, we suppose we are given a Glaeser-stable bundle with nonempty fibers, written in the form

$$\mathcal{H} = (v(x) + H_x^0)_{x \in Q}, \tag{3.54}$$

where $\mathcal{H}^0 = (H_x^0)_{x \in Q}$ is a homogeneous bundle, and

$$v(x) \perp H_x^0 \quad \text{for each } x \in Q. \tag{3.55}$$

(As before, we use the standard inner product on $\mathbb{R}^r$.) Our goal here is to give a formula for a section F of the bundle $\mathcal{H}$. We will take

$$F(x) = \sum_{y \in S(x)} A(x,y)v(y) \in \mathbb{R}^r \text{ for each } x \in Q, \text{ where} \tag{3.56}$$

$$S(x) \subset Q \text{ is a finite set for each } x \in Q \text{ and} \tag{3.57}$$

$$A(x,y) : \mathbb{R}^r \to \mathbb{R}^r \text{ is a linear map, for each } x \in Q, y \in S(x). \tag{3.58}$$

Here, the sets $S(x)$ and the linear maps $A(x,y)$ are determined by $\mathcal{H}^0$; they do not depend on the family of vectors $(v(x))_{x \in Q}$.

We will establish the following result.

Theorem 20 *We can pick the $S(x)$ and $A(x,y)$ so that (3.57), (3.58) hold, and the function $F : Q \to \mathbb{R}^r$, defined by (3.56), is a section of the bundle $\mathcal{H}$. Moreover, that section satisfies:*

(1) $\max_{x \in Q} |F(x)| \leq C \sup_{x \in Q} |v(x)|$, *where C depends only on n and r.*
(2) Furthermore, each of the sets $S(x)$ contains at most d points, where d depends only on n and r.

Note: Since $v(x)$ is the shortest vector in $v(x) + H_x^0$ by (3.55), it follows that $\sup_{x \in Q} |v(x)| = \sup_{x \in Q}$ distance $(0, v(x) + H_x^0) = \| \mathcal{H} \| < \infty$; see our earlier discussion of Michael's theorem.

Proof. Roughly speaking, the idea of our proof is as follows. We partition Q into finitely many "strata," among which we single out the "lowest stratum" E_1. For $x \in E_1$, we simply set $F(x) = v(x)$. To define F on $Q \setminus E_1$, we cover $Q \setminus E_1$ by Whitney cubes Q_ν. Each Q_ν^* fails to meet E_1, by definition, and therefore has fewer strata than Q. Hence, by induction on the number of strata, we can produce a formula for a section F_ν of the bundle $\mathcal{H}$ restricted to Q_ν^*. Patching together the F_ν by using the Whitney partition of unity, we define our section F on $Q \setminus E_1$ and complete the proof of Theorem 20.

Let us begin our proof. For $k = 0, 1, \dots, r$, the kth "stratum" of $\mathcal{H}$ is defined by

$$E(k) = \{x \in Q : \dim\ H_x^0 = k\}. \tag{3.59}$$

The "number of strata" of $\mathcal{H}$ is defined as the number of nonempty $E(k)$; this number is at least 1 and at most $r + 1$. We write E_1 to denote the stratum $E(k_{\min})$, where $k_{\min}$ is the least k such that $E(k)$ is nonempty. We call E_1 the "lowest stratum."

We will prove Theorem 20 by induction on the number of strata, allowing the constants C and d on (1), (2), to depend on the number of strata, as well as on n and r. Since the number of strata is at most $r + 1$, such an induction will yield Theorem 20 as stated.

Thus, we fix a positive integer Λ and assume the inductive hypothesis:

(H1) Theorem 20 holds, with constants $C_{\Lambda-1}, d_{\Lambda-1}$ in (3.8), (3.9), whenever the number of strata is less than Λ.

We will then prove Theorem 20, with constants C_Λ, d_Λ in (1), (2), whenever the number of strata is equal to Λ. Here, C_Λ and d_Λ are determined by $C_{\Lambda-1}, d_{\Lambda-1}, n$ and r. To do so, we start with (3.54), (3.3) and assume that:

(H2) The number of strata of $\mathcal{H}$ is equal to Λ.

We must produce sets $S(x)$ and linear maps $A(x, y)$ satisfying (3.57) $\cdots$ (2), with constants C_Λ, d_Λ depending only on $C_{\Lambda-1}, d_{\Lambda-1}, n, r$. This will complete our induction and establish Theorem 20.

For the rest of the proof of Theorem 20, we write c, C, C', etc. to denote constants determined by $C_{\Lambda-1}, d_{\Lambda-1}, n, r$. These symbols need not denote the same constant in different occurrences.

The following useful remark is a simple consequence of our assumption that the bundle (3.54) is Glaeser stable. Let $x \in E(k)$ and let

$$v_1, \dots, v_{k+1} \in v(x) + H_x^0 \tag{3.60}$$

be the vertices of a nondegenerate affine k-simplex in $\mathbb{R}^r$. Given $\epsilon > 0$, there exists $\delta > 0$ such that for any $y \in Q \cap B(x, \delta)$, there exist $v_1', \dots, v_{k+1}' \in v(y) + H_y^0$ satisfying $|v_i' - v_i| < \epsilon$ for each i. Here, as usual, $B(x, \delta)$ denotes the ball of radius δ about x.

Taking ϵ small enough in (3.60), we conclude that $v_1', \dots, v_{k+1}' \in v(y) + H_y^0$ are the vertices of a nondegenerate affine k-simplex in $\mathbb{R}^r$. Therefore, (3.60) yields at once that if $x \in E(k)$, then $\dim\ H_y^0 \geq k$ for all $y \in Q$ sufficiently close to x. In particular, the lowest stratum E_1 is a nonempty closed subset of Q. Also, for each $k = 0, 1, 2, \dots, r$, (3.60) shows that the map

$$x \mapsto v(x) + H_x^0 \tag{3.61}$$

is continuous from $E(k)$ to the space of all affine k-dimensional subspaces of $\mathbb{R}^r$.

Since each H_x^0 is a vector subspace of $\mathbb{R}^r$, we learn from (3.55) and (3.61) that the map $x \mapsto v(x)$ is continuous on each $E(k)$. In particular,

$$x \mapsto v(x) \text{ is continuous on } E_1. \tag{3.62}$$

Next, we introduce the Whitney cubes $\{Q_\nu\}$ and the Whitney partition of unity $\{\theta_\nu\}$ for the closed set $E_1 \subset Q$. From the previous section, we have the following results. We write δ_ν for the side length of the Whitney cube Q_ν. Note that

$$\delta_\nu \leq \text{dist}(Q_\nu^*, E_1) \leq C\delta_\nu \text{ for each } \nu. \tag{3.63}$$

$$Q_\nu^* \cap E_1 = \phi \quad \text{for each } \nu. \tag{3.64}$$

$$\bigcup_\nu Q_\nu = Q \setminus E_1. \tag{3.65}$$

$$\begin{gathered}\text{Any given } y \in Q \setminus E_1 \text{ has a neighborhood that meets}\\ Q_\nu^* \text{ for at most } C \text{ distinct } Q_\nu.\end{gathered} \tag{3.66}$$

$$\begin{gathered}\text{Each } \theta_\nu \text{ is a nonnegative continuous function on } Q,\\ \text{vanishing outside } Q \cap Q_\nu^*.\end{gathered} \tag{3.67}$$

$$\sum_\nu \theta_\nu(x) = 1 \quad \text{if } x \in Q \setminus E_1, \quad 0 \text{ if } x \in E_1. \tag{3.68}$$

Thanks to (3.19), we can pick points $x_\nu \in E_1$ such that

$$\text{dist}(x_\nu, Q_\nu^*) \leq C\delta_\nu. \tag{3.69}$$

We next prove a continuity property of the fibers $v(x) + H_x^0$.

Lemma 21 *Given $x \in E_1$ and $\epsilon > 0$, there exists $\delta > 0$ for which the following holds. Let Q_ν be a Whitney cube such that distance $(x, Q_\nu^*) < \delta$. Then:*

(1) $|v(x) - v(x_\nu)| < \epsilon$, *and*
(2) $\text{dist}(v(x), v(y) + H_y^0) < \epsilon$ *for all* $y \in Q_\nu^* \cap Q$.

Proof. Fix $x \in E_1$ and $\epsilon > 0$. Let $\delta > 0$ be a small enough number, to be picked later. Let Q_ν be a Whitney cube such that

$$\text{dist}(x, Q_\nu^*) < \delta. \tag{3.70}$$

Then, by (3.19), we have

$$\delta_\nu \leq \text{dist}(E_1, Q_\nu^*) \leq \text{dist}(x, Q_\nu^*) < \delta, \tag{3.71}$$

hence, (3.69) and (3.70) yield the estimates

$$|x - x_\nu| \leq \text{dist}(x, Q_\nu^*) + \text{ diameter } (Q_\nu^*) + \text{dist}(Q_\nu^*, x_\nu) \leq \delta + C\delta_\nu \leq C'\delta. \tag{3.72}$$

Since x and x_ν belong to E_1, (3.72) implies (1), thanks to (3.62), provided we take δ small enough. Also, for any $y \in Q_\nu^* \cap Q$, we learn from (3.70), (3.71) that $|y - x| \leq$ diameter $(Q_\nu^*) + \text{dist}(x, Q_\nu^*) < C\delta_\nu + C\delta \leq C'\delta$.

Since the bundle $(v(z) + H_z^0)_{z \in Q}$ is Glaeser stable, it follows that (3.26) holds, provided we take δ small enough.

We now pick $\delta > 0$ small enough that the above arguments go through. Then (3.25) and (3.26) hold. The proof of Lemma 21 is complete.

We return to the proof of Theorem 20. For each Whitney cube Q_ν, we prepare to apply our inductive hypothesis (H1) to the family of affine subspaces

$$\mathcal{H}_\nu = (v(y) - v(x_\nu) + H_y^0)_{y \in Q_\nu^* \cap Q}. \tag{3.73}$$

Since $Q_\nu^* \cap Q$ is a closed rectangular box, but not necessarily a cube, it may happen that (3.73) fails to be a bundle. The cure is simply to fix an affine map $\rho_\nu : \mathbb{R}^n \to \mathbb{R}^n$, such that $\rho_\nu(Q^o) = Q_\nu^* \cap Q$, where Q^o denotes the unit cube.

The family of affine spaces

$$\check{\mathcal{H}}_\nu = \left(v(\rho_\nu \check{y}) - v(x_\nu) + H^0_{\rho_\nu \check{y}}\right)_{\check{y} \in Q^o} \quad \text{is then a bundle.} \tag{3.74}$$

We write (3.73) in the form

$$\mathcal{H}_\nu = \left(v_\nu(y) + H_y^0\right)_{y \in Q_\nu^* \cap Q}, \text{ where} \tag{3.75}$$

$$v_\nu(y) \perp H_y^0 \text{ for each } y \in Q_\nu^* \cap Q. \tag{3.76}$$

The vector $v_\nu(y)$ is given by

$$v_\nu(y) = \Pi_y v(y) - \Pi_y v(x_\nu) \text{ for } y \in Q_\nu^* \cap Q, \text{ where} \tag{3.77}$$

Π_y denotes the orthogonal projection from $\mathbb{R}^r$ onto the orthocomplement of H_y^0.

Passing to the bundle $\check{\mathcal{H}}_\nu$, we find that

$$\check{\mathcal{H}}_\nu = \left(\check{v}_\nu(\check{y}) + H^0_{\rho_\nu \check{y}}\right)_{\check{y} \in Q^o}, \text{ with} \tag{3.78}$$

$$\check{v}_\nu(\check{y}) \perp H^0_{\rho_\nu \check{y}} \text{ for each } \check{y} \in Q^o. \tag{3.79}$$

Here, $\check{v}_\nu(\check{y})$ is given by

$$\check{v}_\nu(\check{y}) = v_\nu(\rho_\nu \check{y}). \tag{3.80}$$

It is easy to check that $\check{\mathcal{H}}_\nu$ is a Glaeser-stable bundle with nonempty fibers. Moreover, from (3.12) and (21), we see that the function $y \mapsto \dim H_y^0$ takes at most $\Lambda - 1$ values as y ranges over $Q_\nu^* \cap Q$. Therefore, the bundle $\check{\mathcal{H}}_\nu$ has at most $\Lambda - 1$ strata.

Thus, our inductive hypothesis (3.11) applies to the bundle $\check{\mathcal{H}}_\nu$. Consequently, we obtain the following results for the family of affine spaces $\mathcal{H}_\nu$.

We obtain sets

$$S_\nu(x) \subset Q_\nu^* \cap Q \quad \text{for each } x \in Q_\nu^* \cap Q, \tag{3.81}$$

and linear maps

$$A_\nu(x, y) : \mathbb{R}^r \to \mathbb{R}^r \text{ for each } x \in Q_\nu^* \cap Q, y \in S_\nu(x). \tag{3.82}$$

$$\text{The sets } S_\nu(x) \text{ each contain at most } C \text{ points.} \tag{3.83}$$

$$\text{The } S_\nu(x) \text{ and } A_\nu(x, y) \text{ are determined by } (H_z^0)_{z \in Q_\nu^* \cap Q}. \tag{3.84}$$

Moreover, setting

$$F_\nu(x) = \sum_{y \in S_\nu(x)} A_\nu(x, y) v_\nu(y) \quad \text{for} \quad x \in Q_\nu^* \cap Q, \tag{3.85}$$

we find that

$$F_\nu \text{ is continuous on } Q_\nu^* \cap Q, \tag{3.86}$$

$$F_\nu(x) \in v_\nu(x) + H_x^0 = v(x) - v(x_\nu) + H_x^0 \quad \text{for each} \quad x \in Q_\nu^* \cap Q, \tag{3.87}$$

and

$$\max_{x \in Q_\nu^* \cap Q} \left| F_\nu(x) \right| \le C \sup_{y \in Q_\nu^* \cap Q} \left| v_\nu(y) \right|. \tag{3.88}$$

Let us estimate the right-hand side of (3.88). For any Q_ν, formula (3.77) shows that

$$\sup_{y \in Q_\nu^* \cap Q} \left| v_\nu(y) \right| \le 2 \sup_{y \in Q} \left| v(y) \right|. \tag{3.89}$$

Moreover, let $x \in E_1$, $\epsilon > 0$ be given, and let δ be as in Lemma 21. Given any Q_ν such that distance $(x, Q_\nu^*) < \delta$, and given any $y \in Q_\nu^* \cap Q$, Lemma 21 tells us that

$$\left| v(x) - v(x_\nu) \right| < \epsilon \text{ and distance } (v(x), v(y) + H_y^0) < \epsilon.$$

Consequently,

$$\text{dist}(0, v(y) - v(x_\nu) + H_y^0) < 2\epsilon \text{ and } \left| v(x) - v(x_\nu) \right| < \epsilon. \tag{3.90}$$

From (3.73), (3.75), (3.76), we see that $v_\nu(y)$ is the shortest vector in $v(y) - v(x_\nu) + H_y^0$. Hence, (3.90) yields the estimate $\left| v_\nu(y) \right| < 2\epsilon$.

Therefore, we obtain the following result. Let $x \in E_1$ and $\epsilon > 0$ be given. Let δ be as in Lemma 21. Then, for any Q_ν such that distance $(x, Q_\nu^*) < \delta$, we have

$$\sup_{y \in Q_\nu^* \cap Q} \left| v_\nu(y) \right| \leq 2\epsilon, \text{ and } \left| v(x) - v(x_\nu) \right| < \epsilon. \tag{3.91}$$

From (3.88), (3.89), (3.91), we see that

$$\max_{x \in Q_\nu^* \cap Q} \left| F_\nu(x) \right| \leq C \sup_{y \in Q} \left| v(y) \right| \tag{3.92}$$

for each ν and that the following holds. Let $x \in E_1$ and $\epsilon > 0$ be given. Let δ be as in Lemma 21 and let $y \in Q_\nu^* \cap Q \cap B(x, \delta)$. Then

$$\left| F_\nu(y) \right| \leq C\epsilon, \text{ and } \left| v(x) - v(x_\nu) \right| < \epsilon. \tag{3.93}$$

We now define a map $F : Q \to \mathbb{R}^r$, by setting

$$F(x) = v(x) \text{ for } x \in E_1, \text{ and} \tag{3.94}$$

$$F(x) = \sum_\nu \theta_\nu(x) \cdot F_\nu(x) + v(x_\nu) \text{ for } x \in Q \setminus E_1. \tag{3.95}$$

Note that (3.95) makes sense, because the sum contains finitely many nonzero terms and because $\theta_\nu = 0$ outside the set where F_ν is defined.

We will show that F is given in terms of the $(v(y))_{y \in Q}$ by a formula of the form (3.56) and that conditions (3.57) $\cdots$ (20) are satisfied. As we noted just after (H2), this will complete our induction on Λ and establish Theorem 20.

First, we check that our $F(x)$ is given by (3.56), for suitable $S(x)$, $A(x, y)$. We proceed by cases. If $x \in E_1$, then already (3.94) has the form (3.56), with

$$S(x) = \{x\} \text{ and } A(x, y) = \text{ identity}. \tag{3.96}$$

Suppose $x \in Q \setminus E_1$. Then $F(x)$ is defined by (3.95).

Thanks to (3.67), we may restrict the sum in (3.95) to those ν such that $x \in Q_\nu^*$. For each such ν, we substitute (3.77) into (3.85) and then substitute the resulting formula for $F_\nu(x)$ into (3.95). We find that

$$F(x) = \sum_{Q_\nu^* \ni x} \theta_\nu(x) \cdot v(x_\nu) + \sum_{y \in S_\nu(x)} A_\nu(x, y) \cdot \left(\Pi_y v(y) - \Pi_y v(x_\nu)\right) \tag{3.97}$$

which is a formula of the form (3.4).

Thus, in all cases, F is given by a formula (3.4). Moreover, examining (3.96) and (3.97) (and recalling (3.81) $\cdots$ (3.84) as well as (3.20), we see that (3.5)–(3.7) hold and that in our formula (3.4) for F, each $S(x)$ contains at most C points. Thus, (3.9) holds, with a suitable d_Λ in place of d.

It remains to prove (3.8) and to show that our F is a section of the bundle $\mathcal{H}$. Thus, we must establish the following.

$$F : Q \to \mathbb{R}^r \text{ is continuous.} \tag{3.98}$$

$$F(x) \in v(x) + H_x^0 \text{ for each } x \in Q. \tag{3.99}$$

$$|F(x)| \le C \sup_{y \in Q} |v(y)| \text{ for each } x \in Q. \tag{3.100}$$

The proof of Theorem 20 is reduced to proving (3.98)–(3.100).

Let us prove (3.98). Fix $x \in Q$; we show that F is continuous at x. If $x \notin E_1$, then (3.66), (3.67), (3.86), and (3.95) easily imply that F is continuous at x.

On the other hand, suppose $x \in E_1$. To show that F is continuous at x, we must prove that

$$\lim_{y \to x, y \in E_1} v(y) = v(x) \quad \text{and that} \tag{3.101}$$

$$\lim_{y \to x, y \in Q \setminus E_1} \sum_\nu \theta_\nu(y) F_\nu(y) + v(x_\nu) = v(x). \tag{3.102}$$

We obtain (3.101) as an immediate consequence of (3.62). To prove (3.102), we bring in (3.93). Let $\epsilon > 0$ and let $\delta > 0$ arise from ϵ, x as in (3.93). Let $y \in Q \setminus E_1$ and suppose $|y - x| < \delta$. For each ν such that $y \in Q_\nu^*$, (3.93) gives

$$|\theta_\nu(y) \cdot [F_\nu(y) + v(x_\nu) - v(x)]| \le C\epsilon\theta_\nu(y). \tag{3.103}$$

For each ν such that $y \notin Q_\nu^*$, (3.103) holds trivially, since $\theta_\nu(y) = 0$. Thus, (3.103) holds for all ν. Summing on ν, and recalling (3.68), we conclude that

$$\left|\sum_\nu \theta_\nu(y) \cdot F_\nu(y) + v(x_\nu) - v(x)\right| \le C\epsilon.$$

This holds for any $y \in Q \setminus E_1$ such that $|y - x| < \delta$. The proof of (3.102) is complete. Thus, (3.98) is now proven.

To prove (3.99), we again proceed by cases. If $x \in E_1$, then (3.99) holds trivially, by (3.94). On the other hand, suppose $x \in Q \setminus E_1$. Then (3.87) gives $[F_\nu(x) + v(x_\nu)] \in v(x) + H_x^0$ for each ν such that $Q_\nu^* \ni x$.

Since also $\theta_\nu(x) = 0$ for $x \notin Q_\nu^*$, and since $\sum_\nu \theta_\nu(x) = 1$, it follows that

$$\sum_\nu \theta_\nu(x) \cdot [F_\nu(x) + v(x_\nu)] \in v(x) + H_x^0, \text{i.e.,}$$

$F(x) \in v(x) + H_x^0$. Thus, (3.99) holds in all cases.

Finally, we check (3.100). For $x \in E_1$, (3.100) is trivial from the definition (3.94). On the other hand, suppose $x \in Q \setminus E_1$. For each ν such that $Q_\nu^* \ni x$, (3.92) gives

$$|\theta_\nu(x) \cdot [F_\nu(x) + v(x_\nu)]| \le C\theta_\nu(x) \cdot \sup_{y \in Q} |v(y)|. \tag{3.104}$$

Estimate (3.104) also holds trivially for $x \notin Q_\nu^*$, since then $\theta_\nu(x) = 0$. Thus, (3.104) holds for all ν. Summing on ν, we find that

$$\left|F(x)\right| \leq \sum_\nu \left|\theta_\nu(x)\cdot[F_\nu(x)+v(x_\nu)]\right| \leq C \sup_{y\in Q}\left|v(y)\right|\cdot\sum_\nu \theta_\nu(x) = C \sup_{y\in Q}\left|v(y)\right|,$$

thanks to (3.68) and (3.95).

Thus, (3.100) holds in all cases. The proof of Theorem 20 is complete. □

Let $\widetilde{F}$ be any section of the bundle $\mathcal{H}$ in Theorem 20. For each $x \in Q$, we have $|v(x)| \leq |\widetilde{F}(x)|$, since $\widetilde{F}(x) \in v(x) + H_x^0$ and $v(x)\perp H_x^0$. Therefore, the section F produced by Theorem 20 satisfies the estimate $\max_{x\in Q}|F(x)| \leq C\cdot\max_{x\in Q}|\widetilde{F}(x)|$, where C depends only on n, r.

3.5 *Computing the Section of a Bundle*

Here, we combine our results from the last few sections. Let

$$\mathcal{H} = (v(x) + H_x^0)_{x\in Q} \text{ be a bundle, where} \tag{3.105}$$

$$\mathcal{H}^0 = (H_x^0)_{x\in Q} \text{ is a homogeneous bundle, and} \tag{3.106}$$

$$v(x)\perp H_x^0 \quad \text{for each } x \in Q. \tag{3.107}$$

Suppose $\mathcal{H}$ has a section. Then the iterated Glaeser refinements of $\mathcal{H}$ have nonempty fibers and may therefore be written as

$$\mathcal{H}^\ell = (v^\ell(x) + H_x^{0,\ell})_{x\in Q} \text{ where} \tag{3.108}$$

$$\mathcal{H}^{0,\ell} = (H_x^{0,\ell})_{x\in Q} \text{ is a homogeneous bundle, and} \tag{3.109}$$

$$v^\ell(x)\perp H_x^{0,\ell} \quad \text{for each } x \in Q. \tag{3.110}$$

Let $\xi_i^\ell(x) \in \mathbb{R}$, $y_i^{\ell,\nu}(x) \in Q$, $\beta_{ij}^{\ell,\nu}(x) \in \mathbb{R}$, $w_i(x) \in \mathbb{R}^r$ be as in Sect. 3.2. Thus,

$$\xi_i^0(x) = i\text{th component of } v(x), \text{ for } x \in Q; \tag{3.111}$$

$$\xi_i^\ell(x) = \lim_{\nu\to\infty} \sum_{j=1}^r \beta_{ij}^{\ell,\nu}(x)\, \xi_j^{\ell-1}\big(y_i^{\ell,\nu}(x)\big) \tag{3.112}$$

for $x \in Q, 1 \leq \ell \leq 2r+1, \quad 1 \leq i \leq r$, and

$$v^{2r+1}(x) = \sum_{i=1}^r \xi_i^{2r+1}(x) w_i(x) \quad \text{for } x \in Q. \tag{3.113}$$

Recall that $\beta_{ij}^{\ell,\nu}(x)$, $\gamma_i^{\ell,\nu}(x)$ and $w_i(x)$ are determined by the homogeneous bundle $\mathcal{H}^0$, independently of the vectors $(v(z))_{z\in Q}$. The bundle $\mathcal{H}^{2r+1} = (v^{2r+1}(x) + H_x^{0,2r+1})_{x\in Q}$ is Glaeser stable, with nonempty fibers. Hence, the results of Sect. 3.4 apply to $\mathcal{H}^{2r+1}$. Thus, we obtain a section of $\mathcal{H}^{2r+1}$ of the form

$$F(x) = \sum_{y\in S(x)} A(x,y)v^{2r+1}(y) \qquad (\text{all } x \in Q), \tag{3.114}$$

where $S(x) \subset Q$ and $\#(S(x)) \leq d$ for each $x \in Q$ and $A(x,y) : \mathbb{R}^r \to \mathbb{R}^r$ is a linear map, for each $x \in Q$, $y \in S(x)$. Our section F satisfies the estimate

$$\max_{x\in Q} |F(x)| \leq C \max_{x\in Q} |\widetilde{F}(x)|, \text{ for any section } \widetilde{F} \text{ of } \mathcal{H}^{2r+1}. \tag{3.115}$$

Here, d and C depend only on n and r; and the $S(x)$ and $A(x,y)$ are determined by $\mathcal{H}^{0,2r+1}$, independently of the vectors $v^{2r+1}(z)$ $(z \in Q)$.

Recall that the bundles $\mathcal{H}$ and $\mathcal{H}^{2r+1}$have the same sections. Therefore, substituting (3.113) into (3.114), and setting

$$A_i(x,y) = A(x,y)w_i(y) \in \mathbb{R}^r \quad \text{for } x \in Q,\ y \in S(x),\ i = 1,\ldots,r, \tag{3.116}$$

we find that

$$F(x) = \sum_{y\in S(x)} \sum_{1}^{r} \xi_i^{2r+1}(y)A_i(x,y) \text{ for all } x \in Q. \tag{3.117}$$

Moreover, F is a section of $\mathcal{H}$, and

$$\max_{x\in Q} |F(x)| \leq C \max_{x\in Q} |\widetilde{F}(x)| \text{ for any section } \widetilde{F} \text{ of } \mathcal{H}. \tag{3.118}$$

Furthermore, the $A_i(x,y)$ are determined by $\mathcal{H}^0$, independently of the family of vectors $(v(z))_{z\in Q}$.

Thus, we can compute a section of $\mathcal{H}$ by starting with (3.111), then computing the $\xi_i^\ell(x)$ using the recursion (3.112), and finally applying (3.117) once we know the $\xi_i^{2r+1}(x)$. In particular, we guarantee that the limits in (3.112) exist. Here, of course, we make essential use of our assumption that $\mathcal{H}$ has a section.

3.6 Computing a Continuous Solution of Linear Equations

We apply the results of the preceding section, to find continuous solutions of

$$\phi_1 f_1 + \cdots + \phi_r f_r = \phi \text{ on } Q. \tag{3.119}$$

Such a solution $(\phi_1, \ldots, \phi_r)$ is a section of the bundle

$$\mathcal{H} = (H_x)_{x \in Q}, \text{ where} \tag{3.120}$$

$$H_x = \{v = (v_1, \ldots, v_r) \in \mathbb{R}^r : v_1 f_1(x) + \cdots + v_r f_r(x) = \phi(x)\}. \tag{3.121}$$

We write $\mathcal{H}$ in the form

$$\mathcal{H} = (v(x) + H_x^0)_{x \in Q}, \text{ where} \tag{3.122}$$

$$H_x^0 = \{v = (v_1, \ldots, v_r) \in \mathbb{R}^r : v_1 f_1(x) + \cdots + v_r f_r(x) = 0\}, \text{ and} \tag{3.123}$$

$$v(x) = \phi(x) \cdot (\widetilde{\xi}_1(x), \ldots, \widetilde{\xi}_r(x)); \text{ here,} \tag{3.124}$$

$$\widetilde{\xi}_i(x) = \begin{cases} 0 & \text{if} \quad f_1(x) = f_2(x) = \cdots = f_r(x) = 0 \\ f_i(x)/\left(f_1^2(x) + \cdots + f_r^2(x)\right) & \text{otherwise.} \end{cases} \tag{3.125}$$

Note that

$$v(x) \perp H_x^0 \quad \text{for each } x \in Q. \tag{3.126}$$

Specializing the discussion in the preceding section to the bundle (3.108) $\cdots$ (3.112), we obtain the following objects:

- Coefficients $\beta_{ij}^{\ell,\nu}(x) \in \mathbb{R}$, for $x \in Q$, $1 \le \ell \le 2r+1$, $\nu \ge 1, 1 \le i, j \le r$;
- Points $y_i^{\ell,\nu}(x) \in Q$, for $x \in Q$, $1 \le \ell \le 2r+1$, $\nu \ge 1$, $1 \le i \le r$;
- Finite sets $S(x) \subset Q$, for $x \in Q$; and
- Vectors $A_i(x, y) \in \mathbb{R}^r$, for $x \in Q, y \in S(x)$, $1 \le i \le r$.

These objects depend only on the functions $f_1, \ldots, f_r$.

We write $A_{ij}(x, y)$ to denote the ith component of the vector $A_j(x, y)$.

To attempt to solve (3.119), we use the following

Procedure 22 *First, compute $\xi_i^\ell(x) \in \mathbb{R}$, for all $x \in Q$, $0 \le \ell \le 2r+1$, $1 \le i \le r$, by the recursion:*

$$\xi_i^0(x) = \widetilde{\xi}_i(x) \cdot \phi(x) \quad \text{for } 1 \le i \le r; \text{ and} \tag{3.127}$$

$$\xi_i^\ell(x) = \lim_{\nu \to \infty} \sum_{j=1}^r \beta_{ij}^{\ell,\nu}(x) \cdot \xi_j^{\ell-1}(y_i^{\ell,\nu}(x)) \tag{3.128}$$

for $1 \le i \le r, 1 \le \ell \le 2r+1$.

Then define functions $\Phi_1, \ldots, \Phi_r : Q \to \mathbb{R}$, by setting

$$\Phi_i(x) = \sum_{y \in S(x)} \sum_{j=1}^r A_{ij}(x, y) \cdot \xi_j^{2r+1}(y) \quad \text{for} \quad x \in Q, \ 1 \le i \le r \tag{3.129}$$

If, for some $x \in Q$ and $i = 1, \ldots, r$, the limit in (3.128) fails to exist, then our procedure (22) fails. Otherwise, procedure (22) produces functions $\Phi_1, \ldots, \Phi_r : Q \to \mathbb{R}$. These functions may or may not be continuous.

The next result follows at once from the discussion in the preceding section. It tells us that, if (3.105) has a continuous solution, then procedure (22) produces an essentially optimal continuous solution of (3.105).

Theorem 23 *(1) The objects $\widetilde{\xi}_i(x), \beta_{ij}^{\ell,\nu}(x), y_i^{\ell,\nu}(x), S(x)$, and $A_{ij}(x,y)$, used in procedure (22), depend only on $f_1, \cdots, f_r$, and not on the function ϕ.*

(2) For each $x \in Q$, the set $S(x) \subset Q$ contains at most d points, where d depends only on n and r.

(3) Let $\phi : Q \to \mathbb{R}$ and let $\phi_1, \ldots, \phi_r : Q \to \mathbb{R}$ be continuous functions such that $\phi_1 f_1 + \cdots + \phi_r f_r = \phi$ on Q. Then procedure (22) succeeds, the resulting functions $\Phi_1, \ldots, \Phi_r : Q \to \mathbb{R}$ are continuous, and $\Phi_1 f_1 + \cdots + \Phi_r f_r = \phi$ on Q. Moreover,

$$\max_{\substack{x \in Q \\ 1 \le i \le r}} \left|\Phi_i(x)\right| \le C \cdot \max_{\substack{x \in Q \\ 1 \le i \le r}} \left|\phi_i(x)\right|$$

where C depends only on n, r.

For particular functions $f_1, \ldots, f_r$, it is a tedious, routine exercise to go through the arguments in the past several sections and compute the $\xi_i(x), \beta_i^{\ell,\nu}(x), y_i^{\ell,\nu}(x), S(x)$ and $A_{ij}(x,y)$ used in our procedure (22). We invite the reader to carry this out for the case of Hochster's equation 3.4 and to compare the resulting formulas with those given in Sect. 3.

So far, we have dealt with a single equation (3.119) for continuous functions $\phi_1, \ldots, \phi_r$. To handle a system of equations, we simply take $f_1, \ldots, f_r$ and ϕ to be vector valued in (3.119). In place of (3.124), (3.125), and (3.127), we now define $v(x) = \left(\xi_1^0(x), \ldots, \xi_r^0(x)\right)$ to be the shortest vector in $\mathbb{R}^r$ that solves the equation $\sum_i \xi_i^0(x) f_i(x)$ for each fixed x. (If, for some x, this equation has no solution, then (3.119) has no solution.) We can easily compute the $\xi_i^0(x)$ from $f_1(x), \ldots, f_r(x)$ and $\phi(x)$ by linear algebra. Starting from the above $\xi_i^0(x)$, we can repeat the proof of Theorem 23, with trivial changes.

4 Algebraic Geometry Approach

The following simple example illustrates this method.

Example 24 *Which functions ϕ on $\mathbb{R}^2_{xy}$ can be written in the form*

$$\phi = \phi_1 x^2 + \phi_2 y^2 \tag{4.1}$$

where ϕ_1, ϕ_2 are continuous on $\mathbb{R}^2$? (We know that the pointwise tests (3) give an answer in this case, but the following method will generalize better.)

An obvious necessary condition is that ϕ should vanish to order 2 at the origin. This is, however, not sufficient since xy cannot be written in this form.

To see what happens, we blow up the origin. The resulting real algebraic variety $p : B_0\mathbb{R}^2 \to \mathbb{R}^2$ can be covered by two charts: one given by coordinates $x_1 = x/y, y_1 = y$ and the other by coordinates $x_2 = x, y_2 = y/x$. Working in the first chart, pulling back (4.1), we get the equation

$$\phi \circ p = (\phi_1 \circ p) \cdot x_1^2 y_1^2 + (\phi_2 \circ p) \cdot y_1^2. \tag{4.2}$$

The right-hand side is divisible by y_1^2, so we have our first condition

(24.1) First test. Is $(\phi \circ p)/y_1^2$ continuous?

If the answer is yes, then we divide by y_1^2, set $\psi := (\phi \circ p)/y_1^2$, and try to solve

$$\psi = \psi_1 \cdot x_1^2 + \psi_2. \tag{4.3}$$

This always has a continuous solution, but we need a solution where $\psi_i = \phi_i \circ p$ for some ϕ_i. Clearly, the ψ_i have to be constant along the line $(y_1 = 0)$. This is easily seen to be the only restriction. We thus set $y_1 = 0$ and try to solve

$$\psi(x_1, 0) = r_1 x_1^2 + r_2 \quad \text{where } r_i \in \mathbb{R}. \tag{4.4}$$

The original 2-variable problem has been reduced to a 1-variable question. Solvability is easy to decide using either of the following.

(24.2.i) Second test, Wronskian form. The following determinant is identically zero

$$\begin{vmatrix} 1 & 1 & 1 \\ a^2 & b^2 & c^2 \\ \psi(a,0) & \psi(b,0) & \psi(c,0) \end{vmatrix}$$

(24.2.ii) Second test, finite set form. For every $a, b, c \in \mathbb{R}$, there are $r_i := r_i(a,b,c) \in \mathbb{R}$ (possibly depending on a, b, c) such that

$$\psi(a,0) = r_1 a^2 + r_2, \quad \psi(b,0) = r_1 b^2 + r_2 \quad \text{and} \quad \psi(c,0) = r_1 c^2 + r_2.$$

(In principle, we should check what happens on the second chart, but in this case, it gives nothing new.)

Working on $\mathbb{R}^n$, let us now consider the general case

$$\phi = \textstyle\sum_i \phi_i f_i.$$

As in (24), we start by blowing up either the common zero set $Z = (f_1 = \cdots = f_r = 0)$ or, what is computationally easier, the ideal $(f_1, \ldots, f_r)$. We get a real algebraic variety $p : Y \to \mathbb{R}^n$.

Working in various coordinate charts on Y, we get analogs of the first test (24.1) and new equations

$$\psi = \sum_i \psi_i g_i.$$

The solvability again needs to be checked only on an $(n-1)$-dimensional real algebraic subvariety $Y_E \subset Y$. One sees, however, that the second tests (24.2.i–ii) are both equivalent to the pointwise tests (3), thus not sufficient in general.

Instead, we focus on what kind of question we need to solve on Y_E. This leads to the following concept.

Definition 25 *A descent problem is a compound object*

$$\mathbf{D} = \big(p : Y \to X, f : p^*E \to F\big)$$

consisting of a proper morphism of real algebraic varieties $p : Y \to X$, an algebraic vector bundle E on X, an algebraic vector bundle F on Y, and an algebraic vector bundle map $f : p^*E \to F$. (See (31) for the basic notions related to real algebraic varieties.)

Our aim is to understand the image of $f \circ p^* : C^0(X, E) \to C^0(Y, F)$.

We have the following analog of (24.2.ii).

Definition 26 *Let $\mathbf{D} = \big(p : Y \to X, f : p^*E \to F\big)$ be a descent problem and $\phi_Y \in C^0(Y, F)$. We say that ϕ_Y satisfies the finite set test if for every $y_1, \dots, y_m \in Y$, there is a $\phi_X = \phi_{X, y_1, \dots, y_m} \in C^0(X, E)$ (possibly depending on $y_1, \dots, y_m$) such that*

$$\phi_Y(y_i) = f \circ p^*(\phi_X)(y_i) \quad \textit{for } i = 1, \dots, m.$$

Definition 27 *A descent problem $\mathbf{D} = \big(p : Y \to X, f : p^*E \to F\big)$ is called finitely determined if for every $\phi_Y \in C^0(Y, F)$, the following are equivalent:*

(1) $\phi_Y \in \operatorname{im}\big[f \circ p^* : C^0(X, E) \to C^0(Y, F)\big]$.
(2) ϕ_Y satisfies the finite set test.

28 (Outline of the Main Result) *Our Theorem (34) gives an algorithm to decide the answer to Question 1. The precise formulation is somewhat technical to state, so here is a rough explanation of what kind of answer it gives and what we mean by an "algorithm." There are three main parts:*

Part 1. *First, starting with $\mathbb{R}^n$ and $f_1, \dots, f_r$, we construct a finitely determined descent problem $\mathbf{D} = \big(p : Y \to \mathbb{R}^n, f : p^*E \to F\big)$. This is purely algebraic, can be effectively carried out and independent of ϕ.*

Part 2. *There is a partially defined "twisted pull-back" map $p^{(*)} : C^0(\mathbb{R}^n) \dashrightarrow C^0(Y, F)$ (32) which is obtained as an iteration of three kinds of steps:*

(1) We compose a function by a real algebraic map.
(2) We create a vector function out of several functions or decompose a vector function into its coordinate functions.

(3) *We choose local (real analytic) coordinates* $\{y_i\}$ *and ask if a certain function of the form* $\psi_{j+1} := \psi_j \cdot \prod_i y_i^{-m_i}$ *is continuous or not where* $m_i \in \mathbb{Z}$.

If any of the answers is no, then the original ϕ *cannot be written as* $\sum_i \phi_i f_i$ *and we are done. If all the answers are yes, then we end up with* $p^{(*)}\phi \in C^0(Y, F)$.

Part 3. *We show that* $\phi = \sum_i \phi_i f_i$ *is solvable iff* $p^{(*)}\phi \in C^0(Y, F)$ *satisfies the finite set test (26).*

By following the proof, one can actually write down solutions ϕ_i, *but this relies on some artificial choices. The main ingredient that we need is to choose extensions of certain functions defined on closed semialgebraic subsets to the whole* $\mathbb{R}^n$. *In general, there does not seem to be any natural extension, and we do not know if it makes sense to ask for the "best possible" solution or not.*

Negative Aspects. There are two difficulties in carrying out this procedure in any given case. First, in practice, (28) of Part 28 may not be effectively doable. Second, we may need to compose ψ_{j+1} with a real algebraic map r_{j+1} such that ψ_j vanishes on the image of r_{j+1}. Thus, we really need to compute limits and work with the resulting functions. This also makes it difficult to interpret our answer on $\mathbb{R}^n$ directly.

Positive Aspects. On the other hand, just knowing that the answer has the above general structure already has some useful consequences.

First, the general framework works for other classes of functions; for instance, the same algebraic setup also applies in case ϕ and the ϕ_i are Hölder continuous.

Another consequence we obtain is that if $\phi = \sum_i \phi_i f_i$ is solvable and ϕ has certain additional properties, then one can also find a solution $\phi = \sum_i \psi_i f_i$ where the ψ_i also have these additional properties. We list two such examples below; see also (12). For the proof, see (50) and (37).

Corollary 29 *Fix* $f_1, \dots, f_r$ *and assume that* $\phi = \sum_i \phi_i f_i$ *is solvable. Then:*

(1) *If* ϕ *is semialgebraic (31), then there is a solution* $\phi = \sum_i \psi_i f_i$ *such that the* ψ_i *are also semialgebraic.*
(2) *Let* $U \subset \mathbb{R}^n \setminus Z$ *be an open set such that* ϕ *is* C^m *on* U *for some* $m \in \{1, 2, \dots, \infty, \omega\}$. *Then there is a solution* $\phi = \sum_i \psi_i f_i$ *such that the* ψ_i *are also* C^m *on* U.

Examples 30 *The next series of examples shows several possible variants of (29) that fail.*

(1) *Here* ϕ *is a polynomial, but the* ϕ_i *must have very small Hölder exponents.*

For $m \geq 1$, *take* $\phi := x^{2m} + (x^{2m-1} - y^{2m+1})^2$ *and* $f_1 = x^{2m+2} + y^{2m+2}$. *There is only one solution,*

$$\phi_1 = \frac{x^{2m} + (x^{2m-1} - y^{2m+1})^2}{x^{2m+2} + y^{2m+2}}.$$

We claim that it is Hölder with exponent $\frac{2}{2m-1}$. The exponent is achieved along the curve $x^{2m-1} - y^{2m+1} = 0$, parametrized as $\big(t^{(2m+1)/(2m-1)}, t\big)$.

(2) *Here ϕ is C^n, there is a C^0 solution but no Hölder solution.*
On $[-\frac{1}{2}, \frac{1}{2}] \subset \mathbb{R}^1$ set $f = x^n$ and $\phi = x^n/\log|x|$. Then ϕ is C^n and $\phi = \frac{1}{\log|x|} \cdot f$. Note that $\frac{1}{\log|x|}$ is continuous but not Hölder. (These can be extended to $\mathbb{R}^1$ in many ways.)

(3) *Question: If ϕ is C^∞ and there is a C^0 solution, is there always a Hölder solution?*

(4) *Let $g(x)$ be a real-analytic function. Set $f_1 := y$ and $\phi := \sin\big(g(x)y\big)$. Then $\phi_1 := \phi/y$ is also real analytic and $\phi = \phi_1 \cdot f_1$ is the only solution. Note that $|\phi(x, y)| \le 1$ everywhere, yet $\phi_1(x, 0) = g(x)$ can grow arbitrary fast.*

(5) *In general, there is no solution $\phi = \sum_i \psi_i f_i$ such that $\operatorname{Supp}\psi_i \subset \operatorname{Supp}\phi$ for every i. As an example, take $f_1 = x^2 + x^4$, $f_2 = x^2 + y^2$ and*

$$\phi(x, y) = \begin{cases} x^4 - y^2 & \text{if } y^2 \ge x^4 \text{ and} \\ 0 & \text{if } y^2 \le x^4. \end{cases}$$

Note that $\phi = f_1 - \phi_2 f_2$ where

$$\phi_2(x, y) = \begin{cases} 1 & \text{if } y^2 \ge x^4 \text{ and} \\ \frac{x^2+x^4}{x^2+y^2} & \text{if } y^2 \le x^4. \end{cases}$$

Let $\phi = \phi_1 \cdot (x^2 + x^4) + \psi_2 \cdot (x^2 + y^2)$ be any continuous solution. Setting $x = 0$, we get that $-y^2 = \psi_2(0, y) \cdot y^2$; hence, $\psi_2(0, 0) = -1$. Thus, $\operatorname{Supp}\psi_2$ cannot be contained in $\operatorname{Supp}\phi$.

On the other hand, given any solution $\phi = \sum_i \phi_i f_i$, let χ be a function that is 1 on $\operatorname{Supp}\phi$ and 0 outside a small neighborhood of it. Then $\phi = \chi\phi = \sum(\chi\phi_i) f_i$. Thus, we do have solutions whose support is close to $\operatorname{Supp}\phi$.

4.1 Descent Problems and Their Scions

31 (Basic Setup) *From now on, X denotes a fixed real algebraic variety. We always think of X as the real points of a complex affine algebraic variety $X_{\mathbb{C}}$ that is defined by real equations. (All our algebraic varieties are assumed reduced, i.e., a function is zero iff it is zero at every point).*

By a projective variety over X, we mean the real points of a closed subvariety $Y \subset X \times \mathbb{CP}^N$. Every such Y is again the set of real points of a complex affine algebraic variety $Y_{\mathbb{C}} \subset X_{\mathbb{C}} \times \mathbb{CP}^N$ that is defined by real equations. For instance, $X \times \mathbb{RP}^N$ is contained in the affine variety which is the complement of the hypersurface $(\sum y_i^2 = 0)$ where y_i are the coordinates on $\mathbb{P}^N$.

A variety Y over X comes equipped with a morphism $p : Y \to X$ to X, given by the first projection of $X \times \mathbb{CP}^N$. Given such $p_i : Y_i \to X$, a morphism between them is a morphism of real algebraic varieties $\phi : Y_1 \to Y_2$ such that $p_1 = p_2 \circ \phi$.

Given $p_i : Y_i \to X$, their fiber product is

$$Y_1 \times_X Y_2 := \{(y_1, y_2) : p_1(y_1) = p_2(y_2)\} \subset Y_1 \times Y_2.$$

This comes with a natural projection $p : Y_1 \times_X Y_2 \to X$ and $p^{-1}(x) = p_1^{-1}(x) \times p_2^{-1}(x)$ for every $x \in X$. (Note, however, that even if the Y_i are smooth, their fiber product can be very singular.) If X is irreducible, we are frequently interested only in those irreducible components that dominate X, called the dominant components.

$\mathcal{R}(Y)$ denotes the ring of all regular functions on Y. These are locally quotients of polynomials $p(x)/q(x)$ where $q(x)$ is nowhere zero.

By an algebraic vector bundle on Y, we mean the restriction of a complex algebraic vector bundle from $Y_{\mathbb{C}}$ to Y. All such vector bundles can be given by patching trivial bundles on a Zariski open cover $X = \cup_i U_i$ using transition functions in $\mathcal{R}(U_i \cap U_j)$. (Note that the latter condition is not quite equivalent to our definition, but this is not important for us, cf. [BCR98, Chap. 12].)

Note that there are two natural topologies on a real algebraic variety Y, the Euclidean topology and the Zariski topology. The closed sets of the latter are exactly the closed subvarieties of Y. A Zariski closed (resp. open) subset of Y is also Euclidean closed (resp. open).

A closed basic semialgebraic subset of Y is defined by finitely many inequalities $g_i \geq 0$. Using finite intersections and complements, we get all semialgebraic subsets. A function is semialgebraic iff its graph is semialgebraic. See [BCR98, Chap. 2] for a detailed treatment.

We need various ways of modifying descent problems. The following definition is chosen to consist of simple and computable steps yet be broad enough for the proofs to work. (It should become clear that several variants of the definition would also work. We found the present one convenient to use.)

Definition 32 (Scions of Descent Problems) *Let $\mathbf{D} = \big(p : Y \to X, f : p^*E \to F\big)$ be a descent problem. A scion of $\mathbf{D}$ is any descent problem $\mathbf{D}_s = \big(p_s : Y_s \to X, f_s : p_s^*E \to F_s\big)$ that can be obtained by repeated application of the following procedures:*

(1) For a proper morphism $r : Y_1 \to Y$, set

$$r^*\mathbf{D} := \big(p \circ r : Y_1 \to X, r^*f : (p \circ r)^*E \to r^*F\big).$$

As a special case, if $Z \subset X$ is a closed subvariety, then the scion $\mathbf{D}_Z = \big(p_Z : Y_Z \to Z, f_Z : p_Z^*(E|_Z) \to F|_{Y_Z}\big)$ (where $Y_Z := p^{-1}(Z)$) is called the restriction of $\mathbf{D}$ to Z.

(2) Given Y_w, assume that there are several proper morphisms $r_i : Y_w \to Y$ such that the composites $p_w := p \circ r_i$ are all the same. Set

$$(r_1, \ldots, r_m)^*\mathbf{D} := \big(p_w : Y_w \to X, \textstyle\sum_{i=1}^m r_i^* f : p_w^*E \to \sum_{i=1}^m r_i^* F\big)$$

where $\sum_{i=1}^{m} r_i^* f$ *is the natural diagonal map.*

(3) *Assume that* f *factors as* $p^*E \stackrel{q}{\to} F' \stackrel{j}{\hookrightarrow} F$ *where* F' *is a vector bundle and* $\operatorname{rank}_y j = \operatorname{rank}_y F'$ *for all* y *in a Euclidean dense Zariski open subset* $Y^0 \subset Y$. *Then set*

$$\mathbf{D}' := \big(p : Y \to X, f' := q : p^*E \to F'\big).$$

(The choice of Y^0 *is actually a quite subtle point. Algebraic maps have constant rank over a suitable Zariski open subset, and we want this open set to determine what happens with an arbitrary continuous function. This is why* Y^0 *is assumed Euclidean dense, not just Zariski dense. If* Y *is smooth, these are equivalent properties, but not if* Y *is singular. As an example, consider the Whitney umbrella* $Y := (x^2 = y^2 z) \subset \mathbb{R}^3$. *Here,* $Y \setminus (x = y = 0)$ *is Zariski open and Zariski dense. Its Euclidean closure does not contain the "handle"* $(x = y = 0,\ z < 0)$, *so it is not Euclidean dense.)*

Each scion remembers all of its forebears. That is, two scions are considered the "same" only if they have been constructed by an identical sequence of procedures. This is quite important since the vector bundle F_s *on a scion* $\mathbf{D}_s$ *does depend on the whole sequence.*

Every scion comes with a structure map $r_s : Y_s \to Y$.

If $\phi \in C^0(Y, F)$, *then* $r^*\phi \in C^0(Y_1, r^*F)$ *and* $\sum_{i=1}^{m} r_i^*\phi \in C^0(Y_w, \sum_{i=1}^{m} r_i^* F)$ *are well defined. In (32) above,* $j : C^0(Y, F') \to C^0(Y, F)$ *is an injection; hence, there is at most one* $\phi' \in C^0(Y, F')$ *such that* $j(\phi') = \phi$. *Iterating these, for any scion* $\mathbf{D}_s$ *of* $\mathbf{D}$ *with structure map* $r_s : Y_s \to Y$, *we get a partially defined map, called the twisted pull-back,*

$$r_s^{(*)} : C^0(Y, F) \dashrightarrow C^0(Y_s, F_s).$$

We will need to know which functions ϕ *are in the domain of a twisted pull-back map. A complete answer is given in (43).*

The twisted pull-back map sits in a commutative square

$$\begin{array}{ccc} C^0(Y, F) & \stackrel{r_s^{(*)}}{\dashrightarrow} & C^0(Y_s, F_s) \\ \uparrow & & \uparrow \\ C^0(X, E) & = & C^0(X, E). \end{array}$$

If the structure map $r_s : Y_s \to Y$ *is surjective, then* $r^{(*)} : C^0(Y, F) \dashrightarrow C^0(Y_s, F_s)$ *is injective (on its domain). In this case, understanding the image of* $f \circ p^* : C^0(X, E) \to C^0(Y, F)$ *is pretty much equivalent to understanding the image of* $f_s \circ p_s^* : C^0(X, E) \to C^0(Y_s, F_s)$.

We are now ready to state our main result, first in the inductive form.

Proposition 33 *Let* $\mathbf{D} = \big(p : Y \to X, f : p^*E \to F\big)$ *be a descent problem. Then there is a scion* $\mathbf{D}_s = \big(p_s : Y_s \to X, f_s : p_s^*E \to F_s\big)$ *with surjective structure map* $r_s : Y_s \to Y$ *and a closed subvariety* $Z \subset X$ *such that* $\dim Z < \dim X$ *and for every* $\phi \in C^0(Y, F)$, *the following are equivalent:*

(1) $\phi \in \operatorname{im}[f \circ p^* : C^0(X,E) \to C^0(Y,F)]$.
(2) $r_s^{(*)}\phi$ *is defined and* $r_s^{(*)}\phi \in \operatorname{im}[f_s \circ p_s^* : C^0(X,E) \to C^0(Y_s,F_s)]$.
(3) (a) $r_s^{(*)}\phi$ *satisfies the finite set test (26) and*
    (b) $\phi|_{Y_Z} \in \operatorname{im}[f_Z \circ p_Z^* : C^0(Z,E|_Z) \to C^0(Y_Z,F_Z)]$, *where the scion* $\mathbf{D}_Z = (p_Z : Y_Z \to Z, f_Z : p_Z^*(E|_Z) \to F_Z)$ *is the restriction of* $\mathbf{D}_s$ *to* Z *(32.1).*

We can now set $X_1 := Z, \mathbf{D}_1 := \mathbf{D}_Z$ apply (33) to $\mathbf{D}_1$ and get a descent problem $\mathbf{D}_2 := (\mathbf{D}_1)_Z$. Repeating this, we obtain descent problems $\mathbf{D}_i = (p_i : Y_i \to X, f_i : p_i^*E \to F_i)$ such that the dimension of $p_i(Y_i)$ drops at every step. Eventually, we reach the case where $p_i(Y_i)$ consists of points. Then the finite set test (26) gives the complete answer. The disjoint union of all the Y_i can be viewed as a single scion; hence, we get the following algebraic answer to Question 1.

Theorem 34 *Let* $\mathbf{D} = (p : Y \to X, f : p^*E \to F)$ *be a descent problem. Then it has a finitely determined scion* $\mathbf{D}_w = (p_w : Y_w \to X, f_w : p_w^*E \to F_w)$ *with surjective structure map* $r_w : Y_w \to Y$.

That is, for every $\phi \in C^0(Y,F)$, *the following are equivalent:*

(1) $\phi \in \operatorname{im}[f \circ p^* : C^0(X,E) \to C^0(Y,F)]$.
(2) *The twisted pull-back* $r_w^{(*)}\phi$ *is defined, and it is contained in the image of* $f_w \circ p_w^* : C^0(X,E) \to C^0(Y_w,F_w)$.
(3) *The twisted pull-back* $r_w^{(*)}\phi$ *is defined and satisfies the finite set test (26).* □

The proof of (34) works for many subclasses of continuous functions as well. Next, we axiomatize the necessary properties and describe the main examples.

4.2 Subclasses of Continuous Functions

Assumption 35 *For real algebraic varieties* Z, *we consider vector subspaces* $C^*(Z) \subset C^0(Z)$ *that satisfy the following properties:*

(1) *(Local property) If* $Z = \cup_i U_i$ *is an open cover of* Z, *then* $\phi \in C^*(Z)$ *iff* $\phi|_{U_i} \in C^*(U_i)$ *for every* i.
(2) *(*$\mathcal{R}(Z)$*-module) If* $\phi \in C^*(Z)$ *and* $h \in \mathcal{R}(Z)$ *is a regular function (31), then* $h \cdot \phi \in C^*(Z)$.
(3) *(Pull-back) For every morphism* $g : Z_1 \to Z_2$, *composing with* g *maps* $C^*(Z_2)$ *to* $C^*(Z_1)$.
(4) *(Descent property) Let* $g : Z_1 \to Z_2$ *be a proper, surjective morphism,* $\phi \in C^0(Z_2)$ *and assume that* $\phi \circ g \in C^*(Z_1)$. *Then* $\phi \in C^*(Z_2)$.
(5) *(Extension property) Let* $Z_1 \subset Z_2$ *be a closed semialgebraic subset (38). Then the twisted pull-back map* $C^*(Z_2) \to C^*(Z_1)$ *is surjective.*

Since every closed semialgebraic subset is the image of a proper morphism (38), we can unite (35) and (35) and avoid using semialgebraic subsets as follows.

(4+5) *(Strong descent property) Let $g : Z_1 \to Z_2$ be a proper morphism and $\psi \in C^*(Z_2)$. Then $\psi = \phi \circ g$ for some $\phi \in C^*(Z_2)$ iff ψ is constant on every fiber of g.*

The following additional condition comparing 2 classes $C_1^* \subset C_2^*$ is also of interest.

(6) *(Division property) Let $h \in \mathcal{R}(Z)$ be any function whose zero set is nowhere Euclidean dense. If $\phi \in C_1^*(Z)$ and $\phi / h \in C_2^*(Z)$, then $\phi / h \in C_1^*(Z)$.*

Example 36 *Here are some natural examples satisfying the assumptions (35.1–5):*

(1) *$C^0(Z)$, the set of all continuous functions on Z*
(2) *$C^h(Z)$, the set of all locally Hölder continuous functions on Z*
(3) *$S^0(Z)$, the set of continuous semialgebraic functions on Z*

Moreover, the pairs $S^0 \subset C^0$ and $S^0 \subset C^h$ both satisfy (35.6). (By contrast, by (30.2), the pair $C^h \subset C^0$ does not satisfy (35.6).)

37 (Proof of (29.1)) *More generally, consider two classes $C_1^* \subset C_2^*$ that satisfy (35.1–5) and also (35.6). Let **D** be a descent problem and $\phi \in C_1^*(Y, F)$. We claim that if $\phi = f \circ p^*(\phi_X)$ is solvable with $\phi_X \in C_2^*(X, E)$, then it also has a solution $\phi = f \circ p^*(\psi_X)$ where $\psi_X \in C_1^*(X, E)$.*

To see this, let $\mathbf{D}_w$ be a scion as in (34). By our assumption, the twisted pull-back $r_w^{(*)}\phi$ is in $C_2^*(Y_w, F_w)$, and it satisfies the finite set test. For the finite set test, it does not matter what type of functions we work with. Thus, we need to show that $r_w^{(*)}\phi$ is in $C_1^*(Y_w, F_w)$.

In a scion construction, this holds for steps as in (32.1–2) by (35.3). The key question is (32.3). The solution given in (43) shows that it is equivalent to (35.6). □

38 (C^*-Valued Functions over Semialgebraic Sets) *Let $S \subset Z$ be a closed semialgebraic subset. We can think of S as the image of a proper morphism $g : W \to Z$ (cf. [BCR98, Sect. 2.7]). One can define $C^*(S)$ either as the image of $C^*(Z)$ in $C^0(S)$ or as the preimage of $C^*(W)$ under the pull-back by g. By (35.4+5), these two are equivalent.*

We also have the following:

(1) *(Closed patching condition) Let $S_i \subset Z$ be closed semialgebraic subsets. Let $\phi_i \in C^*(S_i)$ and assume that $\phi_i|_{S_i \cap S_j} = \phi_j|_{S_i \cap S_j}$ for every i, j.*

Then there is a unique $\phi \in C^*(\cup_i S_i)$ such that $\phi|_{S_i} = \phi_i$ for every i.

To see this, realize each S_i as the image of some proper morphism $g_i : W_i \to Z$. Let $W := \amalg_i W_i$ be their disjoint union and $g : W \to Z$ the corresponding morphism. Define $\psi \in C^*(W)$ by the conditions $\psi|_{W_i} = \phi_i \circ g_i$.

The patching condition guarantees that ψ is constant on the fibers of g. Thus, by (35.4+5), $\psi = \phi \circ g$ for some $\phi \in C^*(\cup_i S_i)$.

These arguments also show that each $C^*(Z)$ is in fact a module over $S^0(Z)$, the ring of continuous semialgebraic functions.

Definition 39 (**C^*-Valued Sections**) *By Serre's theorems, every vector bundle on a complex affine variety can be written as a quotient bundle of a trivial bundle and also as a subbundle of a trivial bundle. Furthermore, every extension of vector bundles splits.*

Thus, on a real algebraic variety, every algebraic vector bundle can be written as a quotient bundle (and a subbundle) of a trivial bundle and every constant rank map of vector bundles splits.

Let F be an algebraic vector bundle on Z and $Z = \cup_i U_i$ an open cover such that $F|_{U_i}$ is trivial of rank r for every i. Let

$$C^*(Z, F) \subset C^0(Z, F)$$

denote the set of those sections $\phi \in C^0(Z, F)$ such that $\phi|_{U_i} \in C^*(U_i)^r$ for every i. If C^* satisfies the properties (35.1–2), this is independent of the trivializations and the choice of the covering.

If C^* satisfies the properties (35.1–6), then their natural analogs also hold for $C^*(Z, F)$. This is clear for the properties (35.2–4) and (35.6).

In order to check the extension property (35.5), first note that we have the following:

(1) Let $f : F_1 \to F_2$ be a surjection of vector bundles. Then $f : C^*(Z, F_1) \to C^*(Z, F_2)$ is surjective.

Now let $Z_1 \subset Z_2$ be an closed subvariety and F a vector bundle on Z_2. Write it as a quotient of a trivial bundle $\mathbb{C}^N_{Z_2}$. Every section $\phi_1 \in C^*(Z_1, F|_{Z_1})$ lifts to a section in $C^*(Z_1, \mathbb{C}^N_{Z_1})$ which in turn extends to a section in $C^*(Z_2, \mathbb{C}^N_{Z_2})$ by (35.6). The image of this lift in $C^*(Z_2, F|_{Z_2})$ gives the required lifting of ϕ_1.

4.3 Local Tests and Reduction Steps

Next we consider various descent problems whose solution is unique, if it exists.

40 (Pull-Back Test) *Let $g : Z_1 \to Z_2$ be a proper surjection of real algebraic varieties. Let F be a vector bundle on Z_2 and $\phi_1 \in C^*(Z_1, g^*F)$. When can we write $\phi_1 = g^*\phi_2$ for some $\phi_2 \in C^*(Z_2, F)$?*

Answer: By (35.4), such a ϕ_2 exists iff ϕ_1 is constant on every fiber of g. This can be checked as follows.

Take the fiber product $Z_3 := Z_1 \times_{Z_2} Z_1$ with projections $\pi_i : Z_3 \to Z_1$ for $i = 1, 2$. Note that $F_3 := \pi_1^* g^* F$ is naturally isomorphic to $\pi_2^* g^* F$. We see that ϕ_1 is constant on every fiber of g iff

$$\pi_1^*\phi_1 - \pi_2^*\phi_1 \in C^*(Z_3, F_3) \quad \textit{is identically 0.}$$

Note that this solves descent problems $\mathbf{D} = \big(p : Y \to X, f : p^* E \cong F\big)$ *where* f *is an isomorphism. We use two simple cases:*

(1) *Assume that there is a closed subset* $Z \subset X$ *such that* p *induces an isomorphism* $Y \setminus p^{-1}(Z) \to X \setminus Z$ *and* $\phi_Y \in C^0\big(Y, p^*E\big)$ *vanishes along* $p^{-1}(Z)$. *Then there is a* $\phi_X \in C^0\big(X, E\big)$ *such that* $\phi_Y = p^*\phi_X$ *(and* ϕ_X *vanishes along* Z*).*
(2) *Assume that there is a finite group* G *acting on* Y *such that* G *acts transitively on every fiber of* $\big(Y \setminus (\phi_Y = 0)\big) \to X$. *Then there is a* $\phi_X \in C^0\big(X, E\big)$ *such that* $\phi_Y = p^*\phi_X$.

41 (Wronskian Test) *Let* $\phi, f_1, \dots, f_r$ *be functions on a set* Z. *Assume that the* f_i *are linearly independent. Then* ϕ *is a linear combination of the* f_i *(with constant coefficients) iff the determinant*

$$\begin{vmatrix} f_1(\mathbf{z}_1) & \cdots & f_1(\mathbf{z}_r) & f_1(\mathbf{z}_{r+1}) \\ \vdots & & \vdots & \vdots \\ f_r(\mathbf{z}_1) & \cdots & f_r(\mathbf{z}_r) & f_r(\mathbf{z}_{r+1}) \\ \phi(\mathbf{z}_1) & \cdots & \phi(\mathbf{z}_r) & \phi(\mathbf{z}_{r+1}) \end{vmatrix}$$

is identically zero as a function on Z^{r+1}.

Proof. Since the f_i are linearly independent, there are $\mathbf{z}_1, \dots, \mathbf{z}_r \in Z$ such that the upper left $r \times r$ subdeterminant of is nonzero. Fix these $\mathbf{z}_1, \dots, \mathbf{z}_r$ and solve the linear system

$$\phi(\mathbf{z}_i) = \textstyle\sum_j \lambda_j f_j(\mathbf{z}_i) \quad \text{for } i = 1, \dots, r.$$

Replace ϕ by $\psi := \phi - \sum_i \lambda_i f_i$ and let $\mathbf{z}_{r+1}$ vary. Then our determinant is

$$\begin{vmatrix} f_1(\mathbf{z}_1) & \cdots & f_1(\mathbf{z}_r) & f_1(\mathbf{z}_{r+1}) \\ \vdots & & \vdots & \vdots \\ f_r(\mathbf{z}_1) & \cdots & f_r(\mathbf{z}_r) & f_r(\mathbf{z}_{r+1}) \\ 0 & \cdots & 0 & \psi(\mathbf{z}_{r+1}) \end{vmatrix}$$

and it vanishes iff $\psi(\mathbf{z}_{r+1})$ is identically zero. That is, when $\phi \equiv \sum_j \lambda_j f_j$.

42 (Linear Combination Test) *Let* Z *be a real algebraic variety,* F *a vector bundle on* Z *and* $f_1, \dots, f_r$ *linearly independent algebraic sections of* F.

Given $\phi \in C^*\big(Z, F\big)$, *when can we write* $\phi = \sum_i \lambda_i f_i$ *for some* $\lambda_i \in \mathbb{C}$?

Answer: One can either write down a determinantal criterion similar to (41) or reduce this to the Wronskian test as follows.

Consider $q : \mathbb{P}(F) \to X$, *the space of 1-dimensional quotients of* F. *Let* $u : q^*F \to Q$ *be the universal quotient line bundle. Then* $\phi = \sum_i \lambda_i f_i$ *iff*

$$u \circ q^*(\phi) = \textstyle\sum_i \lambda_i \cdot u \circ q^*(f_i).$$

The latter is enough to check on a Zariski open cover of $\mathbb{P}(F)$ *where* Q *is trivial. Thus, we recover the Wronskian test.* □

43 (Membership Test for Sheaf Injections) *Let* Z *be a real algebraic variety,* E, F *algebraic vector bundles, and* $h : E \to F$ *a vector bundle map such that* $\operatorname{rank} h = \operatorname{rank} E$ *on a Euclidean dense Zariski open set* $Z^0 \subset Z$*. Given a section* $\phi \in C^*(Z, F)$*, when is it in the image of* $h : C^*(Z, E) \to C^*(Z, F)$*?*

Answer: Over Z^0*, there is a quotient map* $q : F|_{Z^0} \to Q_{Z^0}$ *where* $\operatorname{rank} Q_{Z^0} = \operatorname{rank} F - \operatorname{rank} E$ *and* $\operatorname{im}(h|_{Z^0}) = \ker q$*. Then the first lifting condition is:*

(1) $q(\phi) = 0$*. Note that, in the local coordinate functions of* ϕ*, this is a linear condition with polynomial coefficients.*

By (39.3), $h|_{Z^0}$ *has an algebraic splitting* $s : F|_{Z^0} \to E|_{Z^0}$*. Note that* s *is not unique on* E *but it is unique on the image of* h*. Thus, the second condition says:*

(2) The section $s(\phi|_{Z^0}) \in C^*(Z^0, E|_{Z^0})$ *extends to a section of* $C^*(Z, E)$*.*

In order to make this more explicit, choose local algebraic trivializations of E *and of* F*. Then* ϕ *is given by coordinate functions* $(\phi_1, \dots, \phi_m)$*, and* s *is given by a matrix* (s_{ij}) *where the* s_{ij} *are rational functions on* Z *that are regular on* Z^0*. We can bring them to common denominator and write* $s_{ij} = u_{ij}/v$ *where* u_{ij} *and* v *are regular on* Z*. Thus,*

$$s(\phi|_{Z^0}) = \Big(\sum_j s_{1j}\phi_j, \dots, \sum_j s_{nj}\phi_j\Big) = \frac{1}{v}\Big(\sum_j u_{1j}\phi_j, \dots, \sum_j u_{nj}\phi_j\Big).$$

Let Φ *denote the vector function in the parenthesis on the right. Then* $\Phi \in C^*(Z, E)$*, and we are asking if* $\Phi/v \in C^*(Z, E)$ *or not. This is exactly one of the questions considered in Part 2 of (28).*

Also, if we are considering two function classes $C_1^* \subset C_2^*$*, then (43.3) and the assumption (35.6) say that a function* $\phi \in C_1^*(Z, F)$ *is in the image of* $h : C_2^*(Z, E) \to C_2^*(Z, F)$ *iff it is in the image of* $h : C_1^*(Z, E) \to C_1^*(Z, F)$*.* □

44 (Resolution of Singularities) *Let* $\mathbf{D} = (p : Y \to X, f : p^*E \to F)$ *be a descent problem. By Hironaka's theorems (see [Kol07, Chap. 3] for a relatively simple treatment), there is a resolution of singularities* $r_0 : Y' \to Y$*. That is,* Y' *is smooth and* r_0 *is proper and birational (i.e., an isomorphism over a Zariski dense open set). Note however, that* r_0 *is not surjective in general. In fact,* $r_0(Y')$ *is precisely the Euclidean closure of the smooth locus* Y^{ns}*. Thus,* $Y \setminus r_0(Y') \subset \operatorname{Sing}(Y)$*.*

We resolve $\operatorname{Sing} Y$ *to obtain* $r_1 : Y_1' \to \operatorname{Sing}(Y)$*. The resulting map* $Y' \amalg Y_1' \to Y$ *is surjective, except possibly along* $\operatorname{Sing}(\operatorname{Sing}(Y))$*. We can next resolve* $\operatorname{Sing}(\operatorname{Sing}(Y))$ *and so on. After at most* $\dim Y$ *such steps, we obtain a smooth, proper morphism* $R : Y^R \to Y$ *such that* Y^R *is smooth and* R *is surjective.* R *is an isomorphism over* Y^{ns} *but it can have many irreducible components that map to* $\operatorname{Sing}(Y)$*.*

We refer to $Y' \subset Y^R$ *as the main components of the resolution.*

Proposition 45 *Let* $\mathbf{D} = (p : Y \to X, f : p^*E \to F)$ *be a descent problem. Assume that* X, Y *are irreducible, the generic fiber of* p *is irreducible and smooth, and* $h(x) : E(x) \to C^0(Y_x, F|_{Y_x})$ *is an injection for general* $x \in p(Y)$. *Then* $\mathbf{D}$ *has a scion* $\mathbf{D}_s = (p_s : Y_s \to X, f_s : p_s^*E \to F_s)$ *with surjective structure map* $r_s : Y_s \to Y$ *such that:*

(1) Y_s *is a disjoint union* $Y_s^h \amalg Y_s^v$.
(2) $\dim p_s(Y_s^v) < \dim X$.
(3) f_s *is an isomorphism over* Y_s^h.

Proof. Set $n = \operatorname{rank} E$ and let Y_X^{n+1} be the union of the dominant components (31) of the $n+1$-fold fiber product of $Y \to X$ with coordinate projections π_i. Let $\tilde{p} : Y_X^{n+1} \to X$ be the map given by any of the $p \circ \pi_i$. Consider the diagonal map

$$\tilde{f} : \tilde{p}^*E \to \sum_{i=1}^{n+1} \pi_i^* F$$

which is an injection over a Zariski dense Zariski open set $Y^0 \subset Y_X^{n+1}$ by assumption. By (32), these define a scion of $\mathbf{D}$ with surjective structure map.

We want to use the local lifting test (43) to replace $\sum_{i=1}^{n+1} \pi_i^* F$ by $\tilde{p}^*E$. For this, we need Y^0 to be also Euclidean dense. To achieve this, we resolve Y_X^{n+1} as in (44) to get Y_s. The main components give Y_s^h but we may have introduced some other components Y_s^v that map to $\operatorname{Sing}(Y)$. Since the general fiber of p is smooth, Y_s^v maps to a lower dimensional subvariety of X.

Proposition 46 *Let* $\mathbf{D} = (p : Y \to X, f : p^*E \to F)$ *be a descent problem. Assume that* X, Y *are irreducible and the generic fiber of* p *is irreducible and smooth. Then there is a commutative diagram*

$$\begin{array}{ccc} \bar{Y} & \stackrel{\tau_Y}{\to} & Y \\ \bar{p} \downarrow & & \downarrow p \\ \bar{X} & \stackrel{\tau_X}{\to} & X \end{array}$$

where τ_X, τ_Y *are proper, birational and there is a quotient bundle* $\tau_X^*E \twoheadrightarrow \bar{E}$ *such that* $\bar{p}^*\tau_X^*E \to \tau_Y^*F$ *factors through* $\bar{p}^*\bar{E}$ *and the descent problem*

$$\bar{\mathbf{D}} = (\bar{p} : \bar{Y} \to \bar{X}, \bar{f} : \bar{p}^*\bar{E} \to \bar{F} := \tau_Y^*F)$$

satisfies the assumptions of (45). That is, $\bar{f}(x) : \bar{E}(x) \to C^0(\bar{Y}_x, \tau_Y^*F|_{\bar{Y}_x})$ *is an injection for general* $x \in \bar{p}(\bar{Y})$.

Moreover, if a finite group G *acts on* $\mathbf{D}$, *then we can choose* $\bar{\mathbf{D}}$ *such that the* G*-action lifts to* $\bar{\mathbf{D}}$.

(Note that, as shown by (48), the conclusions can fail if the general fibers of p are not irreducible.)

Proof. Complexify $p : Y \to X$ to get a complex proper morphism $p_{\mathbb{C}} : Y_{\mathbb{C}} \to X_{\mathbb{C}}$ and set

$$E'_{\mathbb{C}} := \operatorname{im}\big[E_{\mathbb{C}} \to (p_{\mathbb{C}})_* F_{\mathbb{C}}\big].$$

Let $x \in p(Y)$ be a general point. Then Y_x is irreducible and the real points Y_x are Zariski dense in the complex fiber $(Y_{\mathbb{C}})_x$. Thus, $H^0\big((Y_{\mathbb{C}})_x, F_{\mathbb{C}}\big) = H^0\big(Y_x, F\big)$.

So far, $E'_{\mathbb{C}}$ is only a coherent sheaf which is a quotient of $E_{\mathbb{C}}$. Using (47) and then (45), we obtain $\tau_X : \bar{X} \to X$ as desired.

47 *Let X be an irreducible variety $q : E \to E'$, a map of vector bundles on X. In general, we cannot write q as a composite of a surjection of vector bundles followed by an injection, but the following construction shows how to achieve this after modifying X.*

Let $\operatorname{Gr}(d, E) \to X$ *be the universal Grassmann bundle of rank d quotients of E where d is the rank of q at a general point. At a general point, $x \in X$, $q(x) : E(x) \twoheadrightarrow \operatorname{im} q(x) \subset E'(x)$ is such a quotient. Thus, q gives a rational map $X \dashrightarrow \operatorname{Gr}(d, E)$, defined on a Zariski dense Zariski open subset. Let $\bar{X} \subset \operatorname{Gr}(d, E)$ denote the closure of its image and $\tau_X : \bar{X} \to X$ the projection. Then τ_X is a proper birational morphism, and we have a decomposition*

$$\tau_X^* q : \tau_X^* E \overset{s}{\twoheadrightarrow} \bar{E} \overset{j}{\hookrightarrow} \tau_X^* E'$$

where $\bar{E}$ is a vector bundle of rank d on $\bar{X}$, s is a rank d surjection everywhere, and j is a rank d injection on a Zariski dense Zariski open subset.

4.4 Proof of the Main Algebraic Theorem

In order to answer Question 1 in general, we try to create a situation where (46) applies.

First, using (44), we may assume that Y is smooth. Next, take the Stein factorization $Y \to W \to X$; that is, $W \to X$ is finite and all the fibers of $Y \to W$ are connected (hence, general fibers are irreducible).

After some modifications, (45) applies to $Y \to W$; thus, we are reduced to comparing $C^0(W, p_W^* E)$ and $C^0(X, E)$.

This is easy if $W \to X$ is Galois, since then the sections of $p_W^* E$ that are invariant under the Galois group descend to sections of E.

If $p : W \to X$ is a finite morphism of (smooth or at least normal) varieties over $\mathbb{C}$, the usual solution would be to take the Galois closure of the field extension $\mathbb{C}(W)/\mathbb{C}(X)$ and let $W^{\mathrm{Gal}} \to X$ be the normalization of X in it. Then the Galois group G acts on $W^{\mathrm{Gal}} \to X$ and the action is transitive on every fiber.

This does not work for real varieties since in general, W^{Gal} has no real points. (For instance, take $X = \mathbb{R}$ and let $W \subset \mathbb{R}^2$ be any curve given by an irreducible equation of the form $y^m = f(x)$. If $m = 2$, then W/X is Galois, but for $m \geq 3$, the Galois closure W^{Gal} has no real points.) Some other problems are illustrated by the next example.

Example 48 *Let $W \subset \mathbb{R}^2$ be defined by $(y^5 - 5y = x)$ with $p : W \to \mathbb{R}^1_x =: X$ the projection. Set $E = \mathbb{C}^4_X$ and $F = \mathbb{C}_W$ with $f : p^*E \to F$ given by $f(\psi(x)e_i) = y^i\psi(x)|_W$.*

Note that p has degree 5 as a map of (complex) Riemann surfaces, but $p^{-1}(x)$ consists of 3 points for $-1 < x < 1$ and of 1 point if $|x| > 1$. Therefore, the kernel of $f \circ p^*(x) : \mathbb{C}^4 = E(x) \to C^0(W_x, F|_{W_x})$ has rank 1 if $-1 < x < 1$ and rank 3 if $|x| > 1$. Thus, $\ker(f \circ p^*) \subset E$ is a rank 1 subbundle on the interval $-1 < x < 1$ and a rank 3 subbundle on the intervals $|x| > 1$.

These kernels depend only on some of the 5 roots of $y^5 - 5y = x$; hence, they are semialgebraic subbundles but not real algebraic subbundles.

As a replacement of the Galois closure W^{Gal}, we next introduce a series of varieties $W^{(m)}_X \to X$. The $W^{(m)}_X$ are usually reducible, the symmetric group S_m acts on them, but the S_m-action is usually not transitive on every fiber. Nonetheless, all the $W^{(m)}_X$ together provide a suitable analog of the Galois closure.

Definition 49 *Let $s : W \to X$ be a finite morphism of (possibly reducible) varieties and $X^0 \subset X$ the largest Zariski open subset over which p is smooth.*

Consider the m-fold fiber product $W^m_X := W \times_X \cdots \times_X W$ with coordinate projections $\pi_i : W^m_X \to W$. For every $i \neq j$, let $\Delta_{ij} \subset W^m_X$ be the preimage of the diagonal $\Delta \subset W \times_X W$ under the map (π_i, π_j). Let $W^{(m)}_X \subset W^m_X$ be the union of the dominant components in the closure of $W^m_X \setminus \cup_{i \neq j} \Delta_{ij}$ with projection $s^{(m)} : W^{(m)}_X \to X$. The symmetric group S_m acts on $W^{(m)}_X$ by permuting the factors.

If $x \in X^0$ then $\left(s^{(m)}\right)^{-1}(x)$ consists of ordered m-element subsets of $s^{-1}(x)$. Thus, $\left(s^{(m)}\right)^{-1}(x)$ is empty if $|s^{-1}(x)| < m$ and S_m acts transitively on $\left(s^{(m)}\right)^{-1}(x)$ if $|s^{-1}(x)| = m$. If $\left|s^{-1}(x)\right| > m$, then S_m does not act transitively on $\left(s^{(m)}\right)^{-1}(x)$. We obtain a decreasing sequence of semialgebraic subsets

$$s^{(1)}\left(W^{(1)}_X\right) \supset s^{(2)}\left(W^{(2)}_X\right) \supset \cdots .$$

Set

$$X^0_{W,m} := X^0 \cap \left(s^{(m)}\left(W^{(m)}_X\right) \setminus s^{(m+1)}\left(W^{(m+1)}_X\right)\right).$$

The $X^0_{W,m}$ are disjoint, $\bigcup_m X^0_{W,m}$ is a Euclidean dense semialgebraic open subset of $p(Y) \cap X^0$, and the S_m-action is transitive on the fibers of $s^{(m)}$ that lie over $X^0_{W,m}$. Thus, $s^{(m)} : W^{(m)} \to X$ behaves like a Galois extension over $X^0_{W,m}$ and together the $X^0_{W,m}$ cover most of X.

Let now $p : Y \to X$ be a proper morphism of (possibly reducible) normal varieties with Stein factorization $p : Y \stackrel{q}{\to} W \stackrel{s}{\to} X$. Let Y^m_X denote the m-fold fiber product $Y \times_X \cdots \times_X Y$ with coordinate projections $\pi_i : Y^m_X \to Y$.

Let $Y_X^{(m)} \subset Y_X^m$ denote the dominant parts of the preimage of $W_X^{(m)}$ under the natural map $q^m : Y_X^m \to W_X^m$ with projection $p^{(m)} : Y_X^{(m)} \to X$. Note that, for general $x \in X$, $\left(p^{(m)}\right)^{-1}(x)$ is empty if $p^{-1}(x)$ has fewer than m irreducible components and S_m acts transitively on the irreducible components of $\left(p^{(m)}\right)^{-1}(x)$ if $p^{-1}(x)$ has exactly m irreducible components. Thus, we obtain a decreasing sequence of semialgebraic subsets $p^{(1)}\left(Y_X^{(1)}\right) \supset p^{(2)}\left(Y_X^{(2)}\right) \supset \cdots$.

Let F be a vector bundle on Y. Then $\oplus_i \pi_i^* F$ is a vector bundle on Y_X^m. Its restriction to $Y_X^{(m)}$ is denoted by $F^{(m)}$.

Note that the S_m-action on $Y_X^{(m)}$ naturally lifts to an S_m-action on $F^{(m)}$. If E is a vector bundle on X and $f : p^* E \to F$ is a vector bundle map, then we get an S_m-invariant vector bundle map $f^{(m)} : \left(p^{(m)}\right)^* E \to F^{(m)}$. For each m, we get a scion of **D**

$$\mathbf{D}^{(m)} := \left(p^{(m)} : Y_X^{(m)} \to X, f^{(m)} : \left(p^{(m)}\right)^* E \to F^{(m)}\right).$$

Below, we will use all the $\mathbf{D}^{(m)}$ together to get a scion with Galois-like properties.

50 (Proof of (33)) *If $\mathbf{D}_s$ is a scion of* **D** *with surjective structure map $r_s : Y_s \to Y$, then (33.1) $\Leftrightarrow$ (33.2) by definition and (33.2) $\Rightarrow$ (33.3) holds for any choice of Z.*

Assume next that we have a candidate for $\mathbf{D}_s$ and Z such that. How do we check (33.3) $\Rightarrow$ (33.2)?

Pick $\Phi_s \in C^*(Y_s, F_s)$ and assume that there is a section $\phi_Z \in C^*(Z, E|_Z)$ whose pull-back to Y_Z equals the restriction of Φ_s. By (39), we can lift ϕ_Z to a section $\phi_X \in C^*(X, E)$. Consider next

$$\Psi_s := \Phi_s - f_s(p_s^* \phi_X) \in C^*(Y_s, F_s).$$

We are done if we can write $\Psi_s = f_s \circ p_s^*(\psi_X)$ for some $\psi_X \in C^*(X, E)$.

By assumption, Ψ_s satisfies the finite set test (26), but the improvement is that Ψ_s vanishes on Y_Z. As we saw already in (2), this can make the problem much easier. We deal with this case in (51).

Note that by [Whi34], we can choose ϕ_X to be real analytic away from Z and the rest of the construction preserves differentiability properties. Thus, (29.2) holds once the rest of the argument is worked out. □

Proposition 51 *Let $\mathbf{D} = \left(p : Y \to X, f : p^* E \to F\right)$ be a descent problem. Then there is a closed algebraic subvariety $Z \subset X$ with $\dim Z < \dim X$ and a scion $\mathbf{D}_s = \left(p_s : Y_s \to X, f_s : p_s^* E \to F_s\right)$ with surjective structure map $r_s : Y_s \to Y$ such that the following holds.*

Let $\psi_s \in C^0(Y_s, F_s)$ be a section that vanishes on $p_s^{-1}(Z)$ and satisfies the finite set test (26). Then there is a $\psi_X \in C^0(X, E)$ such that ψ_X vanishes on Z and $\psi_s = f_s \circ p_s^*(\psi_X)$.

Proof. We may harmlessly assume that $p(Y)$ is Zariski dense in X. Using (44), we may also assume that Y is smooth.

After we construct $\mathbf{D}_s$, the plan is to make sure that Z contains all of its "singular" points. In the original setting of Question 1, Z was the set where the map $(f_1, \dots, f_r) : \mathbb{C}^r \to \mathbb{C}$ has rank 0. In the general case, we need to include points over which f_s drops rank and also points over which p_s drops rank.

During the proof, we gradually add more and more irreducible components to Z. To start with, we add to Z the lower dimensional irreducible components of X, the locus where X is not normal and the (Zariski closures of) the $p(Y_i)$ where $Y_i \subset Y$ is an irreducible component that does not dominate any of the maximal dimensional irreducible components of X. We can thus assume that X is irreducible and every irreducible component of Y dominates X.

Take the Stein factorization $p : Y \stackrel{q}{\to} W \stackrel{s}{\to} X$ and set $M = \deg(W/X)$. For each $1 \le m \le M$, consider the following diagram

$$\begin{array}{ccccccc}
 & & \left(\bar{\bar{q}}^{(m)}\right)^* \bar{\bar{E}}^{(m)} \cong \bar{F}^{(m)} & & F^{(m)} & & F \\
 & & \downarrow & & \downarrow & & \downarrow \\
\left(t_W^{(m)} \circ s_W^{(m)}\right)^* E \twoheadrightarrow \bar{\bar{E}}^{(m)} & & \bar{Y}_X^{(m)} & \stackrel{t_Y^{(m)}}{\to} & Y_X^{(m)} & \stackrel{\pi_l^{(m)}}{\to} & Y \\
 & \searrow & \downarrow \bar{q}^{(m)} & & \downarrow q^{(m)} & & \downarrow p \\
 & & \bar{W}^{(m)} & \stackrel{t_W^{(m)}}{\to} & W^{(m)} & \stackrel{s^{(m)}}{\to} & X
\end{array} \tag{51.m}$$

where $W^{(m)}$ and its column is constructed in (49) and out of this $\bar{W}^{(m)}$, its column and the vector bundle $\bar{\bar{E}}^{(m)}$ are constructed in (46). Note that the symmetric group S_m acts on the whole diagram.

The $\mathbf{D}_s$ we use will be the disjoint union of the scions

$$\bar{\mathbf{D}}_s^{(m)} := \left(\bar{p}^{(m)} : \bar{Y}_X^{(m)} \to X, \bar{f}^{(m)} : \left(\bar{p}^{(m)}\right)^* E \to \bar{F}^{(m)}\right) \quad \text{for } m = 1, \dots, M.$$

By enlarging Z if necessary, we may assume that $Y_X^{(m)} \to X$ is smooth over $X \setminus Z$ and each $t_W^{(m)}$ is an isomorphism over $X \setminus Z$. Note that, for every m,

$$X_m^0 := p^{(m)}\left(Y_X^{(m)}\right) \setminus \left(Z \cup p^{(m+1)}\left(Y_X^{(m+1)}\right)\right) \subset X$$

is an open semialgebraic subset of $X \setminus Z$ whose boundary is in Z. Furthermore, $p(Y) \setminus Z$ is the disjoint union of the X_m^0 and the fiber Y_x has exactly m irreducible components if $x \in X_m^0$.

Let $\Psi_s \in C^0(Y_s, F_s)$ be a section that vanishes on $p_s^{-1}(Z)$. We can then uniquely write $\Psi_s = \sum_m \Psi_s^{(m)}$ such that each $\Psi_s^{(m)}$ vanishes on $Y_s \setminus p_s^{-1}(X_m^0)$. Moreover, Ψ_s satisfies the finite set test (26) iff all the $\Psi_s^{(m)}$ satisfy it.

Thus, it is sufficient to prove that each $\Psi_s^{(m)}$ is the pull-back of a section $\psi_X^{(m)} \in C^*(X, E)$ that vanishes on $X \setminus X_m^0$. For each m we use the corresponding diagram (51_m).

Each $\Psi_s^{(m)}$ lifts to a section $\bar{\Psi}_s^{(m)}$ of $\left(\bar{q}^{(m)}\right)^* \bar{E}^{(m)}$ that satisfies the pull-back conditions for $\bar{Y}^{(m)} \to \bar{W}^{(m)}$. Thus, $\bar{\Psi}_s^{(m)}$ is the pull-back of a section $\bar{\Psi}_W^{(m)}$ of $\bar{E}^{(m)}$. By construction, $\bar{\Psi}_W^{(m)}$ is S_m-invariant and it vanishes outside $\left(t^{(m)} \circ s^{(m)}\right)^{-1}\left(X_m^0\right)$. Using a splitting of $\left(s_W^{(m)} t_W^{(m)}\right)^* E \twoheadrightarrow \bar{E}^{(m)}$, we can think of $\bar{\Psi}_W^{(m)}$ as an S_m-invariant section of $\left(t_W^{(m)} \circ s_W^{(m)}\right)^* E$. By the choice of Z, $t^{(m)}$ is an isomorphism over X_m^0; hence, $\bar{\Psi}_W^{(m)}$ descends to an S_m-invariant section $\Psi_W^{(m)}$ of $\left(s_W^{(m)}\right)^* E$ that vanishes outside $\left(s_W^{(m)}\right)^{-1}\left(X_m^0\right)$. Therefore, by (40.2), $\Psi_W^{(m)}$ descends to a section $\psi_X^{(m)} \in C^0(X, E)$ that vanishes on $X \setminus X_m^0$.

4.5 Semialgebraic, Real, and p-Adic Analytic Cases

52 (Real-Analytic Case) *It is natural to ask Question 1 when the f_i are real-analytic functions and $\mathbb{R}^n$ is replaced by an arbitrary real-analytic variety. As before, we think of X as the real points of a complex Stein space $X_{\mathbb{C}}$ that is defined by real equations. Our proofs work without changes for descent problems* $\mathbf{D} = \left(p : Y \to X, f : p^*E \to F\right)$ *where Y and f are relatively algebraic over X.*

By definition, this means that Y is the set of real points of a closed (reduced but possibly reducible) complex analytic subspace of some $X_{\mathbb{C}} \times \mathbb{CP}^N$ and that f is assumed algebraic in the $\mathbb{CP}^N$-variables.

This definition may not seem the most natural, but it is exactly the setting needed to answer Question 1 if the f_i are real-analytic functions on a real-analytic space.

53 (Semialgebraic Case) *It is straightforward to consider semialgebraic descent problems* $\mathbf{D} = \left(p : Y \to X, f : p^*E \to F\right)$ *where X, Y are semialgebraic sets; E, F are semialgebraic vector bundles; and p, f are semialgebraic maps. (See [BCR98, Chap. 2] for basic results and definitions.) It is not hard to go through the proofs and see that everything generalizes to the semialgebraic case.*

In fact, some of the constructions could be simplified since one can break up any descent problem $\mathbf{D}$ *into a union of descent problems* $\mathbf{D}_i$ *such that each $Y_i \to X_i$ is topologically a product over the interior of X_i. This would allow one to make some noncanonical choices to simplify the construction of the diagrams (51.m).*

It may be, however, worthwhile to note that one can directly reduce the semialgebraic version to the real algebraic one as follows.

Note first that in the semialgebraic setting it is natural to replace a real algebraic descent problem $\mathbf{D} = \left(p : Y \to X, f : p^*E \to F\right)$ *by its semialgebraic reduction* sa-red($\mathbf{D}$) $:= \left(p : Y \to p(Y), f : p^*\left(E|_{p(Y)}\right) \to F\right)$.

We claim that for every semialgebraic descent problem $\mathbf{D}$, *there is a proper surjection $r : Y_s \to Y$ such that the corresponding scion* $r^*\mathbf{D}$ *is semialgebraically isomorphic to the semialgebraic reduction of a real algebraic descent problem.*

To see this, first, we can replace the semialgebraic X by a real algebraic variety X^a that contains it and extend E to semialgebraic vector bundle over X^a. Not all *semialgebraic vector bundles are algebraic, but we can realize E as a semialgebraic subbundle of a trivial bundle $\mathbb{C}^M$. This in turn gives a semialgebraic embedding of X into $X \times \mathrm{Gr}(\mathrm{rank}\, E, M)$. Over the image, E is the restriction of the algebraic universal bundle on $\mathrm{Gr}(\mathrm{rank}\, E, M)$. Thus, up to replacing X by the Zariski closure of its image, we may assume that X and E are both algebraic. Replacing Y by the graph of p in $Y \times X$, we may assume that p is algebraic. Next write Y as the image of a real algebraic variety. We obtain a scion where now $p : X \to Y, E, F$ are all algebraic. To make f algebraic, we use that f defines a semialgebraic section of $\mathbb{P}(\mathcal{H}om_X(p^*E, F)) \to Y$. Thus, after replacing Y by the Zariski closure of its image in $\mathbb{P}(\mathcal{H}om_X(p^*E, F)) \to Y$, we obtain an algebraic scion with surjective structure map.*

54 (**p-Adic Case**) *One can also consider Question 1 in the p-adic case and the proofs work without any changes. In fact, if we start with polynomials $f_i \in \mathbb{Q}[x_1, \ldots, x_n]$, then in Theorem 34, it does not matter whether we want to work over $\mathbb{R}$ or $\mathbb{Q}_p$; we construct the same descent problems. It is only in checking the finite set test (26) that the field needs to be taken into account: if we work over $\mathbb{R}$, we need to check the condition for fibers over all real points; if we work over $\mathbb{Q}_p$, we need to check the condition for fibers over all p-adic points.*

Acknowledgements We thank M. Hochster for communicating his unpublished example (3.4). We are grateful to B. Klartag and A. Naor for bringing Michael's theorem to our attention at a workshop organized by the American Institute of Mathematics (AIM), to which we are also grateful. Our earlier proof of (6) was unnecessarily complicated. We thank H. Brenner, A. Isarel, K. Luli, R. Narasimhan, A. Némethi, and T. Szamuely for helpful conversations and F. Wroblewski for TeXing several sections of this chapter.

Partial financial support for CF was provided by the NSF under grant number DMS-0901040 and by the ONR under grant number N00014-08-1-0678. Partial financial support for JK was provided by the NSF under grant number DMS-0758275.

References

[BCR98] Jacek Bochnak, Michel Coste, and Marie-Françoise Roy, *Real algebraic geometry*, Ergebnisse der Mathematik und ihrer Grenzgebiete, vol. 36, Springer-Verlag, Berlin, 1998, Translated from the 1987 French original, Revised by the authors. MR1659509 (2000a:14067)

[BL00] Yoav Benyamini and Joram Lindenstrauss, *Geometric nonlinear functional analysis. Vol. 1*, American Mathematical Society Colloquium Publications, vol. 48, American Mathematical Society, Providence, RI, 2000. 1727673 (2001b:46001)

[BMP03] Edward Bierstone, Pierre D. Milman, and Wiesław Pawłucki, *Differentiable functions defined in closed sets. A problem of Whitney*, Invent. Math. **151** (2003), no. 2, 329–352. 1953261 (2004h:58009)

[Bre06] Holger Brenner, *Continuous solutions to algebraic forcing equations*, arXiv.org:0608611, 2006.

[Ehr70] Leon Ehrenpreis, *Fourier analysis in several complex variables*, J. Wiley & Sons, New York-London-Sydney, 1970, reprinted by Dover, 2006

[Fef06] Charles Fefferman, *Whitney's extension problem for C^m*, Ann. of Math. (2) **164** (2006), no. 1, 313–359. 2233850 (2007g:58013)

[Gla58] Georges Glaeser, *Étude de quelques algèbres tayloriennes*, J. Analyse Math. **6** (1958), 1–124; erratum, insert to 6 (1958), no. 2. 0101294 (21 #107)

[Hoc10] Melvin Hochster, *(personal communication)*, 2010.

[Kol07] János Kollár, *Lectures on resolution of singularities*, Annals of Mathematics Studies, vol. 166, Princeton University Press, Princeton, NJ, 2007. MR2289519 (2008f:14026)

[Kol10] János Kollár, *Continuous closure of sheaves*, arXiv.org:1010.5480, 2010.

[Kol11] ———, *Continuous rational functions on real and p-adic varieties*, arXiv.org:1101.3737, 2011.

[Mal67] B. Malgrange, *Ideals of differentiable functions*, Tata Institute of Fundamental Research Studies in Mathematics, No. 3, Tata Institute of Fundamental Research, Bombay, 1967. MR0212575 (35 #3446)

[Mic56] Ernest Michael, *Continuous selections. I*, Ann. of Math. (2) **63** (1956), 361–382. 0077107 (17,990e)

[Nar68] Raghavan Narasimhan, *Analysis on real and complex manifolds*, Advanced Studies in Pure Mathematics, Vol. 1, Masson & Cie, Éditeurs, Paris, 1968. 0251745 (40 #4972)

[Ste70] Elias. M. Stein, *Singular integrals and differentiability properties of functions*, Princeton Univ. Press, Princeton, NJ, 1970.

[Tou72] Jean-Claude Tougeron, *Idéaux de fonctions différentiables*, Springer-Verlag, Berlin, 1972, Ergebnisse der Mathematik und ihrer Grenzgebiete, Band 71. MR0440598 (55 #13472)

[Whi34] Hassler Whitney, *Analytic extensions of differentiable functions defined in closed sets*, Trans. Amer. Math. Soc. **36** (1934), no. 1, 63–89. MR1501735

Recurrence for Stationary Group Actions

Hillel Furstenberg and Eli Glasner

In memory of Leon Ehrenpreis.

Abstract Using a structure theorem from [Furstenberg and Glasner, Contemporary Math. **532**, 1–28 (2010)], we prove a version of multiple recurrence for sets of positive measure in a general stationary dynamical system.

Key words Stationary dynamical systems • Szemerédi theorem • SAT • Multiple recurrence

Mathematics Subject Classification (2000): Primary 22D05, 37A30, 37A50. Secondary 22D40, 37A40

Introduction

The celebrated theorem of E. Szemerédi regarding the existence of long arithmetical progressions in subsets of the integers having positive (upper) density is known to be equivalent to a statement involving "multiple recurrence" in the framework of dynamical systems theory (see, e.g., [2]). Stated precisely, this is the following assertion:

H. Furstenberg (✉)
Department of Mathematics, Hebrew University of Jerusalem, Jerusalem, Israel
e-mail: harry@math.huji.ac.il

E. Glasner
Department of Mathematics, Tel Aviv University, Ramat Aviv, Israel
e-mail: glasner@math.tau.ac.il

H.M. Farkas et al. (eds.), *From Fourier Analysis and Number Theory to Radon Transforms and Geometry*, Developments in Mathematics 28, DOI 10.1007/978-1-4614-4075-8_11,

Theorem 0.1. *Let $(X, \mathcal{B}, \mu, T)$ be a measure preserving dynamical system; i.e., $(X, \mathcal{B}, \mu)$ is a probability space and $T : X \to X$ a measure preserving mapping of X to itself. If $A \in \mathcal{B}$ is a measurable subset with $\mu(A) > 0$, then for any $k = 1, 2, 3, \ldots$, there exists $m \in \mathbb{N} = \{1, 2, 3, \ldots\}$ with*

$$\mu\left(A \cap T^{-m} A \cap T^{-2m} A \cap \cdots \cap T^{-km}\right) > 0.$$

The case $k = 1$ is the "Poincaré recurrence theorem" and is an easy exercise in measure theory. The general case is more recondite (see, e.g., [2]). In principle, recurrence phenomena make sense in the framework of more general group actions, and we can inquire what is the largest domain of their validity. Specifically, if a group G acts on a measure space (X, ν) (we have suppressed the σ-algebra of measurable sets) with $(g, x) \to T_g x$ by non-singular maps $\{T_g\}$, and A is a measurable subset of X with $\nu(A) > 0$, under what conditions can we find for large k an element $g \in G$, $g \neq$ identity, with

$$\nu\left(A \cap T_g^{-1} A \cap T_g^{-2} A \cap \cdots \cap T_g^{-k} A\right) > 0\,?$$

Some conditions along the line of measure preservation will be necessary. Without this, we could take $G = \mathbb{Z}$, $X = \mathbb{Z} \cup \{\infty\}$, $\forall\, t, n \in \mathbb{Z}$, $T_t n = n + t$, $n \neq \infty$, $T_t \infty = \infty$, and $\nu(\{n\}) = \frac{1}{3 \cdot 2^{|n|}}$, $\nu(\{\infty\}) = 0$. Here no $t \neq 0$ will satisfy $T_t(\{n\}) \cap \{n\} \neq 0$.

The present work extends an earlier paper on "stationary" systems [3]. Here we shall show that quite generally, under the hypothesis of "stationarity," which we shall presently define, one obtains a version of multiple recurrence for sets of positive measure.

We recall the basic definitions here, although we will rely on the treatment in [3] for fundamental results. Throughout, G will represent a locally compact, second countable group, and μ, a fixed probability measure on Borel sets of G. We consider measure spaces (X, ν) on which G acts measurably, i.e., the map $G \times X \to X$ which we denote $(g, x) \to gx$ is measurable, and so the convolution of the measure μ on G and $\nu \in \mathcal{P}(X)$, $\mu * \nu$, is defined as the image of $\mu \times \nu$ on X under this map; thus $\mu * \nu$ is again a probability measure on X, an element of $\mathcal{P}(X)$. *We will always assume that G acts on (X, ν) by non-singular* transformations; i.e., $\nu(A) = 0$ implies $\nu(gA) = 0$ for a measurable $A \subset X$ and $g \in G$.

Definition 0.2. When $\mu * \nu = \nu$, we say that (X, ν) is a stationary (G, μ) space.

This can be interpreted as saying that ν is invariant "on the average." It is also equivalent to the statement that for measurable $A \subset X$

$$\nu(A) = \int_G \nu\left(g^{-1} A\right) \, d\mu(g). \tag{0.1}$$

Associated with the space (G, μ), we will consider the probability space

$$(\Omega, P) = (G, \mu) \times (G, \mu) \times \cdots$$

where we will denote the random variables representing the coordinates of a point $\omega \in \Omega$ by $\{\xi_1(\omega), \xi_2(\omega), \ldots, \xi_n(\omega), \ldots\}$. We will draw heavily on the "martingale convergence theorem" which for our purposes can be formulated:

Theorem 0.3 (Martingale Convergence Theorem (MCT)). *Let $\{F_n(\omega)\}_{n\in\mathbb{N}}$ be a sequence of uniformly bounded, measurable, real-valued functions on Ω with F_n measurable with respect to $\xi_1, \xi_2, \ldots, \xi_n$ and such that*

$$F_n(\xi_1, \xi_2, \ldots, \xi_n) = \int_G F_{n+1}(\xi_1, \xi_2, \ldots, \xi_n, \eta)\, \mathrm{d}\mu(\eta). \tag{0.2}$$

(Such a sequence is called a martingale.*) Then with probability one, the sequence $\{F_n(\omega)\}$ converges almost surely to a limit $F(\omega)$ satisfying:*

$$\mathcal{E}(F) = \int F(\omega)\, \mathrm{d}P(\omega) = \int_G F_1(\eta)\, \mathrm{d}\mu(\eta). \tag{0.3}$$

The theory of stationary actions is intimately related to boundary theory for topological groups and the theory of harmonic functions. For details, we refer the reader to [1].

1 Poincaré Recurrence for Stationary Actions

A first application will be a proof of a particular version of the Poincaré recurrence phenomenon for stationary actions.

Theorem 1.1. *Let G be an infinite discrete group and let μ be a probability measure on G whose support $S(\mu)$ generates G as a group. Let (X, ν) be a stationary space for (G, μ) and let $A \subset X$ be a measurable subset with $\nu(A) > 0$. Then there exists $g \in G,\ g \neq$ identity, with $\nu(A \cap g^{-1}A) > 0$.*

We start with a lemma.

Lemma 1.2. *If $\Sigma(\mu)$ is the semigroup in G generated by $S(\mu)$, there exists a sequence of elements $\alpha_1, \alpha_2, \alpha_3, \ldots \in \Sigma(\mu)$ such that no product $\alpha_{i_1}\alpha_{i_2}\cdots\alpha_{i_n}$ with $i_1 < i_2 < i_3 < \cdots < i_n$ equals the identity element of G.*

Proof. The semigroup $\Sigma(\mu)$ is infinite since a finite subsemigroup of a group is a group. We proceed inductively so that having defined $\alpha_1, \alpha_2, \ldots, \alpha_n$ where products do not degenerate, we can find $\alpha_{n+1} \in \Sigma(\mu)$ so that no product

$$\alpha_{i_1}\alpha_{i_2}\cdots\alpha_{i_s}\alpha_{n+1} = \mathrm{id},$$

there being only finitely many values to avoid.

Proof (Proof of the Theorem 1.1:). The proof is based on two ingredients. First, if we define functions on Ω by

$$F_n(\omega) = \nu\left(\xi_n^{-1}\xi_{n-1}^{-1}\cdots\xi_1^{-1}A\right)$$

then by (0.1), the sequence $\{F_n\}$ forms a martingale. The second ingredient is the fact that in almost every sequence $\xi_1(\omega), \xi_2(\omega), \ldots, \xi_n(\omega), \ldots$, every word in the "letters" of $S(\mu)$ appears infinitely far out, and then every element in $\Sigma(\mu)$ appears as a partial product. Now let $f(\omega) = \lim F_n(\omega)$, which by the MCT is defined almost everywhere, then $\mathcal{E}(f) = \int \nu(g^{-1}A)\, d\mu(g) = \nu(A) > 0$. So, if $\delta = \nu(A)/2$, there will be a random variable $n(\omega)$ which is finite with positive probability so that for $n > n(\omega)$

$$\nu\left(\xi_n^{-1}\xi_{n-1}^{-1}\cdots\xi_1^{-1}A\right) > \delta.$$

Now choose $\alpha_1, \alpha_2, \alpha_3, \cdots \in \Sigma(\mu)$ as in the foregoing lemma, and let $N > 1/\delta$. With positive probability, there is $l \geq n(\omega)$ and $0 = r_0 < r_1 < r_2 < \cdots < r_N$ so that in $\Sigma(\mu)$,

$$\xi_{l+r_{i-1}+1}(\omega)\xi_{l+r_{i-1}+2}(\omega)\cdots\xi_{l+r_i}(\omega) = \alpha_i,$$

for $i = 1, 2, \ldots, N-1$. By definition of $n(\omega)$,

$$\nu\left(\alpha_i^{-1}\cdots\alpha_l^{-1}\beta^{-1}A\right) > \delta \qquad i = 1, 2, \ldots, N,$$

where $\beta = \xi_1\xi_2\cdots\xi_{l-1}$. But this yields N sets of measure $> 1/N$ in X, and we conclude that for some $i < j$,

$$\nu\left(\alpha_i^{-1}\cdots\alpha_l^{-1}\beta^{-1}A \cap \alpha_j^{-1}\cdots\alpha_l^{-1}\beta^{-1}A\right) > 0.$$

This however implies that for a conjugate γ of the product $\alpha_j^{-1}\alpha_{j-1}^{-1}\cdots\alpha_{i+1}^{-1}$, we have $\nu(A \cap \gamma A) > 0$. Here $\gamma \neq \mathrm{id}$ since by construction, $\alpha_{i+1}\alpha_{i+2}\cdots\alpha_j \neq \mathrm{id}$.

2 Multiple Recurrence for SAT Actions

Our main result is a multiple recurrence theorem for stationary actions. We proceed step by step proving the theorem first for the special category of actions known as SAT actions. These were introduced by Jaworski in [4].

Definition 2.1. The action of a group G on a probability measure space (X, ν) is SAT (strongly approximately transitive) if for every measurable $A \subset X$ with $\nu(A) > 0$, we can find a sequence $\{g_n\} \subset G$ with $\nu(g_n A) \to 1$.

We now have a second recurrence result:

Theorem 2.2. *If (X, ν) is a probability measure space on which the group G acts by non-singular transformations and the G action is SAT, then for every measurable $A \subset X$ with $\nu(A) > 0$ and any integer $k \geq 1$, there is a $\gamma \in G$, $\gamma \neq \mathrm{id}$ with*

$$\nu(A \cap \gamma^{-1}A \cap \gamma^{-2}A \cap \cdots \cap \gamma^{-k}A) > 0\,. \tag{2.1}$$

Moreover, if F is any finite subset of G, γ can be chosen outside of F.

We use the following basic lemma from measure theory.

Lemma 2.3. *If $\sigma : X \to X$ is a non-singular transformation with respect to a measure ν on X, then for any $\varepsilon > 0$, there exists a $\delta > 0$ so that $\nu(A) < \delta$ implies $\nu(\sigma A) < \varepsilon$.*

Proof. If such a δ did not exist, we could find $B \subset X$ with $\nu(B) = 0$ and $\nu(\sigma B) \geq \varepsilon$.

Proposition 2.4. *Assume G acts on (X, ν) by non-singular transformations and let $\gamma_1, \gamma_2, \ldots, \gamma_k \in G$. There exists $\delta > 0$ so that if $\nu(B) > 1 - \delta$, then*

$$\nu(\gamma_1 B \cap \gamma_2 B \cap \cdots \cap \gamma_k B) > 0.$$

Proof. The desired inequality will take place provided the measure of each $\gamma_i B'$ is less than $1/k$, where $B' = X \setminus B$. By Lemma 2.3, this will hold if $\nu(B')$ is sufficiently small.

Proof (Proof of the Theorem 2.2). Let $\sigma \neq \mathrm{id}$ be any element of G. Apply Proposition 2.4 with $\gamma_0 = \mathrm{id}$, $\gamma_i = \sigma^{-i}$, $i = 1, 2, \ldots, k$ and find $\delta > 0$ so that $\nu(B) > 1 - \delta$ implies

$$\nu(B \cap \sigma B \cap \sigma^2 B \cap \cdots \cap \sigma^k B) > 0\,.$$

Use the SAT property to find $g \in G$ with $\nu(g^{-1}A) > 1 - \delta$. Then

$$\nu(g^{-1}A \cap \sigma g^{-1}A \cap \sigma^2 g^{-1}A \cap \cdots \cap \sigma^k g^{-1}A) > 0.$$

Applying g to the set appearing here, we get:

$$\nu(A \cap g\sigma g^{-1}A \cap g\sigma^2 g^{-1}A \cap \cdots \cap g\sigma^k g^{-1}A) > 0.$$

Letting $\gamma = g\sigma^{-1}g^{=1}$, we obtain the desired result.

We turn now to the last statement of the theorem. One sees easily that if G has a nontrivial SAT action, then G is infinite. Let H be a finite subset of G with greater cardinality than F. Now in the foregoing discussion, we consider a sequence $\{g_n\}$ in G with $\nu(g_n^{-1}A) \to 1$; then for any σ, if n is sufficiently large if we take $\gamma = g_n \sigma g_n^{-1}$ for large n, we will get (2.1). We claim that σ can be chosen so that for an infinite subsequence $\{n_j\}$, we will have $g_{n_j}\sigma g_{n_j}^{-1} \notin F$. For this, we simply consider the sets $\{g_n H g_n^{-1}\}$ each of which has some element outside of F. $\{n_j\}$ is then a sequence for which there is a fixed $\sigma \in H$ with $g_{n_j}\sigma g_{n_j}^{-1} \notin F$. This completes the proof.

3 A Structure Theorem for Stationary Actions

In order to formulate our structure theorem, we will introduce a few definitions and some well-known basic tools from the general theory of dynamical systems.

3.1 Factors and the Disintegration of Measures

Definition 3.1. Let (X, ν) and (Y, ρ) be two (G, μ) spaces. A measurable map $\pi : (X, \nu) \to (Y, \rho)$ is called *a factor map*, or *an extension*, depending on the view point, if it intertwines the group actions: for every $g \in G$, $g\pi(x) = \pi(gx)$ for ν almost every $x \in X$.

Definition 3.2. If (Y, ρ) is a factor of (X, ν), we can decompose the measure ν as $\nu = \int_Y \nu_y d\rho(y)$, where the ν_y are probability measures on X with $\nu_y(\pi^{-1}(y)) = 1$ and the map $y \mapsto \nu_y$ is measurable from Y into the space of probability measures on X, equipped with its natural Borel structure. We say (X, ν) is a *measure preserving extension* of (Y, ρ) if for each $g \in G$, $g\nu_y = \nu_{gy}$ for almost every $y \in Y$. Note that a stationary system (X, ν) is measure preserving (i.e., $g\nu = \nu$ for every $g \in G$) if and only if the extension $\pi : X \to Y$, where the factor (Y, ρ) is the trivial one-point system is a measure preserving extension.

Topological Models. We begin this subsection with some remarks regarding stationary actions of (G, μ) on (X, ν) in the case that X is a compact metric space. We then speak of a topological stationary system. In this case, we can form the measure-valued martingale

$$\theta_n(\omega) = \xi_1 \xi_2 \cdots \xi_n \nu.$$

The martingale convergence theorem is valid also in this context by the separability of $C(X)$, and so we obtain a measure-valued random variable $\theta(\omega) = \lim_{n\to\infty} \theta_n(\omega)$.

Definition 3.3. A topological stationary system (X, ν) is *proximal* if with probability 1, the measure $\theta(\omega)$ is a Dirac measure: $\theta(\omega) = \delta_{z(\omega)}$.

Definition 3.4. A stationary system (X, ν) is *proximal* if every compact metric factor (X', ν') is proximal.

Definition 3.5. Let (X, ν) and (X', ν') be two (G, μ) stationary systems, and suppose that X' is a compact metric space. We say that the stationary system (X', ν') is *a topological model* for (X, ν) if there is an isomorphism of the measure spaces $\phi : (X, \nu) \to (X', \nu')$ which intertwines the G actions.

The following proposition is well known and has several proofs. We will be content here with just a sketch of an abstract construction.

Proposition 3.6. *Every (G, μ) system (X, ν) admits a topological model. Moreover, if $A \subset X$ is measurable, we can find a topological model $\phi : (X, \nu) \to (X', \nu')$ such that the set $A' = \phi(A)$ is a clopen subset of the compact space X'.*

Proof. Choose a sequence of functions $\{f_n\} \subset L_\infty(X,\nu)$ which spans $L_2(X,\nu)$, with $f_1 = \mathbf{1}_A$. Let $G_0 \subset G$ be a countable dense subgroup and let $\mathcal{A}$ be the G_0-invariant closed unital C^*-subalgebra of $L_\infty(X,\nu)$ which is generated by $\{f_n\}$. We let X' be the, compact metric, Gelfand space which corresponds to the G-invariant, separable, C^*-algebra $\mathcal{A}$. Since $f_1^2 = f_1$, we also have $\tilde{f}_1^2 = \tilde{f}_1$, where the latter is the element of $C(X')$ which corresponds to f_1. Since $\tilde{f}_1$ is continuous, it follows that $A' := \{x' : \tilde{f}_1(x') = 1\}$ is indeed a clopen subset of X' with $\tilde{f}_1 = \mathbf{1}_{A'}$. The probability measure ν' is the measure which corresponds, via Riesz' theorem, to the linear functional $\tilde{f} \mapsto \int f \, d\nu$.

Proposition 3.7. *If (X,ν) is a proximal stationary system for (G,μ), then the action of G on (X,ν) is SAT.*

Proof. Let A be a measurable subset of X with $\nu(A) > 0$. There is a topological model (X',ν') of (X,ν) such that A is the pullback of a closed-open set A' with $\nu'(A') = \nu(A)$. As in Sect. 1, we form the martingale $\nu'(\xi_n^{-1}\xi_{n-1}^{-1}\cdots\xi_1^{-1}A')$ which converges to $\theta(\omega)(A) = \delta_{z'(\omega)}(A')$, since by the proximality of (X,ν), the topological factor (X',ν') is proximal. Now the latter limit is 0 or 1, and since the expectation of $\nu'(\xi_n^{-1}\xi_{n-1}^{-1}\cdots\xi_1^{-1}A)$ is $\nu(A) > 0$, there is positive probability that $z'(\omega) \in A'$. When this happens,

$$\nu(\xi_n^{-1}\xi_{n-1}^{-1}\cdots\xi_1^{-1}A) = \nu'(\xi_n^{-1}\xi_{n-1}^{-1}\cdots\xi_1^{-1}A') \to 1.$$

This proves that the action is SAT.

The Structure Theorem. We now reformulate the structure theorem (Theorem 4.3) of [3] to suit our needs. (The theorem in [3] gives more precise information.)

Theorem 3.8. *Every stationary system is a factor of a stationary system which is a measure preserving extension of a proximal system.*

Alternatively, in view of Proposition 3.7:

Theorem 3.9. *If (X,ν) is a stationary action of (G,μ), there is an extension (X^*,ν^*) of (X,ν) which is a measure preserving extension of an SAT action of G on a stationary space (Y,ρ).*

This is the basic structure theorem which we will use to deduce a general multiple recurrence result for stationary actions.

4 Multiple Recurrence for Stationary Actions

We recall the terminology of [3]:

Definition 4.1. A (G,μ) stationary action of on (X,ν) is *standard* if (X,ν) is a measure preserving extension of a proximal action.

Since proximality implies SAT, we can extend this notion and replace "proximal" by SAT. Theorem 3.8 asserts that every stationary action has a standard extension. The nature of recurrence phenomena is such that if such a phenomenon is valid for an extension of a system, it is valid for the system. Precisely, if $\pi : X \to X'$ and $A' \subset X'$ and for the pullback $A = \pi^{-1}(A')$ and a set $g_1, g_2, \ldots, g_k$, we have $\nu(g_1^{-1}A \cap g_2^{-1}A \cap \cdots \cap g_k^{-1}A) > 0$, then $\nu'(g_1^{-1}A' \cap g_2^{-1}A' \cap \cdots \cap g_2^{-1}A') > 0$. It follows now from Theorem 3.8 that for a general multiple recurrence theorem for stationary actions, it will suffice to treat standard actions. Using the definition of a standard action, we will take advantage of the multiple recurrence theorem proved in Sect. 2 for SAT actions and show that this now extends to any standard action. For this, we use a lemma which is based on Szemerédi's theorem. By the latter, there is a function $N(\delta, \ell)$, for $\delta > 0$ and ℓ a natural number, so that for $n \geq N(\delta, \ell)$, if $E \subset \{1, 2, 3, \ldots, n\}$ with $|E| \geq \delta n$, then E contains an ℓ-term arithmetic progression. We now have:

Lemma 4.2. *In any probability space* (Ω, P), *for* $n \geq N(\delta, \ell)$, *if* $A_1, A_2, \ldots, A_n$ *are* n *subsets of* Ω *with* $P(A_i) > \delta$ *for* $i = 1, 2, \ldots, n$, *then there exist* a *and* d *so that*

$$P(A_a \cap A_{a+d} \cap A_{a+d} \cap \cdots \cap A_{a+(\ell-1)d}) > 0.$$

Proof. Set $f_i(x) = 1_{A_i}(x)$, $i = 1, 2, \ldots, n$, and let $E(x) = \{i : f_i(x) = 1\}$. $|E(x)| = \Sigma_{i=1}^n f_i(x)$ and the condition $|E(x)| > \delta n$ is implied by $F(x) = \Sigma f_i(x) > \delta n$. But $\int F(x)\,dP(x) = \Sigma P(A_i) > \delta n$, and so for some set, $B \subset \Omega$ with $P(B) > 0$, $F(x) > \delta n$. Thus, for each $x \in B$, we have $|E(x)| > n\delta$ and there is an ℓ-term arithmetic progression $R_{a,d}(x) \subset E(x)$, so that x lies in the intersection of the A_r, as r ranges over the arithmetic progression $R_{a,d}(x)$. There being only finitely many progressions, we obtain for one of these $P(\bigcap_{r \in R_{a,d}} A_r) > 0$.

We will need an additional hypothesis to obtain a general multiple recurrence theorem.

Definition 4.3. A group G is OU (order unbounded) if for any integer n we have for some $g \in G$, $g^n \neq id$.

For an OU group, we can find, for any given k, elements $\sigma \in G$ so that none of the powers $\sigma, \sigma^2, \ldots, \sigma^k$ give the identity. Note that in our proof of multiple recurrence for SAT actions, Theorem 2.2, we obtain, for any subset $A \subset X$ of positive measure, an element id $\neq \gamma \in G$ with:

$$\nu\left(A \cap \gamma^{-1}A \cap \gamma^{-2}A \cap \cdots \cap \gamma^{-k}A\right) > 0,$$

where for an OU group we can demand that each $\gamma^j \neq id$, $j = 1, 2, \ldots, k$. In fact, in that proof, we show that the element γ can be found within the conjugacy class of any nonidentity element σ of G.

We can now prove:

Theorem 4.4. *Let (X, ν) represent a stationary action of (G, μ) with the elements of G acting on (X, ν) by non-singular transformations and where G is an OU group. Let $A \subset X$ be a measurable set with $\nu(A) > 0$ and let $k \geq 1$ be any integer; then there exists an element γ in G with $\gamma^j \neq id$, $j = 1, 2, \ldots, k$ and with*

$$\nu\left(A \cap \gamma^{-1} A \cap \gamma^{-2} A \cap \cdots \cap \gamma^{-k} A\right) > 0.$$

Proof. We can assume (X, ν) is a measure preserving extension of (Y, ρ) where the action of G on (Y, ρ) is SAT. Let $\pi : X \rightarrow Y$ and decompose $\nu = \int \nu_y d\rho(y)$. Let $A \subset X$ be given and let $\delta > 0$ be such that $B = \{y : \nu_y(A) > \delta\}$ has positive measure. Set $N = N(\delta, k)$ in Theorem 2.2 and find γ with $\gamma^j \neq$ id for $j = 1, 2, \ldots, N$, and with

$$\rho\left(B \cap \gamma^{-1} B \cap \gamma^{-2} B \cap \cdots \cap \gamma^{-N} B\right) > 0.$$

For $y \in B \cap \gamma^{-1} B \cap \cdots \cap \gamma^{-N} B$ and $j = 1, 2, \ldots, N$, we will have

$$\nu_y\left(\gamma^{-j} A\right) = \gamma^j \nu_y(A) = \nu_{\gamma^j y}(A) > \delta.$$

We now use Lemma 4.2 to obtain for each $y \in B \cap \gamma^{-1} B \cap \cdots \cap \gamma^N B$ a k-term arithmetic progression R of powers of γ with $\nu_y\left(\bigcap_{j \in R} \gamma^{-j} A\right) > 0$. In particular, $\bigcap_{j \in R} \gamma^{-j} A \neq \emptyset$ for some arithmetic progression $R = \{a, a + d, a + 2d, \ldots, a + (k-1)d\}$ so that with $\gamma' = \gamma^d$,

$$A \cap \gamma'^{-1} A \cap \gamma'^{-2} A \cap \cdots \cap \gamma'^{-k} \neq \emptyset.$$

Obtaining a nonempty intersection suffices to obtain an intersection of positive measure, and so our theorem is proved.

References

1. H. Furstenberg, *Boundary theory and stochastic processes on homogeneous spaces*, Harmonic analysis on homogeneous spaces (Proc. Sympos. Pure Math., Vol. XXVI, Williams Coll., Williamstown, Mass., 1972), 193–229. Amer. Math. Soc., Providence, R.I., 1973.
2. H. Furstenberg, *Recurrence in ergodic theory and combinatorial number theory*, Princeton university press, Princeton, N.J., 1981.
3. H. Furstenberg and E. Glasner, *Stationary dynamical systems*, Dynamical numbers, AMS, Contemporary Math. **532**, 1–28, Providence, Rhode Island, 2010.
4. W. Jaworski *Strongly approximately transitive group actions, the Choquet-Deny theorem, and polynomial growth*, Pacific J. of Math., **165**, (1994), 115–129.

On the Honda - Kaneko Congruences*

P. Guerzhoy

In memory of Leon Ehrenpreis

Abstract Several years ago, Kaneko experimentally observed certain congruences for the Fourier coefficients of a weakly holomorphic modular form modulo powers of primes. Recently, Kaneko and Honda proved that a special case of these congruences, namely, the congruences modulo single primes. In this chapter, we consider this weakly holomorphic modular form in the framework of the theory of mock modular forms, and prove a limit version of these congruences.

Mathematics Subject Classification (2010): 11F33

Let $q = \exp(2\pi i \tau)$ with $\operatorname{Im}(\tau) > 0$. Consider the Eisenstein series

$$E_4(\tau) = 1 + 240 \sum_{n>0} \left(\sum_{d|n} d^3 \right) q^n$$

$$E_6(\tau) = 1 - 504 \sum_{n>0} \left(\sum_{d|n} d^5 \right) q^n$$

of weights 4 and 6 correspondingly. Let

*The author is thankful to Simons Foundation collabration Grant for the support.

P. Guerzhoy (✉)
Department of Mathematics, University of Hawaii, 2565 McCarthy Mall, Honolulu, HI 96822-2273, USA
e-mail: pavel@math.hawaii.edu

H.M. Farkas et al. (eds.), *From Fourier Analysis and Number Theory to Radon Transforms and Geometry*, Developments in Mathematics 28, DOI 10.1007/978-1-4614-4075-8_12,

$$\eta(\tau) = q^{\frac{1}{24}} \prod_{n>0} (1 - q^n)$$

be Dedekind's eta function.

For a prime p, denote by U Atkin's U_p operator. We say that a function ϕ with a Fourier expansion $\phi = \sum u(n)q^n$ is congruent to zero modulo a power of a prime p,

$$\phi = \sum u(n)q^n \equiv 0 \mod p^w,$$

if all its Fourier expansion coefficients are divisible by this power of the prime; $u(n) \equiv 0 \mod p^w$ for all n.

In this chapter, we prove the following congruences.

Theorem 1. (i) *If $p > 3$ is a prime, then for all integers $l > 0$,*

$$\left(\frac{E_6(6\tau)}{\eta(6\tau)^4}\right) |U^l \equiv 0 \mod p^{3l}.$$

(ii) *Let p be a prime such that $p \equiv 1 \mod 3$. There exists an integer $A_p \geq 0$ such that for all integers $l \geq 0$*

$$\left(\frac{E_4(6\tau)}{\eta(6\tau)^4}\right) |U^l \equiv 0 \mod p^{l-A_p}.$$

(iii) *Let $p > 3$ be a prime such that $p \equiv 2 \mod 3$. There exists an integer $A_p \geq 0$ such that for all integers $l \geq 0$*

$$\left(\frac{E_4(6\tau)}{\eta(6\tau)^4}\right) |U^l \equiv 0 \mod p^{[l/2]-A_p}.$$

These congruences were first observed by Masanobu Kaneko several years ago as a result of numerical experiments. In a recent paper by Honda and Kaneko [6], the congruences of Theorem 1 (i) and (ii) were proved in the case $l = 1$ with $A_p = 0$. Their result in this case is thus sharper than ours. They also conjecture that these congruences are true for all $l \geq 0$ with $A_p = 0$. The techniques which they use in their proof are quite different from ours.

As the author was informed by Professor Kaneko, the function $E_4(6\tau)/\eta(6\tau)^4$ was considered by him in relation to his study with Koike [7–10] of a differential equation of second order that first arose from the work of Kaneko and Zagier [11] on supersingular j-invariants. This differential equation is also related to the classification of 2-dimensional conformal field theories. The current author falls short of being an expert in these areas and knows nothing about possible interpretations of the congruences above. It seems, however, interesting that an understanding and a proof of these congruences are far from being obvious and requiring the theory of *weak harmonic Maass forms* which was developed very recently (see [13] for details and a bibliography).

The author is very grateful to Masanobu Kaneko for many valuable communications. The author would like to thank Marvin Knopp for his deep and interesting comments. The author is thankful to the referee for remarks which allowed the author to improve the presentation.

The three congruences in Theorem 1, although they look similar, are quite different in their nature: Theorem 1 (i) is pretty easy (see Proposition 1 below) while Theorem 1 (ii), (iii) are more involved. In particular, the conditions $p \equiv 1$ or $2 \mod 3$ indeed make a difference and are related to complex multiplication for the elliptic curve $X_0(36)$. Our proof is easily generalizable and indicates that congruences of this type are far from being isolated. Similar congruences may be related to all weakly holomorphic modular forms which may be produced by means of differentiation from the mock modular forms whose shadows are complex multiplication cusp forms (see, e.g., [13] for the basic definitions related to weak harmonic Maass forms and mock modular forms). In particular, let $g = \sum b(n)q^n$ be the weight two normalized cusp form which is the pullback of the holomorphic differential on $X_0(36)$ (incidentally, $g = \eta(6\tau)^4$). If a prime p is inert in the CM field (in this case, the CM field is $\mathbb{Q}(\sqrt{-3})$, and inert primes are the odd primes $p \equiv 2 \mod 3$), then $b(p) = 0$. The congruences of Theorem 1 (ii) are, in a sense, tightly related to this fact. A general theory which indicates how to produce similar congruences is developed in [1]. In this chapter, however, we concentrate only on the case of $X_0(36)$ in order to obtain a clean and specific result.

Denote by D the differentiation

$$D := \frac{1}{2\pi \mathrm{i}} \frac{\mathrm{d}}{\mathrm{d}\tau}.$$

We write $M_s^! = M_s^!(N)$ for the space of weakly holomorphic (i.e., holomorphic in the upper half plane with possible poles at the cusps) modular forms of weight s on $\Gamma_0(N)$ of Hauptypus (i.e., having the trivial character). Bol's identity implies that for an even positive integer k,

$$D^{k-1} : M_{2-k}^! \to M_k^!. \tag{1}$$

For a weakly holomorphic modular form $f \in M_{2-k}^!$, which has rational q-expansion coefficients, the bounded denominator principle allows us to claim the existence of an integer T such that all q-expansion coefficients of Tf are integers. The following fact follows from this observation.

Proposition 1. *Let p be a prime. If $f \in M_{2-k}^!$ has rational q-expansion coefficients, then there exists an integer $A \geq 0$ such that for all integers $l \geq 0$*

$$(D^{k-1} f)|U^l \equiv 0 \mod p^{l(k-1)-A}.$$

In particular, if $f \in M_{2-k}^!$ has p-integral q-expansion coefficients, then $A = 0$.

As an example, we apply Proposition 1 to prove Theorem 1 (i). Indeed, it is easy to check that

$$\frac{E_6(6\tau)}{\eta(6\tau)^4} = -D^3\left(\eta(6\tau)^{-4}\right), \tag{2}$$

and Theorem 1 (i) follows from Proposition 1 and this identity. Being an identity between two modular forms, (2) can be verified, for example, by a straightforward computer calculation of their sufficiently many Fourier coefficients. Note that these are weakly holomorphic modular forms, and their principal parts at all cusps coincide. One cannot prove Theorem 1 (ii), (iii) in a similar way because the map (1) is not surjective. Specifically, $E_4(6\tau)/\eta(6\tau)^4 \in M_2^!(36)$ does not belong to the image of this map.

The rest of the chapter is devoted to the proof of Theorem 1 (ii), (iii). We obtain our results as an application of the theory of *weak harmonic Maass forms*. We refer to [2, 4] for definitions and detailed discussion of their properties. The extension of (1) to the space $H_{2-k} \supset M_{2-k}^!$ of weak harmonic Maass forms $D^{k-1} : H_{2-k} \to M_k^!$ is still not surjective. We, however, have the following proposition.

Proposition 2. *There exists a weak harmonic Maass form M of weight zero (on $\Gamma_0(36)$ of Haupttypus) such that*

$$\frac{E_4(6\tau)}{\eta(6\tau)^4} = D(M) + \gamma\eta(6\tau)^4$$

for some $\gamma \in \mathbb{C}$.

Proof. The existence of M with any given principal parts at cusps and in particular such that the principal parts of $D(M)$ and $E_4(6\tau)/\eta(6\tau)^4$ at all cusps coincide follows from [2, Proposition 3.11]. Note that the constant terms of the Fourier expansion of $E_4(6\tau)/\eta(6\tau)^4$ at all cusps vanish. This follows from the fact that $E_4(\tau)/\eta(\tau)$ has no constant term at infinity combined with the modularity of E_4 and the transformation law of the Dedekind η-function. Since the constant terms of the Fourier expansion of $E_4(6\tau)/\eta(6\tau)^4$ at all cusps vanish, the difference $D(M) - E_4(6\tau)/\eta(6\tau)^4 \in S_2(36)$ is a cusp form. However, $\dim S_2(36) = 1$, and this space is generated by the unique normalized cusp form $\eta(6\tau)^4$.

We will later show that in fact, $\gamma = 0$. But first investigate some properties of M. It is well known (see [13, Sect. 7.2], [2, Sect. 3] for the details) that being a weak harmonic Maass form M has a canonical decomposition

$$M = M^+ + M^-$$

into a sum of a holomorphic function M^+ and a non-holomorphic function M^- (in the case under the consideration that is simply a decomposition of a C^∞ function into the sum of a holomorphic and an anti-holomorphic functions). The holomorphic part M^+ has a Fourier expansion

$$M^+ = \sum_{n \gg -\infty} a(n)q^n.$$

Proposition 3. *The Fourier coefficients $a(n)$ of M^+ are algebraic numbers. More specifically, there is a cyclotomic extension K of $\mathbb{Q}$ such that $a(n) \in K$ for all n.*

Proof. Let $g := \eta(6\tau)^4 \in S_2(\Gamma_0(36))$ be the unique normalized cusp form of weight 2 and level 36. Let $\tau = x + iy$. The differential operator $\xi := 2i\overline{\frac{\partial}{\partial\bar{\tau}}}$ takes weight zero weak harmonic Maass forms to cusp forms. In particular, since $\dim S_2(36) = 1$, we conclude that

$$\xi(M) = tg$$

for some $t \in \mathbb{C}$. Since Fourier coefficients of M at all cusps are rational, we derive from [2, Proposition 3.5] that $t||g||^2 \in \mathbb{Q}$, where $||g||^2$ denotes the Petersson norm of g. At the same time, it follows from [4, Proposition 5.1] that there exists a weak harmonic Maass form M_g which is *good for* g. That means (see [4]) that, in particular, $\xi(M_g) = ||g||^{-2}g$, and M_g has its principal part at $i\infty$ in $\mathbb{Q}[q^{-1}]$ and is bounded at all other cusps. Since the rational linear combination $M - t||g||^2 M_g$ obviously satisfies

$$\xi(M - t||g||^2 M_g) = 0,$$

we conclude that it is a weight zero weakly holomorphic modular form

$$M - t||g||^2 M_g \in M_0^!(36).$$

Since M_g is *good for* g, the modular form $M - t||g||^2 M_g$ has principal parts with rational Fourier expansion at all cusps and, therefore, rational Fourier coefficients at $i\infty$. Proposition 3 now follows from a theorem of Bruinier, Ono, and Rhoades [4, Theorem 1.3] which tells us that the Fourier expansion coefficients of the holomorphic part M_g^+ belong to a cyclotomic field K, because g is a CM-form.

We now prove that in Proposition 2, in fact, $\gamma = 0$.

Proposition 4. *There exists a weak harmonic Maass form M of weight zero (on $\Gamma_0(36)$ of Haupttypus) such that*

$$\frac{E_4(6\tau)}{\eta(6\tau)^4} = D(M).$$

Proof. We begin with an argument which is closely related to the proof of [4, Theorem 1.2] (and could actually be expanded to an alternative proof of our Proposition 3). Following [13, Lemma 7.2], we write the Fourier expansion of $M = M^+ + M^-$ as

$$M^+ = \sum_{n \gg -\infty} a(n)q^n, \qquad M^- = \sum_{n<0} a^-(n)\Gamma(1, 4\pi|n|y)q^n = \sum_{n<0} a^-(n)\exp(2\pi i n\bar{\tau}),$$

where $\Gamma(a, x)$ is the incomplete Γ-function. We thus have that

$$tg = \xi(M) = -4\pi \sum_{n \geq 1} a^{-}(-n) n q^{n}.$$

Since g has complex multiplication by $\mathbb{Q}(\sqrt{-3})$, we conclude that $a^{-}(n) = 0$ if $n \equiv 0, 2 \mod 3$. (Alternatively, this follows, of course, from the definition $g = \eta(6\tau)^4$, which implies immediately that the nonzero Fourier coefficients of g are supported on $n \equiv 1 \mod 6$.) Let $\chi = \left(\frac{-3}{\cdot}\right)$. As in [4], we conclude that the weak harmonic Maass form

$$u := M + M \otimes \chi := M + \sum_{n \gg -\infty} a(n)\chi(n) q^{n} + \sum_{n<0} a^{-}(n)\chi(n) \exp(2\pi i n \bar{\tau})$$

has the property $\xi(u) = 0$ and is therefore a weakly holomorphic weight zero modular form. It follows that the denominators of its Fourier coefficients are bounded. Namely, there exists a nonzero $T \in K^*$ such that the coefficients of

$$Tu = T(M + M \otimes \chi) = T(M^+ + M^+ \otimes \chi) = \sum_{n \gg -\infty} T\left(a(n) + \chi(n)a(n)\right) q^{n}$$

all belong to the ring of integers $\mathcal{O}_K \subset K$. In particular, for a prime $p \equiv 1 \mod 3$, we conclude that the p-adic limit of the coefficients $p^m a(p^m)$ of q^{p^m} of $D(M) = D(M^+)$ as $m \to \infty$ is zero. Since all coefficients of q^{p^m} in $E_4(6\tau)/\eta(6\tau)^4$ are zero, and the coefficients of q^{p^m} in g are not divisible by p, we conclude that $\gamma = 0$.

Remark. There is an alternative way to prove Proposition 4: observe that M^+ is a generalized abelian integral of the second kind (see [12] for a definition), and derive the proposition from the results of Knopp [12].

Proposition 4 allows us to assume further that the Fourier coefficients of M^+ are rational numbers.

We now need the Hecke operators action on M. For a prime p, let $T(p) := U + p^{k-1}V$ be the p-the Hecke operator at weight k. Let

$$g = \sum_{n \geq 1} b(n) q^{n}.$$

The form g is, of course, a Hecke eigenform. Using the same argument as in [3, Lemma 7.4], we have that

$$M|_0 T(p) = p^{-1} b(p) M + R_p,$$

where $R_p \in M_0^!(36)$ is a weakly holomorphic modular form with coefficients in $\mathbb{Q}$. We apply the differential operator D to this identity and use the commutation relation

$$pD\left(H|_0 T(p)\right) = \left(D(H)\right)|_2 T(p),$$

valid for any 1-periodic function H. We obtain that

$$(D(M))|_2 T(p) = b(p)D(M) + pD(R_p). \tag{3}$$

Let β, β' be the roots of equation

$$X^2 - b(p)X + p = 0,$$

such that $ord_p(\beta) \leq ord_p(\beta')$. Note that g is a complex multiplication cusp form, and the complex multiplication field is $\mathbb{Q}(\sqrt{-3})$. In particular, if $p \equiv 1 \mod 3$, then $\beta, \beta' \in \mathbb{Q}_p$ by Hensel's lemma, and $\mathrm{ord}_p(\beta) = 0$ while $\mathrm{ord}_p(\beta') = 1$. If $p \equiv 2 \mod 3$, then $b(p) = 0$, and we have that $\beta = -\beta'$. Thus, $\beta, \beta' \in \mathcal{F} = \mathbb{Q}_p(\sqrt{-p})$ and $ord_p(\beta) = ord_p(\beta') = 1/2$ in this case.

Our next proposition is closely related to calculations made in [1] and [5]. Let $\mathcal{R} \subset \mathcal{F}$ be the ring of p-integers. We consider the topology on $\mathcal{F} \otimes \mathcal{R}[[q]] \supset \mathbb{Q}_p \otimes \mathbb{Z}_p[[q]]$ (the tensor products are taken over $\mathbb{Z}$ throughout) determined by the norm

$$\left| \sum_{n \geq 0} u(n)q^n \right| = p^{-\inf_n(\mathrm{ord}_p(u(n)))}.$$

Proposition 5. (i) *Let* $p \equiv 1 \mod 3$ *be a prime. We have that in* $\mathbb{Q}_p \otimes \mathbb{Z}_p[[q]]$

$$\lim_{l \to \infty} \beta^{-l}(D(M))|U^l = 0.$$

(ii) *Let* $p \equiv 2 \mod 3$ *be a prime. We have that in* $\mathcal{F} \otimes \mathcal{R}[[q]]$ *the limits*

$$\lim_{l \to \infty} \beta^{-2l}(D(M))|U^{2l}$$

and

$$\lim_{l \to \infty} \beta^{-2l-1}(D(M))|U^{2l+1}$$

exist.

Proof. Abbreviate

$$F = D(M), \qquad r_p = pD(R_p),$$

and note that it follows from (3) that the Fourier coefficients of r_p are rational integers since those of F are rational integers. We firstly prove that all limits exist.

We put

$$G(\tau) = F(\tau) - \beta' F(p\tau) \quad \text{and} \quad G'(\tau) = F(\tau) - \beta F(p\tau),$$

and rewrite (3) as

$$(F|U)(\tau) + \beta\beta' F(p\tau) = (\beta + \beta')F(\tau) + r_p.$$

We obtain that

$$G|U = \beta G + r_p, \quad G'|U = \beta' G' + r_p,$$

and

$$F|U = \frac{\beta}{\beta - \beta'}(\beta G + r_p) - \frac{\beta'}{\beta - \beta'}(\beta' G' + r_p).$$

It follows that

$$(\beta - \beta')\beta^{-l} F|U^l = \left(\beta G + r_p + \frac{1}{\beta} r_p \middle| U + \cdots + \frac{1}{\beta^{l-1}} r_p \middle| U^{l-1} \right)$$
$$- (\beta'/\beta)^l \left(\beta' G' + r_p + \frac{1}{\beta'} r_p \middle| U + \cdots + \frac{1}{\beta'^{l-1}} r_p \middle| U^{l-1} \right). \tag{4}$$

The existence of the limit in part (i) follows from (4) since $(\beta'/\beta)^l \to 0$, and the second expression in parenthesis has bounded denominators by Proposition 1, while $\beta^{1-l} r_p|U^{l-1} \to 0$ as $l \to \infty$ again by Proposition 1. In order to prove the existence of the limits in part (ii), we rewrite (4) in this case, taking into the account that $\beta = -\beta'$, as

$$2\beta^{-2l+1} F|U^{2l} = \beta G + \beta G' + 2\frac{1}{\beta} r_p|U + 2\frac{1}{\beta^3} r_p|U^3 + \cdots + 2\frac{1}{\beta^{2l-1}} r_p|U^{2l-1}$$

and

$$2\beta^{-2l} F|U^{2l+1} = \beta G - \beta G' + 2r_p + 2\frac{1}{\beta^2} r_p|U^2 + 2\frac{1}{\beta^4} r_p|U^4 + \cdots + 2\frac{1}{\beta^{2l}} r_p|U^{2l},$$

and Proposition 5 (ii) follows since we still have that $\beta^{-m} r_p|U^m \to 0$ as $m \to \infty$ by Proposition 1.

We now prove that the limit in Proposition 5 (i) is actually zero. Write

$$\lim_{l \to \infty} \beta^{-l}(D(M))|U^l = \sum_{n>0} c(n) q^n.$$

Obviously,

$$\left(\sum_{n>0} c(n) q^n \right) |U = \beta \left(\sum_{n>0} c(n) q^n \right),$$

and we derive from (3), Proposition 1, and the fact that the operators U and $T(m)$ commute for any integer m not divisible by p that

$$\left(\sum_{n>0} c(n) q^n \right) |T(m) = b(m) \left(\sum_{n>0} c(n) q^n \right).$$

A standard inductive argument now allows us to conclude that $\sum c(n)q^n$ must be a multiple of $g(\tau) - \beta' g(p\tau)$. However, $c(1) = 0$ (simply because $F = E_4(6\tau)/\eta(6\tau)^4$, and therefore $c(p^l) = 0$ for all l). Thus, the series $\sum_{n>0} c(n)q^n$ must be a zero multiple of $g(\tau) - \beta' g(p\tau)$.

We are now ready to prove Theorem 1 (ii), (iii).

Proof (Proof of Theorem 1 (ii), (iii)*).* Recall that $F = D(M) = E_4(6\tau)/\eta(6\tau)^4$. Theorem 1 (iii) follows immediately from Proposition 5 (ii) since $\mathrm{ord}_p(\beta) = 1/2$ for $p \equiv 2 \mod 3$.

Assume that $p \equiv 1 \mod 3$. Proposition 1 allows us to pick $A_p \geq 0$ such that

$$\mathrm{ord}_p \left(p^{A_p} r_p | U^m \right) \geq m \tag{5}$$

for all $m \geq 0$. Since F has p-integral Fourier coefficients, so has G', and in view of (5) and the fact that $\mathrm{ord}_p(\beta') = 1$, it now follows from (4) that

$$(\beta - \beta')\beta^{-l} p^{A_p} F \,\Big|\, U^l \equiv p^{A_p} \beta G + p^{A_p} r_p + \frac{p^{A_p}}{\beta} r_p \,\Big|\, U + \cdots + \frac{p^{A_p}}{\beta^{l-1}} r_p \,\Big|\, U^{l-1} \mod p^l.$$

Let $s \geq 1$ be an integer. Pick $l > s$ large enough such that $F|U^l \equiv 0 \mod p^s$, take into the account that both $(\beta - \beta')$ and β are p-adic units, and rewrite the previous congruence as

$$0 \equiv p^{A_p} \frac{\beta - \beta'}{\beta^s} F|U^s + \frac{p^{A_p}}{\beta^s} r_p|U^s + \cdots + \frac{p^{A_p}}{\beta^{l-1}} r_p|U^{l-1} \mod p^s.$$

It now follows from (5) that all terms on the right in this congruence except possibly the first one vanish modulo p^s, and we conclude that $p^{A_p} F|U^s \equiv 0 \mod p^s$ as required.

References

1. Bringmann, K.; Guerzhoy, P.; Kane, B., Mock modular forms as p-adic modular forms. Trans. Amer. Math. Soc. 364 (2012), 2393–2410.
2. Bruinier, Jan Hendrik; Funke, Jens, On two geometric theta lifts, Duke Math. J. 125 (2004), no. 1, 45–90.
3. Bruinier, Jan; Ono, Ken, Heegner divisors, L-functions and harmonic weak Maass forms. Ann. of Math. (2) 172 (2010), no. 3, 2135–2181.
4. Bruinier, Jan Hendrik; Ono, Ken; Rhoades, Robert, Differential operators for harmonic weak Maass forms and the vanishing of Hecke eigenvalues, Math. Ann., 342 (2008), 673–693.
5. Guerzhoy, Pavel; Kent, Zachary A.; Ono, Ken, p-adic coupling of mock modular forms and shadows. Proc. Natl. Acad. Sci. USA 107 (2010), no. 14, 6169–6174.
6. Honda, Yutaro; Kaneko, Masanobu, On Fourier coefficients of some meromorphic modular forms, preprint.

7. Kaneko, Masanobu; Koike, Masao, On modular forms arising from a differential equation of hypergeometric type, Ramanujan J. 7 (2003), no. 1–3, 145–164.
8. Kaneko, Masanobu; Koike, Masao, Quasimodular solutions of a differential equation of hypergeometric type. Galois theory and modular forms, 329–336, Dev. Math., 11, Kluwer Acad. Publ., Boston, MA, 2004.
9. Kaneko, Masanobu; Koike, Masao, On extremal quasimodular forms. Kyushu J. Math. 60 (2006), no. 2, 457–470.
10. Kaneko, Masanobu, On modular forms of weight $(6n + 1)/5$ satisfying a certain differential equation. Number theory, 97–102, Dev. Math., 15, Springer, New York, 2006.
11. Kaneko, M.; Zagier, D., Supersingular j-invariants, hypergeometric series, and Atkin's orthogonal polynomials. Computational perspectives on number theory (Chicago, IL, 1995), 97–126, AMS/IP Stud. Adv. Math., 7, Amer. Math. Soc., Providence, RI, 1998.
12. Knopp, Marvin Isadore, On abelian integrals of the second kind and modular functions. Amer. J. Math. 84 1962 615–628.
13. Ono, Ken, Unearthing the visions of a master: harmonic Maass forms and number theory, Proceedings of the 2008 Harvard-MIT Current Developments in Mathematics Conference, International Press, Somerville, MA, 2009, pages 347–454.

Some Intrinsic Constructions on Compact Riemann Surfaces

Robert C. Gunning

Abstract For any prescribed differential principal part on a compact Riemann surface, there are uniquely determined and intrinsically defined meromorphic abelian differentials with these principal parts, defined independently of any choice of a marking of the surface or of a basis for the holomorphic abelian differentials, and they are holomorphic functions of the singularities. They can be constructed explicitly in terms of intrinsically defined cross-ratio functions on the Riemann surfaces, the classical cross-ratio function for the Riemann sphere, and natural generalizations for surfaces of higher genus.

Key words Riemann surfaces • Abelian differentials • Cross-ratio function

Mathematics Subject Classification (2010): 30F10 (Primary), 30F30, 14H55

1 Introduction

The vector space of holomorphic abelian differentials is intrinsically defined on any compact Riemann surface M; but whether there is an individual uniquely and intrinsically defined holomorphic abelian differential on an arbitrary Riemann surface is a rather different matter. Of course, there is the familiar standard basis for holomorphic abelian differentials on a marked Riemann surface, a surface with a standard homology basis [4, 6], and there is an individual single holomorphic abelian differential on any pointed Riemann surface determined uniquely and intrinsically up to a constant factor, as will be demonstrated here; but in each case, some other normalizing property of the surface is involved. The situation is

R.C. Gunning (✉)
Fine Hall, Princeton University, Princeton, NJ, USA
e-mail: gunning@math.princeton.edu

H.M. Farkas et al. (eds.), *From Fourier Analysis and Number Theory to Radon Transforms and Geometry*, Developments in Mathematics 28, DOI 10.1007/978-1-4614-4075-8_13,

quite different for meromorphic abelian differentials. There is a single uniquely and intrinsically defined meromorphic abelian differential with a specified singularity on any Riemann surface, fully independent of any choice of basis for the homology group or for the space of holomorphic abelian differentials; and the periods and other properties of this differential can be expressed quite simply in terms of any basis for the homology or space of holomorphic abelian differentials on M. This intrinsic differential will be discussed for meromorphic abelian differentials of the second kind in Sect. 2, the simple case in which the integrals of the differentials are well-defined meromorphic functions on the universal covering surface, and for the basic meromorphic abelian differentials of the third kind in Sect. 3, the case in which the integrals of the differentials are well-defined meromorphic functions only on the complements of paths joining the singularities on the universal covering surface. The complications caused by the multiple-valued nature of the integrals are avoided by considering the intrinsic cross-ratio function on M, a uniquely and intrinsically defined basic analytic entity on any Riemann surface. The proofs are perhaps a bit novel, since the point of this note is to show that the constructions involved are quite intrinsic and that all the invariants can be calculated by essentially the same formulas in terms of any bases for the homology and the holomorphic abelian differentials on the surface. Some standard properties of meromorphic abelian differentials and of the cross-ratio function with more standard proofs on marked Riemann surfaces can be found in [9], where the cross-ratio function was introduced but called the prime form for its role in the factorization of meromorphic functions on M; the possibility of intrinsically defined meromorphic abelian differentials was not discussed there.

As background for the discussion here, on a compact Riemann surface M of genus $g > 0$, let $\omega_i(z)$ be a basis for the holomorphic abelian differentials and $\tau_j \in H_1(M)$ be a basis for the first homology group. The intersection matrix of M in terms of these bases is the $2g \times 2g$ skew-symmetric integral matrix P describing the intersection numbers of the homology basis τ_j; so the entries of the matrix P are the integers $p_{jk} = \tau_j \cap \tau_k$. If $\sigma_i(z)$ are the dual differential forms to the homology basis τ_j, the closed differential 1-forms on M for which $\int_{\tau_j} \sigma_i(z) = \delta^i_j$, then equivalently $p_{ij} = \int_M \sigma_i(z) \wedge \sigma_j(z)$. The period matrix of M is the $g \times 2g$ complex matrix Ω with the entries $\omega_{ij} = \int_{\tau_j} \omega_i(z)$ for $1 \le i \le g,\ 1 \le j \le 2g$. Riemann's equality [8, 10] for the period matrix Ω is

$$i \begin{pmatrix} \Omega \\ \overline{\Omega} \end{pmatrix} P \, {}^t\overline{\begin{pmatrix} \Omega \\ \overline{\Omega} \end{pmatrix}} = \begin{pmatrix} H & 0 \\ 0 & -\overline{H} \end{pmatrix} \tag{1}$$

where H is a $g \times g$ complex matrix and ${}^tX$ denotes the transpose of the matrix X; and Riemann's inequality for the period matrix Ω is that the matrix $H = i\Omega P\, \overline{{}^t\Omega}$ is positive-definite Hermitian. It is convenient also to introduce the auxiliary matrices

$$G = {}^tH^{-1} = \overline{H}^{-1} \tag{2}$$

and

$$\begin{pmatrix} \Pi \\ \overline{\Pi} \end{pmatrix} = {}^t\overline{\begin{pmatrix} \Omega \\ \overline{\Omega} \end{pmatrix}}^{-1}, \tag{3}$$

so that the $g \times 2g$ complex matrices Ω and Π satisfy

$$\Omega\,{}^t\Pi = 0, \quad \overline{\Omega}\,{}^t\Pi = \mathrm{I}_{2g}, \quad \text{and} \quad {}^t\Omega\overline{\Pi} + {}^t\overline{\Omega}\Pi = \mathrm{I}_{2g} \tag{4}$$

where I_{2g} is the $2g \times 2g$ identity matrix. It follows from (4) that there is the direct sum decomposition

$$\mathbb{C}^{2g} = {}^t\Omega\overline{\Pi}\mathbb{C}^{2g} \oplus {}^t\overline{\Omega}\Pi\mathbb{C}^{2g} = {}^t\Omega\mathbb{C}^{g} \oplus {}^t\overline{\Omega}\mathbb{C}^{g}. \tag{5}$$

The subspace ${}^t\Omega\mathbb{C}^g \subset \mathbb{C}^{2g}$ in this decomposition can be described more intrinsically as the subspace spanned by the period vectors of the holomorphic abelian differentials, where the period vector of the holomorphic abelian differential $\omega_i(z)$ is the column vector in $\mathbb{C}^{2g}$ with the entries $\int_{\sigma_j} \omega_i(z)$ for $1 \le j \le 2g$; correspondingly, the subspace ${}^t\overline{\Omega}\mathbb{C}^{2g} \subset \mathbb{C}^{2g}$ is the subspace spanned by the period vectors of the complex conjugates of the holomorphic abelian differentials $\omega_i(z)$. A straightforward calculation shows that under changes of bases

$$\omega_i(z) = \sum_{k=1}^{g} c_{ik}\omega_k^*(z) \quad \text{and} \quad \tau_j = \sum_{l=1}^{2g} q_{lj}\tau_l^* \tag{6}$$

for any matrices $C = \{c_{ik}\} \in \mathrm{Gl}(g, \mathbb{C})$ and $Q = \{q_{lj}\} \in \mathrm{Gl}(2g, \mathbb{Z})$, the intersection and period matrices are changed by

$$P = Q^{-1}P^*\,{}^tQ^{-1}, \quad \Omega = C\Omega^*Q \quad \text{and} \quad G = {}^tC^{-1}G^*\overline{C}^{-1}. \tag{7}$$

This note is dedicated to the memory of Leon Ehrenpreis, among whose many mathematical interests were special functions [1, 2] and Riemann surfaces [3], so who might have been amused by these observations.

2 Meromorphic Abelian Differentials of the Second Kind

The singularities of meromorphic abelian differentials on M are described by differential principal parts $\mathfrak{p} = \{\mathfrak{p}_{a_i}\}$ associated to finitely many distinct points $a_i \in M$, where $\mathfrak{p}_{a_i}$ is a finite Laurent expansion of a local meromorphic differential form in an open neighborhood of the point a_i in terms of a local coordinate centered the point a_i. A meromorphic abelian differential on M having the differential principal part $\mathfrak{p}$ is a meromorphic abelian differential $\mu(z)$ on M that differs from the local differential principal part $\mathfrak{p}_{a_l}$ by a local holomorphic differential form in a neighborhood of the point a_l for each $a_l \in M$; so $\mu(z)$ is determined uniquely by its principal part $\mathfrak{p}$ up to the addition of an arbitrary holomorphic abelian differential on M. The general existence theorem for meromorphic abelian differentials [6, 11] asserts that there is a meromorphic abelian differential on M with the differential principal part $\mathfrak{p} = \{\mathfrak{p}_{a_l}\}$ if and only if the sum of the residues of the Laurent expansions $\mathfrak{p}_{a_l}$ at all of the points a_l is zero. A differential principal part of the second kind is a differential principal part $\mathfrak{p} = \{\mathfrak{p}_{a_l}\}$ such that the residue of each Laurent expansion $\mathfrak{p}_{a_l}$ is zero; so any differential principal part of the second kind on M is the differential principal part of a meromorphic abelian differential on M, called a meromorphic abelian differential of the second kind. These differential forms frequently are called just differentials of the second kind, and holomorphic abelian differentials are called differentials of the first kind. A meromorphic abelian differential that is of neither the first nor the second kind is called a meromorphic abelian differential of the third kind, or just a differential of the third kind. A holomorphic abelian differential $\omega(z)$ or a meromorphic abelian differential $\mu(z)$ on M can be identified with a holomorphic or meromorphic abelian differential on the universal covering surface $\widetilde{M}$ of M that is invariant under the action of the covering translation group Γ; the induced differential form on $\widetilde{M}$ generally will be denoted by the same symbol as the differential form on M. The integrals $w(z,a) = \int_a^z \omega$ and $u(z,a) = \int_a^z \mu$ of holomorphic abelian differentials and meromorphic abelian differentials of the second kind are well- defined holomorphic or meromorphic functions of points $z, a \in \widetilde{M}$ and are determined uniquely by the conditions that if $a \in \widetilde{M}$ is viewed as a fixed point, then $dw(z,a) = \omega(z)$ and $du(z,a) = \mu(z)$ while $w(a,a) = 0$ and $u(a,a) = 0$, assuming of course that a is not one of the points a_l. If the base point a is irrelevant, the integral may be denoted just by $w(z)$ or $u(z)$. The period classes of these differentials are the group homomorphisms $\omega \in \operatorname{Hom}(\Gamma, \mathbb{C})$ and $\mu \in \operatorname{Hom}(\Gamma, \mathbb{C})$ from the covering translation group Γ of M to the additive group of complex numbers defined by $\omega(T) = w(Tz,a) - w(z,a)$ and $\mu(T) = u(Tz,a) - u(z,a)$ for any $T \in \Gamma$, where $\omega(T)$ and $\mu(T)$ are constants since $dw(z,a)$ and $du(z,a)$ are Γ-invariant functions of $z \in \widetilde{M}$. Alternatively, since $\mathbb{C}$ is abelian, these period classes can be viewed as homomorphisms $\omega \in \operatorname{Hom}(H_1(M), \mathbb{C})$ and $\mu \in \operatorname{Hom}(H_1(M), \mathbb{C})$ from the homology group $H_1(M)$ of M, the abelianization $H_1(M) = \Gamma/[\Gamma, \Gamma]$ of the covering translation group Γ; and the values of these homomorphisms coincide with the usual period integrals $\omega(\tau) = \int_\tau \omega(z)$ and $\mu(\tau) = \int_\tau \mu(z)$ of these differential forms along closed paths $\tau \in M$ representing the given homology classes, provided of course that the paths τ avoid the singularities of $\mu(z)$. By de Rham's theorem, two closed C^∞ differential 1-forms $\phi(z)$ and $\psi(z)$ on M have the same periods on all 1-cycles $\tau \in H_1(M)$ if and only if they differ by the exterior derivative of a C^∞ function on M; such differential forms are called cohomologous, and that $\phi(z)$ and $\psi(z)$ are cohomologous differential forms will be indicated by writing $\phi(z) \sim \psi(z)$. The following probably quite familiar observations about differential forms are inserted here for convenience of reference in the subsequent discussion.

Lemma 1. *If $\phi'(z), \phi''(z), \psi'(z), \psi''(z)$ are closed C^∞ differential 1-forms on a compact Riemann surface M, where $\phi'(z) \sim \phi''(z)$ and $\psi'(z) \sim \psi''(z)$, then*

$$\int_M \phi'(z) \wedge \psi'(z) = \int_M \phi''(z) \wedge \psi''(z). \tag{8}$$

Proof. If $\phi'(z) \sim \phi''(z)$ so that $\phi'(z) - \phi''(z) = \mathrm{d} f(z)$ for a C^∞ function $f(z)$ on M, then by Stokes's theorem,

$$\begin{aligned}\int_M (\phi'(z) - \phi''(z)) \wedge \psi'(z) &= \int_M \mathrm{d} f(z) \wedge \psi'(z) = \int_M \mathrm{d}\big(f(z)\ \psi'(z)\big) \\ &= \int_{\partial M} f(z)\psi'(z) = 0\end{aligned}$$

since $\partial M = \emptyset$ for the compact manifold M. The obvious iteration of the preceding equation yields (8), and that suffices for the proof.

Lemma 2. *For any bases $\omega_i(z)$ of holomorphic differential forms and $\tau_j \in H_1(M)$ for the first homology group of a compact Riemann surface M,*

$$i \int_M \omega_j(z) \wedge \overline{\omega_k(z)} = h_{jk} \tag{9}$$

where H is the matrix defined in (1).

Proof. That ω_{jm} are the periods of the differential form $\omega_j(z)$ can be restated as the condition that $\omega_j(z) \sim \sum_{m=1}^{g} \omega_{jm}\sigma_m(z)$ where $\sigma_m(z)$ are the dual differential forms to the homology basis τ_j, so

$$\begin{aligned}i \int_M \omega_j(z) \wedge \overline{\omega_k(z)} &= i \sum_{m,n=1}^{g} \omega_{jm}\overline{\omega_{kn}} \int_M \sigma_m(z) \wedge \sigma_n(z) \\ &= i \sum_{m,n=1}^{g} \omega_{jm}\overline{\omega_{kn}}\, p_{mn} = h_{jk}\end{aligned}$$

by definition of the matrix H, and that suffices for the proof.

Theorem 1. (i) *For any differential principal part of the second kind $\mathfrak{p}$ on a compact Riemann surface M of genus $g > 0$, there are a unique meromorphic abelian differential of the second kind $\mu_{\mathfrak{p}}(z)$ and a unique holomorphic abelian differential $\omega_{\mathfrak{p}}(z)$ such that $\mu_{\mathfrak{p}}(z)$ has the differential principal part $\mathfrak{p}$ and has the same period class as the complex conjugate differential $\overline{\omega_{\mathfrak{p}}(z)}$.*

(ii) *The holomorphic abelian differential $\omega_{\mathfrak{p}}(z)$ is characterized by*

$$\int_M \omega(z) \wedge \overline{\omega_{\mathfrak{p}}(z)} = 2\pi \mathrm{i} \sum_{a \in M} \mathrm{res}_a\,(w(z)\ \mathfrak{p}) \tag{10}$$

for all holomorphic abelian differentials $\omega(z)$ *on* M*, where* $\mathrm{d}w(z) = \omega(z)$ *and* $\mathrm{res}_a(w(z)\ \mathfrak{p})$ *is the residue of the local meromorphic differential form* $w(z)\ \mathfrak{p}$ *at the point* $a \in M$*.*

(iii) *In terms of any bases* $\omega_i(z)$ *and* τ_j*, the differential form* $\omega_{\mathfrak{p}}(z)$ *is given by*

$$\omega_{\mathfrak{p}}(z) = -2\pi \sum_{i,j=1}^{g} \sum_{a \in M} g_{ji}\ \overline{\mathrm{res}_a(w_i(z)\ \mathfrak{p})}\omega_j(z) \tag{11}$$

where $G = \{g_{ij}\}$ *is the matrix* (2) *and* $\mathrm{d}w_i(z) = \omega_i(z)$*, and*

(iv) *the period class of the meromorphic abelian differential* $\mu_{\mathfrak{p}}(z)$ *is given by*

$$\mu_{\mathfrak{p}}(T) = -2\pi \sum_{i,j=1}^{g} \sum_{a \in M} g_{ij}\ \ \mathrm{res}_a(w_i(z)\ \mathfrak{p})\ \overline{\omega_j(T)} \tag{12}$$

for any $T \in \Gamma$*.*

Proof. (i) If $\mu(z)$ is an abelian differential of the second kind with the differential principal part $\mathfrak{p}$, then $\mu(z) + \omega(z)$ is an abelian differential of the second kind with the principal part $\mathfrak{p}$ for any holomorphic abelian differential $\omega(z)$, and all the abelian differentials of the second kind with the differential principal part $\mathfrak{p}$ arise in this way. There is a unique holomorphic abelian differential $\omega(z)$ such that the period vector $\{\mu_{\mathfrak{p}}(\tau_j)\} = \{\mu(\tau_j) + \omega(\tau_j)\}$ of the meromorphic differential form $\mu_{\mathfrak{p}}(z) = \mu(z) + \omega(z)$ is contained in the linear subspace $\overline{{}^t\Omega\mathbb{C}^g} \subset \mathbb{C}^{2g}$ in the direct sum decomposition (5), hence such that $\mu_{\mathfrak{p}}(\tau) = \overline{\omega_{\mathfrak{p}}(\tau)}$ for a uniquely determined holomorphic abelian differential $\omega_{\mathfrak{p}}(z)$.

(ii) If the differential principal part is explicitly $\mathfrak{p} = \{\mathfrak{p}_{a_l}\}$, choose points $\tilde{a}_l \in \widetilde{M}$ such that $\pi(\tilde{a}_l) = a_l$ for the covering projection $\pi : \widetilde{M} \longrightarrow M$; the inverse image $\pi^{-1}(a_l) = \Gamma\tilde{a}_l$ then is a Γ-invariant set of points on $\widetilde{M}$. The integral

$$u_{\mathfrak{p}}(z) = \int_{z_0}^{z} \mu_{\mathfrak{p}} \tag{13}$$

for a fixed point $z_0 \in \widetilde{M}$ disjoint from the set $\Gamma\tilde{a}_l$ is a well-defined meromorphic function of the variable $z \in \widetilde{M}$ with poles just at the points $\Gamma\tilde{a}_l$ and $u_{\mathfrak{p}}(Tz) = u_{\mathfrak{p}}(z) + \mu_{\mathfrak{p}}(T)$ for any covering translation $T \in \Gamma$. Choose disjoint coordinate discs Δ_l about each of the points a_l and a connected component $\widetilde{\Delta_l}$ of the inverse image $\pi^{-1}(\Delta_l)$ containing the point $\tilde{a}_l$, so $\pi^{-1}(\Delta_l) = \Gamma\widetilde{\Delta}_l$. Let $\tilde{u}_{\mathfrak{p}}(z)$ be a C^∞ modification of the function $u_{\mathfrak{p}}(z)$ in $\widetilde{\Delta}_l$, the result of multiplying the function $u_{\mathfrak{p}}(z)$ by a C^∞ function in $\widetilde{\Delta}_l$ that vanishes in an open neighborhood of $\tilde{a}_l$ and is identically equal to 1 near the boundary of $\widetilde{\Delta}_l$, and extend this modification to all the discs $\Gamma\widetilde{\Delta}_l$ so that $\tilde{u}_{\mathfrak{p}}(Tz) = \tilde{u}_{\mathfrak{p}}(z) + \mu_{\mathfrak{p}}(T)$ for any covering translation $T \in \Gamma$. Then $\tilde{\mu}_{\mathfrak{p}}(z) = \mathrm{d}\tilde{u}_{\mathfrak{p}}(z)$ is a C^∞ closed Γ-invariant differential form of degree 1 on

$\widetilde{M}$, so it can be viewed as a C^{∞} differential form on M; moreover, $\tilde{\mu}_{\mathfrak{p}}(z)$ is equal to $\mu_{\mathfrak{p}}(z)$ outside the discs $\Gamma\Delta_l$ and has the same periods as $\overline{\omega_{\mathfrak{p}}}(z)$ by (i) of the present theorem, so it follows from Lemma 1 that

$$\int_M \omega(z) \wedge \overline{\omega_{\mathfrak{p}}(z)} = \int_M \omega(z) \wedge \tilde{\mu}_{\mathfrak{p}}(z) \tag{14}$$

for all holomorphic abelian differentials $\omega(z)$. The exterior product $\omega(z) \wedge \tilde{\mu}_{\mathfrak{p}}(z)$ vanishes outside the discs Δ_l since the differential forms $\omega(z)$ and $\tilde{\mu}_{\mathfrak{p}}(z)$ are both holomorphic 1-forms there, and the differential forms $\mu_{\mathfrak{p}}(z)$ and $\tilde{\mu}_{\mathfrak{p}}(z)$ agree on the boundaries $\partial\Delta_l$ of the discs Δ_l. Then if $\mathrm{d}w(z) = \omega(z)$, it follows from Stokes's theorem and the Cauchy integral formula on $\widetilde{M}$ that

$$\begin{aligned}\int_M \omega(z) \wedge \tilde{\mu}_{\mathfrak{p}}(z) &= \sum_l \int_{\Delta_l} \omega(z) \wedge \tilde{\mu}_{\mathfrak{p}}(z) = \sum_l \int_{\widetilde{\Delta}_l} \mathrm{d}\big(w(z)\, \tilde{\mu}_{\mathfrak{p}}(z)\big) \\ &= \sum_l \int_{\partial\widetilde{\Delta}_l} w(z)\, \tilde{\mu}_{\mathfrak{p}}(z) = \sum_l \int_{\partial\widetilde{\Delta}_l} w(z)\, \mu_{\mathfrak{p}}(z) \\ &= 2\pi\mathrm{i} \sum_l \operatorname{res}_{\tilde{a}_l} \big(w(z)\, \mu_{\mathfrak{p}}(z)\big) = 2\pi\mathrm{i} \sum_{a\in M} \operatorname{res}_a \big(w(z)\ \mathfrak{p}\big)\,.\end{aligned} \tag{15}$$

It then follows from (14) and (15) that the differential form $\omega_{\mathfrak{p}}(z)$ satisfies (10). For any choice of bases $\omega_i(z)$ and τ_j and for any holomorphic abelian differential $\omega(z) = \sum_{l=1}^{g} c_l\omega_l(z)$, it then follows from Lemma 2 and (2) that

$$\begin{aligned} i \sum_{k=1}^{g} g_{kj} \int_M \omega_k(z) \wedge \overline{\omega(z)} &= i \sum_{k,l=1}^{g} g_{kj}\overline{c_l} \int_M \omega_k(z) \wedge \overline{\omega_l(z)} \\ &= \sum_{k,l=1}^{g} g_{kj}\overline{c_l}h_{kl} = \sum_{l=1}^{g} \delta_l^j \overline{c_l} = \overline{c_j}; \end{aligned} \tag{16}$$

consequently, (10) fully determines the differential form $\omega_{\mathfrak{p}}(z)$.

(iii) In particular, if $\omega_{\mathfrak{p}}(z) = \sum_{j=1}^{g} c_j\omega_j(z)$ and $\mathrm{d}w_k(z) = \omega_k(z)$, it follows from (16) and (10) that

$$\overline{c}_j = i \sum_{k=1}^{g} g_{kj} \int_M \omega_k(z) \wedge \overline{\omega_{\mathfrak{p}}(z)} = -2\pi \sum_{k=1}^{g} \sum_{a\in M} g_{kj} \operatorname{res}_a(w_k(z)\ \mathfrak{p}) \tag{17}$$

and consequently that

$$\omega_{\mathfrak{p}}(z) = -2\pi \sum_{j,k=1}^{g} \sum_{a\in M} g_{jk}\, \overline{\operatorname{res}_a(w_k(z)\ \mathfrak{p})}\omega_j(z). \tag{18}$$

(iv) Finally, if $\omega_{\mathfrak{p}}(z) = \sum_{j=1}^{g} c_j \omega_j(z)$, then for any homology class $\tau \in H_1(M)$, it follows from (18) that

$$\mu_{\mathfrak{p}}(\tau) = \overline{\omega_{\mathfrak{p}}(\tau)} = \sum_{j=1}^{g} \overline{c}_j\, \overline{\omega_j(\tau)}$$

$$= -2\pi \sum_{j,k=1}^{g} \sum_{a \in M} g_{kj}\ \mathrm{res}_a\,(w_k(z)\ \mathfrak{p})\ \overline{\omega_k(\tau)},$$

which is equivalent to (12), and that suffices to conclude the proof.

It is evident from (12) that

$$\mu_{c_1\mathfrak{p}_1 + c_2\mathfrak{p}_2} = c_1\mu_{\mathfrak{p}_1} + c_2\mu_{\mathfrak{p}_2} \tag{19}$$

for any differential principal parts $\mathfrak{p}_1$ and $\mathfrak{p}_2$ of the second kind and any complex constants c_1, c_2. By construction, the meromorphic abelian differential $\mu_{\mathfrak{p}}(z)$ and the holomorphic abelian differential $\omega_{\mathfrak{p}}(z)$ are determined intrinsically and uniquely by the differential principal part $\mathfrak{p}$, independently of the choice of bases for the holomorphic abelian differentials or the homology of the surface M; it is easy to verify that directly, and it suffices to do so just for the holomorphic abelian differential (18). It is convenient to use matrix notation, so let $\mathbf{w}(z)$ be the column vector with entries $w_k(z)$ and $\boldsymbol{\omega}(z)$ be the column vector with entries $\omega_j(z)$, and then the effect (7) of a change of bases (6) is $\mathbf{w}(z) = C\mathbf{w}^*(z)$ and $\boldsymbol{\omega}(z) = C\boldsymbol{\omega}^*(z)$ while $G = {}^tC^{-1}\, G^*\, \overline{G^{-1}}$. Therefore,

$$\omega_{\mathfrak{p}}(z) = -2\pi \sum_{a \in M} {}^t\boldsymbol{\omega}(z)\, G\, \overline{\mathrm{res}_a\, {}^t(\mathbf{w}(z)\mathfrak{p})}$$

$$= -2\pi \sum_{a \in M} {}^t\boldsymbol{\omega}^*(z)\, {}^tC \cdot {}^tC^{-1} G^* \overline{C^{-1}} \cdot \overline{C\ \mathrm{res}_a\, {}^t(\mathbf{w}^*(z)\mathfrak{p})}$$

$$= -2\pi \sum_{a \in M} {}^t\boldsymbol{\omega}^*(z)\, G^*\ \overline{\mathrm{res}_a\, {}^t(\mathbf{w}^*(z)\mathfrak{p})},$$

exhibiting the invariance of the formula for the holomorphic abelian differential $\omega_{\mathfrak{p}}(z)$ under changes of the bases.

On a pointed Riemann surface M, a Riemann surface with a specified base point $a \in M$, a differential principal part $\mathfrak{p}_a$ having a double pole with zero residue at the point a is determined uniquely up to a constant factor by the point a alone; hence, the meromorphic abelian differential of the second kind $\mu_{\mathfrak{p}_a}(z)$ and the associated holomorphic abelian differential $\omega_{\mathfrak{p}_a}(z)$ are determined uniquely up to a constant factor by the base point a, independently of the choice of bases for the homology or the holomorphic abelian differentials on M, so with that understanding they can be denoted by $\mu_a(z)$ and $\omega_a(z)$. By Theorem 1 (iii), the holomorphic abelian differential $\omega_a(z)$ is

$$\omega_a(z) = -2\pi \sum_{i,j=1}^{g} g_{ji}\overline{w_i'(a)}\omega_j(z) \tag{20}$$

for the matrix $G = \{g_{ij}\}$ in terms of any bases τ_j for the homology of M, where $\omega_i(z) = dw_i(z)$ for the holomorphic abelian differentials on M. The derivative $w_i'(a)$ depends on the choice of a local coordinate near the point a, but only up to a constant factor. The holomorphic abelian differential $\omega_a(z)$ is thus a conjugate holomorphic function of the point $a \in M$, and the mapping that associates to the point $a \in M$ the conjugate holomorphic abelian differential $\overline{\omega_a(z)}$ is a well-defined holomorphic mapping from M to the $(g-1)$-dimensional projective space associated to the space of conjugate holomorphic differentials on M; indeed, since the matrix G is nonsingular, it is evident from (20) that this is equivalent to the canonical mapping of M into the $(g-1)$-dimensional projective space. Some further properties of this special case will be discussed in Sect. 4.

3 Meromorphic Abelian Differentials of the Third Kind

The differential principal part $\mathfrak{p}_{a_+,a_-}$ is defined as having a simple pole at the point $a_+ \in M$ with residue $+1$ and a simple pole at the point $a_- \in M$ with residue -1, so it is described uniquely by the ordered pair of points (a_+, a_-) in M. The residues at these two points are nonzero, so $\mathfrak{p}_{a_+,a_-}$ is a differential principal part of the third kind; but the sum of the residues at these two points is zero, so there are meromorphic abelian differentials on M having the principal part $\mathfrak{p}_{a_+,a_-}$. It is possible to determine one of these differentials uniquely and intrinsically through its period class, just as for the meromorphic abelian differentials of the second kind, but to do so requires a bit of care even to define the period class. If $\nu(z)$ is a meromorphic abelian differential on M with the differential principal part $\mathfrak{p}_{a_+,a_-}$, and if δ is a simple path on M from the point a_- to the point a_+, then $\nu(z)$ when viewed as a Γ-invariant differential form on $\widetilde{M}$ is a holomorphic differential form on the inverse image $\widetilde{M}_\delta = \pi^{-1}(M \sim \delta) \subset \widetilde{M}$ of the complement of the path $\delta \subset M$. The integral of this holomorphic differential form around any closed path in $\widetilde{M}_\delta$ is zero, since the image of any such path on M is a closed path on $M \sim \delta$, hence a path that has the same winding number around the point a_- as around the pole a_+; therefore, for any fixed point $z_0 \in \widetilde{M}_\delta$, the integral $\upsilon(z) = \int_{z_0}^{z} \nu$ is a well- defined holomorphic function on $\widetilde{M}_\delta$. For any covering translation $T \in \Gamma$, the difference $\upsilon(Tz) - \upsilon(z) = \nu(\delta, T)$ is a constant since $d\,\upsilon(z)$ is Γ-invariant; the mapping that associates to the covering translation T the complex number $\nu(\delta, T)$ is a group homomorphism $\nu(\delta) : \Gamma \longrightarrow \mathbb{C}$ that is defined as the period class of the meromorphic abelian differential $\nu(z)$ with respect to the path δ. The period class can be viewed alternatively as a homomorphism $\nu(\delta) : H_1(M) \longrightarrow \mathbb{C}$, and the period $\nu(\delta, \tau)$ for a homology class $\tau \in H_1(M)$ can be identified with the integral $\int_\tau \upsilon(z)$ along any path in $M \sim \delta$ that represents that homology class.

Theorem 2. (i) *For any simple path δ from a point a_- to a point a_+ on a compact Riemann surface M of genus $g > 0$, there are a unique meromorphic abelian differential of the third kind $\nu_\delta(z)$ and a unique holomorphic abelian differential $\omega_\delta(z)$ such that $\nu_\delta(z)$ has the differential principal part $\mathfrak{p}_{a_+,a_-}$ and the period class of $\nu_\delta(z)$ with respect to the path δ is equal to the period class of the complex conjugate differential $\overline{\omega_\delta(z)}$.*

(ii) *The holomorphic abelian differential $\omega_\delta(z)$ is characterized by*

$$\int_M \omega(z) \wedge \overline{\omega_\delta(z)} = 2\pi\mathrm{i} \int_\delta \omega(z) \tag{21}$$

for all holomorphic abelian differentials $\omega(z)$ on M.

(iii) *In terms of any bases $\omega_i(z)$ and τ_j, the differential form $\omega_\delta(z)$ is given by*

$$\omega_\delta(z) = -2\pi \sum_{i,j=1}^{g} g_{ji} \left(\int_\delta \overline{\omega_i(z)} \right) \omega_j(z) \tag{22}$$

where $G = \{g_{ji}\}$ is the matrix (2).

(iv) *The period class of the meromorphic abelian differential $\nu_\delta(z)$ with respect to the path δ is given by*

$$\nu_\delta(\delta, T) = -2\pi \sum_{i,j=1}^{g} g_{ij} \left(\int_\delta \omega_i(z) \right) \overline{\omega_j(T)} \tag{23}$$

for any $T \in \Gamma$.

Proof. (i) If $\nu(z)$ is a meromorphic abelian differential with the differential principal part $\mathfrak{p}_{a_+,a_-}$, then $\nu(z) + \omega(z)$ is a meromorphic abelian differential with the differential principal part $\mathfrak{p}_{a_+,a_-}$ for any holomorphic abelian differential $\omega(z)$, and all the meromorphic abelian differentials with the differential principal part $\mathfrak{p}_{a_+,a_-}$ arise in this way. There is a unique holomorphic abelian differential $\omega(z)$ such that the period vector $\nu_\delta(\delta, \tau_j) = \{\nu(\delta, \tau_j) + \omega(\tau_j)\}$ of the meromorphic differential form $\nu_\delta(z) = \nu(z) + \omega(z)$ with respect to the path δ is contained in the linear subspace $\overline{{}^t\Omega}\mathbb{C}^g \subset \mathbb{C}^{2g}$ in the direct sum decomposition (5), hence such that $\nu_\delta(\delta, \tau_j) = \overline{\omega_\delta(\tau_j)}$ for a uniquely determined holomorphic abelian differential $\omega_\delta(z)$.

(ii) Choose a connected component $\widetilde{\delta} \subset \widetilde{M}$ of the inverse image $\pi^{-1}(\delta) \subset \widetilde{M}$, which must be a simple path from a point $\tilde{a}_- \in \widetilde{M}$ to a point $\tilde{a}_+ \in \widetilde{M}$ where $\pi(\tilde{a}_-) = a_-$ and $\pi(\tilde{a}_+) = a_+$. In addition, choose a contractible open neighborhood U of the path δ in M and let $\widetilde{U} \subset \widetilde{M}$ be that connected component of $\pi^{-1}(U) \subset \widetilde{M}$ containing $\widetilde{\delta}$. Then $\pi^{-1}(\delta) = \Gamma\widetilde{\delta}$ and $\pi^{-1}(U) = \Gamma\widetilde{U}$ are Γ-invariant subsets of $\widetilde{M}$. Let $\tilde{v}_\delta(z)$ be a C^∞ modification of the function $v_\delta(z) = \int_{z_0}^{z} \nu$ in $\widetilde{U}$, the result of multiplying the function $v_\delta(z)$ by a C^∞ function that vanishes in an open neighborhood of $\widetilde{\delta}$ and is identically equal to 1 in an open neighborhood of the boundary of $\widetilde{U}$, and extend this modification to all the subsets $\Gamma\widetilde{U}$ so that $\tilde{v}_\delta(Tz) = \tilde{v}_\delta(z) + \nu(\delta, T)$ for all $T \in \Gamma$. The differential form $\tilde{\nu}_\delta(z) = \mathrm{d}\tilde{v}_\delta(z)$ then is a C^∞ closed Γ-invariant differential 1-form on $\widetilde{M}$, so it can be viewed as a C^∞ differential 1-form on M. This differential form has the same periods as $\nu_\delta(z)$ with respect to δ, so has the same periods as $\overline{\omega_\delta(z)}$, as was demonstrated in the proof of part (i); hence, by Lemma 1,

$$\int_M \omega(z) \wedge \overline{\omega_\delta(z)} = \int_M \omega(z) \wedge \tilde{\nu}_\delta(z) \tag{24}$$

for all holomorphic abelian differentials $\omega(z)$. The exterior product $\omega(z)\wedge\tilde{\nu}_\delta(z)$ vanishes outside the open set $U \subset M$ since the differential forms $\omega(z)$ and $\tilde{\nu}_\delta(z)$ are both holomorphic 1-forms there; and the differential forms $\nu(z)$ and $\tilde{\nu}_\delta(z)$ agree on the boundary ∂U of the set $U \subset M$. Then if $w(z)$ is a holomorphic function on $\widetilde{M}$ such that $\mathrm{d}w(z) = \omega(z)$, it follows from Stokes's theorem and the Cauchy integral formula on $\widetilde{M}$ that

$$\begin{aligned}
\int_M \omega(z) \wedge \tilde{\nu}_\delta(z) &= \int_U \omega(z) \wedge \tilde{\nu}_\delta(z) = \int_{\widetilde{U}} \mathrm{d}\big(w(z)\,\tilde{\nu}_\delta(z)\big) \\
&= \int_{\partial\widetilde{U}} w(z)\,\tilde{\nu}_\delta(z) = \int_{\partial\widetilde{U}} w(z)\,\nu_\delta(z) \\
&= 2\pi\mathrm{i} \sum_{p\in\widetilde{U}} \mathrm{res}_p\big(w(z)\,\mathfrak{p}_{\tilde{a}_+,\tilde{a}_-}\big) = 2\pi\mathrm{i}\big(w(\tilde{a}_+) - w(\tilde{a}_-)\big) \\
&= 2\pi\mathrm{i} \int_{\tilde{\delta}} \omega(z) = 2\pi\mathrm{i} \int_\delta \omega(z).
\end{aligned} \tag{25}$$

It then follows from (24) and (25) that the differential form $\omega_\delta(z)$ satisfies (21). For any choice of bases $\omega_i(z)$ and τ_j and for any holomorphic abelian differential $\omega(z) = \sum_{l=1}^{g} c_l \omega_l(z)$, it follows from Lemma 2 and (2) that

$$\begin{aligned}
i \sum_{k=1}^{g} g_{kj} \int_M \omega_k(z) \wedge \overline{\omega(z)} &= i \sum_{k,l=1}^{g} g_{kj} \overline{c_l} \int_M \omega_k(z) \wedge \overline{\omega_l(z)} \\
&= \sum_{k,l=1}^{g} g_{kj} \overline{c_l} h_{kl} = \sum_{l=1}^{g} \delta_l^j \overline{c_l} = \overline{c_j};
\end{aligned} \tag{26}$$

consequently, (21) fully determines the differential form $\omega_\delta(z)$.

(iii) In particular, if $\omega_\delta(z) = \sum_{j=1}^{g} c_j \omega_j(z)$, it follows from (26) and (21) that

$$\overline{c_j} = i \sum_{k=1}^{g} g_{kj} \int_M \omega_k(z) \wedge \overline{\omega_\delta(z)} = -2\pi \sum_{k=1}^{g} g_{kj} \int_\delta \omega_k(z) \tag{27}$$

and consequently that

$$\omega_\delta(z) = -2\pi \sum_{k,l=1}^{g} g_{jk} \left(\int_\delta \overline{\omega_k(z)} \right) \omega_j(z). \tag{28}$$

(iv) Finally, if $\omega_\delta(z) = \sum_{j=1}^{g} c_j \omega_j(z)$, then for any homology class $\tau \in H_1(M)$, it follows from (28) that

$$\nu_\delta(\delta, \tau) = \overline{\omega_\delta(\tau)} = \sum_{j=1}^{g} \overline{c}_j \overline{\omega_j(\tau)} = -2\pi \sum_{j,k=1}^{g} g_{kj} \left(\int_\delta \omega_k(z) \right) \overline{\omega_j(\tau)},$$

which is equivalent to (23), and that suffices to conclude the proof.

By construction, the meromorphic abelian differential $\nu_\delta(z)$ and the holomorphic abelian differential $\omega_\delta(z)$ are determined intrinsically and uniquely by the differential principal part $\mathfrak{p}_{a_+,a_-}$ and the path δ from a_- to a_+; that can be verified directly, just as for the meromorphic abelian differential of the second kind. These differentials though actually depend only on the homotopy class of the path δ. To see that, for any choice of a point $z_- \in \widetilde{M}$ such that $\pi(z_-) = a_-$, there is a unique choice of a connected component $\widetilde{\delta}$ of $\pi^{-1}(\delta) \subset \widetilde{M}$ that begins at the point z_-, and the path $\widetilde{\delta}$ will end at a point $z_+ \in \widetilde{M}$ for which $\pi(z_+) = a_+$. If $\delta' \in M$ is another path from a_- to a_+ and is homotopic to δ, and if $\widetilde{\delta'}$ is the component of $\pi^{-1}(\delta') \subset \widetilde{M}$ that begins at the point z_-, then as is no doubt quite familiar $\widetilde{\delta'}$ also will end at the point z_+; and conversely, the image under the covering projection π of any path from z_- to z_+ in $\widetilde{M}$ will be a path in M that is homotopic to δ, since $\widetilde{M}$ is simply connected. Thus, a homotopy class of paths from a_- to a_+ is determined uniquely by a pair (z_+, z_-) of points in $\widetilde{M}$ for which $\pi(z_+) = a_+$ and $\pi(z_-) = a_-$. Moreover, for any covering translation $T \in \Gamma$, the pair of points (Tz_+, Tz_-) determines the same homotopy class as the pair of points (z_+, z_-). Altogether then, the set of homotopy classes of paths δ from a_- to a_+ on M can be identified with the set of equivalence classes of pairs (z_+, z_-) of points in $\widetilde{M}$ such that $\pi(z_+) = a_+$ and $\pi(z_-) = a_-$, under the equivalence relation $(z_+, z_-) \sim (Tz_+, Tz_-)$ for any $T \in \Gamma$. With this in mind, it is easy to see that the differentials $\nu_\delta(z)$ and $\omega_\delta(z)$ depend only on the homotopy class of the path δ; indeed, if $w_l(z)$ is any holomorphic function on $\widetilde{M}$ such that $dw_l(z) = \omega_l(z)$ and if the homotopy class of the path $\delta \in M$ is described by the pair of points (z_+, z_-), then $\int_\delta \omega_l(z) = \int_{\widetilde{\delta}} dw_l(z) = w_l(z_+) - w_l(z_-)$ and consequently (22) can be written

$$\omega_\delta(z) = -2\pi \sum_{l,j=1}^{g} g_{jl} \left(\overline{w_l(z_+)} - \overline{w_l(z_-)} \right) \omega_j(z), \tag{29}$$

showing that the holomorphic abelian differential $\omega_\delta(z)$ and consequently the meromorphic abelian differential $\nu_\delta(z)$ both depend only on the homotopy class of the path δ. Since these differentials are determined uniquely and intrinsically by the pair of points (z_+, z_-), they can be denoted alternatively by

$$\nu_\delta(z) = \nu_{z_+, z_-}(z), \quad \text{and} \quad \omega_\delta(z) = \omega_{z_+, z_-}(z); \tag{30}$$

and it is clear from the preceding discussion that

$$\nu_{Tz_+, Tz_-}(z) = \nu_{z_+, z_-}(z) \quad \text{and} \quad \omega_{Tz_+ Tz_-}(z) = \omega_{z_+ z_-}(z) \quad \text{for all} \quad T \in \Gamma. \tag{31}$$

The period class (23) also depends only on the pair of points (z_+, z_-) rather than on the path δ, so it can be denoted alternatively by $\nu_\delta(\delta, T) = \nu_{z_+, z_-}(T)$ and is given by

$$\nu_{z_+, z_-}(T) = -2\pi \sum_{i,j=1}^{g} g_{ij} \Big(w_i(z_+) - w_i(z_-) \Big) \overline{\omega_j(T)}. \tag{32}$$

On the other hand, for any base point $z_0 \in \tilde{M}$ for which $\pi(z_0) \in M \sim (a_+, a_-)$, the integral

$$v_\delta(z, z_0) = \int_{z_0}^{z} \nu_{z_+, z_-} \tag{33}$$

still is defined just in the open subset $\widetilde{M}_\delta \subset \widetilde{M}$, where for any point $z \in \widetilde{M}_\delta$ it is calculated by integration along any path from z_0 to z in $\widetilde{M}$ that is disjoint from $\pi^{-1}(\delta)$, but its exponential is well defined on the entire Riemann surface $\widetilde{M}$, independently of the choice of the path δ.

Theorem 3. (i) *For any choice of distinct points z_0, z_+, z_- in the universal covering space $\widetilde{M}$ of a compact Riemann surface M of genus $g > 0$, the function*

$$q(z, z_0; z_+, z_-) = \exp v_\delta(z, z_0) \tag{34}$$

of the variable $z \in \widetilde{M}_\delta$ extends to a uniquely and intrinsically defined function of the variable $z \in \widetilde{M}$ that has simple zeros at the points Γz_+, simple poles at the points Γz_-, takes the value 1 at the point z_0 and is otherwise holomorphic and nonvanishing on $\widetilde{M}$, and that satisfies

$$q(Tz, z_0; z_+, z_-) = q(z, z_0; z_+, z_-) \exp \nu_{z_+, z_-}(T) \tag{35}$$

for all $T \in \Gamma$.

(ii) *The extended function $q(z, z_0; z_+, z_-)$ is characterized completely by its divisor, its value at z_0, and the functional equations* (35).

Proof. (i) The function $q(z, z_0; z_+, z_-)$ is a uniquely and intrinsically defined holomorphic and nowhere vanishing function in the open subset $\widetilde{M}_\delta \subset \widetilde{M}$, since the integral $v_\delta(z, z_0)$ is a uniquely and intrinsically defined holomorphic function in $\widetilde{M}_\delta$, and $q(z_0, z_0; z_+, z_-) = 1$ since $v_\delta(z_0, z_0) = 0$. For any point,

$z \in \widetilde{M}_\delta$ by definition $v_\delta(z, z_0) = \int_\lambda \nu_{z_+, z_-}$ where λ is a path from z_0 to z that is disjoint from $\pi^{-1}(\delta)$. If λ^* is another path from z_- to z that avoids the singularities $\Gamma z_- \cup \Gamma z_+$ but may not be disjoint from $\pi^{-1}(\delta)$ otherwise, the integral $v_\delta^*(z, z_0) = \int_{\lambda^*} \nu_{z_-, z_+}$ still has a well-defined value. The difference $v_\delta^*(z, z_0) - v_\delta(z, z_0)$ is the integral around a closed path in $\widetilde{M}$, the value of which is $2\pi\mathrm{i}$ times the sum of the residues of the meromorphic abelian differential ν_{z_+, z_-} at the singularities inside that closed path and hence is $2\pi i n$ for some integer n since the differential form ν_{z_-, z_+} has residues ± 1 at each pole; consequently

$$\exp v_\delta^*(z, z_0) = \exp v_\delta(z, z_0), \tag{36}$$

showing that the function $q(z, z_0; z_+, z_-)$ is really independent of the choice of the path of integration defining the function $v_\delta(z, z_0)$ and hence is a well-defined holomorphic and nowhere vanishing function on $\widetilde{M} \sim (\Gamma z_+ \cup \Gamma z_-)$. Since the meromorphic abelian differential $\nu_{z_+, z_+}(z)$ has the periods $\nu_{z_+, z_-}(T)$, it follows that $v_\delta(Tz, z_0) = v_\delta(z, z_0) + \nu_{z_+, z_-}(T)$ for any covering translation $T \in \Gamma$ and consequently that the function $q(Tz, z_0; z_+, z_-)$ satisfies (35). In a local coordinate z in an open neighborhood of a_+ and centered at the point a_+, the differential form $\nu_{a_+, a_-}(z)$ has the differential principal part $z^{-1}\mathrm{d}z$ so its integral $v_\delta(z, z_0)$ differs from the local multiple-valued function $\log z$ by a holomorphic function and consequently $q(z, z_0; z_+, z_-) = \exp v_\delta(z, z_0)$ is holomorphic and has a simple zero at the point a_+. Correspondingly, in a local coordinate z in an open neighborhood of a_- and centered at the point a_-, the differential form $\nu_{a_+, a_-}(z)$ has the differential principal part $-z^{-1}\mathrm{d}z$ so its integral differs from the local multiple-valued function $-\log z$ by a holomorphic function and consequently $q(z, z_0; z_+, z_-) = \exp v_\delta(z, z_0)$ is meromorphic and has a simple pole at the point a_-. It then follows from (35) that $q(z, z_0; z_+, z_-)$ is meromorphic on the Riemann surface $\widetilde{M}$, has simple zeros at the points Γz_+, has simple poles at the points Γz_-, is nonzero at the other points of $\widetilde{M}$, and takes the value 1 at the point z_0.

(ii) If $q^*(z, z_0; z_+, z_-)$ is any meromorphic function of the variable $z \in \widetilde{M}$ that has the same divisor as $q(z, z_0; z_+, z_-)$ and also satisfies (35), then the quotient $q^*(z, z_0; z_+, z_-)/q(z, z_0; z_+, z_-)$ is a holomorphic and nowhere vanishing function on $\widetilde{M}$ that is invariant under the covering translation group and is therefore a nonzero constant, and if $q^*(z_0, z_0; z_+, z_-) = q(z_0, z_0; z_+, z_-) = 1$, that constant is 1. Consequently, the function $q(z, z_0; z_+, z_-)$ is uniquely determined by its divisor, its value at z_0, and the functional equations (35). That suffices for the proof.

The functional equations (35) exhibit the function $q(z, z_0; z_+, z_-)$ as a relatively automorphic function of the variable $z \in \widetilde{M}$ for the action of the covering translation group Γ on the universal covering space $\widetilde{M}$, for a factor of automorphy that has the form of a group homomorphism $T \longrightarrow \exp \nu_{z_+, z_-}(T)$ in $\mathrm{Hom}(\Gamma, \mathbb{C}^*)$, and that factor of automorphy is uniquely and intrinsically defined by the pair of points (z_+, z_-), since the period class $\nu_{z_+, z_-}(T)$ is uniquely and intrinsically defined by those points. If the Riemann surface M is not hyperelliptic (ii) of the preceding theorem can be strengthened to the assertion that the function $q(z,z_0;z_+,z_-)$ of the variable $z \in \widetilde{M}$ is characterized completely as a meromorphic relatively automorphic function for the factor of automorphy $\exp \nu_{z_+,z_-}(T)$ that takes the value 1 at the point z_0 and has as its singularities simple poles on $\widetilde{M}$ that represent a single point on the Riemann surface M, but where that point is not specified; for if there were another function $q^*(z,z_0;z_+,z_-)$ with the same properties, the quotient $q^*(z,z_0;z_+,z_-)/q(z,z_0;z_+,z_-)$ would be a meromorphic function of order 2 on M and hence M would be hyperelliptic. The factor of automorphy $\exp \nu_{z_+,z_-}(T)$ describes divisors of degree 0, or equivalently line bundles of characteristic class 0, on the Riemann surface M. There are other uniquely and intrinsically defined factors of automorphy describing divisors of nonzero degree, or equivalently line bundles of nonzero characteristic class, on M, leading to other uniquely and intrinsically defined functions on compact Riemann surfaces; the classification of these factors of automorphy and their relatively automorphic functions is a somewhat more complicated matter that is discussed in [10].

4 Duality

The holomorphic abelian differential $\omega_{\mathfrak{p}}(z)$ depends rather simply, explicitly, and analytically on the differential principal part $\mathfrak{p}$ as in (11), while the holomorphic abelian differential $\omega_{z_+,z_-}(z)$ depends even more simply, explicitly, and analytically on the points $z_+, z_- \in \widetilde{M}$ as in (29). As might be expected, the meromorphic abelian differentials $\mu_{\mathfrak{p}}(z)$ and $\mu_{z_+,z_-}(z)$ also depend rather simply and analytically, if not so explicitly, on the differential principal part $\mathfrak{p}$ and the pair of points (z_+, z_-), and that can be seen quite directly through natural dualities satisfied by these meromorphic abelian differentials.

Theorem 4. *If δ' is a simple path from a point a'_- to a point a'_+ and δ'' is a disjoint simple path from a point a''_- to a point a''_+ on a compact Riemann surface M of genus $g > 0$, then*

$$\int_{\delta''} \nu_{\delta'}(z) = \int_{\delta'} \nu_{\delta''}(z) \tag{37}$$

and therefore

$$q(a'_+, a'_-; a''_+, a''_-) = q(a''_+, a''_-; a'_+, a'_-). \tag{38}$$

Proof. For a fixed point $z_0 \in \widetilde{M}_{\delta'} \cap \widetilde{M}_{\delta''} \subset \widetilde{M}$, the integral $v_{\delta'}(z,z_0) = \int_{z_0}^{z} \nu_{\delta'}$ is a holomorphic function of the variable $z \in \widetilde{M}_{\delta'}$, and correspondingly, the integral $v_{\delta''}(z,z_0) = \int_{z_0}^{z} \nu_{\delta''}$ is a holomorphic function of the variable $z \in \widetilde{M}_{\delta''}$. Choose connected components $\widetilde{\delta}'$ and $\widetilde{\delta}''$ of the inverse images $\pi^{-1}(\delta')$ and $\pi^{-1}(\delta'')$ in $\widetilde{M}$ and disjoint contractible open neighborhoods U' and U'' of δ' and δ'', and let $\widetilde{U}'$ and $\widetilde{U}''$ be the components of the inverse images $\pi^{-1}(U')$ and $\pi^{-1}(U'')$ in $\widetilde{M}$ that contain the paths $\widetilde{\delta}'$ and $\widetilde{\delta}''$, respectively. Let $\tilde{v}_{\delta'}(z, z_0)$ be a C^∞ modification of the function $v_{\delta'}(z, z_0)$ in $\Gamma\tilde{U}'$ and $\tilde{v}_{\delta''}(z, z_0)$ be a C^∞ modification of the function $v_{\delta''}(z, z_0)$ in $\Gamma\widetilde{U}''$, as in the proof of Theorem 2, and introduce the C^∞ differential forms $\tilde{\nu}_{\delta'}(z) = \mathrm{d}v_{\delta'}(z, z_0)$ and $\tilde{\nu}_{\delta''}(z) = \mathrm{d}v_{\delta''}(z, z_0)$. Both are holomorphic differential forms of degree 1 outside the open subset $\Gamma\widetilde{U}' \cup \Gamma\widetilde{U}''$ of $\widetilde{M}$ so $\tilde{\nu}_{\delta'}(z) \wedge \tilde{\nu}_{\delta''}(z) = 0$ there. Then as in the proof of Theorem 2

$$
\begin{aligned}
\int_M \tilde{\nu}_{\delta'}(z) \wedge \tilde{\nu}_{\delta''}(z) &= \int_{U''} \tilde{\nu}_{\delta'}(z) \wedge \tilde{\nu}_{\delta''}(z) + \int_{U'} \tilde{\nu}_{\delta'}(z) \wedge \tilde{\nu}_{\delta''}(z) \\
&= \int_{\widetilde{U}''} \mathrm{d}\big(\tilde{v}_{\delta'}(z, z_0)\tilde{\nu}_{\delta''}(z)\big) - \int_{\widetilde{U}'} \mathrm{d}\big(\tilde{v}_{\delta''}(z, z_0)\tilde{\nu}_{\delta'}(z)\big) \\
&= \int_{\partial\widetilde{U}''} \tilde{v}_{\delta'}(z, z_0)\tilde{\nu}_{\delta''}(z) - \int_{\partial\widetilde{U}'} \tilde{v}_{\delta''}(z, z_0)\tilde{\nu}_{\delta'}(z) \\
&= \int_{\partial\widetilde{U}''} v_{\delta'}(z, z_0)\ \nu_{\delta''}(z) - \int_{\partial\widetilde{U}'} v_{\delta''}(z, z_0)\ \nu_{\delta'}(z) \\
&= 2\pi\mathrm{i}\Big(v_{\delta'}(z''_+, z_0) - v_{\delta'}(z''_-, z_0)\Big) \\
&\quad - 2\pi\mathrm{i}\Big(v_{\delta''}(z'_+, z_0) - v_{\delta''}(z'_-, z_0)\Big) \\
&= 2\pi\mathrm{i}\int_{\delta''} \nu_{\delta'} - 2\pi\mathrm{i}\int_{\delta'} \nu_{\delta''}
\end{aligned}
\tag{39}
$$

since $v_{\delta'}(z, z_0)$ is a holomorphic function in the contractible set $\widetilde{U}''$ and $v_{\delta''}(z, z_0)$ is a holomorphic function in the contractible set $\widetilde{U}'$. The closed C^∞ differential forms $\tilde{\nu}_{\delta'}(z)$ and $\overline{\omega_{\delta'}(z)}$ have the same periods, as do the closed C^∞ differential forms $\tilde{\nu}_{\delta''}(z)$ and $\overline{\omega_{\delta''}(z)}$; therefore, by Lemma 1

$$
\int_M \tilde{\nu}_{\delta'}(z) \wedge \tilde{\nu}_{\delta''}(z) = \int_M \overline{\omega_{\delta'}(z)} \wedge \overline{\omega_{\delta''}(z)}. \tag{40}
$$

The differential forms $\overline{\omega_{\delta'}(z)}$ and $\overline{\omega_{\delta''}(z)}$ are both conjugate holomorphic differentials, so their wedge product vanishes identically, and consequently,

$$
\int_M \overline{\omega_{\delta'}(z)} \wedge \overline{\omega_{\delta''}(z)} = 0. \tag{41}
$$

It then follows from (39), (40), and (41) that (37) holds. The functions $v_{\delta'}(z, z_)$) and $v_{\delta''}(z, z_0)$ are defined by the integrals (33); hence, it follows from (37) that

$$
v_{\delta'}(a''_+, a''_-) = \int_{\delta''} \nu_{\delta'} = \int_{\delta'} \nu_{\delta''} = v_{\delta''}(a'_-, a'_+) \tag{42}
$$

and consequently, in view of the definition (34), it follows further that

$$q(a'_+, a'_-; a''_+, a''_-) = \exp v_{\delta''}(a'_+, a'_-)$$
$$= \exp v_{\delta'}(a''_+, a''_-) = q(a''_+, a''_-; a'_+, a'_-), \tag{43}$$

which suffices for the proof.

Corollary 1. *The function $q(z_1, z_2; z_3, z_4)$ is a uniquely and intrinsically defined meromorphic function on the complex manifold $\widetilde{M}^4$ such that*

$$q(z_1, z_2; z_3, z_4) = q(z_3, z_4; z_1, z_2) = q(z_2, z_1; z_3, z_4)^{-1}; \tag{44}$$

it has first-order zeros along the subvarieties $z_1 = Tz_3$ and $z_2 = Tz_4$ and first-order poles along the subvarieties $z_1 = Tz_4$ and $z_2 = Tz_3$ for all $T \in \Gamma$ and is otherwise holomorphic and nonvanishing on $\widetilde{M}^4$.

Proof. By Theorem 3, the function $q(z_1, z_2; z_3, z_4)$ is a uniquely and intrinsically defined meromorphic function of the variable $z_1 \in \widetilde{M}$ for any choice of points $z_2, z_3, z_4 \in \widetilde{M}$ that represent distinct points of M. By Theorem 4

$$q(z_1, z_2; z_3, z_4) = q(z_3, z_4; z_1, z_2);$$

and since $v_\delta(z, z_0) = \int_{z_0}^{z} \nu_{z_+, z_-}$ by (33), then $v_\delta(z, z_0) = -\nu_\delta(z_0, z)$, so in view of (34)

$$q(z_1, z_2; z_3, z_4) = q(z_2, z_1; z_3, z_4)^{-1},$$

showing that (44) holds. It follows from these symmetries that $q(z_1, z_2; z_3, z_4)$ is a meromorphic function of each of its variables, so by Rothstein's theorem [12, 13], it is a meromorphic function on the complex manifold $\widetilde{M}^4$. Since this function has the zeros and poles in the variable z_1 as in Theorem 3 (i) and the symmetries (44), it follows that as a meromorphic function of the 4 variables $(z_1, z_2, z_3, z_4) \in \widetilde{M}^4$ it has the zeros and singularities as in the statement of the present corollary, and that suffices for the proof.

The function $q(z_1, z_2; z_3, z_4)$ is called the *intrinsic cross-ratio function* of the Riemann surface M, since it is uniquely and intrinsically defined on M and its analytic properties correspond to those of the classical cross-ratio function

$$q(z_1, z_2; z_3, z_4) = \frac{(z_1 - z_3)(z_2 - z_4)}{(z_1 - z_4)(z_2 - z_3)}$$

on the Riemann sphere $\mathbb{P}^1$. There are other normalizations of this function that are useful in various circumstances but that are not intrinsic to the Riemann surface M; the cross-ratio function with a standard normalization for a marked Riemann surface as in [9] was used by Farkas in [5] and Grant in [7], for instance. The function $q(z_1, z_2; z_3, z_4)$ is in many ways the basic uniquely and intrinsically defined meromorphic function on the Riemann surface M, for the intrinsically defined meromorphic abelian differentials on M described in the preceding discussion can be expressed quite simply in terms of the cross-ratio function. Indeed, it follows immediately from the definition (34) of the cross-ratio function and (33) that

$$\frac{\partial}{\partial z}\log q(z,z_0;z_+,z_-)dz = \frac{\partial}{\partial z}v_{z_+,z_-}(z,z_0)dz = \nu_{z_+,z_-}(z), \tag{45}$$

showing that the meromorphic abelian differential $\nu_{z_+,z_-}(z)$ also is a meromorphic function of the variables $(z_+,z_-)\in\widetilde{M}^2$. The corresponding result holds for differentiation with respect to the other variables z_0,z_+,z_-, in view of the symmetries (44); for instance

$$\frac{\partial}{\partial z_0}\log q(z,z_0;z_+,z_-)dz_0 = \frac{\partial}{\partial z_0}\log q(z_0,z;z_+,z_-)^{-1}dz_0 = -\nu_{z_+,z_-}(z_0). \tag{46}$$

Further results follow from a duality theorem between meromorphic abelian differentials of the second and third kinds.

Theorem 5. *If $\mathfrak{p}=\{\mathfrak{p}_{a_l}\}$ is a differential principal part of the second kind on a compact Riemann surface M and δ is a simple path on M from a point a_- to a point a_+ that avoids the points a_l, then for any point $z_0\in\widetilde{M}$ that is disjoint from the points $\pi^{-1}(a_l)$ and $\pi^{-1}(\delta)$,*

$$\int_\delta \mu_{\mathfrak{p}}(z) = \sum_l \operatorname{res}_{a_l}\left(v_\delta(z,z_0)\,\mathfrak{p}_{a_l}\right). \tag{47}$$

Proof. Choose disjoint coordinate discs Δ_l centered at the points a_l and a contractible open neighborhood U of the path δ that is disjoint from the discs Δ_l and choose points $\tilde{a}_l\in\pi^{-1}(a_l)$ and a connected component $\widetilde{\delta}\subset\pi^{-1}(\delta)$, so $\widetilde{\delta}$ is a path from a point $\tilde{a}_-$ to a point $\tilde{a}_+$ in $\widetilde{M}$ where $\pi(\tilde{a}_-)=a_-$ and $\pi(\tilde{a}_+)=a_+$. Let $\widetilde{\Delta}_l$ be the connected component of $\pi^{-1}(\Delta_l)$ containing the point $\tilde{a}_l$ and $\widetilde{U}$ be the connected component of $\pi^{-1}(U)$ in $\widetilde{M}$ containing $\widetilde{\delta}$. The integral $u_{\mathfrak{p}}(z)=\int_{z_0}^z\mu_{\mathfrak{p}}$ is a well-defined meromorphic function of the variable $z\in\widetilde{M}$ with poles at the points $\pi^{-1}(a_l)$ as in the proof of Theorem 1, and the integral $v_\delta(z)=\int_{z_0}^z\nu_\delta$ is a well-defined holomorphic function of the variable z in the open subset $\widetilde{M}_\delta=\pi^{-1}(M\sim\delta)\subset\widetilde{M}$ defined by integrating from z_0 to z along any path in $\widetilde{M}\sim\delta$, as in the proof of Theorem 1. Let $\tilde{u}_{\mathfrak{p}}(z)$ be a C^∞ modification of the function $u_{\mathfrak{p}}(z)$ in the discs $\Gamma\widetilde{\Delta}_l$ as in the proof of Theorem 1 and let $\tilde{v}_\delta(z)$ be a C^∞ modification of the function $v_\delta(z)$ in the set $\Gamma\widetilde{U}$ as in the proof of Theorem 2. By this construction, $\tilde{\mu}_{\mathfrak{p}}(z)$ has the same periods as $\mu_{\mathfrak{p}}(z)$, which in turn has the same periods as $\overline{\omega_{\mathfrak{p}}(z)}$, and correspondingly, for the differential forms $\tilde{\nu}_{\mathfrak{p}}(z)$, $\nu_{\mathfrak{p}}(z)$ and $\overline{\omega_\delta(z)}$; therefore, it follows from Lemma 1 that

$$\int_M \tilde{\mu}_{\mathfrak{p}}(z)\wedge\tilde{\nu}_\delta(z) = \int_M \overline{\omega_{\mathfrak{p}}(z)}\wedge\overline{\omega_\delta(z)} = 0 \tag{48}$$

since $\omega_{\mathfrak{p}}(z)$ and $\omega_{\delta}(z)$ are holomorphic differential 1-forms on M so their exterior product is identically zero. The exterior product $\tilde{\mu}_{\mathfrak{p}}(z) \wedge \tilde{\nu}_{\delta}(z)$ vanishes identically outside the sets U and Δ_a, where the two differentials are both conjugate holomorphic differential forms, so it follows as before that

$$\begin{aligned}\int_M \tilde{\mu}_{\mathfrak{p}}(z) \wedge \tilde{\nu}_{\delta}(z) &= \int_U \tilde{\mu}_{\mathfrak{p}}(z) \wedge \tilde{\nu}_{\delta}(z) + \sum_l \int_{\Delta_l} \tilde{\mu}_{\mathfrak{p}}(z) \wedge \tilde{\nu}_{\delta}(z) \\ &= \int_{\widetilde{U}} \mathrm{d}\left(\tilde{u}_{\mathfrak{p}}(z)\ \tilde{v}_{\delta}(z)\right) - \sum_l \int_{\widetilde{\Delta}_l} \mathrm{d}\left(\tilde{v}_{\delta}(z)\ \tilde{\mu}_{\mathfrak{p}}(z)\right) \\ &= \int_{\partial\widetilde{U}} \tilde{u}_{\mathfrak{p}}(z)\ \tilde{v}_{\delta}(z) - \sum_l \int_{\partial\widetilde{\Delta}_l} \tilde{v}_{\delta}(z)\, \tilde{\mu}_{\mathfrak{p}}(z) \\ &= \int_{\partial\widetilde{U}} u_{\mathfrak{p}}(z)\ v_{\delta}(z) - \sum_l \int_{\partial\widetilde{\Delta}_l} v_{\delta}(z)\mu_{a_l}(z) \qquad (49)\end{aligned}$$

since the C^{∞} modifications of the differentials and their integrals coincide with the original differentials and their integrals on the boundaries of the sets $\widetilde{U}$ and $\widetilde{\Delta}_l$. The function $u_{\mathfrak{p}}(z)$ is holomorphic in $\widetilde{U}$, while the abelian differential ν_{δ} has the principal part $\mathfrak{p}_{\tilde{a}_+,\tilde{a}_-}$ in $\widetilde{U}$, so

$$\int_{\partial\widetilde{U}} u_{\mathfrak{p}}(z)\, \nu_{\delta}(z) = 2\pi \mathrm{i}\Big(u_{\mathfrak{p}}(\tilde{a}_+) - u_{\mathfrak{p}}(\tilde{a}_-)\Big) = 2\pi \mathrm{i} \int_{\tilde{\delta}} \mu_{\mathfrak{p}}(z) = 2\pi \mathrm{i} \int_{\delta} \mu_{\mathfrak{p}}. \qquad (50)$$

On the other hand, the abelian differential of the second kind $\mu_{\mathfrak{p}}(z)$ is Γ-invariant, while the function $v_{\delta}(z, z_0)$ changes only by an additive constant under the action of the group Γ, so $\mathrm{res}_{\tilde{a}_l}\left(v_{\delta}(z, z_0)\mu_{\mathfrak{p}}(z)\right) = \mathrm{res}_{T\tilde{a}_l}\left(v_{\delta}(z, z_0)\mu_{\mathfrak{p}}(z)\right)$ for any covering translation $T \in \Gamma$ and consequently this residue can be calculated at the point $a_l \in M$; therefore

$$\int_{\partial\widetilde{\Delta}_l} v_{\delta}(z)\mu_{a_l}(z) = 2\pi \mathrm{i}\ \mathrm{res}_{a_l}\left(v_{\delta}(z)\mu_{\mathfrak{p}}(z)\right). \qquad (51)$$

It follows from (48)–(51) that (47) holds, and that suffices for the proof.

For example, if δ is a path in $\widetilde{M}$ from a point z_0 to the point Tz_0 for a covering translation $T \in \Gamma$, the integral $\int_{\delta} \mu_{\mathfrak{p}}(z)$ is just the period $\mu_{\mathfrak{p}}(T)$, so (47) yields an expression for these periods involving the residues of the meromorphic abelian differential $\nu_{\delta}(z)$, as an alternative to (12). The simplest special case of a differential principal part of the second kind is one with a nontrivial double pole at a single point $a \in M$, so has the form

$$\mathfrak{p} = \mathfrak{p}_{a,t} = \frac{1}{t^2}\, \mathrm{d}t \qquad (52)$$

in terms of a local coordinate t in M centered at the point a. The description of this differential principal part specifies not just the point $a \in M$ but also the local coordinate t; a change in the local coordinate changes the differential principal part by a constant factor. For this principal part (11) takes the form

$$\omega_{\mathfrak{p}}(z) = -2\pi \sum_{i,j=1}^{g} g_{ji}\,\overline{w_i'(a)}\omega_j(z), \tag{53}$$

which involves the derivative $w_i'(t)$ of the function $w_i(t)$ with respect to the local coordinate t evaluated at the point a so depends not just on the point a but also on the local coordinate t. The associated differential form $\omega(t) = w'(t)dt$ though is independent of the choice of the local coordinate t, so when the meromorphic abelian differential $\omega_{\mathfrak{p}}(z)$ is viewed as a meromorphic differential form both in the variable $z \in M$ and in the variable $a \in M$, so as the differential 2-form $\omega_t(z) \wedge dt$ on M^2, the preceding equation can be written

$$\omega_t(z) \wedge \overline{dt} = -2\pi \sum_{i,j=1}^{g} g_{ji}\omega_j(z) \wedge \overline{\omega_i(t)}. \tag{54}$$

There is a corresponding interpretation for the duality relation (47); for the differential principal part (52) again and for a path δ from a_- to a_+ that avoids the point a, it follows from (47) that

$$\begin{aligned} u_{\mathfrak{p}}(a_+) - u_{\mathfrak{p}}(a_-) &= \int_{a_-}^{a_+} u_{\mathfrak{p}}'(z)\,dz = \int_{\delta} \mu_{\mathfrak{p}}(z) \\ &= \operatorname{res}_a\Big(v_\delta(z, z_0)\,\mathfrak{p}\Big) = \frac{\partial}{\partial t} v_\delta(t, z_0)\Big|_{t=a}. \end{aligned} \tag{55}$$

When the meromorphic function $u_{\mathfrak{p}}(z)$ of the variable $z \in \widetilde{M}$ is viewed also as a differential form $u_t(z)dt$ in the variable $t \in \widetilde{M}$, this can be rewritten

$$u_t(t_+)\,dt - u_t(t_-)\,dt = \frac{\partial}{\partial t} v_\delta(t, z_0)\,dt = \nu_{t_+,t_-}(t) \tag{56}$$

where the points a_+ and a_- in $\widetilde{M}$ are described by the local coordinates t_+ and t_- in $\widetilde{M}$. Here $u_t(t_+)\,dt$ is a meromorphic function of the variable $t_+ \in \widetilde{M}$ and a meromorphic differential form in the variable $t \in \widetilde{M}$, and consequently, the meromorphic abelian differential $\nu_{t_+,t_-}(t)$ in the variable $t \in \widetilde{M}$ is a meromorphic function of the variables $t_+, t_- \in \widetilde{M}$. What is also interesting is that as function of the variables t_+, t_-, the abelian differential $\nu_{t_+,t_-}(t)$ can be decomposed into the sum of differential forms that are functions of the separate variables t_+ and t_-, as in (56). The exterior derivative of the function $u_t(z)$ of the variable z is the meromorphic abelian differential of the second kind with the differential principal part (52), which can be denoted correspondingly by $\mu_t(z)$; consequently, it follows from (56) that

$$\mu_t(t_+) \wedge dt = \frac{\partial}{\partial t_+}\Big(u_{\mathfrak{p}}(t_+) - u_{\mathfrak{p}}(t_-)\Big)dt_+ \wedge dt = -\frac{\partial}{\partial t_+}\nu_{t_+,t_-}(t) \wedge dt_+, \tag{57}$$

a relation between these two intrinsic meromorphic abelian differentials on M. This can be expressed in terms of the intrinsic cross-ratio function by using (45), so that

$$\mu_t(t_+) \wedge dt = -\frac{\partial}{\partial t_+}\Big(\frac{\partial}{\partial t}\log q(t, t_0; t_+, t_-)\Big)dt \wedge dt_+ \tag{58}$$

or after a change of notation

$$\mu_t(z) \wedge dt = \frac{\partial^2}{\partial z \partial t}\log q(t, t_0; z, z_0)\ dz \wedge dt. \tag{59}$$

From the symmetry (38) of the cross-ratio function, it follows that

$$\mu_t(z) \wedge dt = -\mu_z(t) \wedge dz. \tag{60}$$

This differential form is called the *intrinsic double differential* on the Riemann surface M, since it is uniquely and intrinsically defined as a meromorphic differential form on M^2 with a double pole along the diagonal $D = \{ (z,t) \in M^2 \mid z = t \}$; as a function of one variable for the other variable fixed, it is the meromorphic abelian differential of the second kind with a single double pole, as in Theorem 1. Any meromorphic abelian differential can be written as the sum of basic meromorphic abelian differentials of the third kind and meromorphic abelian differentials of the second kind associated to the singularities, thus providing an intrinsic meromorphic abelian differential with the specified singularities.

The explicit invariant forms of the intrinsic meromorphic abelian differentials suggest an alternative normalization of the holomorphic abelian differentials. A change (6) of the basis $\omega_i(z)$ by a matrix $C \in \mathrm{Gl}(g,\mathbb{C})$ has the effect (7) on the matrix G, so it is possible in this way to choose a basis $\omega_j(z)$ for which G and H are normalized to have the form

$$G = H = \mathrm{I}, \quad \text{the } g \times g \text{ identity matrix;} \tag{61}$$

that normalization is not unique but is preserved by further changes (6) in the basis $\omega_j(z)$ by arbitrary unitary matrices $C \in U(g) \subset \mathrm{Gl}(g,\mathbb{C})$. By Lemma 2, this normalization of the holomorphic abelian differentials amounts to the condition that

$$\int_M \omega_j(z) \wedge \overline{\omega_k(z)} = -i\ \delta^j_k. \tag{62}$$

With this normalization, the explicit formulas derived for the intrinsic meromorphic abelian differentials are simplified by replacing g_{ij} by δ^i_j throughout.

References

1. Leon Ehrenpreis, *Hypergeometric functions*, Algebraic Analysis, vol. I, pp. 85–128, Academic Press, 1988.
2. Leon Ehrenpreis, *Special functions*, Inverse Prob. Imaging, 4 (2010), pp. 639–643.
3. Leon Ehrenpreis, *The Schottky relation on genus 4 curves*, Curves, Jacobians and Abelian varieties, pp. 139- 160. American Mathematical Society, 1992.
4. Hershel Farkas and Irwin Kra, *Riemann Surfaces*, Springer-Verlag, 1980.
5. Hershel Farkas,"The Trisecant Formula and Hyperelliptic Surfaces", in *Curves, Jacobians and Abelian Varieties*, Contemp. Math. 136, American Mathematical Society.
6. Otto Forster, *Lectures on Riemann Surfaces*, Springer-Verlag, 1981.
7. David Grant, "On Gunning's prime form in genus 2," *Canad. Math. Bull*, vol. 45 (2002), pp. 89–96.
8. Phillip Griffiths and Joseph Harris, *Principles of Algebraic Geometry*, Wiley 1978.
9. Robert C. Gunning, *Riemann Surfaces and Generalized Theta Functions*, Springer-Verlag, 1976.
10. Robert C. Gunning, *Function Theory on Compact Riemann Surfaces*, to appear.
11. Rick Miranda, *Algebraic Curves and Riemann Surfaces*, AmericanMathematical Society,1991.
12. W. Rothstein, "Ein neuer Beweis des Hartogsschen Hauptsatzes und eine Ausdehnung auf meromorphe Funktionen," *Math. Zeit.*, vol. 53 (1950), pp. 84–95.
13. Y.-T. Siu, *Techniques of extension of analytic objects*, Lecture notes in pure and applied mathematics, 8, Dekker, 1974.

The Parallel Refractor

Cristian E. Gutiérrez and Federico Tournier

Dedicated to the memory of Leon Ehrenpreis

Abstract Given two homogenous and isotropic media I and II with different refractive indices n_I and n_{II}, respectively, we have a source Ω surrounded by media I and a target screen Σ surrounded by media II. We prove existence of interface surfaces between the media that refract collimated radiation emanating from Ω into Σ with prescribed input and output intensities.

Key words Geometric optics • Optimization • Refraction

Mathematics Subject Classification (2010): 78A05, 35Q60

1 Introduction

The problem considered in this chapter is the following: Suppose we have a domain $\Omega \subset \mathbb{R}^{n-1}$ and a domain Σ contained in an $n-1$-dimensional surface in $\mathbb{R}^n$; Σ is referred to as the target domain or screen to be illuminated (for the practical application, one can think that $n = 3$). Let n_1 and n_2 be the indexes of refraction of two homogeneous and isotropic media I and II, respectively, and suppose that

C.E. Gutiérrez (✉)
Department of Mathematics, Temple University, Philadelphia, PA 19122, USA
e-mail: gutierre@temple.edu

F. Tournier
Instituto Argentino de Matemática, CONICET, Saavedra 15, Buenos Aires (1083), Argentina
e-mail: f-tournier@hotmail.com

H.M. Farkas et al. (eds.), *From Fourier Analysis and Number Theory to Radon Transforms and Geometry*, Developments in Mathematics 28, DOI 10.1007/978-1-4614-4075-8_14,

from the region Ω surrounded by medium I radiation emanates in the direction e_n with intensity $f(x)$ for $x \in \Omega$, and Σ is surrounded by media II. That is, all emanating rays from Ω are collimated. We seek an optical surface $\mathcal{R}$ interface between media I and II, such that all rays refracted by $\mathcal{R}$ into medium II are received at the surface Σ, and the prescribed radiation intensity received at each point $p \in \Sigma$ is $g(p)$. Of course, some conditions on the relative position of Σ and Ω are needed so rays can be refracted to Σ, see conditions (A) and (B) below. Assuming no loss of energy in this process, we have the conservation of energy equation $\int_\Omega f(x)\,\mathrm{d}x = \int_\Sigma g(p)\,\mathrm{d}p$.

The purpose of this chapter is to show the existence of the interface surface $\mathcal{R}$ solving this problem under general conditions on Ω and Σ, and also when g is a Radon measure in D. This implies that one can design a lens refracting a collimated light beam emanating from Ω so that the screen Σ is illuminated in a prescribed way. The lens is bounded by two optical surfaces, the "upper" surface is $\mathcal{R}$ and the "lower" one is a plane perpendicular to e_n.

From the reversibility of the optical paths, we obtain that the surface $\mathcal{R}$ refracts radiation emanating from a surface in $\mathbb{R}^n$ into collimated rays hitting Ω. In particular, we construct an optical surface that refracts radiation emanating from a finite number of sources into a beam of collimated rays.

Our construction uses ideas from [GH09] involving ellipsoids of revolution and where the far field problem is solved when radiation emanates from a source point. However, the method used in the present chapter is different from the mass transportation methods used in [GH09]. We first solve the case when the target is a finite set of points and then construct the solution in the general case by approximation. An essential fact used is that an ellipsoid of revolution separating media I and II, and of eccentricity related to the indices of refraction of the media, refracts all radiation emanating from a focus into a collimated beam parallel to the axis of the ellipsoid. This is a consequence of the Snell law of refraction written in vector form, see [GH09, Sect. 2].

Throughout the chapter, we assume that media II is denser than media I, that is, $\kappa := \dfrac{n_1}{n_2} < 1$. The case when $\kappa > 1$ can be treated in a similar way but the geometry of the surface changes. One needs to use hyperboloids of revolution instead of ellipsoids as it is indicated in detail in [GH09].

2 Definitions and Preliminaries

We work with ellipsoids of the form $|x| = -kx_n + b$ which can be written as

$$\frac{|x'|^2}{\dfrac{b^2}{1-k^2}} + \frac{\left(x_n + \dfrac{kb}{1-k^2}\right)^2}{\dfrac{b^2}{(1-k^2)^2}} = 1,$$

where $x = (x', x_n)$. This is the equation of an ellipsoid of revolution about the x_n-axis with foci $(0,0)$ and $(0, -2\kappa b/(1-\kappa^2))$. If the focus at $(0,0)$ is moved to the point $p = (p', p_n)$, then the corresponding ellipsoid can be written as

$$\frac{|x'-p'|^2}{\frac{b^2}{1-k^2}} + \frac{\left(x_n - \left(p_n - \frac{kb}{1-k^2}\right)\right)^2}{\frac{b^2}{(1-k^2)^2}} = 1; \tag{2.1}$$

let us denote this ellipsoid by $E_{p,b}$.

We consider the lower part of the ellipsoid as the graph of the function $\phi_{p,b}$, that is, we let

$$\phi_{p,b}(x') = p_n - \frac{kb}{1-k^2} - \sqrt{\frac{b^2}{(1-k^2)^2} - \frac{|x'-p'|^2}{1-k^2}}.$$

The reason to look at the lower part of the ellipsoid is that this is the only part that refracts rays parallel to e_n into the point p, see [GH09, Sect. 2.2]. $\phi_{p,b}(x')$ is defined for $|x'-p'| \leq b/\sqrt{1-\kappa^2}$, that is, on the ball $B_{b/\sqrt{1-\kappa^2}}(p')$.

We fix two constants $0 < C_1 < C_2$ and we consider a target set $\Sigma \subseteq \mathbb{R}^n$ such that

$$\Sigma \subseteq \left\{(p', p_n) : C_1 \leq p_n \leq C_2\right\}. \tag{2.2}$$

We also consider a domain $\Omega \subseteq \mathbb{R}^{n-1} = \{(x', x_n) : x_n = 0\}$.

For $p \in \Sigma$, we will consider $\phi_{p,b}$ with $\frac{p_n(1-k^2)}{k} \leq b \leq \frac{C_2(1-k^2)(1+k)^2}{k^3}$.

We make two assumptions regarding Σ and Ω.

(A) We assume that there exists $0 < \delta < 1$ such that $\Omega \subseteq B_{\delta p_n \sqrt{1-\kappa^2}/k}(p')$ for all $p \in \Sigma$. This hypothesis implies that for all $p \in \Sigma$ and $b \geq \frac{p_n(1-k^2)}{k}$, $\phi_{p,b}$ is defined and $\phi_{p,b} \leq 0$ in $\bar{\Omega}$.

(B) This is a visibility condition. Set $M = C_2\left(\frac{1+k}{k}\right)^3 - C_1$. We assume that for all $x \in \bar{\Omega} \times [0, -M]$ and for all $m \in S^{n-1}$, the ray $\{x + tm : t > 0\}$ intersects Σ in at most one point.

We remark that the first condition is equivalent to the assumption that there exists $0 < \beta < 1$ such that $\left\langle -e_n, \frac{x-p}{|x-p|}\right\rangle \geq \beta$ for all $p \in \Sigma$ and for all $x \in \bar{\Omega}$.

We now define a parallel refractor with respect to Σ and Ω.

Definition 2.1. We say a function $u : \bar{\Omega} \longrightarrow \mathbb{R}$ is a parallel refractor if for all $\bar{x} \in \bar{\Omega}$, there exists $p \in \Sigma$ and $b \geq \frac{p_n(1-k^2)}{k}$ such that $\phi_{p,b}(\bar{x}) = u(\bar{x})$ and $\phi_{p,b}(x') \geq u(x')$ for all $x' \in \bar{\Omega}$. That is, $\phi_{p,b}$ touches u from above at $\bar{x}$ in $\bar{\Omega}$. In this case, we say $p \in \mathcal{N}_u(\bar{x})$ or that $\bar{x} \in \mathcal{T}_u(p)$.

We first notice the following:

Lemma 2.2. *If u is a parallel refractor, then u is Lipschitz in $\bar{\Omega}$.*

Proof. Let $x, \bar{x} \in \bar{\Omega}$ and let $p \in \mathcal{N}_u(\bar{x})$. There exists $b \geq \frac{p_n(1-k^2)}{k}$ such that $u(x) \leq \phi_{p,b}(x)$ for all $x \in \bar{\Omega}$ with equality at $\bar{x}$. It follows that

$$\begin{aligned}
u(x) - u(\bar{x}) &\leq \phi_{p,b}(x) - \phi_{p,b}(\bar{x}) \\
&= \sqrt{\frac{b^2}{(1-k^2)^2} - \frac{|x-p'|^2}{1-k^2}} - \sqrt{\frac{b^2}{(1-k^2)^2} - \frac{|\bar{x}-p'|^2}{1-k^2}} \\
&= \frac{|x-p'|^2 - |\bar{x}-p'|^2}{(1-k^2)\left(\sqrt{\frac{b^2}{(1-k^2)^2} - \frac{|x-p'|^2}{1-k^2}} + \sqrt{\frac{b^2}{(1-k^2)^2} - \frac{|\bar{x}-p'|^2}{1-k^2}}\right)} \\
&= \frac{2\langle \xi - p', x - \bar{x}\rangle}{(1-k^2)\left(\sqrt{\frac{b^2}{(1-k^2)^2} - \frac{|x-p'|^2}{1-k^2}} + \sqrt{\frac{b^2}{(1-k^2)^2} - \frac{|\bar{x}-p'|^2}{1-k^2}}\right)} \\
&\leq \frac{2|\xi - p'||x-\bar{x}|}{(1-k^2)\sqrt{\frac{b^2}{(1-k^2)^2} - \frac{|\bar{x}-p'|^2}{1-k^2}}}
\end{aligned}$$

for some $\xi \in [x, \bar{x}]$. By assumption (A), $x, \bar{x} \in B_{\delta p_n \sqrt{1-\kappa^2}/k}(p') \subseteq B_{\delta b/\sqrt{1-k^2}}(p')$ and hence, we have $|\xi - p'| \leq \frac{\delta b}{\sqrt{1-k^2}}$ and also $|\bar{x} - p'|^2 \leq \frac{\delta^2 b^2}{1-k^2}$, and therefore, we get $u(x) - u(\bar{x}) \leq \frac{2\delta}{(1-k^2)\sqrt{1-\delta}}\,|x - \bar{x}|$. Interchanging the roles of x and $\bar{x}$ yields the result.

Definition 2.3. Given a parallel refractor $u(x)$ for $x \in \Omega$, the refractor mapping of u is the multivalued map defined for $x_0 \in \overline{\Omega}$ by

$$\mathcal{N}_u(x_0) = \left\{ p \in \overline{\Sigma} : \ \phi_{p,b} \text{ touches } u \text{ from above at } x_0 \text{ for some } b \geq \frac{p_n(1-k^2)}{k} \right\}.$$

Given $p \in \overline{\Sigma}$, the tracing mapping of u is defined by

$$\mathcal{T}_u(p) = \mathcal{N}_u^{-1}(p) = \{x \in \overline{\Omega} : \ p \in \mathcal{N}_u(x)\}.$$

The singular set of u is defined by

$$S_u = \{x \in \bar{\Omega} : \text{there exist } p, q \in \Sigma \text{ such that } p \neq q \text{ and } p, q \in \mathcal{N}_u(x)\},$$

and as usual, this set has Lebesgue measure zero [Gut01, Lemma 1.1.12]. To see this in the present case, we observe first that if $E_{p,b}$ and $E_{\bar{p},\bar{b}}$ are two ellipsoids given by (2.1) such that $E_{\bar{p},\bar{b}} \subseteq E_{p,b}$, and they touch at some point x, then it follows that $v := \frac{x-p}{|x-p|} = \bar{v} := \frac{x-\bar{p}}{|x-\bar{p}|}$, and hence, p, $\bar{p}$, and x are on a line. Indeed, from the equation of the normals at x, we have that $v + \kappa\, e_n = \lambda\,(\bar{v} + \kappa\, e_n)$ for some $\lambda > 0$. So $v = \lambda \bar{v} + (\lambda - 1)\kappa e_n$, taking norms, and since $\kappa < 1$, we obtain that $\lambda = 1$, and we are done. This together with Lemma 2.2 and the visibility condition (B) yields that $|S_u| = 0$ as desired. Then as in [GH09, Lemma 3.5], this implies that the class of sets $\mathcal{C} = \{F \subset \overline{\Sigma} : \mathcal{T}_u(F) \text{ is Lebesgue measurable}\}$ is a Borel σ-algebra in $\overline{\Sigma}$.

Given a nonnegative $f \in L^1(\Omega)$, we then obtain as in [GH09, Lemma 3.6] that the set function

$$\mathcal{M}_{u,f}(F) = \int_{\mathcal{T}_u(F)} f \, dx$$

is a finite Borel measure defined on $\mathcal{C}$, which we call it the parallel refractor measure associated with u and f.

Lemma 2.4. *Let $G \subseteq \Sigma$ be open and $\bar{G} \subseteq \Sigma$. Assume $u_m \longrightarrow u$ uniformly in $\bar{\Omega}$, where u_m, u are parallel refractors. Then $\mathcal{T}_u(G) \setminus S_u \subseteq \liminf_{m\to\infty} \mathcal{T}_{u_m}(G)$.*

Proof. Suppose not and let $\bar{x} \in \mathcal{T}_u(G) \setminus S_u$ such that $\bar{x} \notin \liminf_{m\to\infty} \mathcal{T}_{u_m}(G)$. Since $\bar{x} \notin S_u$, there exists a unique $\bar{p} \in \mathcal{N}_u(\bar{x})$, $\bar{p} \in G$, and $u \le \phi_{\bar{p},b}$ in $\bar{\Omega}$ with equality at $\bar{x}$ for some b.

Since $\bar{x} \notin \liminf_{m\to\infty} \mathcal{T}_{u_m}(G)$, there is a subsequence m_k such that $\bar{x} \notin \mathcal{T}_{u_{m_k}}(G)$. Hence, $\bar{x} \notin \mathcal{T}_{u_{m_k}}(q)$ for all $q \in G$ or, equivalently, $q \notin \mathcal{N}_{u_{m_k}}(\bar{x})$ for all $q \in G$ and for all m_k's.

Let $p_{m_k} \in \mathcal{N}_{u_{m_k}}(\bar{x})$, then $p_{m_k} \in \Sigma \setminus G$, which is a compact set. Hence, we may assume, passing through a subsequence, that $p_{m_k} \to p_0$, $p_0 \in \Sigma \setminus G$, and we may also assume $b_{m_k} \to b_0$, as $k \to \infty$. But, since $u_m \longrightarrow u$ uniformly in $\bar{\Omega}$, we will have $u \le \phi_{p_0,b_0}$ in $\bar{\Omega}$ with equality at $\bar{x}$. This means that $p_0 \in \mathcal{N}_u(\bar{x})$, but $p_0 \neq \bar{p}$ since $\bar{p} \in G$, a contradiction with the uniqueness of $\bar{p}$.

3 Main Results

We construct in this section the surfaces that refract collimated radiation in a prescribed way.

Lemma 3.1. *Let $p_i \in \Sigma$ be distinct points, $p_i = (p_1^i, \dots, p_n^i) = (p_i', p_n^i)$, and $b_1, \dots, b_N$ be such that $b_i \ge \frac{p_n^i(1-k^2)}{k}$, $i = 1, \dots, N$, and $\Omega \subseteq \bigcap_{i=1}^N B_{\delta p_n^i \sqrt{1-\kappa^2}/\kappa}(p_i')$.*[1] *Define u in Ω by*

$$u(x) = \min_{1 \le i \le N} \phi_{p_i,b_i}(x).$$

[1]This inclusion follows from condition (A).

Then

$$\mathcal{M}_{u,f}(\{p_1,\dots,p_N\}) = \Sigma_{i=1}^N \mathcal{M}_{u,f}(\{p_i\}) = \int_\Omega f(x)\,dx.$$

Proof. Let $S_i = \{x \in \bar{\Omega} : \exists q \neq p_i, q \in \mathcal{N}_u(x)\}$, $1 \leq i \leq N$, and $S_u = \{x \in \bar{\Omega} : \exists\, p, q \in \mathcal{N}_u(x),\ q \neq p\}$. We write $\bar{\Omega} = \bigcup_{i=1}^N \mathcal{T}_u(p_i) = \bigcup_{i=1}^N (\mathcal{T}_u(p_i) \setminus S_i) \bigcup \bigcup_{i=1}^N (\mathcal{T}_u(p_i) \cap S_i)$. We have $\bigcup_{i=1}^N (\mathcal{T}_u(p_i) \cap S_i) \subset S_u$ and $(\mathcal{T}_u(p_i) \setminus S_i) \cap (\mathcal{T}_u(p_j) \setminus S_j)$ for $i \neq j$. The result then follows since $|S_i| = 0$, $i = 1,\dots,N$, and $|S_u| = 0$.

Lemma 3.2. *Let $p_i \in \Sigma$ be distinct points, $p_i = (p_1^i,\dots,p_n^i) = (p_i', p_n^i)$, and $b_1,\dots,b_N$ be such that $b_i \geq \frac{p_n^i(1-k^2)}{k}$, $i = 1,\dots,N$, and $\Omega \subseteq \bigcap_{i=1}^N B_{8p_n^i\sqrt{1-\kappa^2}/\kappa}(p_i')$.*

Let $\epsilon > 0$ and define u and u_ϵ in Ω by

$$u(x) = \min_{1\leq i\leq N} \phi_{p_i,b_i}(x), \quad \textit{and} \quad u_\epsilon(x) = \min\left\{\phi_{p_1,b_1+\epsilon}(x), \phi_{p_i,b_i}(x) : i = 2,\dots,N\right\}.$$

Then $\mathcal{T}_{u_\epsilon}(p_i) \setminus S_{u_\epsilon} \subseteq \mathcal{T}_u(p_i)$ for $i \neq 1$, and $\limsup_{\epsilon\to 0} \mathcal{T}_{u_\epsilon}(p_1) \subseteq \mathcal{T}_u(p_1)$. Similarly, if b_1 is replaced by b_j, then the first conclusion holds for $i \neq j$ and the second for p_j instead of p_1.

Proof. Let $\bar{x} \in \mathcal{T}_{u_\epsilon}(p_i) \setminus S_{u_\epsilon}$, $i \neq 1$, then $u_\epsilon(\bar{x}) = \phi_{p_i,b_i}(\bar{x})$. Since $\phi_{p_1,b_1+\epsilon} \leq \phi_{p_1,b_1}$, we have $u_\epsilon(x) \leq u(x)$, and so $\phi_{p_i,b_i}(\bar{x}) = u(\bar{x})$.

If $\bar{x} \in \limsup_{\epsilon\to 0} \mathcal{T}_{u_\epsilon}(p_1)$, then for all $\epsilon > 0$, there exists $0 < \beta < \epsilon$ such that $\bar{x} \in \mathcal{T}_{u_\beta}(p_1)$. That is, there exists b_β such that $u_\beta(x) \leq \phi_{p_1,b_\beta}(x)$ with equality at $\bar{x}$. Passing through a subsequence $\beta_\beta \to \bar{b} > 0$ as $\beta \to 0$, and so $u(x) \leq \phi_{p_1,\bar{b}}(x)$ with equality at $\bar{x}$, that is, $\bar{x} \in \mathcal{T}_u(p_1)$.

We are now in a position to prove the existence theorem when the target is a set of points.

Theorem 3.3. *Let $p_i \in \Sigma$, $i = 1,\dots,N$ be distinct points as in Lemma 3.2 and $a_i > 0$ such that $\Sigma_{i=1}^N a_i = \int_\Omega f(x)\,dx$.*

Then there exists $u : \bar{\Omega} \to [-M, 0]$ a parallel refractor such that $\mathcal{M}_{u,f}(\{p_i\}) = a_i$ for $i = 1,\dots,N$ and such that if $E \subseteq \Sigma$ and $E \cap \{p_1,\dots,p_N\} = \emptyset$, then $\mathcal{M}_{u,f}(E) = 0$.

Proof. For simplicity in the notation, we write $\mathcal{M}_u$ instead $\mathcal{M}_{u,f}$.

We say $b = (b_1,\dots,b_N)$ is admissible if $b_i \geq \frac{p_n^i(1-k^2)}{k}$ for $i = 1,\dots,N$. For each admissible b define

$$u_b(x) = \min_{1\leq i\leq N} \phi_{p_i,b_i}(x),$$

and set

$$\bar{b}_1 = \frac{1-k^2}{k}\left(p_n^1 + \frac{1}{k}\max_{2\leq i\leq N} p_n^i\right). \tag{3.3}$$

Clearly, $(\bar{b}_1, b_2, \dots, b_N)$ is admissible when $(b_1, b_2, \dots, b_N)$ is admissible. Define the set

$$W = \left\{ (b_2, \dots, b_N) : b_i \geq \frac{p_n^i(1-k^2)}{k},\ \mathcal{M}_{u_b}(\{p_i\}) \leq a_i,\ i = 2, \dots, N \right\},$$

where u_b is defined with $b = (\bar{b}_1, b_2, \dots, b_N)$.

Claim 1: $W \neq \emptyset$.

Indeed, with the choice $b_i = \frac{p_n^i(1-k^2)}{k}$, $i = 2, \dots, N$, we have that $\max\{\phi_{p_1,\bar{b}_1}(x) : x \in \bar{\Omega}\} \leq p_n^1 - \frac{\kappa \bar{b}}{1-\kappa^2} \leq -\frac{b_i}{1-\kappa^2} = \phi_{p_i,b_i}(p_i') = \min\{\phi_{p_i,b_i}(x) : x \in B_{b/\sqrt{1-\kappa^2}}(p_i')\}$ for each $i = 2, \dots, N$. Therefore, $\phi_{p_1,\bar{b}_1}(x) \leq \phi_{p_i,b_i}(x)$ in Ω, and hence, $u_b(x) = \phi_{p_1,\bar{b}_1}(x)$ for all $x \in \bar{\Omega}$, which implies that $\mathcal{M}_{u_b}(\{p_i\}) = 0$ for $i = 2, \dots, N$.

Claim 2: W is bounded

We shall prove that if $b_j \geq \frac{(1-k^2)(1+k)^2 C_2}{k^3}$, for some $2 \leq j \leq N$, where C_2 is the constant in (2.2), then $(b_2, \dots, b_N) \notin W$. We have that

$$\begin{aligned} b_j \geq \frac{(1-k^2)(1+k)^2 C_2}{k^3} &= \frac{1-k^2}{k}\left(C_2 + \frac{(1+k)}{k^2}C_2 + \frac{1}{k}C_2\right) \\ &\geq \left(\frac{1-k^2}{k}\right)\left(p_n^j + \frac{(1+k)}{k^2}\max_{2\leq i\leq N} p_n^i + \frac{1}{k}p_n^1\right), \end{aligned}$$

which implies that

$$\begin{aligned} \max\left\{\phi_{p_j,b_j}(x) : x \in \bar{\Omega}\right\} &\leq p_n^j - \frac{kb_j}{1-k^2} \leq p_n^1 - \frac{(1+k)\bar{b}_1}{1-k^2} \\ &= \phi_{p_1,\bar{b}_1}(p_1') \leq \min\left\{\phi_{p_1,\bar{b}_1}(x) : x \in \bar{\Omega}\right\}. \end{aligned}$$

Therefore, $u_b(x) = \min_{2\leq i\leq N} \phi_{p_i,b_i}(x)$, and so $\mathcal{M}_{u_b}(\{p_1\}) = 0$. Suppose by contradiction that $(b_2, \dots, b_N) \in W$. Then $\mathcal{M}_{u_b}(\{p_i\}) \leq a_i$, for $i = 2, \dots, N$. But, by Lemma 3.1, we have $\int_\Omega f(x)\,dx = \mathcal{M}_{u_b}(\{p_1, \dots, p_N\}) = \Sigma_{i=1}^N \mathcal{M}_{u_b}(\{p_i\}) = \Sigma_{i=2}^N \mathcal{M}_{u_b}(\{p_i\}) \leq \Sigma_{i=2}^N a_i < \int_\Omega f(x)\,dx$, a contradiction.

Claim 3: W is closed

Let $(b_2^m, \dots, b_N^m) \in W$ such that $(b_2^m, \dots, b_N^m) \to (\bar{b}_2, \dots, \bar{b}_N)$ as $m \to \infty$. Set $b_m = (\bar{b}_1, b_2^m, \dots, b_N^m)$ and $\bar{b} = (\bar{b}_1, \bar{b}_2, \dots, \bar{b}_N)$.

We have that $u_{b_m} \longrightarrow u_{\bar{b}}$ uniformly in $\bar{\Omega}$. We claim that $\mathcal{M}_{u_{\bar{b}}}(\{p_i\}) \leq a_i$, for $i = 2, \dots, N$. Without loss of generality, we may assume $i = 2$. Let G be open in Σ such that $p_2 \in G$ and $p_i \notin G$ for $i \neq 2$. Then $\mathcal{M}_{u_{b_m}}(G) = \mathcal{M}_{u_{b_m}}(\{p_2\}) \leq a_2$ for all m. From Lemma 2.4, we have that $\mathcal{T}_{u_{\bar{b}}}(G) \setminus S_{u_{\bar{b}}} \subseteq \liminf_{m\to\infty} \mathcal{T}_{u_{b_m}}(G)$, and so $\mathcal{M}_{u_{\bar{b}}}(\{p_2\}) \leq \mathcal{M}_{u_{\bar{b}}}(G) = \mathcal{M}_{u_{\bar{b}}}(G \setminus S_{u_{\bar{b}}}) \leq \liminf_{m\to\infty} \mathcal{M}_{u_{b_m}}(G) \leq a_2$ and Claim 3 is proved.

Define the function $\psi : W \longrightarrow [0, \infty)$ by $\psi(b_2, \dots, b_N) = b_2 + \cdots + b_N$. Since W is a compact set, ψ attains its maximum at some point $(\bar{b}_2, \dots, \bar{b}_N) \in W$. Set $\bar{b} = (\bar{b}_1, \bar{b}_2, \dots, \bar{b}_N)$, with $\bar{b}_1$ from (3.3). We shall prove that $\mathcal{M}_{u_{\bar{b}}}(\{p_i\}) = a_i$ for $i = 1, 2, \dots, N$.

Since $(\bar{b}_2, \dots, \bar{b}_N) \in W$, we have $\mathcal{M}_{u_{\bar{b}}}(\{p_i\}) \leq a_i$ for $i = 2, \dots, N$. Suppose that for some $i \geq 2$, $\mathcal{M}_{u_{\bar{b}}}(\{p_i\}) < a_i$, say $\mathcal{M}_{u_{\bar{b}}}(\{p_2\}) < a_2$. Let $\bar{b}_\epsilon = (\bar{b}_1, \bar{b}_2 + \epsilon, \dots, \bar{b}_N)$. Then, by the second assertion of Lemma 3.2, $\mathcal{M}_{u_{\bar{b}_\epsilon}}(\{p_2\}) < a_2$ for ϵ sufficiently small. Also from the first assertion of Lemma 3.2, we have $\mathcal{T}_{u_{\bar{b}_\epsilon}}(\{p_i\}) \setminus S_{u_{\bar{b}_\epsilon}} \subseteq \mathcal{T}_{u_{\bar{b}}}(\{p_i\})$ for $i \neq 2$. Therefore, $\mathcal{M}_{u_{\bar{b}_\epsilon}}(\{p_i\}) \leq a_i$ for $i = 2, \dots, N$, and hence, $(\bar{b}_2 + \epsilon, \dots, \bar{b}_N) \in W$, contradicting that ψ has a maximum at $(\bar{b}_2, \dots, \bar{b}_N)$. Therefore, $\mathcal{M}_{u_{\bar{b}}}(\{p_i\}) = a_i$ for $i = 2, \dots, N$. By Lemma 3.1, we have $\Sigma_{i=1}^N \mathcal{M}_{u_{\bar{b}}}(\{p_i\}) = \int_\Omega f(x)\,dx = \Sigma_{i=1}^N a_i$, and therefore, we get $\mathcal{M}_{u_{\bar{b}}}(\{p_1\}) = a_1$. This proves the claim.

We also notice that if $E \subseteq \Sigma$ such that $E \cap \{p_1, \dots, p_N\} = \emptyset$ and $x \in \mathcal{T}_{u_{\bar{b}}}(E)$, then either $x \in \partial\Omega$ or $u_{\bar{b}}$ is not differentiable at x. Since $u_{\bar{b}}$ is Lipschitz in $\bar{\Omega}$, we have that $\mathcal{M}_{u_{\bar{b}}}(E) = 0$.

We also notice that $u_{\bar{b}} \leq 0$ in $\bar{\Omega}$.

Also, recall that from the proof of Claim 2 above, if $b_i \geq \frac{(1-k^2)(1+k)^2 C_2}{k^3}$, for some $2 \leq i \leq N$, then $(b_2, \dots, b_N) \notin W$. Notice that for such b_i, we have that $\min\{\psi_{p_i, b_i}(x) : x \in \bar{\Omega}\} = p_n^i - \frac{(k+1)b_i}{1-k^2} \geq C_1 - \left(\frac{k+1}{k}\right)^3 C_2 = -M$, the constant defined in condition (B) at the outset. Hence, $u_{\bar{b}} \geq -M$ in $\bar{\Omega}$.

For the general case when the distribution of energy to receive is given by a measure, we have the following:

Theorem 3.4. *Let μ be a Borel measure on Σ and $f \in L^1(\Omega)$ such that $\mu(\Sigma) = \int_\Omega f(x)\,dx$. There exists a function $u : \Omega \longrightarrow [-M, 0]$ that is a parallel refractor and $\mathcal{M}_{u,f} = \mu$.*

Proof. Let $\mu_m \to \mu$ weakly such that $\mu_m = \Sigma_{i=1}^{N_m} a_{i_m} \delta_{p_{i_m}}$ and such that $\Sigma_{i=1}^{N_m} a_{i_m} = \int_\Omega f(x)\,dx$ for all m.

From Theorem 3.3, let u_m be a solution of $\mathcal{M}_{u_m, f} = \mu_m$. From Lemma 2.2, the sequence $\{u_m\}$ is uniformly Lipschitz in $\bar{\Omega}$, and $-M \leq u_m \leq 0$ in $\bar{\Omega}$ for all m. Therefore, there exists a subsequence $u_{m_j} \longrightarrow \bar{u}$ uniformly in $\bar{\Omega}$, and hence, $\mu_{m_j} = \mathcal{M}_{u_{m_j}, f} \to \mathcal{M}_{\bar{u}, f}$ weakly, and also $\mu_{m_j} \to \mu$ weakly. Hence, $\mathcal{M}_{\bar{u}, f} = \mu$.

Lemma 3.5. *Let u_b and $u_{\bar{b}}$ be two solutions as in Theorem 3.3 with $b = (b_1, \dots, b_N)$ and $\bar{b} = (\bar{b}_1, \dots, \bar{b}_N)$. If $b_1 \leq \bar{b}_1$, then $b_i \leq \bar{b}_i$ for $i = 2, \dots, N$. Moreover, if $u_b(x_0) = u_{\bar{b}}(x_0)$ at some $x_0 \in \Omega$, then $u_b \equiv u_{\bar{b}}$.*

Proof. Let $J = \{j : b_j > \bar{b}_j\}$ and $I = \{i : b_i \leq \bar{b}_i\}$. Suppose $J \neq \emptyset$. For $j \in J$, we have $\phi_{b_j, p_j} < \phi_{\bar{b}_j, p_j}$ in $\bar{\Omega}$, and for $i \in I$, we have $\phi_{b_i, p_i} \geq \phi_{\bar{b}_i, p_i}$ in $\bar{\Omega}$.

Fix $j \in J$ and let $\bar{x} \in \mathcal{T}_{u_b}(p_j)$. It follows that $u_{\bar{b}}(\bar{x}) = \phi_{\bar{b}_j, p_j}(\bar{x})$. And hence, $\phi_{\bar{b}_j, p_j}(\bar{x}) \leq \phi_{\bar{b}_i, p_i}(\bar{x})$ for all $i \in I$ which implies that $\phi_{b_j, p_j}(\bar{x}) < \phi_{\bar{b}_j, p_j}(\bar{x}) \leq \phi_{\bar{b}_i, p_i}(\bar{x}) \leq \phi_{b_i, p_i}(\bar{x})$ for all $i \in I$. By continuity, there exists $\epsilon > 0$ such that for all

$x \in B_\epsilon(\bar{x})$, $\phi_{b_j,p_j}(x) < \phi_{b_i,p_i}(x)$ for all $i \in I$, and this implies that for $x \in B_\epsilon(\bar{x})$, $u_b(x) = \min\{\phi_{b_j,p_j}(x) : j \in J\}$. This means that $B_\epsilon(\bar{x}) \subseteq \mathcal{T}_{u_b}(\{p_j : j \in J\})$. So we have shown that $\mathcal{T}_{u_{\bar{b}}}(\{p_j : j \in J\}) \subseteq (\mathcal{T}_{u_b}(\{p_j : j \in J\}))^\circ$. Since $\mathcal{T}_{u_{\bar{b}}}(\{p_j : j \in J\})$ is closed, then we obtain that $(\mathcal{T}_{u_b}(\{p_j : j \in J\}))^\circ \setminus \mathcal{T}_{u_{\bar{b}}}(\{p_j : j \in J\})$ is a nonempty open set. Since u_b and $u_{\bar{b}}$ are solutions, we have $\int_{\mathcal{T}_{u_{\bar{b}}}(\{p_j : j\in J\})} f(x), dx = \int_{\mathcal{T}_{u_b}(\{p_j : j\in J\})} f(x)\,dx = \Sigma_{j\in J} a_j$, a contradiction.

If $b_1 = \bar{b}_1$, then $b_j = \bar{b}_j$ for all $j > 1$ from the first part, and we are done. We claim that if $b_1 > \bar{b}_1$, then $b_j > \bar{b}_j$ for all $j > 1$. Indeed, if $b_j = \bar{b}_j$ for some $j \neq 1$, then $b_k = \bar{b}_k$ for all $k \neq j$ by the first part, a contradiction. Therefore, $u_b(x_0) = \min_{1\le i\le N} \phi_{p_j,b_j}(x_0) < \min_{1\le i\le N} \phi_{p_j,\bar{b}_j}(x_0) = u_{\bar{b}}(x_0)$, a contradiction.

Theorem 3.6. *There exists a constant $-\beta < 0$ depending on C_1, C_2, and k such that if $x_0 \in \bar{\Omega}$ and $t \le -\beta$, then there exists a parallel refractor u as in Theorem 3.3 satisfying $u(x_0) = t$.*

Proof. To obtain a solution passing through a given point, we can modify the proof of Theorem 3.3 as follows.

We consider $\bar{b}_1 \ge \dfrac{(1-k^2)}{k}\left(p_n^1 + \frac{1}{k}\max_{2\le i\le N} p_n^i\right)$ and we assume the visibility condition (B) holds on $\bar{\Omega} \times (-\infty, 0]$.

We claim that for each such $\bar{b}_1$, we can obtain a solution denoted $u_{\bar{b}_1}$ with the property that

$$\frac{(1+k)}{k}p_n^1 + \min_{2\le i\le N} p_n^i - \frac{(1+k)}{k}\max_{2\le i\le N} p_n^i - \frac{(1+k)^2}{k(1-k^2)}\bar{b}_1$$
$$\le u_{\bar{b}_1}(x) \le \phi_{p_1,\bar{b}_1}(x)$$

in $\bar{\Omega}$. This follows just as in the proof of Theorem 3.3 defining the set W in the same way and noticing that if $b_i \ge \frac{(1-k^2)}{k}\left(\max_{2\le i\le N} p_n^i - p_n^1 + \frac{(1+k)}{1-k^2}\bar{b}_1\right)$, for $i = 2, \ldots, N$, then $(b_2, \ldots, b_N) \notin W$. Since the solution is of the form u_b with $b = (\bar{b}_1, b_2, \ldots, b_N)$ and $(b_2, \ldots, b_N) \in W$, it follows that $\min_{2\le i\le N} \phi_{p_i,b_i}(x) \le u_b(x) \le \phi_{p_1,\bar{b}_1}(x)$, where $b_i = \frac{(1-k^2)}{k}\left(\max_{2\le i\le N} p_{i_n} - p_{1_n} + \frac{(1+k)}{1-k^2}\bar{b}_1\right)$, and since $\min_{2\le i\le N} \phi_{p_i,b_i}(x) \ge \frac{(1+k)}{k}p_n^1 + \min_{2\le i\le N} p_n^1 - \frac{(1+k)}{k}\max_{2\le i\le N} p_{i_n} - \frac{(1+k)^2}{k(1-k^2)}\bar{b}_1$, the claim follows.

With $\bar{b}_1 = \frac{(1-k^2)}{k}\left(p_n^1 + \frac{1}{k}\max_{2\le i\le N} p_n^i\right)$, let $-\beta = \frac{(1+k)}{k}p_n^1 + \min_{2\le i\le N} p_n^i - \frac{(1+k)}{k}\max_{2\le i\le N} p_n^i - \frac{(1+k)^2}{k(1-k^2)}\bar{b}_1$. That is, $-\beta = -\frac{(1+k)}{k^2}p_n^1 + \min_{2\le i\le N} p_n^i - \frac{(1+k)}{k^2}\max_{2\le i\le N} p_n^i$.

Given a point (x_0, t) with $x_0 \in \bar{\Omega}$ and $t \le -\beta$, we use continuity of the solution $u_{\bar{b}_1}$ in the parameter $\bar{b}_1$ to show that for some $\bar{b}_1 \ge \frac{(1-k^2)}{k}\left(p_n^1 + \frac{1}{k}\max_{2\le i\le N} p_n^i\right)$, we have $u_{\bar{b}_1}(x_0) = t$. Indeed, if $\bar{b}_1 = \frac{(1-k^2)}{k}\left(p_n^1 + \frac{1}{k}\max_{2\le i\le N} p_n^i\right)$, then $u_{\bar{b}_1}(x_0) \ge -\beta \ge t$; while if $\bar{b}_1$ is large enough, then we will have $u_{\bar{b}_1}(x_0) \le \phi_{p_1,\bar{b}_1}(x_0) \le t$.

Acknowledgements The first author was partially supported by NSF grant DMS–0901430. The second author was supported by CONICET, Argentina.

References

[GH09] C. E. Gutiérrez and Qingbo Huang, *The refractor problem in reshaping light beams*, Arch. Rational Mech. Anal. **193** (2009), no. 2, 423–443.

[Gut01] C. E. Gutiérrez, *The Monge–Ampère equation*, Birkhäuser, Boston, MA, 2001.

On a Theorem of N. Katz and Bases in Irreducible Representations

David Kazhdan

Dedicated to the memory of Leon Ehrenpreis

Abstract N. Katz has shown that any irreducible representation of the Galois group of $\mathbb{F}_q((t))$ has unique extension to a *special* representation of the Galois group of $k(t)$ unramified outside 0 and ∞ and tamely ramified at ∞. In this chapter, we analyze the number of not necessarily special such extensions and relate this question to a description of bases in irreducible representations of multiplicative groups of division algebras.

1 A Formula for the Formal Dimension

Let $k = \mathbb{F}_q, q = p^r$ be a finite field, $\bar{k}$ the algebraic closure of k, $F := k((t))$ and $\bar{F}$ be the algebraic closure of F. The restriction to $\bar{k} \subset \bar{F}$ defines a group homomorphism

$$\mathrm{Gal}(\bar{F}/F) \to \mathrm{Gal}(\bar{k}/k) = \hat{\mathbb{Z}},$$

and we define the *Weil group* of the field F as the preimage $\mathcal{G}_0 \subset \mathrm{Gal}(\bar{F}/F)$ of $\mathbb{Z} \subset \hat{\mathbb{Z}}$ under this homomorphism.

We denote by $\underline{\mathbb{P}}^1$ the projective line over k, set $E := k(t)$, and denote by S the set of points of $\underline{\mathbb{P}}^1$. For any $s \in S$, we denote by E_s the completion of E at s. Using the parameter t on $\underline{\mathbb{P}}^1$, we identify the fields E_0 and E_∞ with F and therefore identify $\mathcal{G}_0$ with the Weil groups of the fields E_0 and E_∞.

D. Kazhdan (✉)
Institute of Mathematics, The Hebrew University, Jerusalem, Israel
e-mail: kazhdan@math.huji.ac.il

H.M. Farkas et al. (eds.), *From Fourier Analysis and Number Theory to Radon Transforms and Geometry*, Developments in Mathematics 28, DOI 10.1007/978-1-4614-4075-8_15,

Let $\tilde{E}$ be the maximal extension of the field E unramified outside 0 and ∞ and tamely ramified at ∞. We denote by $\mathcal{G} \subset \mathrm{Gal}(\tilde{E}/E)$ the Weil group corresponding to the extension $\tilde{E}/E$. We have the imbedding

$$\mathcal{G}_0 \hookrightarrow \mathcal{G}, \quad \text{and the homomorphism } \mathcal{G}_\infty \hookrightarrow \mathcal{G}$$

defined up to conjugation. Therefore, for any complex representation ρ of $\mathcal{G}$, the restrictions to $\mathcal{G}_0, \mathcal{G}_\infty$ define representations ρ_0, ρ_∞ of the corresponding local groups. The group $\mathcal{G}$ has a unique maximal quotient $\bar{\mathcal{G}}$ such that the Sylow p-subgroup of $\bar{\mathcal{G}}$ is normal. As shown by Katz [5], the composition $\mathcal{G}_0 \to \bar{\mathcal{G}}$ is an isomorphism.

Remark. A finite-dimensional irreducible representation ρ_0 of $\mathcal{G}$ is called *special* if it factors through a representation of the group $\bar{\mathcal{G}}$. One can restate the theorem of N. Katz by saying that for any irreducible representation ρ_0 of $\mathcal{G}_0$ there exists a unique special representation ρ_{sp} of the group $\mathcal{G}$ whose restriction to $\mathcal{G}_0$ is equivariant to ρ_0.

Let D_0 be a skew-field with center $F, \dim_F D_0 = n^2$, $G_0 := D_0^*$ be the multiplicative group of D_F and ρ_0 be an n-dimensional indecomposable representation of the group $\mathcal{G}_0$.

Definition 1.1. (a) We denote by $\tilde{\sigma}(\rho_0)$ the irreducible discrete series representation of the group $GL_n(F)$ which corresponds to ρ_0 under the local Langlands correspondence (see, e.g., ([3]) and by $\sigma(\rho_0)$ the irreducible representation of the group G_0 which corresponds to $\tilde{\sigma}(\rho_0)$ as in [1].

(b) We denote by $r(\rho_0)$ the formal dimension of the representation $\tilde{\sigma}(\rho_0)$ where the formal dimension is normalized in such a way that the formal dimension of the Steinberg representation is equal to 1. Analogously, for any indecomposable representation ρ_∞ of the group $\mathcal{G}_\infty$, we define an integer $r(\rho_\infty)$.

(c) We denote by $A(\tilde{\rho}_0)$ the set of equivalence classes of n-dimensional irreducible representations ρ of the group $\mathcal{G}$ whose restriction to $\mathcal{G}_0$ is equivalent to ρ_0 and the restriction to $\mathcal{G}_\infty$ is indecomposable.

Theorem 1.2. *For any n-dimensional irreducible $\bar{\mathbb{Q}}_l$-representation of the group $\mathcal{G}_0$, the sum $\sum_{\rho \in A(\rho_0)} r(\rho_\infty)$ is equal to $r(\rho_0)$.*

Proof. Let $\mathbb{A} = \prod'_{s \in S} E_s$ the ring of adeles of E, D be a skew-field with center E unramified outside $\{0, \infty\}$, $D_0 := D \otimes_E E_0$ and $D_\infty := D \otimes_E E_\infty$. Then D_0, D_∞ are local skew-fields. Let $\underline{G}$ be the multiplicative group of D considered as the algebraic E-group.

We denote by $N : D_0 \to F$ the reduced norm and define

$$\mu := \nu \circ N : G_0 \to \mathbb{Z}, K_0 := \mu^{-1}(0),$$

where $\nu : F^* \to \mathbb{Z}$ is the standard valuation. Then $K_0 \subset G_0$ is a maximal compact subgroup. We define the *first congruence subgroup* K_0^1 by

$$K_0^1 := \{k \in K_0 | \mu(k - Id) > 0\}$$

As is well known, K_0^1 is a normal subgroup of D_0^* such that $K_0/K_0^1 = \mathbb{F}_{q^n}^*$ and $D_0^*/K_0^1 = \mathbb{Z} \ltimes \mathbb{F}_{q^n}^*$, where $\mathbb{Z}$ acts on $\mathbb{F}_{q^n}^*$ by $(n, x) \to x^{q^n}$.

For any $s \in S - \{0, \infty\}$, we identify the group G_{E_s} with $GL(n, E_s)$ and define $K_s := GL(n, \mathcal{O}_s)$. We write $G_{\mathbb{A}} := D_\infty^* \times GL_n(\mathbb{A}^\infty)$, where

$$GL_n(\mathbb{A}^\infty) := G_0 \times \prod_{s \in S - \{0,\infty\}} GL(n, E_s),$$

and define

$$K^0 := \prod_{s \in S - \{0,\infty\}} K_s \times K_{E_\infty}, K^1 := \prod_{s \in S - \{0,\infty\}} K_s \times K_{E_\infty}^1,$$

where $K_\infty^1 \subset K_\infty \subset D_\infty^*$ is the first congruence subgroup of G_∞.

Lemma 1.3. *(a) For any irreducible complex representation* $\kappa : D_0^*/K_0^1 \to Aut(W)$ *and any character* $\chi : K_0/K_0^1 \to \mathbb{C}^*$, *we have*

$$\dim(W^\chi) \le 1,$$

where $W^\chi = \{w \in W | \kappa(k)w = \chi(k)w, k \in K_0\}$.

(b) For any irreducible representation π *of the group* G_0, *the formal dimension of* $\tilde{\pi}$ *is equal to the dimension of* π.

Proof. Part a) follows from the isomorphism $G_0/K_0^1 = \mathbb{Z} \ltimes \mathbb{F}_{q^n}^*$.
Part b) follows from [1].

It follows from [6] that we can identify the set $A(\tilde{\rho}_0)$ with the set of automorphic representations $\tilde{\pi} = \otimes'_{s \in S} \tilde{\pi}_s$ of the group $GL_n(\mathbb{A})$ such that the representation $\tilde{\pi}_0$ is equivalent to $\tilde{\sigma}(\rho_0)$ and the representation $\tilde{\pi}_\infty$ is of discrete series. Then it follows from [1] that we can identify the set $A(\tilde{\rho}_0)$ with the set $A(\tilde{\rho}_0)$ of automorphic representations $\pi = \otimes'_{s \in S} \pi_s$ of the group $\underline{G}(\mathbb{A})$ such that the representation π_0 is equivalent to $\sigma(\rho_0)$. The restriction of the representation π_∞ on K_∞^1 is trivial and the representations $\pi_{s,s}$ in $S - \theta, \infty$ are unramified. We will use this identification for the proof of the Theorem 1.2.

We see that the following equality implies the validity of the Theorem 1.2.

Claim 1.4. *For any* n*-dimensional irreducible* $\bar{\mathbb{Q}}_l$*-representation* ρ_0 *of the group* $\mathcal{G}_0$, *the sum* $\sum_{\pi \in A(\rho_0)} \dim(\pi_\infty)$ *is equal to* $\dim(\sigma(\rho_0))$.

The proof of Claim is based on the following result.

Proposition 1.5. *The product map* $G_0 \times K^1 \times G_E \to G_{\mathbb{A}}$ *is a bijection.*

Proof of the Proposition. The surjectivity follows from Lemma 7.4 in [4]. To show the injectivity, it is sufficient to check the equality

$$(G_0 \times K^1) \cap G_E = \{e\}$$

which is obvious.

We denote by $\mathbb{C}(G_{\mathbb{A}}/G_E)$ the space of locally constant functions on $G_{\mathbb{A}}/G_E$ with compact support, by $\mathbb{C}(G_0)$ the space of locally constant functions on G_0 with compact support, and by $L \subset \mathbb{C}(G_{\mathbb{A}}/G_E)$ the subspace of K^1-invariant functions. The group $G_0 \times D^\star_\infty/K^1_\infty$ acts naturally on L.

Let ρ_0 be an indecomposable representation of the group $\mathcal{G}_0$. We denote by $(\sigma(\rho_0), V(\rho_0))$ the corresponding representation of the group G_0 and identify the set $A(\rho_0)$ with the set of automorphic representations $\pi^a = \otimes'_{s\in S}\pi^a_s$ of the group $\underline{G}(\mathbb{A})$ such that the representation π^a_0 is equivalent to $\sigma(\rho_0)$ and the representation π^a_∞ is trivial on K^1_∞. Let

$$\mathcal{H} := \otimes_{s\in S-\{0,\infty\}}\mathcal{H}_s,$$

where $\mathcal{H}_s$ is the spherical Hecke algebra for $G(F_s) = GL(n, F_s)$. By construction, the commutative algebra $\mathcal{H}$ acts on the $G_0 \times D^\star_\infty/K^1$-module L. For any $a \in A(\rho_0)$, we define

$$L_a := \mathrm{Hom}_{G^\infty_{\mathbb{A}}}(\pi^a, \mathbb{C}(G_{\mathbb{A}}/G_E)) = \mathrm{Hom}_{G_0\times\mathcal{H}}(\sigma(\rho_0), L) \subset \mathrm{Hom}_{G_0}(\sigma(\rho_0), L)$$

Lemma 1.6. *(a) The restriction $r : L \to \mathbb{C}(G_0)$ is an isomorphism of G_0-modules where G_0 acts on $\mathbb{C}(G_0)$ by left translation.*
(b) $\mathrm{Hom}_{G_0}(\sigma(\rho_0), L) = V^\vee$ where $V^\vee$ is the dual space to $V(\rho_0)$.
(c) $V^\vee = \oplus L_a, a \in A(\rho_0)$ where the algebra $\mathcal{H}$ acts on $L_a, a \in A(\rho_0)$ by a character $\chi_a : \mathcal{H} \to \bar{\mathbb{Q}}^\star_l$, $\chi_a \neq \chi_{a'}$ for $a \neq a'$, and the representations π^a_∞ of the group $D^\star_\infty/K^1$ on M_a are irreducible.
(d) The representations π^a_∞ are associated with the restriction $\rho(a)_\infty$ by the local Langlands correspondence.

Proof. The Lemma follows immediately from the Proposition and the strong multiplicity one theorem ([1] and [7]).

This Lemma implies the validity of Claim and therefore of Theorem 1.2. Indeed, we have

$$\dim(V) = \dim(V^\vee) = \sum_{a\in A(\rho_0)} \dim(L_a) = \sum_{a\in A(\rho_0)} \dim(\pi^a_\infty) = \sum_{a\in A(\rho_0)} r(\rho(a)_\infty) \qquad \square$$

One can ask whether one can extend Theorem 1.2 to the case of other groups. More precisely, let G be a split reductive group with a connected center and ${}^LG$ be the Langlands dual group. Consider a homomorphism $\rho_0 : \mathcal{G}_0 \to {}^LG$ such that the connected component of the centralizer $Z_{\rho_0} := Z_{^LG}(\mathrm{Im}(\rho_0))$ is unipotent. Let $[Z_\rho]$ be the group of connected components of the centralizer Z_{ρ_0}. Conjecturally, one can associate with ρ_0 an L-packet of irreducible representations $\pi_{\rho_0}(\tau)$ of the group $G_0 := G(F)$ parameterized by irreducible representations τ of $[Z_\rho]$, and there exists an integer $r(\rho_0)$ such that the formal dimension of $\pi_{\rho_0}(\tau)$ is equal to $r(\rho_0)\dim(\tau)$.

We denote by $A^{\mathcal{G}}(\rho_0)$ the set of conjugacy classes of homomorphisms $\rho : \mathcal{G} \to {}^L G$ whose restriction on $\mathcal{G}_0$ is conjugate to ρ_0 and such that the connected component of the centralizer of the restriction on $\mathcal{G}_\infty$ is unipotent.

Question. Is it true that $r(\rho_0) = \sum_{a \in A(\rho_0)} r(\rho_\infty)$, where $r(\rho_\infty)$ is defined in the same way as $r(\rho_0)$?

2 A Construction of a Basis

Let G be a reductive group over a local field. As is well known, one can realize the spherical Hecke algebra $\mathcal{H}$ of G *geometrically*, that is, as the Grothendieck group of the monoidal category of perverse sheaves on the affine Grassmannian. Analogously in the case when G be a reductive group over a global field of positive characteristic, the unramified geometric Langlands conjecture predicts the existence of a geometric realization of the corresponding space of automorphic functions.

Let $\underline{C}$ be a smooth absolutely irreducible $\mathbb{F}_q$-curve, $q = p^m$, S be the set of closed points of $\underline{C}$, $\Gamma := \pi_1(\underline{C})$. For any $s \in S$, we denote by $Fr_s \subset \Gamma$ the conjugacy class of the Frobenius at s.

Let E be the field of rational functions on $\underline{C}$. For any $s \in S$, we denote by E_s the completion of E at s and we denote by $\mathbb{A}$ be the ring of adeles of E. Fix a prime number $l \neq p$.

Let $\underline{G}$ be a split reductive group, and $\hat{K} := \prod_{s \in S} G(\mathcal{O}_s) \subset G(\mathbb{A})$ be the standard maximal compact subgroup. An irreducible representation $(\pi, V) = \otimes'_{s \in S}(\pi_s, V_s)$ of $G(\mathbb{A})$ is *unramified* if $V^{\hat{K}} \neq \{0\}$. In this case, $\dim(V^{\hat{K}}) = 1$. So for any unramified representation (π, V) of the group $G(\mathbb{A})$, there is a special *spherical* vector $v_{sp} \in V$ defined up to a multiplication by a scalar.

Let ${}^L G$ be the Langlands dual group and ρ a homomorphism from Γ to ${}^L G(\bar{\mathbb{Q}}_l)$, such that for any $s \in S$, the conjugacy class $\gamma_s := \rho(Fr_s) \subset {}^L G(\bar{\mathbb{Q}}_l)$ is semisimple. In such a case, we can define unramified representations (π_{γ_s}, V_s) of local groups $G(E_s)$ and the representation $(\pi(\rho), V_\rho) = \otimes_s(\pi_{\gamma_s}, V_s)$ of the adelic group $G(\mathbb{A})$. According to the unramified geometric Langlands conjecture, the homomorphism ρ defines [at least in the case when ρ is tempered] an imbedding

$$i_\rho : V_\rho \to \bar{\mathbb{Q}}_l(K \backslash G(\mathbb{A})/G(E))$$

and a function $f_\rho := i_\rho(v_{sp})$ which is defined up to a multiplication by a scalar.

We can identify the set $K\backslash G(\mathbb{A})/G(E)$ with the set of $\mathbb{F}_q$-points of the stack $\mathcal{B}_G$ of principal G-bundles on $\underline{C}$, and the unramified geometric Langlands correspondence predicts the existence of a perverse Weil sheaf $\mathcal{F}(\rho)$ on $\mathcal{B}_G$ such that the function f_ρ is given by the trace of the Frobenius automorphisms on stalks of $\mathcal{F}(\rho)$. (See [2].)

If one considers ramified automorphic representations $(\pi, V) = \otimes'_{s \in S}(\pi_s, V_s)$ of $G(\mathbb{A})$, then there is no natural way to choose a special vector in V. So on the "geometric" side, one expects not an object $\mathcal{F}(\rho)$ but an abelian category $\mathcal{C}(\rho)$

which is a product of local categories $\mathcal{C}(\rho_s)$ such that the Grothendieck K-group of the category $[\mathcal{C}(\rho_s)]$ coincides with the subspace V_s^0 of the minimal K-type vectors of the space V_s of the local representation. Such geometric realization of the space V_s^0 would define a special basis of vector spaces V_s^0 which would be a non-Archimedean analog of Lusztig's canonical basis. Here, we consider only the case of an anisotropic group when the minimal K-type subspace V_s^0 coincides with the space V_s of the representation of G. Moreover, we will only discuss a slightly weaker data of a *projective basis* where a projective basis in a finite-dimensional vector space T is a decomposition of the space T in a direct sum of one-dimensional subspaces. So one could look for a special basis of vector spaces V_s which would be a non-Archimedean analog of the Lusztig's canonical basis.

Let as before $F := k((t))$, D_0 be a skew-field with center $F, \dim_0 D_0 = n^2, G_0$ be the multiplicative group of D_0 and $\sigma : G_0 \to \mathrm{Aut}(V)$ a complex irreducible continuous representation of the group G_0.

Theorem 2.1. *For any irreducible representation $\tau : D_F^* \to Aut(T)$ of the group D_F^*, there exists a "natural" projective basis $= \oplus_a T_a$ of T.*

Remark 2.2. The construction is global. In particular, I do not know how to define a projective basis in the case when F is a local field of characteristic zero. It would be very interesting to find a local construction of a projective basis.

The construction. As follows from Lemma 1.6(c), we have a decomposition $V^\vee = \sum_{a \in A(\rho_0)} M_a$ where the group $D_\infty^\star / K_\infty^1$ acts irreducibly on M_a. Therefore, the group $\mathbb{F}_{q^n}^\star = K_\infty / K_\infty^1$ acts on M_a, and we have a decomposition of M_a into the sum of eigenspaces for the action of the group $\mathbb{F}_{q^n}^\star$. As follows from Lemma 1.3 these eigenspaces are one dimensional.

Acknowledgements The author acknowledges the support of the European Research Council during the preparation of this paper.

References

1. Deligne, P.; Kazhdan, D.; Vignéras, M.-F. Représentations des alge'bres centrales simples p-adiques. Representations of reductive groups over a local field, 33–117, Travaux en Cours, Hermann, Paris, 1984.
2. Gaitsgory, D. Informal introduction to geometric Langlands. An introduction to Langlands program. 269–281 Burhauser Boston, Boston MA 2003.
3. Henniart, G., Une preuve simple des conjectures de Langlands pour GL(n) sur un corps p-adique, Invent. Math., (2000).
4. Hrushovski, E,; Kazhdan D.; Motivis Poisson summation. Moscow Math. J. 9(2009) no. 3 569–623.
5. Katz, N. Local-to-global extensions of representations of fundamental groups. (French summary) Ann. Inst. Fourier (Grenoble) 36 (1986), no. 4, 69–106.
6. Lafforgue, L. Chtoucas de Drinfeld et correspondance de Langlands. Invent. Math. 147 (2002), no. 1, 1–241.
7. Piatetskii-Shapiro I. Multiplicity one theorems, Proc. Sympos. Pure Math., vol. 33, Part I, Providence, R. I., 1979, pp. 209–212.

Vector-Valued Modular Forms with an Unnatural Boundary

Marvin Knopp[†] **and Geoffrey Mason**

Dedicated to the memory of Leon Ehrenpreis.

Abstract We characterize all logarithmic, holomorphic vector-valued modular forms which can be analytically continued to a region strictly larger than the upper half-plane.

Mathematics Subject Classification (2010): 11F12, 11F99

1 Introduction

Let $\Gamma = SL_2(\mathbb{Z})$ be the modular group with standard generators

$$S = \begin{pmatrix} 0 & -1 \\ 1 & 0 \end{pmatrix}, \ T = \begin{pmatrix} 1 & 1 \\ 0 & 1 \end{pmatrix},$$

and let $\rho : \Gamma \to GL(p, \mathbb{C})$ be a p-dimensional representation of Γ. A holomorphic vector-valued modular form of weight $k \in \mathbb{Z}$ associated to ρ is a holomorphic function $F : \mathfrak{H} \to \mathbb{C}^p$ defined on the upper half-plane $\mathfrak{H}$ which satisfies

$$F|_k\gamma(\tau) = \rho(\gamma)F(\tau) \quad (\gamma \in \Gamma) \tag{1}$$

[†]Marvin Isadore Knopp, born January 4, 1933, passed away unexpectedly on December 24, 2011.

M. Knopp[†]
(Deceased)

G. Mason (✉)
Department of Mathematics, University of California, Santa Cruz, California, USA
e-mail: gem@ucsc.edu

H.M. Farkas et al. (eds.), *From Fourier Analysis and Number Theory to Radon Transforms and Geometry*, Developments in Mathematics 28, DOI 10.1007/978-1-4614-4075-8_16,

and a growth condition at ∞ (see below). As usual, the stroke operator here is defined as

$$F|_k\gamma(\tau) = (c\tau + d)^{-k} F(\gamma\tau) \quad \left(\gamma = \begin{pmatrix} a & b \\ c & d \end{pmatrix} \in \Gamma\right).$$

We generally drop the adjective 'holomorphic' from holomorphic vector-valued modular form unless there is good reason not to. We also refer to the pair (ρ, F) as a vector-valued modular form and call p the *dimension* of (ρ, F). We usually consider F as a vector-valued function[1] $F(\tau) = (f_1(\tau), \ldots, f_p(\tau))^t$ and call the $f_i(\tau)$ the *component functions* of F.

Given a pair of vector-valued modular forms $(\rho, F), (\rho', F')$ of weight k and dimension p, we say that they are *equivalent* if there is an invertible $p \times p$ matrix A such that

$$(\rho, F) = (A\rho' A^{-1}, AF').$$

In particular, the representations ρ, ρ' of Γ are necessarily equivalent in the usual sense.

Suppose that (ρ, F) is a vector-valued modular form. The purpose of this chapter is to investigate whether $F(\tau)$ has a *natural boundary*. If $f(\tau)$ is a nonconstant (scalar) modular form of weight k on a subgroup of finite index in Γ, then it is well known that the real axis is a natural boundary for $f(\tau)$ in the sense that there is *no* real number r such that $f(\tau)$ can be analytically continued to a region containing $\mathfrak{H} \cup \{r\}$. In this chapter, we say that (ρ, F) has the real line as a natural boundary provided that *at least one component* of F does. Note that if we replace (ρ, F) by an equivalent vector-valued modular form, the component functions are replaced by linear combinations of component functions. In particular, the existence of a natural boundary is a property that is shared by any two vector-valued modular forms that are equivalent.

In [KM2], the authors extended the classical result on natural boundaries to the case in which the matrix $\rho(T)$ is *unitary*. Replacing (ρ, F) with an equivalent vector-valued modular form if necessary, we may assume that $\rho(T)$ is both unitary and diagonal. A (ρ, F) such that $\rho(T)$ is unitary and diagonal is called *normal*, and we proved (*loc. cit.*) that a normal vector-valued modular form has the real line as natural boundary. Here, we study the same question for the larger class of *polynomial*, or *logarithmic*, vector-valued modular forms introduced in [KM3], where one assumes only that the eigenvalues of $\rho(T)$ have absolute value 1. This case is more subtle for several reasons, not the least being that the existence of a natural boundary *no longer obtains* in general.

We recall some facts about polynomial vector-valued modular forms (*loc. cit.*). Replacing (ρ, F) by an equivalent vector-valued modular form if necessary, we may, and shall, assume that $\rho(T)$ is in (modified) Jordan canonical form.[2] Let the ith

[1] Superscript t means *transpose* of vectors and matrices.

[2] A minor variant of the usual Jordan canonical form. See [KM3] for details.

Jordan block of $\rho(T)$ have size m_i and label the corresponding component functions of $F(\tau)$ as $\varphi_1^{(i)}(\tau), \ldots, \varphi_{m_i}^{(i)}(\tau)$. By [KM3], they have *polynomial q-expansions*

$$\varphi_l^{(i)}(\tau) = \sum_{s=0}^{l-1} \binom{\tau}{s} h_{l-1-s}^{(i)}(\tau) \quad (1 \leq l \leq m_i), \tag{2}$$

each $h_s^{(i)}(\tau)$ having a left-finite q-series

$$h_s^{(i)}(\tau) = \mathrm{e}^{2\pi \mathrm{i} \mu_i \tau} \sum_{n=\nu_i}^{\infty} a_n(s,i) \mathrm{e}^{2\pi \mathrm{i} n \tau} \quad (0 \leq s \leq m_i - 1, \ \nu_i \in \mathbb{Z}). \tag{3}$$

Here, $\mathrm{e}^{2\pi \mathrm{i} \mu_i}$ is the eigenvalue of $\rho(T)$ determined by the ith block and $0 \leq \mu_i < 1$. (It is here that we are using the assumption that the eigenvalues of $\rho(T)$ have absolute value 1.) $F(\tau)$ is then called a *holomorphic* vector-valued modular form if, for each Jordan block, each q-series $h_s^{(i)}(\tau)$ has only *nonnegative* powers of q, i.e., $a_n(s,i) = 0$ whenever $n + \mu_i < 0$.

Setting $q = \mathrm{e}^{2\pi \mathrm{i} \tau} (\tau \in \mathfrak{H})$ so that $\tau = (2\pi \mathrm{i})^{-1} \log q$, we find from (2) and (3) that $\varphi_l^{(i)}(\tau)$ may alternatively be expressed in the form

$$\varphi_l^{(i)}(\tau) = \sum_{s=0}^{l-1} (\log q)^s g_{l-1-s}^{(i)}(q)$$

with q-series $g_{l-1-s}^{(i)}(q)$. It is this formulation that gives rise to the name *logarithmic* vector-valued modular form. We find it convenient to use the polynomial variation encapsulated by (2) and (3) in this chapter.

The most accessible examples of polynomial vector-valued modular forms that are not normal are as follows (cf. Sect. 2 for more details). If we set

$$C(\tau) = (\tau^{p-1}, \tau^{p-2}, \ldots, 1)^t,$$

then $C(\tau)$ is a vector-valued modular form of weight $1 - p$ associated with a representation σ equivalent to the $(p-1)$th symmetric power $S^{p-1}(\nu)$ of the natural defining representation ν of Γ. The canonical form for $\sigma(T)$ is a single Jordan block, and (σ, C) is equivalent to a vector-valued modular form for which the q-series corresponding to the $h_j^{(i)}(\tau)$ in (3) are constants.

Obviously, (σ, C) is a p-dimensional vector-valued modular form that is analytic throughout the complex plane. The main result of this chapter is that these are essentially the only examples of polynomial vector-valued modular forms whose natural boundary is not the real line. We give two formulations of the main result. As we shall explain, they are essentially equivalent.

Theorem 1. *Suppose that the eigenvalues of $\rho(T)$ have absolute value* 1, *and let (ρ, F) be a nonzero vector-valued modular form of weight k and dimension p. Then the following are equivalent:*

(a) $F(\tau)$ does not have the real line as a natural boundary.
(b) The component functions of $F(\tau)$ span the space of polynomials of degree $l-1$ for some $l \le p$. Moreover, $k = -l$.

Theorem 2. *Suppose that the eigenvalues of $\rho(T)$ have absolute value* 1, *and let (ρ, F) be a vector-valued modular form of weight k and dimension p. Suppose further that the component functions of $F(\tau)$ are linearly independent. Then the following are equivalent:*

(a) $F(\tau)$ does not have the real line as a natural boundary.
(b) (ρ, F) is equivalent to (σ, C) and $k = 1 - p$.

This chapter is organized as follows. In Sect. 2, we consider the basic example (σ, C) introduced above in more detail and explain why Theorems 1 and 2 are equivalent. In Sect. 3, we give the proof of the theorems.

2 The Vector-Valued Modular Form (σ, C)

The space of homogeneous polynomials in variables X, Y is a right Γ-module such that $\gamma = \begin{pmatrix} a & b \\ c & d \end{pmatrix} \in \Gamma$ is an algebra automorphism with

$$\gamma : X \mapsto aX + bY,\ Y \mapsto cX + dY.$$

The subspace of homogeneous polynomials of degree $p-1$ is an irreducible Γ-submodule which we denote by Q_{p-1}. The representation of Γ that it furnishes is the $(p-1)^{\text{th}}$ symmetric power $S^{p-1}(\nu)$ of the defining representation ν.

For $\tau \in \mathfrak{H}$, let $P_{p-1}(\tau)$ be the space of polynomials in τ of degree at most $p-1$. Since

$$\tau^j|_{1-p}\gamma = (c\tau + d)^{p-1}\left(\frac{a\tau + b}{c\tau + d}\right)^j = (a\tau + b)^j (c\tau + d)^{p-1-j},$$

it follows that $P_{p-1}(\tau)$ is a right Γ-module with respect to the stroke operator $|_{1-p}$. Indeed, $P_{p-1}(\tau)$ is isomorphic to Q_{p-1}, an isomorphism being given by

$$X^j Y^{p-1-j} \mapsto \tau^j \ \ (0 \le j \le p-1).$$

Since $1, \tau, \ldots, \tau^{p-1}$ are linearly independent and span a right Γ-module with respect to the stroke operator $|_{1-p}$, we know (cf. [KM1], Sect. 2) that there is a unique representation $\sigma : \Gamma \to GL_p(\mathbb{C})$ such that

$$\sigma(\gamma)C(\tau) = C|_{1-p}\gamma(\tau) \ \ (\gamma \in \Gamma).$$

This shows that (C, σ) is a vector-valued modular form of weight $1 - p$ and that the representation σ is equivalent to $S^{p-1}(\nu)$.

We can now explain why Theorems 1 and 2 are equivalent. Assume first that Theorem 1 holds, and let (ρ, F) be a vector-valued modular form of weight k with linearly independent component functions and such that the real line is not a natural boundary for $F(\tau)$. By Theorem 1, the components of $F(\tau)$ span a space of polynomials of degree no greater than $p - 1$, and by linear independence, they must span the space $P_{p-1}(\tau)$. Moreover, we have $k = 1 - p$. Now there is an invertible $p \times p$ matrix A such that $AF(\tau) = C(\tau)$, whence $(\rho, F(\tau))$ is equivalent to $(A\rho A^{-1}, C(\tau))$. As explained above, we necessarily have $A\rho A^{-1} = \sigma$ in this situation, so that (ρ, F) is equivalent to (σ, C). This shows that (a) $\Rightarrow$ (b) in Theorem 2, in which case Theorem 2 is true.

Now suppose that Theorem 2 holds, and let (ρ, F) be a nonzero vector-valued modular form of dimension p and weight k such that the real line is not a natural boundary for $F(\tau)$. Let $(g_1, \ldots, g_l)$ be a *basis* for the span of the components of F. Setting $G = (g_1, \ldots, g_l)^t$, we again use ([KM1], Sect. 2) to find a representation $\alpha : \Gamma \to GL_l(\mathbb{C})$ such that (α, G) is a vector-valued modular form of weight k. Because the components of G are linearly independent, Theorem 2 tells us that they span the space $P_{l-1}(\tau)$ of polynomials of degree at most l and that $k = -l$. Thus the conclusions of Theorem 1(b) hold, and Theorem 1 is true.

The reader familiar with Eichler cohomology will recognize the Γ-module $P_{p-1}(\tau)$ as a crucial ingredient in that theory. This points to the fact that Eichler cohomology has close connections to the theory of vector-valued modular forms, connections that in fact go well beyond the question of natural boundaries that we treat here. The authors hope to return to this subject in the future.

3 Proof of the Main Theorems

In this section, we will prove Theorem 2. As we have explained, this is equivalent to Theorem 1.

In order to prove Theorem 2, we may replace (ρ, F) by any equivalent vector-valued modular form. Thus we may, and from now on shall, assume without loss that $\rho(T)$ is in (modified) Jordan canonical form. We assume that $\rho(T)$ has t Jordan blocks, which we may, and shall, further assume are ordered in decreasing size $M = m_1 \geq m_2 \geq \cdots \geq m_t$. Thus $m_1 + \cdots + m_t = p$, and we may speak, with an obvious meaning, of the component functions in a block. The ith. block corresponds to an eigenvalue $e^{2\pi i \mu_i}$ of $\rho(T)$, and we let the component functions of $F(\tau)$ in that block be as in (2), (3).

Let

$$\gamma = \begin{pmatrix} a & b \\ c & d \end{pmatrix} \in \Gamma. \tag{4}$$

Because (ρ, F) is a vector-valued modular form of weight k, we have

$$\rho(\gamma)F(\tau) = (c\tau + d)^{-k} F(\gamma\tau).$$

So if $\varphi_u(\tau) = \varphi_v^{(i)}(\tau)$ is the uth component of the ith block of $F(\tau)$ (so that $u = m_1 + \cdots + m_{i-1} + v$), then

$$(c\tau + d)^{-k}\varphi_u(\gamma\tau) = \sum_{j=1}^{t}\sum_{l=1}^{m_j} \alpha_l^{(j)}\varphi_l^{(j)}(\tau),$$

where $(\ldots, \underbrace{\alpha_1^{(j)}, \ldots, \alpha_{m_j}^{(j)}}_{j\text{th block}}, \ldots)$ is the uth row of $\rho(\gamma)$. Using (2), we obtain

$$\begin{aligned}
(c\tau + d)^{-k}\varphi_u(\gamma\tau) &= \sum_{j=1}^{t}\sum_{l=1}^{m_j}\sum_{s=0}^{l-1} \alpha_l^{(j)} \binom{\tau}{s} h_{l-1-s}^{(j)}(\tau) \\
&= \sum_{s=0}^{M-1} \binom{\tau}{s} \left(\sum_{j=1}^{t} \sum_{l=s+1}^{m_j} \alpha_l^{(j)}\, h_{l-1-s}^{(j)}(\tau) \right) \\
&= \sum_{s=0}^{M-1} \binom{\tau}{s} \sum_{j=1}^{t} \left(\alpha_{s+1}^{(j)} h_0^{(j)}(\tau) + \sum_{l=s+2}^{m_j} \alpha_l^{(j)}\, h_{l-1-s}^{(j)}(\tau) \right).
\end{aligned}$$

(Here, $\alpha_{s+1}^{(j)} = 0$ if $s \ge m_j$.)

Because the component functions of (ρ, F) are linearly independent, $\varphi_u(\tau)$ is nonzero and the previous display is not identically zero. So there is a largest integer B in the range $0 \le B \le M - 1$ such that the summand corresponding to $\binom{\tau}{B}$ does *not* vanish. Now note that $\varphi_1^{(j)}(\tau) = h_0^{(j)}(\tau)$. Because the component functions are linearly independent, then in particular the $h_0^{(j)}(\tau)$ are linearly independent, and we can conclude that

$$\begin{aligned}
&\alpha_{s+1}^{(j)} = 0 \;\; (1 \le j \le t,\; s > B + 1), \\
&\alpha_{B+1}^{(j)} \text{ are not all zero } \;(1 \le j \le t).
\end{aligned} \tag{5}$$

It follows that

$$\begin{aligned}
&(c\tau + d)^{-k}\varphi_u(\gamma\tau) \\
&= \binom{\tau}{B} \sum_{j=1}^{t} \alpha_{B+1}^{(j)} h_0^{(j)}(\tau) + \sum_{s=0}^{B-1} \binom{\tau}{s} \sum_{j=1}^{t} \sum_{l=s+2}^{m_j} \alpha_l^{(j)}\, h_{l-1-s}^{(j)}(\tau)
\end{aligned} \tag{6}$$

and the first term on the right-hand side of (6) is nonzero. Incorporating (3), we obtain

$$
\begin{aligned}
&(c\tau+d)^{-k}\varphi_u(\gamma\tau)\\
&=\binom{\tau}{B}\sum_{j=1}^{t}\alpha_{B+1}^{(j)}\mathrm{e}^{2\pi\mathrm{i}\mu_j\tau}\sum_{n=\nu_j}^{\infty}a_n(0,j)\mathrm{e}^{2\pi\mathrm{i}n\tau}\\
&\quad+\sum_{s=0}^{B-1}\binom{\tau}{s}\sum_{j=1}^{t}\sum_{l=s+2}^{m_j}\sum_{n=\nu_j}^{\infty}\alpha_l^{(j)}\mathrm{e}^{2\pi\mathrm{i}\mu_j\tau}a_n(l-1-s,j)\mathrm{e}^{2\pi\mathrm{i}n\tau}.
\end{aligned}\tag{7}
$$

Let $\tilde{\mu}_1,\ldots,\tilde{\mu}_p$ be the *distinct* values among $\mu_1,\ldots,\mu_t$. Then we can rewrite (7) in the form

$$
\begin{aligned}
&(c\tau+d)^{-k}\varphi_u(\gamma\tau)\\
&=\binom{\tau}{B}\sum_{j=1}^{p}\mathrm{e}^{2\pi\mathrm{i}\tilde{\mu}_j\tau}g_B^{(j)}(\tau)+\sum_{s=0}^{B-1}\binom{\tau}{s}\sum_{j=1}^{p}\mathrm{e}^{2\pi\mathrm{i}\tilde{\mu}_j\tau}g_s^{(j)}(\tau),
\end{aligned}\tag{8}
$$

where the first term in (8) is *nonzero* and each $g_m^{(j)}(\tau)$ is a left-finite *pure* q-series, i.e., one with only integral powers of q.

Consider the *nonzero* summands

$$
\begin{aligned}
\mathrm{e}^{2\pi\mathrm{i}\tilde{\mu}_j\tau}g_B^{(j)}(\tau)&=\sum_{n=\nu(j,B)}^{\infty}b_n(j,B)q^{n+\tilde{\mu}_j}\\
&=b_{\nu(j,B)}(j,B)q^{\nu(j,B)+\tilde{\mu}_j}(1+\text{positive integral powers of }q)
\end{aligned}\tag{9}
$$

that occur in the first term on the right-hand side of (8). Let J be the corresponding set of indices j. Because the $\tilde{\mu}_j$ are distinct, there is a *unique* $j_0\in J$ which minimizes the expression

$$\nu(j,B)+\tilde{\mu}_j.$$

Let $J'=J\setminus\{j_0\}$. Hence, there is $y_0>0$ such that for $\Im(\tau)>y_0$, we have

$$\left|\mathrm{e}^{2\pi\mathrm{i}\tilde{\mu}_{j_0}\tau}g_B^{(j_0)}(\tau)\right|>2\left|\sum_{j\in J'}\mathrm{e}^{2\pi\mathrm{i}\tilde{\mu}_j\tau}g_B^{(j)}(\tau)\right|.$$

Taking into account the terms $\mathrm{e}^{2\pi\mathrm{i}\tilde{\mu}_j\tau}g_B^{(j)}(\tau)$ that vanish, we obtain for $\Im(\tau)>y_0$:

$$
\begin{aligned}
\left|\sum_{j=1}^{p}\mathrm{e}^{2\pi\mathrm{i}\tilde{\mu}_j\tau}g_B^{(j)}(\tau)\right|&>\left|\mathrm{e}^{2\pi\mathrm{i}\tilde{\mu}_{j_0}\tau}g_B^{(j_0)}(\tau)\right|-\left|\sum_{j\in J'}\mathrm{e}^{2\pi\mathrm{i}\tilde{\mu}_j\tau}g_B^{(j)}(\tau)\right|\\
&>1/2\left|\mathrm{e}^{2\pi\mathrm{i}\tilde{\mu}_{j_0}\tau}g_B^{(j_0)}(\tau)\right|>0.
\end{aligned}\tag{10}
$$

In (10), for $N \in \mathbb{Z}$, we have $\Im(\tau + N) = \Im(\tau) > y_0$. So (10) holds with τ replaced by $\tau + N$. Because $g_B^{(j)}(\tau + N) = g_B^{(j)}(\tau)$, we see that

$$\left|\sum_{j=1}^{p} \mathrm{e}^{2\pi \mathrm{i}\tilde{\mu}_j(\tau+N)} g_B^{(j)}(\tau)\right| > 1/2 \left|\mathrm{e}^{2\pi \mathrm{i}\tilde{\mu}_{j_0}\tau} g_B^{(j_0)}(\tau)\right| > 0 \;\; (N \in \mathbb{Z},\ \Im(\tau) > y_0). \quad (11)$$

At this point, we return to (8). Replace τ by $\tau + N$ to obtain

$$\begin{aligned}&(c\tau + cN + d)^{-k}\varphi_u(\gamma(\tau + N)) \\ &\quad = \binom{\tau+N}{B}\sum_{j=1}^{p} \mathrm{e}^{2\pi \mathrm{i}\tilde{\mu}_j(\tau+N)} g_B^{(j)}(\tau) + \sum_{s=0}^{B-1}\binom{\tau+N}{s}\sum_{j=1}^{p} \mathrm{e}^{2\pi \mathrm{i}\tilde{\mu}_j(\tau+N)} g_s^{(j)}(\tau).\end{aligned} \quad (12)$$

Set

$$\begin{aligned}\Sigma_1(\tau) = \Sigma_1(\tau;\gamma) &= \sum_{s=0}^{B}\binom{\tau}{s}\sum_{j=1}^{p} \mathrm{e}^{2\pi \mathrm{i}\tilde{\mu}_j(\tau)} g_s^{(j)}(\tau), \\ \Sigma_2(\tau, N) = \Sigma_2(\tau, N;\gamma) &= \Sigma_1(\tau + N).\end{aligned}$$

Thus (12) reads

$$\varphi_u(\gamma(\tau + N)) = (c\tau + cN + d)^k \Sigma_2(\tau, N). \quad (13)$$

Next, we examine the powers of N that appear in $\Sigma_2(\tau, N)$. Now

$$\begin{aligned}\binom{\tau+N}{s} &= \frac{1}{s!}(\tau + N)(\tau + N - 1)\ldots(\tau + N - s + 1) \\ &= \frac{N^s}{s!} + O(N^{s-1}), \;\; N \to \infty.\end{aligned}$$

Therefore, the highest power of N occurring with nonzero coefficient in $\Sigma_2(\tau, N)$ is N^B, the coefficient in question being

$$\frac{1}{B!}\sum_{j=1}^{p} \mathrm{e}^{2\pi \mathrm{i}\tilde{\mu}_j(\tau+N)} g_B^{(j)}(\tau). \quad (14)$$

Hence, we obtain

$$\Sigma_2(\tau, N) = \frac{N^B}{B!}\sum_{j=1}^{p} \mathrm{e}^{2\pi \mathrm{i}\tilde{\mu}_j(\tau+N)} g_B^{(j)}(\tau) + O(N^{B-1}), \;\; N \to \infty, \quad (15)$$

with nonzero leading coefficient (14).

So far, the component function $\varphi_u(\tau)$ of $F(\tau)$ has been arbitrary. Now we claim that there is at least one component such that the integer B occurring in (15), and thereby also in (13), is equal to $M - 1$. Indeed, because the first block has size

M, the Mth component $\varphi_M^{(1)}(\tau)$ of $F(\tau)$ has the polynomial $\binom{\tau}{M-1}$ occurring in its logarithmic q-expansion (2) with *nontrivial* coefficient $h_0^{(1)}(\tau)$. Because $\rho(\gamma)$ is nonsingular, at least one row of $\rho(\gamma)$, say the uth, has a nonzero value $\alpha_M^{(1)}$ in the Mth column. Thanks to (5), we must have $B+1=M$, as asserted.

With $\varphi_u(\tau)$ as in the last paragraph, we have for $N\to\infty$,

$$\begin{aligned}&\varphi_u(\gamma(\tau+N))\\&\quad=(c\tau+cN+d)^k\left(\frac{N^{M-1}}{(M-1)!}\sum_{j=1}^{p}\mathrm{e}^{2\pi\mathrm{i}\tilde{\mu}_j(\tau+N)}g_{M-1}^{(j)}(\tau)+O(N^{M-2})\right).\end{aligned}\tag{16}$$

Lemma 3.1. *Let $\varphi_u(\tau)$ be as before, and suppose that there exists a rational number a/c, $((a,c)=1, c\neq 0)$ at which $\varphi_u(\tau)$ is continuous from above. Then $k\leq 1-M$.*

Proof. First note that

$$\gamma(\tau+N)=\frac{a+b/(\tau+N)}{c+d(\tau+N)}\to a/c \text{ as } N\to\infty \text{ within } \mathfrak{H}.$$

By the continuity assumption of the Lemma, $\varphi_u(\tau)$ remains bounded as $N\to\infty$. Choosing y_0 large enough, we see from (11) that

$$\left|\sum_{j=1}^{p}\mathrm{e}^{2\pi\mathrm{i}\tilde{\mu}_j(\tau+N)}g_{M-1}^{(j)}(\tau)\right|$$

is bounded away from 0 as $N\to\infty$. Because $c\neq 0$, we deduce that the right-hand side of (16) is $\approx\alpha(N)N^{k+M-1}$ with $\alpha(N)\neq 0$. If $k>1-M$, this implies that the right-hand side is unbounded as $N\to\infty$. This contradiction proves the Lemma.

Lemma 3.2. *Let $\varphi_u(\tau)$ be as in (16), and suppose that $\varphi_u(\tau)$ is holomorphic in a region containing $\mathfrak{H}\cup\mathfrak{I}$ with $\mathfrak{I}$ a nonempty open interval in $\mathbb{R}$. Then $k\geq 1-M$.*

Proof. Choose rational a/c as in the last Lemma so that $a/c\in\mathfrak{I}$. The argument of the previous Lemma shows that the right-hand side of (16) is $\approx\alpha(N)N^{k+M-1}$ with $\alpha(N)\neq 0$. Indeed, we easily see from (16) that $\alpha(N)$ has an upper bound independent of N for $N\to\infty$. Then if $k<1-M$, the right-hand side of (16) $\to 0$ as $N\to\infty$.

On the other hand, the left-hand side of (16) $\to\varphi_u(a/c)$ as $N\to\infty$. We conclude that $\varphi_u(a/c)=0$, and because this holds for all rationals in $\mathfrak{I}$, then $\varphi_u(\tau)$ is identically zero, thanks to the regularity assumption on $\varphi_u(\tau)$. Because the components of $F(\tau)$ are linearly independent, $\varphi_u(\tau)$ cannot vanish, and this contradiction shows that $k\geq 1-M$, as required.

Proposition 3.3. *Assume that the regularity assumption of Lemma 3.2 applies to all components of $F(\tau)$. Then $k = 1 - M$ and each component is a polynomial of degree at most $M - 1$.*

Proof. Because of the regularity assumptions of the present proposition, we may apply Lemmas 3.1 and 3.2 to find that $k = 1-M$. If, for some component $\varphi_u(\tau)$, the integer B that occurs in (15) is *less* than $M - 1$, the argument of Lemma 3.1 yields a contradiction. Since, in any case, we have $B \leq M - 1$, then in fact $B = M - 1$ for all components. As a result, (16) holds for *every* component $\varphi(\tau)$ of $F(\tau)$.

Differentiate (8) (now with $B = M - 1$) M times and apply the well-known identity of Bol [B]:

$$D^{(M)}\left((c\tau + d)^{M-1}\varphi(\gamma\tau)\right) = (c\tau + d)^{-1-M}\varphi^{(M)}(\gamma\tau).$$

We obtain (using Leibniz's rule) for $|\tau| \to \infty$ that

$$\begin{aligned}
&\varphi^{(M)}(\gamma\tau)\\
&= (c\tau + d)^{M+1} D^{(M)}\left(\sum_{s=0}^{M-1}\binom{\tau}{s}\sum_{j=1}^{p} \mathrm{e}^{2\pi \mathrm{i}\tilde{\mu}_j\tau} g_s^{(j)}(\tau)\right)\\
&= (c\tau + d)^{M+1}\left(\frac{\tau^{M-1}}{(M-1)!} D^{(M)}\left(\sum_{j=1}^{p} \mathrm{e}^{2\pi \mathrm{i}\tilde{\mu}_j\tau} g_{M-1}^{(j)}(\tau)\right) + O(|\tau|^{M-2})\right).
\end{aligned}$$

Therefore, for $N \to \infty$,

$$\begin{aligned}
&\varphi^{(M)}(\gamma(\tau + N))\\
&= (c\tau + cN + d)^{M+1}\left(\frac{(\tau + N)^{M-1}}{(M-1)!} D^{(M)}\left(\sum_{j=1}^{p} \mathrm{e}^{2\pi \mathrm{i}\tilde{\mu}_j(\tau+N)} g_{M-1}^{(j)}(\tau)\right)\right.\\
&\qquad \left. + O(N^{M-2})\right). \qquad (17)
\end{aligned}$$

Take $c \neq 0$ with γ as in (4) and $a/c \in \mathfrak{J}$, and apply the regularity assumption to $\varphi(\tau)$. Then the left-hand side of (17) has a limit $\varphi^{(M)}(a/c)$ for $N \to \infty$. On the other hand, we know that $\left|\sum_{j=1}^{p} \mathrm{e}^{2\pi \mathrm{i}\tilde{\mu}_j(\tau+N)} g_{M-1}^{(j)}(\tau)\right|$ is bounded away from zero. So if $D^{(M)}\left(\sum_{j=1}^{p} \mathrm{e}^{2\pi \mathrm{i}\tilde{\mu}_j(\tau+N)} g_{M-1}^{(j)}(\tau)\right)$ does not vanish identically, then the right-hand side of (17) is $\approx \alpha(N)N^{2M}$ for $N \to \infty$, with $\alpha(N)$ bounded away from zero. This contradiction shows that in fact

$$D^{(M)}\left(\sum_{j=1}^{p} \mathrm{e}^{2\pi \mathrm{i}\tilde{\mu}_j(\tau+N)} g_{M-1}^{(j)}(\tau)\right) \equiv 0$$

From (9), we have

$$D^{(M)}\left(\sum_{j=1}^{p} \mathrm{e}^{2\pi \mathrm{i}\tilde{\mu}_j(\tau+N)} g_{M-1}^{(j)}(\tau)\right)$$
$$= \sum_{j=1}^{p} e^{2\pi \mathrm{i}\tilde{\mu}_j N} \sum_{n=\nu(j,M-1)}^{\infty} b_n(j, M-1)(2\pi \mathrm{i}(n+\tilde{\mu}_j))^M q^{n+\tilde{\mu}_j},$$

so that

$$\sum_{j=1}^{p} \mathrm{e}^{2\pi \mathrm{i}\tilde{\mu}_j N} b_n(j, M-1)(n+\tilde{\mu}_j)^M = 0 \quad (n \geq \nu(j, M-1)).$$

This implies that $b_n(j, M-1) = 0$ whenever $n + \tilde{\mu}_j \neq 0$. Because $b_{\nu(j,M-1)}(j, M-1) \neq 0$, we must have

$$\tilde{\mu}_j = 0 \quad (1 \leq j \leq p), \tag{18}$$

which amounts to the assertion that all $\mu_j = 0$ $(1 \leq j \leq t)$. Moreover, $b_n(j, M-1) = 0$ for $n \neq 0$, so that

$$g_{M-1}^{(j)}(\tau) = b_0(j, M-1)$$

is constant. Now (8) reads

$$(c\tau + d)^{M-1}\varphi(\gamma\tau) = \binom{\tau}{M-1}\sum_{j=1}^{p} b_0(j, M-1) + \sum_{s=0}^{M-2}\binom{\tau}{s}\sum_{j=1}^{p} g_s^{(j)}(\tau). \tag{19}$$

We now repeat the argument $M-1$ times, starting with (19) in place of (8). We end up with an identity of the form

$$(c\tau + d)^{M-1}\varphi(\gamma\tau) = \sum_{s=0}^{M-1}\binom{\tau}{s}\sum_{j=1}^{p} b_0(j, s),$$

where of course the right-hand side is a polynomial $p(\tau)$ of degree at most $M-1$. Then

$$\varphi(\tau) = (c\gamma^{-1}\tau + d)^{1-M} p\left(\frac{d\tau - b}{-c\tau + a}\right)$$
$$= (c\tau + d)^{M-1} p\left(\frac{d\tau - b}{-c\tau + a}\right)$$

is itself a polynomial of degree at most $M-1$. This completes the proof of Proposition 3.3.

It is now easy to complete the proof of Theorem 2. It is only necessary to establish the implication (a) $\Rightarrow$ (b). Assuming (a) means that Proposition 3.3 is applicable, so that we have $k = 1 - M$ and the components of $F(\tau)$ are polynomials of degree at most $M - 1$. Because the components are linearly independent, it must be the case that the maximal block size M is equal to the dimension p of the representation ρ. Thus $k = 1 - p$, and the component functions *span* the space of polynomials of degree at most $p - 1$. The fact that (ρ, F) is equivalent to (σ, C), then follows from the discussion in Sect. 2.

Acknowledgements Supported by the NSF and NSA.

References

B. Bol, G., Invarianten linearer Differentialgleichungen, Abh. Math. Sem. Univ. Hamburg **16** Nos. 3–4 (1949), 1–28.

KM1. Knopp, M., and Mason, G., On vector-valued modular forms and their Fourier coefficients, Acta Arith. **110.2** (2003), 117–124.

KM2. Knopp, M. and Mason, G., Vector-valued modular forms and Poincaré series, Ill. J. Math. **48** No. 4 (2004), 1345–1366.

KM3. Knopp, M. and Mason, G., Logarithmic vector-valued modular forms, to appear in Acta Arith. 147.3(2011), 264–282.

Loss of Derivatives

J.J. Kohn

Dedicated to the memory of Leon Ehrenpreis

Revised November 30, 2011
Abstract In 1957, Hans Lewy (see Lewy [L]) obtained a remarkable result. Namely, he found a first-order partial differential operator L such that there exists a function $f \in C^\infty(\mathbb{R}^3)$ so that the equation $Lu = f$ does not have any distribution solutions u on any open set, equivalently the associated laplacian $Eu = LL^*u = f$ does not have any distribution solution. This operator comes from the study of the induced Cauchy-Riemann equation on the sphere in $\mathbb{C}^2$. Roughly speaking, nonexistence of distribution solutions means that no derivative of u can be uniformly estimated by some derivatives of f, that is, "E loses infinitely many derivatives." In Kohn (Ann. Math. 162:943–986, 2005), the operator E was approximated by a sequence of operators $\{E_k\}$, each of which loses exactly $k - 1$ derivatives but nevertheless is locally solvable and hypoelliptic. Here we study these phenomena for operators of the form $\sum X_i^* X_i$, where the X_i are complex-valued vector fields and the corresponding approximating operators lose a finite number of derivatives.

1 Introduction

Operators on boundaries of domains in $\mathbb{C}^n$, associated with the Cauchy-Riemann equations, sometimes exhibit the kind of behavior studied here. The loss of derivatives phenomenon has been studied by Parenti and Parmeggiani; see [PP]. In [KR], H. Rossi and I studied the induced tangential Cauchy-Riemann equations and the associated laplacians on $(0, q)$-forms, with $q > 0$, strongly pseudoconvex domains

J.J. Kohn (✉)
Department of Mathematics, Princeton University, Princeton, NJ, USA
e-mail: kohn@princeton.edu

H.M. Farkas et al. (eds.), *From Fourier Analysis and Number Theory to Radon Transforms and Geometry*, Developments in Mathematics 28, DOI 10.1007/978-1-4614-4075-8_17,

in $\mathbb{C}^n$ with $n > 2$. In those cases, in contrast with the case $n = 2$, the laplacians are hypoelliptic. The Lewy operator and its associated laplacian on the sphere in $\mathbb{C}^2$ motivate the paper [K]. This work was generalized to operators on boundaries of certain weakly pseudoconvex domains in $\mathbb{C}^2$ in [BDKT]. Further generalizations of these operators were treated by G. Zampieri and various coauthors; see, for example, [KPZ].

Let S be the hypersurface in $\mathbb{C}^2$ given by

$$S = \left\{(z_1, z_2) \in \mathbb{C}^2 \mid \mathrm{Re}(z_2) = |z_1|^2\right\}.$$

Let L denote the $(1,0)$ vector field tangent to S given by

$$L = \frac{\partial}{\partial z_1} - 2\bar{z}_1 \frac{\partial}{\partial z_2}.$$

We identify S with $\mathbb{R}^3 \sim \mathbb{C}\times\mathbb{R}^2$ by means of the coordinates $z = z_1$ and $t = Im(z_2)$; in terms of these coordinates, we have

$$L = \frac{\partial}{\partial z} - i\bar{z}\frac{\partial}{\partial t}.$$

This is the Lewy operator. The same nonexistence theorem holds for the operator $E = L\bar{L}^* = -L\bar{L}$. The operators E_k, defined below, near the origin are approximations of $-L\bar{L}$. The E_k do have local existence and are hypoelliptic, but in order to estimate s derivatives of u, we need $s + k - 1$ derivatives of $E_k u$. The operator $-L\bar{L}$ locally is the limit, as $k \to \infty$, of the operators E_k defined by

$$E_k = -L\bar{L} - \bar{L}|z|^{2k} L,$$

in the sense that the limits of the coefficients E_k are the corresponding coefficients of E when $|z| < 1$. In [K], it is shown that the operators E_k "lose" $k - 1$ derivatives. That is, if $f \in H^s_{\mathrm{loc}}(\mathbb{R}^3)$ and $Lu = f$, then $u \in H^{s-k+1}_{\mathrm{loc}}(\mathbb{R}^3)$ and further that there exists $f \in H^s_{\mathrm{loc}}(\mathbb{R}^3)$ such that there does not exist a solution $u \in H^{s'}_{\mathrm{loc}}(\mathbb{R}^3)$ when $s' > s - k + 1$. In [BDKT], this result is generalized, and it is proved that if L, for positive m, is defined by

$$L = \frac{\partial}{\partial z} - i\bar{z}|z|^{2(m-1)}\frac{\partial}{\partial t},$$

then the operators E_k "lose" $\frac{k-1}{m}$ derivatives. These results are proved by means of the following optimal estimates, for each $s \in \mathbb{R}$ and $k \in \mathbb{Z}^+$, there exists $C = C(k,s)$ such that

$$\|u\|_{s-\frac{k-1}{m}} \leq C\|E_k u\|_s,$$

for all $u \in C_0^\infty(\mathbb{R}^n)$. The estimates are optimal in the sense that for any s, k, and a bounded $U \subset \mathbb{R}^3$, there exists a sequence $\{u^\nu\}$ with $u^\nu \in C_0^\infty(U)$, $\|E_k u^\nu\|_s = 1$, and when $\nu \to \infty$ then $\|u^\nu\|_{s'} \to \infty$, whenever $s' > s - \frac{k-1}{m}$.

To discuss this problem in a more general setting, let $X_1, \ldots, X_l$ be complex vector fields on $\mathbb{R}^n$, that is,

$$X_i = \sum_{j=1}^{n} a_i^j \frac{\partial}{\partial x_j},$$

for $i = 1, \ldots, l$ with a_i^j complex valued and $a_i^j \in C^\infty(\mathbb{R}^n)$.

Definition. The vector fields $X_1, \ldots, X_l$ satisfy the *bracket condition* at $x_0 \in \mathbb{R}^n$ if the Lie algebra generated by $X_1, \ldots, X_l$ evaluated at x_0 equals $\mathbb{C}T_{x_0}(\mathbb{R}^n)$.

Let $\mathcal{F}^1(X_1, \ldots, X_l)$ denote the vector space consisting of all vector fields that are linear combinations, with C^∞ coefficients, of the $X_1, \ldots, X_l$. Inductively, we define $\mathcal{F}^p(X_1, \ldots, X_l)$ by

$$\mathcal{F}^j(X_1, \ldots, X_l) = \mathcal{F}^{j-1}(X_1, \ldots, X_l) + [\mathcal{F}^1(X_1, \ldots, X_l), \mathcal{F}^{j-1}(X_1, \ldots, X_l)].$$

Note that the bracket condition at x_0 is equivalent to the condition that there exists m such that

$$\mathcal{F}^m(X_1, \ldots, X_l)|_{x_0} = \mathbb{C}T_{x_0}(\mathbb{R}^n). \tag{1}$$

Definition. If $\{X_1, \ldots, X_l\}$ satisfy the bracket condition at x_0, then the *type* of $\{X_1, \ldots, X_l\}$ at x_0 is the smallest m for which (1) holds.

Here we are concerned with existence and hypoellipticity for the operator:

$$E = \sum_{i=1}^{l} X_i^* X_i.$$

Suppose that the vector fields $\{X_1, \ldots, X_l\}$ are real, then Hörmander has obtained the following result (see [H]).

Theorem. *Suppose that the $\{X_1, \ldots, X_l\}$ are real and satisfy the bracket condition at x_0. Then, there exists a neighborhood U of x_0 and $\varepsilon > 0$ such that if f and u are distributions on U such that $Eu = f$ and if $f|_V \in H^s(V)$, then $u|_V \in H^{s+2\varepsilon}_{\text{loc}}(V)$.*

Hörmander's proof depends on the following *subelliptic energy estimate*.

Theorem (Hörmander). *Suppose that the $\{X_1, \ldots, X_l\}$ are real and satisfy the bracket condition at x_0. Then, there exists a neighborhood U of x_0 and constants C and $\varepsilon > 0$ such that*

$$\|u\|_\varepsilon^2 \leq C \sum_{i=1}^{l} \|X_i u\|^2, \tag{2}$$

for all $u \in C_0^\infty(U)$.

Existence

The estimate (2) for $\varepsilon > 0$ implies that for any $s \in \mathbb{R}$, there exists C_s such that

$$\|u\|^2_{s+2\varepsilon} \leq C_s \left\| \sum_{i=1}^{l} X_i^* X_i u \right\|_s^2, \tag{3}$$

for all $u \in C_0^\infty(U)$. Then, (3) implies that if $f \in H^s_{\text{loc}}(U)$, then there exists a $u \in H^{s+2\varepsilon}_{\text{loc}}(U)$ such that $\sum_{i=1}^{l} X_i^* X_i u = f$.

When (2) holds with $\varepsilon > 0$, then (3) is proved by substituting $\Lambda^{s+\varepsilon}u$ for u in (2). This substitution is justified, despite the fact that $\Lambda^{s+\varepsilon}u$ is not supported in U, as follows. If $\zeta \in C_0^\infty(U)$ with $\zeta = 1$ in a neighborhood of supp(u), then

$$\Lambda^{s+\varepsilon}u = \Lambda^{s+\varepsilon}\zeta u = \zeta\Lambda^{s+\varepsilon}u - [\zeta, \Lambda^{s+\varepsilon}]u.$$

Thus, the symbol of $[\zeta, \Lambda^{s+\varepsilon}]u$ is zero on supp(u), and therefore, the operator $[\zeta, \Lambda^{s+\varepsilon}]$ is of order $-\infty$ on all u with $\zeta = 1$ in a neighborhood of supp(u). When $\varepsilon > 0$, the derivation of (3) proceeds in a straight forward way. The operators $[X_i, \Lambda^{s+\varepsilon}]$ are of order $s + \varepsilon$, and when $\varepsilon > 0$, we have

$$\|u\|_{s+\varepsilon} \leq \text{ small const.} \|u\|_{s+2\varepsilon},$$

when supp(u) has small diameter. Thus, error terms of the form $\|[X_i, \Lambda^{s+\varepsilon}]u\|$ can be absorbed in the left-hand side. This is not the case when (2) holds with $\varepsilon \leq 0$. Here we overcome this difficulty by restricting ourselves only to the special complex vector fields defined below, although (2) holds in greater generality.

Definition. The complex vector fields $X_1, \ldots, X_l$ on $\mathbb{R}^{n+1} = \mathbb{R}^n_x \times \mathbb{R}_t$ are called *special vector fields* if they satisfy the following:

- • $X_i^* = -\bar{X}_i$.
- •• $X_i = \sum_{j=1}^n a_i^j(x)\frac{\partial}{\partial x_j} + b_i(x)\frac{\partial}{\partial t}$ with $a_i^j, b_i \in C^\infty(\mathbb{R}^n_x)$.
- ••• For each x, the l vectors $(a_i^1(x), \ldots, a_i^n(x))$, with $i = 1, \ldots, l$, span an n-dimensional vector space over $\mathbb{C}$.

In order to study the estimates (2) and (3) for the special vector fields, we will microlocalize as follows. Denote by $\{\xi_1, \ldots, \xi_n, \tau\}$, the coordinates dual to $\{x_1, \ldots, x_n, t\}$. Let $\gamma, \gamma_0 \in C^\infty(\mathbb{R}^{n+1}$ with $\gamma(c\xi, c\tau) = \gamma(\xi, \tau)$ and $\gamma_0(c\xi, c\tau) = \gamma_0(\xi, \tau)$ when $|\xi|^2 + |\tau|^2 \geq 1$ and $c \geq 1$. Further, assume that $a, b \in (0, 1)$ and that $\gamma(\xi, \tau) = 0$ when $|\xi| > a|\tau|$ and $|\xi|^2 + |\tau|^2 \geq 1$ and that $\gamma_0(\xi, \tau) = 0$ when $|\xi| < b|\tau|$ and $|\xi|^2 + |\tau|^2 \geq 1$. Define Γu and $\Gamma_0 u$ by

$$\widehat{\Gamma u}(\xi, \tau) = \gamma(\xi, \tau)\hat{u}(\xi, \tau)$$

and

$$\widehat{\Gamma_0 u}(\xi, \tau) = \gamma_0(\xi, \tau)\hat{u}(\xi, \tau),$$

respectively. Denote by $\mathfrak{G}$ and by $\mathfrak{G}_0$ the sets of symbols γ and γ_0. The sets of the corresponding operators will be denoted by $\mathcal{G}$ and by $\mathcal{G}_0$, respectively. Then, $|\tau\gamma_0| \le b^{-1}|\xi|$ when $|\xi|^2 + |\tau|^2 \ge 1$ so that

$$\left(\sum_{i=1}^{n} |\xi_i|^2 + |\tau|^2 + 1\right)^{\frac{1}{2}} \gamma_0(\xi,\tau) \le C\left(\sum_{i=1}^{n} |\xi_i|^2 + 1\right)^{\frac{1}{2}} \gamma_0(\xi,\tau),$$

when $|\xi|^2 + |\tau|^2 \ge 1$. Then, since the $\frac{\partial}{\partial x_i}$ are combinations of the X_i and $\frac{\partial}{\partial t}$, we have

$$\|\Gamma_0 u\|_1^2 \le C \sum \|X_i \Gamma_0 u\|^2 + \|Ru\|^2,$$

where R is a pseudodifferential operator whose symbol has compact support and hence is of order $-\infty$. Therefore,

$$\|\Gamma_0 u\|_1^2 \le C \sum \|X_i \Gamma_0 u\|^2 + O(\|u\|_{-\infty}^2),$$

and for any $s \in \mathbb{R}$,

$$\|\Gamma_0 u\|_{s+1}^2 \le C \sum \|X_i \Gamma_0 u\|_s^2 + O(\|u\|_{-\infty}^2).$$

Let $\Gamma_0' \in \mathcal{G}_0$ be an operator with symbol $\gamma_0'(\xi,\tau) = 1$ in a neighborhood of *supp* $\gamma \cap \{(\xi,\tau) \mid |\xi|^2 + |\tau|^2 \ge 1\}$. Then, since the operator $[X_i, \Gamma_0](1 - \Gamma_0')$ is of order $-\infty$, we have

$$\|[X_i, \Gamma_0]u\|_s^2 \le C\|\Gamma_0' u\|_s^2 + O(\|u\|_{-\infty}^2) \le C\|\Gamma_0'' u\|_{s-1}^2 + O(\|u\|_{-\infty}^2),$$

where the symbol $\gamma_0''(\xi,\tau) = 1$ in a neighborhood of supp $\gamma_0' \cap \{(\xi,\tau) \mid |\xi|^2 + |\tau|^2 \ge 1\}$. Hence, proceeding inductively, we obtain

$$\|\Gamma_0 u\|_{s+1}^2 \le C \sum \|X_i \Gamma_0 u\|_s^2 + O(\|u\|_{-\infty}^2) \le C \sum \|X_i u\|_s^2 + O(\|u\|_{-\infty}^2). \quad (4)$$

Lemma. *If $X_1, \dots, X_l$ are special vector fields as above and if (2) holds for some $\varepsilon \in \mathbb{R}$, then (3) holds.*

Proof. Choose $\Gamma \in \mathcal{G}$ and $\Gamma_0 \in \mathcal{G}_0$ such that $\gamma(\xi,\tau) + \gamma_0(\xi,\tau) = 1$ when $|\xi|^2 + |\tau|^2 \ge 1$. Then, substituting $s + \varepsilon - 1$ for s and $\Lambda^\varepsilon u$ for u, we get

$$\begin{aligned}\|\Gamma_0 \Lambda^\varepsilon u\|_{s+\varepsilon}^2 u &\le C \sum \|X_i \Gamma_0 \Lambda^\varepsilon u\|_{s-1}^2 + O(\|u\|_{-\infty}^2) \\ &\le C\left|\left(\sum \Lambda^s \Gamma_0 \bar{X}_i X_i u, \Gamma_0 \Lambda^{s+2\varepsilon-2} u\right)\right| + O(\|u\|_{-\infty}^2),\end{aligned}$$

and since

$$\|\Gamma_0 \Lambda^{s+2\varepsilon-2} u\| \le \|\Gamma_0 u\|_{s+2\varepsilon-2} + \|\Gamma_0' \Lambda^\varepsilon u\|_{s+\varepsilon-2} + O(\|u\|_{-\infty}),$$

we obtain by induction

$$\|\Gamma_0 u\|_{s+2\varepsilon}^2 \le C\|\sum \bar{X}_i X_i u\|_s^2 + O(\|u\|_{-\infty}^2).$$

Let $\Gamma \in \mathcal{G}$ with $\gamma(\xi.\tau) = 1$ in a neighborhood of $\{(0,\tau) \mid \tau \ge 1\}$. Let Λ_t denote the operator with symbol $(1+|\tau|^2)^{\frac{1}{2}}$ and substitute $\zeta\Lambda_t^{s+\varepsilon}\Gamma u$ for u in (2), where $\zeta \in C_0^\infty(U)$ with $\zeta = 1$ on a neighborhood of supp(u). Then, we obtain

$$\|\zeta\Lambda_t^{s+\varepsilon}\Gamma u\|_\varepsilon^2 \le C\sum \|X_i\zeta\Lambda_t^{s+\varepsilon}\Gamma u\|^2.$$

Since the symbol of $[\zeta, \Lambda_t^{s+\varepsilon}]\Gamma$ is zero on supp(u), we have

$$\|\zeta\Lambda_t^{s+\varepsilon}\Gamma u\|_\varepsilon^2 \equiv \|\Gamma u\|_{s+2\varepsilon}^2 + O(\|u\|_{-\infty}^2).$$

Next, we have

$$[X_i, \zeta\Lambda_t^{s+\varepsilon}\Gamma] = [X_i, \zeta\Lambda_t^{s+\varepsilon}]\Gamma + \zeta\Lambda_t^{s+\varepsilon}[X_i, \Gamma].$$

The symbol of the first operator on the right is zero on supp(u). Let $W \subset \mathbb{R}^{n+1}$ be the interior of the set on which $\gamma = 1$, where γ is the symbol of Γ. Let $\tilde{\gamma}_0 = 1$ on a neighborhood of $\mathbb{R}^{n+1} - W$ and $\tilde{\Gamma}_0 \in \mathcal{G}_0$. Then, if $\sigma(\zeta\Lambda_t^{s+\varepsilon}[X_i, \Gamma])$ is the symbol of $\zeta\Lambda_t^{s+\varepsilon}[X_i, \Gamma]$, we have

$$\sigma(\zeta\Lambda_t^{s+\varepsilon}[X_i, \Gamma]) = \sigma(\zeta\Lambda_t^{s+\varepsilon}[X_i, \Gamma])\tilde{\gamma}_0.$$

Hence,

$$\begin{aligned}\|\zeta\Lambda_t^{s+\varepsilon}[X_i, \Gamma])u\|_\varepsilon^2 &= \|\zeta\Lambda_t^{s+\varepsilon}[X_i, \Gamma])\tilde{\Gamma}_0 u\|_\varepsilon^2 + O(\|u\|_{-\infty}^2)\\ &\le C\|\tilde{\Gamma}_0 u\|_{s+2\varepsilon}^2 + O(\|u\|_{-\infty}^2)\\ &\le C\|\sum \bar{X}_i X_i u\|_s^2 + O(\|u\|_{-\infty}^2).\end{aligned}$$

Then, we obtain

$$\|\Gamma u\|_{s+2\varepsilon}^2 \le C\left\|\sum \bar{X}_i X_i u\right\|_s^2 + O(\|u\|_{-\infty}^2).$$

Therefore, since

$$\|u\|_{s+2\varepsilon} \le \|\Gamma u\|_{s+2\varepsilon} + \|\Gamma_0 u\|_{s+2\varepsilon},$$

we conclude that (3) holds, thus completing the proof.

Theorem 1. *If $X_1, \ldots, X_l$ are special vector fields as above and if (3) holds on $U \subset \mathbb{R}^{n+1}$ for some $\varepsilon \in \mathbb{R}$, then if $f \in H^s(\mathbb{R}^n)$ there exists a $u \in H^{s+2\varepsilon}(\mathbb{R}^{n+1})$ such that on U we have $Eu = f$.*

Proof. Let $\tilde{E}^{\varepsilon}$ denote the operator on $C_0^{\infty}(U)$ defined by

$$(\tilde{E}^{\varepsilon}v, w) = (v, E\Lambda^{-2\varepsilon}w),$$

for all distributions w on $\mathbb{R}^{n+1}$. So that $\tilde{E}^{\varepsilon} = \Lambda^{-2\varepsilon}E$. Then, substituting $-s-2\varepsilon$ for s and v for u in (3), we get

$$\|v\|_{-s} \le C\|Ev\|_{-s-2\varepsilon} = C\|\tilde{E}^{\varepsilon}\|_{-s},$$

for all $v \in C_0^{\infty}(U)$. Let $\mathcal{S}$ be the subspace of $H^{-s}(\mathbb{R}^n)$ given by

$$\mathcal{S} = \{g \in C^{\infty}(\mathbb{R}^n) \mid \exists v \in C_0^{\infty}(U) \, \textit{such that}\, g = \tilde{E}^{\varepsilon}v\}.$$

Given $f \in H^s(\mathbb{R}^n)$ let $T : \mathcal{S} \to \mathbb{C}$ be the linear functional defined by $T(\tilde{E}^{\varepsilon}v) = (v, f)$. Then, we have

$$|T(\tilde{E}^{\varepsilon}v)| = |(v, f)| \le \|v\|_{-s}\|f\|_s \le C\|\tilde{E}^{\varepsilon}v\|_{-s}.$$

Hence, T is bounded in the $-s$ norm and hence can be extended to a bounded functional on $H^{-s}(\mathbb{R}^{n+1})$ so that there exists $w \in H^s(\mathbb{R}^{n+1})$ such that $T(\tilde{E}^{\varepsilon}v) = (\tilde{E}^{\varepsilon}v, w)$. Therefore, $(v, E\Lambda^{-2\varepsilon}w) = (v, f)$ for all $v \in C_0^{\infty}(U)$ so setting $u = \Lambda^{-2\varepsilon}w$, we have $u \in H^{s+2\varepsilon}(\mathbb{R}^{n+1})$ and $Eu = f$ on U. Concluding the proof of the theorem.

The Energy Estimate

When the vector fields $X_1, \dots, X_l$ are real, Hörmander in [H] proved that a necessary and sufficient condition for the energy estimate (2), with $\varepsilon > 0$, is that the bracket condition hold for $X_1, \dots, X_l$. In the case of complex special vector fields and $\varepsilon \in \mathbb{R}$, the bracket condition is still sufficient (2), but it is not necessary. To see this, consider the vector field $X = \frac{\partial}{\partial \bar{z}}$ on $\mathbb{C} \sim \mathbb{R}^2$. We have here $\|Xu\| \sim \|u\|_1$, for $u \in C_0^{\infty}(U)$, even though the bracket condition does not hold.

Observe that the subelliptic energy estimate (2) holds for complex vector fields whenever it holds for the vector fields $X_1, \dots, X_l, X_1^*, \dots, X_l^*$ and if

$$\sum_1^l \|X_i^* v\|^2 \le C\left(\sum_1^l \|X_i v\|^2 + \|v\|^2\right), \tag{5}$$

for all $v \in C_0^{\infty}(U)$.

Theorem 2. *If $X_1, \dots, X_l, \bar{X}_1, \dots, \bar{X}_l$ are complex vector fields satisfying the bracket condition at x_0 and if $[X_i, \bar{X}_i] \in \mathcal{F}^2(X_1, \dots, X_l)$ in a neighborhood of x_0, then there exists $U \ni x_0$ and $\varepsilon > 0$ such that the subelliptic energy estimate (2) holds.*

Proof. We have

$$[X_i, X_i^*] = \sum a_{ij}^{hk}[X_h, X_k] + \sum b_{ij}^h X_h + c_{ij}.$$

Then,

$$\begin{aligned}\|X_i^* v\|^2 &= \|X_i v\|^2 + ([X_i, X_i^*]v, v)\\ &= \|X_i v\|^2 + \sum (a_{ij}^{hk}[X_h, X_k]v, v) + \sum (b_{ij}^h X_h v, v) + (c_{ij} v, v)\end{aligned}$$

and

$$\begin{aligned}|(a_{ij}^{hk}[X_h, X_k]v, v)| &= |(a_{ij}^{hk}(X_h X_k v, v)| + |(a_{ij}^{hk} X_k X_h v, v)|\\ &\quad + O\left(\sum \|X_h v\|^2 + \|v\|^2\right).\end{aligned}$$

Furthermore,

$$|(a_{ij}^{hk}(X_h X_k v, v)| \le l.c.\|X_k v\|^2 + s.c.\|\bar{X}_h v\|^2 + s.c.\|v\|^2.$$

Hence, combining the above and summing, we obtain

$$\sum \|\bar{X}_i v\|^2 \le C\left(\sum \|X_i v\|^2 + \|v\|^2\right) + s.c. \sum \|\bar{X}_i v\|^2.$$

Thus, choosing the small constant $l.c.$ small enough, we obtain (5) concluding the proof.

Conjecture. Suppose that $X_1, \ldots, X_l$ are complex vector fields on $\mathbb{R}^{n+1}$. Suppose that $X_1, \ldots, X_l, \bar{X}_1, \ldots, \bar{X}_l$ satisfy the bracket condition of type m, that is, m is the least integer such that

$$\mathbb{C}T_0(\mathbb{R}^{n+1}) = \mathcal{F}_0^m(X_1, \ldots, X_l, \bar{X}_1, \ldots, \bar{X}_l).$$

Further suppose that q is an integer such that

$$[X_i, \bar{X}_i]_0 \in \mathcal{F}^q(X_1, \ldots, X_l),$$

for $i = 1, \ldots, l$. Then, there exist a neighborhood U of 0 and $C > 0$ such that

$$\|u\|^2_{\frac{3-q}{m+1}} \le C \sum_1^l \|X_i u\|^2$$

when $q \ge 2$ and for all $u \in C_0^\infty(U)$.

Note that when the X are real, then when can take $q = 2$, and the conjecture gives $\varepsilon = \frac{1}{m+1}$ which was proved by Rothschild and Stein (see [RS]). For the deformations of the Lewy operators discussed at the beginning of this chapter, we have $q = k + 2$. The conjecture holds for the special complex vector fields defined above. Here we will prove it for the following vector fields.

In $\mathbb{R}^2$, let

$$X_1 = \frac{\partial}{\partial x} + \mathrm{i}x^{m-1}\frac{\partial}{\partial t} \qquad \text{and} \qquad \mathrm{X}_2 = \mathrm{x}^{\mathrm{k}}\bar{\mathrm{X}}_1. \tag{6}$$

When m is odd then, as proved in [M], both operators X_1 and E are subelliptic with gains of $\frac{1}{m}$ and $\frac{2}{m}$, respectively. However, when m is even, the estimate proved below (i.e., with a loss of $\frac{2(k-1)}{m}$ derivatives for E) is optimal as proved in the following section.

Note that we have

$$\mathbb{C}T_0 = \mathcal{F}^m(X_1, X_2, \bar{X}_1, \bar{X}_2)$$

and

$$[X_1, \bar{X}_1] \in \mathcal{F}^{k+2}(X_1, X_2)$$

so that $q = k + 2$.

The subelliptic estimate proved in the lemma below is a special case of the result proved by Rothschild and Stein in [RS] mentioned above.

Lemma. *If U is a neighborhood of the origin, then there exists C such that*

$$\|u\|^2_{\frac{1}{m}} \le C(\|X_1u\|^2 + \|\bar{X}_1u\|^2), \tag{7}$$

for all $u \in C_0^\infty(U)$. Furthermore,

$$\left\|x^{m-2}\frac{\partial u}{\partial t}\right\|^2_{-\frac{1}{m}} \le C(\|X_1u\|^2 + \|\bar{X}_1u\|^2), \tag{8}$$

for all $u \in C_0^\infty(U)$.

Proof. Since $\bar{X}_1 = -\frac{\partial}{\partial x} + \mathrm{i}x^{m-1}\frac{\partial}{\partial t}$, we have

$$\frac{\partial}{\partial x} = \frac{1}{2}(X_1 - \bar{X}_1)$$

and

$$x^{m-1}\frac{\partial}{\partial t} = \frac{-\mathrm{i}}{2}(X_1 + \bar{X}_1).$$

Thus, $\mathcal{F}^h(X_1, \bar{X}_1) = \mathcal{F}^h(\frac{\partial}{\partial x}, x^{m-1}\frac{\partial}{\partial t})$. Hence, for $0 \le h \le m-1$, $\mathcal{F}^h(X_1, \bar{X}_1)$ is spanned by $\frac{\partial}{\partial x}$ and $x^{m-h-1}\frac{\partial}{\partial t}$. In particular, when $h = 0$, we have

$$\|X_1u\|^2 + \|X_1^*u\|^2 \sim \left\|\frac{\partial u}{\partial x}\right\|^2 + \left\|x^{m-1}\frac{\partial u}{\partial t}\right\|^2. \tag{9}$$

Thus, when $m = 1$, the right sides of (7) and (8) equal $\|u\|_1^2$, and the lemma is proved. When $m > 1$, suppose that $\varepsilon \geq 0$ and that

$$\|u\|_\varepsilon^2 \leq C(\|X_1 u\|^2 + \|\bar{X}_1 u\|^2),$$

for all $u \in C_0^\infty(U)$. Then,

$$\begin{aligned}\|u\|_\varepsilon^2 &= \left(\frac{\partial x}{\partial x}\Lambda^\varepsilon u, \Lambda^\varepsilon u\right) = -\left(x\frac{\partial}{\partial x}\Lambda^\varepsilon u, \Lambda^\varepsilon u\right) - \left(x\Lambda^\varepsilon u, \frac{\partial}{\partial x}\Lambda^\varepsilon u\right)\\ &\leq 2\left|\left(x\Lambda^\varepsilon u, \frac{\partial}{\partial x}\Lambda^\varepsilon u\right)\right| \leq \|x\Lambda^{2\varepsilon}u\|^2 + \left\|\frac{\partial u}{\partial x}\right\|^2 + l.c.\|[\Lambda^\varepsilon, x]\Lambda^\varepsilon u\|^2 + s.c.\|u\|_\varepsilon^2.\end{aligned}$$

The pseudodifferential operator $[\Lambda^\varepsilon, x]\Lambda^\varepsilon$ is of order $2\varepsilon - 1$ so that if $\varepsilon < 1$ and if the diameter of U is sufficiently small, we have

$$\|[\Lambda^\varepsilon, x]\Lambda^\varepsilon u\|^2 \leq C\ \|u\|_{2\varepsilon-1}^2 \leq s.c.\ \|u\|_\varepsilon^2.$$

Hence, inductively, we obtain

$$\begin{aligned}\|u\|_\varepsilon^2 &\leq C\ \left\|x\Lambda^{2\varepsilon}u\right\|^2 + \left\|\frac{\partial u}{\partial x}\right\|^2\\ &\leq C\left(\left|(x^2\Lambda^{3\varepsilon}u, \Lambda^\varepsilon u)\right| + \left\|[x^2, \Lambda^\varepsilon]\Lambda^{2\varepsilon}u\right\|^2 + \left\|\frac{\partial u}{\partial x}\right\|^2\right)\\ &\leq C\ \left\|x^2\Lambda^{3\varepsilon}u\right\|^2 + \left\|\frac{\partial u}{\partial x}\right\|^2 + \text{“error}_2\text{”}\\ &\vdots \qquad\qquad \vdots\\ &\leq C\left(\left\|x^{m-2}\Lambda^{(m-1)\varepsilon}u\right\|^2 + \left\|\frac{\partial u}{\partial x}\right\|^2\right) + \text{“error}_{m-2}\text{”}\\ &\leq C\left(\left\|x^{m-1}\Lambda^{m\varepsilon}u\right\|^2 + \left\|\frac{\partial u}{\partial x}\right\|^2\right) + \text{“error}_{m-1}\text{”},\end{aligned}$$

where

$$\text{“error}_h\text{”} \leq C\|x^{h-1}\Lambda^{(h+1)\varepsilon-1}u\|^2 \leq s.c.\|x^{h-1}\Lambda^{h\varepsilon}u\|^2,$$

when $\varepsilon < 1$. Therefore, if $\varepsilon = \frac{1}{m}$, we have

$$\|u\|_{\frac{1}{m}}^2 \leq C\left(\left\|x^{m-1}\Lambda^1 u\right\|^2 + \left\|\frac{\partial u}{\partial x}\right\|^2\right) \leq C\left(\|x^{m-1}u\|_1^2 + \left\|\frac{\partial u}{\partial x}\right\|^2 + s.c.\|u\|^2\right).$$

Then,

$$\begin{aligned}\left\|x^{m-1}u\right\|_1^2 &= \left\|x^{m-1}u\right\|^2 + \left\|\frac{\partial}{\partial x}(x^{m-1}u)\right\|^2 + \left\|\frac{\partial}{\partial t}(x^{m-1}u)\right\|^2\\ &\le \|u\|^2 + \left\|\frac{\partial u}{\partial x}\right\|^2 + \left\|x^{m-1}\frac{\partial u}{\partial t}\right\|^2.\end{aligned}$$

which proves (7) since $\|u\|^2 \le s.c.\|u\|^2_{\frac{1}{m}}$ when the diameter of U is small. Similarly, we obtain

$$\|x^h \Lambda^{\frac{h+1}{m}} u\|^2 = \|x^h u\|_1^2 + O(\|u\|^2).$$

Then, since $x^h \frac{\partial}{\partial t} \in \mathcal{F}^{m-h-1}(X_1, \bar{X}_1)$, we obtain (8) with the same argument as the proof of (7) completing the proof of the lemma.

Note that if $\alpha > 0$ and if

$$\|\bar{X}_1 u\|^2_{-\alpha} \lesssim \sum_{j=1,2} \|X_j u\|^2 + \|u\|^2_{-\alpha}$$

then

$$\|u\|^2_{-\alpha+\frac{1}{m}} \lesssim \sum_{j=1,2} \|X_j u\|^2.$$

Proposition. *If U has small enough diameter so that the above lemmas hold and if $s \in \mathbb{R}$, then*

$$\|u\|^2_{s-\frac{k-1}{m}} \le C \sum_{j=1,2} \|X_j u\|^2_s, \tag{10}$$

for all $u \in C_0^\infty(U)$.

Proof. Since $\bar{X}_1 = \frac{\partial}{\partial x} - \mathrm{i}x^{m-1}\frac{\partial}{\partial t}$, we have

$$x^{m-1}\frac{\partial}{\partial t} = \frac{\mathrm{i}}{2}(\bar{X}_1 - X_1).$$

Hence,

$$\begin{aligned}\|\bar{X}_1 u\|^2_{-\alpha} &\le \left\|x^{m-1}\frac{\partial u}{\partial t}\right\|^2_{-\alpha} + \|X_1 u\|^2_{-\alpha} \le |\left(\Lambda^{-\alpha} x^m \frac{\partial u}{\partial t}, \Lambda^{-\alpha} x^{m-2}\frac{\partial u}{\partial t}\right)| + \|X_1 u\|^2_{-\alpha}\\ &\le \left|\left(\Lambda^{-\alpha} x\bar{X}_1 u, \Lambda^{-\alpha} x^{m-2}\frac{\partial u}{\partial t}\right)\right| + \|X_1 u\|^2_{-\alpha}\\ &\le \left|\left(\Lambda^{-\alpha+\frac{1}{m}} x\bar{X}_1 u, \Lambda^{-\alpha-\frac{1}{m}} x^{m-2}\frac{\partial u}{\partial t}\right)\right| + \|X_1 u\|^2_{-\alpha}\end{aligned}$$

$$\leq l.c.\|\Lambda^{-\alpha+\frac{1}{m}}\, x\bar{X}_1 u\|^2 + s.c.\left\|\Lambda^{-\alpha-\frac{1}{m}}\, x^{m-2}\frac{\partial u}{\partial t}\right\|^2 + \|X_1 u\|^2_{-\alpha}.$$

From (8), we get

$$\left\|x^{m-2}\frac{\partial u}{\partial t}\right\|^2_{-\alpha-\frac{1}{m}} \leq C(\|X_1 u\|^2_{-\alpha} + \|\bar{X}_1 u\|^2_{-\alpha}).$$

Therefore, using the Schwarz inequality and induction, we obtain

$$\begin{aligned}\|\bar{X}_1 u\|^2_{-\alpha} &\leq C(\|x\bar{X}_1 u\|^2_{-\alpha+\frac{1}{m}} + \|X_1 u\|^2_{-\alpha})\\ &\leq C(\|x^2\bar{X}_1 u\|^2_{-\alpha+\frac{2}{m}} + \|X_1 u\|^2_{-\alpha}) \leq \cdots\\ &\leq C(\|x^k\bar{X}_1 u\|^2_{-\alpha+\frac{k}{m}} + \|X_1 u\|^2_{-\alpha}) \leq C(\|X_2 u\|^2_{-\alpha+\frac{k}{m}} + \|X_1 u\|^2_{-\alpha}).\end{aligned}$$

Then, setting $\alpha = \frac{k}{m}$, we get

$$\|\bar{X}_1 u\|^2_{-\frac{k}{m}} \leq C(\|X_1 u\|^2_{-\frac{k}{m}} + \|X_2 u\|^2).$$

Hence,

$$\|u\|^2_{s-\frac{k-1}{m}} \leq C(\|X_1 u\|^2_{s-\frac{k}{m}} + \|X_2 u\|^2_s) \leq C\sum_{j=1,2}\|X_j u\|^2_s.$$

This concludes the proof of the proposition.

Note that the proposition implies

$$\|u\|_{s-2\left(\frac{k-1}{m}\right)} \leq C\|Eu\|_s, \tag{11}$$

for all $u \in C_0^\infty(U)$.

Optimal Estimates

For X_1, X_2 defined by (6), when m is even, then the estimate (11) is optimal, as proved below. However, if m is odd, then E is subelliptic with a gain of $\frac{2}{m}$ derivatives as is shown in [M].

Proposition. *Given a small neighborhood, U of $(0,0)$ and $C, s, s' \in \mathbb{R}$ such that*

$$\|u\|^2_{s'} \leq C\sum_{j=1}^{2}\|X_j u\|^2_s, \tag{12}$$

for all $u \in C_0^\infty(U)$. When m is even, we have $s \leq s' - \frac{k-1}{m}$ for all s.

Proof. When m is even, we define for each $\lambda >> 0$ the function $u_\lambda \in C_0^m(U)$ as follows. Let $\varphi \in C_0^\infty(U)$ such that $\varphi = 1$ in a neighborhood of the origin and then set

$$u_\lambda = \varphi e^{-\lambda h_m},$$

where h_m is given by

$$h_m(x,t) = \frac{1}{m}x^m + \mathrm{i}t - \left(\frac{1}{m}x^m + \mathrm{i}t\right)^2 \tag{13}$$

Then, $X_1 h_m = 0$, and for $|x|$ small, we have

$$\mathrm{Re}(h_m) \sim |x|^m + t^2.$$

Hence, when $|\alpha| > 0$ we have $(D^\alpha \varphi)u_\lambda = O(\lambda^{-N})$ for every N, since $D^\alpha \varphi$ vanishes in a neighborhood of the origin. Let $(x,t;\xi,\tau)$ be coordinates of the cotangent space and let Γ, as above, have support in a cone containing the τ axis but not the ξ axis. Then, in the support of Γ, we have $1 + |\xi|^2 + |\tau|^2 \leq \text{const.}(1 + \|\tau\|^2)$. Hence,

$$\|\Gamma v\|_s \sim \|\Lambda_t^s \Gamma v\|,$$

for all $v \in C_0^\infty(U)$, where Λ_t^s is the operator with symbol $(1 + |\tau|^2)^{\frac{s}{2}}$. Using the change of variables

$$\begin{cases} x' = \lambda^{\frac{1}{m}} x \\ t' = \lambda t \\ \xi' = \lambda^{-\frac{1}{m}} \xi \\ \tau' = \lambda^{-1} \tau \end{cases}$$

we obtain $\lambda^s \leq \text{const.}\lambda^{s' - \frac{k-1}{m}}$ for large λ so that $s \leq s' - \frac{k-1}{m}$, which concludes the proof.

Hypoellipticity

Definition. An operator E is *hypoelliptic* if for any distributions u and f, satisfying $Eu = f$ with $f|_U \in C^\infty(U)$ then $u|_U \in C^\infty(U)$.

This definition implies the following estimate. If $\zeta, \tilde{\zeta} \in C_0^\infty(U)$ with $\tilde{\zeta} = 1$ on a neighborhood of supp(ζ), then for each s there exist s' and $C \in \mathbb{R}$ such that

$$\|\zeta u\|_s \leq C\|\tilde{\zeta} E u\|_{s'} + O(\|u\|_{-\infty}), \tag{14}$$

for all $u \in C^\infty(U)$. Conversely, if for each $x_0 \in U$ there exist $\zeta, \tilde{\zeta} \in C_0^\infty(U)$ such that $\zeta = 1$ in a neighborhood of x_0 and $\tilde{\zeta} = 1$ on a neighborhood of supp(ζ), then (14) implies hypoellipticity whenever there exists an appropriate smoothing operator as described below.

Next, we will outline the proof of hypoellipticity when the vector fields are given by (6) and when m is even. To simplify matters, we will replace m by $2m$ and (6) by

$$X_1 = \frac{\partial}{\partial x} + \mathrm{i}x^{2m-1}\frac{\partial}{\partial t} \qquad \text{and} \qquad \mathrm{X}_2 = \mathrm{x}^{\mathrm{k}}\bar{\mathrm{X}}_1. \tag{15}$$

Here we outline the proof of the following:

Theorem 3. *The operator $E = -\bar{X}_1X_1 - \bar{X}_2X_2$, where X_1 and X_2 are given by (15), is hypoelliptic.*

The main step is the a priori estimate

$$\|\zeta u\|_s \le C\,\|\tilde{\zeta} Eu\|_{s+\frac{k-1}{2m}} + O(\|u\|_{-\infty}), \tag{16}$$

The proof involves an additional microlocalization as follows. Let

$$\mathfrak{G}^+ = \{\gamma \in \mathfrak{G} \mid \gamma(\xi,\tau) = 0, \text{ when } |\xi|^2 + |\tau|^2 \ge 1 \text{ and } \tau \le 0$$

and let $\mathfrak{G}^- = \mathfrak{G} - \mathfrak{G}^+$. The corresponding sets of operators are then denoted by $\mathcal{G}^+$ and $\mathcal{G}^-$, respectively. If $\Gamma^- \in \mathcal{G}^-$ then, since $\tau = -|\tau|$ on $\mathrm{supp}(\gamma^-)$

$$(\mathrm{i}x^{2(m-1)}\Gamma^-\left(\frac{\partial w}{\partial t}\right), \Gamma^- w) \sim (-\Lambda_t x^{2(m-1)}\Gamma^-\left(\frac{\partial w}{\partial t}\right), \Gamma^- w) \sim -\|x^{m-1}\Gamma^- w\|^2_{\frac{1}{2}};$$

hence, since

$$\begin{aligned}\|X_1\Gamma^- w\|^2 &= (-[\bar{X}_1, X_1]\Gamma^- w, \Gamma^- w) + \|\bar{X}_1\gamma^- w\|^2\\ &= \left(-2\mathrm{i}x^{2(m-1)}\Gamma^-\left(\frac{\partial w}{\partial t}\right), \Gamma^- w\right) + \|\bar{X}_1\gamma^- w\|^2,\end{aligned}$$

then

$$\|x^{m-1}\Gamma^- w\|^2_{\frac{1}{2}} + \|\bar{X}_1\Gamma^- w\|^2 \lesssim \|X_1\Gamma^- w\|^2$$

and since

$$\|\Gamma^- w\|^2_{\frac{1}{2m}} \lesssim \|X_1\Gamma^- w\|^2 + \|\bar{X}_1\Gamma^- w\|^2$$

we have

$$\|\Gamma^- w\|^2_{\frac{1}{2m}} + \|\bar{X}_1\Gamma^- w\|^2 \lesssim \|X_1\Gamma^- w\|^2 \tag{17}$$

and similarly

$$\|\Gamma^+ w\|^2_{\frac{1}{2m}} + \|X_1\Gamma^+ w\|^2 \lesssim \|\bar{X}_1\Gamma^+ w\|^2, \tag{18}$$

for all $w \in C_0^\infty(U)$. Then, it follows that

$$\|\Gamma^- w\|^2_{\frac{1}{m}} \lesssim \|Ew\|^2 + \|\Gamma_0 w\|^2;$$

hence, replacing w by $\Lambda_t^{s-\frac{k-2}{m}} \zeta u$, we get, inductively, the following estimate that "gains" $\frac{1}{m}$ derivatives

$$\|\Gamma^- \zeta u\|^2_{s-\frac{k-1}{m}} \lesssim \|\Gamma^- \tilde{\zeta} E u\|^2_{s-\frac{k-2}{m}} + \|\tilde{\zeta}\Gamma_0 u\|^2_{s-\frac{k-2}{m}} + O(\|w\|^2_{-\infty}),$$

therefore

$$\|\Gamma^- \zeta u\|^2_{s-\frac{k-1}{m}} \lesssim \|\tilde{\zeta} E u\|^2_s + \|\tilde{\zeta}\Gamma_0 u\|^2_{s-\frac{k-2}{m}} + O(\|u\|^2_{-\infty}), \tag{19}$$

Note that in a neighborhood of (x,t) with $x \neq 0$, the operator E is elliptic; hence, it suffices to treat the case $x_0 = (0, t_0)$. Let $\varphi, \tilde{\varphi}, \theta, \tilde{\theta} \in C_0^\infty(\mathbb{R})$ with $\varphi(x) = 1$ in a neighborhood of 0, $\theta(t) = 1$ in a neighborhood of t_0, let $\tilde{\varphi}(x) = 1$ in a neighborhood of $\mathrm{supp}(\varphi)$, $\tilde{\theta}(t) = 1$ in a neighborhood of $\mathrm{supp}(\theta)$, let $\zeta(x,t) = \varphi(x)\theta(t)$, and let $\tilde{\zeta}(x,t) = \tilde{\varphi}(x)\tilde{\theta}(t)$. Then,

$$X_1(\zeta u)(x,t) = \varphi'(x)\theta(t)u(x,t) + \mathrm{i}x^{m-1}\varphi(x)\theta'(t)u(x,t) + \zeta(x,t)X_1(u)(x,t)$$

and

$$X_2(\zeta u)(x,t) = \varphi'(x)\theta(t)u(x,t) + \mathrm{i}x^{k+m-1}\varphi(x)\theta'(t)u(x,t) + \zeta(x,t)X_2(u)(x,t).$$

First, we observe that $\varphi' = 0$ in a neighborhood of the t-axis and $\tilde{\varphi} = 1$ in a neighborhood of $\mathrm{supp}(\varphi)$ then, since E is elliptic in $\mathrm{supp}(\varphi')$, we have

$$\|\varphi' u\|_s \leq C\|E(\varphi u)\|_{s-2} \leq C(\|\varphi E u\|_{s-2} + \|\tilde{\varphi}' u\|_{s-1} + \|u\|_{-\infty},$$

where $\tilde{\varphi}'$ is a function that vanishes in a neighborhood of the t-axis and $\tilde{\varphi}' = 1$ in a neighborhood of $\mathrm{supp}(\theta')$. Then, applying the above to the last term recursively, we obtain

$$\|\varphi' u\|_s \lesssim \|\tilde{\varphi} E u\|_{s-2} + \|u\|_{-\infty},$$

Similarly, since E is elliptic on $\Gamma_0 u$, we obtain

$$\|\zeta\Gamma_0 u\|_s \lesssim \|\tilde{\zeta} E u\|_{s-2} + \|u\|_{-\infty}. \tag{20}$$

Thus, to prove (16), it suffices to prove it in the case when u is replaced by $u^+ = \Gamma^+(\varphi u)$ and ζ by θ. The main difficulty is the localization in space; one cannot have a term with the cutoff function θ between u and X_1, or $\bar{X}_1$, unless the terms also contains suitable powers of x. We will give brief description of the method used in [K], a variant is given in [BDKT]. Substituting $\theta\Lambda_t^s X_1 u^+$ for $\Gamma^+ w$ in (18), we have

$$\|X_1\theta\Lambda_t^s X_1 u^+\|^2 + \|\theta\Lambda_t^s X_1 u^+\|^2_{\frac{1}{2m}} \lesssim \|\bar{X}_1\theta\Lambda_t^s X_1 u^+\|^2 + \|\theta\Lambda_t^s X_1 u^+\|^2 + \|u\|^2_{-\infty},$$

so that

$$\begin{aligned}&\|\theta\bar{X}_1 X_1 u^+\|_s^2 + \|\theta X_1^2 u^+\|_s^2 + \|\theta X_1 u^+\|^2_{s+\frac{1}{2m}} \\ &\qquad \lesssim |(\theta X_1\bar{X}_1 X_1 u^+, \theta X_1 u^+)_s| + \|\theta' X_1 u^+\|_s^2 + \|\tilde{\theta} u^+\|_{s-1}^2 + \|u\|^2_{-\infty}.\end{aligned}$$

Then,

$$X_1\bar{X}_1 X_1 = -X_1 E - X_1^2 x^{2k}\bar{X}_1$$

and using integration by parts, as in [K] pp. 971–972, we obtain

$$\begin{aligned}&\|\theta\bar{X}_1 X_1 u^+\|_s^2 + \|\theta X_1 u^+\|^2_{s+\frac{1}{2m}} \\ &\qquad \lesssim \|\theta' E u^+\|_s^2 + \|x^{2k-2}\theta' u^+\|_s^2 + \|x^{2k-1}\theta' u^+\|^2_{s+\frac{1}{2m}} + \|u\|^2_{-\infty}.\end{aligned}$$

The estimate (16) is then obtained by following the arguments in [K] pp. 973–978. Alternately, a somewhat different derivation of this estimate is given in [BDKT] pp. 4–10.

To prove that (16) implies that E is hypoelliptic, we will show that if u is a distribution solution of $Eu = f$ in U, if ζ and $\tilde{\zeta}$ are in $C_0^\infty(U)$ with $\tilde{\zeta} = 1$ in a neighborhood of supp(ζ), and if $\tilde{\zeta} f \in H^{s+\frac{k-1}{m}}$, then $\zeta u \in H^s$. To do this, we will use smoothing operators S_δ and S_δ^+ having the property that for any distribution u, we have $S_\delta u$, $S_\delta^+\Gamma^+ u \in C^\infty$ and that if $\|\zeta S_\delta u\|_s \le C$, with C independent of δ, then $\zeta u \in H^s$. Similarly, if $\|\zeta S_\delta^+\Gamma^+ u\|_s \le C$, with C independent of δ, then $\zeta\Gamma^+ u \in H^s$. Since E is elliptic on $\Gamma^0 u$, it follows, using a standard smoothing operator S_δ, that if $\tilde{\zeta} f \in H^{s-2}$ and $\tilde{\zeta}\Gamma u \in H^s$, then $\Gamma^0 u \in H^s$. Similarly, since E is subelliptic on $\Gamma^- u$, then if $\tilde{\zeta} f \in H^{s-\frac{1}{m}}$ and $\tilde{\zeta}\Gamma^+ u \in H^s$, then $\Gamma^0 u \in H^s$. Let $\chi \in C_0^\infty(\mathbb{R})$ such that $\chi(0) = 1$ and let S_δ^+ be a pseudodifferential operator whose symbol $\sigma(S_\delta^+)$ satisfies

$$\gamma^+\sigma(S_\delta^+)(\xi,\tau) = \gamma^+(\xi,\tau)\chi(\delta\tau).$$

Then, the support of the symbols of $[X_i, \Gamma^+ S_\delta]$ lies in the support of some $\gamma^0 \in \mathfrak{G}^0$. Substituting $S_\delta^+\Gamma^+ u$ for u in (16), we obtain

$$\|\zeta S_\delta^+\Gamma^+ u\|_s \lesssim \|\tilde{\zeta} f\|_{s+\frac{k-1}{m}} + \|\tilde{\zeta}\Gamma^0 u\|_{s+\frac{k-1}{m}} + O(\|u\|_{-\infty}),$$

and by the above, the left-hand side is bounded independently of δ. Therefore, $\zeta\Gamma^+u$, $\zeta\Gamma^-u$, and $\zeta\Gamma^0u$ are in H^s. It then follows that $\zeta u \in H^s$ which concludes the proof of hypoellipticity.

Remark. M. Christ in [C] has proved the following. If $X_1, \ldots, X_l$ are complex-valued vector fields on $\mathbb{R}^n$ and the corresponding operator $E = \sum X_i^* X_i$ is hypoelliptic with loss of derivatives, then the operator

$$E' = E - \frac{\partial^2}{\partial x_{n+1}^2}$$

on $\mathbb{R}^{n+1}$ is not necessarily hypoelliptic. For more examples of this phenomenon, see [BMT].

References

[BDKT] Bove, A., Derridj, M., Kohn, J.J. and Tartakoff, D.S.: "Sums of squares of complex vector fields and (analytic-) hypoellipticity." *Math. Res. Lett.* 13 (2006), no. 5-6, 683–701.

[BMT] Bove, A., Mughetti, M., and Tartakoff, D.S.: "Hypoellipticity and non-hypoellipticity for sums of squares of complex vector fields," preprint (2011) to appear in Analysis and PDE

[C] Christ, M.: "A remark on sum of squares of complex vector fields," preprint posted in the archive math.CV/0503506.

[H] Hörmander, L.,: "Hypoelliptic second order differential equations," *Acta Math.* 119 (1967), 147–171.

[KPZ] Khanh, T.V., Pinton, S., and Zampieri, G.,: "Loss of derivatives for systems of complex vector fields and sums of squares," *arXiv:*submit/0102171 (2010).

[K] Kohn, J.J.,: "Hypoellipticity and loss of derivatives," *Annals of Math.* 162(2005), 943–986.

[KR] Kohn, J.J. and Rossi, H.,: "On the extension of holomorphic functions on the boundary of a manifold," *Annals of Mathematics*, 81, (1965), 451–472.

[L] Lewy, H.: "An example of a smooth linear partial differential equation without solution", *Annals of Mathematics*, 66 (1957): 155–158.

[M] Menikoff, A.: "Some examples of hypoelliptic partial differential equations," *Math. Ann.* 221 (1976), 167–181.

[PP] Parenti, C. and Parmeggiani, A.,: "On the hypoellipticity with a big loss of derivatives," *Kyushu J. Math.* 59 (2005), 155–230

[RS] Rothschild, L. P. and Stein, E. M.,: "Hypoelliptic differential operators and nilpotent groups," *Acta Math.*, 137 (1976), 247–320.

On an Oscillatory Result for the Coefficients of General Dirichlet Series

Winfried Kohnen and Wladimir de Azevedo Pribitkin

In memory of Leon Ehrenpreis, so strong and full of brightness

Key words General Dirichlet series • Oscillatory sequence

Mathematics Subject Classification: Primary 11M41, 30B50.

1 Introduction

Let $(a_n)_{n\geq 1}$ be a sequence of real numbers. Then we say that $(a_n)_{n\geq 1}$ is oscillatory if there exist infinitely many n with $a_n > 0$ and infinitely many n with $a_n < 0$.

Recall that a general Dirichlet series is a series of the form

$$\sum_{n\geq 1} a_n \mathrm{e}^{-\lambda_n s},$$

where the a_n $(n \geq 1)$ are complex numbers, the exponent sequence $(\lambda_n)_{n\geq 1}$ is real and strictly increasing to ∞, and s is a complex variable.

In [3] the second author proved the following general result.

W. Kohnen
Mathematisches Institut der Universität, INF 288, 69120 Heidelberg, Germany
e-mail: winfried@mathi.uni-heidelberg.de

W. de A. Pribitkin (✉)
Department of Mathematics, College of Staten Island, CUNY, 2800 Victory Boulevard, Staten Island, NY 10314, USA
e-mail: Wladimir.Pribitkin@csi.cuny.edu; w_pribitkin@msn.com

The second author's work was supported (in part) by The City University of New York PSC-CUNY Research Award Program (grant #62571-00 40 and grant #63516-00 41).

H.M. Farkas et al. (eds.), *From Fourier Analysis and Number Theory to Radon Transforms and Geometry*, Developments in Mathematics 28, DOI 10.1007/978-1-4614-4075-8_18,

Theorem. *Let* $(a_n)_{n\geq 1}$ *be a sequence of real numbers, not all of the* a_n *being zero, and suppose that the general Dirichlet series*

$$F(s) = \sum_{n\geq 1} a_n \mathrm{e}^{-\lambda_n s}$$

is convergent for $\sigma := \Re(s) > \sigma_0$. *Suppose further that* $F(s)$ *has an analytic continuation to an open connected subset of* $\mathbf{C}$ *containing the real line and that* $F(s)$ *has infinitely many real zeros. Then* $(a_n)_{n\geq 1}$ *is oscillatory.*

In [3] this result was deduced by combining two classical results in number theory, namely Landau's well-known theorem on general Dirichlet series with nonnegative coefficients (see, e.g., [1]) and secondly Laguerre's rule concerning the sign changes of coefficients of general Dirichlet series [2].

From the above theorem one can establish immediately the oscillatory behavior of the coefficients of many important families of Dirichlet series occurring in number theory and the theory of automorphic forms. For a variety of such examples, see [4].

The purpose of this note is to point out another very simple proof of the above theorem, based only on Landau's result coupled with some completely elementary arguments.

2 A Very Simple Proof

First note that $F(\sigma) \neq 0$ for $\sigma \in \mathbf{R}$, $\sigma \gg 0$. Indeed, this is well known, but we want to give the short argument for the reader's convenience. Let n_0 be the smallest index $n \geq 1$ with $a_n \neq 0$ and write

$$F(s) = \mathrm{e}^{-\lambda_{n_0} s} \sum_{n\geq n_0} a_n \mathrm{e}^{-(\lambda_n - \lambda_{n_0})s} \quad (\sigma > \sigma_0). \tag{1}$$

Suppose that $F(\sigma_\nu) = 0$ for a sequence $(\sigma_\nu)_{\nu\geq 1}$ of real numbers with $\sigma_\nu \to \infty$ and $\sigma_\nu > \sigma_0$ for all ν. On the one hand, we see from (1) that

$$\sum_{n\geq n_0} a_n \mathrm{e}^{-(\lambda_n - \lambda_{n_0})\sigma_\nu} = 0 \qquad (\forall \nu = 1, 2, \ldots). \tag{2}$$

On the other hand, because of uniform convergence on $\{\sigma \in \mathbf{R} : \sigma > \sigma_0\}$ and the fact that $\lambda_n > \lambda_{n_0}$ for $n > n_0$, the left-hand side of (2) has the limit $a_{n_0} \neq 0$ as $\nu \to \infty$, a contradiction.

Since $F(s)$ has an analytic continuation to an open connected set D (containing $\mathbf{R}$), the set of zeros of $F(s)$ in D is discrete in D, i.e., contains no accumulation point of D. By hypothesis $F(s)$ has infinitely many real zeros; therefore we see that there is a sequence $(r_\nu)_{\nu\geq 1}$ of negative real numbers with $r_\nu \to -\infty$ and $F(r_\nu) = 0$ for all ν.

Now assume that $(a_n)_{n\geq 1}$ is not oscillatory. Then by Landau's theorem, either $F(s)$ must have a singularity at the real point of its abscissa of convergence or must converge for all $s \in \mathbf{C}$. By hypothesis $F(s)$ is analytic on $\mathbf{R}$; hence the first alternative cannot hold.

Therefore, in particular $F(s)$ must converge at r_ν for all ν, and we obtain

$$\sum_{n\geq 1} a_n \mathrm{e}^{-\lambda_n r_\nu} = 0 \qquad (\forall \nu = 1, 2, \ldots). \tag{3}$$

Now choose N large enough so that $a_N \neq 0$ and either $a_n \geq 0$ for $n \geq N+1$ or $a_n \leq 0$ for $n \geq N+1$. Then (3) implies that

$$\sum_{n\geq N+1} a_n \mathrm{e}^{(\lambda_n - \lambda_N)|r_\nu|} = -\sum_{n=1}^{N-1} a_n \mathrm{e}^{(\lambda_n - \lambda_N)|r_\nu|} - a_N. \tag{4}$$

On the one hand, the right-hand side of (4) has the limit $-a_N \neq 0$ as $\nu \to \infty$. On the other hand, either the left-hand side of (4) is identically zero (if $a_n = 0$ for $n \geq N+1$) or it grows without bound as $\nu \to \infty$. This gives a contradiction and concludes the proof of the theorem.

3 Two Remarks

1. It is also possible to give a Laguerre-style proof of the following: If the general Dirichlet series $F(s)$ converges everywhere and has infinitely many real zeros, then its coefficient sequence is oscillatory. (As before, we assume that the coefficients are real but not all zero.) This follows largely from Rolle's theorem, as did Laguerre's original rule. The key observation here is that if $F(s)$ has infinitely many real zeros, then so does the first derivative of $\mathrm{e}^{\lambda_1 s} F(s)$. By using this idea repeatedly, we may annihilate as many terms as we please. So if we suppose from the outset that the coefficient sequence is not oscillatory and if we restrict our attention to real values of s, then we are faced eventually with the absurdity that a sum of functions, all positive or all negative, has infinitely many real zeros. Although this approach circumvents the use of the uniqueness theorem for real analytic functions, the notion of analyticity is obviously required for the application of Landau's result.
2. Let $(a_n)_{n\geq 1}$ be a sequence of complex numbers, not all of the a_n being zero. In [3] such a sequence is called oscillatory if, for each real number $\phi \in [0, \pi)$, either the sequence $(\Re(\mathrm{e}^{-\mathrm{i}\phi} a_n))_{n\geq 1}$ is oscillatory or all of its terms are zero. Then the theorem proved above remains valid in this broader setting. In fact, as demonstrated in [3], the case of complex-valued coefficients follows without difficulty from the real-valued case.

Acknowledgement The authors thank the referee for making useful suggestions.

References

1. G.H. Hardy and M. Riesz: The General Theory of Dirichlet's Series, Dover, New York, 2005.
2. E.N. Laguerre: Sur la théorie des équations numériques, J. Math. Pures Appl. 9 (1883), 99–146.
3. W. Pribitkin: On the sign changes of coefficients of general Dirichlet series, Proc. Amer. Math. Soc. 136 (2008), 3089–3094.
4. W. Pribitkin: On the oscillatory behavior of certain arithmetic functions associated with automorphic forms, J. Number Theory 131 (2011), 2047–2060.

Representation Varieties of Fuchsian Groups

Michael Larsen and Alexander Lubotzky

Dedicated to the memory of Leon Ehrenpreis

Abstract We estimate the dimension of varieties of the form $Hom(\Gamma, G)$ where Γ is a Fuchsian group and G is a simple real algebraic group, answering along the way a question of I. Dolgachev.

1 Introduction

Let G be an almost simple real algebraic group, i.e., a non-abelian linear algebraic group over $\mathbb{R}$ with no proper normal $\mathbb{R}$-subgroups of positive dimension. Let Γ be a finitely generated group. The set of representations $Hom(\Gamma, G(\mathbb{R}))$ coincides with the set of real points of the representation variety $X_{\Gamma,G} := Hom(\Gamma, G)$. (We note here, that by a *variety*, we mean an affine scheme of finite type over $\mathbb{R}$; in particular, we do not assume that it is irreducible or reduced.)

Let $X^{\mathrm{epi}}_{\Gamma,G}$ denote the Zariski closure in $X_{\Gamma,G}$ of the set of *Zariski-dense homomorphisms* $\Gamma \to G(\mathbb{R})$, i.e., homomorphisms with Zariski-dense image. Our goal is to estimate the dimension of $X^{\mathrm{epi}}_{\Gamma,G}$ when Γ is a cocompact Fuchsian group. Our main results assert that in most cases, this dimension is roughly $(1 - \chi(\Gamma)) \dim G$, where $\chi(\Gamma)$ is the Euler characteristic of Γ.

M. Larsen
Department of Mathematics, Indiana University, Bloomington, IN 47405, USA
e-mail: mjlarsen@indiana.edu

A. Lubotzky (✉)
Einstein Institute of Mathematics, Edmond J. Safra Campus, Givat Ram,
The Hebrew University of Jerusalem, Jerusalem 91904, Israel
e-mail: alexlub@math.huji.ac.il

H.M. Farkas et al. (eds.), *From Fourier Analysis and Number Theory to Radon Transforms and Geometry*, Developments in Mathematics 28, DOI 10.1007/978-1-4614-4075-8_19,

To formulate our results more precisely, we need some notation and definitions. A cocompact oriented Fuchsian group Γ (and all Fuchsian groups in this chapter will be assumed to be cocompact and oriented without further mention) always admits a presentation of the following kind: Consider nonnegative integers m and g and integers $d_1, \ldots, d_m$ greater than or equal to 2, such that

$$2 - 2g - \sum_{i=1}^{m}(1 - d_i^{-1}) \tag{1.1}$$

is negative. For some choice of m, g, and d_i, Γ has a presentation

$$\begin{aligned}\Gamma := \langle x_1, \ldots, x_m, y_1, \ldots, y_g, z_1, \ldots, z_g \mid x_1^{d_1}, \ldots, x_m^{d_m}, \\ x_1 \cdots x_m [y_1, z_1] \cdots [y_g, z_g] \rangle,\end{aligned} \tag{1.2}$$

and its Euler characteristic $\chi(\Gamma)$ is given by (1.1). If $g = 0$ in the presentation (1.2), we sometimes denote Γ by $\Gamma_{d_1,\ldots,d_m}$. If, in addition, $m = 3$, Γ is called a *triangle group*, and its isomorphism class does not depend on the order of the subscripts. Note that the parameter g and the multiset $\{d_1, \ldots, d_m\}$ are determined by the isomorphism class of Γ. Every nontrivial element of Γ of finite order is conjugate to a power of one of the x_i, which is an element of order exactly d_i.

Definition 1.1. Let H be an almost simple algebraic group. We say that a Fuchsian group Γ is *H-dense* if and only if there exists a homomorphism $\phi\colon \Gamma \to H(\mathbb{R})$ such that $\phi(\Gamma)$ is Zariski dense in H and ϕ is injective on all finite cyclic subgroups of Γ (equivalently, $\phi(x_i)$ has order d_i for all i).

We can now state our main theorems.

Theorem 1.2. *For every Fuchsian group Γ and every integer $n \geq 2$,*

$$\dim X^{\text{epi}}_{\Gamma,\mathrm{SU}(n)} = (1 - \chi(\Gamma)) \dim \mathrm{SU}(n) + O(1),$$

where the implicit constant depend only on Γ.

In particular, this answers a question of Igor Dolgachev, proving the existence in sufficiently high degree, of uncountably many absolutely irreducible, pairwise nonconjugate, representations.

Theorem 1.3. *For every Fuchsian group Γ and every split simple real algebraic group G,*

$$\dim X^{\text{epi}}_{\Gamma,G} = (1 - \chi(\Gamma)) \dim G + O(rankG),$$

where the implicit constant depend only on Γ.

Theorem 1.4. *For every* SO(3)*-dense Fuchsian group Γ and every compact simple real algebraic group G,*

$$\dim X^{\text{epi}}_{\Gamma,G} = (1 - \chi(\Gamma)) \dim G + O(rankG),$$

where the implicit constant depend only on Γ.

Let us mention here that all but finitely many Fuchsian groups are SO(3)-dense (see Proposition 6.2 for the complete list of exceptions).

The proof of the theorems is based on deformation theory. It is a well-known result of Weil [We] that the Zariski tangent space to $X_{\Gamma,G}$ at any point $\rho \in X_{\Gamma,G}(\mathbb{R})$ is equal to the space of 1-cocycles $Z^1(\Gamma, \mathrm{Ad} \circ \rho)$, where $\mathrm{Ad} \circ \rho$ is the representation of Γ on the Lie algebra $\mathfrak{g}$ of G determined by ρ. (For brevity, we often denote $\mathrm{Ad} \circ \rho$ by $\mathfrak{g}$, where the action of Γ is understood.) In general, the dimension of the tangent space to $X_{\Gamma,G}$ at ρ can be strictly larger than the dimension of a component of $X_{\Gamma,G}$ containing ρ, thanks to obstructions in $H^2(\Gamma, \mathrm{Ad} \circ \rho)$. Weil showed that if the coadjoint representation $(\mathrm{Ad} \circ \rho)^*$ has no Γ-invariant vectors, then ρ is a non-singular point of $X_{\Gamma,G}$, i.e., it lies on a unique component of $X_{\Gamma,G}$ whose dimension is given by $\dim Z^1(\Gamma, \mathrm{Ad} \circ \rho)$, the dimension of the Zariski tangent space to $X_{\Gamma,G}$ at ρ. Computing this dimension is easy; the difficulty is to find ρ for which the obstruction space vanishes. A basic technique is to find a subgroup H of G for which the homomorphisms $\Gamma \to H$ are better understood and to choose ρ to factor through H. To this end, we make particular use of the homomorphisms from $H = \mathsf{A}_n$ to $G = \mathrm{SO}(n-1)$ and of the principal homomorphisms from $H = \mathrm{PGL}(2)$ and $H = \mathrm{SO}(3)$ to various groups G—see Sects. 3 and 4, respectively.

It is interesting to compare our results (Theorems 1.2–1.4) to the results of Liebeck and Shalev [LS2]. They also estimate $\dim X_{\Gamma,G}$ (and implicitly $\dim X^{\mathrm{epi}}_{\Gamma,G}$), but their methods work only for genus $g \geq 2$, while the most difficult (and interesting) case is $g = 0$. On the other hand, their methods work in arbitrary characteristic, while our methods appear to break down when the characteristic of the field divides the order of some generator x_i. A striking difference is that they deduce their information about $X_{\Gamma,G}$ from deep results on the finite quotients of Γ, while we work directly with $X^{\mathrm{epi}}_{\Gamma,G}$ and can deduce that various families of finite groups of Lie type can be realized as quotients of Γ (see [LLM]).

It may also be worth comparing our results to those of Benyash-Krivatz, Chernousov, and Rapinchuk [BCR], who consider X_{Γ,SL_n} where Γ is a surface group. They not only compute the dimension but prove a strong rationality result. It would be interesting to know if similar rationality results hold for more general semisimple groups G.

The material is organized as follows. In Sect. 2, we give a uniform proof of the upper bound in Theorems 1.2–1.4. This requires estimating the dimensions of suitable cohomology groups and boils down to finding lower bounds on dimensions of centralizers.

To prove the lower bounds of these three theorems, we present in each case a representation of Γ which is "good" in the sense that it is a non-singular point of the representation variety to which it belongs. We then compute the dimension of the tangent space at the good point. In Sect. 3, we explain how one can go from a good representation of Γ into a smaller group H to a good representation into a larger group G. The initial step of this kind of induction is via a representation of

Γ into an alternating group, SO(3), or $\mathrm{PGL}_2(\mathbb{R})$. We discuss the alternating group strategy in Sect. 3, where we prove Theorem 1.2 and begin the proof of Theorem 1.3. In Sect. 4, we discuss the principal homomorphism strategy, treating the remaining cases of Theorem 1.3, proving Theorem 1.4, and proving the existence of dense homomorphisms from SO(3)-dense Fuchsian groups to exceptional compact Lie groups (Proposition 5.3).

Proposition 6.2 in Sect. 5 shows that there are only six Fuchsian groups which are not SO(3)-dense. We do not have a good strategy for finding dense homomorphisms from these groups to compact simple Lie groups, since the methods of Sect. 3 are not effective. Y. William Yu found explicit surjective homomorphisms, described in the Appendix, from these groups to small alternating groups, which may serve as base cases for inductively constructing dense homomorphisms $\Gamma \to G(\mathbb{R})$ for these groups. We are grateful to him for his help.

All *Fuchsian groups* in this chapter are assumed to be cocompact and oriented. A *variety* is an affine scheme of finite type over $\mathbb{R}$. Its *dimension* is understood to mean its Krull dimension. *Points* are $\mathbb{R}$-points, and *non-singular* points should be understood scheme-theoretically; i.e., a point x is non-singular if and only if it lies in only one irreducible component X, and the dimension of X equals the dimension of the Zariski tangent space at x. An *algebraic group* will mean a linear algebraic group over $\mathbb{R}$. Unless otherwise stated, all topological notions will be understood in the sense of the Zariski topology. In particular, a *closed subgroup* is taken to be Zariski-closed. Note, however, that an algebraic group G is *compact* if $G(\mathbb{R})$ is so in the real topology.

We would like to thank the referee for a quick and thorough reading and a number of very helpful comments.

This work is dedicated to the memory of Leon Ehrenpreis who was a leading figure in Fuchsian groups and was an inspiration in several other directions—not only mathematically.

2 Upper Bounds

We recall some results from [We]. For every finitely generated group Γ, the Zariski tangent space to $\rho \in X_{\Gamma,G}(\mathbb{R})$ is equal to $Z^1(\Gamma, \mathrm{Ad} \circ \rho)$ where $\mathrm{Ad} : G \to \mathrm{Aut}(\mathfrak{g})$ is the adjoint representation of G on its Lie algebra. We will often write this more briefly as $Z^1(\Gamma, \mathfrak{g})$. Note that $\dim Z^1(\Gamma, \mathfrak{g})$ is always at least as great as the dimension of any component of $X_{\Gamma,G}$ in which ρ lies. Moreover, if Γ is a Fuchsian group and the coadjoint representation $\mathfrak{g}^* = (\mathrm{Ad} \circ \rho)^*$ has no Γ-invariant vectors, then ρ is a non-singular point of $X_{\Gamma,G}$.

If V denotes any finite-dimensional real vector space V on which Γ acts, then

$$\dim Z^1(\Gamma, V) := (2g-1)\dim V + \dim(V^*)^\Gamma + \sum_{j=1}^{m}(\dim V - \dim V^{\langle x_j\rangle}).$$

$$= (1-\chi(\Gamma))\dim V + \dim(V^*)^\Gamma + \sum_{j=1}^{m}\Big(\frac{\dim V}{d_j} - \dim V^{\langle x_j\rangle}\Big). \tag{2.1}$$

The following proposition essentially gives the upper bounds in Theorems 1.2–1.4, since for every irreducible component C of $X^{\mathrm{epi}}_{\Gamma,G}$, there exists a representation in $C(\mathbb{R})$, $\rho\colon \Gamma \to G(\mathbb{R})$ with Zariski-dense image; $\dim Z^1(\Gamma, \mathfrak{g})$ is at least as great as the dimension of any irreducible component of $X_{\Gamma,G}$ to which ρ belongs and therefore at least as great as $\dim C$.

Proposition 2.1. *For every Fuchsian group Γ, every reductive $\mathbb{R}$-algebraic group G with a Lie algebra $\mathfrak{g}$ and every representation $\rho : \Gamma \to G(\mathbb{R})$ with Zariski-dense image, we have*

$$\dim Z^1(\Gamma, \mathfrak{g}) \le (1-\chi(\Gamma))\dim G + (2g + m + \operatorname{rank} G) + \frac{3}{2}m \operatorname{rank} G,$$

where g and m are as in (1.2).

Proof. By Weil's formula (2.1),

$$\dim Z^1(\Gamma, \mathfrak{g}) = (1-\chi(\Gamma))\dim G + \dim(\mathfrak{g}^*)^\Gamma + \sum_{j=1}^{m}\left(\frac{\dim G}{d_j} - \dim \mathfrak{g}^{\langle x_j\rangle}\right). \tag{2.2}$$

Note that if $\mathfrak{g}$ is the real Lie algebra of G, then $\mathfrak{g} \otimes_{\mathbb{R}} \mathbb{C}$ is the complex Lie algebra of G. By abuse of notation, we will also denote it by $\mathfrak{g}$. Of course, they have the same dimensions over $\mathbb{R}$ and $\mathbb{C}$, respectively.

We have the following dimension estimates.

Lemma 2.2. *Under the above assumptions,*

$$\dim(\mathfrak{g}^*)^\Gamma \le 2g + m + \operatorname{rank} G.$$

Let us say that an automorphism α of G of order k is a *pure outer automorphism* of G if α^l is not inner for any l satisfying $1 \le l < k$.

For inner or pure automorphisms, we have

Lemma 2.3. *Let α be either an inner or a pure outer automorphism of G of order k. Then,*

$$\dim \mathrm{Fix}_G(\alpha) \ge \frac{\dim G}{k} - \operatorname{rank} G. \tag{2.3}$$

Lemma 2.4. *If G is a complex reductive group and α any automorphism of G of order k, then*

$$\dim \mathrm{Fix}_G(\alpha) \geq \frac{\dim G}{k} - \frac{3}{2}\mathrm{rank}\, G,$$

where $\mathrm{Fix}_G(\alpha)$ *denotes the subgroup of the fixed points of* α.

Plugging the results of Lemmas 2.2 and 2.4 into (2.1), and noting that $\dim \mathfrak{g}^{\langle x_j \rangle}$ is equal to $\dim \mathrm{Fix}_G(x_j)$, we have

$$\dim Z^1(\Gamma, \mathfrak{g}) \leq (1 - \chi(\Gamma)) \dim G + (2g + m + \mathrm{rank}\, G) + \frac{3}{2} m \,\mathrm{rank}\, G.$$

Proof (Proof of Lemma 2.2). The dimension of the Γ-invariants on $\mathfrak{g}^*$, $\dim(\mathfrak{g}^*)^\Gamma$, is equal to the dimension of the Γ-coinvariants on $\mathfrak{g}$. As Γ is Zariski-dense in G, this is equal to the dimension of the coinvariants of G acting on $\mathfrak{g}$ via Ad. Letting G^0 act first, we deduce that the space of G-coinvariants is a quotient space of $\mathfrak{g}/[\mathfrak{g}, \mathfrak{g}]$. More precisely, it is equal to the coinvariants of $\mathfrak{g}/[\mathfrak{g}, \mathfrak{g}]$ acted upon by the finite group G/G^0. As $\mathfrak{g}/[\mathfrak{g}, \mathfrak{g}]$ is a characteristic zero vector space, the dimension of the coinvariants is the same as that of the G/G^0-invariant subspace. Now, the space of linear maps $\mathrm{Hom}(\mathfrak{g}/[\mathfrak{g}, \mathfrak{g}], \mathbb{R})$ corresponds to the homomorphisms from G^0 to $\mathbb{R}$, and the G/G^0-invariants are those which can be extended to G. So, altogether $\dim(\mathfrak{g}^*)^\Gamma$ is bounded by $\dim \mathrm{Hom}(G, \mathbb{R})$. Now

$$\dim \mathrm{Hom}(G, \mathbb{R}) = \dim G^{\mathrm{ab}},$$

where $G^{\mathrm{ab}} = G/[G, G]$, and

$$G^{\mathrm{ab}} = U \times T \times A,$$

where U is a unipotent group, T a torus, and A a finite group. So $\dim G^{\mathrm{ab}} = \dim U + \dim T$. As Γ is Zariski-dense in G, its image is Zariski-dense in U, and hence,

$$\dim U \leq d(\Gamma) \leq 2g + m,$$

where $d(\Gamma)$ denotes the number of generators of Γ. Now, T, being a quotient of G, satisfies $\dim T \leq \mathrm{rank}\, G$. Altogether,

$$\dim(\mathfrak{g}^*)^\Gamma \leq 2g + m + \mathrm{rank}\, G,$$

as claimed. This completes the proof of Lemma 2.2.

Proof (Proof of Lemma 2.3.). Without loss of generality, we can assume G is connected. Let $\mathfrak{g}$ be the Lie algebra of G. Then α acts also on $\mathfrak{g}$, and $\dim \mathrm{Fix}_G(\alpha) = \dim \mathfrak{g}^\alpha$, so we can work at the level of Lie algebras. As α respects the decomposition of $\mathfrak{g}$ into $[\mathfrak{g}, \mathfrak{g}] \oplus \mathfrak{z}$ where $\mathfrak{z}$ is the Lie algebra of the central torus, and $\mathrm{rank}\mathfrak{g} = \mathrm{rank}[\mathfrak{g}, \mathfrak{g}] + \dim \mathfrak{z}$, we can restrict α to $[\mathfrak{g}, \mathfrak{g}]$ and assume $\mathfrak{g}$ is semisimple.

Moreover, we can write $\mathfrak{g}$ as a direct sum $\mathfrak{g} = \bigoplus_{i=1}^{s} \mathfrak{g}_i$ where each $\mathfrak{g}_i$ is itself a direct sum of isomorphic simple Lie algebras such that for each i, α acts transitively on the simple components. As both sides of the inequality are additive on a direct sum of α-invariant subalgebras, we can assume $\mathfrak{g}$ is a sum of t isomorphic simple algebras, $t|k$, and α acts transitively on the summands. If α is inner, then $t = 1$. If α is pure outer, it is equivalent to an action of the form

$$\alpha(x_1, \ldots, x_t) = (\beta(x_t), x_1, \ldots, x_{t-1}),$$

where β is a pure outer automorphism of a simple factor $\mathfrak{h}$, of order k/t. Thus,

$$\dim \mathfrak{g}^{\alpha} = \dim\{(x, x, \ldots, x) \mid x \in \mathfrak{h}^{\beta}\} = \dim \mathfrak{h}^{\beta}.$$

Thus, for the outer case, it suffices to prove the result when $t = 1$. If $k = 1$, the result is trivial. The possibilities for $(\mathfrak{g}, \mathfrak{h})$ are well-known (see, e.g., [He, Chap. X, Table 1]). For $k = 2$, they are $(\mathfrak{sl}(2n), \mathfrak{sp}(2n))$, $(\mathfrak{sl}(2n+1), \mathfrak{so}(2n+1))$, $(\mathfrak{so}(2n), \mathfrak{so}(2n-1))$, and $(\mathfrak{e}_6, \mathfrak{f}_4)$, and for $k = 3$, there is the unique case $(\mathfrak{so}(8), \mathfrak{g}_2)$.

Now, assume α is inner. Here, (2.3) follows from work of Lawther [Lw]. We thank the referee for suggesting this reference. For type A, a stronger estimate than (2.3) holds, namely,

$$\dim \mathrm{Fix}_G(\alpha) \geq \frac{\dim G}{k} - 1.$$

This will be needed for the upper bound in Theorem 1.2 and is easy to see. Namely, for $x \in G = \mathrm{SL}_n$ of order k, let a_j denote the multiplicity of $e^{2\pi i k/j}$ as an eigenvalue of x. By the Cauchy–Schwartz inequality,

$$\dim Z_G(x) + 1 = \sum_{j=0}^{k-1} a_j^2 \geq \frac{\left(\sum_{j=0}^{k-1} a_j\right)^2}{k} = n^2/k > \frac{\dim G}{k}. \tag{2.4}$$

Proof (Proof of Lemma 2.4.). To prove the statement, we still need to handle the case where α is neither an inner nor a pure outer automorphism. This means that for some l dividing k, with $1 < l < k$, α^l is inner while α is not. Let $H = Z_G(\alpha^l) = \mathrm{Fix}_G(\alpha^l)$. As α^l is an inner automorphism of order k/l, Lemma 2.3 implies that

$$\dim H \geq \frac{\dim G}{k/l} - \mathrm{rank}\, G.$$

Now α acts on the reductive group H as a pure outer automorphism of order at most l. Thus, again by Lemma 2.3,

$$\begin{aligned}\dim \mathrm{Fix}_G(\alpha) &= \dim \mathrm{Fix}_H(\alpha)\\ &\geq \frac{\dim H}{l} - \mathrm{rank}\, H\end{aligned}$$

$$\geq \frac{1}{l}\left(\frac{\dim G}{k/l} - \operatorname{rank} G\right) - \operatorname{rank} G$$
$$\geq \frac{\dim G}{k} - \left(1 + \frac{1}{l}\right)\operatorname{rank} G.$$

As $l > 1$, we get

$$\dim \operatorname{Fix}_G(\alpha) \geq \frac{\dim G}{k} - \frac{3}{2}\operatorname{rank} G,$$

completing the proof of Lemma 2.4.

In summary, we have proved the upper bounds for Theorems 1.2–1.4. For Theorems 1.3 and 1.4, the bounds follow immediately from Proposition 2.1, while the bound for Theorem 1.2 requires the better estimate proved in (2.4).

3 A Density Criterion

The results in this section are valid for general finitely generated groups Γ. The main result is Theorem 3.4, which gives a criterion for an irreducible component C of $X_{\Gamma,G}$ to be contained in $X^{\mathrm{epi}}_{\Gamma,G}$, i.e., to have the property that there exists a Zariski-dense subset of $C(\mathbb{R})$ consisting of representations ρ such that $\rho(\Gamma)$ is Zariski-dense in G. We begin with the technical results needed in the proof of Theorem 3.4.

Proposition 3.1. *Let G be a linear algebraic group over $\mathbb{R}$ and $H \subset G$ a closed subgroup such that $G(\mathbb{R})/H(\mathbb{R})$ is compact. Let C denote an irreducible component of $X_{\Gamma,H}$. The condition on $\rho \in X_{\Gamma,G}(\mathbb{R})$ that ρ is not contained in any $G(\mathbb{R})$-conjugate of $C(\mathbb{R})$ is open in the real topology.*

Proof. The conjugation map $H \times X_{\Gamma,H} \to X_{\Gamma,H}$ restricts to a map

$$H^\circ \times C \to X_{\Gamma,H}.$$

As H° and C are irreducible, the image of this morphism lies in an irreducible component of $X_{\Gamma,H}$, which must therefore be C.

The proposition can be restated as follows: the condition on ρ that ρ *is* contained in some $G(\mathbb{R})$-conjugate of $C(\mathbb{R})$ is closed in the real topology. To prove this, consider a sequence $\rho_i \in X_{\Gamma,G}(\mathbb{R})$ converging to ρ. Suppose that for each ρ_i there exists $g_i \in G(\mathbb{R})$ such that $\rho_i \in g_i C(\mathbb{R}) g_i^{-1}$. Let $\bar{g}_i$ denote the image of g_i in $G(\mathbb{R})/H^\circ(\mathbb{R})$. As this set is compact, there exists a subsequence which converges to some $\bar{g} \in G(\mathbb{R})/H^\circ(\mathbb{R})$. Passing to this subsequence, we may assume that $\bar{g}_1, \bar{g}_2, \ldots$ converges to $\bar{g}$. If $g \in G(\mathbb{R})$ represents the coset $\bar{g}$, we claim that $\rho \in gC(\mathbb{R})g^{-1}$. The claim implies the proposition.

By the implicit function theorem, there exists a continuous section $s\colon G(\mathbb{R})/H^\circ(\mathbb{R}) \to G(\mathbb{R})$ in a neighborhood of $\bar{g}$, and we may normalize so that $s(\bar{g}) = g$. For i sufficiently large, $s(\bar{g}_i)$ is defined, and $g_i = s(\bar{g}_i)h_i$ for some $h_i \in H^\circ(\mathbb{R})$. As conjugation by elements of $H^\circ(\mathbb{R})$ preserves C, we may assume without loss of generality that $g_i = s(\bar{g}_i)$ for all i sufficiently large. As $\lim_{i\to\infty} g_i = g$ and $C(\mathbb{R})$ is closed in the real topology in $X_{\Gamma,G}(\mathbb{R})$,

$$g^{-1}\rho g = \lim_{i\to\infty} g_i^{-1}\rho_i g_i \in C(\mathbb{R}).$$

The following proposition is surely well-known, but for lack of a precise reference, we give a proof.

Proposition 3.2. *Let G be an almost simple real algebraic group. There exists a finite set $\{H_1,\dots,H_k\}$ of proper closed subgroups of G such that every proper closed subgroup is contained in some group of the form gH_ig^{-1}, where $g \in G(\mathbb{R})$.*

Proof. The theorem is proved for $G(\mathbb{R})$ compact in [La, 1.3], so we may assume henceforth that G is not compact.

First, we prove that every proper closed subgroup K is contained in a maximal closed subgroup of positive dimension. If $\dim K > 0$, then for every infinite ascending chain $K_1 = K \subsetneq K_2 \subsetneq \cdots \subset G$ of closed subgroups of dimension $\dim K$, there exists a proper subgroup L of G which contains every K_i and for which $\dim L > \dim K$. Indeed, we can take $L := N_G(K^\circ)$, which contains all K_i, since $K_i^\circ = K^\circ$. It cannot equal G since G is almost simple, and if $\dim K = \dim L$, then $L^\circ = K^\circ$, and there are only finitely many groups between K and L. Thus, every proper subgroup of G of positive dimension is either contained in a maximal subgroup of G of the same dimension or in a proper subgroup of higher dimension. It follows that each such proper subgroup is contained in a maximal subgroup. For finite subgroups K, we can embed K in a maximal compact subgroup of G, which lies in a conjugacy class of proper closed subgroups of positive dimension since G itself is not compact, and maximal compact subgroups are maximal subgroups.

We claim that every maximal closed subgroup H of positive dimension is either parabolic or the normalizer of a connected semisimple subgroup or the normalizer of a maximal torus. Indeed, H is contained in the normalizer of its unipotent radical U. If U is nontrivial, this normalizer is contained in a parabolic P [Hu, 30.3, Cor. A], so $H = P$. If U is trivial, H is reductive and is contained in the normalizer of the derived group of its identity component H°. If this is nontrivial, H is the normalizer of a semisimple subgroup. If not, H° is a torus T. Then H is contained in the normalizer of the derived group of $Z_G(T)^\circ$, which is again the normalizer of a semisimple subgroup unless $Z_G(T)^\circ$ is a torus. In this case, it is a maximal torus, and H is the normalizer of this torus. Since a real semisimple group has finitely many conjugacy classes of parabolics and maximal tori, we need only consider the normalizers of semisimple subgroups. There are finitely many conjugacy classes of these by a theorem of Richardson [Ri].

The proof of Proposition 3.2 gives some additional information, which we employ in the following lemma:

Lemma 3.3. *If H is a maximal proper subgroup of a split almost simple algebraic group G over $\mathbb{R}$, then either H is parabolic or* $\dim H \leq \frac{9}{10} \dim G$.

Proof. For exceptional groups, all proper subgroups have dimension $\leq \frac{9}{10} \dim G$. Indeed, if G is an exceptional group over a finite fields $\mathbb{F}_q$ and H is a closed subgroup over $\mathbb{F}_q$, then the action of $G(\mathbb{F}_{q^n})$ on the set of $H(\mathbb{F}_{q^n})$-cosets gives a nontrivial complex representation of degree $G(\mathbb{F}_{q^n})/H(\mathbb{F}_{q^n})$. As $|H(\mathbb{F}_{q^n})| = O(q^{n \dim H})$, the Landazuri–Seitz estimates for the minimal degree of a nontrivial complex representation of $G(\mathbb{F}_q)$ [LZ] now imply $\dim H \leq \frac{9}{10} \dim G$. The same result follows in characteristic zero by a specialization argument.

We therefore consider only the case that G is of type A, B, C, or D. Also, we can ignore isogenies and assume that G is either SL_n, a split orthogonal group, or a split symplectic group. Let V be the natural representation of G. If $\dim V = n$, then $\dim G$ is $n^2 - 1$, $n(n-1)/2$, or $n(n+1)/2$, depending on whether G is linear, orthogonal, or symplectic.

It suffices to consider the case that H is the normalizer of a semisimple subgroup $K \subset G$. The action of H must preserve the decomposition of V into K-irreducible factors. Therefore, H lies in a parabolic subgroup unless all factors have equal dimension. If all factors have equal dimension and there are at least three factors, then $\dim H \leq n^2/3$, so the theorem holds in such cases. If H° respects a decomposition $V = W_1 \oplus W_2$ where $\dim W_i = n/2$, then either G is linear and $\dim H < (1/2) \dim G + 1$, G is orthogonal and $\dim H \leq (n/2)^2$, or G is symplectic and $\dim H \leq (n/2)(n/2+1)$. If $V \otimes \mathbb{C}$ is reducible, it decomposes into two factors of degree $n/2$, and the same estimates apply.

We have therefore reduced to the case that K is semisimple and $V \otimes \mathbb{C}$ is irreducible, so we may and do extend scalars to $\mathbb{C}$ for the remainder of the proof. If K is not almost simple, then any element of G which normalizes K must respect a nontrivial tensor decomposition, and therefore H respects such a decomposition. This implies

$$\dim H \leq m^2 + (n/m)^2 - 1 \leq 3 + n^2/4.$$

We may therefore assume that K is almost simple and V is associated to a dominant weight of K. It is easy to deduce from the Weyl dimension formula that every nontrivial irreducible representation of a simple Lie algebra L of rank r, other than the natural representation and its dual, has dimension at least $(r^2 + r)/2$; we need only consider the case that V is a natural representation. As $H \subsetneq G$, we need only consider the inclusions $\mathrm{SO}(n) \subset \mathrm{SL}_n$ and $\mathrm{Sp}(n) \subset \mathrm{SL}_n$. In all cases, we have $\dim H \leq \frac{2}{3} \dim G$.

We recall that $X^{\mathrm{epi}}_{\Gamma,G}$ is the Zariski closure in $X_{\Gamma,G}$ of the set of Zariski-dense homomorphisms $\Gamma \to G(\mathbb{R})$.

Theorem 3.4. *Let Γ be a finitely generated group, G an almost simple real algebraic group, and $\rho_0 \in Hom(\Gamma, G(\mathbb{R}))$ a non-singular $\mathbb{R}$-point of $X_{\Gamma,G}$. For every closed subgroup H of G such that $\rho_0(\Gamma) \subset H(\mathbb{R})$, let t_H denote the dimension of the Zariski tangent space of $X_{\Gamma,H}$ at ρ_0 (i.e., $t_H = \dim Z^1(\Gamma, \mathfrak{h})$, where $\mathfrak{h}$ is the Lie algebra of $H(\mathbb{R})$ with the adjoint action of Γ.) We assume*

(1) If H is any maximal closed subgroup such that $\rho_0(\Gamma) \subset H(\mathbb{R})$, then

$$t_G - \dim G > t_H - \dim H.$$

(2) If H is any maximal closed subgroup such that $G(\mathbb{R})/H(\mathbb{R})$ is not compact, then

$$t_G - \dim G > \dim X_{\Gamma,H} - \dim H.$$

Then, $X^{\mathrm{epi}}_{\Gamma,G}$ contains the irreducible component of $X_{\Gamma,G}$ to which ρ_0 belongs.

Proof. Let C denote the irreducible component of $X_{\Gamma,G}$ containing ρ_0, which is unique since ρ_0 is a non-singular point of $X_{\Gamma,G}$. Again, since ρ_0 is a non-singular point, there is an open neighborhood U of ρ_0 in $C(\mathbb{R})$ which is diffeomorphic to $\mathbb{R}^n$, where $n := \dim C = t_G$.

Let $\{H_1, \ldots, H_k\}$ represent the conjugacy classes of maximal proper closed subgroups of G given by Lemma 3.2. Let $C_{i,j}$ denote the irreducible components of X_{Γ,H_i}. For each component, we consider the conjugation morphism $\chi_{i,j}: G \times C_{i,j} \to X_{\Gamma,G}$. We claim that the fibers of this morphism have dimension at least $\dim H_i$. Indeed, the action of H_i° on $G \times C_{i,j}$ given by

$$h.(g, \rho_0) = (gh^{-1}, h\rho_0 h^{-1})$$

is free, and $\chi_{i,j}$ is constant on the orbits of the action. Thus, the closure of the image of $\chi_{i,j}$ has dimension at most $\dim C_{i,j} + \dim G - \dim H_i$. Condition (2) guarantees that if $G(\mathbb{R})/H_i(\mathbb{R})$ is not compact, then a nonempty Zariski-open subset of C lies outside the image of $\chi_{i,j}$ for all j. Condition (1) guarantees the same thing if $G(\mathbb{R})/H_i(\mathbb{R})$ is compact, and some conjugate of ρ_0 lies in $C_{i,j}(\mathbb{R})$. Note that $\dim C_{i,j} \leq t_{H_i}$ if $\rho_0 \in C_{i,j}(\mathbb{R})$.

Finally, we consider components $C_{i,j}$ for which $G(\mathbb{R})/H_i(\mathbb{R})$ is compact, but no conjugate of ρ_0 lies in $C_{i,j}(\mathbb{R})$. By Proposition 3.1, the $G(\mathbb{R})$-orbit of each such $C_{i,j}(\mathbb{R})$ meets $C(\mathbb{R})$ in a set which is closed in the real topology. Since ρ_0 belongs to none of these sets, there is a neighborhood U of ρ_0 consisting of homomorphisms ρ such that no conjugate of ρ lies in any such $C_{i,j}$. The intersection of U with any nonempty Zariski-open subset of $C(\mathbb{R})$ is therefore Zariski-dense in C, and for every ρ in this set, $\rho(\Gamma)$ is Zariski-dense in $G(\mathbb{R})$. It follows that $X^{\mathrm{epi}}_{\Gamma,G}$ contains C.

Note that if G is compact, condition (2) is vacuous.

Corollary 3.5. *If G is a compact almost simple algebraic group over $\mathbb{R}$, H is a connected maximal proper closed subgroup of G with finite center, and $\rho_0: \Gamma \to H(\mathbb{R})$ has dense image, then $t_G - \dim G > t_H - \dim H$ implies $X^{\mathrm{epi}}_{\Gamma,G}$ contains the irreducible component of $X_{\Gamma,G}$ to which ρ_0 belongs.*

Proof. To apply the theorem, we need only prove that ρ_0 is a non-singular point of $X_{\Gamma,G}$. As H is maximal, the product $Z_G(H)H$ must equal H, which means $Z_G(H) = Z(H)$ is finite. Thus, $\mathfrak{g}^\Gamma = \mathfrak{g}^H = \{0\}$, and since $\mathfrak{g}$ is a self-dual $G(\mathbb{R})$-representation, this implies $(\mathfrak{g}^*)^\Gamma = \{0\}$, which implies that ρ_0 is a non-singular point of $X_{\Gamma,G}$.

4 The Alternating Group Method

In this section, Γ is any (cocompact, oriented) Fuchsian group. We first consider $G = \mathrm{SO}(n)$.

Proposition 4.1. *For Γ a Fuchsian group and $G = \mathrm{SO}(n)$, we have*

$$\dim X^{\mathrm{epi}}_{\Gamma,\mathrm{SO}(n)} = (1 - \chi(\Gamma)) \dim \mathrm{SO}(n) + O(n)$$

where the implicit constant depends only on Γ.

Proof. Proposition 2.1 gives the upper bound, so it suffices to prove

$$\dim X^{\mathrm{epi}}_{\Gamma,\mathrm{SO}(n)} \geq (1 - \chi(\Gamma)) \dim \mathrm{SO}(n) + O(n).$$

Let $d_1, \ldots, d_m$ be defined as in (1.2). For large n, denote C_i, for $i = 1, \ldots, m$, the conjugacy class in the alternating group A_{n+1} which consists of even permutations of $\{1, 2, \ldots, n+1\}$ with only d_i-cycles and 1-cycles and with as many d_i-cycles as possible. Thus, any element of C_i has at most $2d_i - 1$ fixed points. Theorem 1.9 of [LS1] ensures that for large enough n, there exist epimorphisms ρ_0 from Γ onto A_{n+1}, sending x_i to an element of C_i for $i = 1, \ldots, m$ and x_i as in (1.2).

Now, $\mathsf{A}_{n+1} \subset \mathrm{SO}(n)$ and moreover the action of A_{n+1} on the Lie algebra $\mathfrak{so}(n)$ of $\mathrm{SO}(n)$ is the restriction to A_{n+1} of the irreducible S_{n+1} representation associated to the partition $(n-1)+1+1$ [FH, Ex. 4.6]. If $n \geq 5$, this partition is not self-conjugate, so the restriction to A_{n+1} is irreducible. By (2.2),

$$\dim Z^1(\Gamma, \mathrm{Ad} \circ \text{æ}_0) = (1 - \chi(\Gamma)) \dim \mathfrak{so}(n) + \sum_{i=1}^{m} \left(\frac{\dim \mathfrak{so}(n)}{d_i} - \dim \mathfrak{so}(n)^{\langle x_i \rangle} \right).$$

Now, $\dim \mathfrak{so}(n)^{\langle x_i \rangle}$ is equal to the multiplicity of the eigenvalue 1 of $x = \rho_0(x_i)$ acting via Ad on $\mathfrak{so}(n)$. Note that the multiplicity of every d_ith root of unity as an eigenvalue for our element $x = \rho_0(x_i)$, when acting on the natural n-dimensional representation, is of the form $\frac{n}{d_i} + O(1)$, where the implied constant depends only on d_i. Thus, using the same arguments as in the proof of Lemma 2.3 (see (2.4)), we can deduce that

$$\left|\frac{\dim \mathfrak{so}(n)}{d_i} - \dim \mathfrak{so}(n)^{\langle x_i \rangle}\right| = O(n),$$

where again the constant depends only on d_i.

As $\mathfrak{so}(n)^*$ has no A_{n+1}-invariants, $X_{\Gamma,\mathrm{SO}(n)}$ is non-singular at ρ_0. By Theorem 3.4, as long n is large enough that

$$\begin{aligned} t_{\mathrm{SO}(n)} &= \dim Z^1(\Gamma, \mathrm{Ad} \circ æ_0) \\ &> \dim \mathrm{SO}(n) - \dim \mathsf{A}_{n+1} + t_{\mathsf{A}_{n+1}} \\ &= \dim \mathrm{SO}(n), \end{aligned}$$

$X^{\mathrm{epi}}_{\Gamma,\mathrm{SO}(n)}$ contains the component of $X_{\Gamma,\mathrm{SO}(n)}$ to which ρ_0 belongs, and this has dimension $t_{\mathrm{SO}(n)} = (1 - \chi(\Gamma)) \dim \mathrm{SO}(n) + O(n)$.

We remark that in this case, there is a more elementary alternative argument. The condition on $X_{\Gamma,\mathrm{SO}(n)}$ of irreducibility on $\mathfrak{so}(n)$ is open. It is impossible that all representations in a neighborhood of ρ_0 have finite image and those with infinite image should have Zariski-dense image (since the Lie algebra of the connected component of the Zariski closure is $\rho(\Gamma)$-invariant).

We can now prove Theorem 1.2.

Proof. The upper bound has already been proved in Sect. 1. It therefore suffices to prove

$$\dim X^{\mathrm{epi}}_{\Gamma,\mathrm{SU}(n)} \geq (1 - \chi(\Gamma)) \dim \mathrm{SU}(n) + O(1).$$

Throughout the argument, we may always assume that n is sufficiently large,

We begin by defining ρ_0 as in the proof of Proposition 4.1. Let C denote the irreducible component of $X_{\Gamma,\mathrm{SO}(n)}$ to which ρ_0 belongs. We may choose $\rho_0' \in C(\mathbb{R})$ such that $\rho_0'(\Gamma)$ is Zariski-dense in $\mathrm{SO}(n)$. As there are finitely many conjugacy classes of order d_i in $\mathrm{SO}(n)$, the conjugacy class of $\rho(x_i)$ does not vary as ρ ranges over the irreducible variety C, so $\rho_0(x_i)$ is conjugate to $\rho_0'(x_i)$ in $\mathrm{SO}(n)$.

We have no further use for ρ_0 and now redefine ρ_0 to be the composition of ρ_0' with the inclusion $\mathrm{SO}(n) \hookrightarrow \mathrm{SU}(n)$. The eigenvalues of $\rho_0(x_i)$ are d_ith roots of unity, and each appears with multiplicity $n/d_i + O(1)$, where the implicit constant may depend on d_i but does not depend on n. The representation $\mathrm{SO}(n) \to \mathrm{SU}(n)$ is irreducible, so $(\mathfrak{su}(n))^{\mathrm{SO}(n)} = \{0\}$. As $\mathfrak{su}(n)$ is a self-dual representation of $\mathrm{SU}(n)$, it is a self-dual representation of $\mathrm{SO}(n)$, so as $\rho_0(\Gamma)$ is dense in $\mathrm{SO}(n)$,

$$(\mathfrak{su}(n)^*)^{\Gamma} = (\mathfrak{su}(n)^*)^{\mathrm{SO}(n)} = \{0\}.$$

It follows that $X_{\Gamma,\mathrm{SU}(n)}$ is non-singular at ρ_0. Since each eigenvalue of $\rho_0(x_i)$ has multiplicity $n/d_i + O(1)$,

$$t_{\mathrm{SU}(n)} = \dim Z^1(\Gamma, \mathrm{Ad} \circ \rho_0) = (1 - \chi(\Gamma)) \dim \mathrm{SU}(n) + O(1).$$

We claim that $\mathrm{SO}(n)$ is contained in a unique maximal closed subgroup of $\mathrm{SU}(n)$. Indeed, if G is any intermediate group, the Lie algebra $\mathfrak{g}$ of G must be an $\mathrm{SO}(n)$-subrepresentation of $\mathfrak{su}(n)$ which contains $\mathfrak{so}(n)$. Since $\mathfrak{su}(n)/\mathfrak{so}(n)$ is an irreducible $\mathrm{SO}(n)$-representation (namely, the symmetric square of the natural representation of $\mathrm{SO}(n)$), it follows that $\mathfrak{g} = \mathfrak{su}(n)$ or $\mathfrak{g} = \mathfrak{so}(n)$. In the former case, $G = \mathrm{SU}(n)$. In the latter case, G is contained in $N_G(\mathrm{SO}(n))$. This is therefore the unique maximal proper closed subgroup of $\mathrm{SU}(n)$ containing $\mathrm{SO}(n)$, or (equivalently) $\rho_0(\Gamma)$. The theorem now follows from Theorem 3.4 together with the upper bound estimate Proposition 2.1 applied to $N_G(\mathrm{SO}(n))$.

We can also deduce Theorem 1.3 for G of types A and D from Proposition 4.1.

Proof. If $G_1 \to G_2$ is an isogeny, the morphism $X_{\Gamma,G_1} \to X_{\Gamma,G_2}$ is quasi-finite, and so

$$\dim X_{\Gamma,G_2} \geq \dim X_{\Gamma,G_1}.$$

Likewise, the composition of a homomorphism with dense image with an isogeny still has dense image, so

$$\dim X^{\mathrm{epi}}_{\Gamma,G_2} \geq \dim X^{\mathrm{epi}}_{\Gamma,G_1}.$$

In particular, to prove our dimension estimate for an adjoint group, it suffices to prove it for any covering group. We begin by proving it for $G = \mathrm{SL}_n$, which also gives it for PGL_n.

Let ρ_0 now denote a homomorphism $\Gamma \to \mathrm{SO}(n) \subset \mathrm{SL}_n(\mathbb{R})$ with dense image and such that every eigenvalue of $\rho_0(x_i)$ has multiplicity $n/d_i + O(1)$. Such a homomorphism exists by the proof of Proposition 4.1. It is well-known that $\mathrm{SO}(n)$ is a maximal closed subgroup of SL_n, and $\mathfrak{g}^{\mathrm{SO}(n)} = \{0\}$. Thus ρ_0 is a non-singular point of $X_{\Gamma,G}(\mathbb{R})$. Let C denote the unique irreducible component to which it belongs. In applying Theorem 3.4, we do not need to consider parabolic subgroups at all since $\rho_0(\Gamma)$ is not contained in any and $G(\mathbb{R})/H(\mathbb{R})$ is compact when H is parabolic. All other maximal subgroups are reductive, and we may therefore apply Proposition 2.1 to get an upper bound

$$\dim X_{\Gamma,H} \leq (1 - \chi(\Gamma)) \dim H + 2g + m + (3m/2 + 1)n$$

By Lemma 3.3, $\dim H < \frac{9}{10}(n^2 - 1)$, so for n sufficiently large,

$$\dim X_{\Gamma,H} - \dim H < \dim X_{\Gamma,G} - \dim G.$$

Thus, condition (2) of Theorem 3.4 holds, and so the component C of $X_{\Gamma,G}$ to which ρ_0 belongs lies in $X^{\mathrm{epi}}_{\Gamma,G}$. It is therefore a non-singular point of C, and it follows that

$$\dim X^{\mathrm{epi}}_{\Gamma,G} \geq \dim C = \dim Z^1(\Gamma, \mathfrak{g}) = (1 - \chi(\Gamma)) \dim \mathrm{SL}_n + O(n).$$

The argument for type D is very similar. Here we work with $G = \mathrm{SO}(n,n)$, which is a double cover of the split adjoint group of type D_n over $\mathbb{R}$. Our starting point is a homomorphism $\rho_0\colon \Gamma \to \mathrm{SO}(n) \times \mathrm{SO}(n)$ with dense image and such that the eigenvalues of

$$\rho(x_i) \in \mathrm{SO}(n) \times \mathrm{SO}(n) \subset \mathrm{SO}(n,n) \subset \mathrm{GL}_{2n}(\mathbb{C})$$

have multiplicity $(2n)/d_i + O(1)$. Such a ρ_0 is given by a pair (σ, τ) of dense homomorphisms $\Gamma \to \mathrm{SO}(n)$ satisfying a balanced eigenvalue multiplicity condition and the additional condition that σ and τ do not lie in the same orbit under the action of $\mathrm{Aut}(\mathrm{SO}(n))$ on $X_{\Gamma,\mathrm{SO}(n)}$. This additional condition causes no harm, since $\dim \mathrm{Aut}\,\mathrm{SO}(n) = \dim \mathrm{SO}(n)$, while the components of $\dim X^{\mathrm{epi}}_{\Gamma,\mathrm{SO}(n)}$ constructed above (which satisfy the balanced eigenvalue condition) have dimension greater than $\dim \mathrm{SO}(n)$ for large n. Given a pair (σ, τ) as above, the closure H of $\rho_0(\Gamma)$ is a subgroup of $\mathrm{SO}(n) \times \mathrm{SO}(n)$ which maps onto each factor but which does not lie in the graph of an isomorphism between the two factors. By Goursat's lemma, $H = \mathrm{SO}(n) \times \mathrm{SO}(n)$. From here, one passes from H to $G = \mathrm{SO}(n,n)$ just as in the case of groups of type A.

5 Principal Homomorphisms

It is a well-known theorem of de Siebenthal [dS] and Dynkin [D1] that for every (adjoint) simple algebraic group G over $\mathbb{C}$, there exists a conjugacy class of *principal* homomorphisms $\mathrm{SL}_2 \to G$ such that the image of any nontrivial unipotent element of $\mathrm{SL}_2(\mathbb{C})$ is a regular unipotent element of $G(\mathbb{C})$. The restriction of the adjoint representation of G to SL_2 via the principal homomorphism is a direct sum of V_{2e_i}, where $e_1, \ldots, e_r$ is the sequence of exponents of G and V_m denotes the mth symmetric power of the 2-dimensional irreducible representation of SL_2, which is of dimension $m+1$ [Ko]. In particular,

$$\dim G = \sum_{i=1}^{r} (2e_i + 1),$$

where r denotes $\mathrm{rank} G$. As each V_{2e_i} factors through PGL_2, the same is true for the homomorphism $\mathrm{SL}_2 \to \mathrm{Ad}(G)$. More generally, if G is defined and split over any field K of characteristic zero, the principal homomorphism can be defined over K.

The following proposition is due to Dynkin:

Proposition 5.1. *Let G be an adjoint simple algebraic group over $\mathbb{C}$ of type A_1, A_2, B_n ($n \geq 4$), C_n ($n \geq 2$), E_7, E_8, F_4, or G_2. Let H denote the image of a principal homomorphism of G. Let K be a closed subgroup of G whose image in the adjoint representation of G is conjugate to that of H. Then K is a maximal subgroup of G.*

Proof. As K is conjugate to H in $\mathrm{GL}(\mathfrak{g})$, in particular, the number of irreducible factors of $\mathfrak{g}$ restricted to H and to K is the same. By [Ko], this already implies that H and K are conjugate in G. The fact that H is maximal is due to Dynkin. The classical and exceptional cases are treated in [D3] and [D2], respectively.

As SL_2 is simply connected, the principal homomorphism $\mathrm{SL}_2 \to G$ lifts to a homomorphism $\mathrm{SL}_2 \to H$ if H is a split semisimple group which is simple modulo its center. Again, this is true for split groups over any field of characteristic zero. We also call such homomorphisms principal.

If G is an adjoint simple group over $\mathbb{R}$ with $G(\mathbb{R})$ compact and $\phi\colon \mathrm{PGL}_{2,\mathbb{C}} \to G_{\mathbb{C}}$ is a principal homomorphism over $\mathbb{C}$, ϕ maps the maximal compact subgroup $\mathrm{SO}(3) \subset \mathrm{PGL}_2(\mathbb{C})$ into a maximal compact subgroup of $G(\mathbb{C})$. Thus, ϕ can be chosen to map $\mathrm{SO}(3)$ to $G(\mathbb{R})$, and such a homomorphism will again be called principal. Likewise, if H is almost simple and $H(\mathbb{R})$ is compact, a principal homomorphism $\phi\colon \mathrm{SL}_{2,\mathbb{C}} \to H_{\mathbb{C}}$ can be chosen so that $\phi(\mathrm{SU}(2)) \subset H(\mathbb{R})$.

Proposition 5.2. *Let G be an adjoint compact simple real algebraic group of type A_1, A_2, B_n $(n \geq 4)$, C_n $(n \geq 2)$, E_7, E_8, F_4, or G_2, and let Γ be an* SO(3)*-dense Fuchsian group. Let $\rho_0\colon \Gamma \to G$ denote the composition of the map $\Gamma \to \mathrm{SO}(3)$ and the principal homomorphism $\phi\colon \mathrm{SO}(3) \to G$. If*

$$-\chi(\Gamma)\dim G + \sum_{j=1}^{m} \frac{\dim G}{d_j} - \sum_{j=1}^{m}\sum_{i=1}^{r}(1 + 2\lfloor e_i/d_j\rfloor)$$
$$> -\chi(\Gamma)\dim \mathrm{SO}(3) + \sum_{j=1}^{m} \frac{\dim \mathrm{SO}(3)}{d_j} - m,$$

then

$$\dim X_{\Gamma,G}^{\mathrm{epi}} \geq (1-\chi(\Gamma))\dim G + \sum_{j=1}^{m} \frac{\dim G}{d_j} - \sum_{j=1}^{m}\sum_{i=1}^{r}(1 + 2\lfloor e_i/d_j\rfloor). \tag{5.1}$$

Proof. Let x_j denote the jth generator of finite order in the presentation (1.2). If $\phi(x_j)$ lifts to an element of SU(2) whose eigenvalues are $\zeta^{\pm 1}$, where ζ is a primitive $2d_j$-root of unity, the eigenvalues of the image of x_j in $\mathrm{Aut}(\mathfrak{g})$ are

$$\zeta^{-2e_1}, \zeta^{2-2e_1}, \zeta^{4-2e_1}, \ldots, 1, \ldots, \zeta^{2e_1}, \zeta^{-2e_2}, \ldots, \zeta^{2e_2}, \ldots, \zeta^{-2e_r}, \ldots, \zeta^{2e_r}.$$

The multiplicity of 1 as eigenvalue is therefore $\sum_{i=1}^{r}(1 + 2\lfloor e_i/d_j\rfloor)$. By (2.2), the left-hand side of (5.1) is $\dim Z^1(\Gamma, \mathfrak{g})$. By Corollary 3.5, we need only check that

$$t_G - \dim G = -\chi(\Gamma)\dim G + \sum_{j=1}^{m} \frac{\dim G}{d_j} - \sum_{j=1}^{m}\sum_{i=1}^{r}(1 + 2\lfloor e_i/d_j\rfloor).$$

is greater than

$$t_{\mathrm{SO}(3)} - \dim \mathrm{SO}(3) = -\chi(\Gamma) \dim \mathrm{SO}(3) + \sum_{j=1}^{m} \frac{\dim \mathrm{SO}(3)}{d_j} - \sum_{j=1}^{m} 1,$$

which is true by hypothesis.

We can now prove Theorem 1.4.

Proof. The upper bound was proved in Sect. 2. Recall that if $G_1 \to G_2$ is an isogeny, we can prove the lower bound of the theorem for G_1 and immediately deduce it for G_2. Theorem 1.2 and Proposition 4.1 therefore cover groups of type A, B, and D. This leaves only the symplectic case, where Proposition 5.2 applies. Note that

$$\begin{aligned}
&\sum_{j=1}^{m} \frac{\dim G}{d_j} - \sum_{j=1}^{m} \sum_{i=1}^{r} (1 + 2\lfloor e_i/d_j \rfloor) \\
&= \sum_{j=1}^{m} \sum_{i=1}^{r} \frac{1+2e_i}{d_j} - \sum_{j=1}^{m} \sum_{i=1}^{r} (1 + 2\lfloor e_i/d_j \rfloor) \\
&= \sum_{i=1}^{r} \sum_{j=1}^{m} \left(\frac{1+2e_i}{d_j} - 1 + 2\lfloor e_i/d_j \rfloor \right).
\end{aligned}$$

As

$$-1 < 2x + 1/d_j - 1 - 2\lfloor x \rfloor < 1,$$

the error term is at most mr in absolute value.

The following proposition illustrates the fact that the methods of this section are not only useful in the large rank limit. We make essential use of the technique illustrated below in [LLM].

Proposition 5.3. *Every* SO(3)*-dense Fuchsian group is also* $F_4(\mathbb{R})$*-dense,* $E_7(\mathbb{R})$*-dense, and* $E_8(\mathbb{R})$*-dense, where* F_4*,* E_7*, and* E_8 *denote the compact simple exceptional real algebraic groups of absolute rank* 4, 7, *and* 8 *respectively.*

Proof. Let G be one of F_4, E_7, and E_8. Let E denote the set of exponents of G, other than 1, which is the only exponent of SO(3). We map Γ to $G(\mathbb{R})$ via the principal homomorphism $\mathrm{SO}(3) \to G$ and apply Corollary 3.5. To show that there exists a homomorphism from Γ to $G(\mathbb{R})$ with dense image, we need only check that

$$t_G - \dim G > t_{\mathrm{SO}(3)} - \dim \mathrm{SO}(3).$$

The proof of Theorem 3.4 proceeds by deforming the composed homomomorphism $\Gamma \to \mathrm{SO}(3) \to G(\mathbb{R})$, and under continuous deformation, the order of the image of a torsion element remains constant. We therefore obtain more, namely, that Γ is $G(\mathbb{R})$-dense.

By replacing t_G and $t_{\mathrm{SO}(3)}$ by the middle expression in (2.1) for $V = \mathfrak{g}$ and $V = \mathfrak{so}(3)$, respectively, the desired inequality can be rewritten

$$(2g-2+m)(\dim G - \dim \mathrm{SO}(3)) - \sum_{j=1}^{m}\sum_{e\in E}(1+2\lfloor e/d_j\rfloor) > 0. \tag{5.2}$$

The summand is nonincreasing with each d_j. In particular,

$$\sum_{j=1}^{m}\sum_{e\in E}(1+2\lfloor e/d_j\rfloor) \le \sum_{j=1}^{m}\sum_{e\in E}(1+2\lfloor e/2\rfloor) < \sum_{j=1}^{m}\sum_{e\in E}(1+2e)$$
$$= \dim G - \dim \mathrm{SO}(3).$$

Therefore, if $g \ge 1$, the expression (5.2) is positive. For $g = 0$, $(d_1, \ldots, d_m)$ is dominated by $(2, 2, \ldots, 2)$ for $m \ge 5$, $(2, 2, 2, 3)$ for $m = 4$, and $(2, 3, 7)$, $(2, 4, 5)$, or $(3, 3, 4)$ for $m = 3$.

The following table presents the value of

$$\sum_{i=1}^{r}\left((1+2\lfloor d_i/n\rfloor) - \frac{2d_i+1}{n}\right)$$

for each root system of exceptional type and for each $n \le 7$.

n	A_1	E_6	E_7	E_8	F_4	G_2
2	$-1/2$	-1	$-7/2$	-4	-2	-1
3	0	-2	$-4/3$	$-8/3$	$-4/3$	$-2/3$
4	$1/4$	$1/2$	$-1/4$	-2	-1	$1/2$
5	$2/5$	$2/5$	$2/5$	$-8/5$	$8/5$	$6/5$
6	$1/2$	-1	$-7/6$	$-4/3$	$-2/3$	$-1/3$
7	$4/7$	$6/7$	0	$4/7$	$4/7$	0

By (2.2), the relevant values of $t_G - \dim G$ are given in the following table:

d_i vector	A_1	E_6	E_7	E_8	F_4	G_2
$(2, 2, 2, 3)$	2	18	34	56	16	6
$(2, 3, 7)$	0	4	8	12	4	2
$(2, 4, 5)$	0	4	10	20	4	0
$(3, 3, 4)$	0	10	14	28	8	2

For $(\underbrace{2, \ldots, 2}_{m})$, $m \ge 5$, the values of $t_G - \dim G$ for A_1, E_6, E_7, E_8, F_4, G_2 are $2m-6$, $40m-136$, $70m-266$, $128m-496$, $28m-104$, $8m-28$, respectively. In all cases except $(2, 4, 5)$ for G_2, the desired inequality holds.

We conclude by proving Theorem 1.3 in the remaining cases, i.e., for adjoint groups G of type B or C.

Proof. We begin with a Zariski-dense homomorphism $\rho_0 \colon \Gamma \to \mathrm{PGL}_2(\mathbb{R})$. Such a homomorphism always exists since Γ is Fuchsian. We now embed PGL_2 via the principal homomorphism in a split adjoint group G of type B_n or C_n. Assuming $n \geq 4$, the image is a maximal subgroup, and we can apply Theorem 3.4 as in the A and D cases.

6 SO(3)-Dense Groups

In this section, we show that almost all Fuchsian groups are SO(3)-dense and classify the exceptions.

Lemma 6.1. *Let $d \geq 2$ be an integer.*

(1) *If $d \neq 6$, there exists an integer a relatively prime to d such that*

$$\frac{1}{4} \leq \frac{a}{d} \leq \frac{1}{2},$$

with equality only if $d \in \{2, 4\}$.

(2) *If $d \notin \{4, 6, 10\}$, then a can be chosen such that*

$$\frac{1}{3} \leq \frac{a}{d} \leq \frac{1}{2},$$

with equality only if $d \in \{2, 3\}$.

(3) *If $d \notin \{2, 3, 18\}$, there exists a such that*

$$\frac{1}{12} < \frac{a}{d} < \frac{4}{15},$$

with equality only if $d = 12$.

Proof. For (1) and (2), let

$$a = \begin{cases} \frac{d-1}{2} \text{ if } d \equiv 1 \pmod 2, \\ \frac{d-4}{2} \text{ if } d \equiv 2 \pmod 4, \\ \frac{d-2}{2} \text{ if } d \equiv 0 \pmod 4. \end{cases}$$

As long as $d > 12$, these fractions satisfy the desired inequalities, and for $d \leq 12$, this can be checked by hand.

For (3), let $a = \frac{d-b}{6}$, where b depends on d (mod 36) and is given as follows:

b	d (mod 4)	d (mod 9)
-12	2	3
-6	0	6
-4	2	2, 5, 8
-3	1, 3	3
-2	0	1, 4, 7
-1	1, 3	2, 5, 8
1	1, 3	1, 4, 7
2	0	2, 5, 8
3	1, 3	0, 6
4	2	1, 4, 7
6	0	0, 3
12	2	0, 6

As long as $d > 24$, these fractions satisfy the desired inequalities, and the cases $d \le 24$ can be checked by hand.

Proposition 6.2. *A cocompact oriented Fuchsian group is* SO(3)*-dense if and only if it does not belong to the set*

$$\{\Gamma_{2,4,6}, \Gamma_{2,6,6}, \Gamma_{3,4,4}, \Gamma_{3,6,6}, \Gamma_{2,6,10}, \Gamma_{4,6,12}\}. \tag{6.1}$$

Proof. We recall that every proper closed subgroup of SO(3) is contained in a subgroup of SO(3) isomorphic to O(2), A_5, or S_4. The set of homomorphisms $\mathrm{O}(2) \to \mathrm{SO}(3)$, $A_5 \to \mathrm{SO}(3)$, and $S_4 \to \mathrm{SO}(3)$ have dimension 2, 3, and 3 respectively. Furthermore, $\dim X_{\Gamma,\mathrm{O}(2)} \le 2g+m$, while $\dim X_{\Gamma,S_4} = \dim X_{\Gamma,A_5} = 0$.

Every nontrivial conjugacy class in SO(3) has dimension 2. As the commutator map $\mathrm{SO}(3) \times \mathrm{SO}(3) \to \mathrm{SO}(3)$ is surjective and every fiber has dimension at least 3, if $g \ge 1$, we have $\dim X_{\Gamma,\mathrm{SO}(3)} \ge 3 + 3(2g-2) + 2m$. For $g \ge 2$ or $g = 1$ and $m \ge 2$, the dimension of $\dim X_{\Gamma,\mathrm{SO}(3)}$ exceeds the dimension of the space of all homomorphisms whose image lies in a proper closed subgroup, so there exists a homomorphism with dense image with $\rho(x_i)$ of order d_i for all i. If $g = m = 1$, and $\rho(\Gamma) \subset \mathrm{O}(2)$, then the commutator $\rho([y_1, z_1])$ lies in SO(2), so $\rho(x_1) \in \mathrm{SO}(2)$. The set of elements of order d_1 in SO(2) is finite, so $\dim X_{\Gamma,\mathrm{O}(2)} \le 2$, and the set of elements of $X_{\Gamma,\mathrm{SO}(3)}$ which can be conjugated into a fixed O(2) has dimension ≤ 4; again, there exists ρ with dense image and with $\rho(x_i)$ of order d_i for all i.

This leaves the case $g = 0$, $m \ge 3$. By (1.1), $\sum 1/d_i < m - 2$. We claim that unless we are in one of the cases of (6.1), there exist elements $\bar{x}_1, \dots, \bar{x}_m \in \mathrm{SO}(3)$ of orders $d_1, \dots, d_m$, respectively, such that $\bar{x}_1 \cdots \bar{x}_m = e$ and the elements $\bar{x}_i$ generate a dense subgroup of SO(3). For $m = 3$, the order of terms in the sequence d_1, d_2, d_3 does not matter since $\bar{x}_1\bar{x}_2\bar{x}_3 = e$ implies $\bar{x}_2\bar{x}_3\bar{x}_1 = e$ and $\bar{x}_3^{-1}\bar{x}_2^{-1}\bar{x}_1^{-1} = e$.

Without loss of generality, we may therefore assume that $d_1 \leq d_2 \leq d_3$ when $m = 3$. If the base case $m = 3$ holds whenever d_3 is sufficiently large, the higher m cases follow by induction, since one can replace the $m+1$-tuple $(d_1, \ldots, d_{m+1})$ by the m-tuple $(d_1, \ldots, d_{m-1}, d)$ and the triple (d_m, d_{m+1}, d), where d is sufficiently large.

If $\alpha_1, \alpha_2, \alpha_3 \in (0, \pi]$ satisfy the triangle inequality, by a standard continuity argument, there exists a nondegenerate spherical triangle whose sides have angles α_i. If α_1, α_2, and α_3 are of order exactly d_1, d_2, and d_3 in the group $\mathbb{R}/2\pi$, respectively, then there exists a homomorphism from the triangle group Γ_{d_1,d_2,d_3} to SO(3) such that the generators x_i map to elements of order d_i, and these elements do not commute. We claim that except in the cases $(2,4,6)$, $(2,6,6)$, $(3,6,6)$, $(2,6,10)$, and $(4,6,12)$, there always exist positive integers $a_i \leq d_i/2$ such that a_i is relatively prime to d_i and a_i/d_i satisfy the triangle inequality. We can therefore set $\alpha_i = 2a_i\pi/d_i$.

Every nondecreasing triple from the interval $[1/4, 1/2]$ except for $1/4, 1/4, 1/2$ satisfies the triangle inequality. As (d_1, d_2, d_3) cannot be $(2,4,4)$, Lemma 6.1 (1) implies the claim unless at least one of d_1, d_2, d_3 equals 6. We therefore assume that at least one of the d_i is 6. As $1/6$ and any two elements of $[1/3, 1/2]$ other than $1/3$ and $1/2$ satisfy the triangle inequality and as $(d_1, d_2, d_3) \neq (2,3,6)$, Lemma 6.1 (2) implies the claim except if one of the d_i is 4, one of the d_i is 10, or two of the d_i are 6. By Lemma 6.1 (3), the remaining a_i/d_i can then be chosen to lie in $(1/12, 4/15)$ unless this $d_i \in \{2, 3, 12, 18\}$. If a_i/d_i is in this interval, the triangle inequality follows. Examination of the remaining 12 cases reveal five exceptions: $(2,4,6)$, $(2,6,6)$, $(2,6,10)$, $(3,6,6)$, and $(4,6,12)$.

Assuming that we are in none of these cases, there exist non-commuting elements $\bar{x}_i$ in SO(3) of order d_1, d_2, and d_3, such that $\bar{x}_1\bar{x}_2\bar{x}_3 = e$. They cannot all lie in a common SO(2). In fact, they cannot all lie in a common O(2), since any element in the nontrivial coset of O(2) has order 2, $d_3 \geq d_2 > 2$, and if three elements multiply to the identity, it is impossible that exactly two lie in SO(2). If Γ maps to S_4 or A_5, then $\{d_1, d_2, d_3\}$ is contained in $\{2,3,4\}$ or $\{2,3,5\}$ respectively. The possibilities for (d_1, d_2, d_3) are therefore $(2,5,5)$, $(3,3,5)$, $(3,5,5)$, $(5,5,5)$, $(3,4,4)$, $(3,3,4)$, and $(4,4,4)$. The realization of $\Gamma_{a,b,b}$ as an index-2 subgroup of $\Gamma_{2,2a,b}$ implies the proposition for $\Gamma_{2,5,5}$, $\Gamma_{3,3,5}$, $\Gamma_{3,5,5}$, $\Gamma_{5,5,5}$, $\Gamma_{3,3,4}$, and $\Gamma_{4,4,4}$. The only remaining case is $\Gamma_{3,4,4}$.

Lastly, we show that none of the groups in (6.1) are SO(3)-dense. Suppose there exist elements x_1, x_2, x_3 of orders d_1, d_2, d_3, respectively, such that $x_1x_2x_3$ equals the identity and $\langle x_1, x_2, x_3 \rangle$ is dense in SO(3). These elements can be regarded as rotations through angles $2\pi a_1$, $2\pi a_2$, $2\pi a_3$, respectively, where the a_i can be taken in $[0, 1/2)$, and no two axes of rotation coincide. Choosing a point P on the great circle of vectors perpendicular to the axis of rotation of x_1, the three points $P, x_2^{-1}(P), x_1(P) = x_3^{-1}x_2^{-1}(P)$ satisfy the strict spherical triangle inequality, so $a_1 < a_2 + a_3$. Likewise, $a_2 < a_3 + a_1$ and $a_3 < a_1 + a_2$. However, one easily verifies in each of the cases (6.1) that one cannot find rational numbers $a_1, a_2, a_3 \in (0, 1/2]$ with denominators d_1, d_2, d_3, respectively, such that a_1, a_2, a_3 satisfy the strict triangle inequality.

7 Appendix by Y. William Yu

The following triples of permutations, which evidently multiply to 1, have been checked by machine to generate the full alternating groups in which they lie:

- $\Gamma_{2,4,6} \to \mathsf{A}_{14}$:

$$x_1 = (1\ 2)(3\ 4)(5\ 6)(7\ 8)(9\ 10)(11\ 12)$$
$$x_2 = (1\ 10\ 9\ 8)(2\ 14\ 13\ 3)(4\ 5)(6\ 7\ 12\ 11)$$
$$x_3 = (1\ 3\ 5\ 11\ 7\ 9)(2\ 8\ 6\ 4\ 13\ 14)$$

- $\Gamma_{2,6,6} \to \mathsf{A}_{14}$:

$$x_1 = (1\ 2)(3\ 4)(5\ 6)(7\ 8)(9\ 10)(11\ 12)$$
$$x_2 = (1\ 14\ 8\ 7\ 4\ 2)(3\ 5\ 13\ 11\ 9\ 6)$$
$$x_3 = (1\ 4\ 6\ 3\ 7\ 14)(5\ 9\ 10\ 11\ 12\ 13)$$

- $\Gamma_{3,6,6} \to \mathsf{A}_{12}$:

$$x_1 = (1\ 2\ 3)(4\ 5\ 6)(7\ 8\ 9)(10\ 11\ 12)$$
$$x_2 = (1\ 12\ 11\ 6\ 2\ 3)(4\ 10\ 8\ 9\ 5\ 7)$$
$$x_3 = (1\ 2\ 3\ 6\ 9\ 10)(4\ 11)(5\ 7\ 8)$$

- $\Gamma_{3,4,4} \to \mathsf{A}_{14}$:

$$x_1 = (1\ 2\ 3)(4\ 5\ 6)(7\ 8\ 9)(10\ 11\ 12)$$
$$x_2 = (1\ 14\ 11\ 12)(2\ 3\ 4\ 5)(7\ 10\ 13\ 9)(6\ 8)$$
$$x_3 = (1\ 2\ 12\ 14)(3\ 5)(4\ 8\ 9\ 6)(7\ 13\ 10\ 11)$$

- $\Gamma_{2,6,10} \to \mathsf{A}_{12}$:

$$x_1 = (1\ 2)(3\ 4)(5\ 6)(7\ 8)(9\ 10)(11\ 12)$$
$$x_2 = (1\ 8\ 6\ 7\ 5\ 3)(4\ 10\ 11)(9\ 12)$$
$$x_3 = (1\ 2\ 3\ 11\ 9\ 4\ 5\ 8\ 6\ 7)(10\ 12)$$

- $\Gamma_{4,6,12} \to \mathsf{A}_{12}$:

$$x_1 = (1\ 4\ 3\ 2)(5\ 8\ 7\ 6)(9\ 10)(11\ 12)$$
$$x_2 = (1\ 2\ 5\ 9\ 10\ 3)(4\ 7\ 11\ 8\ 6\ 12)$$
$$x_3 = (2\ 10\ 5\ 8)(3\ 12\ 7\ 11\ 6\ 4)$$

In each case, one can use (2.1) to compute that

$$\dim Z^1(\Gamma, \mathfrak{so}(n)) - \dim \mathrm{SO}(n) > 0.$$

The reasoning of Proposition 4.1 therefore applies to give a homomorphism $\Gamma \to \mathrm{SO}(n)$ either for $n = 11$ or for $n = 13$, with dense image.

Acknowledgments ML was partially supported by the National Science Foundation and the United States-Israel Binational Science Foundation. AL was partially supported by the European Research Council and the Israel Science Foundation.

References

[BCR] Rapinchuk, A. S.; Benyash-Krivetz, V. V.; Chernousov, V. I.: Representation varieties of the fundamental groups of compact orientable surfaces. *Israel J. Math.* **93** (1996), 29–71.
[D1] Dynkin, E. B.: Some properties of the system of weights of a linear representation of a semisimple Lie group. (Russian) *Doklady Akad. Nauk SSSR (N.S.)* **71** (1950), 221–224.
[D2] Dynkin, E. B.: Semisimple subalgebras of semisimple Lie algebras. *Amer. Math. Soc. Transl. (Ser. 2)* **6** (1957), 111–245.
[D3] Dynkin, E. B.: Maximal subgroups of the classical groups. *Amer. Math. Soc. Transl. (Ser. 2)* **6** (1957), 245–378.
[FH] Fulton, William; Harris, Joe: Representation theory. A first course. Graduate Texts in Mathematics, 129. Readings in Mathematics. Springer-Verlag, New York, 1991.
[He] Helgason, Sigurdur: Differential Geometry, Lie Groups, and Symmetric Spaces. Academic Press, Orlando, Florida, 1978.
[Hu] Humphreys, James E.: Linear algebraic groups. Graduate Texts in Mathematics, No. 21. Springer-Verlag, New York-Heidelberg, 1975.
[Ko] Kostant, Bertram: The principal 3-dimensional subgroup and the Betti numbers of a complex simple Lie group. *Amer. J. Math.* **81** (1959), 973–1032.
[LZ] Landazuri, Vicente; Seitz, Gary M.: On the minimal degrees of projective representations of the finite Chevalley groups. *J. Algebra* **32** (1974), 418–443.
[La] Larsen, Michael: Rigidity in the invariant theory of compact groups, arXiv: math.RT/0212193.
[LLM] Larsen, Michael; Lubotzky, Alex; Marion, Claude: Deformation theory and finite simple quotients of triangle groups I, preprint.
[Lw] Lawther, Ross: Elements of specified order in simple algebraic groups. *Trans. Amer. Math. Soc.* **357** (2005), no. 1, 221–245.
[LS1] Liebeck, Martin W.; Shalev, Aner: Fuchsian groups, coverings of Riemann surbfaces, subgroup growth, random quotients and random walks. *J. Algebra* **276** (2004), no. 2, 552–601.
[LS2] Liebeck, Martin; Shalev, Aner: Fuchsian groups, finite simple groups and representation varieties. *Invent. Math.* **159** (2005), no. 2, 317–367.
[Ri] Richardson, R. W., Jr.: A rigidity theorem for subalgebras of Lie and associative algebras. *Illinois J. Math.* **11** (1967) 92–110.
[dS] de Siebenthal, Jean: Sur certaines sous-groupes de rang un des groupes de Lie clos. *C. R. Acad. Sci. Paris* 230 (1950), 910–912.
[We] Weil, André: Remarks on the cohomology of groups. *Annals of Math.* **80** (1964), 149–157.

Two Embedding Theorems

Gerardo A. Mendoza

To Leon Ehrenpreis, in memoriam

Abstract We first consider pairs $(\mathcal{N}, \mathcal{T})$ where $\mathcal{N}$ is a closed connected smooth manifold and $\mathcal{T}$ a nowhere vanishing smooth real vector field on $\mathcal{N}$ that admits an invariant metric and shows that there is an embedding $F : \mathcal{N} \to S^{2N-1} \subset \mathbb{C}^N$ for some N mapping $\mathcal{T}$ to a vector field of the form $\mathcal{T}' = \mathrm{i}\sum_{j=1}^{N} \tau_j \big(z^j \frac{\partial}{\partial z^j} - \overline{z}^j \frac{\partial}{\partial \overline{z}^j}\big)$ for some $\tau_j \neq 0$. We further consider pairs $(\mathcal{N}, \mathcal{T})$ with the additional datum of an involutive subbundle $\overline{\mathcal{V}} \subset \mathbb{C}T\mathcal{N}$ such that $\mathcal{V} + \overline{\mathcal{V}} = \mathbb{C}T\mathcal{N}$ and $\mathcal{V} \cap \overline{\mathcal{V}} = \operatorname{span}_{\mathbb{C}} \mathcal{T}$ for which there is a section β of the dual bundle of $\overline{\mathcal{V}}$ such that $\langle \beta, \mathcal{T} \rangle = -\mathrm{i}$ and

$$X\langle \beta, Y\rangle - Y\langle \beta, X\rangle - \langle \beta, [X,Y]\rangle = 0 \quad \text{whenever } X, Y \in C^\infty(\mathcal{N}; \overline{\mathcal{V}}).$$

Then $\overline{\mathcal{K}} = \ker \beta$ is a CR structure, and we give necessary and sufficient conditions for the existence of a CR embedding of $\mathcal{N}$ (with a possibly different, but related, CR structure) into S^{2N-1} mapping $\mathcal{T}$ to $\mathcal{T}'$. The first result is an analogue of the fact that for any line bundle $L \to \mathcal{B}$ over a compact base, there is an embedding $f : \mathcal{B} \to \mathbb{CP}^{N-1}$ such that L is isomorphic to the pullback by f of the tautological line bundle $\Gamma \to \mathbb{CP}^{N-1}$. The second is an analogue of the statement in complex differential geometry that a holomorphic line bundle over a compact complex manifold is positive if and only if one of its tensor powers is very ample.

Key words Classifying space • Complex manifolds • CR manifolds • Embedding theorems • Kodaira embedding theorem • Line bundles

Mathematics Subject Classification (2010): Primary 32V30, 57R40. Secondary 32V10, 32V20, 32Q15, 32Q40

G.A. Mendoza (✉)
Department of Mathematics, Temple University, Philadelphia, PA 19122, USA
e-mail: gmendoza@temple.edu

H.M. Farkas et al. (eds.), *From Fourier Analysis and Number Theory to Radon Transforms and Geometry*, Developments in Mathematics 28, DOI 10.1007/978-1-4614-4075-8_20,

1 Introduction

Let $\mathscr{F}$ be the family of pairs $(\mathcal{N}, \mathcal{T})$ where $\mathcal{N}$ is a closed connected smooth manifold and $\mathcal{T}$ is a smooth nowhere vanishing real vector field on $\mathcal{N}$ admitting an invariant metric. An example of such a pair is the sphere $S^{2N-1} \subset \mathbb{C}^N$ with the vector field $\mathcal{T}'$ in formula (1.2) below.

We will show:

Theorem 1.1. *Let $(\mathcal{N}, \mathcal{T}) \in \mathscr{F}$. Then there is a positive integer N, an embedding $F : \mathcal{N} \to \mathbb{C}^N$ with image contained in the sphere S^{2N-1}, and positive numbers τ_j such that $F_*\mathcal{T}$ is the vector field*

$$\mathcal{T}' = \mathrm{i}\sum_{j=1}^{N} \tau_j \left(z^j \frac{\partial}{\partial z^j} - \overline{z}^j \frac{\partial}{\partial \overline{z}^j} \right). \tag{1.2}$$

Furthermore, no component function of F is flat at any point of $\mathcal{N}$.

An element $(\mathcal{N}, \mathcal{T}) \in \mathscr{F}$ is like the circle bundle of a complex line bundle over a closed manifold $\mathcal{B}$ (with $\mathcal{T}$ being the infinitesimal generator of the circle action), and the theorem is like the basic ingredient in the classification theorem for line bundles. In our general setting, the orbits of $\mathcal{T}$ need not be compact.

Theorem 1.1 was stated without proof in [14] as Theorem 3.11. The fact that no component of F is flat was used there in an argument involving the Malgrange preparation theorem. The complete proof is given here in Sect. 2.

The statement of our second result requires us recalling some terminology and a few facts. Associated with any involutive subbundle $\mathcal{W}$ of $T\mathcal{N}$ or its complexification $\mathbb{C}T\mathcal{N}$, there is a first-order differential cochain complex on the exterior powers of its dual,

$$\cdots \to C^\infty(\mathcal{N}; {\textstyle\bigwedge}^q \mathcal{W}^*) \to C^\infty(\mathcal{N}; {\textstyle\bigwedge}^{q+1} \mathcal{W}^*) \to \cdots$$

where the coboundary operator is given by Cartan's formula for the differential. We review this in more detail below. The complex is elliptic if and only if $\mathcal{W} + \overline{\mathcal{W}} = \mathbb{C}T\mathcal{N}$ (or $= T\mathcal{N}$ if $\mathcal{W} \subset T\mathcal{N}$), in which case, $\mathcal{W}$ is referred to as an elliptic structure.

Let $\mathscr{F}_{\mathrm{ell}}$ be the set of triples $(\mathcal{N}, \mathcal{T}, \overline{\mathcal{V}})$ such that $(\mathcal{N}, \mathcal{T}) \in \mathscr{F}$, $\overline{\mathcal{V}} \subset \mathbb{C}T\mathcal{N}$ is an elliptic structure with $\mathcal{V} \cap \overline{\mathcal{V}} = \mathrm{span}_{\mathbb{C}}\mathcal{T}$, and there is a closed section β of $\overline{\mathcal{V}}^*$ such that $\langle \beta, \mathcal{T} \rangle = -\mathrm{i}$. Closed means in the sense of the associated complex, that is, $\overline{\mathbb{D}}\beta = 0$, where $\overline{\mathbb{D}}$ refers to the coboundary operator of the induced complex:

$$V\langle \beta, W \rangle - W\langle \beta, V \rangle - \langle \beta, [V, W] \rangle = 0 \quad \text{for all } V,\ W \in C^\infty(\mathcal{N}; \overline{\mathcal{V}}). \tag{1.3}$$

If β, $\beta' \in C^\infty(\mathcal{N}; \overline{\mathcal{V}}^*)$ are two sections as described, we say that β and β' are equivalent if $\beta' - \beta = \overline{\mathbb{D}}u$ with a real-valued function u and write $\boldsymbol{\beta}$ for the class of β. Here $\overline{\mathbb{D}}u$ means the restriction of du to $\overline{\mathcal{V}}$. Observe that necessarily $\mathcal{T}u = 0$.

Suppose that $(\mathcal{N}, \mathcal{T}, \overline{\mathcal{V}}) \in \mathscr{F}_{\mathrm{ell}}$ and that β is a section of $\overline{\mathcal{V}}^*$ as just described. Let

$$\overline{\mathcal{K}}_\beta = \{v \in \overline{\mathcal{V}} : \langle \beta, v\rangle = 0\}$$

Then $\overline{\mathcal{K}}_\beta$ is a CR structure of codimension 1. Let θ_β be the real 1-form that satisfies $\langle \theta_\beta, \mathcal{T}\rangle = 1$ and whose restriction to $\overline{\mathcal{K}}_\beta$ vanishes. Define

$$\mathrm{Levi}_{\theta_\beta}(v, w) = -\mathrm{i} d\theta_\beta(v, \overline{w}), \quad v,\ w \in \mathcal{K}_{\beta,p},\ p \in \mathcal{N}; \tag{1.4}$$

$\mathcal{K}_\beta$ is the conjugate of $\overline{\mathcal{K}}_\beta$.

A map $F : \mathcal{N} \to \mathbb{C}^N$ will be called equivariant if $F_* \mathcal{T} = \mathcal{T}'$ for some $\mathcal{T}'$ of the form (1.2).

Theorem 1.5. *Suppose that $(\mathcal{N}, \mathcal{T}, \overline{\mathcal{V}}) \in \mathscr{F}_{\mathrm{ell}}$ with* $\dim \mathcal{N} \geq 5$. *Fix a class $\boldsymbol{\beta}$ as described above. The following are equivalent:*

(1) There is $\beta \in \boldsymbol{\beta}$ and an equivariant CR immersion of $\mathcal{N}$ with the CR structure $\overline{\mathcal{K}}_\beta$ into $\mathbb{C}^N$ for some N.
(2) There is $\beta' \in \boldsymbol{\beta}$ and an equivariant CR immersion of $\mathcal{N}$ with the CR structure $\overline{\mathcal{K}}_{\beta'}$ into $\mathbb{C}^N$ for some N with image in S^{2N-1}.
(3) There is $\beta' \in \boldsymbol{\beta}$ such that the CR structure $\overline{\mathcal{K}}_{\beta'}$ is definite.
(4) There is $\beta' \in \boldsymbol{\beta}$ and an equivariant CR embedding of $\mathcal{N}$ with the CR structure $\overline{\mathcal{K}}_{\beta'}$ into $\mathbb{C}^N$ with image in S^{2N-1} for some N.

The implication (3) $\Longrightarrow$ (4) is like Kodaira's embedding theorem of Kähler manifolds with integral fundamental form into complex projective space. This is explained in some detail the paragraphs following Example 1.7. The proof of the implication relies on Boutet de Monvel's construction in [4] of an embedding under the same condition, strict pseudoconvexity; this is the only reason for the restriction on the dimension of $\mathcal{N}$ in the hypothesis of the theorem.

Concrete models of manifolds $\mathcal{N}$ with the structure described above are the following.

Example 1.6. Let $\mathcal{N} = S^{2n+1} \subset \mathbb{C}^{n+1}$, let

$$\mathcal{T} = \mathrm{i} \sum_{j=1}^{n+1} \tau_j \left(z^j \frac{\partial}{\partial z^j} - \overline{z}^j \frac{\partial}{\partial \overline{z}^j} \right).$$

Then $(\mathcal{N}, \mathcal{T}) \in \mathscr{F}$, since $\mathcal{T}$ preserves the standard metric of S^{2n+1}. Suppose all τ_j have the same sign. Let $\overline{\mathcal{K}}$ be the standard CR structure of S^{2n+1} (as a subbundle of $T^{0,1}\mathbb{C}^{n+1}$ along S^{2n+1}). Then $\mathcal{T}$ is transverse to $\overline{\mathcal{K}}$ and $\overline{\mathcal{V}} = \mathcal{K} \oplus \mathrm{span}_{\mathbb{C}} \mathcal{T}$ is involutive. Let θ be the unique real 1-form on S^{2n+1} which vanishes on $\overline{\mathcal{K}}$ and satisfies $\langle \theta, \mathcal{T}\rangle = 1$ and let $\beta = -\mathrm{i} \jmath^* \theta$ where $\jmath : \overline{\mathcal{K}} \to \mathbb{C}TS^{2n+1}$ is the inclusion map. Then (1.3) holds. Indeed, if V and W are CR vector fields, then so is $[V, W]$ since $\overline{\mathcal{K}}$ is involutive, and if V is CR, then $[V, \mathcal{T}]$ is also CR, so (1.3) holds if V is

CR and $W = \mathcal{T}$. Since $\langle\beta, \mathcal{T}\rangle = -\mathrm{i}$, $(\mathcal{N}, \mathcal{T}, \overline{\mathcal{V}}) \in \mathscr{F}_{\mathrm{ell}}$. The set of closures of the orbits of $\mathcal{T}$ is a Hausdorff space, an analogue of complex projective space.

Example 1.7. Let $\mathcal{B}$ be a complex manifold, let $E \to \mathcal{B}$ a Hermitian holomorphic line bundle, and let $\rho : \mathcal{N} \to \mathcal{B}$ be its circle bundle. Define

$$\overline{\mathcal{V}} = \{v \in \mathbb{C}T\mathcal{N} : \rho_* v \in T^{0,1}\mathcal{B}\}. \tag{1.8}$$

Then $(\mathcal{N}, \mathcal{T}, \overline{\mathcal{V}}) \in \mathscr{F}_{\mathrm{ell}}$; the vector field $\mathcal{T}$ is the infinitesimal generator of the standard circle action on $\mathcal{N}$. Identifying $\mathcal{N}$ with the bundle of oriented orthonormal bases of the real bundle underlying E, let θ be the connection form of the Hermitian holomorphic connection, a real smooth 1-form with $\langle\theta, \mathcal{T}\rangle = 1$ and $\mathcal{L}_{\mathcal{T}}\theta = 0$, where $\mathcal{L}_{\mathcal{T}}$ is Lie derivative. Let $\iota : \overline{\mathcal{V}} \to \mathbb{C}T\mathcal{N}$ be the inclusion map. Using the dual map $\iota^* : \mathbb{C}T^*\mathcal{N} \to \overline{\mathcal{V}}^*$, let $\beta = -\mathrm{i}\iota^*\theta$. Then $\mathrm{Im}\langle\beta, \mathcal{T}\rangle = -1$ and $\overline{\mathbb{D}}\beta = 0$; that β is $\overline{\mathbb{D}}$-closed and is equivalent to the statement that θ corresponds to a holomorphic connection. Adding $\overline{\mathbb{D}}u$ to β with u real valued and $\mathcal{T}u = 0$ corresponds to a change of the Hermitian metric.

In the context of Example 1.7, let $\overline{\mathcal{K}} = \ker\beta$; this is a CR structure. The statement that Levi_θ (as defined in (1.4)) is positive definite is equivalent to the statement that the line bundle $E \to \mathcal{B}$ is negative (Grauert [6], see also Kobayashi [8, p. 87]), that is, the form ω on $\mathcal{B}$ such that $\rho^*\omega = -\mathrm{i}d\theta$ is the fundamental form of a Kähler metric on $\mathcal{B}$.

Kodaira's embedding theorem [9] asserts that if $\mathcal{B}$ is compact and admits a Kähler metric whose fundamental form is in the image of an integral class, then $\mathcal{B}$ admits an embedding into a projective space. The line bundle $E \to \mathcal{B}$ associated to such fundamental form is, by definition, negative, and its circle bundle with the induced CR structure, strictly pseudoconvex. For any integer m, let $\mathfrak{H}(\mathcal{B}, E^{\otimes m})$ be the space of holomorphic sections of $E^{\otimes m}$. The proof of Kodaira's existence theorem consists of showing that for a suitable m (a negative number here), the map sending the point $b \in \mathcal{B}$ to the kernel of the map

$$\mathfrak{H}(\mathcal{B}, E^{\otimes m}) \ni \phi \mapsto \phi(b) \in E_b^{\otimes m}$$

defines an embedding $\Psi : \mathcal{B} \to \mathbb{P}\mathfrak{H}(\mathcal{B}, E^{\otimes m})^*$. We describe an interpretation of this along the lines of the last assertion in Theorem 1.5. Fix a Hermitian metric on E and use it to induce metrics on each of the tensor powers of E. For each integer $m \neq 0$, define $\wp_m : SE \to SE^{\otimes m}$ by

$$\wp_m(p) = \begin{cases} p \otimes \cdots \otimes p & \text{if } m > 0 \\ p^* \otimes \cdots \otimes p^* & \text{if } m < 0, \end{cases}$$

($|m|$ factors in either case) with $p^* \in E^*_{\rho(p)}$ such that $\langle p^*, p\rangle = 1$. A section ϕ of $E^{\otimes(-m)}$ is a map $E^{\otimes m} \to \mathbb{C}$ which in turn gives a map $SE \to \mathbb{C}$ by way of the formula

$$SE \ni p \mapsto f_\phi(p) = \langle \phi, \wp_m(p) \rangle \in \mathbb{C}.$$

This map has the property that

$$\frac{\mathrm{d}}{\mathrm{d}t} f_\phi(\mathrm{e}^{\mathrm{i}t} p) = \mathrm{i}m f_\phi(\mathrm{e}^{\mathrm{i}t} p).$$

Conversely, any $f : SE \to E$ with this property determines a section ϕ of $E^{\otimes(-m)}$ such that $f_\phi = f$. It is not hard to see that f_ϕ is a CR function if and only if ϕ is a holomorphic section. Suppose $E \to \mathcal{B}$ is a negative line bundle. Suppose m is so large that the map Ψ described above is an embedding. Let $\phi_1, \dots, \phi_N$ be a basis of $\mathfrak{H}(\mathcal{B}, E^{-m})$. Then the map $F : SE \to \mathbb{C}^N$ with components f_{ϕ_j} is an equivariant CR embedding, the assertion in part (1.5) of Theorem 1.5. In this case, since $F(\mathrm{e}^{\mathrm{i}t} p) = \mathrm{e}^{\mathrm{i}mt} F(p)$, the numbers τ_j are all equal to m (here a positive number). Kodaira's embedding map consists of sending the point $b \in \mathcal{B}$ to the complex line containing $F(SE_b)$.

Theorems 1.1 and 1.5 are generalization of classical theorems about line bundles. Other generalizations of classical results about line bundles to the contexts of these theorems were given in [14] (generalizing classification by the first Chern class) and [15] (concerning a kind of Gysin sequence). We point out, however, that Theorem 1.5 applied to line bundles does not quite give Kodaira's embedding theorem because one cannot guarantee that the vector field $\mathcal{T}'$ alluded to in the statement about the embedding being equivariant has all τ_j equal to each other. A similar remark applies to Theorem 1.1.

The proof of Theorem 1.1, contained in Sect. 2, exploits an idea used by Bochner [3] to prove analytic embeddability in $\mathbb{R}^N$ of real analytic compact manifolds with analytic Riemannian metric. The rest of this chapter is devoted to the proof of Theorem 1.5. In Sect. 3, we recall some basic facts about involutive structures and their associated complexes, including some aspects of elliptic structures (of which the subbundles $\overline{\mathcal{V}}$ in the definition of $\mathscr{F}_{\mathrm{ell}}$ are examples). In Sect. 4, we discuss the complexes relevant to this work. The presentation here is motivated by earlier work on complex b-structures; see [12, 13] and [14, Sect. 1]. Section 5 is a preliminary analysis of the structure of the space of CR functions on $\mathcal{N}$ for a given β. This is used in Sect. 6 to prove that (1) $\Longrightarrow$ (2) (Proposition 6.9) and that (2) $\Longrightarrow$ (3) (Proposition 6.11) in Theorem 1.5. The implication (3) $\Longrightarrow$ (4) is proved in Sect. 7 (Theorem 7.1). This last section includes a result (Proposition 7.5) about a decomposition of the space of L^2 CR functions into eigenspaces of $\mathcal{L}_{\mathcal{T}}$. This can be interpreted as giving a global version of the Baouendi–Treves approximation theorem [1]; see Remark 7.15. The implication (4) $\Longrightarrow$ (1) is immediate.

2 Real Embeddings

Suppose $(\mathcal{N}, \mathcal{T}) \in \mathscr{F}$ and fix some $\mathcal{T}$-invariant Riemannian metric g on $\mathcal{N}$. Let Δ denote the Laplace–Beltrami operator. Since $\mathcal{L}_{\mathcal{T}} g = 0$, Δ commutes with $\mathcal{T}$. It is of course well known that the eigenspaces of Δ are finite dimensional and consist of smooth functions. Since Δ commutes with $\mathcal{T}$, these eigenspaces are invariant under $-\mathrm{i}\mathcal{T}$. The latter operator acts on these finite-dimensional spaces as a self-adjoint operator (with the inner product of the L^2 space defined by the Riemannian density), in particular with real eigenvalues. Let

$$\mathcal{E}_{\tau,\lambda} = \{\phi \in C^\infty(\mathcal{N}) : -\mathrm{i}\mathcal{T}\phi = \tau\phi,\ \Delta\phi = \lambda\phi\}$$

and let

$$\operatorname{spec}(-\mathrm{i}\mathcal{T}, \Delta) = \{(\tau, \lambda) : \mathcal{E}_{\tau,\lambda} \neq 0\}.$$

The latter set, the joint spectrum of Δ and $-\mathrm{i}\mathcal{T}$, is a discrete subset of $\mathbb{R}^2$. Since Δ is a real operator (that is, $\Delta\overline{\phi} = \overline{\Delta\phi}$),

$$(\tau, \lambda) \in \operatorname{spec}(-\mathrm{i}\mathcal{T}, \Delta) \implies (-\tau, \lambda) \in \operatorname{spec}(-\mathrm{i}\mathcal{T}, \Delta). \tag{2.1}$$

Note that the map F satisfies $F_*\mathcal{T} = \mathcal{T}'$ with $\mathcal{T}'$ given by (1.2) if and only if its component functions f^j satisfy $\mathcal{T} f^j = \mathrm{i}\tau_j f^j$. This justifies using functions in the spaces $\mathcal{E}_{\tau,\lambda}$ as building blocks for the components of F. For each $(\tau, \lambda) \in \operatorname{spec}(-\mathrm{i}\mathcal{T}, \Delta)$, let $\phi_{\tau,\lambda,j}$, $j = 1, \dots, N_{\tau,\lambda}$, be an orthonormal basis of $\mathcal{E}_{\tau,\lambda}$, so

$$\{\phi_{\tau,\lambda,j} : (\tau, \lambda) \in \operatorname{spec}(-\mathrm{i}\mathcal{T}, \Delta),\ j = 1, \dots, N_{\tau,\lambda}\}$$

is an orthonormal basis of $L^2(\mathcal{N})$. To construct F, we will take advantage of the following two properties of the $\phi_{\tau,\lambda,j}$:

1. For all $p_0 \in \mathcal{N}$, $\mathbb{C}T^*_{p_0}\mathcal{N} = \operatorname{span}\{d\phi_{\tau,\lambda,j}(p_0) : (\tau, \lambda) \in \operatorname{spec}(-\mathrm{i}\mathcal{T}, \Delta),\ j = 1, \dots, N_{\tau,\lambda}\}$.
2. The functions $\phi_{\tau,\lambda,j}$, $(\tau, \lambda) \in \operatorname{spec}(-\mathrm{i}\mathcal{T}, \Delta)$, $j = 1, \dots, N_{\tau,\lambda}$, separate points of $\mathcal{N}$.

To prove the first assertion, suppose that the span of the $d\phi_{\tau,\lambda,j}(p_0)$ is a proper subspace W of $\mathbb{C}T^*_{p_0}\mathcal{N}$, and let $f : \mathcal{N} \to \mathbb{C}$ be a smooth function such that $df(p_0) \notin W$. By standard results from the theory of elliptic self-adjoint operators on compact manifolds, the Fourier series of f,

$$f = \sum_{(\tau,\lambda)\in\Sigma} \sum_{j=1}^{N_{\tau,\lambda}} f_{\tau,\lambda,j}\phi_{\tau,\lambda,j} \tag{2.2}$$

converges to f in $C^\infty(\mathcal{N})$; here we used Σ to denote $\operatorname{spec}(-\mathrm{i}\mathcal{T}, \Delta)$. So

$$\mathrm{d}f(p_0) = \sum_{(\tau,\lambda)\in\Sigma} \sum_{j=1}^{N_{\tau,\lambda}} f_{\tau,\lambda,j} \mathrm{d}\phi_{\tau,\lambda,j}(p_0)$$

with uniform convergence of the series. The terms of the series belong to W, a complete space because it is finite dimensional, so the convergence takes place in W. But $\mathrm{d}f(p_0) \notin W$, a contradiction. Thus in fact $W = \mathbb{C}T^*_{p_0}\mathcal{N}$ as claimed.

The second assertion is proved by using the pointwise convergence of the series (2.2) for a smooth function f separating two distinct points p_0 and p_1 to contradict the supposition that $\phi_{\tau,\lambda,j}(p_0) = \phi_{\tau,\lambda,j}(p_1)$ for all values of the indices.

It follows from property (1) that there are (τ_k, λ_k, j_k), $k = 1, \dots, \dim\mathcal{N}$ such that the differentials at p_0 of the functions $f^k = \phi_{\tau_k,\lambda_k,j_k}$ span $\mathbb{C}T^*_{p_0}\mathcal{N}$. Then, if v is a real tangent vector at p_0, the condition $\mathrm{d}f^k(v) = 0$ for all k implies $v = 0$. The same property is true if some or all of the functions f^k are replaced by their conjugates. So replacing f^k by $\overline{f}^k$ if $\tau_k < 0$, we get that the map

$$p \mapsto (f^1(p), \dots, f^{\dim\mathcal{N}}(p))$$

has injective differential at p_0 (hence in a neighborhood of p_0) and components that satisfy $\mathcal{T} f^k = \mathrm{i}\tau_k f^k$ with $\tau_k > 0$; see (2.1).

By the compactness of $\mathcal{N}$, there are smooth functions $\tilde{f}^1, \dots, \tilde{f}^{\tilde{N}}$ such that $\mathcal{T}\tilde{f}^k = \mathrm{i}\tau_k\tilde{f}^k$ for each k with $\tau_k > 0$ and such that the map $\tilde{F} : \mathcal{N} \to \mathbb{C}^{\tilde{N}}$ with components f^k is an immersion. The origin of $\mathbb{C}^{\tilde{N}}$ is not in the image of $\tilde{F}$. Indeed, if there is p_0 such that $\tilde{f}^k(p_0) = 0$ for all k, then $\mathcal{T}\tilde{f}^k(p_0) = \mathrm{i}\tau_k\tilde{f}(p_0) = 0$ for all k, so $\mathcal{T}(p_0)$ belongs to the kernel of $\mathrm{d}\tilde{F}(p_0)$, a contradiction.

Since $\|\tilde{F}(p)\| \neq 0$ for all p, the map $p \mapsto \|\tilde{F}(p)\|^{-1}\tilde{F}(p)$ is smooth and has image in $S^{2\tilde{N}-1}$. However, it may not be an immersion, since the differential of the radial projection $\mathbb{C}^{\tilde{N}}\backslash 0 \to S^{2\tilde{N}-1}$ has nontrivial kernel at every point: the kernel at $z \in \mathbb{C}^{\tilde{N}}\backslash 0$ is the radial vector $R = \sum_\ell z^\ell \partial_{z^\ell} + \overline{z}^\ell \partial_{\overline{z}^\ell}$. To fix this problem, we augment $\tilde{F}$ by adjoining the functions $(\tilde{f}^k)^2$: redefine $\tilde{F}$ to be

$$\tilde{F} = (\tilde{f}^1, \dots, \tilde{f}^{\tilde{N}}, (\tilde{f}^1)^2, \dots, (\tilde{f}^{\tilde{N}})^2).$$

Then $\tilde{F}$ is again an immersion. Moreover, for all $p \in \mathcal{N}$, $R(\tilde{F}(p)) \notin \operatorname{rg}\mathrm{d}\tilde{F}(p)$. To see this, suppose $v \in T_{p_0}\mathcal{N}$ is such that

$$\mathrm{d}\tilde{F}(v) = cR(\tilde{F}(p_0))$$

for some c. Then

$$\langle \mathrm{d}\tilde{f}^k, v\rangle = c\tilde{f}^k(p_0) \text{ and } \langle \mathrm{d}(\tilde{f}^k)^2, v\rangle = c(\tilde{f}^k(p_0))^2,\ k = 1, \dots, \tilde{N}.$$

Using the first set of equations in the second, we get

$$c(\tilde{f}^k(p_0))^2 = \langle \mathrm{d}(\tilde{f}^k)^2, v\rangle = 2\tilde{f}^k(p_0)\langle \mathrm{d}\tilde{f}^k, v\rangle = 2c(\tilde{f}^k(p_0))^2 \text{ for all } k$$

so, since $\tilde{f}^k(p_0) \neq 0$ for some k, $c = 2c$, hence $c = 0$. Thus the composition of $\tilde{F}$ with the radial projection on $S^{4\tilde{N}-1}$,

$$F_0(p) = \frac{1}{\|\tilde{F}(p)\|}\tilde{F}(p),$$

is an immersion. Let $N_0 = 2\tilde{N}$ and let $f^1, \dots, f^{N_0}$ denote the components of F_0. Since $\mathcal{T}|\tilde{f}^j|^2 = 0$ (because $\mathcal{T}\tilde{f}^j = i\tau_j\tilde{f}^j$ and τ_j is real), $\mathcal{T}f^j = i\tau_j f^j$ with $\tau_j > 0$.

We will now augment F_0 so as to obtain an injective map. Let

$$Z = \{(p_0, p_1) \in \mathcal{N} \times \mathcal{N} : p_0 \neq p_1,\ f^k(p_0) = f^k(p_1) \text{ for all } k\}.$$

Since F_0 is an immersion, the diagonal in $\mathcal{N} \times \mathcal{N}$ has a neighborhood U on which the condition

$$(p_0, p_1) \in U \text{and } F_0(p_0) = F_0(p_1) \implies p_0 = p_1$$

holds. Thus Z is a closed set. Suppose $(p_0, p_1) \in Z$. By the second property of the functions $\phi_{\tau,\lambda,j}$, there is f smooth such that $\mathcal{T}f = i\tau f$ and $f(p_0) \neq f(p_1)$. If the latter happens, then also $\overline{f}(p_0) \neq \overline{f}(p_1)$, so we may assume $\tau > 0$. With such f, the map

$$F_1 : p \mapsto \frac{1}{\sqrt{1 + |f(p)|^2}}(F(p), f(p)),$$

which has image in the unit sphere in $\mathbb{C}^{N_0+1}$, separates p_0 and p_1. Indeed, if

$$\frac{F(p_0)}{\sqrt{1 + |f(p_0)|^2}} = \frac{F(p_1)}{\sqrt{1 + |f(p_1)|^2}} \quad \text{and}$$
$$\frac{f(p_0)}{\sqrt{1 + |f(p_0)|^2}} = \frac{f(p_1)}{\sqrt{1 + |f(p_1)|^2}},$$

then, since $F(p_0) = F(p_1)$ (because $(p_0, p_1) \in Z$), $\sqrt{1 + |f(p_0)|^2} = \sqrt{1 + |f(p_1)|^2}$, so $f(p_0) = f(p_1)$ contradicting the choice of f. So $F_1(p_0) \neq F_1(p_1)$, and (p_0, p_1) has a neighborhood U such that $(p, p') \in U \implies F_1(p) \neq F_1(p')$. Using the compactness of Z, we get a finite number of maps $F_1, \dots, F_L$, each mapping $\mathcal{N}$ into the unit sphere in $\mathbb{C}^{N_0+1}$, such that $(p_0, p_1) \in Z$ implies $F_\ell(p_0) \neq F_\ell(p_1)$ for some ℓ. Then, with $N = N_0 + (N_0 + 1)L + 1$,

$$F = \frac{1}{\sqrt{L + 1}}(F_0, F_1, \dots, F_L) : \mathcal{N} \to S^{2N+1}$$

is an embedding whose components f^j satisfy $\mathcal{T} f^j = \mathrm{i}\tau_j f^j$ with $\tau_j > 0$, hence $F_*\mathcal{T} = T'$ with $\mathcal{T}'$ given by (1.2) as claimed.

That no component of the map F just constructed is flat at any point of $\mathcal{N}$ is a consequence of the fact that these functions are constructed out of eigenfunctions of a second-order elliptic real operator (see [7, Theorem 17.2.6]). In particular, the set $\{p \in \mathcal{N} : \forall j\ F^j(p) \neq 0\}$ is dense in $\mathcal{N}$.

Remark 2.1. The last assertion of Theorem 1.1 was an essential component in the proof of Proposition 3.7 used in [14].

3 Involutive Structures

Let $\mathcal{M}$ be a smooth manifold. An involutive structure on $\mathcal{M}$ is a subbundle of the complexification $\mathbb{C}T\mathcal{M}$ of the tangent bundle of $\mathcal{M}$. We will briefly review some facts in connection with such structures here and then discuss particularities in the context of Theorem 7.1. For a detailed account of various aspects of such structures, see Treves [18–20].

Associated to any involutive structure $\mathcal{W} \subset \mathbb{C}T\mathcal{M}$, there is a complex based on the exterior powers of the dual bundle:

$$\cdots \to C^\infty(\mathcal{M}; \textstyle\bigwedge^q \mathcal{W}^*) \xrightarrow{\mathfrak{D}} C^\infty(\mathcal{M}; \textstyle\bigwedge^{q+1} \mathcal{W}^*) \to \cdots. \tag{3.1}$$

Namely, if $\eta \in C^\infty(\mathcal{M}; \bigwedge^q \mathcal{W}^*)$ and $V_0, \dots, V_q$ are smooth sections of $\mathcal{W}$, then

$$\begin{aligned}(q+1)\mathfrak{D}\eta(V_0,\dots,V_q) &= \sum_j (-1)^j V_j \eta(V_0,\dots,\hat{V}_j,\dots,V_q) \\ &\quad + \sum_{j<k} (-1)^{j+k} \eta([V_j,V_k],V_1,\dots,\hat{V}_j,\dots,\hat{V}_k,\dots,V_q).\end{aligned} \tag{3.2}$$

These satisfy

$$\mathfrak{D}^2 = 0$$

and

$$\mathfrak{D}(\eta \wedge \zeta) = \mathfrak{D}(\eta) \wedge \zeta + (-1)^q \eta \wedge \mathfrak{D}(\zeta) \tag{3.3}$$

if $\eta \in C^\infty(\mathcal{M}; \bigwedge^q \mathcal{W}^*)$ and $\zeta \in C^\infty(\mathcal{M}; \bigwedge^{q'} \mathcal{W}^*)$. For a function f, we have $\mathfrak{D} f = \iota^* \mathrm{d} f$, where $\iota^* : \mathbb{C}T^*\mathcal{M} \to \mathcal{W}^*$ is the dual of the inclusion homomorphism $\iota : \mathcal{W} \to \mathbb{C}T\mathcal{M}$. This just means that

$$\langle \mathfrak{D} f, v \rangle = v f \tag{3.4}$$

if $v \in \mathcal{W}$.

The structure $\mathcal{W}$ is said to be elliptic if $\mathcal{W} + \overline{\mathcal{W}} = \mathbb{C}T\mathcal{M}$, the reason being that the complex (3.1) is elliptic if and only if $\mathcal{W}$ is. If $\mathcal{W}$ is an elliptic structure, $\mathcal{W} \cap \overline{\mathcal{W}}$ is the complexification of a subbundle of $T\mathcal{M}$; its integral manifolds are called the real leaves of the structure.

Suppose $\mathcal{W}$ is an elliptic structure. By a theorem of Nirenberg [17] (a consequence of the Newlander–Nirenberg theorem [16]), every point $p_0 \in \mathcal{M}$ has a neighborhood U on which there are local coordinates

$$x^1, \dots, x^{2n}, t^1, \dots, t^\kappa$$

such that, with $z^\mu = x^\mu + \mathrm{i}x^{\mu+n}$, $\mathcal{W}|_U$ is the span of the vector fields

$$\partial_{\overline{z}^1}, \dots, \partial_{\overline{z}^n}, \partial_{t^1}, \dots, \partial_{t^\kappa}. \tag{3.5}$$

Such a local chart $(z^1, \dots, z^n, t^1, \dots, t^\kappa)$ is called a hypoanalytic chart (Baouendi-Chang-Treves [2], Treves [20]). The intersection of the real leaves and U are the level sets of the function $p \mapsto (z^1(p), \dots, z^n(p))$. If U is connected and $\zeta : U \to \mathbb{C}$ satisfies $\mathfrak{D}\zeta = 0$, then ζ is constant on the connected components of the real leaves in U and a holomorphic function of the z^μ.

Lemma 3.6. *Suppose that $\mathcal{M}$ is connected, let $\mathcal{W} \subset \mathbb{C}T^*\mathcal{M}$ be an elliptic structure, and let $\beta \in C^\infty(\mathcal{M}; \mathcal{W}^*)$ be $\mathfrak{D}$-closed. If $\zeta : \mathcal{M} \to \mathbb{C}$ is not identically zero and $\mathfrak{D}\zeta + \zeta\beta = 0$, then the set $\{p \in \mathcal{M} : \zeta(p) = 0\}$ has empty interior.*

Proof. Let $p_0 \in \zeta^{-1}(0)$ and let $(z^1, \dots, z^n, t^1, \dots, t^\kappa)$ be a hypoanalytic chart centered at p_0, mapping its domain U onto $B \times C$ where B is a ball in $\mathbb{C}^n$ with center 0 and C is the cube $(-1, 1)^\kappa \subset \mathbb{R}^\kappa$. We will show in a moment that there is $f : U \to \mathbb{C}$ such that $\mathfrak{D}f = \beta$ in U. Assuming this, we have

$$\mathfrak{D}(\mathrm{e}^f\zeta) = \mathrm{e}^f(\mathfrak{D}\zeta + \zeta\mathfrak{D}f) = \mathrm{e}^f(-\zeta\beta + \zeta\beta) = 0$$

so $\mathrm{e}^f\zeta$ is a holomorphic function of the z^μ. Thus if the set $\zeta^{-1}(0) \cap U$ does not have empty interior, then ζ vanishes on U. A simple argument using the connectedness of $\mathcal{M}$ then leads to the conclusion that if the interior of $\zeta^{-1}(0)$ is not empty, then ζ is identically 0.

To complete the proof, we show that β is exact on U using a well-known argument. Over U, the sections $\mathfrak{D}\overline{z}^\mu$, $\mathfrak{D}t^j$ of $\mathcal{W}^*$ form the frame dual to the frame (3.5) of $\mathcal{W}$. Writing

$$\beta = \sum_{\mu=1}^n \beta_\mu \mathfrak{D}\overline{z}^\mu + \sum_{j=1}^\kappa \beta_j \mathfrak{D}t^j$$

we have

$$\mathfrak{D}\beta = \sum_{\mu<\nu} \left(\frac{\partial\beta_\nu}{\partial\overline{z}^\mu} - \frac{\partial\beta_\mu}{\partial\overline{z}^\nu} \right) \mathfrak{D}\overline{z}^\mu \wedge \mathfrak{D}\overline{z}^\nu + \sum_{\mu=1}^n \sum_{j=1}^\kappa \left(\frac{\partial\beta_j}{\partial\overline{z}^\mu} - \frac{\partial\beta_\mu}{\partial t^j} \right) \mathfrak{D}\overline{z}^\mu \wedge \mathfrak{D}t^j$$
$$+ \sum_{j<k} \left(\frac{\partial\beta_k}{\partial t^j} - \frac{\partial\beta_j}{\partial t^k} \right) \mathfrak{D}t^j \wedge \mathfrak{D}t^k.$$

From the condition $\mathfrak{D}\beta = 0$, we derive the existence of a smooth function g such that $\partial g/\partial t^j = \beta_j$ for each j. Then

$$\beta' = \beta - \mathfrak{D}g = \sum_{\mu=1}^{n} \left(\beta_\mu - \frac{\partial g}{\partial \overline{z}^\mu} \right) \mathfrak{D}\, \overline{z}^\mu$$

is again $\mathfrak{D}$-closed, and consequently the coefficients of β' are independent of the t^j. We may then view β' as a $(0,1)$-form, and as such it is $\overline{\partial}$-closed. Since B is a ball, there is $h(z)$ such that $\overline{\partial} h = \beta'$, and it follows that $\beta = \mathfrak{D}(g+h)$ in U.

We end our discussion of general elliptic structures with the following observation:

Lemma 3.7. *Suppose that $\mathcal{M}$ is compact and connected. If $\zeta : \mathcal{M} \to \mathbb{C}$ solves $\mathfrak{D}\zeta = 0$, then ζ is constant.*

Proof. Let p_0 be an extremal point of $|\zeta|$. Fix a hypoanalytic chart (z,t) for $\overline{\mathcal{V}}$ centered at p_0. Since $\mathfrak{D}\zeta = 0$, $\zeta(z,t)$ is independent of t and $\partial_{\overline{z}^\nu}\zeta = 0$. So there is a holomorphic function Z defined in a neighborhood of 0 in $\mathbb{C}^n$ such that $\zeta = Z \circ z$. Then $|Z|$ has a maximum at 0, so Z is constant near 0. Therefore ζ is constant, say $\zeta(p) = c$, near p_0. Let $C = \{p : \zeta(p) = c\}$, a closed set. Let $p_1 \in C$. Since p_1 is also an extremal point of ζ, the above argument gives that ζ is constant near p_1, therefore equal to c. Thus C is open, and consequently ζ is constant on $\mathcal{M}$.

4 Underlying Complexes

Fix $(\mathcal{N},\mathcal{T},\overline{\mathcal{V}}) \in \mathscr{F}_{\text{ell}}$, that is, $(\mathcal{N},\mathcal{T}) \in \mathscr{F}$, $\overline{\mathcal{V}} \subset \mathbb{C}T\mathcal{N}$ is an involutive elliptic subbundle with $\mathcal{V} \cap \overline{\mathcal{V}} = \operatorname{span}_{\mathbb{C}}\mathcal{T}$, and there is a global section $\beta \in C^\infty(\mathcal{N};\overline{\mathcal{V}}^*)$ such that

$$\begin{aligned} &\text{(a) } \langle \beta, \mathcal{T} \rangle = -\mathrm{i};\\ &\text{(b) } \overline{\mathbb{D}}\beta = 0 \end{aligned} \tag{4.1}$$

where $\overline{\mathbb{D}}$ refers to the coboundary operator of the induced differential complex on $\mathcal{V}^*$:

$$\cdots \to C^\infty(\mathcal{N};\textstyle\bigwedge^q\overline{\mathcal{V}}^*) \xrightarrow{\overline{\mathbb{D}}} C^\infty(\mathcal{N};\textstyle\bigwedge^{q+1}\overline{\mathcal{V}}^*) \to \cdots. \tag{4.2}$$

In addition to the complex (4.2), which exists independently of β, there is another complex on $\mathcal{N}$ induced by β, namely, let

$$\overline{\mathcal{K}}_\beta = \{v \in \overline{\mathcal{V}} : \langle \beta, v \rangle = 0\}.$$

Indeed, $\mathcal{K}_\beta$ is involutive: For if V and W are sections of $\mathcal{K}_\beta$, then by (3.2),

$$\langle \beta, [V,W] \rangle = -2\overline{\mathbb{D}}\beta(V,W) + V\langle \beta, W \rangle - W\langle \beta, V \rangle$$

which vanishes by property (b) above and because $\langle\beta, V\rangle = \langle\beta, W\rangle = 0$. Now, $\overline{\mathcal{K}}_\beta$ is a CR structure: $\overline{\mathcal{K}}_\beta \cap \mathcal{K}_\beta = 0$. To see this, suppose $v \in \overline{\mathcal{K}}_\beta \cap \mathcal{K}_\beta$. Then in particular, $v \in \overline{\mathcal{V}} \cap \mathcal{V}$, so $v = c\mathcal{T}$ for some c. Thus $0 = \langle\beta, v\rangle = \langle\beta, c\mathcal{T}\rangle = ic$, hence $v = 0$. We will write $\overline{\partial}_b$ for the coboundary operators of the complex

$$\cdots \to C^\infty(\mathcal{N}; \textstyle\bigwedge^q \overline{\mathcal{K}}_\beta^*) \to C^\infty(\mathcal{N}; \textstyle\bigwedge^{q+1} \overline{\mathcal{K}}_\beta^*) \to \cdots.$$

Occasionally, there will be two such complexes involved, determined by sections β and β'. We will not distinguish this in the notation.

There is a third complex associated with $\overline{\mathcal{V}}$ and β, in which the terms of the cochain complex are those in (4.2), but the coboundary operator is

$$\overline{\mathcal{D}}_q(\sigma)\phi = \overline{\mathbb{D}}\phi + i\sigma\beta \wedge \phi, \quad \phi \in C^\infty(\mathcal{N}; \textstyle\bigwedge^q \overline{\mathcal{V}}^*) \tag{4.3}$$

with a fixed $\sigma \in \mathbb{C}$. That $\overline{\mathcal{D}}_{q+1}(\sigma)\overline{\mathcal{D}}_q(\sigma) = 0$ follows immediately form the corresponding property for $\overline{\mathbb{D}}$ together with b) in (4.1). This complex is, again, elliptic. Write $H^q_{\overline{\mathcal{D}}(\sigma)}(\mathcal{N})$ for the cohomology groups and let

$$\operatorname{spec}^q(\overline{\mathcal{D}}) = \{\sigma : H^q_{\overline{\mathcal{D}}(\sigma)}(\mathcal{N}) \neq 0\}.$$

Lemma 4.4. *The cohomology groups $H^q_{\overline{\mathcal{D}}(\sigma)}(\mathcal{N})$ are finite dimensional for each $\sigma \in \mathbb{C}$. For each q, the set $\operatorname{spec}^q(\overline{\mathcal{D}})$ is closed and discrete; in fact,*

$$\{\sigma \in \operatorname{spec}^q(\overline{\mathcal{D}}) : -a \leq \operatorname{Im}\sigma \leq a\}$$

is finite for each $a > 0$.

Proof. Fix a $\mathcal{T}$-invariant metric g on $\mathcal{N}$ for which $g(\mathcal{T}, \mathcal{T}) = 1$. It determines a metric on $\overline{\mathcal{V}}$, hence on the various exterior powers of $\overline{\mathcal{V}}^*$. We use these metrics and the Riemannian measure to give an L^2 inner product to each of the spaces $C^\infty(N; \bigwedge^q \overline{\mathcal{V}}^*)$. Let

$$\overline{\mathcal{D}}_q^\star(\sigma) : C^\infty(\mathcal{N}; \textstyle\bigwedge^{q+1} \overline{\mathcal{V}}^*) \to C^\infty(\mathcal{N}; \textstyle\bigwedge^q \overline{\mathcal{V}}^*)$$

denote the formal adjoint of $\overline{\mathcal{D}}_q(\overline{\sigma})$; it depends holomorphically on σ. Define

$$\Box_q(\sigma) = \overline{\mathcal{D}}_q^\star(\sigma)\overline{\mathcal{D}}_q(\sigma) + \overline{\mathcal{D}}_{q-1}(\sigma)\overline{\mathcal{D}}_{q-1}^\star(\sigma).$$

This is a family of elliptic operators depending holomorphically on σ. Since $\Box_q(\sigma)$ is elliptic (because the complex is) and $\mathcal{N}$ is compact, this is a Fredholm family. Furthermore, if $\boldsymbol{\sigma}(\Box_q(\sigma))$ denotes the principal symbol of $\Box_q(\sigma)$ and $\|\beta\|$ denotes the pointwise norm of β, we have that

$$(\boldsymbol{\sigma}\Box_q(\sigma))(\boldsymbol{\xi}) + \sigma^2 \|\beta\|^2 I$$

is invertible at every point $\boldsymbol{\xi} \in T^*\mathcal{N}$ and every $\sigma \in \mathbb{C}$ with estimates

$$\left\| \left(\boldsymbol{\sigma} \left(\Box_q(\sigma) \right) (\boldsymbol{\xi}) + \sigma^2 \, \|\beta\|^2 I \right)^{-1} \right\| \leq \frac{C}{\|\boldsymbol{\xi}\|^2 + |\sigma|^2}$$

with uniform C for arbitrary $\boldsymbol{\xi}$ and σ such that $|\operatorname{Im}\sigma| \leq a$ (C depends on a) because $\|\beta\|$ is nowhere zero. This estimate implies that for each $a > 0$ there is b such that $\Box_q(\sigma)$ is invertible if $|\operatorname{Im}\sigma| \leq a$ and $|\Re\sigma| > b$. Since $\Box_q(\sigma)$ is a holomorphic Fredholm family, the intersection of

$$\Sigma_q = \{\sigma \in \mathbb{C} : \Box_q(\sigma) \text{ is invertible}\}$$

with any horizontal strip $\{\sigma \in \mathbb{C} : |\operatorname{Im}\sigma| \leq a\}$ is finite. We now show that the analogous statement holds for $\operatorname{spec}^q(\overline{\mathcal{D}})$. Let

$$\mathcal{G}_q(\sigma) : L^2(\mathcal{N}; E^q_{\mathcal{N}}) \to H^2(\mathcal{N}; E^q_{\mathcal{N}})$$

be the inverse of $\Box_q(\sigma)$, $\sigma \notin \Sigma_q$. The map $\sigma \mapsto \mathcal{G}_q(\sigma)$ is meromorphic with poles in Σ_q. The operators $\Box_q(\sigma)$ are the Laplacians of the complex (4.2) with the coboundary operators (4.3) when σ is real. Thus for $\sigma \in \mathbb{R}\backslash(\operatorname{spec}_{b,\mathcal{N}}(\Box_q) \cup \operatorname{spec}_{b,\mathcal{N}}(\Box_{q+1}))$, we have

$$\overline{\mathcal{D}}_q(\sigma)\mathcal{G}_q(\sigma) = \mathcal{G}_{q+1}(\sigma)\overline{\mathcal{D}}_q(\sigma), \quad \overline{\mathcal{D}}_q(\sigma)^\star \mathcal{G}_{q+1}(\sigma) = \mathcal{G}_q(\sigma)\overline{\mathcal{D}}_q(\overline{\sigma})^\star$$

by standard Hodge theory. Since all operators depend holomorphically on σ, the same equalities hold for $\sigma \in \mathfrak{R} = \mathbb{C}\backslash(\Sigma_q \cup \Sigma_{q+1})$. It follows that

$$\overline{\mathcal{D}}_q^\star(\sigma)\overline{\mathcal{D}}_q(\sigma)\mathcal{G}_q(\sigma) = \mathcal{G}_q(\sigma)\overline{\mathcal{D}}_q^\star(\sigma)\overline{\mathcal{D}}_q(\sigma)$$

in $\mathfrak{R}$. By analytic continuation, the equality holds on all of $\mathbb{C}\backslash\Sigma_q$. Thus if $\sigma_0 \notin \Sigma_q$ and ϕ is a $\overline{\mathcal{D}}_q(\sigma_0)$-closed section, $\overline{\mathcal{D}}_q(\sigma_0)\phi = 0$, then the formula

$$\phi = [\overline{\mathcal{D}}_q^\star(\sigma_0)\overline{\mathcal{D}}_q(\sigma_0) + \overline{\mathcal{D}}_{q-1}(\sigma_0)\overline{\mathcal{D}}_{q-1}^\star(\sigma_0)]\mathcal{G}_q(\sigma_0)\phi$$

leads to

$$\phi = \overline{\mathcal{D}}_{q-1}(\sigma_0)[\overline{\mathcal{D}}_{q-1}^\star(\sigma_0)\mathcal{G}_q(\sigma_0)\phi].$$

Therefore $\sigma_0 \notin \operatorname{spec}^q(\overline{\mathcal{D}})$. Thus $\operatorname{spec}^q(\overline{\mathcal{D}}) \subset \Sigma_q$.

Remark 4.5. The argument concerning the poles of the inverse of $\Box_q(\sigma)$ is extracted from a related problem in the analysis of elliptic operators on b-manifolds; see Melrose [11].

Later, we will allow replacing the section β by an equivalent section in the following sense.

Definition 4.6. Two smooth sections β, β' of $\overline{\mathcal{V}}^*$ satisfying (4.1) are equivalent if $\beta' - \beta = \overline{\mathbb{D}}u$ for some real-valued function u. The class of β is denoted $\boldsymbol{\beta}$.

Lemma 4.7. *Suppose β, β' are equivalent, let $\overline{\mathcal{D}}(\sigma)$, $\overline{\mathcal{D}}'(\sigma)$ be the associated operators. Then*

$$H^q_{\overline{\mathcal{D}}(\sigma)}(\mathcal{N}) \approx H^q_{\overline{\mathcal{D}}'(\sigma)}(\mathcal{N})$$

for all q and σ. Consequently, $\operatorname{spec}^q(\overline{\mathcal{D}})$ *depends only on the class of β.*

Proof. There is u real valued such that $\beta' = \beta + \overline{\mathbb{D}}u$. Using (3.3) and $\overline{\mathbb{D}}e^{i\sigma u} = i\sigma e^{i\sigma u}\overline{\mathbb{D}}u$, we see that

$$\begin{array}{ccccccc}
\cdots \longrightarrow & C^\infty(\mathcal{N};\bigwedge^q\overline{\mathcal{V}}^*) & \xrightarrow{\overline{\mathcal{D}}_\beta(\sigma)} & C^\infty(\mathcal{N};\bigwedge^{q+1}\overline{\mathcal{V}}^*) & \longrightarrow \cdots \\
& \Big\downarrow e^{i\sigma u} & & \Big\downarrow e^{i\sigma u} & \\
\cdots \longrightarrow & C^\infty(\mathcal{N};\bigwedge^q\overline{\mathcal{V}}^*) & \xrightarrow{\overline{\mathcal{D}}_{\beta'}(\sigma)} & C^\infty(\mathcal{N};\bigwedge^{q+1}\overline{\mathcal{V}}^*) & \longrightarrow \cdots
\end{array} \tag{4.8}$$

is a cochain isomorphism for any σ.

5 CR Functions

We continue our discussion with a fixed element $(\mathcal{N}, \mathcal{T}, \overline{\mathcal{V}})$ of $\mathscr{F}_{\mathrm{ell}}$ and section β of $\overline{\mathcal{V}}^*$ satisfying (4.1). The section β gives a CR structure $\overline{\mathcal{K}}_\beta = \ker\beta$ and operators $\overline{\mathcal{D}}(\sigma)$ defined in (4.3).

The one-parameter group of diffeomorphisms generated by $\mathcal{T}$ will be denoted by $t \mapsto \mathfrak{a}_t$. We write $\mathcal{O}_p$ for the orbit of $\mathcal{T}$ through p. The integral curves of $\mathcal{T}$ need not be periodic, i.e., the orbits need not be closed.

Lemma 5.1. *A distribution $\zeta \in C^{-\infty}(\mathcal{N})$ solves $\overline{\mathcal{D}}(\sigma)\zeta = 0$ if and only if it is a CR function and satisfies*

$$\mathcal{T}\zeta + \sigma\zeta = 0 \tag{5.2}$$

If $\overline{\mathcal{D}}(\sigma)\zeta = 0$, then ζ is smooth.

Proof. Since $\overline{\mathcal{V}} = \overline{\mathcal{K}}_\beta \oplus \operatorname{span}\mathcal{T}$, the statement that $\overline{\mathbb{D}}\zeta + i\sigma\beta\zeta$ vanishes is equivalent to

$$\langle \overline{\mathbb{D}}\zeta + i\sigma\zeta\beta, v\rangle = 0 \ \forall v \in \overline{\mathcal{K}}_\beta \quad \text{and} \quad \langle \overline{\mathbb{D}}\zeta + i\sigma\beta\zeta, \mathcal{T}\rangle = 0.$$

In view of part (a) of (4.1), and since $\langle\beta, v\rangle = 0$ and $\langle\overline{\mathbb{D}}\zeta, v\rangle = \langle\overline{\partial}_b\zeta, v\rangle$ if $v \in \overline{\mathcal{K}}_\beta$, these statements are equivalent, respectively, to

$$\overline{\partial}_b\zeta = 0 \quad \text{and} \quad \mathcal{T}\zeta + \sigma\zeta = 0$$

as claimed. That ζ is smooth if $\overline{\mathcal{D}}(\sigma)\zeta = 0$ is a consequence of the complex (4.2) being elliptic (the principal symbol of $\overline{\mathcal{D}}(\sigma)$ on functions is injective).

The space of smooth CR functions, $C^\infty_{\mathrm{CR}}(\mathcal{N}) = C^\infty(\mathcal{N}) \cap \ker \overline{\partial}_b$, is a ring. We will see that $C^\infty_{\mathrm{CR}}(\mathcal{N})$ decomposes as a direct sum of the spaces $\ker \overline{\mathcal{D}}_0(\sigma)$, $\sigma \in \operatorname{spec}^0(\overline{\mathcal{D}})$.

Lemma 5.3. *The set* $\operatorname{spec}^0(\overline{\mathcal{D}}) \subset \mathbb{C}$ *is a subset of the imaginary axis and an additive discrete semigroup with identity. If* $\operatorname{spec}^0(\overline{\mathcal{D}})$ *is not a group, then* $\operatorname{spec}^0(\overline{\mathcal{D}})\backslash 0$ *is contained in a single component of* $\mathrm{i}\mathbb{R}\backslash\{0\}$.

Proof. That $\operatorname{spec}^0(\overline{\mathcal{D}})$ is discrete is a consequence of Lemma 4.4. Suppose $\sigma_0 \in \operatorname{spec}^0(\overline{\mathcal{D}})$ is not zero and let ζ be a nonzero function that satisfies $\overline{\mathcal{D}}(\sigma_0)\zeta = 0$; such ζ exists precisely because $\sigma_0 \in \operatorname{spec}^0(\overline{\mathcal{D}})$. Furthermore, ζ is bounded because it is smooth and $\mathcal{N}$ is compact. By Lemma 5.1, $\zeta(\mathfrak{a}_t p) = \mathrm{e}^{-\sigma_0 t}\zeta(p)$. So $|\mathrm{e}^{-\sigma_0 t}\zeta(p)|$ is bounded. Since ζ is not identically zero, there is p such that $\zeta(p) \neq 0$. Thus $|\mathrm{e}^{-\sigma_0 t}|$ is bounded, hence $\Re\sigma_0 = 0$.

Since $\overline{\mathbb{D}}1 = 0$, $0 \in \operatorname{spec}^0(\overline{\mathcal{D}})$. Let $\sigma_1, \sigma_2 \in \operatorname{spec}^0(\overline{\mathcal{D}})$, and pick nonvanishing elements $\zeta^1 \in H^0_{\overline{\mathcal{D}}(\sigma_1)}(\mathcal{N})$, $\zeta^2 \in H^0_{\overline{\mathcal{D}}(\sigma_2)}(\mathcal{N})$. Since

$$\overline{\mathbb{D}}(\zeta^1\zeta^2) = \zeta^2\overline{\mathbb{D}}\zeta^1 + \zeta^1\overline{\mathbb{D}}(\zeta^2) = -\mathrm{i}(\sigma_1 + \sigma_2)\zeta^1\zeta^2\beta,$$

$\zeta^1\zeta^2 \in H^0_{\overline{\mathcal{D}}(\sigma_1+\sigma_2)}(\mathcal{N})$ which by Lemma 3.6 is not identically 0 (since neither of ζ^1, ζ^2 is). Thus $\sigma_1 + \sigma_2 \in \operatorname{spec}^0(\overline{\mathcal{D}})$.

Suppose now that $\operatorname{spec}^0(\overline{\mathcal{D}})$ has elements in both components of $\mathbb{C}\backslash\mathbb{R}$ and let σ_+ be the element with smallest modulus among the elements of $\operatorname{spec}^0(\overline{\mathcal{D}})$ with positive imaginary part, and let σ_- be the analogous element with negative imaginary part. If $\sigma = \sigma_+ + \sigma_- \neq 0$, then either $\operatorname{Im}\sigma > 0$ and $|\sigma| < |\sigma_+|$ or $\operatorname{Im}\sigma < 0$ and $|\sigma| < |\sigma_-|$. Either way, we get a contradiction, since $\sigma \in \operatorname{spec}^0(\overline{\mathcal{D}})$. So $\sigma_- = -\sigma_+$. In particular, $m\sigma_+ \in \operatorname{spec}^0(\overline{\mathcal{D}})$ for every $m \in \mathbb{Z}$. If $\sigma \in \operatorname{spec}^0(\overline{\mathcal{D}})$ is arbitrary, then there is $m \in \mathbb{Z}$ such that $|\sigma - m\sigma_+| < |\sigma_+|$. Consequently, $\sigma - m\sigma_+ = 0$. Thus $\operatorname{spec}^0(\overline{\mathcal{D}}) = \sigma_+\mathbb{Z}$, a group. Therefore, if $\operatorname{spec}^0(\overline{\mathcal{D}})$ is not a group, then $\operatorname{spec}^0(\overline{\mathcal{D}})\backslash 0$ is contained in a single component of $\mathbb{C}\backslash\mathbb{R}$.

Thus the space

$$\bigoplus_{\sigma\in\operatorname{spec}^0(\overline{\mathcal{D}})} H^0_{\overline{\mathcal{D}}(\sigma)}(\mathcal{N}),$$

is a subring of $C^\infty_{\mathrm{CR}}(\mathcal{N})$ graded by $\operatorname{spec}^0(\overline{\mathcal{D}})$.

The spaces $H^0_{\overline{\mathcal{D}}(\sigma)}(\mathcal{N})$ are particularly simple when $\operatorname{spec}^0(\overline{\mathcal{D}})$ is a group.

Proposition 5.4. *Suppose that* $\operatorname{spec}^0(\overline{\mathcal{D}})$ *is a group. Then all cohomology groups* $H^0_{\overline{\mathcal{D}}(\sigma)}(\mathcal{N})$, $\sigma \in \operatorname{spec}^0(\overline{\mathcal{D}})$, *are one dimensional, and all their nonzero elements are nowhere vanishing functions.*

Proof. The dimension of $H^0_{\overline{\mathcal{D}}(0)}(\mathcal{N}) = H^0_{\overline{\mathbb{D}}}(\mathcal{N})$ is 1, since this space contains the constant functions, and only the constant functions by Lemma 3.7. If $\operatorname{spec}^0(\overline{\mathcal{D}}) = \{0\}$, there is nothing more to prove. So suppose $\operatorname{spec}^0(\overline{\mathcal{D}}) \neq \{0\}$. Pick a generator σ_1 of $\operatorname{spec}^0(\overline{\mathcal{D}})$ and a nonzero element $\eta \in H^0_{\overline{\mathcal{D}}(\sigma_1)}(\mathcal{N})$. If $\eta' \in H^0_{\overline{\mathcal{D}}(-\sigma_1)}(\mathcal{N})$ is a nonzero element, then $\eta\eta'$ is not identically zero (Lemma 3.6) and belongs to $H^0_{\overline{\mathcal{D}}(0)}(\mathcal{N})$. Therefore $\eta\eta'$ is a nonzero constant. Thus η vanishes nowhere and η^k belongs to $H^0_{\overline{\mathcal{D}}(k\sigma_1)}(\mathcal{N})$ for each $k \in \mathbb{Z}$. If $\zeta \in H^0_{\overline{\mathcal{D}}(k\sigma_1)}(\mathcal{N})$, then $\zeta\eta^{-k}$ is a constant c, so $\zeta = c\eta^k$. Thus each group $H^0_{\overline{\mathcal{D}}(k\sigma_1)}(\mathcal{N})$ is one dimensional, and its nonzero elements vanish nowhere.

As a consequence of the proof, we get

Corollary 5.5. *Suppose that* $\operatorname{spec}^0(\overline{\mathcal{D}})$ *is a group. If* $\zeta^j \in H^0_{\overline{\mathcal{D}}(\sigma_j)}(\mathcal{N})$, $j = 1, 2$, *then* $d\zeta^1$ *and* $d\zeta^2$ *are everywhere linearly dependent.*

If $\dim \mathcal{N} = 1$, then $\mathcal{N}$ is a circle and $\operatorname{spec}^0(\overline{\mathcal{D}})$ is a group. Somewhat less trivially:

Example 5.6. Let $\mathcal{B}$ be a compact complex manifold, let $E \to \mathcal{B}$ be a flat line bundle; the holomorphic structure is the one for which the local flat section is holomorphic. Pick a Hermitian metric and let $\mathcal{N}$ be the circle bundle, with the usual structure as in Example 1.7. If some power E^m, $m \neq 0$, is holomorphically trivial, then with the smallest such power, m_0, we get that $\operatorname{spec}^0(\overline{\mathcal{D}}) = \mathrm{i}m_0\mathbb{Z}$. If no such m exists, then $\operatorname{spec}^0(\overline{\mathcal{D}}) = \{0\}$.

6 CR Maps into $\mathbb{C}^N$

We now analyze maps $\mathcal{N} \to \mathbb{C}^N \backslash 0$.

Proposition 6.1. *Suppose that there is a map* $F : \mathcal{N} \to \mathbb{C}^N \backslash 0$ *whose components* ζ^j *satisfy*

$$\overline{\mathbb{D}}\zeta^j + \mathrm{i}\sigma_j \zeta^j \beta = 0 \tag{6.2}$$

with all the σ_j *in one component of* $\mathrm{i}\mathbb{R}\backslash\{0\}$. *Then there is* $u : \mathcal{N} \to \mathbb{R}$ *smooth such that the map* $\tilde{F} : \mathcal{N} \to \mathbb{C}^N \backslash 0$ *with components* $\tilde{\zeta}^j = e^{-\mathrm{i}\sigma_j u}\zeta^j$ *has image in* S^{2N-1}.

Proof. Let $s_j = -\operatorname{Im}\sigma_j$ and define $g : \mathbb{R}_+ \times (\overline{\mathbb{R}}^N_+ \backslash 0) \to \mathbb{R}$ by

$$g(\rho, y^1, \ldots, y^N) = \sum_{j=1}^{N} \rho^{-2s_j} y^j.$$

Since all s_j have the same sign, $\partial_\rho g(\rho, y)$ does not vanish. If the s_j are positive, then for fixed y, $g(\rho, y) \to \infty$ as $\rho \to 0$ and $g(\rho, y) \to 0$ as $\rho \to \infty$. An analogous statement holds if all s_j are negative. So for each $y \in \overline{\mathbb{R}}^N_+\backslash 0$, there is a unique positive $\rho(y)$ such that $g(\rho(y), y) = 1$, and $\rho(y)$ depends smoothly on y. Define $f : \mathcal{N} \to \mathbb{R}$ by $f = \rho(|\zeta^1|^2, \dots, |\zeta^N|^2)$. The function f is well defined because the ζ^j do not vanish simultaneously, is positive everywhere, and satisfies

$$\sum_{j=1}^{N} f^{-2s_j} |\zeta^j|^2 = 1. \tag{6.3}$$

By Lemma 5.3, $\sigma_j = \mathrm{i}s_j$ for some real number s_j. The identity $\mathcal{T}\zeta^j = -\sigma_j\zeta^j$ gives $\zeta^j(\mathfrak{a}_t p) = \mathrm{e}^{-\mathrm{i}s_j t}\zeta^j(p)$ so

$$\mathcal{T}|\zeta^j|^2 = 0. \tag{6.4}$$

Applying $\mathcal{T}$ to both members of (6.3) and using (6.4) gives

$$-2f^{-1}\sum_{j=1}^{N} s_j f^{-2s_j}|\zeta^j|^2 \mathcal{T} f = 0.$$

The function $\sum_{j=1}^{N} s_j f^{-2s_j}|\zeta^j|^2$ vanishes nowhere, since all the s_j have the same sign. Thus $\mathcal{T} f = 0$, and with $u = \log f$, we also have

$$\mathcal{T} u = 0. \tag{6.5}$$

Define $\tilde{\zeta}^j = \mathrm{e}^{-\mathrm{i}\sigma_j u}\zeta^j$. Then $|\tilde{\zeta}^j|^2 = f^{-2s_j}|\zeta^j|^2$, so by (6.3) the map $\tilde{F} : \mathcal{N} \to \mathbb{C}^N\backslash 0$ with components $\tilde{\zeta}^j$ has image in S^{2N-1}.

With the notation of the proof, let $\beta' = \beta + \overline{\mathbb{D}}u$. Thus $\overline{\mathbb{D}}\beta' = 0$ and because of (6.5), also $\langle \beta', \mathcal{T}\rangle = -\mathrm{i}$. Thus β' is an admissible section of $\overline{\mathcal{V}}^*$. The functions $\tilde{\zeta}^j$ satisfy

$$\overline{\mathbb{D}}\tilde{\zeta}^j + \mathrm{i}\sigma_j\tilde{\zeta}^j\beta' = 0. \tag{6.6}$$

Therefore, by Lemma 5.1, they are CR functions with respect to the CR structure $\overline{\mathcal{K}}_{\beta'}$. Thus $\tilde{F} : \mathcal{N} \to S^{2N-1}$ is a CR map (as was F but for the CR structure defined by β).

Using (6.5) in (6.6) gives

$$\mathcal{T}\tilde{\zeta}^j - \mathrm{i}\tau_j\tilde{\zeta}^j = 0$$

with $\tau_j = -s_j$ (they all have the same sign). Therefore

$$\tilde{F}_*\mathcal{T}(p) = \mathcal{T}'(\tilde{F}(p))$$

where

$$\mathcal{T}' = \mathrm{i}\sum_{j=1}^{N} \tau_j (w^j \partial_{w^j} - \overline{w}^j \partial_{\overline{w}^j}) \tag{6.7}$$

using $w^1, \dots, w^N$ as coordinates in $\mathbb{C}^N$.

Suppose that ζ vanishes nowhere and solves $\overline{\mathbb{D}}\zeta + \mathrm{i}\sigma_0\zeta\beta$ with $\sigma_0 \neq 0$. Applying Proposition 6.1, we may assume that $|\zeta| = 1$ after a suitable change of β (in this case, this just means that ζ is replaced by $\zeta/|\zeta|$ and β is changed accordingly). The following is analogous to the situation of the circle bundle of a flat line bundle; see Example 5.6.

Proposition 6.8. *Suppose $\zeta : \mathcal{N} \to S^1$ solves $\overline{\mathbb{D}}\zeta + \mathrm{i}\sigma_0\zeta\beta = 0$ with $\sigma_0 \neq 0$. Then ζ is a submersion whose fibers are complex manifolds.*

Proof. Since ζ vanishes nowhere and $\sigma_0 \neq 0$, $\mathcal{T}\zeta \neq 0$ everywhere. Thus ζ is a submersion. Since ζ is CR with respect to $\overline{\mathcal{K}}_\beta$, $v\zeta = 0$ if $v \in \overline{\mathcal{K}}_\beta$. Since ζ is nowhere 0, also $v(1/\zeta) = 0$ if $v \in \overline{\mathcal{K}}_\beta$. But $1/\zeta = \overline{\zeta}$. Thus v is tangent to the fibers of ζ: the CR structure $\overline{\mathcal{K}}_{\beta'}$ is tangent to the fibers of ζ and can be viewed as the $(0, 1)$-tangent bundle of a complex structure.

The case $\dim \mathcal{N} = 1$ is trivially included in Proposition 6.8. On the other hand, we have

Proposition 6.9. *Suppose that* $\dim \mathcal{N} = 2n + 1 > 1$ *and that* $F : \mathcal{N} \to \mathbb{C}^N$ *is a map whose components* ζ^j *satisfy* (6.2) *with* $\sigma_j \neq 0$. *Suppose further that at every* $p \in \mathcal{N}$, $n + 1$ *of the differentials* $\mathrm{d}\zeta^j(p)$ *are independent. Then*

(1) $\mathrm{spec}^0(\overline{\mathcal{D}})\backslash 0$ *is contained in one component of* $\mathrm{i}\mathbb{R}\backslash\{0\}$.
(2) 0 *is not in the image of* F.
Let $\tilde{F} : \mathcal{N} \to S^{2N-1}$ *be the map in Proposition 6.1.*
(3) Then for each p, $n + 1$ *of the differentials* $\mathrm{d}\tilde{\zeta}^j(p)$ *are independent.*

Proof. Since $\dim \mathrm{span} \mathrm{d}\zeta^j > 1$, Corollary 5.5 gives that $\mathrm{spec}^0(\overline{\mathcal{D}})$ is not a group, so $\mathrm{spec}^0(\overline{\mathcal{D}})\backslash 0$ is contained in one component of $\mathrm{i}\mathbb{R}\backslash\{0\}$ by Lemma 5.3.

To show that the image of F does not contain 0, we show that for every $p \in \mathcal{N}$, there is j_0 such that $\mathcal{T}\zeta^{j_0}(p) \neq 0$. Since $\mathcal{T}\zeta^{j_0} = -\sigma_{j_0}\zeta^{j_0}(p)$ and $\sigma_{j_0} \neq 0$, we conclude from $\mathcal{T}\zeta^{j_0}(p) \neq 0$ that $\zeta^{j_0}(p) \neq 0$.

Let then $p \in \mathcal{N}$ and suppose that the differentials $\mathrm{d}\zeta^j(p)$, $j = 1, \dots, n + 1$, are independent. The restrictions to the fiber $\overline{\mathcal{K}}_{\beta,p}$ of these differentials vanish, so they give $n + 1$ independent linear functions on the $(n + 1)$-dimensional vector space $\mathbb{C}T_p\mathcal{N}/\overline{\mathcal{K}}_{\beta,p}$. The image of $\mathcal{T}(p)$ in this quotient is not 0, so for some j_0, $\mathcal{T}\zeta^{j_0}(p) \neq 0$. Thus the image of F does not contain 0. This and the fact that the σ_j lie in one component of $\mathrm{i}\mathbb{R}\backslash\{0\}$ allow us to apply Proposition 6.1.

Let then $u : \mathcal{N} \to \mathbb{R}$ be the function in Proposition 6.1. Suppose again that $\mathrm{d}\zeta^1, \dots, \mathrm{d}\zeta^{n+1}$ are independent at p and $\mathcal{T}\zeta^{n+1}(p) \neq 0$. Let $\tilde{\zeta}^j = \mathrm{e}^{-\mathrm{i}\sigma_j u}\zeta^j$. Then also $\mathcal{T}\tilde{\zeta}^{n+1}(p) \neq 0$. The 1-forms

$$\mathrm{d}\zeta^j - \frac{\mathcal{T}\zeta^j}{\mathcal{T}\zeta^{n+1}}\mathrm{d}\zeta^{n+1}, \; j = 1, \dots n$$

are independent at p, and a brief calculation gives that

$$\mathrm{d}\tilde{\zeta}^j - \frac{\mathcal{T}\tilde{\zeta}^j}{\mathcal{T}\tilde{\zeta}^{n+1}}\mathrm{d}\tilde{\zeta}^{n+1} = \mathrm{e}^{-\mathrm{i}\sigma_j u}\left(\mathrm{d}\zeta^j - \frac{\mathcal{T}\zeta^j}{\mathcal{T}\zeta^{n+1}}\mathrm{d}\zeta^{n+1}\right), \; j = 1, \dots n \tag{6.10}$$

so these n differential forms are also independent at p. They all vanish when paired with $\mathcal{T}$. Furthermore, since $\tilde{\zeta}^{n+1}(p) \neq 0$, $\mathcal{T}\tilde{\zeta}^{n+1}(p) \neq 0$. So the differential forms (6.10) together with $\mathrm{d}\tilde{\zeta}^{n+1}$ are independent at p.

The differentials of the component functions of both F and $\tilde{F}$ are independent over $\mathbb{C}$. Since they are CR function, this is equivalent to F and $\tilde{F}$ being immersions.

The following result is similar to the statement in complex geometry asserting that very ample holomorphic line bundles are ample.

Proposition 6.11. *Let $F : \mathcal{N} \to \mathbb{C}^N$ be an immersion with image in S^{2N-1}, $N > 1$, and components ζ^j that satisfy* (6.2). *Then the Levi form of the CR structure $\overline{\mathcal{K}}_\beta$ is definite.*

Proof. Let $\theta_\beta \in C^\infty(\mathcal{N}; T^*\mathcal{N})$ be the 1-form which vanishes on $\mathcal{K}_\beta \oplus \overline{\mathcal{K}}_\beta$ and satisfies $\langle \theta_\beta, \mathcal{T} \rangle = 1$. Define the Levi form with respect to θ_β as

$$\mathrm{Levi}_{\theta_\beta}(v, w) = -\mathrm{i}\mathrm{d}\theta_\beta(v, \overline{w}), \quad v,\ w \in \mathcal{K}_{\beta,p},\ p \in \mathcal{N}.$$

In this definition, we switched to the conjugate of $\overline{\mathcal{K}}_\beta$ to adapt to the traditional setup. Give S^{2N-1} the standard CR structure $\overline{\mathcal{K}}$ as in Example 1.6, let $\mathcal{T}'$ be the vector field in (6.7) and let θ' be real 1-form which vanishes on $\overline{\mathcal{K}}$ and satisfies $\langle \theta', \mathcal{T}' \rangle = 1$. Then $F^*\theta' = \theta_\beta$, since F is a CR map and $F_*\mathcal{T} = \mathcal{T}'$. The Levi form $\mathrm{Levi}_{\theta'}$ is positive (negative) definite if the τ_j are positive (negative). Let $v, w \in \mathcal{K}_{\beta,p}$. Then $-\mathrm{i}\mathrm{d}\theta_\beta(v, \overline{w}) = -\mathrm{i}\mathrm{d}\theta'(F_*v, \overline{F_*w})$. Since F is an immersion, $(v, w) \mapsto -\mathrm{i}\mathrm{d}\theta'(F_*v, \overline{F_*w})$ is nondegenerate with the same signature as $\mathrm{Levi}_{\theta'}$.

Propositions 6.9 and 6.11 give (1) $\implies$ (2) and (2) $\implies$ (3) in Theorem 1.5.

7 CR Embeddings

Boutet de Monvel [4] showed that if $\mathcal{N}$ is a compact strictly pseudoconvex CR manifold of dimension ≥ 5, then there is a CR embedding $F : \mathcal{N} \to \mathbb{C}^N$ for some N. The proof of the following theorem, a version of the assertion that ample line bundles are very ample, takes advantage of this and, as mentioned already, an idea of Bochner [3].

Theorem 7.1. *Suppose that* $(\mathcal{N},\mathcal{T},\overline{\mathcal{V}}) \in \mathscr{F}_{\mathrm{ell}}$ *with* $\dim\mathcal{N} \geq 5$ *and that* β *is a smooth* $\overline{\mathbb{D}}$*-closed section of* $\overline{\mathcal{V}}^*$ *such that* $\overline{\mathcal{K}}_\beta$ *has definite Levi form. Then there is* $\beta' \in \boldsymbol{\beta}$ *(see Definition 4.6) and a CR embedding* $F : \mathcal{N} \to S^{2N-1} \subset \mathbb{C}^N$ *of* $\mathcal{N}$ *with the CR structure* $\overline{\mathcal{K}}_{\beta'}$ *such that, with* $w^1,\dots,w^N$ *denoting the standard coordinates in* $\mathbb{C}^N$

$$F_*\mathcal{T} = \mathrm{i}\sum_j \tau_j (w^j\,\partial_{w^j} - \overline{w}^j\,\partial_{\overline{w}^j}) \tag{7.2}$$

for some numbers τ_j, $j = 1,\dots,N$. *The* τ_j *are all positive or all negative depending on the signature of* $\mathrm{Levi}_{\theta_\beta}$.

Let $\mathscr{H}^0_{\overline{\partial}_b}(\mathcal{N})$ be the subspace of $L^2(\mathcal{N})$ consisting of CR functions. If the Levi form of $\mathcal{K}_\beta$ is definite, as in the theorem, the space $\mathscr{H}^0_{\overline{\partial}_b}(\mathcal{N}) \cap C^\infty(\mathcal{N})$ is infinite dimensional. Boutet de Monvel's proof of his embedding theorem consists essentially on proving that

(a) For all $p_0 \in \mathcal{N}$, $\operatorname{span}\{\mathrm{d}f(p_0) : f \in \mathscr{H}^0_{\overline{\partial}_b}(\mathcal{N}) \cap C^\infty(\mathcal{N})\}$ is the annihilator of $\overline{\mathcal{K}}_\beta$ in $\mathbb{C}T^*_{p_0}\mathcal{N}$.
(b) The functions in $\mathscr{H}^0_{\overline{\partial}_b}(\mathcal{N}) \cap C^\infty(\mathcal{N})$ separate points of $\mathcal{N}$.

The embedding map is then constructed, taking advantage of these properties. In the present case, we also wish (7.2) to hold, so in addition the component functions ζ^j of F should satisfy $\mathcal{L}_\mathcal{T}\zeta^j = \mathrm{i}\tau_j\zeta_j$ with all τ_j of the same sign. We will therefore prepare for the proof of Theorem 7.1 by exhibiting a decomposition of $\mathscr{H}^0_{\overline{\partial}_b}(\mathcal{N})$, more generally without assumptions on the Levi form, a decomposition of the $\overline{\partial}_b$ cohomology spaces in any degree, into eigenspaces of $-\mathrm{i}\mathcal{T}$.

We begin with the following two lemmas whose proofs are elementary.

Lemma 7.3. *If* α *is a smooth section of the annihilator of* $\overline{\mathcal{V}}$ *in* $\mathbb{C}T^*\mathcal{N}$, *then* $(\mathcal{L}_\mathcal{T}\alpha)|_{\overline{\mathcal{V}}} = 0$. *Consequently, for each* $p \in \mathcal{N}$ *and* $t \in \mathbb{R}$, $\mathrm{d}\mathfrak{a}_t : \mathbb{C}T_p\mathcal{N} \to \mathbb{C}T_{\mathfrak{a}_t(p)}\mathcal{N}$ *maps* $\overline{\mathcal{V}}_p$ *onto* $\overline{\mathcal{V}}_{\mathfrak{a}_t(p)}$.

It follows that there is a well-defined smooth bundle homomorphism $\mathfrak{a}_t^* : \bigwedge^q\overline{\mathcal{V}}^* \to \bigwedge^q\overline{\mathcal{V}}^*$ covering $\mathfrak{a}_{-t}$. In particular, one can define the Lie derivative $\mathcal{L}_\mathcal{T}\phi$ with respect to $\mathcal{T}$ of an element in $\phi \in C^\infty(\mathcal{N};\bigwedge^q\overline{\mathcal{V}}^*)$. The usual formula holds:

Lemma 7.4. *If* $\phi \in C^\infty(\mathcal{N};\bigwedge^q\overline{\mathcal{V}}^*)$, *then* $\mathcal{L}_\mathcal{T}\phi = \mathbf{i}_\mathcal{T}\overline{\mathbb{D}}\phi + \overline{\mathbb{D}}\mathbf{i}_\mathcal{T}\phi$, *where* $\mathbf{i}_\mathcal{T}$ *denotes interior multiplication by* $\mathcal{T}$. *Consequently, for each* t *and* $\phi \in C^\infty(\mathcal{N};\bigwedge^q\overline{\mathcal{V}}^*)$, $\overline{\mathbb{D}}\mathfrak{a}_t^*\phi = \mathfrak{a}_t^*\overline{\mathbb{D}}\phi$.

In particular, it follows from (4.1) that $\mathcal{L}_\mathcal{T}\beta = 0$. Let θ_β and the Levi form $\mathrm{Levi}_{\theta_\beta}$ be defined as at the beginning of the proof of Proposition 6.11. If $\mathrm{Levi}_{\theta_\beta}$ is either positive or negative definite (as in the hypothesis of Theorem 7.1), we may use it to define a Hermitian metric on $\overline{\mathcal{K}}_\beta$ and extend it to $\overline{\mathcal{V}}$ so that $\mathcal{T}$ is a unit vector field orthogonal to $\overline{\mathcal{K}}_\beta$. Lemma 7.4 gives that $\mathcal{L}_\mathcal{T}\beta = 0$, so $\overline{\mathcal{K}}_\beta$, θ, hence also $\mathrm{Levi}_{\theta_\beta}$ are all $\mathcal{T}$-invariant, h is $\mathcal{T}$-invariant. This metric gives an obvious metric on

$\mathcal{K}_\beta \oplus \overline{\mathcal{K}}_\beta \oplus \operatorname{span}_{\mathbb{C}}\mathcal{T}$ which in turn gives a $\mathcal{T}$-invariant Riemannian metric on $\mathcal{N}$ giving a $\mathcal{T}$-invariant positive density on $\mathcal{N}$.

In the general case where there is no assumption on the behavior of $\mathrm{Levi}_{\theta_\beta}$, we first construct a $\mathcal{T}$-invariant Hermitian metric on $\overline{\mathcal{K}}_\beta$ as follows. Fix some $\mathcal{T}$-invariant metric $\tilde{g}$ on $\mathcal{N}$, let $\mathcal{H} = (\mathcal{K}_\beta + \overline{\mathcal{K}}_\beta) \cap T\mathcal{N}$ and define

$$g(v, w) = \frac{1}{2}(\tilde{g}(u, v) + \tilde{g}(Ju, Jv)), \quad u, v \in \mathcal{H}_p,\ p \in \mathcal{N},$$

where J is the complex structure on $\mathcal{H}$ for which the $(0, 1)$ subbundle of $\mathbb{C}\mathcal{H}$ is $\overline{\mathcal{K}}_\beta$. Since $g(Ju, Jv) = g(u, v)$, there is an induced Hermitian metric h on $\overline{\mathcal{K}}_\beta$. Now define the rest of the object as was done in the previous paragraph.

Use the metric h (extended to each exterior power $\bigwedge^q \overline{\mathcal{V}}^*$) and the Riemannian density to define $L^2(\mathcal{N}; \bigwedge^q \overline{\mathcal{V}}^*)$ and the formal adjoint operators $\overline{\partial}_b^\star$. With these, construct the Laplacian $\Box_{b,q}$ in each degree. This operator commutes with $\mathcal{L}_\mathcal{T}$.

Let $\mathscr{H}^q_{\overline{\partial}_b}(\mathcal{N})$ be the kernel of $\Box_{b,q}$ in $L^2(\mathcal{N}; \bigwedge^q \overline{\mathcal{V}}^*)$,

$$\mathscr{H}^q_{\overline{\partial}_b}(\mathcal{N}) = \{\phi \in L^2(\mathcal{N}; \textstyle\bigwedge^q \overline{\mathcal{K}}^*) : \Box_{b,q}\phi = 0\}.$$

In each degree, the operator $-\mathrm{i}\mathcal{L}_\mathcal{T}$, viewed initially as acting on distributional sections, gives by restriction an operator on $\mathscr{H}^q_{\overline{\partial}_b}(\mathcal{N})$ with values in distributional sections in the kernel of $\Box_{b,q}$. Let

$$\operatorname{Dom}(\mathcal{L}_\mathcal{T}) = \{\phi \in \mathscr{H}^q_{\overline{\partial}_b}(\mathcal{N}) : -\mathrm{i}\mathcal{L}_\mathcal{T}\phi \in L^2(\mathcal{N}; \textstyle\bigwedge^q \overline{\mathcal{V}}^*)\}$$

Thus $\mathcal{L}_\mathcal{T}\phi \in \mathscr{H}^q_{\overline{\partial}_b}(\mathcal{N})$ if $\phi \in \mathscr{H}^q_{\overline{\partial}_b}(\mathcal{N})$

Proposition 7.5. *The operator*

$$-\mathrm{i}\mathcal{L}_\mathcal{T} : \operatorname{Dom}(\mathcal{L}_\mathcal{T}) \subset \mathscr{H}^q_{\overline{\partial}_b}(\mathcal{N}) \to \mathscr{H}^q_{\overline{\partial}_b}(\mathcal{N}), \tag{7.6}$$

is Fredholm self-adjoint with compact resolvent. Hence, $\operatorname{spec}(-\mathrm{i}\mathcal{L}_\mathcal{T})$ *is a closed discrete subset of* $\mathbb{R}$ *and there is an orthogonal decomposition*

$$\mathscr{H}^q_{\overline{\partial}_b}(\mathcal{N}) = \bigoplus_{\tau \in \operatorname{spec}(-\mathrm{i}\mathcal{L}_\mathcal{T})} \mathscr{H}^q_{\overline{\partial}_b,\tau}(\mathcal{N})$$

where

$$\mathscr{H}^q_{\overline{\partial}_b,\tau}(\mathcal{N}) = \{\phi \in \mathscr{H}^q_{\overline{\partial}_b}(\mathcal{N}) : -\mathrm{i}\mathcal{L}_\mathcal{T}\phi = \tau\phi\}.$$

It is immediate that (7.6) is densely defined.

Proof. The operator $\Box_{b,q} - \mathcal{L}^2_\mathcal{T}$ is a nonnegative symmetric operator when viewed in the space of smooth sections. Furthermore, it is elliptic. To see this, let $\iota : \overline{\mathcal{K}}_\beta \to$

$\mathbb{C}T\mathcal{N}$ be the inclusion map. The kernel of dual map $\iota^* : \mathbb{C}T^*\mathcal{N} \to \overline{\mathcal{V}}^*$ intersects $T^*\mathcal{N}$ (the real covectors) in exactly the characteristic variety of $\overline{\mathcal{K}}_\beta$, the span of the form θ_β. The principal symbol of $\overline{\partial}_b$ at $\boldsymbol{\xi} \in T^*\mathcal{N}$ is $\boldsymbol{\varpi}(\overline{\partial}_b)(\boldsymbol{\xi})(\phi) = \mathrm{i}(\iota^*\boldsymbol{\xi}) \wedge \phi$, so just as for the standard Laplacian, $\boldsymbol{\varpi}(\Box_{b,q})(\boldsymbol{\xi}) = \|\iota^*(\boldsymbol{\xi})\|^2 I$ where the norm is the one induced on $\overline{\mathcal{V}}^*$ by that of $\overline{\mathcal{V}}$. So $\boldsymbol{\varpi}\big(\overline{\partial}_b\big)(\boldsymbol{\xi})$ is nonnegative and vanishes to exactly order 2 on $\operatorname{Char}\overline{\mathcal{K}}_\beta$. The principal symbol of $-\mathcal{L}_\mathcal{T}^2$ is $\boldsymbol{\varpi}(-\mathcal{L}_\mathcal{T}^2) = (\langle\boldsymbol{\xi}, \mathcal{T}\rangle)^2 I$, hence $\boldsymbol{\varpi}(-\mathcal{L}_\mathcal{T}^2)$ is positive when $\boldsymbol{\xi}$ is nonzero and proportional to θ. Thus

$$(\boldsymbol{\varpi}\Box_{b,q} - \mathcal{L}_\mathcal{T}^2)(\boldsymbol{\xi})$$

is invertible for any $\boldsymbol{\xi} \in T^*\mathcal{N}\backslash 0$. This analysis also leads to the conclusion that $\operatorname{Dom}(\mathcal{L}_\mathcal{T})$ is a subspace of the Sobolev space $H^1(\mathcal{N}; \bigwedge^q \overline{\mathcal{V}}^*)$.

Using ellipticity and that $\Box_{b,q} - \mathcal{L}_\mathcal{T}^2$ is symmetric, we deduce the existence of a parametrix B so that

$$B(\Box_{b,q} - \mathcal{L}_\mathcal{T}^2) = (\Box_{b,q} - \mathcal{L}_\mathcal{T}^2)B = I - \Pi_q$$

where Π_q is the orthogonal projection on $H = \ker(\Box_{b,q} - \mathcal{L}_\mathcal{T}^2)$, a finite-dimensional space consisting of smooth sections. The operator

$$B : L^2(\mathcal{N}; \textstyle\bigwedge^q \overline{\mathcal{V}}^*) \to L^2(\mathcal{N}; \textstyle\bigwedge^q \overline{\mathcal{V}}^*)$$

is pseudodifferential of order -2, self-adjoint, and commutes with $\mathcal{L}_\mathcal{T}$, hence with $\Box_{b,q}$. In particular, it maps $\ker \Box_{b,q}$ into $\ker \Box_{b,q}$, that is, $\mathscr{H}^q_{\overline{\partial}_b}(\mathcal{N})$ into itself. If $\phi \in H$, then

$$\|\overline{\partial}_b \phi\|^2 + \|\overline{\partial}_b^\star \phi\|^2 + \|\mathcal{L}_\mathcal{T}\phi\|^2 = 0,$$

so $\phi \in \mathscr{H}^q_{\overline{\partial}}(\mathcal{N})$ and $\mathcal{L}_\mathcal{T}\phi = 0$. In particular, $H \subset \mathscr{H}^q_{\overline{\partial}_b}(\mathcal{N})$ and we may view the restriction of Π to $\mathscr{H}^q_{\overline{\partial}_b}(\mathcal{N})$ as a finite rank projection $\mathscr{H}^q_{\overline{\partial}_b}(\mathcal{N}) \to \mathscr{H}^q_{\overline{\partial}_b}(\mathcal{N})$ (mapping into $\operatorname{Dom}(\mathcal{L}_\mathcal{T})$). Suppose $\phi \in \operatorname{Dom}(\mathcal{L}_\mathcal{T})$. Then

$$[B(-\mathrm{i}\mathcal{L}_\mathcal{T})](-\mathrm{i}\mathcal{L}_\mathcal{T})\phi = -B\mathcal{L}_\mathcal{T}^2\phi = B(\Box_{b,q} - \mathcal{L}_\mathcal{T}^2)\phi = \phi - \Pi\phi \tag{7.7}$$

using that $\operatorname{Dom}(\mathcal{L}_\mathcal{T}) \subset \mathscr{H}^q_{\overline{\partial}_b}(\mathcal{N})$. Since B commutes with $\mathcal{L}_\mathcal{T}$, we may write the equality of the left and rightmost terms also as

$$[-\mathrm{i}\mathcal{L}_\mathcal{T} B](-\mathrm{i}\mathcal{L}_\mathcal{T})\phi = \phi - \Pi\phi, \quad \phi \in \operatorname{Dom}(\mathcal{L}_\mathcal{T}). \tag{7.8}$$

If $\phi \in \mathscr{H}^q_{\overline{\partial}_b}(\mathcal{N})$, then $B\phi \in \mathscr{H}^q_{\overline{\partial}_b}(\mathcal{N}) \cap H^2(\mathcal{N}; \bigwedge^q \overline{\mathcal{V}}^*)$, so

$$B\mathcal{L}_\mathcal{T}\phi \in \mathscr{H}^q_{\overline{\partial}_b}(\mathcal{N}) \cap H^1(\mathcal{N}; \textstyle\bigwedge^q \overline{\mathcal{V}}^*) \subset \operatorname{Dom}(\mathcal{L}_\mathcal{T})$$

Thus if

$$\pi : L^2(\mathcal{N}; \textstyle\bigwedge^q \overline{\mathcal{V}}) \to L^2(\mathcal{N}; \textstyle\bigwedge^q \overline{\mathcal{V}}), \quad \iota : \mathscr{H}^q_{\overline{\partial}_b}(\mathcal{N}) \to L^2(\mathcal{N}; E)$$

are, respectively, the orthogonal projection on $\mathscr{H}^q_{\overline{\partial}_b}(\mathcal{N})$ and the inclusion map, then (7.7), (7.8) give that $S = -\mathrm{i}\pi\mathcal{L}_\mathcal{T} B\iota$ is a parametrix for (7.6), compact because $\mathcal{L}_\mathcal{T} B$ is of order -1.

We now show that $\mathrm{Dom}(\mathcal{L}_\mathcal{T})$ is dense in $\mathscr{H}^q_{\overline{\partial}_b}(\mathcal{N})$. Let $\psi \in \mathscr{H}^q_{\overline{\partial}_b}(\mathcal{N})$ be orthogonal to $\mathrm{Dom}(\mathcal{L}_\mathcal{T})$. If $\phi \in \mathscr{H}^q_{\overline{\partial}_b}(\mathcal{N})$, then $B\mathcal{L}_\mathcal{T}\phi \in \mathrm{Dom}(\mathcal{L}_\mathcal{T})$, so $(B\mathcal{L}_\mathcal{T}\phi, \psi) = 0$. Since

$$(B\mathcal{L}_\mathcal{T}\phi, \psi) = (\phi, \mathcal{L}_\mathcal{T} B\psi)$$

and ϕ is arbitrary, we conclude that $\mathcal{L}_\mathcal{T} B\psi = 0$. Thus also $\mathcal{L}_\mathcal{T}^2 B\psi = 0$, hence $(\Box_{b,q} - \mathcal{L}_\mathcal{T}^2)B\psi = 0$. Consequently, $\psi = \Pi\psi$, hence $\psi \in \mathrm{Dom}(\mathcal{L}_\mathcal{T})$. Therefore $\psi = 0$. Thus (7.6) is a densely defined operator.

Finally, to prove self-adjointness of (7.6), we only need to verify that its deficiency indices vanish. This can be accomplished as follows. Suppose

$$B(\lambda)(\Box_{b,q} - \mathcal{L}_\mathcal{T}^2 - \lambda^2) = I.$$

This formula can be viewed as holding in the Sobolev space $H^1(\mathcal{N}; \bigwedge^q \overline{\mathcal{V}}^*)$, and gives

$$B(\lambda)(-\mathcal{L}_\mathcal{T}^2 - \lambda^2)\phi = \phi, \quad \phi \in \mathrm{Dom}(\mathcal{L}_\mathcal{T})$$

since $\mathrm{Dom}(\mathcal{L}_\mathcal{T}) \subset H^1(\mathcal{N}; \bigwedge^q \overline{\mathcal{V}}^*)$. Writing this as

$$[B(\lambda)(-\mathrm{i}\mathcal{L}_\mathcal{T} + \lambda)](-\mathrm{i}\mathcal{L}_\mathcal{T} - \lambda)\phi = \phi, \quad \phi \in \mathrm{Dom}(\mathcal{L}_\mathcal{T})$$

and using that $[B(\lambda)(-\mathrm{i}\mathcal{L}_\mathcal{T} + \lambda)]$ commutes with $(-\mathrm{i}\mathcal{L}_\mathcal{T}^2 - \lambda)$, we see that the resolvent set of (7.6) contains $\mathbb{C}\backslash\mathbb{R}$.

This completes the proof of the proposition.

The proof of Theorem 7.1 will also require a rough Weyl estimate. The main ingredient is:

Lemma 7.9. *Let $\{\phi_j\}_{j\in J}$ be an orthonormal basis of $\mathscr{H}^q_{\overline{\partial}_b}(\mathcal{N})$ consisting of eigenvectors of $-\mathrm{i}\mathcal{L}_\mathcal{T}$, $\phi_j \in \mathscr{H}^q_{\overline{\partial}_b,\tau_j}(\mathcal{N})$. Then there are positive constants C and μ such that*

$$\|\phi_j(p)\| \le C(1 + |\tau_j|)^\mu \quad \textit{for all } p \in \mathcal{N},\ j \in J. \tag{7.10}$$

If $\psi \in C^\infty(\mathcal{N}; \bigwedge^q \overline{\mathcal{V}}^*)$, then for each positive integer N there is C_N (depending on ψ) such that

$$(\psi, \phi_j) \le C_N(1 + |\tau_j|)^{-N} \quad \textit{for all } j. \tag{7.11}$$

Proof. The proof is similar to that of the analogous statement for elliptic self-adjoint operators. The ellipticity of $\Box_{b,q} - \mathcal{L}_\mathcal{T}^2$ gives the a priori estimates

$$\|\phi\|_{s+m}^2 \le C_{s+m}(\|\Box_{b,q}\phi - \mathcal{L}_\mathcal{T}^2\phi\|_s^2 + \|\phi\|_s^2), \quad \phi \in H^{s+m}(\mathcal{N}; E)$$

for any s. Replacing ϕ_j for ϕ gives

$$\|\phi_j\|_{s+2}^2 \le C_{s+2}(\|\tau_j^2\phi_j\|_s^2 + \|\phi_j\|_s^2),$$

that is,

$$\|\phi_j\|_{s+2}^2 \le C_{s+2}(1 + |\tau_j|^4)\|\phi_j\|_s^2.$$

By induction, there is, for each $k \in \mathbb{N}$, a constant C_k' such that

$$\|\phi_j\|_{2k}^2 \le C_k'(1 + |\tau_j|^4)^k \|\phi_j\|_0^2$$

With k large enough, the Sobolev embedding theorem gives

$$\|\phi_j\|_{L^\infty}^2 \le C(1 + |\tau_j|^4)^k \|\phi_j\|_0^2 \quad \text{for all } j \in J \tag{7.12}$$

with some constant C. This proves (7.10), since $\|\phi_j\|_0 = 1$. To prove the second statement, let $\psi \in C^\infty(\mathcal{N}; E)$ and pick an integer N. Then

$$|\tau_j|^N |(\phi_j, \psi)| = |(\mathcal{L}_T^N \phi_j, \psi)| = |(\phi_j, \mathcal{L}_T^N \psi)| \le \|\phi_j\|_0 \|\mathcal{L}_T^N \psi\|.$$

Then (7.11) follows, since $\|\phi_j\|_0 = 1$.

The estimates (7.12) can be used as in an argument of W. Allard presented in Gilkey [5, Lemma 1.6.3], see also [10, Proposition 1.4.7], to prove:

Lemma 7.13. *There are positive constants C and μ such that*

$$\dim \bigoplus_{\substack{\tau_0 \in \operatorname{spec}_0(-i\mathcal{L}_T) \\ |\tau_0| < \tau}} \mathcal{E}_{\tau_0} \le C\tau^\mu.$$

This and the estimates (7.11) give:

Lemma 7.14. *Let $\{\phi_j\}_{j \in J}$ be an orthonormal basis of $\mathscr{H}^q_{\overline{\partial}_b}(\mathcal{N})$ consisting of eigenvectors of $-i\mathcal{L}_T$. If $\psi \in \mathscr{H}^q_{\overline{\partial}_b}(\mathcal{N}) \cap C^\infty(\mathcal{N}; E)$, then the Fourier series*

$$\psi = \sum_{j \in J} (\psi, \phi_j)\phi_j$$

converges in $C^\infty(\mathcal{N}; E)$.

Of course, these lemmas are of interest only when $\mathscr{H}^q_{\overline{\partial}_b}(\mathcal{N})$ is infinite dimensional.

Remark 7.15. Suppose $\mathcal{N}$ with the CR structure $\mathcal{K}_\beta$ is nondegenerate. Let $\{\phi_\ell\}_{\ell=0}^\infty$ is an orthonormal basis of $\mathscr{H}^0_{\overline{\partial}_b}(\mathcal{N})$ consisting of eigenfunctions of the operator

(7.6). Using an invariant positive density to trivialize the bundle of densities, we identify generalized functions and densities. If u is a CR distribution, then

$$u = \sum \langle u, \overline{\phi}_\ell \rangle \phi_\ell$$

with convergence in the space of generalized functions. This may be interpreted as a global version of the Baouendi–Treves approximation formula [1] when written as

$$u = \lim_{L\to\infty} \sum_{\ell=0}^{L} \langle u, \overline{\phi}_\ell \rangle \phi_\ell .$$

Proof (Proof of Theorem 7.1). Since Levi_θ is definite, the space $\mathscr{H}^0_{\overline{\partial}_b}(\mathcal{N}) \cap C^\infty(\mathcal{N})$ is infinite dimensional. Let $\{\phi_\ell\}_{\ell=0}^\infty$ be an orthonormal basis of $\mathscr{H}^0_{\overline{\partial}_b}(\mathcal{N})$ as in Example 7.15. Then properties (a–b) on page 418 imply

1. For all $p_0 \in \mathcal{N}$, $\operatorname{span}\{d\phi_\ell(p_0) : \ell = 0, 1, \ldots\}$ is the annihilator of $\overline{\mathcal{K}}_\beta$ in $\mathbb{C}T^*_{p_0}\mathcal{N}$.
2. The functions ϕ_ℓ, $\ell = 1, 2, \ldots$ separate points of $\mathcal{N}$.

This is proved in the same way as the analogous two statements in the proof of Theorem 1.1, taking advantage of the fact that if $f \in \mathscr{H}^0_{\overline{\partial}_b}(\mathcal{N}) \cap C^\infty(\mathcal{N})$, then the Fourier series

$$f = \sum_\ell (f, \phi_\ell)\phi_\ell$$

converges in $C^\infty(\mathcal{N})$; see Lemma 7.14. As in the proof of Theorem 1.1, we conclude that there is an embedding

$$F : \mathcal{N} \to \mathbb{C}^N$$

whose components ζ^j are CR functions with respect to $\overline{\mathcal{K}}_\beta$ and satisfy $-\mathrm{i}\mathcal{T}\zeta^j = \tau_j\zeta^j$. We assume, making full use of (2), that the differentials of these component functions span the annihilator of $\mathcal{K}_\beta$ at each $p \in \mathcal{N}$. By Lemma 5.1,

$$\overline{\mathbb{D}}\zeta^j + \mathrm{i}\sigma_j\zeta^j\beta = 0, \quad j = 1, \ldots, N$$

with $\sigma_j = -\mathrm{i}\tau_j$. The map $\tilde{F}$ in constructed from F as in Proposition 6.1 then has components which are CR with respect to $\beta' = \beta + \overline{\mathbb{D}}u$ and maps into S^{2N-1}. By Proposition 6.9, $\tilde{F}$ is an immersion. However, while F is injective, $\tilde{F}$ may not be. We will correct this by increasing the number of components of F.

Let $w^1, \ldots, w^N$ be the complex coordinates in $\mathbb{C}^N$. The vector fields

$$R = \sum_j \tau_j (w^j \partial_{w^j} + \overline{w}^j \partial_{\overline{w}^j}), \quad \mathcal{T}' = \mathrm{i}\sum_j \tau_j (w^j \partial_{w^j} - \overline{w}^j \partial_{\overline{w}^j})$$

on $\mathbb{C}^N$ are real and commute, so give a foliation $\mathcal{F}$ of $\mathbb{C}^N\backslash 0$ by real two-dimensional submanifolds. Since $JR = T$, the leaves are one-dimensional complex (immersed) submanifolds of $\mathbb{C}^N\backslash 0$. The leaves are parametrized by their intersection with S^{2N-1}, each intersection being an orbit of $\mathcal{T}'$ in the sphere (the leaves are analogues of the complex lines forming $\mathbb{CP}^{N-1}$). For $\varrho \in \mathbb{C}$ and $w \in \mathbb{C}^N\backslash 0$ define

$$\varrho \cdot w = (\mathrm{e}^{\tau_1 \varrho} w^1, \dots, \mathrm{e}^{\tau_N \varrho} w^N).$$

For each $\varrho \in \mathbb{C}$ and $w \in \mathbb{C}^N\backslash 0$, $\varrho \cdot w$ belongs to the leaf passing through w. Since $\mathcal{T}\zeta^j = \mathrm{i}\tau_j \zeta^j$, $F_*\mathcal{T} = \mathcal{T}'$, so F maps orbits to orbits. In particular, F maps orbits of $\mathcal{T}$ into leaves of the foliation. Since the components of $\tilde{F}$ are $\mathrm{e}^{-\mathrm{i}\sigma_j u}\zeta^j = \mathrm{e}^{-\tau_j u}\zeta^j$,

$$\tilde{F}(p) = -u(p) \cdot F(p),$$

which means that $\tilde{F}(p)$ lies in the intersection of the leaf containing $F(p)$ and the unit sphere. Using that F is injective, it is easy to see that the restriction of $\tilde{F}$ to any orbit of $\mathcal{T}$ is injective. But it may happen that points $p_0, p_1 \in \mathcal{N}$ on different orbits of $\mathcal{T}$ are mapped by F to the same leaf of $\mathcal{F}$, so the two orbits are mapped to the same orbit by $\tilde{F}$ with the effect that $\tilde{F}$ is not injective. To solve this problem, we will increase the number of components of the original map F.

Let $Z = \{(p_0, p_1) \in \mathcal{N} \times \mathcal{N} : p_0 \neq p_1,\ \tilde{F}(p_0) = \tilde{F}(p_1)\}$. We show that this is a closed set. Suppose $\{(p_{0,k}, p_{1,k})\}$ is a sequence in Z that converges in $\mathcal{N} \times \mathcal{N}$ to some point (p_0, p_1). By continuity, $\tilde{F}(p_0) = \tilde{F}(p_1)$. We will show that $p_0 \neq p_1$, and thus we conclude $(p_0, p_1) \in Z$. Suppose, to the contrary, that $p_0 = p_1$. Since $\tilde{F}$ is an immersion, p_0 has a neighborhood U with the property that $(p_0', p_1') \in U \times U$ and $\tilde{F}(p_0') = \tilde{F}(p_1')$ imply $p_0' = p_1'$. This contradicts the existence of a sequence in Z converging to (p_0, p_0). Thus no point on the diagonal in $\mathcal{N} \times \mathcal{N}$ belongs to Z, hence Z is indeed closed.

More generally, Z contains no pair (p_0, p_1) such that $p_1 \in \mathcal{O}_{p_0}$, the orbit of $\mathcal{T}$ through p_0. For if the latter relation holds for $(p_0, p_1) \in Z$, then $\tilde{F}(p_0) = \tilde{F}(p_1)$ gives $u(p_0) \cdot F(p_0) = u(p_1) \cdot F(p_1)$, but since $u(p_0) = u(p_1)$ (because $\mathcal{T}u = 0$), $F(p_0) = F(p_1)$. Since F is injective, $p_0 = p_1$, but we have already concluded that W contains no point of the diagonal of $\mathcal{N} \times \mathcal{N}$.

If $(p_0, p_1) \in Z$, then $\tilde{F}(p_0) = \tilde{F}(p_1)$, so $F(p_0)$ and $F(p_1)$ belong to the same leaf of $\mathcal{F}$: Therefore, there is $\varrho \in \mathbb{C}$ such that $F(p_0) = \varrho \cdot F(p_1)$, that is,

$$\zeta^j(p_0) = \mathrm{e}^{\tau_j \varrho} \zeta^j(p_1), \quad j = 1, \dots, N. \tag{7.16}$$

If the real part of ϱ vanishes, then $F(p_1)$ and $F(p_0)$ belong to the same orbit of $\mathcal{T}'$, so p_0 and p_1 belong to the same orbit of $\mathcal{T}$ since F is injective. But then $p_0 = p_1$, contradicting $(p_0, p_1) \in Z$. So $\Re\varrho \neq 0$. We will show later that

3. If $\Re\varrho \neq 0$, then $\{(p_0, p_1) : \phi_\ell(p_0) = \mathrm{e}^{\tau_\ell \varrho}\phi_\ell(p_1)$ for all $\ell\}$ is empty.

Granted this, we proceed as follows. Pick $(p_0, p_1) \in Z$. Associated with this pair, there is a number $\varrho(p_0, p_1)$ with $\Re\varrho(p_0, p_1) \neq 0$ such that (7.16) holds. Pick ℓ such that

$$\phi_\ell(p_0) \neq e^{\tau_\ell \varrho(p_0, p_1)} \phi_\ell(p_1) \tag{7.17}$$

taking advantage of (7). Fix some j_0 such that $\zeta^{j_0}(p_1) \neq 0$. Such j_0 exists because of Part (6.9) of Proposition 6.9. There is a neighborhood U of (p_0, p_1) in $\mathcal{N} \times \mathcal{N}$ in which there is a unique continuous function $\varrho : U \to \mathbb{C}$ such that

$$\zeta^j(q_0) = e^{\tau_j \varrho(q_0, q_1)} \zeta^{j_0}(q_1), \quad (q_0, q_1) \in U$$

By continuity and because of (7.17), we may assume

$$\phi_\ell(q_0) \neq e^{\tau_\ell \varrho(q_0, q_1)} \phi_\ell(q_1), \quad (q_0, q_1) \in U$$

shrinking U if necessary. Then, if F is augmented with the function ϕ_ℓ, (7.16) will cease to hold for $(q_0, q_1) \in U$ and all the component functions of the augmented map. Since Z is compact, we can cover it with finitely many such open sets and augment the map F to a map $F' : \mathcal{N} \to \mathbb{C}^{N'}$ for which the construction of Proposition 6.1 gives an injective map $\tilde{F}' : \mathcal{N} \to S^{2N'-1}$, hence an embedding. Indeed, if $\tilde{F}'(p_0) = \tilde{F}'(p_1)$, then, if p_0 and p_1 lie in the same orbit of $\mathcal{T}$ then $p_0 = p_1$, and if p_0 and p_1 lie in different orbits, then (7.16) holds with $j = 1, \ldots, N'$, in particular $j = 1, \ldots, N$, with some ϱ with nonzero real part (determined by F, p_0 and p_1). So $(p_0, p_1) \in Z$, hence for some ζ^j with $j > N$ we must have $\zeta^j(p_0) \neq e^{\tau_j \varrho} \zeta^j(p_1)$, contradicting (7.16).

To complete the proof, we show the validity of (7) (see page 424). Let $\varrho \in \mathbb{C}$ be such that $\Re\varrho \neq 0$. We will assume that there is (p_0, p_1) such that

$$\forall \ell : \phi_\ell(p_0) = e^{\tau_\ell \varrho} \phi_\ell(p_1) \tag{7.18}$$

and derive a contradiction. We first note that $p_0 \neq p_1$, since there is ℓ such that $\phi_\ell(p_0) \neq 0$ (and $\Re\varrho \neq 0$). If $\Re\varrho > 0$, exchange p_0 and p_1, so we may assume that (7.18) holds with $\Re\varrho < 0$. By Part (6.9) of Proposition 6.9, all τ_ℓ have the same sign. Changing $\mathcal{T}$ to $-\mathcal{T}$ (and β to $-\beta$ for the sake of consistency) if necessary, we may assume that all τ_j are positive; this is already the case if $\mathrm{Levi}_{\theta_\beta}$ is positive definite, but we do not need this fact in our proof.

The estimate (7.10) applied to $\phi_\ell(p_1)$ gives

$$|\phi_\ell(p_0)| \leq C e^{\tau_\ell \Re\varrho/2} \tag{7.19}$$

for some $C > 0$. Suppose $u \in \mathscr{H}^0_{\overline{\partial}_b}(\mathcal{N})$. Then u has a restriction to the orbit through p_0. Let $\iota : \mathbb{R} \to \mathcal{N}$ be the map $\iota(t) = \mathfrak{a}_t(p_0)$. Let $W = \mathrm{Char}\,\square_{b,0}$. The Fourier series $u = \sum_\ell u_\ell \phi_\ell$, $u_\ell = (u, \phi_\ell)$, converges in $C^{-\infty}_{\overline{W}}(\mathcal{N})$ because $\square_{b,0} \sum_{\ell=0}^k u_\ell \phi_\ell = 0$ for all k and $\square_{b,0}$ is elliptic off of W. So, since ι^* :

$C^{-\infty}_{\overline{W}}(\mathcal{N}) \to C^{-\infty}(\mathbb{R})$ is continuous, $\iota^* u = \sum u_\ell e^{i\tau_\ell t}\phi_\ell(p_0)$ in $C^{-\infty}(\mathbb{R})$. Let $\chi \in C^\infty_c(\mathbb{R})$. The Fourier transform of $\chi\iota^* u$ is

$$\sum_\ell u_\ell \widehat{\chi}(\tau - \tau_\ell)\phi_\ell(p_0)$$

and the estimates (7.19) imply that $(\chi\iota^* u)\widehat{\ }(\tau)$ is rapidly decreasing in τ (since $\Re_Q < 0$). Thus $\iota^* u$ is smooth.

We will now show that there is $u \in \mathscr{H}^0_{\overline{\partial}_b}(\mathcal{N})$ such that $\iota^* u$ is not smooth using a support function for the CR structure at p_0 and a well-known trick used in the study of hypoelliptic operators. Let (z, t) be a hypoanalytic chart for the structure $\overline{\mathcal{V}}$ centered at p_0, mapping its domain U to $B \times I$ where B is an open ball in $\mathbb{C}^n$ centered at 0 and $I \subset \mathbb{R}$ is an open interval around 0. The vector fields $\partial_{\overline{z}^\mu}$, $\mu = 1, \dots, n$, ∂_t, form a frame of $\overline{\mathcal{V}}$ over U with dual frame $\overline{\mathbb{D}}\overline{z}^\mu$, $\overline{\mathbb{D}}t$, and

$$\beta = \sum_{\mu=1}^n \beta_\mu \overline{\mathbb{D}}\overline{z}^\mu - i\overline{\mathbb{D}}t.$$

Since $\overline{\mathbb{D}}\beta = 0$, the coefficients β_μ are independent of t. Let

$$t' = t + 2\Re\left[i\left(\sum_{\mu=1}^n \beta_\mu(p_0)\overline{z}^\mu + \frac{1}{2}\sum_{\mu,\nu=1}^n \frac{\partial\beta_\mu}{\partial\overline{z}^\nu}(p_0)\overline{z}^\mu\overline{z}^\nu \right)\right].$$

Since $\partial_{\overline{z}^\nu}\beta_\mu = \partial_{\overline{z}^\mu}\beta_\nu$ (because $\overline{\mathbb{D}}\beta = 0$),

$$i\beta - \overline{\mathbb{D}}t' = i\sum_{\mu=1}^n \left(\beta_\mu - \beta_\mu(p_0) - \sum_{\nu=1}^n \frac{\partial\beta_\mu}{\partial\overline{z}^\nu}(p_0)\overline{z}^\nu \right)\overline{\mathbb{D}}\overline{z}^\mu.$$

The right-hand side is $\overline{\mathbb{D}}$-closed, since the left-hand side is, and since the right-hand side is independent of t and $\overline{\mathbb{D}}t$, the form

$$b = i\sum_{\mu=1}^n \left(\beta_\mu - \beta_\mu(p_0) - \sum_{\nu=1}^n \frac{\partial\beta_\mu}{\partial\overline{z}^\nu}(p_0)\overline{z}^\nu \right)d\overline{z}^\mu$$

is $\overline{\partial}$-closed. Let α solve $\overline{\partial}\alpha = b$ in B and let

$$g = \alpha + t' - \alpha(p_0) - \sum_{\mu=1}^n \frac{\partial\alpha}{\partial z^\mu}(p_0)z^\mu - \frac{1}{2}\sum_{\mu,\nu=1}^n \frac{\partial^2\alpha}{\partial z^\mu\partial z^\nu}(p_0)z^\mu z^\nu$$

Then

$$\overline{\mathbb{D}}g = \mathrm{i}\beta,$$

so $\overline{\mathbb{D}}g$ vanishes on $\overline{\mathcal{K}}_\beta$: g is a CR function.

It is easily verified that

$$g = t' + \mathrm{i}\sum_{\mu,\nu} \frac{\partial \beta_\mu}{\partial z_\nu}(p_0) z^\nu \overline{z}^\mu + \mathcal{O}(|z|^3).$$

On the other hand, the form θ_β is given by

$$\theta_\beta = \mathrm{d}t + \mathrm{i}\sum_{\mu=1}^{n} \beta_\mu \mathrm{d}\overline{z}^\mu - \mathrm{i}\sum_{\mu=1}^{n} \overline{\beta}_\mu \mathrm{d}z^\mu,$$

and

$$-\mathrm{i}\mathrm{d}\theta_\beta = \sum_{\mu,\nu=1}^{n} \left[\frac{\partial \beta_\mu}{\partial z^\nu} - \frac{\overline{\beta}_\mu}{\partial \overline{z}^\nu} \right] \mathrm{d}z^\nu \wedge \mathrm{d}\overline{z}^\mu$$

using $\partial_{\overline{z}^\nu} \beta_\mu = \partial_{\overline{z}^\mu} \beta_\nu$. The vector fields

$$L_\nu = \frac{\partial}{\partial z^\nu} + \mathrm{i}\overline{\beta}_\nu \frac{\partial}{\partial t}, \quad \nu = 1, \ldots, n$$

form a frame for $\mathcal{K}_\beta$ in U, and by hypothesis $\mathrm{Levi}_{\theta_\beta}$ is positive definite. So the matrix with coefficients

$$-\mathrm{i}\mathrm{d}\theta_\beta(L_\nu, \overline{L}_\mu) = \frac{\partial \beta_\mu}{\partial z^\nu} - \frac{\overline{\beta}_\mu}{\partial \overline{z}^\nu}$$

is positive definite. It follows that the quadratic part of

$$\operatorname{Im} g = -\frac{i}{2} \sum_{\mu,\nu=1}^{n} \left(\frac{1}{\beta_0} \frac{\partial \beta_\mu}{\partial z^\nu} - \frac{1}{\overline{\beta}_0} \frac{\overline{\beta}_\mu}{\partial \overline{z}^\nu} \right) z^\nu \overline{z}^\mu + \mathcal{O}(|z|^3)$$

at p_0 is positive definite. Thus shrinking B, we may assume that

$$\operatorname{Im} g \geq c|z|^2 \text{ for some } c > 0.$$

Define

$$u_0 = \int_0^\infty \mathrm{e}^{\mathrm{i}\tau g}(1 + \tau^2)^{-1} \mathrm{d}\tau.$$

in U. The function u_0 is CR (since g is) and in L^2_{loc}, but not in $C^\infty(U)$. In fact, $\mathrm{WF}(u_0) = \{\tau\theta_\beta(p_0) \in T^*_{p_0}\mathcal{N} : \tau > 0\}$. Let $\chi \in C^\infty_c(U)$ be equal to 1 near p_0 and let G be Green's operator for $\Box_{b,1}$. The operator G, being a pseudodifferential operator of type $(1/2, 1/2)$, preserves wavefront set. Therefore, since $\overline{\partial}_b \chi u_0$ is smooth, so is $\overline{\partial}_b^\star G \overline{\partial}_b \chi u_0$. Let

$$u = \chi u_0 - \overline{\partial}_b^\star G \overline{\partial}_b \chi u_0.$$

The pullback of $\overline{\partial}_b^\star G \overline{\partial}_b \chi u_0$ to the orbit through p_0 is smooth. The orbit through p_0 intersects U on sets $z = $ const.; in particular, $\{(z,t) : z = 0\}$ is part of the orbit. On the latter set, $g = t$; therefore the pullback of χu_0 is equal to

$$\int_0^\infty \mathrm{e}^{\mathrm{i}\tau t}(1+\tau^2)^{-1}\mathrm{d}\tau$$

near $t = 0$, which is not smooth. Thus for no pair (p_0, p_1) does (7.18) hold.

References

1. Baouendi, M. S., Treves, F., *A property of the functions and distributions annihilated by a locally integrable system of complex vector fields*, Ann. of Math. (2) **113** (1981), 387–421.
2. Baouendi, M. S., Chang, C. H., Treves, F., *Microlocal hypo-analyticity and extension of CR functions*, J. Differential Geom. **18** (1983), 331–391.
3. Bochner, S., *Analytic mapping of compact Riemann spaces into Euclidean space*, Duke Math. J. **3** (1937), 339–354.
4. Boutet de Monvel, L., *Intégration des équations de Cauchy-Riemann induites formelles*, Séminaire Goulaouic-Lions-Schwartz 1974–1975, Exp. No. 9.
5. Gilkey, P. B., *Invariance theory, the heat equation, and the Atiyah-Singer index theorem.* Mathematics Lecture Series, 11. Publish or Perish, Wilmington, 1984.
6. Grauert, H., *Über Modifikationen und exzeptionelle analytische Mengen* (German), Math. Ann. **146** (1962), 331–368.
7. ———, *The analysis of linear partial differential operators. III. Pseudodifferential operators.* Grundlehren der Mathematischen Wissenschaften, **274**, Springer-Verlag, Berlin, 1985.
8. Kobayashi, S., *Differential geometry of complex vector bundles*, Publications of the Mathematical Society of Japan, 15. Kan Memorial Lectures, 5. Princeton University Press, Princeton, NJ and Iwanami Shoten, Tokyo, 1987.
9. Kodaira, K. On Kähler varieties of restricted type (an intrinsic characterization of algebraic varieties). Ann. of Math. (2) **60** (1954), 28–48.
10. Lesch, M., *Operators of Fuchs type, conical singularities, and asymptotic methods*, Teubner-Texte zur Math. Vol. 136, B.G. Teubner, Stuttgart, Leipzig, 1997.
11. ———, *The Atiyah-Patodi-Singer index theorem*, Research Notes in Mathematics, A. K. Peters, Ltd., Wellesley, MA, 1993.
12. Mendoza, G., *Strictly pseudoconvex b-CR manifolds*, Comm. Partial Differential Equations **29** (2004) 1437–1503.
13. ———, *Boundary structure and cohomology of b-complex manifolds.* In "Partial Differential Equations and Inverse Problems", C. Conca et al. (eds.), Contemp. Math., vol. 362 (2004), 303–320.
14. ———, *Characteristic classes of the boundary of a complex b-manifold*, in *Complex Analysis* (Trends in Mathematics), 245–262, P. Ebenfelt et al. (eds.), Birkhäuser, Basel, 2010.

15. ———, *A Gysin sequence for manifolds with* $\mathbb{R}$*-action*, to appear in *Proceedings of the Workshop on Geometric Analysis of Several Complex Variables and Related Topics*, S. Berhanu, A. Meziani, N. Mir, R. Meziani, Y. Barkatou (eds.), Contemp. Math.
16. Newlander, A., Nirenberg, L., *Complex analytic coordinates in almost complex manifolds*, Ann. of Math. **65** (1957), 391–404.
17. L. Nirenberg, *A complex Frobenius theorem*, Seminar on analytic functions I, Princeton, (1957) 172–189.
18. Treves, F., *Approximation and representation of functions and distributions annihilated by a system of complex vector fields*, Centre Math. Ecole Polytechnique, Paliseau, France (1981).
19. Treves, F., *Hypoanalytic structures*, in "Microlocal analysis", M. S. Baouendi et al., eds., Contemp. Math., vol. 27 (1984), 23–44 .
20. Treves, F., *Hypo-analytic structures. Local theory*, Princeton Mathematical Series, **40**, Princeton University Press, Princeton, NJ, 1992.

Cubature Formulas and Discrete Fourier Transform on Compact Manifolds

Isaac Z. Pesenson and Daryl Geller†

Dedicated to Leon Ehrenpreis

Abstract The goal of this chapter is to describe essentially optimal cubature formulas on compact Riemannian manifolds which are exact on spaces of band-limited functions.

Key words Cubature formulas • Discrete Fourier transform on compact manifolds • Eigenspaces • Laplace operator • Plancherel–Polya inequalities

Mathematics Subject Classification (2010): Primary: 42C99, 05C99, 94A20; Secondary: 94A12

1 Introduction

Analysis on two-dimensional surfaces and in particular on the sphere S^2 found many applications in computerized tomography, statistics, signal analysis, seismology, weather prediction, and computer vision. During the last years, many problems of classical harmonic analysis were developed for functions on manifolds and especially for functions on spheres: splines, interpolation, approximation, different

I.Z. Pesenson (✉)
Department of Mathematics, Temple University, Philadelphia, PA 19122, USA
e-mail: pesenson@temple.edu.

D. Geller†
(Deceased)

H.M. Farkas et al. (eds.), *From Fourier Analysis and Number Theory to Radon Transforms and Geometry*, Developments in Mathematics 28, DOI 10.1007/978-1-4614-4075-8_21,

aspects of Fourier analysis, continuous and discrete wavelet transform, quadrature formulas. Our list of references is very far from being complete [1–5], [7–10, 12–15], [17–29, 31–33]. More references can be found in monographs [11, 18].

The goal of this chapter is to describe three types of cubature formulas on general compact Riemannian manifolds which require *essentially optimal number of nodes*. Cubature formulas introduced in Sect. 3 are *exact on subspaces of band-limited functions*. Cubature formulas constructed in Sect. 4 are *exact on spaces of variational splines* and, at the same time, *asymptotically exact on spaces of band-limited functions*. In Sect. 5, we prove existence of cubature formulas with *positive weights* which are exact on spaces of band-limited functions.

In Sect. 7, we prove that on homogeneous compact manifolds the product of two band-limited functions is also band-limited. This result makes our findings about cubature formulas relevant to Fourier transform on homogeneous compact manifolds and allows *exact* computation of Fourier coefficients of band-limited functions on compact *homogeneous* manifolds.

It is worth to note that all results of the first four sections hold true even for non-compact Riemannian manifolds of bounded geometry. In this case, one has properly define spaces of band-limited functions on non-compact manifolds [24].

Let $\mathbf{M}$ be a compact Riemannian manifold and $\mathcal{L}$ is a differential elliptic operator which is self-adjoint in $L_2(\mathbf{M}) = L_2(\mathbf{M}, \mathrm{d}x)$, where $\mathrm{d}x$ is the Riemannian measure. The spectrum of this operator, say $0 = \lambda_0 < \lambda_1 \leq \lambda_2 \leq \ldots$, is discrete and approaches infinity. Let $u_0, u_1, u_2, \ldots$ be a corresponding complete system of real-valued orthonormal eigenfunctions, and let $\mathbf{E}_\omega(\mathcal{L})$, $\omega > 0$, be the span of all eigenfunctions of $\mathcal{L}$, whose corresponding eigenvalues are not greater than ω. For a function $f \in L_2(\mathbf{M})$, its Fourier transform is the set of coefficients $\{c_j(f)\}$, which are given by formulas

$$c_j(f) = \int_{\mathbf{M}} f u_j \mathrm{d}x. \tag{1.1}$$

By a discrete Fourier transform, we understand a discretization of the above formula. Our goal in this chapter is to develop cubature formulas of the form

$$\int_{\mathbf{M}} f \approx \sum_{x_k} f(x_k) w_k, \tag{1.2}$$

where $\{x_k\}$ is a discrete set of points on $\mathbf{M}$ and $\{w_k\}$ is a set of weights. When creating such formulas, one has to address (among others) the following problems:

1. To make sure that there exists a relatively large class of functions on which such formulas are exact.
2. To be able to estimate accuracy of such formulas for general functions.
3. To describe optimal sets of points $\{x_k\}$ for which the cubature formulas exist.
4. To provide "constructive" ways for determining optimal sets of points $\{x_k\}$.
5. To provide "constructive" ways of determining weights w_k.
6. To describe properties of appropriate weights.

In the first five sections of this chapter, we construct cubature formulas on general compact Riemannian manifolds and general elliptic second-order differential operators. Namely, we have two types of cubature formulas: formulas which are *exact* on spaces $\mathbf{E}_\omega(\mathcal{L})$ (see Sect. 3), i.e.,

$$\int_{\mathbf{M}} f = \sum_{x_k} f(x_k) w_k \tag{1.3}$$

and formulas which are *exact* on spaces of variational splines (see Sect. 4). Moreover, the cubature formulas in Sect. 4 are also asymptotically exact on the spaces $\mathbf{E}_\omega(\mathcal{L})$. For both types of formulas, we address first five issues from the list above. However, in the first four sections, we do not discuss the issue 6 from the same list.

In Sect. 5, we construct another set of cubature formulas which are exact on spaces $\mathbf{E}_\omega(\mathcal{L})$ which have *positive weights of the "right" size*. Unfortunately, for this set of cubatures, we are unable to provide constructive ways of determining weights w_k.

If one considers integrals of the form (1.1), then in the general case we do not have any criterion to determine whether the product $f u_j$ belongs to the space $\mathbf{E}_\omega(\mathcal{L})$ in order to have an *exact* relation

$$\int_{\mathbf{M}} f u_j = \sum_{x_k} f(x_k) u_j(x_k) w_k, \tag{1.4}$$

for cubature rules described in Sect. 1–4. However, if $\mathbf{M}$ is a compact homogeneous manifolds, i.e., $\mathbf{M} = G/K$, where G is a compact Lie group and K is its closed subgroup and $\mathcal{L}$ is the second-order Casimir operator (see (6.2) below), then we can show that for $f, g \in \mathbf{E}_\omega(\mathcal{L})$, their product fg is in $\mathbf{E}_{4d\omega}(\mathcal{L})$, where $d = \dim G$ (see Sect. 7).

2 Plancherel–Polya-Type Inequalities

Let $B(x, r)$ be a metric ball on $\mathbf{M}$ whose center is x and radius is r. The following important lemma can be found in [24, 27].

Lemma 2.1. *There exists a natural number $N_{\mathbf{M}}$, such that for any sufficiently small $\rho > 0$, there exists a set of points $\{y_\nu\}$ such that*

(1) *The balls $B(y_\nu, \rho/4)$ are disjoint.*
(2) *The balls $B(y_\nu, \rho/2)$ form a cover of $\mathbf{M}$.*
(3) *The multiplicity of the cover by balls $B(y_\nu, \rho)$ is not greater than $N_{\mathbf{M}}$.*

Definition 1. Any set of points $M_\rho = \{y_\nu\}$ which is as described in Lemma 2.1 will be called a metric ρ-lattice.

To define Sobolev spaces, we fix a cover $B = \{B(y_\nu, r_0)\}$ of $\mathbf{M}$ of finite multiplicity $N_\mathbf{M}$ (see Lemma 2.1)

$$\mathbf{M} = \bigcup B(y_\nu, r_0), \tag{2.1}$$

where $B(y_\nu, r_0)$ is a ball centered at $y_\nu \in \mathbf{M}$ of radius $r_0 \leq \rho_\mathbf{M}$, contained in a coordinate chart, and consider a fixed partition of unity $\Psi = \{\psi_\nu\}$ subordinate to this cover. The Sobolev spaces $H^s(\mathbf{M}), s \in \mathbf{R}$, are introduced as the completion of $C^\infty(\mathbf{M})$ with respect to the norm

$$\|f\|_{H^s(\mathbf{M})} = \left(\sum_\nu \|\psi_\nu f\|^2_{H^s(B(y_\nu, r_0))}\right)^{1/2}. \tag{2.2}$$

Any two such norms are equivalent. Note that spaces $H^s(\mathbf{M}), s \in \mathbf{R}$, are domains of operators $A^{s/2}$ for all elliptic differential operators A of order 2. It implies, that for any $s \in \mathbf{R}$, there exist positive constants $a(s)$, $b(s)$ (which depend on Ψ, A) such that

$$\|f\|_{H^s(\mathbf{M})} \leq a(s)\left(\|f\|^2_{L_2(\mathbf{M})} + \|A^{s/2} f\|_{L_2(\mathbf{M})}\right)^{1/2} \leq b(s)\|f\|_{H^s(\mathbf{M})} \tag{2.3}$$

for all $f \in H^s(\mathbf{M})$.

We are going to keep notations from the introduction. Since the operator $\mathcal{L}$ is of order two, the dimension $\mathcal{N}_\omega$ of the space $\mathbf{E}_\omega(\mathcal{L})$ is given asymptotically by Weyl's formula

$$\mathcal{N}_\omega(\mathbf{M}) \asymp C(\mathbf{M})\omega^{n/2}, \tag{2.4}$$

where $n = \dim\mathbf{M}$.

The next two theorems were proved in [24, 28], for a Laplace–Beltrami operator in $L_2(\mathbf{M})$ on a Riemannian manifold $\mathbf{M}$ of bounded geometry, but their proofs go through for any elliptic second-order differential operator in $L_2(\mathbf{M})$. In what follows, the notation $n = \dim \mathbf{M}$ is used.

Theorem 2.2. *There exist constants $C_1 > 0$ and $\rho_0 > 0$, such that for any natural $m > n/2$, any $0 < \rho < \rho_0$, and any ρ-lattice $M_\rho = \{x_k\}$, the following inequality holds:*

$$\left(\sum_{x_k \in M_\rho} |f(x_k)|^2\right)^{1/2} \leq C_1 \rho^{-n/2} \|f\|_{H^m(\mathbf{M})}, \tag{2.5}$$

for all $f \in H^m(\mathbf{M})$.

Theorem 2.3. *There exist constants $C_2 > 0$, and $\rho_0 > 0$, such that for any natural $m > n/2$, any $0 < \rho < \rho_0$, and any ρ-lattice $M_\rho = \{x_k\}$, the following inequality holds:*

$$\|f\|_{H^m(\mathbf{M})} \leq C_2 \left\{ \rho^{n/2} \left(\sum_{x_k \in M_\rho} |f(x_k)|^2 \right)^{1/2} + \rho^m \|\mathcal{L}^{m/2} f\|_{L_2(\mathbf{M})} \right\}. \tag{2.6}$$

As one can easily verify, the norm of $\mathcal{L}$ on the subspace $\mathbf{E}_\omega(\mathcal{L})$ (the span of eigenfunctions whose eigenvalues $\leq \omega$) is exactly ω. In particular, one has the following Bernstein-type inequality:

$$\|\mathcal{L}^s f\|_{L_2(\mathbf{M})} \leq \omega^s \|f\|_{L_2(\mathbf{M})}, \quad s \in \mathbf{R}_+, \tag{2.7}$$

for all $f \in \mathbf{E}_\omega(\mathcal{L})$. This fact and the previous two theorems imply the following Plancherel–Polya-type inequalities. Such inequalities are also known as Marcinkiewicz–Zygmund inequalities.

Theorem 2.4. *Set* $m_0 = \left[\frac{n}{2}\right] + 1$. *If* $C_1, \; C_2$ *are the same as above,* $a(m_0)$ *is from (2.3), and* $c_0 = \left(\frac{1}{2} C_2^{-1}\right)^{1/m_0}$ *then for any* $\omega > 0$, *and for every metric* ρ*-lattice* $M_\rho = \{x_k\}$ *with* $\rho = c_0 \omega^{-1/2}$, *the following Plancherel–Polya inequalities hold:*

$$C_1^{-1} a(m_0)^{-1} (1+\omega)^{-m_0/2} \left(\sum_k |f(x_k)|^2 \right)^{1/2} \leq \rho^{-n/2} \|f\|_{L_2(\mathbf{M})}$$
$$\leq (2C_2) \left(\sum_k |f(x_k)|^2 \right)^{1/2}, \tag{2.8}$$

for all $f \in \mathbf{E}_\omega(\mathcal{L})$ *and* $n = \dim \mathbf{M}$.

Proof. Since $\mathcal{L}$ is an elliptic second-order differential operator on a compact manifold which is self-adjoint and positive definite in $L_2(\mathbf{M})$, the norm on the Sobolev space $H^{m_0}(\mathbf{M})$ is equivalent to the norm $\|f\|_{L_2(\mathbf{M})} + \|\mathcal{L}^{m_0/2} f\|_{L_2(\mathbf{M})}$. Thus, the inequality (2.5) implies

$$\left(\sum_{x_k \in \mathbf{M}_\rho} |f(x_k)|^2 \right)^{1/2} \leq C_1 a(m_0) \rho^{-n/2} \left(\|f\|_{L_2(\mathbf{M})} + \|\mathcal{L}^{m_0/2} f\|_{L_2(\mathbf{M})} \right).$$

The Bernstein inequality shows that for all $f \in \mathbf{E}_\omega(\mathcal{L})$ and all $\omega \geq 0$,

$$\|f\|_{L_2(\mathbf{M})} + \|\mathcal{L}^{m_0/2} f\|_{L_2(\mathbf{M})} \leq (1+\omega)^{m_0/2} \|f\|_{L_2(\mathbf{M})}.$$

Thus, we proved the inequality

$$C_1^{-1}a(m_0)^{-1}(1+\omega)^{-m_0/2}\left(\sum_{x_k\in M_\rho}|f(x_k)|^2\right)^{1/2}\leq \rho^{-n/2}\|f\|_{L_2(\mathbf{M})},\quad f\in \mathbf{E}_\omega(\mathcal{L}). \tag{2.9}$$

To prove the opposite inequality, we start with inequality (2.6) where $m_0 = \left[\frac{n}{2}\right]+1$. Applying the Bernstein inequality (2.7), we obtain

$$\|f\|_{L_2(\mathbf{M})}\leq C_2\rho^{n/2}\left(\sum_{x_k\in M_\rho}|f(x_k)|^2\right)^{1/2}+C_2\rho^{m_0}\omega^{m_0/2}\|f\|_{L_2(\mathbf{M})}, \tag{2.10}$$

where $f\in \mathbf{E}_\omega(\mathcal{L})$. Now we fix the following value for ρ:

$$\rho=\left(\frac{1}{2}C_2^{-1}\right)^{1/m_0}\omega^{-1/2}=c_0\omega^{-1/2},\quad c_0=\left(\frac{1}{2}C_2^{-1}\right)^{1/m_0}.$$

With such ρ, the factor in the front of the last term in (2.10) is exactly $1/2$. Thus, this term can be moved to the left side of the formula (2.10) to obtain

$$\frac{1}{2}\|f\|_{L_2(\mathbf{M})}\leq C_2\rho^{n/2}\left(\sum_{x_k\in M_\rho}|f(x_k)|^2\right)^{1/2}. \tag{2.11}$$

In other words, we obtain the inequality

$$\rho^{-n/2}\|f\|_{L_2(\mathbf{M})}\leq 2C_2\left(\sum_{x_k\in M_\rho}|f(x_k)|^2\right)^{1/2}.$$

The theorem is proved.

It is interesting to note that our ρ-lattices (appearing in the previous theorems) always produce sampling sets with essentially optimal number of sampling points. In other words, the number of points in a sampling set for $\mathbf{E}_\omega(\mathcal{L})$ is "almost" the same as the dimension of the space $\mathbf{E}_\omega(\mathcal{L})$ which is given by the Weyl's formula (2.4).

Theorem 2.5. *If the constant $c_0>0$ is the same as above, then for any $\omega>0$ and $\rho=c_0\omega^{-1/2}$, there exist positive a_1, a_2 such that the number of points in any ρ-lattice M_ρ satisfies the following inequalities:*

$$a_1\omega^{n/2}\leq |M_\rho|\leq a_2\omega^{n/2}; \tag{2.12}$$

Proof. According to the definition of a lattice M_ρ, we have

$$|M_\rho| \inf_{x\in M} \mathrm{Vol}(B(x,\rho/4)) \le \mathrm{Vol}(\mathbf{M}) \le |M_\rho| \sup_{x\in M} \mathrm{Vol}(B(x,\rho/2))$$

or

$$\frac{\mathrm{Vol}(\mathbf{M})}{\sup_{x\in\mathbf{M}} \mathrm{Vol}(B(x,\rho/2))} \le |M_\rho| \le \frac{\mathrm{Vol}(\mathbf{M})}{\inf_{x\in\mathbf{M}} Vol\,(B(x,\rho/4))}.$$

Since for certain $c_1(\mathbf{M}), c_2(\mathbf{M})$, all $x \in \mathbf{M}$ and all sufficiently small $\rho > 0$, one has a double inequality

$$c_1(\mathbf{M})\rho^n \le \mathrm{Vol}(B(x,\rho)) \le c_2(\mathbf{M})\rho^n,$$

and since $\rho = c_0\omega^{-1/2}$, we obtain the inequalities (2.12) for certain $a_1 = a_1(\mathbf{M}),\ a_2 = a_2(\mathbf{M})$.

3 Cubature Formulas on Manifolds Which are Exact on Band-Limited Functions

Theorem 2.4 shows that if x_k is in a ρ lattice M_ρ and ϑ_k is the orthogonal projection of the Dirac measure δ_{x_k} on the space $\mathbf{E}_\omega(\mathcal{L})$ (in a Hilbert space $H^{-n/2-\varepsilon}(\mathbf{M}),\ \varepsilon > 0$), then there exist constants $c_1 = c_1(\mathbf{M},\mathcal{L},\omega) > 0,\ c_2 = c_2(\mathbf{M},\mathcal{L}) > 0$, such that the following frame inequality holds for all $f \in \mathbf{E}_\omega(\mathcal{L})$

$$c_1\left(\sum_k |\langle f,\vartheta_k\rangle|^2\right)^{1/2} \le \rho^{-n/2}\|f\|_{L_2(\mathbf{M})} \le c_2\left(\sum_k |\langle f,\vartheta_k\rangle|^2\right)^{1/2}, \tag{3.1}$$

where

$$\langle f,\vartheta_k\rangle = f(x_k),\quad f \in \mathbf{E}_\omega(\mathcal{L}).$$

From here by using the classical ideas of Duffin and Schaeffer about dual frames [6], we obtain the following reconstruction formula.

Theorem 3.1. *If M_ρ is a ρ-lattice in Theorem 2.4 with $\rho = c_0\omega^{-1/2}$, then there exists a frame $\{\Theta_j\}$ in the space $\mathbf{E}_\omega(\mathcal{L})$ such that the following reconstruction formula holds for all functions in $\mathbf{E}_\omega(\mathcal{L})$*

$$f = \sum_{x_k\in M_\rho} f(x_k)\Theta_k. \tag{3.2}$$

This formula implies that for any linear functional F on the space $\mathbf{E}_\omega(\mathcal{L})$, one has

$$F(f) = \sum_{x_k \in M_\rho} f(x_k) F(\Theta_k), \quad f \in \mathbf{E}_\omega(\mathcal{L}).$$

In particular, we have the following exact cubature formula.

Theorem 3.2. *If M_ρ is a ρ-lattice in Theorem 2.4 with $\rho = c_0 \omega^{-1/2}$ and*

$$\nu_k = \int_{\mathbf{M}} \Theta_k,$$

then for all $f \in \mathbf{E}_\omega(\mathcal{L})$, the following holds:

$$\int_{\mathbf{M}} f = \sum_{x_k \in M_\rho} f(x_k)\nu_k, \quad f \in \mathbf{E}_\omega(\mathcal{L}). \tag{3.3}$$

Thus, we have a cubature formula which is exact on the space $\mathbf{E}_\omega(\mathcal{L})$. Now, we are going to consider general functions $f \in L_2(\mathbf{M})$. Let f_ω be orthogonal projection of f onto space $\mathbf{E}_\omega(\mathcal{L})$. As it was shown in [30], there exists a constant $C_{k,m}$ that the following estimate holds for all $f \in L_2(\mathbf{M})$:

$$\|f - f_\omega\|_{L_2(\mathbf{M})} \leq \frac{C_{k,m}}{\omega^k} \Omega_{m-k}\left(\mathcal{L}^k f, 1/\omega\right), \quad k, m \in \mathbf{N}. \tag{3.4}$$

Here the modulus of continuity is defined as

$$\Omega_r(g, s) = \sup_{|\tau| \leq s} \|\Delta_\tau^r g\|, \quad g \in L_2(\mathbf{M}), \quad r \in \mathbf{N}, \tag{3.5}$$

where

$$\Delta_\tau^r g = (-1)^{r+1} \sum_{j=0}^{r} (-1)^{j-1} C_r^j e^{j\tau(i\mathcal{L})} g, \quad \tau \in \mathbf{R}, \quad r \in \mathbf{N}. \tag{3.6}$$

Thus, by combining (3.3) and (3.4), we obtain the following theorem.

Theorem 3.3. *There exists a $c_0 = c_0(\mathbf{M}, \mathcal{L})$, and for any $0 \leq k \leq m, k, m \in \mathbb{N}$, there exists a constant $C_{k,m} > 0$ such that if $M_\rho = \{x_k\}$ is a ρ-lattice with $0 < \rho \leq c_0\omega^{-1}$, then for the same weights $\{\nu_j\}$ as in (3.3)*

$$\left| \int_{\mathbf{M}} f - \sum_{x_j} f_\omega(x_j)\nu_j \right| \leq \frac{C_{k,m}}{\omega^k} \Omega_{m-k}\left(\mathcal{L}^k f, 1/\omega\right), \tag{3.7}$$

where f_ω is the orthogonal projection of $f \in L_2(\mathbf{M})$ onto $\mathbf{E}_\omega(\mathcal{L})$.

Note (see [30]) that $f \in L_2(\mathbf{M})$ belongs to the Besov space $\mathbf{B}^{\alpha}_{2,\infty}(\mathbf{M})$ if and only if

$$\Omega_m(f, 1/\omega) = O(\omega^{-\alpha}),$$

when $\omega \longrightarrow \infty$. Thus, we obtain that for functions in $\mathbf{B}^{\alpha}_{2,\infty}(\mathbf{M})$, the following relation holds:

$$\left| \int_{\mathbf{M}} f - \sum_{x_j} \nu_j f_{\omega}(x_j) \right| = O(\omega^{-\alpha}), \quad \omega \longrightarrow \infty. \tag{3.8}$$

4 Cubature Formulas on Compact Manifolds Which are Exact on Variational Splines

Given a ρ lattice $M_\rho = \{x_\gamma\}$ and a sequence $\{z_\gamma\} \in l_2$, we will be interested to find a function $s_k \in H^{2k}(\mathbf{M})$, where k is large enough, such that

1. $s_k(x_\gamma) = z_\gamma, x_\gamma \in M_\rho$.
2. Function s_k minimizes functional $g \to \|\mathcal{L}^k g\|_{L_2(\mathbf{M})}$.

We already know (2.5), (2.6) that for $k \geq d$ the norm $H^{2k}(\mathbf{M})$ is equivalent to the norm

$$C_1(\rho)\|f\|_{H^{2k}(\mathbf{M})} \leq \|\mathcal{L}^k f\|_{L_2(\mathbf{M})} + \left(\sum_{x_\gamma \in M_\rho} |f(x_\gamma)|^2 \right)^{1/2} \leq C_2(\rho)\|f\|_{H^{2k}(\mathbf{M})}.$$

For the given sequence $\{z_\gamma\} \in l_2$, consider a function f from $H^{2k}(\mathbf{M})$ such that $f(x_\gamma) = z_\gamma$. Let Pf denote the orthogonal projection of this function f in the Hilbert space $H^{2k}(\mathbf{M})$ with the inner product

$$< f, g >= \sum_{x_\gamma \in M_\rho} f(x_\gamma) g(x_\gamma) + < \mathcal{L}^{k/2} f, \mathcal{L}^{k/2} g >$$

on the subspace $U^{2k}(M_\rho) = \left\{ f \in H^{2k}(\mathbf{M}) | f(x_\gamma) = 0 \right\}$ with the norm generated by the same inner product. Then the function $g = f - Pf$ will be the unique solution of the above minimization problem for the functional $g \to \|\mathcal{L}^k g\|_{L_2(\mathbf{M})}, k \geq d$.

Different parts of the following theorem can be found in [29].

Theorem 4.1. *The following statements hold:*

(1) *For any function f from $H^{2k}(\mathbf{M})$, $k \geq d$, there exists a unique function $s_k(f)$ from the Sobolev space $H^{2k}(\mathbf{M})$, such that $f|_{M_\rho} = s_k(f)|_{M_\rho}$; and this function $s_k(f)$ minimizes the functional $u \to \|\mathcal{L}^k u\|_{L_2(\mathbf{M})}$.*

(2) *Every such function $s_k(f)$ is of the form*

$$s_k(f) = \sum_{x_\gamma \in M_\rho} f(x_\gamma) L_\gamma^{2k}$$

where the function $L_\gamma^{2k} \in H^{2k}(\mathbf{M})$, $x_\gamma \in M_\rho$ minimizes the same functional and takes value 1 at the point x_γ and 0 at all other points of M_ρ.

(3) *Functions L_γ^{2k} form a Riesz basis in the space of all polyharmonic functions with singularities on M_ρ, i.e., in the space of such functions from $H^{2k}(\mathbf{M})$ which in the sense of distributions satisfy equation*

$$\mathcal{L}^{2k} u = \sum_{x_\gamma \in M_\rho} \alpha_\gamma \delta(x_\gamma),$$

where $\delta(x_\gamma)$ is the Dirac measure at the point x_γ.

(4) *If in addition the set M_ρ is invariant under some subgroup of diffeomorphisms acting on M, then every two functions L_γ^{2k}, L_μ^{2k} are translates of each other.*

The crucial role in the proof of the above Theorem 4.1 belongs to the following lemma which was proved in [24].

Lemma 4.2. *A function $f \in L_2(\mathbf{M})$ satisfies equation*

$$\mathcal{L}^{2k} f = \sum_{x_\gamma \in M_\rho} \alpha_\gamma \delta(x_\gamma),$$

where $\{\alpha_\gamma\} \in l_2$ if and only if f is a solution to the minimization problem stated above.

Next, if $f \in H^{2k}(\mathbf{M}), k \geq d$, then $f - s_k(f) \in U^{2k}(M_\rho)$, and we have for $k \geq d$,

$$\|f - s_k(f)\|_{L_2(\mathbf{M})} \leq (C_0\rho)^k \|\mathcal{L}^{k/2}(f - s_k(f))\|_{L_2(\mathbf{M})}.$$

Using minimization property of $s_k(f)$, we obtain the inequality

$$\left\| f - \sum_{x_\gamma \in M_\rho} f(x_\gamma) L_{x_\gamma} \right\|_{L_2(\mathbf{M})} \leq (c_0\rho)^k \|\mathcal{L}^{k/2} f\|_{L_2(\mathbf{M})}, k \geq d, \tag{4.1}$$

and for $f \in \mathbf{E}_\omega(\mathcal{L})$, the Bernstein inequality gives for any $f \in \mathbf{E}_\omega(\mathcal{L})$

$$\left\| f - \sum_{x_\gamma \in M_\rho} f(x_\gamma) L_{x_\gamma} \right\|_{L_2(\mathbf{M})} \leq (c_0\rho\sqrt{\omega})^k \|f\|_{L_2(\mathbf{M})}, \tag{4.2}$$

for $k \geq d$. The last inequality shows, in particular, that for any $f \in \mathbf{E}_\omega(\mathcal{L})$ one has the following reconstruction algorithm.

Theorem 4.3. *There exists a* $c_0 = c_0(M)$ *such that for any* $\omega > 0$ *and any* M_ρ *with* $\rho = c_0\omega^{-1}$, *the following reconstruction formula holds in* $L_2(M)$*-norm*

$$f = \lim_{l\to\infty} \sum_{x_j \in M_\rho} f(x_j) L_{x_j}^{(k)}, \; k \geq d, \tag{4.3}$$

for all $f \in \mathbf{E}_\omega(\mathcal{L})$.

To develop a cubature formula, we introduce the notation

$$\lambda_\gamma^{(k)} = \int_{\mathbf{M}} L_{x_\gamma}^{(k)}(x)\mathrm{d}x, \tag{4.4}$$

where $L_{x_\gamma} \in S^k(M_\rho)$ is the Lagrangian spline at the node x_γ.

Theorem 4.4. (1) *For any* $f \in H^{2k}(M)$, *one has*

$$\int_{\mathbf{M}} f\mathrm{d}x \approx \sum_{x_j \in M_\rho} \lambda_j^{(k)} f(x_j), \; k \geq d, \tag{4.5}$$

and the error given by the inequality

$$\left| \int_{\mathbf{M}} f\mathrm{d}x - \sum_{x_\gamma \in M_\rho} \lambda_\gamma^{(k)} f(x_\gamma) \right| \leq \mathrm{Vol}(\mathbf{M})(c_0\rho)^k \|\mathcal{L}^{k/2} f\|_{L_2(\mathbf{M})}, \tag{4.6}$$

for $k \geq d$. *For a fixed function* f *the right-hand side of (4.6) goes to zero as long as* ρ *goes to zero.*

(2) *The formula (4.5) is exact for any variational spline* $f \in S^k(M_\rho)$ *of order* k *with singularities on* M_ρ.

By applying the Bernstein inequality, we obtain the following theorem. This result explains our term "asymptotically correct cubature formulas."

Theorem 4.5. *For any* $f \in \mathbf{E}_\omega(\mathcal{L})$, *one has*

$$\left| \int_{\mathbf{M}} f\mathrm{d}x - \sum_{x_\gamma \in M_\rho} \lambda_\gamma^{(k)} f(x_\gamma) \right| \leq \mathrm{Vol}(\mathbf{M})(c_0\rho\sqrt{\omega})^k \|f\|_{L_2(\mathbf{M})}, \tag{4.7}$$

for $k \geq d$. *If* $\rho = c_0\omega^{-1/2}$, *the right-hand side in (4.7) goes to zero for all* $f \in \mathbf{E}_\omega(\mathcal{L})$ *as long as* k *goes to infinity.*

5 Positive Cubature Formulas on Compact Manifolds

Let $M_\rho = \{x_k\}$, $k = 1, \ldots, N(M_\rho)$, be a ρ-lattice on $\mathbf{M}$. We construct the Voronoi partition of $\mathbf{M}$ associated to the set $M_\rho = \{x_k\}$, $k = 1, \ldots, N(M_\rho)$. Elements of this partition will be denoted as $\mathcal{M}_{k,\rho}$. Let us recall that the distance from each point in $\mathcal{M}_{j,\rho}$ to x_j is less than or equal to its distance to any other point of the family $M_\rho = \{x_k\}$, $k = 1, \ldots, N(M_\rho)$. Some properties of this cover of $\mathbf{M}$ are summarized in the following Lemma. which follows easily from the definitions.

Lemma 5.1. *The sets $\mathcal{M}_{k,\rho}$, $k = 1, \ldots, N(M_\rho)$, have the following properties:*

1. *They are measurable.*
2. *They are disjoint.*
3. *They form a cover of $\mathbf{M}$.*
4. *There exist positive a_1, a_2, independent of ρ and the lattice $M_\rho = \{x_k\}$, such that*

$$a_1 \rho^n \leq \mu\left(\mathcal{M}_{k,\rho}\right) \leq a_2 \rho^n. \tag{5.1}$$

In what follows, we are using partition of unity $\Psi = \{\psi_\nu\}$ which appears in (2.2). Our next goal is to prove the following fact.

Theorem 5.2. *Say $\rho > 0$, and let $\{\mathcal{M}_{k,\rho}\}$ be the disjoint cover of $\mathbf{M}$ which is associated with a ρ-lattice M_ρ. If ρ is sufficiently small, then for any sufficiently large $K \in \mathbb{N}$, there exists a $C(K) > 0$ such that for all smooth functions f the following inequality holds:*

$$\left|\sum_\nu \sum_{x_k \in M_\rho} \psi_\nu f(x_k)\, \mu \mathcal{M}_{k,\rho} - \int_{\mathbf{M}} f(x) \mathrm{d}x\right|$$

$$\leq C(K) \sum_{|\beta|=1}^{K} \rho^{n/2+|\beta|} \|(I + \mathcal{L})^{|\beta|/2} f\|_{L_2(\mathbf{M})}, \tag{5.2}$$

where $C(K)$ is independent of ρ and the ρ-lattice M_ρ.

Proof. We start with the Taylor series

$$\psi_\nu f(y) - \psi_\nu f(x_k) = \sum_{1 \leq |\alpha| \leq m-1} \frac{1}{\alpha!} \partial^\alpha (\psi_\nu f)(x_k)(x_k - y)^\alpha$$

$$+ \sum_{|\alpha|=m} \frac{1}{\alpha!} \int_0^\tau t^{m-1} \partial^\alpha \psi_\nu f(x_k + t\theta) \theta^\alpha \mathrm{d}t, \tag{5.3}$$

where $f \in C^\infty(\mathbb{R}^d)$, $y \in B(x_k, \rho/2)$, $x = (x^{(1)}, \ldots, x^{(d)})$, $y = (y^{(1)}, \ldots, y^{(d)})$, $\alpha = (\alpha_1, \ldots, \alpha_d)$, $(x - y)^\alpha = (x^{(1)} - y^{(1)})^{\alpha_1} \ldots (x^{(d)} - y^{(d)})^{\alpha_d}$, $\tau = \|x - x_i\|$, $\theta = (x - x_i)/\tau$.

We are going to use the following inequality, which is essentially the Sobolev imbedding theorem:

$$|(\psi_\nu f)(x_k)| \leq C_{n,m} \sum_{0\leq j\leq m} \rho^{j-n/p} \|(\psi_\nu f)\|_{W_p^j(B(x_k,\rho))}, \quad 1 \leq p \leq \infty, \tag{5.4}$$

where $m > n/p$ and the functions $\{\psi_\nu\}$ form the partition of unity which we used to define the Sobolev norm in (2.2). Using (5.4) for $p = 1$, we obtain the following inequality:

$$\left| \sum_{1\leq|\alpha|\leq m-1} \frac{1}{\alpha!} \partial^\alpha (\psi_\nu f)(x_k)(x_k - y)^\alpha \right| \\ \leq C(n,m)\rho^{|\alpha|} \sum_{1\leq|\alpha|\leq m} \sum_{0\leq|\gamma|\leq m} \rho^{|\gamma|-n} \|\partial^{\alpha+\gamma}(\psi_\nu f)\|_{L_1(B(x_k,\rho))}, \quad m > n, \tag{5.5}$$

for some $C(n,m) \geq 0$. Since, by the Schwarz inequality,

$$\|\partial^\alpha(\psi_\nu f)\|_{L_1(B(x_k,\rho))} \leq C(n)\rho^{n/2} \|\partial^\alpha(\psi_\nu f)\|_{L_2(B(x_k,\rho))} \tag{5.6}$$

we obtain the following estimate, which holds for small ρ:

$$\sup_{y\in B(x_k,\rho)} \left| \sum_{1\leq|\alpha|\leq m-1} \frac{1}{\alpha!} \partial^\alpha (\psi_\nu f)(x_k)(x_k - y)^\alpha \right| \\ \leq C(n,m) \sum_{1\leq|\beta|\leq 2m} \rho^{|\beta|-n/2} \|\partial^\beta(\psi_\nu f)\|_{L_2(B(x_k,\rho))}, \quad m > n. \tag{5.7}$$

Next, using the Schwarz inequality and the assumption that $m > n = \dim \mathbf{M}$, $|\alpha| = m$, we obtain

$$\left| \int_0^\tau t^{m-1} \partial^\alpha \psi_\nu f(x_k + t\theta)\theta^\alpha \mathrm{d}t \right| \\ \leq \int_0^\tau t^{m-n/2-1/2} |t^{n/2-1/2} \partial^\alpha \psi_\nu f(x_k + t\theta)| \mathrm{d}t \\ \leq C \left(\int_0^\tau t^{2m-n-1} \right)^{1/2} \left(\int_0^\tau t^{n-1} |\partial^\alpha \psi_\nu f(x_k + t\theta)|^2 \mathrm{d}t \right)^{1/2} \\ \leq C\tau^{m-n/2} \left(\int_0^\tau t^{n-1} |\partial^\alpha \psi_\nu f(x_k + t\theta)|^2 \mathrm{d}t \right)^{1/2}, \quad m > n.$$

We square this inequality, and integrate both sides of it over the ball $B(x_k, \rho/2)$, using the spherical coordinate system (τ, θ). We find

$$\int_{B(x_k,\rho)} \left| \int_0^\tau t^{m-1} \partial^\alpha \psi_\nu f(x_k + t\theta)\theta^\alpha \mathrm{d}t \right|^2 \tau^{n-1} \mathrm{d}\theta \mathrm{d}\tau$$

$$\leq C(m,n) \int_0^{\rho/2} \tau^{2m-n} \int_0^{2\pi} \left| \int_0^\tau t^{n-1} \partial^\alpha (\psi_\nu f)(x_k + t\theta)\theta^\alpha \mathrm{d}t \right|^2 \tau^{n-1} \mathrm{d}\theta \mathrm{d}\tau$$

$$\leq C(m,n) \int_0^{\rho/2} t^{n-1} \left(\int_0^{2\pi} \int_0^{\rho/2} \tau^{2m-n} \left| \partial^\alpha (\psi_\nu f)(x_k + t\theta) \right|^2 \tau^{n-1} \mathrm{d}\tau \mathrm{d}\theta \right) \mathrm{d}t$$

$$\leq C_{m,n} \rho^{2|\alpha|} \| \partial^\alpha (\psi_\nu f) \|^2_{L_2(B(x_k,\rho))},$$

where $\tau = \|x - x_k\| \leq \rho/2,\ m = |\alpha| > n$. Let $\{\mathcal{M}_{k,\rho}\}$ be the Voronoi cover of $\mathbf{M}$ which is associated with a ρ-lattice M_ρ (see Lemma 5.1). From here, we obtain

$$\int_{\mathcal{M}_k} |\psi_\nu f(y) - \psi_\nu f(x_k)| \, \mathrm{d}x$$

$$\leq C(n,m) \sum_{1 \leq |\beta| \leq 2m} \rho^{|\beta|+n/2} \| \partial^\beta (\psi_\nu f) \|_{L_2(B(x_k,\rho))}$$

$$+ \sum_{|\alpha|=m} \frac{1}{\alpha!} \int_{B(x_k,\rho)} \left| \int_0^\tau t^{m-1} \partial^\alpha \psi_\nu f(x_k + t\theta)\theta^\alpha \mathrm{d}t \right|$$

$$\leq C(n,m) \sum_{1 \leq |\beta| \leq 2m} \rho^{|\beta|+n/2} \| \partial^\beta (\psi_\nu f) \|_{L_2(B(x_k,\rho))}$$

$$+ \rho^{n/2} \sum_{|\alpha|=m} \frac{1}{\alpha!} \left(\int_{B(x_k,\rho)} \left| \int_0^\tau t^{m-1} \partial^\alpha \psi_\nu f(x_k + t\theta)\theta^\alpha \mathrm{d}t \right|^2 \tau^{n-1} \mathrm{d}\tau \mathrm{d}\theta \right)^{1/2}$$

$$\leq C(n,m) \sum_{1 \leq |\beta| \leq 2m} \rho^{|\beta|+n/2} \| \partial^\beta (\psi_\nu f) \|_{L_2(B(x_k,\rho))}. \tag{5.8}$$

Next, we have the following inequalities:

$$\sum_\nu \sum_{x_k \in M_\rho} \psi_\nu f(x_k)\, \mu\mathcal{M}_{k,\rho} - \int_{\mathbf{M}} f(x) \mathrm{d}x$$

$$= -\sum_\nu \left(\sum_k \int_{\mathcal{M}_{k,\rho}} \psi_\nu f(x) \mathrm{d}x - \sum_k \psi_\nu f(x_k)\, \mu\mathcal{M}_{k,\rho} \right)$$

$$\leq \sum_{\nu}\sum_{k}\left|\int_{\mathcal{M}_{k,\rho}} \psi_\nu f(x) - \psi_\nu f(x_k)\ \mu\mathcal{M}_{k,\rho}\mathrm{d}x\right|$$

$$\leq C(n,m)\rho^{n/2}\sum_{\nu}\sum_{x_k\in M_\rho}\sum_{1\leq|\beta|\leq 2m}\rho^{|\beta|}\|\partial^\beta(\psi_\nu f)\|_{L_2(B(x_k,\rho))}, \tag{5.9}$$

where $m > n$. Using the definition of the Sobolev norm and elliptic regularity of the operator $I + \mathcal{L}$, where I is the identity operator on $L_2(\mathbf{M})$, we obtain the inequality (5.2).

Now we are going to prove existence of cubature formulas which are exact on $\mathbf{E}_\omega(\mathbf{M})$ and have positive coefficients of the "right" size.

Theorem 5.3. *There exists a positive constant a_0, such that if $\rho = a_0(\omega+1)^{-1/2}$, then for any ρ-lattice $M_\rho = \{x_k\}$, there exist strictly positive coefficients $\mu_{x_k} > 0$, $x_k \in M_\rho$, for which the following equality holds for all functions in $\mathbf{E}_\omega(\mathcal{L})$:*

$$\int_{\mathbf{M}} f\,\mathrm{d}x = \sum_{x_k\in M_\rho}\mu_{x_k} f(x_k). \tag{5.10}$$

Moreover, there exists constants c_1, c_2, such that the following inequalities hold:

$$c_1\rho^n \leq \mu_{x_k} \leq c_2\rho^n,\ n = \dim \mathbf{M}. \tag{5.11}$$

Proof. By using the Bernstein inequality, and our Plancherel–Polya inequalities (2.8), and assuming that

$$\rho < \frac{1}{2\sqrt{\omega+1}}, \tag{5.12}$$

we obtain from (5.2) the following inequality:

$$\left|\sum_{\nu}\sum_{x_k\in M_\rho}\psi_\nu f(x_k)\ \mu\mathcal{M}_{k,\rho} - \int_{\mathbf{M}} f(x)\mathrm{d}x\right| \leq C_1\rho^{n/2}\sum_{|\beta|=1}^{K}\left(\rho\sqrt{1+\omega}\right)^{|\beta|}\|f\|_{L_2(\mathbf{M})}$$

$$\leq C_2\rho^n\left(\rho\sqrt{1+\omega}\right)\left(\sum_{x_k\in M_\rho}|f(x_k)|^2\right)^{1/2}, \tag{5.13}$$

where C_2 is independent of $\rho \in \left(0, \left(2\sqrt{\omega+1}\right)^{-1}\right)$ and the ρ-lattice M_ρ.

Let $R_\omega(\mathcal{L})$ denote the space of real-valued functions in $\mathbf{E}_\omega(\mathcal{L})$. Since the eigenfunctions of $\mathcal{L}$ may be taken to be real, we have $\mathbf{E}_\omega(\mathcal{L}) = R_\omega(\mathcal{L}) + \mathrm{i}R_\omega(\mathcal{L})$, so it is enough to show that (5.10) holds for all $f \in R_\omega(\mathcal{L})$.

Consider the sampling operator

$$S : f \to \{f(x_k)\}_{x_k \in M_\rho},$$

which maps $R_\omega(\mathcal{L})$ into the space $\mathbb{R}^{|\mathcal{M}_\rho|}$ with the ℓ^2 norm. Let $V = S(R_\omega(\mathcal{L}))$ be the image of $R_\omega(\mathcal{L})$ under S. V is a subspace of $\mathbb{R}^{|\mathcal{M}_\rho|}$, and we consider it with the induced ℓ^2 norm. If $u \in V$, denote the linear functional $y \to (y, u)$ on V by ℓ_u. By our Plancherel–Polya inequalities (2.8) , the map

$$\{f(x_k)\}_{x_k \in M_\rho} \to \int_{\mathbf{M}} f \mathrm{d}x$$

is a well-defined linear functional on the finite dimensional space V, and so equals ℓ_v for some $v \in V$, which may or may not have all components positive. On the other hand, if w is the vector with components $\{\mu(\mathcal{M}_{k,\rho})\}$, $x_k \in M_\rho$, then w might not be in V, but it has all components positive and of the right size

$$a_1 \rho^n \leq \mu\left(\mathcal{M}_{k,\rho}\right) \leq a_2 \rho^n,$$

for some positive a_1, a_2, independent of ρ and the lattice $M_\rho = \{x_k\}$. Since, for any vector $u \in V$, the norm of u is exactly the norm of the corresponding functional ℓ_u, inequality (5.13) tells us that

$$\|Pw - v\| \leq \|w - v\| \leq C_2 \rho^n \left(\rho\sqrt{1+\omega}\right), \tag{5.14}$$

where P is the orthogonal projection onto V. Accordingly, if z is the real vector $v - Pw$, then

$$v + (I - P)w = w + z, \tag{5.15}$$

where $\|z\| \leq C_2 \rho^n \left(\rho\sqrt{1+\omega}\right)$. Note, that all components of the vector w are of order $O(\rho^n)$, while the order of $\|z\|$ is $O(\rho^{n+1})$. Accordingly, if $\rho\sqrt{1+\omega}$ is sufficiently small, then $\mu := w + z$ has all components positive and of the right size. Since $\mu = v + (I - P)w$, the linear functional $y \to (y, \mu)$ on V equals ℓ_v. In other words, if the vector μ has components $\{\mu_{x_k}\}$, $x_k \in M_\rho$, then

$$\sum_{x_k \in M_\rho} f(x_k)\mu_{x_k} = \int_{\mathbf{M}} f \mathrm{d}x$$

for all $f \in R_\omega(\mathcal{L})$, and hence for all $f \in \mathbf{E}_\omega(\mathcal{L})$, as desired.

We obviously have the following result.

Theorem 5.4. (1) *There exists a* $c_0 = c_0(\mathbf{M}, \mathcal{L})$, *and for any* $0 \leq k \leq m, k, m \in \mathbb{N}$, *there exists a constant* $C_{k,m} > 0$ *such that if* $M_\rho = \{x_k\}$ *is a* ρ*-lattice with* $0 < \rho \leq c_0 \omega^{-1}$, *then for the same weights* $\{\mu_{x_j}\}$ *as in (5.10)*

$$\left| \int_{\mathbf{M}} f - \sum_{x_j} f_\omega(x_j)\mu_{x_j} \right| \leq \frac{C_{k,m}}{\omega^k} \Omega_{m-k}\left(\mathcal{L}^k f, 1/\omega\right), \tag{5.16}$$

(2) For functions in $\mathbf{B}^{\alpha}_{2,\infty}(\mathbf{M})$, the following relation holds:

$$\left| \int_{\mathbf{M}} f - \sum_{x_j} f_\omega(x_j)\mu_{x_j} \right| = O(\omega^{-\alpha}), \quad \omega \longrightarrow \infty, \tag{5.17}$$

where f_ω is the orthogonal projection of $f \in L_2(\mathbf{M})$ onto $\mathbf{E}_\omega(\mathcal{L})$.

6 Harmonic Analysis on Compact Homogeneous Manifolds

We review some very basic notions of harmonic analysis on compact homogeneous manifolds [16], Chap. II.

Let $\mathbf{M}$, $\dim \mathbf{M} = n$, be a compact connected C^∞-manifold. One says that a compact Lie group G effectively acts on $\mathbf{M}$ as a group of diffeomorphisms if

1. Every element $g \in G$ can be identified with a diffeomorphism

$$g : \mathbf{M} \to \mathbf{M}$$

of $\mathbf{M}$ onto itself and

$$g_1 g_2 \cdot x = g_1 \cdot (g_2 \cdot x), \; g_1, g_2 \in G, \; x \in \mathbf{M},$$

where $g_1 g_2$ is the product in G and $g \cdot x$ is the image of x under g.
2. The identity $e \in G$ corresponds to the trivial diffeomorphism

$$e \cdot x = x. \tag{6.1}$$

3. For every $g \in G$, $g \neq e$, there exists a point $x \in \mathbf{M}$ such that $g \cdot x \neq x$.

A group G acts on $\mathbf{M}$ *transitively* if in addition to (1)–(3) the following property holds:

4) For any two points $x, y \in \mathbf{M}$, there exists a diffeomorphism $g \in G$ such that

$$g \cdot x = y.$$

A *homogeneous* compact manifold $\mathbf{M}$ is a C^∞-compact manifold on which a compact Lie group G acts transitively. In this case, $\mathbf{M}$ is necessarily of the form G/K, where K is a closed subgroup of G. The notation $L_2(\mathbf{M})$ is used for the usual Banach spaces $L_2(\mathbf{M}, \mathrm{d}x)$, where $\mathrm{d}x$ is an invariant measure.

Every element X of the (real) Lie algebra of G generates a vector field on $\mathbf{M}$, which we will denote by the same letter X. Namely, for a smooth function f on $\mathbf{M}$, one has

$$Xf(x) = \lim_{t\to 0} \frac{f(\exp tX \cdot x) - f(x)}{t}$$

for every $x \in \mathbf{M}$. In the future, we will consider on $\mathbf{M}$ only such vector fields. The translations along integral curves of such vector fields X on $\mathbf{M}$ can be identified with a one-parameter group of diffeomorphisms of $\mathbf{M}$, which is usually denoted as $\exp tX, -\infty < t < \infty$. At the same time, the one-parameter group $\exp tX, -\infty < t < \infty$, can be treated as a strongly continuous one-parameter group of operators acting on the space $L_2(\mathbf{M})$. These operators act on functions according to the formula

$$f \to f(\exp tX \cdot x),\ t \in \mathbb{R},\ f \in L_2(\mathbf{M}),\ x \in \mathbf{M}.$$

The generator of this one-parameter group will be denoted by D_X, and the group itself will be denoted by

$$e^{tD_X} f(x) = f(\exp tX \cdot x),\ t \in \mathbb{R},\ f \in L_2(\mathbf{M}),\ x \in \mathbf{M}.$$

According to the general theory of one-parameter groups in Banach spaces, the operator D_X is a closed operator on every $L_2(\mathbf{M})$.

If $\mathbf{g}$ is the Lie algebra of a compact Lie group G, then ([16], Chap. II,) it is a direct sum $\mathbf{g} = \mathbf{a} + [\mathbf{g}, \mathbf{g}]$, where $\mathbf{a}$ is the center of $\mathbf{g}$ and $[\mathbf{g}, \mathbf{g}]$ is a semi-simple algebra. Let Q be a positive-definite quadratic form on $\mathbf{g}$ which, on $[\mathbf{g}, \mathbf{g}]$, is opposite to the Killing form. Let $X_1, \dots, X_d$ be a basis of $\mathbf{g}$, which is orthonormal with respect to Q. Since the form Q is $Ad(G)$-invariant, the operator

$$-X_1^2 - X_2^2 - \cdots - X_d^2,\ d = \dim G$$

is a bi-invariant operator on G. This implies in particular that the corresponding operator on $L_2(\mathbf{M})$

$$\mathcal{L} = -D_1^2 - D_2^2 - \cdots - D_d^2, \quad D_j = D_{X_j},\ d = \dim G, \tag{6.2}$$

commutes with all operators $D_j = D_{X_j}$. This operator $\mathcal{L}$, which is usually called the Laplace operator, is elliptic, and is involved in most of the constructions and results of our chapter.

In the rest of this chapter, the notation $\mathbb{D} = \{D_1, \dots, D_d\}, \quad d = \dim G$, will be used for the differential operators on $L_2(\mathbf{M})$, which are involved in the formula (6.2).

There are situations in which the operator $\mathcal{L}$ is, or is proportional to, the Laplace–Beltrami operator of an invariant metric on $\mathbf{M}$. This happens, for example, if $\mathbf{M}$ is a n-dimensional torus, a compact semi-simple Lie group, or a compact symmetric space of rank one.

7 On the Product of Eigenfunctions of the Casimir Operator $\mathcal{L}$ on Compact Homogeneous Manifolds

In this section, we will use the assumption that $\mathbf{M}$ is a compact homogeneous manifold and that $\mathcal{L}$ is the operator of (6.2), in an essential way.

Theorem 7.1. *If $\mathbf{M} = G/K$ is a compact homogeneous manifold and $\mathcal{L}$ is defined as in (6.2), then for any f and g belonging to $\mathbf{E}_\omega(\mathcal{L})$, their product fg belongs to $\mathbf{E}_{4d\omega}(\mathcal{L})$, where d is the dimension of the group G.*

Proof. First, we show that if for an $f \in L_2(\mathbf{M})$ and a positive ω there exists a constant $C(f,\omega)$ such that the following inequalities hold:

$$\|\mathcal{L}^k f\|_{L_2(\mathbf{M})} \le C(f,\omega)\omega^k \|f\|_{L_2(\mathbf{M})} \tag{7.1}$$

for all natural k, then $f \in \mathbf{E}_\omega(\mathcal{L})$. Indeed, assume that

$$\lambda_m \le \omega < \lambda_{m+1}$$

and

$$f = \sum_{j=0}^{\infty} c_j u_j, \tag{7.2}$$

$$c_j(f) =< f, u_j >= \int_{\mathbf{M}} f(x)\overline{u_j(x)}\mathrm{d}x.$$

Then by the Plancherel Theorem

$$\lambda_{m+1}^{2k} \sum_{j=m+1}^{\infty} |c_j|^2 \le \sum_{j=m+1}^{\infty} |\lambda_j^k c_j|^2 \le \|\mathcal{L}^k f\|^2_{L_2(\mathbf{M})}$$
$$\le C^2\omega^{2k}\|f\|^2_{L_2(\mathbf{M})}, \quad C = C(f,\omega),$$

which implies

$$\sum_{j=m+1}^{\infty} |c_j|^2 \le C^2 \left(\frac{\omega}{\lambda_{m+1}}\right)^{2k} \|f\|^2_{L_2(\mathbf{M})}.$$

In the last inequality, the fraction ω/λ_{m+1} is strictly less than 1, and k can be any natural number. This shows that the series (7.2) does not contain terms with $j \ge m+1$, i.e., the function f belongs to $\mathbf{E}_\omega(\mathcal{L})$.

Now, since every smooth vector field on $\mathbf{M}$ is a differentiation of the algebra $C^\infty(\mathbf{M})$, one has that for every operator $D_j, 1 \le j \le d$, the following equality holds for any two smooth functions f and g on $\mathbf{M}$:

$$D_j(fg) = fD_j g + gD_j f, \quad 1 \le j \le d. \tag{7.3}$$

Using formula (6.2), one can easily verify that for any natural $k \in \mathbb{N}$, the term $\mathcal{L}^k(fg)$ is a sum of d^k, $(d = \dim G)$, terms of the following form:

$$D_{j_1}^2 \dots D_{j_k}^2(fg), \ 1 \le j_1, \dots, j_k \le d. \tag{7.4}$$

For every D_j, one has

$$D_j^2(fg) = f(D_j^2 g) + 2(D_j f)(D_j g) + g(D_j^2 f).$$

Thus, the function $\mathcal{L}^k(fg)$ is a sum of $(4d)^k$ terms of the form

$$(D_{i_1} \dots D_{i_m} f)(D_{j_1} \dots D_{j_{2k-m}} g).$$

This implies that

$$\left|\mathcal{L}^k(fg)\right| \le (4d)^k \sup_{0 \le m \le 2k} \sup_{x,y \in \mathbf{M}} |D_{i_1} \dots D_{i_m} f(x)| \left|D_{j_1} \dots D_{j_{2k-m}} g(y)\right|. \tag{7.5}$$

Let us show that the following inequalities hold:

$$\|D_{i_1} \dots D_{i_m} f\|_{L_2(\mathbf{M})} \le \omega^{m/2} \|f\|_{L_2(\mathbf{M})} \tag{7.6}$$

and

$$\|D_{j_1} \dots D_{j_{2k-m}} g\|_{L_2(\mathbf{M})} \le \omega^{(2k-m)/2} \|g\|_{L_2(\mathbf{M})} \tag{7.7}$$

for all $f, g \in \mathbf{E}_\omega(\mathcal{L})$. First, we note that the operator

$$-\mathcal{L} = D_1^2 + \dots + D_d^2$$

commutes with every D_j (see the explanation before the formula (6.2)). The same is true for $\mathcal{L}^{1/2}$. But then

$$\|\mathcal{L}^{1/2} f\|_{L_2(\mathbf{M})}^2 = <\mathcal{L}^{1/2} f, \mathcal{L}^{1/2} f> = <\mathcal{L} f, f>$$
$$= -\sum_{j=1}^d <D_j^2 f, f> = \sum_{j=1}^d <D_j f, D_j f> = \sum_{j=1}^d \|D_j f\|_{L_2(\mathbf{M})}^2,$$

and also

$$\|\mathcal{L} f\|_{L_2(\mathbf{M})}^2 = \|\mathcal{L}^{1/2}\mathcal{L}^{1/2} f\|_{L_2(\mathbf{M})}^2 = \sum_{j=1}^d \|D_j \mathcal{L}^{1/2} f\|_{L_2(\mathbf{M})}^2$$
$$= \sum_{j=1}^d \|\mathcal{L}^{1/2} D_j f\|_{L_2(\mathbf{M})}^2 = \sum_{j,k=1}^d \|D_j D_k f\|_{L_2(\mathbf{M})}^2.$$

From here, by induction on $s \in \mathbb{N}$, one can obtain the following equality:

$$\|\mathcal{L}^{s/2} f\|^2_{L_2(\mathbf{M})} = \sum_{1 \le i_1,\dots,i_s \le d} \|D_{i_1} \dots D_{i_s} f\|^2_{L_2(\mathbf{M})}, \; s \in \mathbb{N}, \tag{7.8}$$

which implies the estimates (7.6) and (7.7). For example, to get (7.6), we take a function f from $\mathbf{E}_\omega(\mathcal{L})$, an $m \in \mathbb{N}$ and do the Following

$$\|D_{i_1} \dots D_{i_m} f\|_{L_2(\mathbf{M})} \le \left(\sum_{1 \le i_1,\dots,i_m \le d} \|D_{i_1} \dots D_{i_m} f\|^2_{L_2(\mathbf{M})} \right)^{1/2}$$
$$= \|\mathcal{L}^{m/2} f\|_{L_2(\mathbf{M})} \le \omega^{m/2} \|f\|_{L_2(\mathbf{M})}. \tag{7.9}$$

In a similar way, we obtain (7.7).

The formula (7.5) along with the formula (7.9) implies the estimate

$$\|\mathcal{L}^k(fg)\|_{L_2(\mathbf{M})} \le (4d)^k \sup_{0 \le m \le 2k} \|D_{i_1} \dots D_{i_m} f\|_{L_2(\mathbf{M})} \|D_{j_1} \dots D_{j_{2k-m}} g\|_\infty$$
$$\le (4d)^k \omega^{m/2} \|f\|_{L_2(\mathbf{M})} \sup_{0 \le m \le 2k} \|D_{j_1} \dots D_{j_{2k-m}} g\|_\infty. \tag{7.10}$$

Using the Sobolev embedding theorem and elliptic regularity of $\mathcal{L}$, we obtain for every $s > \frac{\dim \mathbf{M}}{2}$

$$\|D_{j_1} \dots D_{j_{2k-m}} g\|_\infty \le C(\mathbf{M}) \|D_{j_1} \dots D_{j_{2k-m}} g\|_{H^s(\mathbf{M})}$$
$$\le C(\mathbf{M}) \left\{ \|D_{j_1} \dots D_{j_{2k-m}} g\|_{L_2(\mathbf{M})} \right.$$
$$\left. + \|\mathcal{L}^{s/2} D_{j_1} \dots D_{j_{2k-m}} g\|_{L_2(\mathbf{M})} \right\}, \tag{7.11}$$

where $H^s(\mathbf{M})$ is the Sobolev space of s-regular functions on $\mathbf{M}$. Since the operator $\mathcal{L}$ commutes with each of the operators D_j, the estimate (7.9) gives the following inequality:

$$\|D_{j_1} \dots D_{j_{2k-m}} g\|_\infty \le C(\mathbf{M}) \left\{ \omega^{k-m/2} \|g\|_{L_2(\mathbf{M})} + \omega^{k-m/2+s} \|g\|_{L_2(\mathbf{M})} \right\}$$
$$\le C(\mathbf{M}) \omega^{k-m/2} \left\{ \|g\|_{L_2(\mathbf{M})} + \omega^{s/2} \|g\|_{L_2(\mathbf{M})} \right\}$$
$$= C(\mathbf{M}, g, \omega, s) \omega^{k-m/2}, \quad s > \frac{\dim \mathbf{M}}{2}. \tag{7.12}$$

Finally, we have the following estimate:

$$\|\mathcal{L}^k(fg)\|_{L_2(\mathbf{M})} \le C(\mathbf{M}, f, g, \omega, s)(4d\omega)^k, \quad s > \frac{\dim \mathbf{M}}{2}, \quad k \in \mathbb{N}, \tag{7.13}$$

which leads to our result. The theorem is proved.

Acknowledgments The first author was supported in part by the National Geospatial-Intelligence Agency University Research Initiative (NURI), grant HM1582-08-1-0019.

References

1. P. Alfeld, M. Neamtu, L. L. Schumaker, *Fitting scattered data on sphere-like surfaces using spherical splines*, J. Comput. Appl. Math., 73, (1996), 5–43.
2. S. B. Damelin; J. Levesley; D. L. Ragozin; X. Sun, *Energies, group-invariant kernels and numerical integration on compact manifolds*, J. Complexity 25 (2009), no. 2, 152–162.
3. O. Davydov, ; L. L. Schumaker, *Interpolation and scattered data fitting on manifolds using projected Powell-Sabin splines*, IMA J. Numer. Anal. 28 (2008), no. 4, 785–805.
4. F. Dai, Z. Ditzian, *Jackson inequality for Banach spaces on the sphere*, Acta Math. Hungar. 118 (2008), no. 1–2, 171–195.
5. J.R. Driscoll and D.M. Healy, *Computing Fourier Transforms and Convolutions on the 2-Sphere*, Adv. in Applied Math. 15(1994), 202–250.
6. R. Duffin, A. Schaeffer, *A class of nonharmonic Fourier series*, Trans. AMS, 72, (1952), 341–366.
7. N. Dyn, F.J. Narcovich, and J.D. Ward, *Variational Principles and Sobolev-Type Estimates for generalized Interpolation on a Riemannian Manifold*, Constr. Approx., 15(1999), 175–208.
8. G. E. Fasshauer; L. L. Schumaker, *Scattered data fitting on the sphere. Mathematical methods for curves and surfaces, II* (Lillehammer, 1997), 117–166, Innov. Appl. Math., Vanderbilt Univ. Press, Nashville, TN, 1998.
9. G. Fasshauer, L. L. Schumaker, *Scattered data fitting on the sphere. Mathematical Methods for Curves and Surfaces II* , (M. Dhlen, T. Lyche, L. L. Schumaker eds). Nashville, TN: Vanderbilt University Press, (1998)pp. 117–166.
10. F. Filbir, H. Mhaskar, *A quadrature formula for diffusion polynomials corresponding to a generalized heat kernel*, J. Fourier Anal. Appl. 16 (2010), no. 5, 629–657.
11. W. Freeden, T. Gervens, M. Schreiner, *Constructive approximation on the spheres. With applications to geomathematics*, Numerical Mathematics and Scientific Computation, The Claredon Press, Oxford University Press, New York, 1998.
12. D. Geller and A. Mayeli, *Nearly Tight Frames and Space-Frequency Analysis on Compact Manifolds* (2009), Math. Z. **263** (2009), 235–264.
13. D. Geller and I. Pesenson, *Band-limited localized Parseval frames and Besov spaces on compact homogeneous manifolds*, J. Geom. Anal. 21 (2011), no. 2, 334–371.
14. D. Geller and I. Pesenson, *Kolmogorov and Linear Widths of Balls in Sobolev and Besov Norms on Compact Manifolds*, arXiv:1104.0632v1 [math.FA]
15. T. Hangelbroek, ; F. J. Narcowich; X. Sun; J. D. Ward, *Kernel approximation on manifolds II: the L_∞ norm of the L_2 projector*, SIAM J. Math. Anal. 43 (2011), no. 2, 662–684.
16. S. Helgason, *Differential Geometry and Symmetric Spaces*, Academic, N.Y., 1962.
17. K. Hesse, H.N. Mhaskar, I.H. Sloan, *Quadrature in Besov spaces on the Euclidean sphere*, J. Complexity 23 (2007), no. 4–6, 528–552.
18. Lai, Ming-Jun; Schumaker, Larry L. Spline functions on triangulations. Encyclopedia of Mathematics and its Applications, 110. Cambridge University Press, Cambridge, 2007. xvi+592 pp.
19. J. Levesley; D. L. Ragozin, *Radial basis interpolation on homogeneous manifolds: convergence rates*, Adv. Comput. Math. 27 (2007), no. 2, 237–246.
20. H.N. Mhaskar, F.J. Narcowich, J.D. Ward, *Spherical Marcinkiewicz–Zygmund inequalities and positive quadrature*, Math. Comp. 70 (2001), no. 235, 1113–1130.
21. H. N. Mhaskar; F. J. Narcowich; J. Prestin; J. D. Ward, *Lp-Bernstein estimates and approximation by spherical basis functions*, Math. Comp. 79 (2010), no. 271, 1647–1679.

22. F.J. Narcowich, J.D. Ward, *Scattered data interpolation on spheres: Error estimates and locally supported basis functions*, SIAM J. Math. Anal. 33 (2002) 1393–1410.
23. I. Pesenson, *The Bernstein Inequality in the Space of Representation of a Lie group*, Dokl. Acad. Nauk USSR **313** (1990), 86–90; English transl. in Soviet Math. Dokl. **42** (1991).
24. I. Pesenson, *A sampling theorem on homogeneous manifolds*, Trans. Amer. Math. Soc. **352** (2000), no. 9, 4257–4269.
25. I. Pesenson, E. Grinberg, *Invertion of the spherical Radon transform by a Poisson type formula*, Contemp. Math., 278(2001), 137–147.
26. I. Pesenson, *Variational splines on Riemannian manifolds with applications to integral geometry*, Adv. in Appl. Math. 33 (2004), no. 3, 548–572.
27. I. Pesenson, *An approach to spectral problems on Riemannian manifolds,* Pacific J. of Math. Vol. 215(1), (2004), 183–199.
28. I. Pesenson, *Poincare-type inequalities and reconstruction of Paley–Wiener functions on manifolds,* J. of Geometric Analysis, (4), 1, (2004), 101–121.
29. I. Pesenson, *Bernstein-Nikolski inequality and Riesz interpolation Formula on compact homogeneous manifolds,* J. Approx. Theory, 150, (2008). no. 2, 175–198.
30. I. Pesenson, *Paley–Wiener approximations and multiscale approximations in Sobolev and Besov spaces on manifolds,* J. of Geometric Analysis, 4, (1), (2009), 101–121.
31. E. Saucan; E. Appleboim; Y. Y. Zeevi, *Image projection and representation on* S^n. J. Fourier Anal. Appl. 13 (2007), no. 6, 711–727.
32. G. Wahba, *Surface fitting with scattered noisy data on Euclidean d-space and on the sphere,* Rocky Mountain J. Math., 14(1984), 281–299.
33. W. zu Castell, F. Filbir, *Radial basis functions and corresponding zonal series expansions on the sphere*, J. Approx. Theory, 134 (2005), pp. 65–79.

The Moment Zeta Function and Applications

Igor Rivin

Abstract Motivated by a probabilistic analysis of a simple game (itself inspired by a problem in computational learning theory), we introduce the *moment zeta function* of a probability distribution and study in depth some asymptotic properties of the moment zeta function of those distributions supported in the interval [0, 1]. One example of such zeta functions is Riemann's zeta function (which is the moment zeta function of the uniform distribution in [0, 1]. For Riemann's zeta function, we are able to show particularly sharp versions of our results.

Key words Asymptotics • Learning theory • Zeta functions

Mathematics Subject Classification (2010): 60E07, 60F15, 60J20, 91E40, 26C10

Introduction

Consider the following setup: (Ω, μ) is a space with a probability measure μ, and $\omega_1, \ldots, \omega_n$ is a collection of measurable subsets of Ω, with $\mu(\omega_i) = p_i$. We play a game as follows: The jth step consists of picking a point $x_j \in \Omega$ at random, so that after k steps we have the set $X_k = \{x_1, \ldots, x_k\}$. The game is considered to be over when

$$\forall i \leq n, \quad X_k \cap \omega_i \neq X_k.$$

I. Rivin (✉)
Mathematics Department, Temple University, Philadelphia, PA 19122, USA

H.M. Farkas et al. (eds.), *From Fourier Analysis and Number Theory to Radon Transforms and Geometry*, Developments in Mathematics 28, DOI 10.1007/978-1-4614-4075-8_22,

We consider the duration of our game to be a random variable $T = T(p_1, \dots, p_n)$, and wish to compute the expectation $E(p_1, \dots, p_n)$ of T. This cannot, in general, be done without knowing the measures $p_{i_1 i_2 \dots i_k} = \mu(\omega_{i_1} \cap \omega_{i_2} \cap \dots \cap \omega_{i_k})$, and in the sequel we will introduce the

Independence Hypothesis:

$$p_{i_1 i_2 \dots i_k} = p_{i_1} \times \dots \times p_{i_k}.$$

Estimates without using the independence hypothesis are shown in the companion paper [8].

We now assume further that we do not actually know the measures $p_1, \dots, p_n$, but know that they themselves are a sample from some (known) probability distribution $\mathcal{F}$, of necessity supported in $[0, 1]$. We consider $E(p_1, \dots, p_n) \stackrel{\text{def}}{=} E((p))$ as our random variable, and we wish to compute *its* expectation (over the space of all n-element samples from $\mathcal{F}$), and in particular we are interested in the limiting situation when n is large.

Under the independence assumption, it turns out that we can write (Lemma 1.3):

$$E(\mathbf{p}) = \sum_{\mathbf{s} \subseteq \{1, \dots, n\}} (-1)^{|s|-1} \left(\frac{1}{1 - p_{\mathbf{s}}} - 1 \right), \tag{1}$$

where if $\mathbf{s} = \{i_1, \dots, i_k\}$, we write $p_{\mathbf{s}} = p_{i_1} \times \dots \times p_{i_k}$. To use (1) to understand the statistical behavior of T, we must introduce the *moment zeta function* of the probability distribution $\mathcal{F}$, defined as follows:

Definition A *Let $m_k = \int_0^1 x^k \mathrm{d}\mathcal{F}$ be the kth moment of $\mathcal{F}$. Then*

$$\zeta_{\mathcal{F}}(s) = \sum_{k=1}^{\infty} m_k^s;$$

The sum in the definition above obviously converges only in some half-plane $\Re s > s_0$; the function can be analytically continued, but in the sequel, we will be interested in asymptotic results for s a large real number, so this will not use complex variable methods at all.

The relevance of this to our questions comes from Lemma 2.2, which we restate for convenience as

Lemma B *Let $\mathcal{F}$ be a probability distribution as above, and let $x_1, \dots, x_n$ be independent random variables with common distribution $\mathcal{F}$. Then*

$$\mathbb{E}\left(\frac{1}{1 - x_1 \dots x_n} \right) = \zeta_{\mathcal{F}}(n). \tag{2}$$

In particular, the expectation is undefined whenever the zeta function is undefined.

Now, we can write (using Lemma B) the following *formal* identity:

$$\mathbb{E}(T) = -\sum_{k=1}^{n}(-1)^k \binom{n}{k} \zeta_{\mathcal{F}}(k). \tag{3}$$

The identity is only formal, because $\zeta_{\mathcal{F}}(k)$ is not necessarily defined for all positive integers k. It *is* defined for all positive integers k when $\mathcal{F}([1-x,1]) \sim x^{\alpha}$, for $\alpha > 1$—this case is analyzed in Sect. 3. If $\alpha = 1$ (we will not deal with the case $\alpha < 1$ in this chapter; see [8]), we write

$$T = T_1 - T',$$

where

$$T_1 = \sum_{i=1}^{n} \frac{1}{1-p_i}.$$

T_1 has infinite expectation, but as n goes to ∞, T_1/n does converge in distribution to a stable law of exponent 1 (see [4] and [3] for many related results). The variable T' does possess a finite expectation, given by

$$\mathbb{E}(T') = \sum_{k=2}^{n}(-1)^k \binom{n}{k} \zeta_{\mathcal{F}}(k). \tag{4}$$

The expressions given by (3) and (4) are analyzed in Sects. 3 and 4, and the following theorems are shown:

Theorem C (Thm. 3.5) *Let $\mathcal{F}$ be a continuous distribution supported on* $[0,1]$, *and let f be the density of $\mathcal{F}$. Suppose further that*

$$\lim_{x\to 1} \frac{f(x)}{(1-x)^{\beta}} = c,$$

for $\beta, c > 0$. Then,

$$\begin{aligned}\lim_{n\to\infty} n^{-\frac{1}{1+\beta}} &\left[\sum_{k=1}^{n} \binom{n}{k} (-1)^k \zeta_{\mathcal{F}}(k)\right] \\ &= -\int_0^{\infty} \frac{1-\exp\left(-c\Gamma(\beta+1)u^{1+\beta}\right)}{u^2}\, du \\ &= -\left(c\Gamma(\beta+1)\right)^{\frac{1}{\beta+1}} \Gamma\left(\frac{\beta}{\beta+1}\right).\end{aligned}$$

and

Theorem D (Thm. 4.8) *Let $\mathcal{F}$ be a continuous distribution supported on* $[0, 1]$, *and let f be the density of $\mathcal{F}$. Suppose further that*

$$\lim_{x\to 1} \frac{f(x)}{(1-x)} = c > 0.$$

Then,

$$\sum_{k=2}^{n} \binom{n}{k} (-1)^k \zeta_{\mathcal{F}}(k) \sim cn \log n.$$

To get error estimates, we need stronger assumption on the function f than the weakest possible assumption made in Theorem 4.8. The proof of the below follows by modifying slightly the proof of Lemma 4.7:

Theorem E (Thm. 4.9) *Let $\mathcal{F}$ be a continuous distribution supported on* $[0, 1]$, *and let f be the density of $\mathcal{F}$. Suppose further that*

$$f(x) \sim c(1-x) + O\left((1-x)^{\delta}\right),$$

where $\delta > 0$. Then,

$$\sum_{k=2}^{n} \binom{n}{k} (-1)^k \zeta_{\mathcal{F}}(k) \sim cn \log n + O(n).$$

Our original probabilistic problem is thus completely resolved, but the sums given by (3) and (4) are interesting in and of itself, and, with some more work (Sect. 5), we can considerably strengthen the results above as follows for the Riemann zeta function and its scaling:

Theorem F (Thm. 5.1)

$$\sum_{k=2}^{n} \binom{n}{k} (-1)^k \zeta(k) \sim n \log n + (2\gamma - 1)n + O\left(\frac{1}{n}\right),$$

where ζ is the Riemann zeta function and γ is Euler's constant.

Theorem G (Thm. 5.2) *Let $s > 1$, and then*

$$\sum_{k=1}^{n} \binom{n}{k} (-1)^k \zeta(sk) \sim \Gamma\left(1 - \frac{1}{s}\right) n^{\frac{1}{s}}.$$

It should be remarked that using the methods of Sect. 5, higher-order terms in the asymptotics can be obtained, if desired, but they seem to be of more limited interest.

1 A Formula for the Winning Time T

An application of the inclusion–exclusion principle gives us the following:

Lemma 1.1. *The probability l_k that we have won after k steps is given by*

$$l_k = \prod_{i=1}^{n}(1 - p_i^k).$$

Note that the probability s_k of winning the game *on the kth step* is given by $s_k = l_k - l_{k-1} = (1 - l_{k-1}) - (1 - l_k)$. Since the expected number of steps T is given by

$$E(T) = \sum_{k=1}^{\infty} k s_k,$$

we immediately have

$$T = \sum_{k=1}^{\infty}(1 - l_k).$$

Lemma 1.2.

$$E(T) = \sum_{k=1}^{\infty}\left(1 - \prod_{i=1}^{n}(1 - p_i^k)\right). \tag{5}$$

Since the sum above is absolutely convergent, we can expand the products and interchange the order of summation to get the formula (6) for $E(T)$:

Notation. Below, we identify subsets of $\{1, \ldots, n\}$ with multindexes (in the obvious way), and if $s = \{i_1, \ldots, i_l\}$, then

$$p_s \stackrel{\text{def}}{=} p_{i_1} \cdots p_{i_l}.$$

Lemma 1.3. *The expression (5) can be rewritten as*

$$E(T) = \sum_{s \subseteq \{1,\ldots,n\}} (-1)^{|s|-1}\left(\frac{1}{1 - p_s} - 1\right). \tag{6}$$

Proof. With notation as above,

$$\prod_{i=1}^{m}(1 - p_i^k) = \sum_{s \subseteq \{1,\ldots,n\}} (-1)^{|s|} p_s^k,$$

so

$$\begin{aligned} E(T) &= \sum_{k=1}^{\infty}\left(1-\prod_{i=1}^{n}(1-p_i^k)\right) \\ &= \sum_{k=1}^{\infty}\left(1-\sum_{s\subseteq\{1,\dots,n\}}(-1)^{|s|}p_s^k\right) \\ &= \sum_{s\subseteq\{1,\dots,n\}}(-1)^{|s|-1}\sum_{k=1}^{\infty}p_s^k \\ &= \sum_{s\subseteq\{1,\dots,n\}}(-1)^{|s|-1}\left(\frac{1}{1-p_s}-1\right), \end{aligned}$$

where the change in the order of summation is permissible since all sums converge absolutely.

Formula (6) is useful in and of itself, but we now use it to analyze the statistical properties of the time of success T under our distribution and independence assumptions. For this, we shall need to study the *moment zeta function* of a probability distribution, introduced below.

2 Moment Zeta Function

Definition 2.1. Let $\mathcal{F}$ be a probability distribution on a (possibly infinite) interval I, and let $m_k(\mathcal{F}) = \int_I x^k \mathcal{F}(\mathrm{d}x)$ be the kth moment of $\mathcal{F}$. Then the *moment zeta function of* $\mathcal{F}$ is defined to be

$$\zeta_{\mathcal{F}}(s) = \sum_{k=1}^{\infty} m_k^s(\mathcal{F}),$$

whenever the sum is defined.

The definition is, in a way, motivated by the following:

Lemma 2.2. *Let $\mathcal{F}$ be a probability distribution as above, and let $x_1, \dots, x_n$ be independent random variables with common distribution $\mathcal{F}$. Then*

$$\mathbb{E}\left(\frac{1}{1-x_1\dots x_n}\right) = \zeta_{\mathcal{F}}(n). \tag{7}$$

In particular, the expectation is undefined whenever the zeta function is undefined.

Proof. Expand the fraction in a geometric series and apply Fubini's theorem.

Example 2.3. For $\mathcal{F}$ the uniform distribution on [0, 1], $\zeta_{\mathcal{F}}$ is the familiar Riemann zeta function.

Our first observation is that for distributions supported in [0, 1], the asymptotics of the moments are determined by the local properties of the distribution at $x = 1$. To show this, first recall that the *Mellin transform* of f is defined to be

$$\mathcal{M}(f)(s) = \int_0^1 f(x)x^{s-1}\mathrm{d}x.$$

Mellin transform is closely related to thc Laplace transform. Making the substitution $x = \exp(-u)$, we see that

$$\mathcal{M}(f) = \int_0^\infty f(\exp(-u))\exp(-su)\,\mathrm{d}u,$$

so the Mellin transform of f is equal to the Laplace transform of $f \circ \exp$, where $\circ$ denotes functional composition.

The following observation is both obvious and well-known:

Lemma 2.4. $m_k(\mathcal{F}) = \mathcal{M}(f)(k+1)$.

It follows that computing the asymptotic behavior of the kth moment of $\mathcal{F}$ as a function of k reduces to calculating the large s asymptotics of the Mellin transform, which is tantamount to computing the asymptotics of the Laplace transform of $f \circ \exp$.

Theorem 2.5. *Let $\mathcal{F}$ be a continuous distribution supported in* [0, 1], *let f be the density of the distribution $\mathcal{F}$, and suppose that $f(1-x) = cx^\beta + O(x^{\beta+\delta})$, for some $\delta > 0$. Then the kth moment of $\mathcal{F}$ is asymptotic to $Ck^{-(1+\beta)}$, for $C = c\Gamma(\beta+1)$.*

Proof. The asymptotics of the Laplace transform are easily computed by Laplace's method, and in the case we are interested in, Watson's lemma (see, e.g., [1]) tells us that if $f(x) \asymp c(1-x)^\beta$, then $\mathcal{M}(f)(s) \asymp c\Gamma(\beta+1)s^{-(\beta+1)}$.

Corollary 2.6. *Under the assumptions of Theorem 2.5, $\zeta_{\mathcal{F}}(s)$ is defined for $s > 1/(1+\beta)$.*

We will need another observation:

Lemma 2.7. *For $\mathcal{F}$ supported in* [0, 1], *$m_k(\mathcal{F})$ is monotonically decreasing as a function of k.*

Proof. Immediate.

Below, we shall analyze three cases. In the sequel, we set $\alpha = \beta + 1$.

3 $\alpha > 1$

In this case, we use our assumptions to rewrite (6) as

$$\mathbb{E}(T) = -\sum_{k=1}^{n} \binom{n}{k} (-1)^k \zeta_{\mathcal{F}}(k). \tag{8}$$

This, in turn, can be rewritten (by expanding the definition of zeta) as

$$\mathbb{E}(T) = -\sum_{j=1}^{\infty} \left[(1 - m_j(\mathcal{F}))^n - 1\right] = \sum_{j=1}^{\infty} \left[1 - (1 - m_j(\mathcal{F}))^n\right]. \tag{9}$$

Since the terms in the sum are monotonically decreasing (as a function of j) by Lemma 2.7, the sum in (9) can be approximated by an integral of *any* monotonic interpolation m of the sequence $m_j(\mathcal{F})$—we will interpolate by $m(x) = \mathcal{M}(f)(x+1))$. The error of such an approximation is bounded by the first term, which, in turn, is bounded in absolute value by 2, to get

$$T = -\int_1^{\infty} \left[(1 - m(x))^n - 1\right] \mathrm{d}x + O(1), \tag{10}$$

where the error term is bounded above by 2. We shall write

$$T_0 = -\int_1^{\infty} \left[(1 - m(x))^n - 1\right] \mathrm{d}x.$$

Now, let us assume that

$$\lim_{x \to \infty} x^{\alpha} m(x) = L, \tag{11}$$

for some $\alpha > 1$. We substitute $x = n^{1/\alpha}/u$, to get

$$T_0 = n^{\frac{1}{\alpha}} \int_0^{n^{\frac{1}{\alpha}}} \frac{\left[1 - \left(1 - m(n^{1/\alpha}/u)\right)^n\right]}{u^2}\, du = n^{\frac{1}{\alpha}} \left[I_1(n) + I_2(n)\right],$$

where

$$I_1(n) = \int_0^{n^{\frac{1}{3\alpha}}} \frac{\left[1 - (1 - m(n^{1/\alpha}/u)^n\right]}{u^2}\, du,$$

and

$$I_2(n) = \int_{n^{\frac{1}{3\alpha}}}^{n^{\frac{1}{\alpha}}} \frac{\left[1 - (1 - m(n^{1/\alpha}/u)^n\right]}{u^2}\, du.$$

We will need the following:

Lemma 3.1. *Let $f_n(x) = (1 - x/n)^n$, and let $0 \le x < 1/2$.*

$$f_n(x) = \exp(-x)\left[1 - \frac{x^2}{2n} + O\left(\frac{x^3}{n^2}\right)\right].$$

Proof. Note that

$$\log f_n(x) = n\log(1 - x/n) = -x - \sum_{k=2}^{\infty} \frac{x^k}{kn^{k-1}}.$$

The assertion of the lemma follows by exponentiating the two sides of the above equation.

Lemma 3.2.

$$\lim_{n\to\infty} n^{\frac{1}{\alpha}} I_2(n) = 0.$$

Proof. The integrand of $I_2(n)$ is monotonically decreasing, and so

$$I_2(n) \le n^{-\frac{2}{3\alpha}}\left[1 - \left(1 - m\left(n^{\frac{2}{3\alpha}}\right)\right)^n\right].$$

By our assumption (11) and by Lemma 3.1, we see that the right-hand side goes to zero (exponentially fast).

Lemma 3.3.

$$\lim_{n\to\infty} I_1(n) = \int_0^{\infty} \frac{1 - \exp(-Lu^{\alpha})}{u^2}\, du.$$

Proof. Immediate from (11) and Lemma 3.1. Note that the integral converges when α is greater than 1.

Remark 3.4.

$$\int_0^{\infty} \frac{1 - \exp(-Lu^{\alpha})}{u^2}\, du = L^{\frac{1}{\alpha}}\Gamma\left(\frac{\alpha - 1}{\alpha}\right).$$

Proof.

$$\int_0^{\infty} \frac{1 - \exp(-Lu^{\alpha})}{u^2}\, du = \lim_{\epsilon\to 0}\left[\frac{1}{\epsilon} - \int_{\epsilon}^{\infty} \frac{\exp(-Lu^{\alpha})}{u^2}\, du\right].$$

To prove the remark, we need to analyze the behavior of the integral above as $\epsilon \to 0$. First, we change variables: $v = Lu^{\alpha}$. Then,

$$\int_{\epsilon}^{\infty} \frac{\exp(-Lu^{\alpha})}{u^2}\, du = \frac{L^{1/\alpha}}{\alpha}\int_{L\epsilon^{\alpha}}^{\infty} \exp(-v)v^{-(1+1/\alpha)} dv.$$

Integrating by parts, get

$$\int_{L\epsilon^\alpha}^{\infty} \exp(-v)v^{-(1+1/\alpha)}\mathrm{d}v = -\alpha \exp(-v)v^{1/\alpha}\Big|_{L\epsilon^\alpha}^{\infty} - \alpha \int_{L\epsilon^\alpha} \exp(-v)v^{-1/\alpha}.$$

Since $1/\alpha < 1$, $\int_0^\infty \exp(-v)v^{-1/\alpha}\mathrm{d}v = \Gamma(1 - 1/\alpha)$, from which the assertion of the remark follows.

We summarize as follows:

Theorem 3.5. *Let $\mathcal{F}$ be a continuous distribution supported on* $[0, 1]$, *and let f be the density of $\mathcal{F}$. Suppose further that*

$$\lim_{x\to 1} \frac{f(x)}{(1-x)^\beta} = c,$$

for $\beta, c > 0$. Then,

$$\begin{aligned}
&\lim_{n\to\infty} n^{-\frac{1}{1+\beta}} \left[\sum_{k=1}^{n} \binom{n}{k}(-1)^k \zeta_{\mathcal{F}}(k)\right] \\
&= -\int_0^\infty \frac{1-\exp\left(-c\Gamma(\beta+1)u^{1+\beta}\right)}{u^2}\,du \\
&= -(c\Gamma(\beta+1))^{\frac{1}{\beta+1}}\,\Gamma\left(\frac{\beta}{\beta+1}\right).
\end{aligned}$$

Proof. The assertion follows from Lemmas 3.3 and 3.2 together with Theorem 2.5 and Remark 3.4.

4 $\alpha = 1$

In this case,

$$f(x) = L + o(1) \tag{12}$$

as x approaches 1, and so Theorem 2.5 tells us that

$$\lim_{j\to\infty} j m_j(\mathcal{F}) = L. \tag{13}$$

It is not hard to see that $\zeta_{\mathcal{F}}(n)$ is defined for $n \geq 2$. We break up the expression in (6) as

$$T = \sum_{j=1}^{n} \frac{1}{1-p_j} - 1 + \sum_{s\subseteq\{1,\dots,n\},\quad |s|>1} (-1)^{|s|-1}\left(\frac{1}{1-p_s} - 1\right). \tag{14}$$

Let

$$T_1 = \sum_{j=1}^{n} \frac{1}{1 - p_j} - 1,$$

$$T_2 = \sum_{s \subseteq \{1,\dots,n\}, \quad |s|>1} (-1)^{|s|-1} \left(\frac{1}{1 - p_s} - 1 \right).$$

The first sum T_1 has infinite expectation; however, T_1/n does have a stable distribution centered on $c \log n + c_2$. We will keep this in mind, but now let us look at the second sum T_2. It can be rewritten as

$$T_2(n) = -\sum_{j=1}^{\infty} \left[\left(1 - m_j(\mathcal{F})\right)^n - 1 + n m_j(\mathcal{F}) \right]. \tag{15}$$

Lemma 4.1. *The quantity* $y_j = \left(1 - m_j(\mathcal{F})\right)^n - 1 + n m_j(\mathcal{F})$ *is a monotonic function of* j.

Proof. We know that $m_j(\mathcal{F})$ is a monotonically decreasing positive function of j, and that $m_0(\mathcal{F}) = 1$. It is sufficient to show that the function $g_n(x) = (1-x)^n + nx$ is monotonic for $x \in (0, 1]$. We compute

$$\frac{\mathrm{d}g_n(x)}{\mathrm{d}x} = n \left(1 - (1 - x)^{n-1}\right) > 0,$$

for $x \in (0, 1)$.

Lemma 4.1 allows us to use the same method as in Sect. 3 under the assumption that the kth moment is asymptotic to k^α (this time for $\alpha \le 1$). Since the term y_j is bounded above by a constant times n, we can write

$$T_2(n) = S_2(n) + O(n), \tag{16}$$

where

$$S_2(n) = n \int_0^n \frac{[1 - nm(n/u) - (1 - m(n/u)^n]}{u^2} \,\mathrm{d}u. \tag{17}$$

Remark 4.2. The error term in (16) above can be improved in the case where $\mathcal{F}$ is the uniform distribution on $[0, 1]$, in which case $m_j = 1/j$. In that case $T_2(n) = S_2(n) - \gamma n + O(1)$, where γ is Euler's constant.

Proof. In this case, we write

$$T_2(n) = \lim_{k\to\infty} -\sum_{j=1}^{k} \left[(1-m_j(\mathcal{F}))^n - 1 + nm_j(\mathcal{F})\right]$$

$$= -\sum_{j=1}^{k} \left[(1-m_j(\mathcal{F}))^n - 1\right] - n\sum_{j=1}^{k} m_j.$$

The terms in the first sum are decreasing, so the first sum can be approximated by an integral with total error $O(1)$. As for the second sum, since $m_j = 1/j$, it is well-known (e.g., Euler–Maclaurin summation) that

$$\sum_{j=1}^{k} \frac{1}{j} = \int_1^k \frac{\mathrm{d}x}{x} + \gamma + O\left(\frac{1}{k}\right),$$

from which the assertion of the remark follows.

To understand the asymptotic behavior of $S_2(n)$, we write

$$S_2(n) = n\left[I_1(n) + I_2(n) + I_3(n) + I_4(n)\right],$$

where

$$I_1(n) = \int_0^1 \frac{\left[1 - nm(n/u) - \left(1 - m\left(\frac{n}{u}\right)\right)^n\right]}{u^2}\, du, \tag{18}$$

$$I_2(n) = \int_1^{n^{\frac{1}{3}}} \frac{\left[1-\left(1-m\left(\frac{n}{u}\right)\right)^n\right]}{u^2}\, du, \tag{19}$$

$$I_3(n) = \int_{n^{\frac{1}{3}}}^{n} \frac{\left[1-\left(1-m\left(\frac{n}{u}\right)\right)^n\right]}{u^2}\, du, \tag{20}$$

$$I_4(n) = -n\int_1^n \frac{m(n/u)}{u^2}\, du. \tag{21}$$

Lemma 4.3.

$$\lim_{n\to\infty} I_1(n) = \int_0^1 \frac{1 - \exp(-Lu) - Lu}{u^2}\, du.$$

Proof. Immediate from the estimate (13) and Lemma 3.1.

Lemma 4.4.

$$\lim_{n\to\infty} I_2(n) = \int_1^\infty \frac{1 - \exp(-Lu)}{u^2}\, du.$$

Proof. Again, immediate from (13) and Lemma 3.1.

Remark 4.5.

$$\int_0^1 \frac{1-\exp(-Lu)-Lu}{u^2} + \int_1^\infty \frac{1-\exp(-Lu)}{u^2} = L(1-\gamma-\log L),$$

where γ is Euler's constant.

Proof.

$$\begin{aligned}
&\int_0^1 \frac{1-\exp(-Lu)-Lu}{u^2}\,du + \int_1^\infty \frac{1-\exp(-Lu)}{u^2}\,du \\
&= \lim_{\epsilon\to 0}\left\{-\int_\epsilon^\infty \left[\frac{\exp(-Lu)}{u^2}+\frac{1}{u^2}\right] du - L\int_\epsilon^1 \frac{du}{u}\right\} \\
&= \lim_{\epsilon\to 0}\left\{\frac{1}{\epsilon} + L\log\epsilon - \int_\epsilon^\infty \frac{\exp(-Lu)}{u^2}\,du\right\}. \qquad (22)
\end{aligned}$$

To evaluate the last limit, we need to compute the expansion as $\epsilon \to 0$ of the last integral. Changing variables $v = Lu$, we get

$$\begin{aligned}
\int_\epsilon^\infty \frac{\exp(-Lu)}{u^2}\,du &= L\int_{L\epsilon}^\infty \frac{\exp(-v)}{v^2}dv \\
&= L\left[-\left.\frac{\exp(-v)}{v}\right|_{L\epsilon}^\infty - \int_{L\epsilon}^\infty \frac{\exp(-v)}{v}dv\right] \\
&= L\left[\frac{\exp(-v)}{L\epsilon} - \exp(-v)\log(v)|_{L\epsilon}^\infty - \int_{L\epsilon}^\infty \exp(-v)\log(v)dv\right] \\
&= \frac{\exp(-L\epsilon)}{\epsilon} + L\exp(-L\epsilon)\log(\epsilon) \\
&\quad + L\log L\exp(-L\epsilon) - L\int_{L\epsilon}^\infty \exp(-v)\log(v)dv.
\end{aligned}$$

Substituting into (22), we get

$$\begin{aligned}
&\int_0^1 \frac{1-\exp(-Lu)-Lu}{u^2}\,du + \int_1^\infty \frac{1-\exp(-Lu)}{u^2}\,du \\
&= \lim_{\epsilon\to 0}\left\{\frac{1-\exp(-L\epsilon)}{\epsilon} + L(1-\exp(-L\epsilon))\log\epsilon \right.\\
&\qquad\left. - L\log L\exp(-L\epsilon) + \int_{L\epsilon}^\infty \exp(-v)\log v dv\right\} \\
&= L\left(1-\log L + \int_{L\epsilon}^\infty \exp(-v)\log v dv\right).
\end{aligned}$$

Since $\int_0^\infty \log(x)\exp(-x)dx = -\gamma$, the result follows.

Lemma 4.6.

$$\lim_{n\to\infty} nI_3(n) = 0.$$

Proof. See the proof of Lemma 3.2.

Lemma 4.7.

$$\lim_{n\to\infty} -\frac{I_4(n)}{\log n} = L.$$

Proof. We shall show that the limit in question lies between $(1-\epsilon)L$ and $(1+\epsilon)L$, for any $\epsilon > 0$, from which the conclusion of the lemma obviously follows. To do that, pick C, such that

$$1-\epsilon/4 \leq xm(x) \leq 1+\epsilon/4$$

for $x > C$. Now, write

$$\int_1^n \frac{m(n/u)}{u^2}\,\mathrm{d}u = J_1(n) + J_2(n),$$

where

$$J_1(n) = \int_1^{\frac{n}{C}} \frac{m(n/u)}{u^2}\,\mathrm{d}u \tag{23}$$

$$J_2(n) = \int_{\frac{n}{C}}^n \frac{m(n/u)}{u^2}\,\mathrm{d}u. \tag{24}$$

Observe that

$$0 < J_2(n) = \frac{1}{n}\int_1^C m(x)\mathrm{d}x \leq \frac{C-1}{n},$$

while

$$\frac{1-\epsilon/4}{n}\int_1^{\frac{n}{C}} \frac{\mathrm{d}u}{u} \leq J_1(n) \leq \frac{1+\epsilon/4}{n}\int_1^{\frac{n}{C}} \frac{\mathrm{d}u}{u},$$

so

$$\frac{(1-\epsilon/4)}{n}(\log n - \log C) \leq J_1(n) \leq \frac{(1+\epsilon/4)}{n}(\log n - \log C).$$

If we now pick $N = C^{4/\epsilon}$, it is clear that for $n > N$,

$$(1-\epsilon/2)\log n \leq J_1(n) \leq (1+\epsilon/2)\log n,$$

while J_2 is bounded above in absolute value by $C^{1-4/\epsilon}$.

The above lemmas can be summarized in the following:

Theorem 4.8. *Let $\mathcal{F}$ be a continuous distribution supported on* $[0, 1]$, *and let f be the density of $\mathcal{F}$. Suppose further that*

$$\lim_{x \to 1} \frac{f(x)}{(1-x)} = c > 0.$$

Then,

$$\sum_{k=2}^{n} \binom{n}{k} (-1)^k \zeta_{\mathcal{F}}(k) \sim cn \log n.$$

To get error estimates, we need stronger assumption on the function f than the (weakest possible) assumption made in Theorem 4.8. The proof of the below follows by modifying slightly the proof of Lemma 4.7:

Theorem 4.9. *Let $\mathcal{F}$ be a continuous distribution supported on* $[0, 1]$, *and let f be the density of $\mathcal{F}$. Suppose further that*

$$f(x) \sim c(1-x) + O\left((1-x)^{\delta}\right),$$

where $\delta > 0$. Then,

$$\sum_{k=2}^{n} \binom{n}{k} (-1)^k \zeta_{\mathcal{F}}(k) \sim cn \log n + O(n).$$

5 Riemann Zeta Function

The proof of the key Lemma 4.7 is trivial in the case where $f(x) = 1$, and so $\zeta_{\mathcal{F}}$ is the Riemann zeta function. In that case, however, we get the following much stronger result:

Theorem 5.1.

$$\sum_{k=2}^{n} \binom{n}{k} (-1)^k \zeta(k) \sim n \log n + (2\gamma - 1)n + O\left(\frac{1}{n}\right),$$

where ζ is the Riemann zeta function and γ is Euler's constant.

It should also be noted that the results of Sect. 3 immediately imply the following:

Theorem 5.2. *Let $s > 1$, then*

$$\sum_{k=1}^{n} \binom{n}{k} (-1)^k \zeta(sk) \sim \Gamma\left(1 - \frac{1}{s}\right) n^{\frac{1}{s}}.$$

To prove Theorem 5.1, we need to sharpen some of the estimates of the preceding section. First:

Lemma 5.3. *Let the notation be as in the preceding section. When* $m(x) = \frac{1}{x}$,

$$I_1(n) = \int_0^1 \frac{1 - \exp(-u) - u}{u^2}\, du + \frac{1}{2n} \int_0^1 \exp(-u)\, du + O\left(\frac{1}{n^2}\right), \quad (25)$$

$$I_2(n) = \int_1^\infty \frac{1 - \exp(-u)}{u^2}\, du + \frac{1}{2n} \int_1^\infty \exp(-u)\, du. \quad (26)$$

Proof. Immediate from the expansion in Lemma 3.1.

We can also sharpen the statement of Lemma 4.6:

Lemma 5.4.

$$\lim_{n \to \infty} n^k I_3(n) = 0,$$

for any k.

Proof. This statement holds in general, and no change in argument is necessary.

In the case where $m(x) = 1/x$, Lemma 4.7 is immediate, and has no error term:

Lemma 5.5.

$$I_4(n) = -\log(n).$$

Proof. Immediate.

We now have the following:

Theorem 5.6.

$$\sum_{k=2}^{n} \binom{n}{k} (-1)^k \zeta(k) \sim n \log n + (2\gamma - 1)n + O(1),$$

Proof. Lemmas 5.3–5.5, combined with Remark 4.2.

Remark 5.7. A statement of a similar flavor can be found in [7, 262.1–2]

To improve the error term from that in Theorem 5.6, it is necessary to sharpen the estimate in Remark 4.2 to the following:

Theorem 5.8. *With the notation of Remark 4.2,*

$$T_2(n) = S_2(n) - \gamma n - \frac{1}{2} + O\left(\frac{1}{n}\right).$$

Proof. The theorem will follow immediately from Lemma 5.9 and the results of Sect. 5.1.

Lemma 5.9.

$$\lim_{N\to\infty}\sum_{j=1}^{N}\frac{1}{j}-\log(n)=\gamma.$$

Proof. Well-known.

5.1 A Sum and an Integral

Let

$$\begin{aligned}S_n(N)&=\sum_{j=1}^{N}\left(1-\frac{1}{j}\right)^n,\\ I_n(N)&=\int_1^N\left(1-\frac{1}{x}\right)^n\mathrm{d}x,\\ D_n(N)&=S_n(N)-I_n(N),\\ D_n&=\lim_{N\to\infty}D_n(N).\end{aligned}$$

In this section, we shall prove the following result:

Theorem 5.10.

$$D_n=\frac{1}{2}+o\left(\frac{1}{n}\right).$$

We will need the following preliminary results:

Lemma 5.11. *Let f be a C^1 function defined on $[0,\infty)$. Then*

$$\sum_{k=0}^{N}f(k)=\frac{1}{2}\left[f(0)+f(N)\right]+\int_0^N f(t)\mathrm{d}t+\int_0^N\left(\{t\}-\frac{1}{2}\right)f'(t)\mathrm{d}t.$$

Proof. Integration by parts—see Exercises for Sect. 6.7 of [1].

Lemma 5.12. *Let f be a C^2 function defined on $[0,\infty)$, such that f'' is bounded, and $f''(x)=O(1/x^2)$. Then*

$$\left|\sum_{k=0}^{\infty}\int_{\frac{k}{n}}^{\frac{k+1}{n}}\left(x-\frac{k+\frac{1}{2}}{n}\right)f(x)\mathrm{d}x-\frac{1}{4n^3}\sum_{k=0}^{\infty}f'\left(\frac{k+\frac{1}{2}}{n}\right)\right|=o\left(\frac{1}{n^4}\right).$$

Proof. On the interval $[k/n, (k+1)/n]$, we can write

$$f(x) = f\left(\frac{k+\frac{1}{2}}{n}\right) + f'\left(\frac{k+\frac{1}{2}}{n}\right)\left(x - \frac{k+\frac{1}{2}}{n}\right) + R_2(x), \qquad (27)$$

where, by Taylor's theorem, $|R_2(x)| \leq x^2 \max_{x\in[k/n,(k+1)/n]} f''(x)$. The assertion of the lemma then follows by integration of (27).

Lemma 5.13. *Under the assumptions of Lemma 5.12, together with the assumption that f and all of its derivatives vanish at* 0

$$\left|\sum_{k=0}^{\infty} f'\left(\frac{k+\frac{1}{2}}{n}\right)\right| = O\left(\frac{1}{n}\right).$$

Proof. Let $g(y) = f'((x+1/2)/n)$. Then,

$$\begin{aligned}
\sum_{k=0}^{\infty} f'\left(\frac{k+\frac{1}{2}}{n}\right) &= \sum_{k=0}^{\infty} g(k) \\
&= \frac{1}{2}g(0) + \int_0^{\infty} g(x)\mathrm{d}x + \int_0^{\infty}\left(\{x\} - \frac{1}{2}\right) g'(x)\mathrm{d}x \\
&= \frac{1}{2}f'\left(\frac{1}{2n}\right) + \int_0^{\infty} f'\left(\frac{x+\frac{1}{2}}{n}\right)\mathrm{d}x + O\left(\frac{1}{n}\right) \\
&= n\int_{\frac{1}{2n}}^{\infty} f'(x)\mathrm{d}x + O\left(\frac{1}{n}\right) \\
&= O\left(\frac{1}{n}\right).
\end{aligned}$$

Now we proceed to the proof of Theorem 5.10. First:

Lemma 5.14.

$$D_n = \frac{1}{2} + n\int_1^{\infty} \frac{\left(\{x\} - \frac{1}{2}\right)\left(1 - \frac{1}{x}\right)^{n-1}}{x^2}\mathrm{d}x.$$

Proof. Immediate corollary of Lemma 5.11.

Proof (Proof of Theorem 5.10). By Lemma 5.14, it remains to analyze the asymptotic behavior of

$$J_n = (n+1)\int_1^{\infty} \frac{\left(\{x\} - \frac{1}{2}\right)\left(1 - \frac{1}{x}\right)^{n}}{x^2}\mathrm{d}x.$$

(the expression occurring in Lemma 5.14 is actually J_{n-1}; we have changed the variable for notational convenience). First, we make the substitution $x = ny$, to get

$$J_n = \frac{(n+1)}{n} \underbrace{\int_{\frac{1}{n}}^{\infty} \frac{\left(\{ny\} - \frac{1}{2}\right)\left(1 - \frac{1}{ny}\right)^n}{y^2} \mathrm{d}x}_{K_n},$$

where clearly $J_n \sim K_n$. We now write

$$K_n = \left[\underbrace{\int_{\frac{1}{n}}^{n^{-\frac{1}{3}}}}_{K_n'} + \underbrace{\int_{n^{-\frac{1}{3}}}^{\infty}}_{K_n''} \right] \frac{\left(\{ny\} - \frac{1}{2}\right)\left(1 - \frac{1}{ny}\right)^n}{y^2} \mathrm{d}x.$$

The integrand of K_n' is bounded above by

$$\left(1 - n^{-\frac{2}{3}}\right)^n,$$

while the interval of integration is polynomial in length, which implies that K_n' decreases faster than any power of n, and so can be ignored for our purposes. On the other hand, Lemma 3.1 implies that

$$\begin{aligned} K_n'' &\sim \int_0^{\infty} \left(\{ny\} - \frac{1}{2}\right) \frac{\exp\left(-\frac{1}{y}\right)}{y^2} \mathrm{d}y \\ &= \sum_{k=0}^{\infty} \int_{\frac{k}{n}}^{\frac{k+1}{n}} \left[ny - \frac{1}{2} - k\right] \frac{\exp\left(-\frac{1}{y}\right)}{y^2} \mathrm{d}y \\ &= n \sum_{k=0}^{\infty} \int_{\frac{k}{n}}^{\frac{k+1}{n}} \left[y - \frac{k + \frac{1}{2}}{n}\right] \frac{\exp\left(-\frac{1}{y}\right)}{y^2} \mathrm{d}y. \end{aligned}$$

We can now apply Lemmas 5.12 and 5.13 with

$$f(x) = \frac{\exp\left(-\frac{1}{y}\right)}{y^2};$$

it is easy to check that $f(x)$ satisfies the assumptions. Theorem 5.10 follows.

Acknowledgements The author would like to thank the EPSRC,the NSF, and the Berlin Mathematics School for support, Technische Universität Berlin for its hospitality during the revision of this paper and Natalia Komarova Ilan Vardi, and Jeff Lagarias for useful conversations. Lagarias, in particular, has pointed out the connection of this work to Li's work on the Riemann hypothesis, see [2, 5, 6].

References

1. C.M. Bender and S.A. Orszag. *Advanced mathematical methods for scientists and engineers: Asymptotic methods and perturbation theory*, volume 1. Springer Verlag, 1978.
2. E. Bombieri and J.C. Lagarias. Complements to Li's criterion for the riemann hypothesis. *Journal of Number Theory*, 77(2):274–287, 1999.
3. N. Komarova and I. Rivin. Mathematics of learning. *Arxiv preprint math/0105235*, 2001.
4. N.L. Komarova and I. Rivin. Harmonic mean, random polynomials and stochastic matrices. *Advances in Applied Mathematics*, 31(2):501–526, 2003.
5. J.C. Lagarias. Li coefficients for automorphic l-functions. *Arxiv preprint math/0404394*, 2004.
6. X.J. Li. The positivity of a sequence of numbers and the riemann hypothesis, 1. *Journal of Number Theory*, 65(2):325–333, 1997.
7. G. Pólya, G. Szegö, C.E. Billigheimer, and G. Pólya. *Problems and theorems in analysis*, volume II. Springer, 1998.
8. I. RIVIN. The performance of the batch learning algorithm. *Arxiv preprint cs.LG/0201009*, 2002.

A Transcendence Criterion for CM on Some Families of Calabi–Yau Manifolds

Paula Tretkoff and Marvin D. Tretkoff

Dedicated to the memory of Leon Ehrenpreis, my Ph.D. advisor and friend for 50 years (Marvin Tretkoff)

Abstract In this chapter, we give some examples of the validity of a special case of a recent conjecture of Green et al. (Ann. Math. Studies, no. 183, Princeton University Press, 2012). This special case is an analogue of a celebrated theorem of Schneider (Math. Annalen 113:1–13, 1937) on the transcendence of values of the elliptic modular function and its generalization in Cohen (Rocky Mountain J. Math. 26:987–1001, 1996) and Shiga and Wolfart (J. Reine Angew. Math. 463:1–25, 1995). Related techniques apply to all the examples of CMCY families in the work of Rohde (Lecture Notes in Mathematics 1975, Springer, Berlin, 2009), and this is the subject of a paper in preparation by the author (Tretkoff, Transcendence and CM on Borcea-Voisin towers of Calabi-Yau manifolds).

Key words Calabi–Yau manifolds • Complex multiplication • Transcendence

Mathematics Subject Classification(2010): 11J81 (Primary), 14C30 (Secondary)

P. Tretkoff (✉)
Department of Mathematics, Texas A&M University, College Station, TX 77842–3368, USA

CNRS, UMR 8524, Université de Lille 1, Cité Scientifique, 59655 Villeneuve d'Ascq CEDEX, FRANCE
e-mail: paulatretkoff@neo.tamu.edu

M.D. Tretkoff
Department of Mathematics, Texas A&M University, College Station, TX 77842–3368, USA
e-mail: tretkoff@math.tamu.edu

H.M. Farkas et al. (eds.), *From Fourier Analysis and Number Theory to Radon Transforms and Geometry*, Developments in Mathematics 28, DOI 10.1007/978-1-4614-4075-8_23,

1 Introduction

In a recent monograph [7], Green, Griffiths, and Kerr propose a general theory of Mumford–Tate domains in order to examine new problems on arithmetic, geometry, and representation theory, generalizing the well-established results of the theory of Shimura varieties. In the last section of this monograph, they formulate an algebraic independence conjecture for points in period domains. This conjecture has its genesis in Grothendieck's period conjecture for algebraic varieties (see [1,6,10]).

Transcendence and linear independence properties of periods of 1-forms on abelian varieties defined over number fields are well understood, even though only a few more general algebraic independence results have been established. Using the linear independence properties, we can deduce results about the transcendence of automorphic functions at algebraic points.

The first important result of this type is due to Schneider [13] in 1937. Let $\mathcal{H}$ be the upper half plane, namely, the complex numbers with positive imaginary part. Let $j(\tau)$, $\tau \in \mathcal{H}$, be the elliptic modular function, which is the unique function, automorphic with respect to $\mathrm{PSL}(2,\mathbb{Z})$, holomorphic with a simple pole at infinity, and with Fourier series of the form

$$j(\tau) = \mathrm{e}^{-2\pi i\tau} + 744 + \sum_{n=1}^{\infty} a_n e^{2\pi i n\tau}, \qquad a_n \in \mathbb{C}.$$

Th. Schneider proved that

$$\left\{\tau \in \mathcal{H} \cap \overline{\mathbb{Q}} : j(\tau) \in \overline{\mathbb{Q}}\right\} = \{\tau \in \mathcal{H} : [\mathbb{Q}(\tau) : \mathbb{Q}] = 2\}.$$

Therefore, $j(\tau)$ is a transcendental number for all $\tau \in \mathcal{H} \cap \overline{\mathbb{Q}}$ which are not imaginary quadratic, that is, are not complex multiplication (CM) points. We view this as a transcendence criterion for complex multiplication, not only because it is equivalent to a statement about transcendence of special values of automorphic functions but, more importantly, because the proof uses techniques from transcendental number theory. The analogous result for Shimura varieties of PEL type is due to the author, jointly with Shiga and Wolfart [4, 15]. There, the key transcendence technique is the Analytic Subgroup Theorem of Wüstholz [23]. Recall that to every polarized abelian variety A of complex dimension g, we can associate a normalized period matrix τ_A in the Siegel upper half space $\mathcal{H}_g$ of genus g, consisting of the $g \times g$ symmetric matrices with positive definite imaginary part. Then, the results of [4, 13, 15] are equivalent to the statement that A is defined over $\overline{\mathbb{Q}}$ as an algebraic variety, and the entries of the matrix τ_A are algebraic numbers if and only if A has complex multiplication (CM). Of course, the matrix τ_A is only defined up to the action on $\mathcal{H}_g$ of the integer points of a symplectic group, but this does not affect the statement.

The simplest case of Conjecture (VIII.A.8) of [7] asks for similar results for variations of Hodge structure of weight $n \geq 1$ (the Shimura variety case is of weight 1). In this chapter, we prove such results for certain examples, namely, for families of

Calabi–Yau threefolds shown by Borcea [2] and Viehweg–Zuo [19] to have Zariski dense sets of complex multiplication fibers. We also indicate how to treat the first step of a tower construction of Calabi–Yau manifolds due to Borcea [3] and Voisin [20]. Similar considerations in [18], where full details will be given, enable us to treat all the examples of Rohde in [11].

We thank Colleen Robles for her informative series of lectures on the monograph [7], given at Texas A&M University in Fall 2010. We also thank the EPFL, Lausanne, and the ETH, Zürich, in particular G. Wüstholz, for their hospitality and the opportunity to lecture on the content of this chapter and [18].

2 The Problem and the Main Results

In this section, we describe the problem we are studying. We then mention briefly the families of Calabi–Yau manifolds, proved by Rohde [11] to have dense sets of CM fibers, for which the problem can be solved [18]. After that, we focus for the rest of this chapter on the examples of Borcea [2], of Viehweg–Zuo [19], and the first step of what Rohde calls a "Borcea–Voisin" tower [3, 20].

As defined in [11], a Calabi–Yau n-fold X is a complex compact Kähler manifold with $H^{k,0}(X) = \{0\}$, $k = 1, \ldots, n-1$, and a nowhere vanishing holomorphic n-form.

For the convenience of the reader, we first recall some basic definitions from Hodge theory. They are well-documented in literature spanning many years and can be found in [7]. For a $\mathbb{Q}$-vector space V and a field $k \supseteq \mathbb{Q}$, we denote $V_k = V \otimes_{\mathbb{Q}} k$ and $\mathrm{GL}(V)_k = \mathrm{GL}(V_k)$. A Hodge structure of weight $n \in \mathbb{Z}$ is a finite dimensional $\mathbb{Q}$-vector space V, endowed with the following three equivalent things:

- A decomposition of vector spaces $V_{\mathbb{C}} = \oplus_{p+q=n} V^{p,q}$, with $V^{p,q} = \overline{V^{q,p}}$.
- A filtration $F^n \subset F^{n-1} \subset \cdots \subset F^0 = V_{\mathbb{C}}$, with $F^p \oplus \overline{F}^{n-p+1} \simeq V_{\mathbb{C}}$.
- A homomorphism of $\mathbb{R}$-algebraic groups

$$\varphi : \mathbb{U}(\mathbb{R}) \to \mathrm{SL}(V)_{\mathbb{R}}$$

with specified weight n and $\varphi(-\mathrm{Id}_{\mathbb{U}}) = (-1)^n \mathrm{Id}_V$. Here $\mathbb{U}$ is the group whose k-points, where $k \supseteq \mathbb{Q}$ is a field, are

$$\mathbb{U}(k) = \left\{ \begin{pmatrix} a & -b \\ b & a \end{pmatrix} : a^2 + b^2 = 1,\ a, b \in k \right\}.$$

We do not specify the weight if it is clear from the context. For $z \in \mathbb{C}$ with $|z| = 1$, we have $\varphi(z) = z^{p-q}$ on $V^{p,q}$, where $z = a + ib$, $a, b \in \mathbb{R}$, is identified with the matrix $\begin{pmatrix} a & -b \\ b & a \end{pmatrix} \in \mathbb{U}(\mathbb{R})$. The endomorphism $C = \varphi(i)$ is called the Weil operator. The $\mathbb{Q}$-vector space $V = \mathbb{Q}$ is assumed to have the trivial Hodge structure

φ_{triv} of weight 0 which maps $\mathbb{U}(\mathbb{R})$ to Id_V. A Hodge structure (V, φ) is polarized if there is a bilinear nondegenerate map

$$Q : V \otimes V \to \mathbb{Q},$$

with

$$Q(u, v) = (-1)^n Q(v, u) \tag{1}$$

satisfying the Hodge–Riemann (HR) relations

$$Q(F^p, F^{n-p+1}) = 0, \qquad \text{(HR1)},$$
$$Q(u, C\,\overline{u}) > 0, \quad u \neq 0, \quad u \in V_{\mathbb{C}}, \qquad \text{(HR2)}.$$

Let $G = \mathrm{Aut}(V, Q)$, and denote by $G(k)$, $\mathbb{Q} \subseteq k$ a field, the k-points of G. Usually there will be a lattice $V_{\mathbb{Z}}$ with $V = V_{\mathbb{Z}} \otimes \mathbb{Q}$, so that $G(\mathbb{Z})$ is the arithmetic subgroup of G preserving $V_{\mathbb{Z}}$. All our Hodge structures will be polarized, although we often do not explicitly refer to the polarization.

The Mumford-Tate group (MT) M_φ of a Hodge structure (V, φ) is the smallest $\mathbb{Q}$-algebraic subgroup of $\mathrm{SL}(V)$ whose real points contain $\varphi(\mathbb{U}(\mathbb{R}))$. Here, we have used the terminology of [7], rather than calling this the Hodge group or special Mumford–Tate group. A Hodge structure (V, φ) is called a CM (complex multiplication) Hodge structure if and only if its Mumford-Tate group is abelian. We just say φ is CM, if the intended V is clear from the context, or just say V is CM, if the intended φ is clear from the context. We refer to $(\mathbb{Q}, \varphi_{\mathrm{triv}})$ as the trivial CM Hodge structure.

Let a $\mathbb{Q}$-vector space V and a nondegenerate bilinear form Q satisfying (1) be given. Furthermore, for all integers p, q with $p + q = n$, let integers $h^{p,q} \geq 0$ summing to dim V with $h^{p,q} = h^{q,p}$ also be given. The $h^{p,q}$ are called the Hodge numbers. We define the period domain D to be the set of polarized Hodge structures (V, Q, φ) with $\dim(V^{p,q}) = h^{p,q}$. Therefore, each Hodge structure satisfies both HR relations for Q. The period domain is a homogeneous space. If we fix a Hodge structure φ_0 with isotropy group H_0 in $G(\mathbb{R})$, then $D = G(\mathbb{R})/H_0$. *For all the examples we consider, there exists a CM Hodge structure in D. Therefore we may, and we will, assume that φ_0 is a fixed CM Hodge structure.* We have a bijection (with g ranging over $G(\mathbb{R})$),

$$\{g\varphi_0 g^{-1} = \varphi_g : \mathbb{U}(\mathbb{R}) \to G(\mathbb{R})\} \simeq G(\mathbb{R})/H_0$$

$$g\varphi_0 g^{-1} \to gH_0.$$

In order to introduce the analogue of Schneider's theorem, we need the context of variations of Hodge structure since, in general, there may not exist suitable $G(\mathbb{Z})$-invariant functions on D. From now on, we do not use the abstract setting, as our examples are geometric. Indeed, all the examples we consider in this chapter, and in [18], are smooth proper algebraic families defined over $\overline{\mathbb{Q}}$:

$$\pi : \mathcal{X} \to S.$$

In particular, the map π is surjective and proper. The base S is a quasi-projective variety defined over $\overline{\mathbb{Q}}$. Moreover, the fibers $\mathcal{X}_s$, $s \in S$, are smooth projective varieties, with $\mathcal{X}_s(\mathbb{C})$ a compact Kähler n-fold. When $s \in S(\overline{\mathbb{Q}})$, the fiber $\pi^{-1}(s) = \mathcal{X}_s$ is defined over $\overline{\mathbb{Q}}$ as an algebraic variety. Let b be a fixed base point in S and let $V = H^n(\mathcal{X}_b, \mathbb{Q})_{\text{prim}}$, the primitive cohomology, with its usual polarization Q (see [11], p.14, or [22]), given by

$$Q(v, w) = \int_{\mathcal{X}_b} v \wedge w. \tag{2}$$

When X is a curve, or a Calabi–Yau threefold, we have $H^n(X, \mathbb{Q})_{\text{prim}} = H^n(X, \mathbb{Q})$, $n = \dim X$.

For $s \in S$, the filtration associated to the usual Hodge decomposition, namely, $H^n(\mathcal{X}_s, \mathbb{C}) = \oplus_{p+q=n} H^{p,q}(\mathcal{X}_s)$, can be pulled back to a filtration of $V_{\mathbb{C}}$ with Hodge numbers independent of s. We denote either by $\varphi_{\mathcal{X}_s}$ or by $H^n(\mathcal{X}_s, \mathbb{Q}_{\mathcal{X}_s})$ the corresponding Hodge structure on V. The induced map from S to the corresponding period domain D is multivalued when S has nontrivial fundamental group, but its image in $\Gamma\backslash D$ is well defined, where $\Gamma \subseteq G(\mathbb{Z})$ is the image of the monodromy representation ([5], Chap. 4, [21], Chap. 1). Therefore, we have a well-defined period map

$$\Phi : S \to \Gamma\backslash D.$$

Let $\rho : D \to \Gamma\backslash D$ be the natural projection. We can now state the analogue of Schneider's problem on the j-function in this context (it is a special case of Conjecture (VIII.A.8) of [7]).

Problem. Let $s \in S(\overline{\mathbb{Q}})$ and suppose that $\varphi \in D$ satisfies $\rho(\varphi) = \Phi(s)$. Show that $\varphi = g\varphi_0 g^{-1}$ for $g \in G(\overline{\mathbb{Q}})$ if and only if (V, φ) has CM.

When the pair (V, Q) is clear from the context, we just say "φ is conjugate over $\overline{\mathbb{Q}}$ to a CM Hodge structure" instead of "$\varphi = g\varphi_0 g^{-1}$ for $g \in G(\overline{\mathbb{Q}})$."

The "if" part of the above statement is immediate in the examples we consider, the only work being in the "only if" part. Notice that once one choice of $\varphi \in D$ with $\rho(\varphi) = \Phi(s)$ is conjugate in $G(\overline{\mathbb{Q}})$ to φ_0, then every $\varphi \in D$ with $\rho(\varphi) = \Phi(s)$ is conjugate in $G(\overline{\mathbb{Q}})$ to φ_0.

Using the well-known description of the Siegel upper half space $\mathcal{H}_g$ of genus g in terms of complex structures on $\mathbb{R}^{2g}$ (see, e.g., [12], Sect. 3), we have:

Proposition. Let $\pi : \mathcal{X} \to S$ be a family of smooth projective algebraic curves of genus g satisfying the above assumptions. Then, we may take $D = \mathcal{H}_g$ and $\Gamma \subseteq \mathrm{PSp}(2g, \mathbb{Z})$, and the statement of the problem is true by [4, 13, 15].

In [18], we show the following:

Claim. The statement of the problem is true for all the families of Calabi–Yau manifolds with dense sets of CM fibers constructed by Rohde in [11] (and called CMCY families in that same reference).

In this chapter, we focus on two examples of families of Calabi–Yau threefolds with dense sets of CM fibers, studied respectively by Borcea and Viehweg–Zuo, and the first step in a tower of Calabi–Yau manifolds that starts with these two examples. We use the fact that the Hodge structures associated to each fiber of our families are sub-Hodge structures of ones involving direct sums and tensor products of Hodge structures on curves and various CM Hodge structures. The CM criterion on a curve is then the one from [4, 15]. Similar considerations allow one to deal with all the examples of [11]. Indeed, this is directly related to the proofs that these families have dense sets of CM fibers. The definition of a CMCY family in [11], Chap. 7, p.143, involves a stronger CM condition. Namely, a family of Calabi–Yau n-manifolds over a quasi-projective base space, which contains a Zariski dense set of fibers X such that the Mumford–Tate group of $H^k(X, \mathbb{Q}_X)$ is a torus for all k, is defined to be a CMCY family. All the examples we consider satisfy the stronger CMCY condition. We say that a variety X (Calabi–Yau or not) such that the Mumford–Tate group of $H^k(X, \mathbb{Q}_X)$ is a torus for all k "has CM for all levels."

3 The Main Lemmas

In this section, we collect, for the convenience of the reader, the main lemmas that we use from other references.

Lemma 1. *[2, 19]*

(i) *Let (V_1, φ_1) and (V_2, φ_2) be two Hodge structures of weight n and $\varphi_1 \oplus \varphi_2$ the induced Hodge structure on $V_1 \oplus V_2$. Then,*

$$M_{\varphi_1 \oplus \varphi_2} \subset M_{\varphi_1} \times M_{\varphi_2} \subset \mathrm{SL}(V_1) \times \mathrm{SL}(V_2) \subset \mathrm{SL}(V_1 \oplus V_2),$$

and the projections

$$M_{\varphi_1 \oplus \varphi_2} \to M_{\varphi_1}, \qquad M_{\varphi_1 \oplus \varphi_2} \to M_{\varphi_2}$$

are surjective.

(ii) *The Mumford-Tate group does not change under Tate twists.*

(iii) *The Mumford-Tate group of a Hodge structure concentrated in bidegree (p, p), $p \in \mathbb{Z}$, is trivial.*

(iv) *Let $\varphi_1 \otimes \varphi_2$ be the induced Hodge structure on $V_1 \otimes V_2$. Then $\varphi_1 \otimes \varphi_2$ has CM if and only if both φ_1 and φ_2 have CM.*

Lemma 2. *[11, 22]. Let X_1 and X_2 be compact Kähler manifolds. Then, for any integers $k, r, s \geq 0$, we have*

$$H^k(X_1 \times X_2, \mathbb{Q}) = \oplus_{i+j=k} H^i(X_1, \mathbb{Q}) \otimes H^j(X_2, \mathbb{Q})$$

and

$$H^{r,s}(X_1 \times X_2) = \oplus_{p+p'=r, q+q'=s} H^{p,q}(X_1) \otimes H^{p',q'}(X_2).$$

Lemma 3. *[11, 22]. Let X be an algebraic manifold of dimension n and let $\widehat{X}$ be the blowup of X along a submanifold Z of codimension 2 in X. Then, for all k, we have an isomorphism of Hodge structures*

$$H^k(X, \mathbb{Q}_X) \oplus H^{k-2}(Z, \mathbb{Q}_Z)(-1) \simeq H^k(\widehat{X}, \mathbb{Q}_{\widehat{X}}),$$

where $H^{k-2}(Z, \mathbb{Q}_Z)(-1)$ is $H^{k-2}(Z, \mathbb{Q}_Z)$ shifted by $(1, 1)$ in bidegree. Therefore, the Mumford–Tate group of $H^k(\widehat{X}, \mathbb{Q}_{\widehat{X}})$ is commutative if and only if the Mumford–Tate groups of both $H^k(X, \mathbb{Q}_X)$ and $H^{k-2}(Z, \mathbb{Q}_Z)$ are commutative. What's more, if X is a smooth surface and Z is a point of X, then the Mumford–Tate groups of $H^2(X, \mathbb{Q}_X)$ and $H^2(\widehat{X}, \mathbb{Q}_{\widehat{X}})$ are isomorphic.

4 The Borcea Family as a Two-Step Tower

Let

$$M_1 = \{x = (x_i)_{i=1}^4 \in \mathbb{P}_1^4 : x_i \neq x_j, i \neq j\}/\mathrm{Aut}(\mathbb{P}_1),$$

where $\mathrm{Aut}(\mathbb{P}_1)$ acts diagonally. It is noncanonically isomorphic to

$$\Lambda = \mathbb{P}_1 \setminus \{0, 1, \infty\}.$$

Consider three families $\mathcal{E}_i$, $i = 1, 2, 3$, of elliptic curves (Calabi–Yau onefolds)

$$\mathcal{E}_i \to \Lambda$$

with fiber $\mathcal{E}_{\lambda_i}$ of $\mathcal{E}_i$ at $\lambda_i \in \Lambda$ given by

$$y^2 = x(x-1)(x-\lambda_i), \qquad i = 1, 2, 3.$$

By the theorem of Schneider [13] mentioned in Sect. 1, the statement of the problem in Sect. 2 is true for these families of elliptic curves.

Each elliptic curve $\mathcal{E}_{\lambda_i}$ carries an involution $\iota_i : (x, y) \mapsto (x, -y)$ fixing the group $\mathcal{E}_{\lambda_i}[2]$ of 2-torsion points, which has four elements, and reversing the sign of the holomorphic 1-form dx/y. The product involution $\iota = \iota_2 \times \iota_3$ on the abelian surface $\mathcal{A}_{\lambda_2,\lambda_3} = \mathcal{E}_{\lambda_2} \times \mathcal{E}_{\lambda_3}$ sends a point to its group inverse and has $4 \times 4 = 16$ fixed points. Blowing up these 16 points, we get a surface $\widehat{\mathcal{A}}_{\lambda_2,\lambda_3}$ with an involution $\widehat{\iota}$, induced by ι, whose ramification locus is the 16 exceptional divisors. The quotient $\mathcal{K}_{\lambda_2,\lambda_3} = \widehat{\mathcal{A}}_{\lambda_2,\lambda_3}/\widehat{\iota}$ is smooth and is a $K3$-surface (hence a Calabi–Yau twofold), called the Kummer surface of $\mathcal{A}_{\lambda_2,\lambda_3}$. The surface $\mathcal{K}_{\lambda_2,\lambda_3}$ is isomorphic to one obtained by resolving the 16 singular double points of the quotient $\mathcal{E}_{\lambda_2} \times \mathcal{E}_{\lambda_3}/\iota_2 \times \iota_3$. On this last quotient surface, the maps $\iota_2 \times \mathrm{Id}$ and $\mathrm{Id} \times \iota_3$ define the same involution, which in turn induces an involution σ on $\mathcal{K}_{\lambda_2,\lambda_3}$. The involutions $\widehat{\iota}$ and σ exist by the universal property of blowing up (see [8], II, Corollary 7.15). The ramification locus R_σ of σ has 8 connected components consisting of smooth rational curves given by the union of the image, under the degree 2 rational map $\mathcal{A}_{\lambda_2,\lambda_3} \to \mathcal{K}_{\lambda_2,\lambda_3}$, of $\mathcal{E}_{\lambda_2}[2]\times\mathcal{E}_{\lambda_3}$ and of $\mathcal{E}_{\lambda_2}\times\mathcal{E}_{\lambda_3}[2]$. The involution σ reverses the sign of any (nonzero) holomorphic 2-form on $\mathcal{K}_{\lambda_2,\lambda_3}$. This construction is a first step in a tower: we build a Calabi–Yau twofold, with involution reversing the sign of any holomorphic 2-form, from two Calabi–Yau onefolds with involution reversing the sign of any holomorphic 1-form. What's more, the rational Hodge structure $\varphi_{\lambda_2,\lambda_3}$ of level 2 on $\mathcal{K}_{\lambda_2,\lambda_3}$ is the $\iota_2 \times \iota_3$-invariant part of the weight 2 Hodge structure on $\mathcal{A}_{\lambda_2,\lambda_3}$. This is just the tensor product of the rational Hodge structure φ_{λ_2} of level 1 on $\mathcal{E}_{\lambda_2}$ with that, φ_{λ_3}, on $\mathcal{E}_{\lambda_3}$. This is a CM Hodge structure if and only if both $\mathcal{E}_{\lambda_2}$ and $\mathcal{E}_{\lambda_3}$ have CM by [2], Proposition 1.2 (Lemma 1(iv), Sect. 3). Suppose $\varphi_{\lambda_2,\lambda_3}$ is conjugate over $\overline{\mathbb{Q}}$ to a CM Hodge structure φ_0. We can write φ_0 as a tensor product $\varphi_{0,2} \otimes \varphi_{0,3}$ of weight 1 CM Hodge structures on the elliptic curves. By [9], Proposition 2.5, p.563, it follows that φ_{λ_2} and φ_{λ_3} are both also conjugate over $\overline{\mathbb{Q}}$ to a CM Hodge structure. Applying Th. Schneider's theorem [13], we deduce that if λ_2 and λ_3 are also algebraic numbers, then $\mathcal{E}_{\lambda_2}$ and $\mathcal{E}_{\lambda_3}$ are both CM and hence that $\varphi_{\lambda_2,\lambda_3}$ is CM. We therefore have:

Theorem 1. *The statement of the problem of Sect. 2 holds for the family*

$$\mathcal{K} \to \Lambda^2$$

of Calabi–Yau twofolds constructed above, which has a dense set of CM fibers.

The next step in the tower applies a construction similar to the above, but now to $(\mathcal{E}_1, \iota_1)$ and $(\mathcal{K}, \sigma)$ (see [3]). Let $(\lambda_1, \lambda_2, \lambda_3) \in \Lambda^3$ and blow up the product $\mathcal{T}_{\lambda_1,\lambda_2,\lambda_3} = \mathcal{E}_{\lambda_1} \times \mathcal{K}_{\lambda_2,\lambda_3}$ along the connected components of the codimension 2 ramification divisor $\mathcal{E}_{\lambda_1}[2] \times R_\sigma$ of the involution $\iota_1 \times \sigma$. Consider the induced involution $\widehat{\iota_1 \times \sigma}$ on this blowup $\widehat{\mathcal{T}_{\lambda_1,\lambda_2,\lambda_3}}$. The quotient

$$\mathcal{Y}_{\lambda_1,\lambda_2,\lambda_3} = \widehat{\mathcal{T}_{\lambda_1,\lambda_2,\lambda_3}}/\widehat{\iota_1 \times \sigma}$$

is a Calabi–Yau threefold with involution γ induced by $\mathrm{Id} \times \sigma = \iota_1 \times \mathrm{Id}$ on $\mathcal{E}_{\lambda_1} \times \mathcal{K}_{\lambda_1,\lambda_2}/\iota_1 \times \sigma$ of which $\mathcal{Y}_{\lambda_1,\lambda_2,\lambda_3}$ is a resolution. It is also a minimal resolution of

$$\mathcal{E}_{\lambda_1} \times \mathcal{E}_{\lambda_2} \times \mathcal{E}_{\lambda_3}/H$$

where H is the group of order 4 generated by $\iota_1 \times \iota_2 \times \mathrm{Id}$ and $\mathrm{Id} \times \iota_2 \times \iota_3$. The singularities of this last quotient lie along a configuration of 48 rational curves with 4^3 intersection points. The ramification locus R_γ of γ consists of the image under the degree 4 rational map

$$\mathcal{E}_{\lambda_1} \times \mathcal{E}_{\lambda_2} \times \mathcal{E}_{\lambda_3} \to \mathcal{Y}_{\lambda_1,\lambda_2,\lambda_3}$$

of the union of

$$\mathcal{E}_{\lambda_1}[2] \times \mathcal{E}_{\lambda_2} \times \mathcal{E}_{\lambda_3}, \quad \mathcal{E}_{\lambda_1} \times \mathcal{E}_{\lambda_2}[2] \times \mathcal{E}_{\lambda_3}, \quad \mathcal{E}_{\lambda_1} \times \mathcal{E}_{\lambda_2} \times \mathcal{E}_{\lambda_3}[2].$$

Moreover, γ reverses the sign of any holomorphic 3-form on $\mathcal{Y}_{\lambda_1,\lambda_2,\lambda_3}$. The Hodge structure $\varphi_{\lambda_1,\lambda_2,\lambda_3}$ given by $H^3(\mathcal{Y}_{\lambda_1,\lambda_2,\lambda_3}, \mathbb{Q}_{\mathcal{Y}_{\lambda_1,\lambda_2,\lambda_3}})$ is $\varphi_{\lambda_1} \otimes \varphi_{\lambda_2} \otimes \varphi_{\lambda_3}$, where φ_{λ_i} is the level 1 rational Hodge structure of the elliptic curve $\mathcal{E}_{\lambda_i}$ (for details, see [2]). Therefore, again by Th. Schneider's theorem [13], and the fact that $\varphi_{\lambda_1} \otimes \varphi_{\lambda_2} \otimes \varphi_{\lambda_3}$ is CM if and only if each φ_i, $i = 1, 2, 3$ is CM, we deduce easily that

Theorem 2. *The statement of the problem of Sect. 2 holds for the family*

$$\mathcal{Y} \to \Lambda^3$$

of Calabi–Yau threefolds constructed above, which has a dense set of CM fibers.

5 The Viehweg–Zuo Family

Viehweg and Zuo [19] have constructed iterated cyclic covers of degree 5 which give a family of Calabi–Yau threefolds (which we call the VZCY family) with a dense set of CM fibers. The fibers of the family are smooth quintics in $\mathbb{P}_4$. The study of this family is taken up again in [11], Sect. 7.3. For a projective hypersurface $X \subset \mathbb{P}_4$, only the Mumford-Tate group of the Hodge structure on $H^3(X, \mathbb{Q})$ can be nontrivial, so the CM condition in the CMCY definition is just the usual one. Consider the parameter space

$$M_2 = \left\{(x_i)_{i=1}^5 \in \mathbb{P}_1^5 : x_i \neq x_j, \quad i \neq j\right\} / \mathrm{Aut}(\mathbb{P}_1)$$

which is noncanonically isomorphic to

$$S = \{u, v \in \mathbb{P}_1(\mathbb{C}) : u \neq 0, 1, \infty, \quad v \neq 0, 1, \infty, \quad u \neq v\}.$$

Explicitly, the VZCY family is given by

$$\pi : \mathcal{X} \to S$$

with fiber $\mathcal{X}_{(u,v)}$ the projective variety with equation,

$$x_4^5 + x_3^5 + x_2^5 + x_1(x_1 - x_0)(x_1 - ux_0)(x_1 - vx_0)x_0 = 0, \qquad (3)$$

in homogeneous coordinates $[x_0 : x_1 : x_2 : x_3 : x_4] \in \mathbb{P}_4$. The fibers $\mathcal{X}_{(u,v)}$ are smooth hypersurfaces of degree 5 in $\mathbb{P}_4$. They are therefore Calabi–Yau threefolds, by the well-known fact that any smooth hypersurface of degree $d + 1$ in $\mathbb{P}_d$ is a Calabi–Yau $(d - 1)$-fold. As in Sect. 2, fix a base point $b \in S$ and let $V = H^3(\mathcal{X}_b, \mathbb{Q})$. The VZCY family is an example of an iterated cyclic cover. Indeed, consider the family of smooth algebraic curves of genus 6 in $\mathbb{P}_2$ given by the following family $\mathcal{C} \to S$ of cyclic covers of $\mathbb{P}_1$ of degree 5:

$$x_2^5 + x_1(x_1 - x_0)(x_1 - ux_0)(x_1 - vx_0)x_0 = 0, \qquad (u, v) \in S. \qquad (4)$$

The fibers of this family are the ramification loci of the family $\mathcal{S} \to S$ of cyclic covers of $\mathbb{P}_2$ of degree 5 given by the family of smooth surfaces in $\mathbb{P}_3$:

$$x_3^5 + x_2^5 + x_1(x_1 - x_0)(x_1 - ux_0)(x_1 - vx_0)x_0 = 0. \tag{5}$$

Iterating again, the fibers of this last family are the ramification loci of the family of cyclic covers of $\mathbb{P}_3$ of degree 5 given by the VZCY family.

Let $\mathcal{F}_5$ be the Fermat curve of degree 5 given by $x^5 + y^5 + z^5 = 0$. The usual Hodge structure $(H^1(\mathcal{F}_5, \mathbb{Q}), \varphi_{\mathcal{F}_5})$ associated to the Hodge decomposition $H^1(\mathcal{F}_5, \mathbb{C}) = H^{(1,0)}(\mathcal{F}_5) \oplus H^{(0,1)}(\mathcal{F}_5)$ has CM, since it is well known that the Jacobian of every Fermat curve is of CM type.

Let $s = (u, v) \in S$, with $u, v \in \overline{\mathbb{Q}}$. Suppose, in addition, that the usual Hodge decomposition on $H^3(\mathcal{X}_s, \mathbb{C})$ gives a representative homomorphism

$$\varphi_s : \mathbb{U}(\mathbb{R}) \to \mathrm{SL}(V)_{\mathbb{R}}$$

satisfying $\varphi_s = g\varphi_0 g^{-1}$ for $g \in G(\overline{\mathbb{Q}})$, where $G = \mathrm{Aut}(V, Q)$ with Q as in Sect. 2, (2), and φ_0 is a fixed CM Hodge structure.

By the argument following Claim 8.6 of [19], p. 525, the Hodge structure $(H^3(\mathcal{X}_b, \mathbb{Q}), \varphi_s)$ is a sub-Hodge structure of the Hodge structure given by

$$[\varphi_s^1 \otimes \varphi_{\mathcal{F}_5} \otimes \varphi_{\mathcal{F}_5}] \oplus [\varphi_{\mathcal{F}_5} \otimes \mathrm{Id}_W] \oplus [\varphi_s^1(-1)]$$

on

$$[H^1(\mathcal{C}_b, \mathbb{Q}) \otimes H^1(\mathcal{F}_5, \mathbb{Q}) \otimes H^1(\mathcal{F}_5, \mathbb{Q})] \oplus [H^1(\mathcal{F}_5, \mathbb{Q}) \otimes W] \oplus [H^1(\mathcal{C}_b, \mathbb{Q})(-1)],$$

where (-1) denotes the Tate twist and W is a $\mathbb{Q}$-vector space with a constant $(1, 1)$ Hodge structure. For each $s \in S$, the homomorphism φ_s^1 is associated to the usual Hodge decomposition $H^1(\mathcal{C}_s, \mathbb{C}) = H^{(1,0)}(\mathcal{C}_s) \oplus H^{(0,1)}(\mathcal{C}_s)$. It is now easy to see that if $\varphi_s = g\varphi_0 g^{-1}$ for $g \in G(\overline{\mathbb{Q}}) \subseteq \mathrm{SL}(V)_{\overline{\mathbb{Q}}}$, then we have $\varphi_s^1 = h\varphi_1 h^{-1}$ for $(H^1(\mathcal{C}_0, \mathbb{Q}), \varphi_1)$ CM and $h \in \mathrm{Sp}(12, \overline{\mathbb{Q}})$. Therefore, by the proposition of Sect. 2, we have that φ_s^1 is CM. Now, as φ_s is therefore a sub-Hodge structure of a Hodge structure built up of tensor products and direct sums of CM Hodge structures, by Lemma 8.1 of [19] (see also the lemmas of our Sect. 3), it follows that φ_s has CM as required. We therefore have

Theorem 3. *The statement of the problem of Sect. 2 holds for the VZCY family of Calabi–Yau threefolds constructed above, which has a dense set of CM fibers.*

On each fiber $\mathcal{X}_s$, $s \in S$, we have the involution

$$\delta : ([x_0 : x_1 : x_2 : x_3 : x_4] \mapsto [x_0 : x_1 : x_2 : x_4 : x_3]$$

which leaves the smooth divisor $D_s : x_3 = x_4$ invariant. Moreover, D_s is isomorphic to $\mathcal{S}_s$ of (5), which is CMCY for a dense set of $s \in S$ (see [11],

p. 151). Moreover, by [19], p. 525, $H^2(\mathcal{S}_s, \mathbb{Q}_{\mathcal{S}_s})$ is a sub-Hodge structure of the tensor product of $H^1(\mathcal{C}_s, \mathbb{Q}_{\mathcal{C}_s})$ and $H^1(\mathcal{F}_5, \mathbb{Q}_{\mathcal{F}_5})$, so, using arguments similar to the above, the statement of the problem of Sect. 2 holds for the family $\mathcal{S} \to S$.

The fibers of the VZCY family isomorphic to the Fermat quintic threefold have CM (see [11], p. 151, [19]). The periods of the holomorphic 3-forms defined over $\overline{\mathbb{Q}}$, and their transcendence, are discussed in the Appendix, authored by Marvin D. Tretkoff.

6 The First Step of a Borcea-Voisin Tower

In this section, we indicate how to prove the claim of Sect. 2 for the Borcea–Voisin towers of CMCY manifolds constructed by Rohde [11], by summarizing the ideas for one step in such a tower using the families of Sect. 4 and Sect. 5. In Sect. 4, we already saw examples of such a construction. Full details for the general case will be given in [18].

Using the CMCY families with involution of Sects. 4 and 5, we build a CMCY family with involution of higher dimension using the construction in [3] and in [11], Proposition 7.2.5, and show that the statement of the problem of Sect. 2 holds for this new family.

Let $(\mathcal{Y}, \gamma)$ be the Borcea family of Calabi–Yau threefolds with involution constructed in Sect. 4, and $(\mathcal{X}, \delta)$ be the VZCY family of Calabi–Yau threefolds with involution from Sect. 5. Let $\mathcal{Y}_{1,2,3}$ be the fiber of $\mathcal{Y}$ at $(\lambda_1, \lambda_2, \lambda_3) \in \Lambda^3$ and $\mathcal{X}_s$ be the fiber of $\mathcal{X}$ at $s \in S$. The ramification divisors $R_\gamma = D_{1,2,3} \subset \mathcal{Y}_{1,2,3}$ of γ and $D_s \subset \mathcal{X}_s$ of δ consist of smooth nontrivial disjoint hypersurfaces. From Sect. 4, the divisor $D_{1,2,3}$ is CM for all levels, as $\mathbb{P}_1$ carries the trivial CM Hodge structure for all levels. As noted at the end of Sect. 5, the divisor D_s is isomorphic to $\mathcal{S}_s$ of (5) for which the statement of the problem of Sect. 2 holds for all levels.

Let $\widehat{\mathcal{Y}_{1,2,3} \times \mathcal{X}_s}$ be the blowup of $\mathcal{Y}_{1,2,3} \times \mathcal{X}_s$ with respect to $D_{1,2,3} \times D_s$ and $\widehat{\gamma \times \delta}$ be the involution on the blowup induced by $\gamma \times \delta$. Then

$$\mathcal{Z}_{1,2,3,s} = \widehat{\mathcal{Y}_{1,2,3} \times \mathcal{X}_s} / \widehat{\gamma \times \delta}$$

is a Calabi–Yau sixfold with involution ε generating $(\mathbb{Z}/2 \times \mathbb{Z}/2)/\langle \gamma \times \delta \rangle$, whose fixed points are a smooth nontrivial divisor, and which reverses the sign of any holomorphic 6-form on $\mathcal{Z}_{1,2,3,s}$ [3, 11, 20].

Following [11], Chap. 7, and [22], using Lemmas 1–3 of Sect. 3, we express the Hodge structure on $H^*(\mathcal{Z}_{1,2,3,s}, \mathbb{Q})$ in terms of tensor products and direct sums of the Hodge structures on $H^*(\mathcal{Y}_{1,2,3}, \mathbb{Q})$, $H^*(\mathcal{X}_s, \mathbb{Q})$, $H^*(D_{1,2,3}, \mathbb{Q})$, and $H^*(D_s, \mathbb{Q})$, where several levels $*$ may intervene. By using this analysis of the Hodge structure, we deduce that the statement of the problem of Sect. 2 holds for the family $\mathcal{Z}$, since it holds for the family $\mathcal{Y}$ of Sect. 4 (Theorem 2), for the family

$\mathcal{X}$ of Sect. 5 (Theorem 3), and for the ramification divisors. The fact that several levels of Hodge structure are involved is not an obstacle, as all relevant fibers of our families turn out to have CM for all levels.

For example, the Hodge structure of level 6 on $\mathcal{Z}_{1,2,3,s}$ is given by the Hodge substructure of level 6 on $\widehat{\mathcal{Y}_{1,2,3} \times \mathcal{X}_s}$ invariant under $\widehat{\gamma \times \delta}$. Using Lemma 3 of Sect. 3, this, in turn, can be expressed in terms of the Hodge structure of level 6 on $\mathcal{Y}_{1,2,3} \times \mathcal{X}_s$ and the Hodge structure of level 4 on $D_{1,2,3} \times D_s$. We then use Lemmas 1 and 2 of Sect. 3 to express these Hodge structures in terms of tensor products and direct sums of the Hodge structures, of various levels, on $\mathcal{Y}_{1,2,3}$, $\mathcal{X}_s$, $D_{1,2,3}$, D_s. If the Hodge structure $H^6(\mathcal{Z}_{1,2,3,s}, \mathbb{Q}_{\mathcal{Z}_{1,2,3,s}})$ is conjugate over $\overline{\mathbb{Q}}$ to a CM Hodge structure, the same is true of the Hodge structures on $\mathcal{Y}_{1,2,3}$, $\mathcal{X}_s$, $D_{1,2,3}$, D_s. As the statement of the problem of Sect. 2 applies to them, again using Lemmas 1–3 of Sect. 3, we deduce that the statement of the problem of Sect. 2 holds for the family $\mathcal{Z} \to \Lambda^3 \times S$.

Appendix: Transcendence of the Periods on Calabi–Yau-Fermat Hypersurfaces

A famous transcendence theorem of Th. Schneider (see Schneider [14], Siegel [16]) asserts that if ω is a holomorphic 1-form on a compact Riemann surface of genus at least 1, then there is a 1-cycle γ on that Riemann surface such that $\int_\gamma \omega$ is a transcendental number. Here, the Riemann surface and the 1-form ω are both supposed to be defined over the same algebraic number field. The possibility of generalizing Schneider's theorem to higher dimensional hypersurfaces is a natural question.

Let V denote the Fermat hypersurface defined in affine coordinates by the equation

$$z_1^r + \cdots + z_{n+1}^r = 1.$$

In [17], we explicitly determined the periods of the n-forms on V for *all* values of n and r. When $r = n + 2$, V is a Calabi–Yau manifold because on it there is a nowhere vanishing holomorphic n-form, ω, given by

$$\omega = z_{n+1}^{-(n+1)} \mathrm{d}z_1 \ldots \mathrm{d}z_n .$$

In order that Schneider's theorem generalize to these Calabi–Yau manifolds, it is necessary and sufficient that at least one period of ω be transcendental.

For these hypersurfaces, the formula for the periods of ω obtained in [17] simplifies to

$$\text{(I)} \qquad \int_\gamma \omega = \alpha(\gamma)\ (\Gamma(1/(n+2)))^{n+1} / \Gamma((n+1)/(n+2)),$$

where $\Gamma(u)$ is the classical gamma function applied to u. Here the n-cycle γ is any member of the basis constructed in [17] for the group of primitive n-cycles on V, and $\alpha(\gamma)$ is an *algebraic number that depends on* γ.

Using the classical identity

$$\Gamma(u)\Gamma(1-u) = \pi \csc(\pi u)$$

and the fact that $\sin(\frac{\pi}{m})$ is an algebraic number for all positive integers m, we can restate (I) as

$$\text{(II)} \qquad \int_\gamma \omega = \beta(\gamma)\frac{1}{\pi}\left(\Gamma\left(\frac{1}{n+2}\right)\right)^{n+2},$$

where $\beta(\gamma)$ is an algebraic number depending on γ.

It follows that we have the

Theorem. *Schneider's theorem extends to the n-dimensional Fermat hypersurfaces of degree $n+2$ if and only if either*

$$(*) \qquad (\Gamma(1/(n+2)))^{n+1} \,/\, \Gamma((n+1)/(n+2))$$

or

$$(**) \qquad \frac{1}{\pi}\left(\Gamma\left(\frac{1}{n+2}\right)\right)^{n+2}$$

is a transcendental number.

Of course, $(*)$ is transcendental if and only if $(**)$ is transcendental. When $n = 1$, Schneider's theorem [14] implies that $\frac{1}{\pi}(\Gamma(\frac{1}{3}))^3$ is transcendental.

Although the Fermat curves of degree $r > 3$ are *not* Calabi–Yau manifolds, the results in [17] allow us to determine their periods explicitly. For example, $\omega = dz/w^3$ is a holomorphic differential on the Fermat quartic curve

$$z^4 + w^4 = 1.$$

With respect to the basis for $H_1(V)$ given in [17], each period of ω is of the form

$$\beta(\gamma)\frac{\Gamma\left(\frac{1}{4}\right)\Gamma\left(\frac{1}{4}\right)}{\Gamma\left(\frac{1}{2}\right)},$$

with $\beta(\gamma)$ an algebraic number. Because $\Gamma(\frac{1}{2}) = \sqrt{\pi}$, Schneider's Theorem implies that $\frac{1}{\sqrt{\pi}}(\Gamma(\frac{1}{4}))^2$ is transcendental. Of course, $\frac{1}{\pi}(\Gamma(\frac{1}{4}))^4$ is therefore transcendental.

Recall that the Fermat quartic *surface* V is a K3 surface and, as such, its group of primitive 2-cycles is free abelian of rank 21.

A basis for this group is given in [17]. Now, V is also a Calabi–Yau manifold and the period of the nonvanishing holomorphic 2-form, ω, along each 2-cycle, γ, belonging to this basis is of the form $\beta(\gamma)\frac{1}{\pi}(\Gamma(\frac{1}{4}))^4$, where $\beta(\gamma)$ is an algebraic number. Therefore, each of these periods is transcendental and we have the

Theorem. *Schneider's theorem extends to the Fermat quartic surface defined by*

$$x^4 + z^4 + w^4 = 1.$$

Finally, we turn to the Fermat quintic threefold, V, defined in affine coordinates by the equation

$$x^5 + y^5 + z^5 + w^5 = 1.$$

A nowhere vanishing holomorphic 3-form on V is given by

$$\omega = w^{-4} dx dy dz.$$

Now, let

$$A(x, y, z, w) = (\zeta x, y, z, w), \quad B(x, y, z, w) = (x, \zeta y, z, w),$$

$$C(x, y, z, w) = (x, y, \zeta z, w), \quad D(x, y, z, w) = (x, y, z, \zeta w),$$

with ζ a primitive 5th root of unity, be automorphisms of the ambient 4-space. Clearly, V is left fixed by A, B, C, D and the group ring $\mathbb{Z}[A, B, C, D]$ acts on the group of 3-cycles on V. It is shown in [17] that there is a 3-cycle γ on V for which we have the following result.

Theorem. (a) *The images*

$$\gamma(i, j, k, \ell) = A^{(i-1)} B^{(j-1)} C^{(k-1)} D^{(\ell-1)} \gamma$$

span a cyclic $\mathbb{Z}[A, B, C, D]$-module and a subset of them forms a basis for the group of 3-cycles on V.

(b) *The 3-form ω can be evaluated explicitly along the $\gamma(i, j, k, \ell)$. In fact,*

$$\int_{\gamma(i,j,k,\ell)} \omega = \frac{1}{5^3} \zeta^{i+j+k+\ell} (1 - \zeta)^4 \Gamma(1/5)^4 \Gamma(4/5)^{-1}.$$

Therefore, each period of ω is the product of a nonzero algebraic number and $\Gamma(1/5)^4\Gamma(4/5)^{-1}$. The algebraic number depends on the 3-cycle in question.

It follows that Schneider's theorem generalizes to the Fermat quintic threefold if and only if $\Gamma(1/5)^4\Gamma(4/5)^{-1}$ is transcendental.

Apparently the transcendence of this number is unknown.

Finally, we note that our formula for the periods of n-forms on Fermat hypersurfaces of degree $r \neq n+2$ is substantially more complicated than that for the Calabi–Yau-Fermat hypersurfaces treated in the present note. Namely, in [17] we show that

$$\int_{\gamma(i_1,\dots,i_{n+1})} z_1^{a_1-1} z_2^{a_2-1} \dots z_n^{a_n-1} z_{n+1}^{a_{n+1}-r} \mathrm{d}z_1 \dots \mathrm{d}z_n$$

$$= \frac{1}{r^n} \zeta^{a_1 i_1 + \dots + a_{n+1} i_{n+1}} (1-\zeta^{a_1}) \dots (1-\zeta^{a_{n+1}}) \left(\frac{\Gamma(\frac{a_1}{r})\Gamma(\frac{a_2}{r}) \dots \Gamma(\frac{a_{n+1}}{r})}{\Gamma(\frac{a_1+\dots+a_{n+1}}{r})} \right),$$

where ζ is a primitive rth root of unity and $a_1, a_2, \dots a_{n+1}$, $i_1, \dots, i_{n+1}$ are appropriate integers between 1 and $r-1$. See [17] for the details. Therefore, we conclude that *Schneider's theorem extends to these Fermat hypersurfaces if and only if the numbers*

$$\frac{\Gamma(\frac{a_1}{r})\Gamma(\frac{a_2}{r}) \dots \Gamma(\frac{a_{n+1}}{r})}{\Gamma(\frac{a_1+\dots+a_{n+1}}{r})}$$

are transcendental.

Acknowledgements The author is supported by NSF grant number DMS–0800311 and NSA grant 1100362.

References

1. André, Y., *Galois theory, motives, and transcendental numbers*, arXiv: 0805.2569
2. Borcea, C., *Calabi–Yau threefolds and complex multiplication*, In: Essays on Mirror Manifolds, Int. Press, Hong Kong, 1992, 489–502.
3. Borcea, C., $K3$ surfaces with involution and mirror pairs of Calabi–Yau manifolds, In: Mirror Symmetry II, Studies in Advanced Mathematics 1, AMS/Int. Press, Providence, RI, 1997, 717–743.
4. Cohen, P.B., *Humbert surfaces and transcendence properties of automorphic functions*, Rocky Mountain J. Math. 26 (1996), 987–1001.
5. Carlson, J., Müller–Stach, S., Peters, C., Period mappings and Period Domains, Camb. Studies in Adv. Math. 85, 2003.
6. Deligne, P., Milne, J.S., Ogus, A., Shih, K.-Y., *Hodge cycles, motives, and Shimura varieties*, Lecture Notes in Mathematics, vol. 900, Springer-Verlag, Berlin, 1982.
7. Green, M., Griffiths P, Kerr M, *Mumford-Tate groups and Domains: their geometry and arithmetic*, Annals of Math. Studies, no. 183, Princeton University Press, 2012.
8. Hartshorne, R., Algebraic Geometry, Graduate Texts in Math. 52, 1977.
9. Lang, S., Algebra, Second Edition, Addison-Wesley, 1984.
10. Präve, A, *Über den Transzendenzgrad von Periodenquotienten*, Dissertation zur Erlangen des Doktorgrades, Fachbereich Mathematik Univ. Frankfurt am Main, 1997.
11. Rohde, J.C., *Cyclic Coverings, Calabi–Yau Manifolds and Complex Multiplication*, Lect. Notes Math 1975, Springer 2009.

12. Runge, B., *On Algebraic Families of Polarized Abelian Varieties*, Abh. Math. Sem. Univ. Hamburg 69 (1999), 237–258.
13. Schneider, Th. *Arithmetische Untersuchungen elliptischer Integrale*, Math. Annalen 113 (1937), 1–13.
14. Schneider, Th., Zur Theorie der Abelschen Funktionen und Integrale, J. reine angew. Math. 183 (1941), 110–128.
15. Shiga, H., Wolfart, J., *Criteria for complex multiplication and transcendence properties of automorphic functions*, J. Reine Angew. Math. 463 (1995), 1–25.
16. Siegel, C.L., Transcendental Numbers, Annals of Math. Studies, Princeton University Press, 1949.
17. Tretkoff, M.D., The Fermat hypersurfaces and their periods, in preparation.
18. Tretkoff, P., Transcendence and CM on Borcea-Voisin towers of Calabi-Yau manifolds, in preparation.
19. Viehweg, E., Zuo, K., *Complex multiplication, Griffiths-Yukawa couplings, and rigidity for families of hypersurfaces*, J. Algebraic Geom. 14 (2005), 481–528.
20. Voisin, C., Miroirs et involutions sur les surfaces $K3$, Journées de géom. alg. d'Orsay, Astérisque 218 (1993), 273–323.
21. Voisin, C., Variations of Hodge structure of Calabi–Yau threefolds, Lezioni Lagrange Roma 1996, Scuola Normale Superiore Pub. Cl. Scienze.
22. Voisin, C., Théorie de Hodge et géométrie algébrique complexe, Cours spécialisé 10, SMF (2002) France.
23. Wüstholz, G., *Algebraische Punkte auf analytischen Untergruppen algebraischer Gruppen*, Annals of Math. 129 (1989), 501–517.

Ehrenpreis and the Fundamental Principle

François Treves

Abstract This chapter outlines the underpinnings and the proof of the Fundamental Principle of Leon Ehrenpreis, according to which every solution of a system (in general, overdetermined) of homogeneous partial differential equations with constant coefficients can be represented as the integral with respect to an appropriate Radon measure over the complex "characteristic variety" of the system.

Key words Fourier transform • Overdetermined systems • PDE with constant coefficients

Mathematics Subject Classification (2010): Primary 35E20, Secondary 35C15, 35E10

1 Introduction

A preliminary version of the Fundamental Principle was first announced by Leon Ehrenpreis at a Functional Analysis conference in Jerusalem in 1960 (see [Ehrenpreis 1954]),with a more detailed version, and an outline of the steps of a potential proof, provided at a Harmonic Analysis conference at Stanford in August 1961. The essence of its statement is that every distribution solution of a system of homogeneous PDE with constant coefficients in a convex open subset Ω of $\mathbb{R}^n$,

$$\sum_{k=1}^{q} P_{j,k}(D)\, h_k = 0,\ j = 1, \ldots, p, \tag{1.1}$$

F. Treves (✉)
Department of Mathematics, Rutgers University, Hill Center for the Mathematical Sciences, Piscataway, NJ, USA
e-mail: treves.jeanfrancois@gmail.com

H.M. Farkas et al. (eds.), *From Fourier Analysis and Number Theory to Radon Transforms and Geometry*, Developments in Mathematics 28, DOI 10.1007/978-1-4614-4075-8_24,

can be represented as an integral of exponential-polynomial solutions with respect to Radon measures on the "null variety" (properly defined) of the system.

A statement of this kind was strikingly bolder and deeper than the results on the existence and approximation of solutions to *scalar* PDE with constant coefficients proved by Ehrenpreis and B. Malgrange (see [Ehrenpreis 1954, Malgrange 1955]) in the early 1950s. At that time, such depth was practically unattainable (even in the scalar case), considering the tools then available to analysts. Needed were the conceptual and technical arms of the Oka-Cartan-Serre theory of analytic sheaves as well as those of Homological Algebra, which were being perfected precisely around that time (see [Oka 1936–1953, Serre 1955]). Not only was the Theorem B of H. Cartan needed but a version of it with fairly precise enumerations and bounds had to be devised.

In the period 1962–1964 the theory of coherent analytic sheaves and Homological Algebra had fully matured (see [Gunning and Rossi 1965]), enabling Malgrange (see [Malgrange 1955]) and V. P. Palamodov (see [Palamodov 1963]) to establish firmly most of the statements on which the proof of the Fundamental Principle was to be based. Among other things Palamodov constructed an example showing that the Fundamental Principle as initially stated could not be valid (see [Hörmander 1966], p. 228); to restore it Palamodov introduced his "Noetherian operators" (Subsection 3.1 in this article). At a 1965 conference in Erevan, Palamodov presented a complete proof of the corrected statement. This proof and much additional material about systems of PDE with constant coefficients make up the content of his 1967 book (in Russian; English translation: [Palamodov 1970]).

In the middle 1960s, a renewal of interest in the questions surrounding the Fundamental Principle was sparked by a series of lectures by J. E. Björk at a summer school in Sweden. The 1960s proofs of the F. P. were substantially simplified in Chap. 8 of the monograph [Björk 1979] and in Hansen's "habilitation" thesis [Hansen 1982]. Finally, Hörmander's L^2 estimates for $\bar{\partial}$ made it possible to go directly to bounds in the cohomology of coherent analytic sheaves and rid the proof of any nonlinearity (as used in the proofs of Cartan's theorems A and B).

This note has no pretention to any originality whatsoever. Its publication is only justified by the sense that a volume in honor of Leon Ehrenpreis ought to contain at least a mention, however imperfect, of his most famous theorem. It is essentially a brief outline of the proof of the Fundamental Principle as provided in Sects. 7.6 and 7.7 of [Hörmander 1966], to which the reader is referred for all fine points and technicalities (only the simplest of proofs are included here). I have also been guided by the tutoring of Otto Liess, whom I wish to thank warmly.

2 The Classical Problems

2.1 Existence and Approximation of Solutions

We suppose given a rectangular matrix $\boldsymbol{P} = \left(P_{j,k}\right)_{1\leq j\leq p, 1\leq k\leq q}$ with polynomial entries $P_{j,k} \in \mathbb{C}\left[\zeta_1,\ldots,\zeta_n\right]$; it is assumed that the set $\{\zeta \in \mathbb{C}^n;\ \boldsymbol{P}(\zeta) = 0\}$ is a proper subvariety. Using the notation $D_j = \frac{1}{\sqrt{-1}}\frac{\partial}{\partial x_j}$ and $D = (D_1,\ldots,D_n)$, we consider the system of *inhomogeneous* linear PDE with constant coefficients

$$\boldsymbol{P}(D)\mathbf{u} = \mathbf{f}, \tag{2.1}$$

or, more explicitly,

$$\sum_{k=1}^{q} P_{j,k}(D)\,u_k = f_j,\ j = 1,\ldots,p, \tag{2.2}$$

where the right-hand sides f_j are functions or distributions in an open set $\Omega \subset \mathbb{R}^n$ and the solutions u_k belong to the same or related function or distribution spaces.[1] The problem is to show that, under the right hypotheses on the data $\mathbf{f}$, the systems (2.1)–(2.2) have solutions in the "natural" classes, primarily $\mathcal{C}^\infty$ or $\mathcal{D}'$ (i.e., smooth functions or distributions). In view of the P-convexity condition in the scalar case $(p = q = 1)$, necessary and sufficient for the surjectivity $P(D)\,\mathcal{C}^\infty(\Omega) = \mathcal{C}^\infty(\Omega)$ as established in Malgrange's thesis [Malgrange 1955], the only domains Ω for which the problem makes sense, in its "grand" generality, are the convex ones. Indeed, when $P(D) = \sum_{j=1}^{n} c_j D_j$ is a real vector field, the P-convexity of Ω means that the intersection of Ω with every orbit of P (a straight line in $\mathbb{R}^n$) is a segment.

A parallel problem concerns the system of *homogeneous* equations

$$\boldsymbol{P}(D)\mathbf{h} = \mathbf{0}. \tag{2.3}$$

The latter have distinguished solutions, the so-called *exponential-polynomial solutions* (sometimes simply called *exponential solutions*), linear combinations of solutions of the type

$$\mathbf{h}(x) = \mathbf{g}(x)\exp \mathrm{i}\,\langle \zeta \cdot x\rangle \tag{2.4}$$

where $\mathbf{g}$ is a q-vector (below we often omit the arrows) with polynomial components and $\zeta \in \mathbb{C}^n$ belongs to a suitably defined algebraic variety $\boldsymbol{V}_{\boldsymbol{P}}$. What is important [also for the solution of (2.1)] is a Runge-type theorem: *Every solution of* (2.3) *in*Ω *is the limit of a sequence of exponential-polynomial solutions.*

---

[1]Most of the time in the sequel the arrows indicating vector values will be omitted.

2.2 *The State of Affairs in the Scalar Case ca 1954*

When $p = q = 1$, some of the questions raised here were settled in the earlier work of Ehrenpreis and Malgrange ([Ehrenpreis 1954, Malgrange 1955]). The Runge result was first proved in [Malgrange 1955].

It is perhaps worthwhile to sketch the proof of the existence of solutions in $\mathcal{C}^\infty(\Omega)$ of the scalar equation

$$P(D)\,u = f \tag{2.5}$$

with $P \in \mathbb{C}[\zeta] = \mathbb{C}[\zeta_1, \ldots, \zeta_n]$. As a first step, (2.5) is solved for compactly supported f, i.e., $f \in \mathcal{C}_c^\infty(\Omega)$. This can be done by making use of a fundamental solution, i.e., a solution of the equation $P(D)\,E = \delta$ (δ : Dirac measure) and by taking the convolution $u = E * f$ as the solution. The existence of a fundamental solution of every nonzero polynomial P was first proved by Ehrenpreis in 1952 and soon after, by a different (and very simple) method, by Malgrange (see [Malgrange 1955]). Another way of approaching (2.5) is by using Fourier transform to transform (2.5) into a *division* problem:

$$P(\xi)\,\hat{u}(\xi) = \hat{f}(\xi) \tag{2.6}$$

where the right-hand side belongs to the Schwartz space $\mathcal{S}(\mathbb{R}^n)$ and can be extended as an entire function $\hat{f}(\zeta)$ of exponential type [Paley–Wiener theorem; we write $\hat{f} \in \boldsymbol{Exp}(\mathbb{C}^n)$].[2] Here $\hat{u}$ is sought as some meromorphic function whose (inverse) Fourier transform defines a smooth function in $\mathbb{R}^n$. This is exactly the approach in [Ehrenpreis 1954] and what was to be the start of his approach to the Fundamental Principle. In the scalar case, this settles the solvability problem in $\mathbb{R}^n$ for compactly supported right-hand sides (the same approach can be followed when $\mathcal{C}^\infty$ is replaced by $\mathcal{D}'$ and many other distribution spaces).

In the case of $f \in \mathcal{C}^\infty(\Omega)$ not compactly supported, one represents f as the limit of a sequence of $f_\nu \in \mathcal{C}_c^\infty(\Omega)$ with $f_{\nu+1} = f_\nu$ in convex open sets $\Omega_\nu \subset\subset \Omega$, $\Omega_\nu \nearrow \Omega$. The preceding reasoning yields $u_\nu \in \mathcal{C}^\infty(\mathbb{R}^n)$ verifying $P(D)\,u_\nu = f_\nu$ in $\mathbb{R}^n$ for each $\nu = 1, 2, \ldots$, whence $P(D)\,(u_{\nu+1} - u_\nu) = 0$ in Ω_ν. The approximation theorem (proved in [Malgrange 1955])provides an exponential-polynomial solution h_ν of (2.3) such that $u_{\nu+1} - u_\nu - h_\nu$ is "very small" in the complete metric space $\mathcal{C}^\infty(\Omega_\nu)$. Then the standard Mittag-Leffler argument applies: as $N \to +\infty$, the limit of

$$u_1 + \sum_{\nu=1}^{N} (u_{\nu+1} - u_\nu - h_\nu) = u_{N+1} - \sum_{\nu=1}^{N} h_\nu \in \mathcal{C}^\infty(\Omega_N)$$

defines a solution $u \in \mathcal{C}^\infty(\Omega)$ of (2.5).

---

[2]The notations $\boldsymbol{P}, \boldsymbol{Q}, \ldots$ will always stand for matrix-valued polynomials. The corresponding differential operators will always be denoted by $\boldsymbol{P}(D), \boldsymbol{Q}(D), \ldots$

3 Solution of the Classical Problems

3.1 Simple Algebra in the General Case

Back to (2.2) assuming $p+q \geq 3$. On the Fourier transform side, we are faced with the division problem:

$$\sum_{k=1}^{q} P_{j,k}(\xi)\,\hat{u}_k(\xi) = \hat{f}_j(\xi)\,,\; j = 1,\ldots,p, \tag{3.1}$$

where $\hat{f}_j(\zeta) \in \boldsymbol{Exp}(\mathbb{C}^n)$ [and $\hat{f}_j(\xi) \in \mathcal{S}(\mathbb{R}^n)$ or $\hat{f}_j(\xi) \in \mathcal{S}'(\mathbb{R}^n)$]. The homogeneous equations (2.3) translate into

$$\sum_{k=1}^{q} P_{j,k}(\xi)\,\hat{h}_k(\xi) = 0,\; j = 1,\ldots,p. \tag{3.2}$$

The following $\mathbb{C}[\zeta]$-modules of vector-valued polynomials are obviously relevant to the problems under discussion:

1. $\mathfrak{M}_{\boldsymbol{P}}$: submodule of $\mathbb{C}[\zeta]^q$ generated by the "row" polynomials $\mathbf{P}_j = (P_{j,k})_{k=1,\ldots,q}$, $j = 1,\ldots,p$.
2. $\mathfrak{R}_{\boldsymbol{P}}$, the set of vectors $\mathbf{R} = (R_j)_{j=1,\ldots,p} \in \mathbb{C}[\zeta]^p$ such that $\mathbf{R}\boldsymbol{P} = 0$, i.e.,

$$\sum_{j=1}^{p} R_j P_{j,k} = 0,\, k = 1,\ldots,q$$

($\mathfrak{R}_{\boldsymbol{P}}$ is often referred to as the *set of relations* of $\mathfrak{M}_{\boldsymbol{P}}$).
3. $\mathfrak{S}_{\boldsymbol{P}}$, the set of vectors $\mathbf{S} = (S_k)_{k=1,\ldots,q} \in \mathbb{C}[\zeta]^q$ such that $\boldsymbol{P}\mathbf{S} = 0$.

Since $\mathbb{C}[\zeta_1,\ldots,\zeta_n]$ is Noetherian, every submodule of $\mathbb{C}[\zeta]^N$ $(1 \leq N \in \mathbb{Z}_+)$ is finitely generated: we can select a finite set of generators $\mathbf{R}_i = (R_{i,j})_{j=1,\ldots,p}$ $(i = 1,\ldots,m)$ of $\mathfrak{R}_{\boldsymbol{P}}$ and a finite set of generators $\mathbf{S}_\ell = (S_{k,\ell})_{k=1,\ldots,q} \in \mathbb{C}[\zeta]^q$ $(k = 1,\ldots,r)$ of $\mathfrak{S}_{\boldsymbol{P}}$. The $m \times p$ matrix $\boldsymbol{R} = (R_{i,j})_{1\leq i\leq m,1\leq j\leq p}$ provides the *compatibility conditions* for the inhomogeneous equations (2.1): for (3.1) to be solvable, it is necessary that

$$\sum_{j=1}^{p} R_{i,j}(\xi)\,\hat{f}_j(\xi) = 0,\, i = 1,\ldots,m. \tag{3.3}$$

The $q \times r$ matrices $\boldsymbol{S} = (S_{k,\ell})_{1\leq k\leq q,1\leq \ell\leq r}$ provides solutions of (3.2):

$$\sum_{k=1}^{q}\sum_{\ell=1}^{r} P_{j,k}(\xi)\,S_{k,\ell}(\xi)\,\psi_\ell(\xi) = 0,\; j = 1,\ldots,p, \tag{3.4}$$

whatever the functions ψ_ℓ. In the linear algebra sequence

$$\mathbb{C}[\zeta]^r \xrightarrow{S} \mathbb{C}[\zeta]^q \xrightarrow{P} \mathbb{C}[\zeta]^p \xrightarrow{R} \mathbb{C}[\zeta]^m \tag{3.5}$$

we have $\ker \boldsymbol{P} = \mathfrak{S}_{\boldsymbol{P}}$, the range of $\boldsymbol{S}$; the range of the map $\boldsymbol{P}$ is equal to $\mathfrak{M}_{\boldsymbol{P}} \subset \ker \boldsymbol{R}$. But we do not necessarily have $\mathfrak{M}_{\boldsymbol{P}} = \ker \boldsymbol{R}$: for instance, in the scalar case $\mathfrak{M}_{\boldsymbol{P}} = P\mathbb{C}[\zeta]$, the ideal generated by the polynomial P and $RP = 0$ entails $R = 0$.

3.2 Analytic Sheaf Theory to the Rescue

The roles of the multipliers $\boldsymbol{R}$ and $\boldsymbol{S}$ are, in a sense, *generic*: at a particular point $\xi \in \mathbb{R}^n$ or more generally $\xi \in \mathbb{C}^n$, there could be much "richer" relations among the $P_{j,k}(\xi)$ than those expressed by $\boldsymbol{R}(\xi)\boldsymbol{P}(\xi) = 0$ or $\boldsymbol{P}(\xi)\boldsymbol{S}(\xi) = 0$: as an extreme example, think of a root ξ of $\boldsymbol{P}(\xi) = 0$. But we do need some genericity (or stability) to construct solutions $\mathbf{u}, \mathbf{h}$ of (2.1) and (2.3), respectively, of the desired function or distribution class. Thus, even at the local level, we need something "more" than the elementary algebra of (3.5). Fortunately, by 1960, the Oka-Cartan-Serre theory of coherent algebraic (or analytic) sheaves had been satisfactorily constructed (see [Serre 1955]).

Let $\mathcal{O}_{\zeta^\circ}$ stand for the ring of germs of holomorphic functions at $\zeta^\circ \in \mathbb{C}^n$ (or, equivalently, the ring of convergent series in the powers of $\zeta - \zeta^\circ$); the sheaf $\mathcal{O}$ is the disjoint union of the "stalks" $\mathcal{O}_\zeta$ as ζ ranges over $\mathbb{C}^n$, equipped with its natural "sheaf" topology: if $U \subset \mathbb{C}^n$ is open and if $h \in \mathcal{O}(U)$, then $U \ni \zeta \longrightarrow h_\zeta \in \mathcal{O}_\zeta$ is a homeomorphism of U onto an open subset of $\mathcal{O}$, its inverse being the base projection; every open subset of $\mathcal{O}$ is a union of such sections. The definition of the powers $\mathcal{O}^N$ $(N = 1, 2, \ldots)$ is self-evident; $\mathbb{C}[\zeta_1, \ldots, \zeta_n]^N$ and its submodules can be identified to subsheafs of $\mathcal{O}^N$.

Going back to our matrix-valued polynomial $\boldsymbol{P}$, we consider the sheaf map

$$\mathcal{O}^q \xrightarrow{P} \mathcal{O}^p \tag{3.6}$$

with $\boldsymbol{P}$ acting multiplicatively on each stalk. The range of (3.6) is the sheaf $\mathcal{M}_{\boldsymbol{P}}$ generated over the sheaf of rings $\mathcal{O}$ by the submodule $\mathfrak{M}_{\boldsymbol{P}}$.

Remark 1. Keep in mind that $\mathbf{v} \in \mathcal{M}_{\boldsymbol{P}}$ means that $\mathbf{v} = \sum_{j=1}^{p} \mathbf{P}_j g_j = \boldsymbol{P}^\top \mathbf{g}$ for some $\mathbf{g} \in \mathcal{O}^q$ ($\boldsymbol{P}^\top$: transpose of the matrix $\boldsymbol{P}$).

A basic result of analytic sheaf theory (see, e.g., [Gunning and Rossi 1965], p. 130) is that the kernel and cokernel of a map such as (3.6) are coherent, meaning that, locally, they are both finitely generated and so are their sheaves of relations. This means that there is an exact sequence of sheaf maps

$$\mathcal{O}^r \xrightarrow{\varphi} \mathcal{O}^q \xrightarrow{P} \mathcal{O}^p \xrightarrow{\psi} \mathcal{O}^p/\mathcal{M}_{\boldsymbol{P}} \longrightarrow 0, \tag{3.7}$$

where ψ is the quotient map. Actually, a celebrated theorem of Oka tells us much more about the possible choice of the map $\boldsymbol{\varphi}$: it can be taken to be *algebraic*.

Theorem 1. *The kernel $\mathcal{K}_{\boldsymbol{P}}$ of the sheaf map (3.6) is generated (over the sheaf of rings $\mathcal{O}$) by a number $r < +\infty$ of vector-valued polynomials $S_\ell = (S_{k,\ell})_{k=1,\ldots,q} \in \mathbb{C}[\zeta_1, \ldots, \zeta_n]^q$.*

The proof consists in showing (by Oka's argument, see [Hörmander 1966], Lemma 7.6.3) that the elements of $\mathbb{C}[\zeta_1, \ldots, \zeta_n]^q$ belonging to $\mathcal{K}_{\boldsymbol{P}}$ generate $\mathcal{K}_{\boldsymbol{P}}$. Then the matrix $\boldsymbol{S}$ in (3.5) can still be made use of. Thus, the short sequence of sheaf maps

$$\mathcal{O}^r \xrightarrow{S} \mathcal{O}^q \xrightarrow{P} \mathcal{O}^p \tag{3.8}$$

is exact, and we can take φ =(multiplication by)$\boldsymbol{S}$ in (3.7). The exactness of (3.8) means that a convergent power series $\mathbf{h}(\zeta) = \sum_{\alpha\in\mathbb{Z}^n} \mathbf{h}_\alpha (\zeta - \zeta^\circ)^\alpha$ ($\mathbf{h}_\alpha \in \mathbb{C}^q$) satisfies the multiplication equation $\boldsymbol{P}\mathbf{h} = \mathbf{0}$ if and only if there is a convergent power series $\mathbf{g}(\zeta) = \sum_{\alpha\in\mathbb{Z}^n} \mathbf{g}_\alpha (\zeta - \zeta^\circ)^\alpha$ ($\mathbf{g}_\alpha \in \mathbb{C}^r$) such that $\boldsymbol{S}\mathbf{g} = \mathbf{h}$.

Let us denote by $\boldsymbol{P}^\top$ the transpose of the matrix $\boldsymbol{P} = \left(P_{j,k}\right)_{1\le j\le p, 1\le k\le q}$ and define $\boldsymbol{P}^\flat(\zeta) = \boldsymbol{P}^\top(-\zeta)$; $\boldsymbol{P}^\flat(\zeta)$ is the "total symbol" of the transpose of the differential operator $\boldsymbol{P}(D)$, $\boldsymbol{P}(D)^\top = \boldsymbol{P}^\top(-D)$. We can apply Theorem 1 with $\boldsymbol{P}^\flat$ in the place of $\boldsymbol{P}$:

Theorem 2. *The kernel $\mathcal{K}_{\boldsymbol{P}^\flat}$ of the sheaf map $\mathcal{O}^p \xrightarrow{P^\flat} \mathcal{O}^q$ is generated (over the sheaf of rings $\mathcal{O}$) by a number $m < +\infty$ of vector-valued polynomials $T_\ell = (T_{k,\ell})_{k=1,\ldots,p} \in \mathbb{C}[\zeta_1, \ldots, \zeta_n]^p$.*

The matrix $\boldsymbol{R}(\zeta)$ in (3.5) can be made use of, by taking $\boldsymbol{T} = \boldsymbol{R}^\flat$, i.e., $T_{k,\ell}(\zeta) = R_{\ell,k}(-\zeta)$. We get the exact short sequence of sheaf maps

$$\mathcal{O}^m \xrightarrow{R^\flat} \mathcal{O}^p \xrightarrow{P^\flat} \mathcal{O}^q. \tag{3.9}$$

Remark 2. In the scalar case, when $p = q = 1$, multiplication of convergent series by the polynomial P is an injective map of $\mathcal{O}$ onto the proper ideal $P\mathcal{O}$; in the sequences (3.5), (3.8), and (3.9), we must take $R = 0$ and $S = 0$. This shows that we cannot transpose (3.9) and glue the result to (3.8) to obtain an exact sequence $\mathcal{O}^r \xrightarrow{S} \mathcal{O}^q \xrightarrow{P} \mathcal{O}^p \xrightarrow{R} \mathcal{O}^m$.

3.3 Estimates and Their Exploitation

Below we use the notation

$$\Delta^n = \left\{z \in \mathbb{C}^n;\ \left|z_j\right| < 1,\ j = 1, \ldots, n\right\}.$$

Let φ be a plurisubharmonic function in $\mathbb{C}^n$ satisfying the condition

$$\forall (z, \zeta) \in \Delta^n \times \mathbb{C}^n, \ |\varphi(z+\zeta) - \varphi(\zeta)| \leq C_\circ. \tag{3.10}$$

Theorem 3. *Given the system* $\boldsymbol{P}$, *there is an integer* N *such that to each* $\mathbf{u} \in \mathcal{O}(\mathbb{C}^n)^q$, *there is* $\mathbf{v} \in \mathcal{O}(\mathbb{C}^n)^q$ *verifying* $\boldsymbol{P}\mathbf{v} = \boldsymbol{P}\mathbf{u}$ *and*

$$\int_{\mathbb{R}^{2n}} |\mathbf{v}(\zeta)|^2 e^{-\varphi(\zeta)} \left(1+|\zeta|^2\right)^{-N} d\xi d\eta \leq C_1 \int_{\mathbb{R}^{2n}} |\boldsymbol{P}(\zeta)\mathbf{u}(\zeta)|^2 e^{-\varphi(\zeta)} d\xi d\eta \tag{3.11}$$

where $C_1 > 0$ *depends solely on* $C_\circ$, *the constant in (3.10).*

Theorem 3 is a special case of Theorem 7.6.11 in [Hörmander 1966];it enters under the heading of "cohomology with bounds" for the sheaf of modules $\mathcal{M}_{\boldsymbol{P}}$ generated by $\mathfrak{M}_{\boldsymbol{P}}$: one needs a Cartan Theorem B with bounds. If $f \in \Gamma(\mathbb{C}^n, \mathcal{O}^p, k+1)$ is a $(k+1)$-cocycle (meaning that $\boldsymbol{R}f = 0$), the classical Theorem B tells us that there is $u \in \Gamma(\mathbb{C}^n, \mathcal{O}^q, k)$ such that $\boldsymbol{P}u = f$. We now require f to satisfy suitable estimates

$$\int_U |f(\zeta)|^2 e^{-\varphi(\zeta)} d\xi d\eta < +\infty,$$

for every element U of the open covering of $\mathbb{C}^N$ used to define the cochains (U is commonly taken to be a suitably small hypercube). One must find a k-cochain $v \in \Gamma(\mathbb{C}^n, \mathcal{O}^q, k)$ which satisfies the equation $\boldsymbol{P}v = \boldsymbol{P}u = f$ as well as an estimate of the type (3.11) but with domain of integration U. The latter is achieved by taking a closer look at the Weierstrass preparation theorem and devising lower bounds for $|\boldsymbol{P}u|$ in U.

The same type of argument, combined with the exactness of the sequence (3.8), leads to

Theorem 4. *Given the system* $\boldsymbol{P}$, *there is* $N \in \mathbb{Z}$ *such that to each* $\mathbf{h} \in \mathcal{O}(\mathbb{C}^n)^q$ *verifying* $\boldsymbol{P}\mathbf{h} = 0$, *there is* $\mathbf{v} \in \mathcal{O}(\mathbb{C}^n)^r$ *verifying* $\mathbf{h} = \boldsymbol{S}\mathbf{v}$ *and*

$$\int_{\mathbb{R}^{2n}} |\mathbf{v}(\zeta)|^2 e^{-\varphi(\zeta)} \left(1+|\zeta|^2\right)^{-N} d\xi d\eta \leq C_1 \int_{\mathbb{R}^{2n}} |\mathbf{h}(\zeta)|^2 e^{-\varphi(\zeta)} d\xi d\eta$$

where $C_1 > 0$ *depends solely on* $C_\circ$, *the constant in (3.10).*

The next two lemmas are important consequences of Theorems 3 and 4. In every one of the remaining statements in this section, Ω will be an arbitrary convex open subset of $\mathbb{R}^n$.

Lemma 1. *If* $f \in C_c^\infty(\Omega)^p$ *satisfies the compatibility conditions* $\boldsymbol{R}(D)f = 0$ *[cf. (3.9)], then there exists* $g \in C_c^\infty(\Omega)^q$ *such that* $f = \boldsymbol{P}(D)g$.

Proof. One must show that, under the hypotheses of the statement, the linear functional $\lambda_f : \psi = \boldsymbol{P}^\flat(D)\chi \to \langle f, \chi \rangle$ is continuous for the topology induced on

$\boldsymbol{P}^{\flat}(D)\,\mathcal{D}'(\Omega)^p$ by $\mathcal{D}'(\Omega)^q$. Indeed, if this is true then the Hahn–Banach theorem allows one to extend λ_f as a continuous linear functional $\mathcal{D}'(\Omega)^q \ni \phi \to \langle g, \phi\rangle$ with $g \in C_c^{\infty}(\Omega)^q$, and then obviously $\langle f, \chi\rangle = \left\langle g, \boldsymbol{P}^{\flat}(D)\,\chi\right\rangle = \langle \boldsymbol{P}(D)\,g, \chi\rangle$ for all $\chi \in \mathcal{D}'(\Omega)^p$. Actually it suffices to deal with distributions $\chi \in \mathcal{E}'(K)^p$, meaning $\chi \in \mathcal{D}'(\mathbb{R}^n)^p$ and $\operatorname{supp}\chi \subset K$, with K an arbitrary convex compact subset of Ω whose interior contains $\operatorname{supp} f$. Since K is convex, the function $\mathbb{C}^n \ni \zeta = \xi + i\eta \longrightarrow H_K(\eta) = \max_{x\in K}(x\cdot\eta) \in \mathbb{R}$ is plurisubharmonic and satisfies (3.10) for a suitably large $C_\circ$; the same is true of $2H_K(\eta) + N\log(1+|\zeta|^2)$ if $N \geq 0$. By the Paley–Wiener theorem, we have, for a suitable $N_\circ \in \mathbb{Z}_+$,

$$\|\hat{\chi}\|^2_{K,N_\circ} = \int_{\mathbb{R}^{2n}} |\hat{\chi}(\zeta)|^2\, \mathrm{e}^{-2H_K(\eta)} \left(1+|\zeta|^2\right)^{-N_\circ} \mathrm{d}\xi \mathrm{d}\eta < +\infty \tag{3.12}$$

as well as $\|\hat{\psi}\|^2_{K,N_\circ+\kappa} < +\infty$ (κ: degree of $\boldsymbol{P}$).

First, we apply Theorem 3 with $\boldsymbol{P}^{\flat}$ in the place of $\boldsymbol{P}$: there is $v \in \mathcal{O}(\mathbb{C}^n)^p$ such that $\boldsymbol{P}^{\flat}(v - \hat{\chi}) = 0$ and $\|v\|^2_{K,N_\circ+N_1} \leq C\,\|\hat{\psi}\|^2_{K,N_\circ+\kappa}$ for suitably large positive constant N_1, C, independent of χ (but not of K). From the Paley–Wiener-Schwartz theorem, it follows that $v = \hat{\chi}_1$, $\chi_1 \in \mathcal{E}'(\mathbb{R}^n)^p$ and $\operatorname{supp}\chi_1 \subset K$.

Next, we apply Theorem 4 with $\boldsymbol{P}^{\flat}$ in the place of $\boldsymbol{P}$: there is $w \in \mathcal{O}(\mathbb{C}^n)^{\kappa}$ verifying $\boldsymbol{R}^{\flat} w = \hat{\chi} - \hat{\chi}_1$ and $\|w\|^2_{K,N_\circ+N_2} \leq C'\,\|\hat{\chi} - \hat{\chi}_1\|^2_{K,N_1}$ for suitable constants $N_2 > N_1, C' > 0$ (independent of $\hat{\chi} - v$). From the Paley–Wiener–Schwartz theorem, it follows that $w = \hat{\mu}$, $\mu \in \mathcal{E}'(\mathbb{R}^n)^{\kappa}$ and $\operatorname{supp}\mu \subset K$. Since $\boldsymbol{R}(D)\,f = 0$, we derive $\langle f, \chi\rangle = \langle f, \chi_1\rangle$. This proves that the map $\mathcal{E}'(\Omega)^q \ni \psi \longrightarrow \chi_1 \in \mathcal{E}'(\Omega)^p$ is continuous, whence the claim.

The proof actually shows that to each sufficiently large order $m_1 \in \mathbb{Z}_+$, there is $m_2 > m_1$ such that if $f \in C_c^{m_1}(\Omega)^q$ satisfies the compatibility conditions $\boldsymbol{R}(D)\,f = 0$, then there exists $g \in C_c^{m_2}(\Omega)^p$ such that $f = \boldsymbol{P}(D)\,g$. Using this observation and standard techniques of the theory of constant coefficients PDE, it is possible to prove

Lemma 2. *If $f \in \mathcal{E}'(\Omega)^q$ satisfies the compatibility conditions $\boldsymbol{R}(D)\,f = 0$, then there exists $g \in \mathcal{E}'(\Omega)^p$ such that $f = \boldsymbol{P}(D)\,g$.*

Theorem 5. *The closure in $C^{\infty}(\Omega)^q$ of the exponential-polynomial solutions of (2.3) is the subspace of $C^{\infty}(\Omega)^q$ of all solutions of (2.2).*

Proof. By the Hahn–Banach theorem, the claim is proved if we show that the hypothesis that $f = (f_1, \ldots, f_q) \in \mathcal{E}'(\Omega)^q$ is orthogonal to all exponential-polynomial solutions of (2.3), implies that f is orthogonal to all of solutions of (2.3) in $C^{\infty}(\Omega)^q$. It is not difficult to prove that this hypothesis implies

$$\forall \zeta \in \mathbb{C}^n,\ \hat{f}(\zeta)\cdot h(-\zeta) = 0$$

for all $h \in \mathcal{O}(\mathbb{C}^n)^q$ such that $\boldsymbol{P}h \equiv 0$. According to Theorem 2, the latter means that there is $w \in \mathcal{O}(U)^r$ such that $h = \boldsymbol{S}\boldsymbol{w}$ whence $\boldsymbol{S}(-\zeta)^{\top}\hat{f} = 0$, i.e., $\boldsymbol{S}^{\flat}(D)f = 0$. The latter are the compatibility conditions for $\boldsymbol{P}^{\flat}(D)$. Then Lemma 2 applied with $\boldsymbol{P}^{\flat}$ in place of $\boldsymbol{P}$ entails that there is $g \in \mathcal{E}'(\Omega)^p$ verifying $\boldsymbol{P}(-D)^{\top}g = f$. The sought conclusion ensues directly from this last fact.

Theorem 5 is the Runge-type theorem in the general case. After the proof that (2.1) can be solved for compactly supported right-hand sides satisfying the appropriate compatibility conditions [obtained through estimates like (3.12) for $\boldsymbol{P}^{\flat}$ in the place of $\boldsymbol{P}$], Theorem 5 is used via the Mittag-Leffler correction (just as in the scalar case) to prove the existence of solutions:

Theorem 6. *To each $f \in \mathcal{C}^{\infty}(\Omega)^q$ such that $\boldsymbol{R}(D)f = 0$, there is a solution $u \in \mathcal{C}^{\infty}(\Omega)^p$ of the system of equations $\boldsymbol{P}(D)u = f$.*

Proof. Lemma 1 shows that we can solve (2.1) if $f \in \mathcal{C}_c^{\infty}(\Omega)^q$. Theorem 5 enables us to duplicate the Mittag-Leffler procedure described, in the scalar case, at the end of Sect. 2.

Keeping track of the integers N in the estimates of type (3.11) enables one to get some precision about the loss of derivatives in solving (2.2).

At this stage, the classical theorems of the early 1950s in the scalar case (at least for convex domains) have been generalized to all systems (2.2).

4 To the Fundamental Principle

4.1 Noetherian Operators

The proof of the Fundamental Principle demands that we further refine the description of the solutions in the multiplicative equation (3.2). By the exactness of the sequence (3.8), we know that they belong to the kernel of the sheaf map $\boldsymbol{Q}$. But this cannot be the whole story, as the case $p = q = 1$ shows: in this case, Q vanishes identically and (3.8) adds nothing to our knowledge (that multiplication by P is injective). Note that we know much more in the one-variable case: think of the ODE $\left(\frac{\mathrm{d}}{\mathrm{d}x} - \zeta\right)^m h = 0$ and of its solutions $h = x^k \mathrm{e}^{\zeta x}$, $k = 0, 1, \ldots, m-1$.

In the general case $\boldsymbol{P} = \left(P_{j,k}\right)_{1 \le j \le p, 1 \le k \le q}$, the first step is to identify the analogue of the null variety in the scalar case, $\overset{-1}{P}(0)$. Returning to the submodule $\mathfrak{M}_{\boldsymbol{P}}$ of $\mathbb{C}[\zeta_1, \ldots, \zeta_n]^q$ generated by the row-vectors $\mathbf{P}_j = \left(P_{j,k}\right)_{1 \le k \le q}$, $j = 1, \ldots, p$, we introduce its *primary decomposition*

$$\mathfrak{M}_{\boldsymbol{P}} = \mathfrak{M}_1 \cap \cdots \cap \mathfrak{M}_{\mathrm{I}} \tag{4.1}$$

where each submodule $\mathfrak{M}_{\iota}$ is proper and primary in the following sense:

Definition 1. A submodule $\mathfrak{M}$ of $\mathbb{C}[\zeta_1,\ldots,\zeta_n]^q$ is proper and primary if $\{0\} \neq \mathfrak{M} \neq \mathbb{C}[\zeta_1,\ldots,\zeta_n]^q$ and if the following condition is satisfied, whatever $F \in \mathbb{C}[\zeta_1,\ldots,\zeta_n]$:

(Pry): *If there is* $M \in \mathbb{C}[\zeta_1,\ldots,\zeta_n]^q \setminus \mathfrak{M}$ *such that* $FM \in \mathfrak{M}$, *then there is* $s \in \mathbb{Z}_+$ *such that* $F^s\mathbb{C}[\zeta_1,\ldots,\zeta_n]^q \subset \mathfrak{M}$.

Condition **(Pry)** says that that either multiplication by $F \in \mathbb{C}[\zeta_1,\ldots,\zeta_n]$ sends into $\mathfrak{M}$ only the vector-valued polynomials that already belong to $\mathfrak{M}$ or else, for large enough $s \in \mathbb{Z}_+$, multiplication by F^s sends all vector-valued polynomials into $\mathfrak{M}$. When $\mathfrak{M}$ is primary, the polynomials F such that $F^s\mathbb{C}[\zeta_1,\ldots,\zeta_n]^q \subset \mathfrak{M}$ form a *prime* ideal $\mathfrak{p}$ of $\mathbb{C}[\zeta_1,\ldots,\zeta_n]$; the algebraic subvariety

$$V_{\mathfrak{p}} = \{\zeta \in \mathbb{C}^n;\ \forall F \in \mathfrak{p},\ F(\zeta) = 0\}$$

is irreducible, equivalent to the fact that the regular part of $V_{\mathfrak{p}}$ is connected (it is dense in $V_{\mathfrak{p}}$).

We denote by $\mathfrak{p}_\iota$ the prime ideal associated with the proper primary submodule $\mathfrak{M}_\iota$. The union

$$V_P = V_{\mathfrak{p}_1} \cup \cdots \cup V_{\mathfrak{p}_I} \tag{4.2}$$

will play the role played by the null variety in the scalar case—in which case $\mathfrak{M}_P$ and $\mathfrak{M}_\iota$ are simply ideals in $\mathbb{C}[\zeta_1,\ldots,\zeta_n]$.

We return to an arbitrary proper and primary submodule $\mathfrak{M}$ of $\mathbb{C}[\zeta_1,\ldots,\zeta_n]^q$ with associated prime ideal $\mathfrak{p}$. The local structure of $V_{\mathfrak{p}}$ is well known; in $\mathbb{C}^n$, a *global* statement is valid:

Lemma 3. *Possibly after an affine change of variables in* $\mathbb{C}^n$, *the following properties hold:*

(1) *There is* $\nu \in \mathbb{Z}_+$, $\nu < n$, *such that* $\mathfrak{p} \cap \mathbb{C}[\zeta_1,\ldots,\zeta_\nu] = \{0\}$ *[*ν *will be the largest such integer and we write* $\zeta' = (\zeta_1,\ldots,\zeta_\nu)$*].*

(2) *For each* $j = 1,\ldots,n-\nu$, *there is an irreducible polynomial* $\Phi_j \in \mathfrak{p} \cap \mathbb{C}\left[\zeta_1,\ldots,\zeta_\nu,\zeta_{\nu+j}\right]$ *of degree* $d_j \geq 1$, *of the form*

$$\Phi_j\left(\zeta',\zeta_{\nu+j}\right) = \zeta_{\nu+j}^{d_j} + \sum_{k=1}^{d_j} a_{j,k}\left(\zeta'\right)\zeta_{\nu+j}^{d_j-k}. \tag{4.3}$$

The discriminant of $\Phi_1(\zeta',\cdot)$, $D(\zeta') \in \mathbb{C}[\zeta_1,\ldots,\zeta_\nu]$, *does not vanish identically.*

(3) *For each* $j = 2,\ldots,n-\nu$, *there are polynomials* $\Psi_j(\zeta',\zeta_{\nu+1}) \in \mathbb{C}[\zeta_1,\ldots,\zeta_\nu,\zeta_{\nu+1}]$ *such that*

$$T_j\left(\zeta',\zeta_{\nu+1},\zeta_{\nu+j}\right) = D\left(\zeta'\right)\zeta_{\nu+j} - \Psi_j\left(\zeta',\zeta_{\nu+1}\right) \in \mathfrak{p}. \tag{4.4}$$

(4) *If* $V_D = \{\zeta \in V_P;\ D(\zeta') = 0\}$, *then* $V_{\mathfrak{p}} \backslash V_D$ *is a connected complex-analytic submanifold of* $\mathbb{C}^n$ *and the projection* $\pi : V_{\mathfrak{p}} \backslash V_D \ni \zeta \longrightarrow \zeta' \in \mathbb{C}^{\nu} \backslash \overset{-1}{D}(0)$ *is a local biholomorphism that turns* $V_{\mathfrak{p}} \backslash V_D$ *into a* d_1*-sheeted covering of* $\mathbb{C}^{\nu} \backslash \overset{-1}{D}(0)$. *The closure of* $V_{\mathfrak{p}} \backslash V_D$ *is equal to* $V_{\mathfrak{p}}$.

(5) *For some* $C > 0$ *and all* $\zeta \in V_{\mathfrak{p}}$, *then* $|\zeta| \le C\,(1 + |\zeta'|)$.

Property #5 is a statement about the degrees of the coefficients $a_{j,k}(\zeta')$ in (4.3).

In the sequel, s shall denote the smallest positive integer such that **(Pry)** is true. Consider then the differential operators acting on polynomials $F \in \mathbb{C}[\zeta_1, \dots, \zeta_n]^q$ of the kind

$$\mathcal{L} = \sum_{|\alpha| < s} c_\alpha(\zeta) \cdot D^{\alpha}_{\zeta''}, \tag{4.5}$$

where $D^{\alpha}_{\zeta''} = D^{\alpha_1}_{\zeta_{\nu+1}} \cdots D^{\alpha_{n-\nu}}_{\zeta_n}$ and $c_\alpha(\zeta) \in \mathbb{C}[\zeta_1, \dots, \zeta_n]^q$ ($\cdot$ stands for the scalar product in $\mathbb{C}[\zeta_1, \dots, \zeta_n]^q$). We denote by $\boldsymbol{Diff}_s(\mathfrak{M})$ the set of operators (4.5) such that $\mathcal{L}F \in \mathfrak{p}$ whatever $F \in \mathfrak{M}$.

The proof of the next lemma is based on the description of $V_{\mathfrak{p}}$ in Lemma 3.

Lemma 4. *For* $F \in \mathbb{C}[\zeta_1, \dots, \zeta_n]^q$ *to belong to* $\mathfrak{M}$, *it is (necessary and) sufficient that* $\mathcal{L}F \in \mathfrak{p}$ *for all* $\mathcal{L} \in \boldsymbol{Diff}_s(\mathfrak{M})$.

It is clear that $\boldsymbol{Diff}_s(\mathfrak{M})$ is a finitely generated $\mathbb{C}[\zeta_1, \dots, \zeta_n]$-module; it is essentially characterized by the property in Lemma 4 together with the property that $[\mathcal{L}, \zeta_j] \in \boldsymbol{Diff}_s(\mathfrak{M})$ for every $j = 1, \dots, n$. The differential operators $\mathcal{L} \in \boldsymbol{Diff}_{s_\iota}(\mathfrak{M}_\iota)$ were called *Noetherian* by Palamodov ([Palamodov 1963], Chap. 4, Sects. 3 and 4; this concept is relative to the variety $V_{\mathfrak{p}}$).

Let $\mathcal{M}$ denote the subsheaf of $\mathcal{O}^q$ generated by $\mathfrak{M}$. If $U \subset \mathbb{C}^n$ is open, we call $\mathcal{M}(U)$ the set of continuous sections of $\mathcal{M}$ over U; $\mathcal{M}(U)$ is an $\mathcal{O}(U)$-submodule of $\mathcal{O}(U)^q$. The preceding lemmas lead to the following description of the elements of $\mathcal{M}(U)$ when U is Stein (the proof exploits the coherence of the analytic subsheaves of $\mathcal{O}^q$):

Theorem 7. *If* $U \subset \mathbb{C}^n$ *is a domain of holomorphy, then, for* $f \in \mathcal{O}(U)^q$ *to belong to* $\mathcal{M}(U)$, *it is (necessary and) sufficient that* $\mathcal{L}f \equiv 0$ *on* $V_{\mathfrak{p}} \cap U$ *for all* $\mathcal{L} \in \boldsymbol{Diff}_s(\mathfrak{M})$.

Returning to the primary decomposition (3.10), we can state

Proposition 1. *If* $U \subset \mathbb{C}^n$ *is a domain of holomorphy, then*

$$\mathcal{M}_P(U) = \mathcal{M}_1(U) \cap \cdots \cap \mathcal{M}_{\mathbf{I}}(U). \tag{4.6}$$

Let s_ι denote the *smallest* positive integer such that $\mathfrak{M}_\iota$ satisfies **(Pry)** with $s = s_\iota$. We select finitely many generators $\mathcal{L}_{\iota,\nu}$ ($\nu = 1, \dots, n_\iota$) of $\boldsymbol{Diff}_{s_\iota}(\mathfrak{M}_\iota)$ for each $\iota = 1, \dots, \mathbf{I}$, The following direct consequence of Theorem 7 is of great importance in what follows:

Theorem 8. *For $f \in \mathcal{O}(\mathbb{C}^n)^q$ to belong to $\mathcal{M}_{\boldsymbol{P}}(\mathbb{C}^n)$, it is necessary and sufficient that $\mathcal{L}_{\iota,\nu} f \equiv 0$ on $\boldsymbol{V}_{\mathfrak{p}_\iota}$ for all $\iota = 1,\ldots,\mathbf{I}$, $\nu = 1,\ldots,n_\iota$.*

The exactness of the sequence (3.5) entails, then:

Corollary 1. *Let $f \in \mathcal{O}(\mathbb{C}^n)^q$ be arbitrary. If $\mathcal{L}_{\iota,\nu} f \equiv 0$ on $\boldsymbol{V}_{\mathfrak{p}_\iota}$ for all $\iota = 1,\ldots,\mathbf{I}$, $\nu = 1,\ldots,n_\iota$ then $\boldsymbol{R} f \equiv 0$.*

If we write

$$\mathcal{L}_{\iota,\nu}\left(\zeta, D_\zeta\right) f(\zeta) = \sum_{|\alpha|<s_\iota} c_{\iota,\nu;\alpha}(\zeta) \cdot D^{\alpha}_{\zeta''} f(\zeta), \; f \in \mathcal{O}(\mathbb{C}^n)^q, \tag{4.7}$$

we can define the "symbol"

$$\widehat{\mathcal{L}}_{\iota,\nu}(\zeta, z) = \sum_{|\alpha|<s_\iota} c_{\iota,\nu;\alpha}(\zeta)\, z''^{\alpha}. \tag{4.8}$$

For each $z \in \mathbb{C}^n$, $\widehat{\mathcal{L}}_{\iota,\nu}(\zeta, z) \in \mathbb{C}[\zeta_1,\ldots,\zeta_n]^q$.

Proposition 2. *If $\zeta^\circ \in V_{\boldsymbol{P}}$, then $\widehat{\mathcal{L}}_{\iota,\nu}(\zeta^\circ, x)\,\mathrm{e}^{\mathrm{i}\langle\zeta^\circ,x\rangle}$ is an exponential-polynomial solution of (2.3).*

Proof. We have, for each $j = 1,\ldots,p$,

$$\begin{aligned}
&P_j(D_x)\left(\widehat{\mathcal{L}}_{\iota,\nu}(\zeta^\circ, x)\,\mathrm{e}^{\mathrm{i}\langle\zeta^\circ,x\rangle}\right)\\
&= \mathrm{e}^{\mathrm{i}\langle\zeta^\circ,x\rangle} \sum_{|\alpha|<s_\iota} \sum_{\beta\preceq\alpha} \frac{1}{\beta!} c_{\iota,\nu;\alpha}\left(\zeta^\circ\right) D^\beta\left(x''^{\alpha}\right) P_j^{(\beta)}\left(\zeta^\circ\right)\\
&= \mathrm{e}^{\mathrm{i}\langle\zeta^\circ,x\rangle} \sum_{|\alpha|<s_\iota} c_{\iota,\nu;\alpha}\left(\zeta^\circ\right) \sum_{\beta\preceq\alpha} \frac{\alpha!}{\beta!\,(\alpha-\beta)!} x''^{\alpha-\beta} P_j^{(\beta)}\left(\zeta^\circ\right)\\
&= \sum_{|\alpha|<s_\iota} c_{\iota,\nu;\alpha}(\zeta) \sum_{\beta\preceq\alpha} \frac{\alpha!}{\beta!\,(\alpha-\beta)!} D_\zeta^{\alpha-\beta}\left(\mathrm{e}^{\mathrm{i}\langle\zeta,x\rangle}\right) P_j^{(\beta)}(\zeta)\Bigg|_{\zeta=\zeta^\circ}\\
&= \mathcal{L}_{\iota,\nu}\left(\zeta, D_\zeta\right)\left(\mathrm{e}^{\mathrm{i}\langle\zeta^\circ,x\rangle} P_j(\zeta)\right)\Big|_{\zeta=\zeta^\circ} = 0
\end{aligned}$$

by Theorem 8.

Let $f \in \mathcal{E}'(\mathbb{R}^n)^q$ be arbitrary and let $\mathcal{L}_{\iota,\nu}\left(\zeta, D_\zeta\right)$ act on

$$\hat{f}(-\zeta) = \int \mathrm{e}^{\mathrm{i}\langle x,\zeta\rangle} f(x)\,\mathrm{d}x \tag{4.9}$$

(the integral $\int \cdot \mathrm{d}x$ stands for the duality bracket between compactly supported distributions and $\mathcal{C}^\infty$ functions); we get directly

$$\mathcal{L}_{\iota,\nu}\left(\zeta, D_\zeta\right) \hat{f}(-\zeta) = \int \mathrm{e}^{\mathrm{i}\langle x,\zeta\rangle} \widehat{\mathcal{L}}_{\iota,\nu}(\zeta, x) f(x)\, \mathrm{d}x. \tag{4.10}$$

Proposition 3. *Let $f \in \mathcal{E}'(\mathbb{R}^n)^q$ be such that $\mathcal{L}_{\iota,\nu}\left(\zeta, D_\zeta\right) \hat{f}(-\zeta) = 0$ for all $\iota = 1, \ldots, \mathbf{I}$, $\nu = 1, \ldots, n_\iota$ and all $\zeta \in V_{\boldsymbol{P}}$. There is $w \in \mathcal{E}'(\mathbb{R}^n)^q$ such that $f = \boldsymbol{P}^\flat(D)\, w$.*

Proof. Taking Remark 1 into account, combine Corollary 1 with Lemma 2.

4.2 Extension with Bounds: Final Statement

The next step is to obtain an extension theorem with bounds for functions. We first state the central result for an arbitrary proper and primary submodule $\mathfrak{M}$ of $\mathbb{C}[\zeta_1, \ldots, \zeta_n]^q$. We select finitely many generators $\mathcal{L}_j$ $(j = 1, \ldots, \rho_\circ)$ of $\boldsymbol{Diff}_s(\mathfrak{M})$.

Theorem 9. *Let φ be a plurisubharmonic function in $\mathbb{C}^n$ satisfying the following condition* [cf. (3.1)]*:*

(Temp): *For some positive constants $C_\circ$, C_1, κ and all $(z, \zeta) \in \mathbb{C}^{2n}$,*

$$|z| \le C_1 (1 + |\zeta|)^{-\kappa} \Longrightarrow |\varphi(\zeta + z) - \varphi(\zeta)| \le C_\circ.$$

Then, for a suitable choice of the positive constants C and N, to each $f \in \mathcal{O}(\mathbb{C}^n)$ there is $g \in \mathcal{O}(\mathbb{C}^n)$ such that $f - g \in \mathcal{M}(\mathbb{C}^n)$ and such, moreover, that

$$\sup_{\zeta \in \mathbb{C}^n} (1 + |\zeta|)^{-N} \mathrm{e}^{-\varphi(\zeta)} |g(\zeta)| \le C \sup_{\zeta \in V_{\mathfrak{p}}} \mathrm{e}^{-\varphi(\zeta)} \left|\mathcal{L}_j f(\zeta)\right|. \tag{4.11}$$

For a proof, see pp. 242–243, [Hörmander 1966]. In generalizing Theorem 9 to the submodule $\mathfrak{M}_{\boldsymbol{P}}$ in (4.1), we make use of the generators $\mathcal{L}_{\iota,j}$ $(j = 1, \ldots, n_\iota)$ of $\boldsymbol{Diff}_{s_\iota}(\mathfrak{M}_\iota)$ introduced above $(\iota = 1, \ldots, \mathbf{I})$.

Theorem 10. *Let the plurisubharmonic function φ in $\mathbb{C}^n$ satisfy **(Temp)**. Then, for a suitable choice of the positive constants C and N, to each $f \in \mathcal{O}(\mathbb{C}^n)$, there is $g \in \mathcal{O}(\mathbb{C}^n)$ such that $f - g \in \mathcal{M}_{\boldsymbol{P}}(\mathbb{C}^n)$ and such, moreover, that*

$$\sup_{\zeta \in \mathbb{C}^n} (1 + |\zeta|)^{-N} \mathrm{e}^{-\varphi(\zeta)} |g(\zeta)| \le C \max_{\iota=1,\ldots,\mathbf{I}} \left(\max_{\nu=1,\ldots,\rho_i} \left(\sup_{\zeta \in V_{\mathfrak{p}_\iota}} \mathrm{e}^{-\varphi(\zeta)} |\mathcal{L}_{\iota,\nu} f(\zeta)| \right)\right). \tag{4.12}$$

In what follows, $\Omega \subset \mathbb{R}^n$ shall once again denote a convex open set and $h = (h_1, \ldots, h_q) \in \mathcal{C}^\infty(\Omega)^q$ will be a solution of (2.3). We have, for every $w = (w_1, \ldots, w_q) \in \mathcal{E}'(\Omega)^q$,

$$\int h \cdot \left(\boldsymbol{P}^\flat(D)\, w\right) \mathrm{d}x = \int w \cdot (\boldsymbol{P}(D)\, h)\, \mathrm{d}x = 0 \tag{4.13}$$

where $\boldsymbol{P}^\flat(\zeta) = \boldsymbol{P}^\top(-\zeta)$ and the integral $\int \cdot \mathrm{d}x$ stands for the duality bracket between $\mathcal{C}^\infty(\Omega)^q$ and $\mathcal{E}'(\Omega)^q$.

Let K denote a convex compact subset of Ω and H_K the supporting function of K. By the Paley–Wiener–Schwartz theorem, to say that $\operatorname{supp} w \subset K$ is equivalent to saying that there are $\kappa \in \mathbb{Z}$ and $C > 0$ such that

$$\forall \zeta \in \mathbb{C}^n,\ \left|\hat{w}_j(\zeta)\right| \le C \mathrm{e}^{H_K(\operatorname{Im}\zeta)} (1 + |\zeta|)^\kappa,\ j = 1, \ldots, q. \tag{4.14}$$

Denote by $\mathbb{E}(K)$ the space of vector-valued functions $V \in \mathcal{O}(\mathbb{C}^n)^q$ such that, for some $N \in \mathbb{Z}$,

$$\sup_{\zeta \in \mathbb{C}^n} |V(\zeta)|\, \mathrm{e}^{-H_K(-\operatorname{Im}\zeta)} (1 + |\zeta|)^{-N} < +\infty. \tag{4.15}$$

Let $v(x) \in \mathcal{E}'(\Omega)^q$ be such that $V(\zeta) = \hat{v}(-\zeta)$. The solution h of (2.3) defines a continuous linear functional on $\mathbb{E}(K)$:

$$\lambda(V) = \int h(x)\, v(x)\, \mathrm{d}x. \tag{4.16}$$

If there is $w \in \mathcal{E}'(\Omega)^q$ such that $v = \boldsymbol{P}^\flat(D)\, w$, i.e., $V(\zeta) = \boldsymbol{P}^\top(\zeta)\, \hat{w}(-\zeta)$, then $\lambda(V) = 0$ by (4.13).

We apply Theorem 10: there is $G \in \mathcal{O}(\mathbb{C}^n)^q$ and $N, N_1 \in \mathbb{Z}$ such that $G - V \in \mathcal{M}_{\boldsymbol{P}}(\mathbb{C}^n)$ and

$$\sup_{\zeta \in \mathbb{C}^n} (1 + |\zeta|)^{-N} \mathrm{e}^{-H_K(-\operatorname{Im}\zeta)} |G(\zeta)|$$
$$\le C \max_{\iota=1,\ldots,\mathrm{I}} \left(\max_{\nu=1,\ldots,\rho_i} \left(\sup_{\zeta \in V_{\mathfrak{p}_\iota}} (1 + |\zeta|)^{-N_1} \mathrm{e}^{-H_K(-\operatorname{Im}\zeta)} |\mathcal{L}_{\iota,\nu} V(\zeta)| \right)\right). \tag{4.17}$$

We can write $G - V = \boldsymbol{P}^\top \Phi$ for some $\Phi \in \mathcal{O}(\mathbb{C}^n)^p$ (see Remark 1) and then apply Theorem 3 with $\boldsymbol{P}^\top$ in place of $\boldsymbol{P}$: there is $\Psi \in \mathcal{O}(\mathbb{C}^n)$ such that $\boldsymbol{P}^\top \Psi = \boldsymbol{P}^\top \Phi$ and, for suitably large $N_2 \in \mathbb{Z}_+$, $C_1 > 0$,

$$\sup_{\zeta \in \mathbb{C}^n} \mathrm{e}^{-H_K(-\operatorname{Im}\zeta)} \left(1 + |\zeta|^2\right)^{-N_2} |\Psi(\zeta)|$$
$$\le C_1 \sup_{\zeta \in \mathbb{C}^n} \mathrm{e}^{-H_K(-\operatorname{Im}\zeta)} (1 + |\zeta|)^{-N} |G(\zeta) - V(\zeta)| < +\infty.$$

We have $0 = \lambda\left(P^{\top}\Psi\right) = \lambda(G) - \lambda(V)$ and, by (4.17),

$$|\lambda(V)| \le C_2 \max_{\iota=1,\dots,\mathrm{I}}\left(\max_{\nu=1,\dots,\rho_\iota}\left(\sup_{\zeta\in V_{\mathfrak{p}_\iota}}(1+|\zeta|)^{-N_3}\,\mathrm{e}^{-H_K(-\mathrm{Im}\,\zeta)}\,|\mathcal{L}_{\iota,\nu}V(\zeta)|\right)\right) \tag{4.18}$$

again for suitably large $N_3 \in \mathbb{Z}_+$, $C_2 > 0$ independent of V satisfying (4.15).

Let now $\mathbb{F}(K, V_P)$ denote the (Banach) space of continuous complex functions F in $V_P = V_{\mathfrak{p}_1} \cup \cdots \cup V_{\mathfrak{p}_\lambda}$ such that

$$\|F\|_{K,V_P} = \sup_{\zeta\in V_P}\left((1+|\zeta|)^{-N_2}\,\mathrm{e}^{-H_K(-\mathrm{Im}\,\zeta)}\,|F(\zeta)|\right) < +\infty.$$

The dual $\mathbb{F}^*(K, V_P)$ of $\mathbb{F}(K, V_P)$ is the space of Radon measures $\mathrm{d}m$ on the locally compact space V_P such that $(1+|\zeta|)^{N_2}\,\mathrm{e}^{H_K(-\mathrm{Im}\,\zeta)}\mathrm{d}m$ is a bounded measure $\mathrm{d}\mu$ on V_P. Thus, if $F \in \mathbb{F}(K, V_P)$ and $\mathrm{d}m = (1+|\zeta|)^{-N_2}\,\mathrm{e}^{-H_K(-\mathrm{Im}\,\zeta)}\mathrm{d}\mu \in \mathbb{F}^*(K, V_P)$, then the duality bracket is

$$\langle F, \mathrm{d}m\rangle = \int_{V_P} F(\zeta)\,(1+|\zeta|)^{-N_2}\,\mathrm{e}^{-H_K(-\mathrm{Im}\,\zeta)}\mathrm{d}\mu. \tag{4.19}$$

By the Hahn–Banach theorem, the inequality (4.18) implies that there are bounded measures $\mathrm{d}\mu_{\iota,\nu}$ on V_P such that

$$\lambda(V) = \sum_{\iota=1}^{\mathrm{I}}\sum_{\nu=1}^{n_\iota}\int_{V_{\mathfrak{p}_\iota}}(1+|\zeta|)^{-N_2}\,\mathrm{e}^{-H_K(-\mathrm{Im}\,\zeta)}\mathcal{L}_{\iota,\nu}V(\zeta)\,\mathrm{d}\mu_{\iota,\nu}(\zeta) \tag{4.20}$$

Recalling that $V(\zeta) = \widehat{v}(-\zeta)$, we see that

$$\mathcal{L}_{\iota,\nu}\left(\zeta, D_\zeta\right)V(\zeta) = \int \mathrm{e}^{\mathrm{i}\langle x,\zeta\rangle}\widehat{\mathcal{L}}_{\iota,\nu}(\zeta, x)\,v(x)\,\mathrm{d}x. \tag{4.21}$$

Recalling (4.16), we see that, for arbitrary $v \in \mathcal{E}'(\Omega)^q$,

$$\langle h, v\rangle = \sum_{\iota=1}^{\mathrm{I}}\sum_{\nu=1}^{n_\iota}\int_{\zeta\in V_{\mathfrak{p}_\iota}} \times \int_{x\in\mathbb{R}^n}\mathrm{e}^{\mathrm{i}\langle x,\zeta\rangle}\,(1+|\zeta|)^{-N_2}\,\mathrm{e}^{-H_K(-\mathrm{Im}\,\zeta)}\widehat{\mathcal{L}}_{\iota,\nu}(\zeta, x)\,v(x)\,\mathrm{d}\mu_{\iota,\nu}(\zeta)\,\mathrm{d}x,$$

which means that, for all x in the interior of K,

$$h(x) = \sum_{\iota=1}^{\mathrm{I}}\sum_{\nu=1}^{n_\iota}\int_{\zeta\in V_{\mathfrak{p}_\iota}}\mathrm{e}^{\mathrm{i}\langle x,\zeta\rangle - H_K(-\mathrm{Im}\,\zeta)}\widehat{\mathcal{L}}_{\iota,\nu}(\zeta, x)\,(1+|\zeta|)^{-N_2}\,\mathrm{d}\mu_{\iota,\nu}(\zeta). \tag{4.22}$$

The integral at the right is absolutely, uniformly convergent and remains so after a number of differentiations—number depending on N_2. Having assumed, here, that h is $\mathcal{C}^\infty$ we are at liberty to select N_2 as large as we wish. Letting $K \nearrow \Omega$ (4.22) gives us a representation of all solutions $h \in \mathcal{C}^\infty(\Omega)^q$ of (2.3). This is the Fundamental Principle.

Reference

[Björk 1979] Björk, J. E., Rings of differential operators, North Holland Publ. Co., Amsterdam 1979.

[Ehrenpreis 1954] Ehrenpreis, L., *Solutions on some problems of division I.* Am. J. of Math. **76** (1954), 883–903.

[Ehrenpreis 1954] Ehrenpreis, L., *A fundamental principle for systems of linear partial differential equations with constant coefficients and some of its applications*, Proceedings Intern. Symp. on Linear Spaces, Jerusalem 1961, 161–174.

[Ehrenpreis 1970] Ehrenpreis, L., Fourier analysis in several complex variables, Pure and Appl. Math. Wiley-Interscience Publ., New York 1970.

[Gunning and Rossi 1965] Gunning, R. C. and Rossi, H., Analytic Functions of Several Complex Variables, Princeton University Press Princeton N. J. 1965.

[Hansen 1982] Hansen, S., *Das Fundamentalprinzip für Systeme partieller Differentialgleichungen mit konstanten Koeffizienten,* Paderboirn Habiliationsschrift 1982.

[Hörmander 1966] Hörmander, L., An introduction to complex analysis in several variables, D. Van Nostrand Co., Princeton, N. J. 1966, 3rd Edition, North Holland Publ. Co., Amsterdam 1994.

[Malgrange 1955] Malgrange, B., *Existence et approximation des solutions des équations aux dérivées partielles et des équations de convolution,* Ann. Inst. Fourier **6** (1955), 271–354.

[Malgrange 1963] Malgrange, B., *Sur les systèmes différentiels à coefficients constants.* Coll. C. N. R. S. Paris 1963, 113–122.

[Oka 1936–1953] Oka, K., *Sur les fonctions analytiques de plusieurs variables*, Iwanami Shoten Tokyo 1961.

[Palamodov 1963] Palamodov, V. P., *The general theorems on the systems of linear equations with constant coefficients.* Outlines of the joint Soviet-American symp. on PDE (1963), 206–213.

[Palamodov 1970] Palamodov, V. P., Linear differential operators with constant coefficients, Grundl. d. Math. Wiss. **168**, Springer-Verlag, Berlin 1970.

[Serre 1955] Serre, J. P., *Faisceaux algébriques cohérents*, Ann. Math. **61**(1955), 197–278.

Minimal Entire Functions

Benjamin Weiss

Abstract Consider the space of nonconstant entire functions $\mathcal{E}$ with the topology of uniform convergence on compact subsets of $\mathbb{C}$ and with the action of $\mathbb{C}$ by translation. A minimal entire function is a nonconstant entire function f with the property that for any $g \in \mathcal{E}$ which is a limit of translates of f, in turn, f is a limit of translates of g. Thus, X, the orbit closure of f is a minimal closed invariant set. It is not clear a priori that there exist such functions with X including functions that are not translates of f. I will show that many such functions can be constructed and that their orbit closures can be quite large and interesting from a dynamical point of view. The main example is based on the construction of a particular compact minimal action of $\mathbb{R}^2$ with rather special properties.

Key words Entire functions • Minimal actions

Mathematics Subject Classification: 30D99, 37B99

1 Introduction

The complex plane $\mathbb{C}$ acts by translation on the space of non-constant entire functions $\mathcal{E}$ by sending $f(z)$ to $f(z+c)$ for $c \in \mathbb{C}$. With the topology of uniform convergence on compact sets, $\mathcal{E}$ becomes a Polish space and it is natural to consider the dynamical aspects of this action. This was first done by G.D. Birkhoff who in [B] constructed an entire function f whose orbit under this action is dense in $\mathcal{E}$. This shows that this action is topologically transitive. In an earlier paper [W2], in response to a question that had been raised by G. Mackey, I showed that there is

B. Weiss (✉)
Institute of Mathematics, Hebrew University of Jerusalem, Jerusalem 91904, Israel
e-mail: weiss@math.huji.ac.il

H.M. Farkas et al. (eds.), *From Fourier Analysis and Number Theory to Radon Transforms and Geometry*, Developments in Mathematics 28, DOI 10.1007/978-1-4614-4075-8_25,

an abundance of non-trivial ergodic invariant probability measures for this action. In fact, for any free probability preserving action of $\mathbb{C}$, $(X, \mathcal{B}, \mu, T_c)$, there is a measurable function $F : X \to \mathbb{C}$ such that for μ-a.e. $x \in X$, the function $f_x(z) = F(T_z(x))$ is a nonconstant entire function. In this note, dedicated to the memory of one of my first teachers, Leon Ehrenpreis, I intend to demonstrate the abundance of ***minimal entire functions***. These are nonconstant entire functions f with the property that for any $g \in \mathcal{E}$ which is a limit of translates of f, in turn, f is a limit of translates of g. In other words, the closure of the orbit of f in $\mathcal{E}$ is a minimal set. Needless to say, these minimal sets are not compact as is customary in topological dynamics.

Nonetheless, they can be quite close to classical minimal actions of $\mathbb{C}$. In fact, the first example of a measurable entire function in [W2] is based on the action on $\mathbb{R}^2$ on a compact group, a two-dimensional solenoid, and defines a minimal entire function. The example there was based on the lattices $L_k = 3^k\mathbb{Z}^2$ with the compact group being the inverse limit of the tori $\mathbb{R}^2/L_k$.

My intent here is to show that more general minimal actions of $\mathbb{R}^2$ can serve as the basis for minimal entire functions. I will illustrate this with one example, which, while close to the solenoid, has quite different properties. In the next section, I will describe in detail a particular minimal action of $\mathbb{R}^2$ with some special properties. In the following one, I will explain how to use it to define a minimal entire function. The final section will contain some further remarks on the minimal system and on extensions of the construction of entire functions based on more general minimal actions. In conclusion, I would like thank Hillel Furstenberg for permission to include the example of the flickering circles.

2 The Flickering Circles

In this section, I shall describe the construction of a minimal action of $\mathbb{R}^2$ on a compact metric space X which will serve as the basis for the minimal entire functions. This construction was carried out many years ago together with Hillel Furstenberg, and I thank him for his allowing me to include it in this note. Our original motivation was to show that $\mathbb{R}^2$ can act minimally without any of its one-parameter subgroups doing so.

The space X will be the closure of the translates of one closed subset $F \subset \mathbb{R}^2$ in the topology induced by the Hausdorff metric on compact subsets of $\mathbb{R}^2$. More explicitly, I mean that F_n converges to F if for every compact subset K, $F_n \cap K$ converges to $F \cap K$ in the Hausdorff metric on subsets of K. It is a standard fact that this space is compact. Let $\{l_i\}$ be a sequence of integers and $\{r_i\}$ a sequence of positive numbers and denote by L_i the lattice $l_i\mathbb{Z}^2$. Furthermore, let C_i denote the circle centered at the origin with radius r_i. The set F will have the form

$$F = \bigcup_{i=1}^{\infty}(M_i + C_i)$$

with $M_i \subset L_i$ suitably defined when the sequences of l_i and r_i satisfy some simple growth properties.

The tricky aspect of the construction is that we want the circles defining F to be disjoint, but nonetheless, we want the M_i to be sufficiently regular so as to guarantee the minimality of the resulting orbit closure. To begin with, we will require that each successive l_{i+1} is a multiple of the preceding l_i so that the basic lattices L_i are nested, i.e., $L_i \supset L_{i+1}$ for all i. Next, we would like the circles $L_i + C_i$ for a fixed i to be disjoint, so we demand that for all i

$$\frac{r_i}{l_i} < \frac{1}{10},$$

and since we would like the new circles to enclose many circles of the previous levels, we demand that for all i

$$\frac{r_{i+1}}{l_i} > 10.$$

The sets M_i will be defined as the limits of a triangular array M_i^n for $i \leq n$. By $d(x, E)$, I mean the distance from the point x to the closed set E. The array M_i^n is defined for each n by a downward induction as follows:

(a) $M_n^n = L_n$.
(b) $M_j^n = \{x \in L_j : d(x, \bigcup_{j<i\leq n}(M_i^n + C_i)) > 2r_j\}$.

Using this array, we define the n-th approximation to F by the formula:

$$F_n = \bigcup_{i=1}^{n}(M_i^n + C_i)$$

For a particular $a \in L_j$, the circle $a + C_j$ appears for the first time in F_j and then may disappear and reappear in subsequent F_n's (hence the name "flickering circles"). Since $a + C_j$ is contained inside C_N for all sufficiently large N, this process eventually stabilizes, and thus, there is a well-defined limit of the M_j^n as n tends to infinity, which is denoted by M_i, and this defines F. Its properties will follow from the properties of the F_n.

The next lemma makes precise the fact that the interior of any one of the translates of C_i in F_n is the same independently of $n \geq i$. Its proof is a straight forward consequence of the fact that the lattices L_i are nested. To state the lemma, I will denote by D_i the disk of radius r_i centered at the origin.

Lemma 1. *For all $i \leq n \leq m$, $a \in M_i^n$, and $b \in M_i^m$, we have:*

$$F_n \cap (a + D_i) - a = F_m \cap (b + D_i) - b.$$

It follows from the lemma that for the limit sets M_i and any a and b in M_i, we have:

$$F \cap (a + D_i) - a = F \cap (b + D_i) - b.$$

For the minimality of the orbit closure of F, it therefore suffices to check that for each i the set M_i is syndetic, i.e., the distance $d(z, M_i)$ is uniformly bounded for $z \in \mathbb{C}$. This in turn follows from the fact that for any $a \in L_i$, there cannot be more than one circle of a higher order whose distance to a is less than $2r_i$. For circles of the same order, this is clear, while the nature of the construction precludes the presence of higher order circles that are too close on the scale of r_i.

Let us denote the orbit closure of F by X and the action by translation by T_c. This minimal system has a natural mapping, π, onto the solenoidal group, S, which is defined as the inverse limit of the tori $\mathbb{R}^2/L_n$. This follows from the fact that the sets M_i are syndetic, and in any limit of translates of F, one sees limits of finite portions of $M_i + D_i$ so that the position of the lattices L_i can be determined in any limit of translates of F.

However, there are also points in X that contain infinite straight lines. These are obtained by translating F so that larger and larger circles pass through a fixed point of $\mathbb{C}$. Clearly, one can obtain a straight line in any desired direction, and then translating in that direction will preserve the straight line so that any fixed one-parameter subgroup of $\mathbb{R}^2$ has proper closed subsets. This is how one sees that no one-parameter subgroup of $\mathbb{R}^2$ acts minimally on X. This phenomenon is in contrast to what happens with ergodicity. If $\mathbb{R}^2$ acts ergodically on a probability space, then with a countable number of exceptions, the one-parameter subgroups of $\mathbb{R}^2$ act ergodically (see [PS]).

A further property that can be established for this example is that X is a ***proximal*** extension of its solenoidal factor. This means that if x and y are two points with the same projection on S, then one can find a sequence $c_k \in \mathbb{C}$ such that the distance between $T_{c_k}(x)$ and $T_{c_k}(y)$ tends to zero. On the other hand, it is also easy to show that for any point $s \in S$, the fiber $\pi^{-1}(s)$ is infinite. In particular, this gives examples of proximal extensions of equicontinuous actions which are not ***almost automorphic***, which are extensions of equicontinuous actions for which the generic fiber consists of only one point (cf. [GW1]).

3 Minimal Entire Functions

In this section, I shall show how to use the structure of the set F together with the classical theorem of C. Runge on approximation of holomorphic functions by polynomials to construct a minimal entire function. Let me begin by recalling Runge's theorem:

Theorem 1. *If K is a compact subset of $\mathbb{C}$ with a connected complement and f is holomorphic in some neighborhood of K, then for any $\epsilon > 0$, there is a polynomial $p(z)$ such that*

$$sup_{z \in K}|f(z) - p(z)| < \epsilon.$$

The construction of the entire function will be carried out in a sequence of steps utilizing the graded structure of F. As before, we denote by D_i the closed desk centered at the origin with radius r_i. For the first step, define $f_1(a+z) = z$ for all $a \in M_1$ and $z \in D_1$. For the second step, set K_2 equal to the union of those translates of D_1 that lie inside D_2 on which f_1 was defined and apply Runge's theorem to find a polynomial p_2 such that

$$\sup_{z \in K_2} |f_1(z) - p_2(z)| < \frac{1}{10}.$$

We set $f_2(a+z) = p_2(z)$ for all $a \in M_2$ and $z \in D_2$, while for those points that are in $M_1 + D_1$ and do not lie in $M_2 + D_2$, we keep $f_2(z) = f_1(z)$. Note that while f_1 looked the same on unit disks centered at elements of M_1, this is no longer the case for f_2. However, it does look the same up to an error that is at most $\frac{1}{5}$. On the other hand, for the disks D_2 centered at elements of M_2, the function f_2 is the same. In general, we will have defined f_n on

$$E_n = \bigcup_{i=1}^{n} (M_i + D_i)$$

to be some polynomial on each of the disks that constitute E_n, and f_n will have the properties:

(A_n) $\quad |f_{n-1}(z) - f_n(z)| < \frac{1}{10^{n-1}}$ for all $z \in E_n$ where f_{n-1} is defined .
(A_n) $\quad f_n(z) = f_n(a+z)$ for all $a \in M_n$ and $z \in D_n$.
(C_i^n) $\quad |f_n(a+z) - f_n(b+z)| < \sum_{j=i}^{n-1} \frac{2}{10^j}$ for all $a, b \in M_i$ and $z \in D_i$.

for all $1 \leq i < n$. In order to define f_{n+1}, we set $K_{n+1} = E_n \cap D_{n+1}$ and apply Runge's theorem to find a polynomial p_{n+1} that satisfies:

$$\sup_{z \in K_{n+1}} |f_n(z) - p_{n+1}(z)| < \frac{1}{10^n}.$$

We set $f_{n+1}(a+z) = p_{n+1}(z)$ for all $a \in M_{n+1}$ and $z \in D_{n+1}$, while for those points that are in E_n and do not lie in $M_{n+1} + D_{n+1}$, we keep $f_{n+1}(z) = f_n(z)$. This defines f_{n+1} on

$$E_{n+1} = \bigcup_{i=1}^{n+1} (M_i + D_i)$$

and it is easy to check that properties A_{n+1}, B_{n+1}, and C_i^{n+1} will hold for all $1 \leq i < n+1$.

The properties A_n imply that the f_n converge to an entire function f. By passing to a limit, one obtains from the properties C_i^n that f will satisfy the properties:

(C_i) $\quad |f(a+z) - f(b+z)| \leq \sum_{j=i}^{\infty} \frac{2}{10^j}$ for all $a, b \in M_i$ and $z \in D_i$.

These properties together with the fact that the sets M_n are syndetic clearly imply that f is a minimal entire function.

The set of entire functions g that we will get as limits of the orbit of f under translation will be denoted by $Y \subset \mathcal{E}$. It is clear from the construction that $f(c_k + z)$ will converge to an entire function $g(z)$ only when the sets $c_k + F$ converge to a limiting set in X, the space of the minimal action defined in the preceding section. Thus, there is a mapping $\pi : Y \to X$ which is of course equivariant with respect to the actions of $\mathbb{R}^2$ on these spaces. This mapping, π, is not onto. This follows easily from the fact that f is not bounded, and thus there is a sequence c_k along which $f(c_k)$ tends to ∞. On the other hand, by the compactness of X, we may assume that $F + c_k$ converges to a limit. However, if the radii r_i are chosen so that the sum $\sum_{i=1}^{\infty} \frac{r_i}{l_i}$ diverges, then it is possible to show that $\pi(Y)$ is big in the following sense.

Define the measures μ_n on X by the formula:

$$\int \phi \mathrm{d}\mu_n = \int_{u+\mathrm{i}v \in D_n} \phi(F + (u + iv))\mathrm{d}u\mathrm{d}v,$$

for continuous functions ϕ on X. Let μ by any invariant measure on X that is a weak*-cluster point of the measures μ_n. Note that since the solenoid is uniquely ergodic, the measure μ necessarily projects onto the Haar measure of the solenoidal factor of X. The image $\pi(Y)$ has μ-measure one. To see this, observe that once $z \in a + D_i$ for $a \in M_i$ and i sufficiently large, $f(z - a)$ is very close to the polynomial $p_i(z)$. Denote by $\hat{D}_i$ the disk centered at the origin but with radius $r_i - d_i$ where d_i is a sequence that tends slowly to ∞ and consider the set:

$$G_k = \{z \in \mathbb{C} : z \in M_i + \hat{D}_i \text{ for some } i \geq k\}.$$

The assumption about the divergence of the series $\sum_{i=1}^{\infty} \frac{r_i}{l_i}$ implies that for any fixed k, the density of G_k in D_n tends to one as n tends to ∞. Thus, the μ measure of the points corresponding to $\cap G_k$ will be one, and corresponding to such points, there will be some limiting entire function. We can summarize the results in the following theorem:

Theorem 2. *There is a minimal entire function f such that if Y denotes the closure of its translates in $\mathcal{E}$, there is an equivariant mapping π from Y to the flickering circle minimal system (X, T_u) with an invariant measure μ such that $\mu(\pi(Y) = 1$. This system (X, T_u) is a proximal extension of a two-dimensional solenoid, and it has the property that every one-parameter subgroup of $\mathbb{R}^2$ has a nontrivial invariant subset.*

4 Concluding Remarks

1. The basic idea used in the construction of the flickering circle system was used again several times both to construct minimal systems as in [GW2] and in finding uniquely ergodic models as in [W1] and the later papers of Alain Rosenthal (e.g., [R]). Most recently, I used it show that any free action of a countable group has a minimal model [W3].

2. A more natural example of an $\mathbb{R}^2$ minimal action such that no one-parameter subgroup acts minimally is given by the product of the horocycle flow with itself. Explicitly, let Γ be a co-compact subgroup of $G = SL(2,\mathbb{R})$ and let $X = G/\Gamma$. The horocycle flow is defined by the action of the group

$$h_t = \begin{pmatrix} 1 & t \\ 0 & 1 \end{pmatrix}$$

and it is minimal as was shown by G. Hedlund in [H]. Clearly, the action of $\mathbb{R}^2$ on $X \times X$ defined by $T_{(s,t)}(x,y) = (h_s(x), h_t(y))$ is also minimal. The geodesic action is defined by the group:

$$g_u = \begin{pmatrix} u & 0 \\ 0 & u^{-1} \end{pmatrix}$$

and it satisfies the commutation relation $h_t g_u = g_u h_{u^{-2}t}$. It follows that for the one-parameter subgroup of $\mathbb{R}^2$, generated by $(1,a)$, the graph

$$\{(x, g_{\sqrt{a}}(x)) : x \in X\}$$

gives a closed invariant set for the direct product of the horocycle flows. Hillel Furstenberg observed many years ago (private communication) that for this example, it is nonetheless true that for any positive ϵ there is some one-parameter subgroup of $\mathbb{R}^2$ such that **all** of its orbits in $X \times X$ are ϵ-dense. In fact, for any $\mathbb{R}^2$ minimal action which is formed as the direct product of two minimal $\mathbb{R}$ actions, $(X, S_u), (Y, T_v)$, given $\epsilon > 0$ if one chooses m to be sufficiently large, then the onc-parameter subgroup, $R_t = S_t \times T_{mt}$,will have the property that the orbit of every point in $X \times Y$ under R_t is ϵ-dense. In the flickering circle example, this is not the case. There is an open set U which contains in its complement entire orbits of every one-parameter subgroup of $\mathbb{R}^2$. Thus, its lack of "one-dimensional minimality" is in some sense stronger.
3. There were two properties of the lattices L_i that were crucial in the construction of the flickering circle system. The first is that they are syndetic, and the second is that for any two points $a, b \in L_i$ for any $j < i$, the sets $(a + D_i) \cap L_j - a = (b + D_i) \cap L_j - b$. If we have a sequence of sets in the plane with these two properties, we can carry out a construction quite similar to the one in Sect. 2. If X is a compact space and T_z is a minimal free action of $\mathbb{C}$ on X and we fix a point $x_0 \in X$ and let U_i be a decreasing sequence of neighborhoods of x_0, then defining $\hat{L}_i = \{z \in \mathbb{C} : T_z(x_0) \in U_i\}$, the minimality implies that the $\hat{L}_i$ are syndetic, while if the neighborhoods decrease sufficiently rapidly, we will have something quite close to the second property. Now, one can use these sets to replace the regular lattices L_i and construct a minimal entire function that is based on the recurrence patterns of the point x_0.
4. At the time that I wrote [W2], I was unaware of Birkhoff's paper [B], which was recently pointed out to me by Eli Glasner. Using the techniques of [W2],

it is not hard to guarantee that closed support of the ergodic measures μ that are constructed there on $\mathcal{E}$ is all of $\mathcal{E}$. The ergodic theorem now shows that μ-a.e. $f \in \mathcal{E}$ has a dense orbit. This is a strengthening of Birkhoff's result.

References

[B] Birkhoff, G.D. *Démonstration d'un theoréorème élémentaire sur les fonctions entières* Comptes rendus des seances de l'Acad. des Sciences **189** (1929), 473–475.

[GW1] Glasner, S. and Weiss, B. *On the construction of minimal skew products* Israel J. Math. **34** (1979), 321–336.

[GW2] Glasner, S. and Weiss, B. *Interpolation sets for subalgebras of* $l^\infty(\mathbb{Z})$ Israel J. Math. **44** (1983), 345–360.

[H] Hedlund, G. *Fuchsian groups and transitive horocycles* Duke Math. J. **2** (1936), 530–542.

[PS] Pugh, C. and Shub, M. *Ergodic elements of ergodic actions* Compositio Math. **23** (1971), 115–122.

[R] Rosenthal, A. *Finite uniform generators for ergodic, finite entropy, free actions of amenable groups* Probab. Theory Related Fields **77** (1988), 147–166.

[W1] Weiss, B. *Strictly ergodic models for dynamical systems* Bull. Amer. Math. Soc. **13** (1985), 143–146.

[W2] Weiss, B. *Measurable entire functions* Ann. Numer. Math. **4** (1997), 599–605.

[W3] Bower, L. Grigorchuk, R. and Vorobets, Y., *Dynamical systems and Group actions ed.*, Contemporary Math. AMS. **567**, (2012), 249–264.

A Conjecture by Leon Ehrenpreis About Zeroes of Exponential Polynomials

Alain Yger

Dedicated to the memory of Leon Ehrenpreis

Abstract Leon Ehrenpreis proposed in his 1970 monograph *Fourier Analysis in several complex variables* the following conjecture: the zeroes of an exponential polynomial $\sum_0^M b_k(z)e^{i\alpha_k z}$, $b_k \in \overline{\mathbb{Q}}[X]$, $\alpha_k \in \overline{\mathbb{Q}} \cap \mathbb{R}$ are well separated with respect to the Paley–Wiener weight. Such a conjecture remains essentially open (besides some very peculiar situations). But it motivated various analytic developments carried by C.A. Berenstein and the author, in relation with the problem of deciding whether an ideal generated by Fourier transforms of differential delayed operators in n variables with algebraic constant coefficients, as well as algebraic delays, is closed or not in the Paley–Wiener algebra $\widehat{\mathcal{E}}(\mathbb{R}^n)$. In this survey, I present various analytic approaches to such a question, involving either the Schanuel-Ax formal conjecture or $\mathcal{D}$-modules technics based on the use of Bernstein–Sato relations for several functions. Nevertheless, such methods fail to take into account the intrinsic rigidity which arises from arithmetic hypothesis: this is the reason why I also focus on the fact that Gevrey arithmetic methods, that were introduced by Y. André to revisit the Lindemann–Weierstrass theorem, could also be understood as an indication for rigidity constraints, for example, in Ritt's factorization theorem of exponential sums in one variable. The objective of this survey is to present the state of the art with respect to L. Ehrenpreis's conjecture, as well as to suggest how methods from transcendental number theory could be combined with analytic ideas, in order precisely to take into account such rigidity constraints inherent to arithmetics.

Key words Bernstein–Sato relations • Differential-difference operator • Exponential polynomial

---

A. Yger (✉)
Institut de Mathématiques de Bordeaux, Université Bordeaux 1, Talence, France
e-mail: yger@math.u-bordeaux1.fr

H.M. Farkas et al. (eds.), *From Fourier Analysis and Number Theory to Radon Transforms and Geometry*, Developments in Mathematics 28, DOI 10.1007/978-1-4614-4075-8_26,

Mathematics Subject Classification (2010): 42A75 (Primary), 65Q10, 11K60, 11J81, 39A70, 14F10 (Secondary)

1 The Conjecture, Various Formulations

In [36], page 322, Leon Ehrenpreis formulated the following conjecture.

Conjecture 1.1 (original form, incorrect). *If $F_1,\dots,F_N$ are N exponential polynomials in n variables with purely imaginary algebraic frequencies, namely,*

$$F_j(z_1,\dots,z_n)=\sum_{k=0}^{M_j} b_{jk}(z)\,\mathrm{e}^{\mathrm{i}\langle \alpha_{jk},z\rangle}\,,\ b_{jk}\in\mathbb{C}[X_1,\dots,X_n]\,,\ \alpha_{jk}\in\overline{\mathbb{Q}}^n\cap\mathbb{R}^n\,,$$
$$j=1,\dots,N,$$

then the ideal $(F_1,\dots,F_N)$ they generate in the Paley–Wiener algebra $\widehat{\mathcal{E}'(\mathbb{R}^n)}$ is slowly decreasing with respect to the Paley–Wiener weight $p(z)=\log|z|+|\mathrm{Im}\,z|$. As a consequence,[1] *this ideal is closed in $\widehat{\mathcal{E}'(\mathbb{R}^n)}$. It coincides with the ideal $[I(F_1,\dots,F_N)]_{\mathrm{loc}}$, which consists of elements in $\widehat{\mathcal{E}'(\mathbb{R}^n)}$ that belong locally to the ideal generated by $F_1,\dots,F_N$ in the algebra of entire functions in n variables.*

This conjecture, in a slightly modified form (see Conjecture 1.2), has been the inspiration for the joint work of C.A. Berenstein and the author since 1985. It is a challenging and fascinating question, one that is closely connected with other open questions in number theory and analytic geometry. In this note, I will point out many of these connections, detail some of the progress that has been made on the problem, and, hopefully, inspire others to continue the work.

As it stands, Conjecture 1.1 would imply, in the one variable setting, the following : *if*

$$f(z)=\sum_{k=0}^{M} b_k(z)\mathrm{e}^{\mathrm{i}\alpha_k z}\,,\quad b_k\in\mathbb{C}[X]\,,\quad \alpha_k\in\overline{\mathbb{Q}}\cap\mathbb{R} \tag{1.1}$$

is an exponential polynomial in one variable with algebraic frequencies and all simple zeroes, then the ideal (f,f') is a non proper ideal in $\widehat{\mathcal{E}'(\mathbb{R})}$ which would imply

$$|f(z)|+|f'(z)|\geq c\,\frac{\mathrm{e}^{-A|\mathrm{Im}\,z|}}{(1+|z|)^p} \tag{1.2}$$

[1]This follows from Theorem 11.2 in [36].

for some $c, A > 0$ *and* $p \in \mathbb{N}$. Unfortunately, such an assertion is false if one does not set any condition of arithmetic nature on the polynomial coefficients b_k. Take, for example,

$$f(z) = f_\gamma(z) = \sin(z-\gamma) - \sin(\sqrt{2}(z-\gamma)),$$

where $2\gamma/\pi$ has excellent approximations belonging to $(2\mathbb{Z}+1) \oplus \sqrt{2}(2\mathbb{Z}+1)$; then some zeroes of f_γ of the form

$$\frac{2l\pi}{1-\sqrt{2}},\ l \in \mathbb{Z},$$

will approach extremely well other zeroes of f_γ of the form

$$\frac{2\alpha + (2l'+1)\pi}{1+\sqrt{2}},\ l' \in \mathbb{Z},$$

and thus the ideal (f_γ, f'_γ) fails to be closed in $\widehat{\mathcal{E}'(\mathbb{R})}$. So Conjecture 1.1 needs to be reformulated as follows.

Conjecture 1.2 (revised form). *If* $F_1, \dots, F_N$ *are exponential polynomials in* n *variables with both algebraic coefficients and purely imaginary algebraic frequencies, namely*

$$F_j(z_1,\dots,z_n) = \sum_{k=0}^{M_j} b_{jk}(z)\, e^{i\langle \alpha_{jk}, z\rangle},\ b_{jk} \in \overline{\mathbb{Q}}[X_1,\dots,X_n],\ \alpha_{jk} \in \overline{\mathbb{Q}}^n \cap \mathbb{R}^n,$$
$$j = 1,\dots,N, \tag{1.3}$$

then the ideal $(F_1, \dots, F_N)$ *they generate in the Paley–Wiener algebra* $\widehat{\mathcal{E}'(\mathbb{R}^n)}$ *is slowly decreasing with respect to the Paley–Wiener weight* $p(z) = \log|z| + |\mathrm{Im}\, z|$. *As a consequence, this ideal is closed in* $\widehat{\mathcal{E}'(\mathbb{R}^n)}$*, and thus coincides with the set of elements in* $\widehat{\mathcal{E}'(\mathbb{R}^n)}$ *which belong locally to the ideal generated by* $F_1, \dots, F_N$ *in the algebra of entire functions in* n *variables.*

Such a conjecture appears to be stronger than the following one.

Conjecture 1.3 (weaker revised form). *If* $F_1, \dots, F_N$ *are exponential polynomials in* n *variables as in Conjecture 1.2, namely*

$$F_j(z_1,\dots,z_n) = \sum_{k=0}^{M_j} b_{jk}(z)\, e^{i\langle \alpha_{jk}, z\rangle},\ b_{jk} \in \overline{\mathbb{Q}}[X_1,\dots,X_n],\ \alpha_{jk} \in \overline{\mathbb{Q}}^n \cap \mathbb{R}^n,$$
$$j = 1,\dots,N,$$

then the closure of the ideal $(F_1, \ldots, F_N)$ *they generate in the Paley–Wiener algebra* $\widehat{\mathcal{E}'(\mathbb{R}^n)}$ *coincides with the set of elements in* $\widehat{\mathcal{E}'(\mathbb{R}^n)}$ *which belong locally to the ideal generated by* $F_1, \ldots, F_N$ *in the algebra of entire functions in* n *variables.*

The conjecture is equivalent to the assertion that the underlying system of difference-differential equations $\mu_1 * f = \cdots = \mu_N * f = 0$ satisfies the spectral synthesis property.

With C.A. Berenstein, we have been developing since [16] a long-term joint research program originally devoted to various attempts to tackle Conjecture 1.2. Such attempts lead to an approach based on multidimensional analytic residue theory that relies on techniques of analytic continuation in one or several complex variables [19, 20]. Conjecture 1.3 seems harder to deal with since it does not fit so well with the search for explicit division formulas in $\widehat{\mathcal{E}'(\mathbb{R}^n)}$ that resolve Ehrenpreis's fundamental principle as studied in [36]. (See also [18] or, more recently, [2]). What is known as the Ehrenpreis-Montgomery conjecture is the particular case of Conjecture 1.2, when $n = 1$. Thanks to Ritt's theorem [51], Conjecture 1.2 in the case $n = 1$ reduces to the following.

Conjecture 1.4 (Ehrenpreis-Montgomery conjecture). *Let*

$$f(z) = \sum_{k=0}^{M} b_k(z)\, \mathrm{e}^{\mathrm{i}\alpha_k z}, \; b_k \in \overline{\mathbb{Q}}[X], \; \alpha_k \in \overline{\mathbb{Q}} \cap \mathbb{R} \tag{1.4}$$

be an exponential polynomial with both algebraic coefficients and frequencies. Then, there are constant $c, A > 0$, $p \in \mathbb{N}$ *(depending on* f*) such that*

$$\Big(f(z) = f(z') = 0 \; and \; z \neq z'\Big) \Longrightarrow |z - z'| \geq c\, \frac{\mathrm{e}^{-A|\mathrm{Im}\, z|}}{(1 + |z|)^p}. \tag{1.5}$$

A possible reason for the terminology is the relation between Conjecture 1.4 and the following conjecture by H. Shapiro (1958) mentioned by H.L. Montgomery in a colloquium in Number Theory (Bolyai Janos ed.), see [56, 57].

Conjecture 1.5 (Montgomery-Shapiro conjecture). *Let* f, g *be two exponential polynomials that have an infinite number of common zeroes. Then, there is an exponential polynomial* h *that divides both* f *and* g *and has also an infinite number of zeroes.*

Unfortunately, I failed to find a precise reference in H. L. Montgomery's work. There seems to be an oral contribution by H. L. Montgomery linking Conjecture 1.4 and Conjecture 1.5. In 1973, Carlos J. Moreno quoted in the introduction of [47] an unpublished manuscript devoted to his work toward such a conjecture. His thesis (New York University, 1972), under the supervision of L. Ehrenpreis, was centered around it. The idea there was to prove Conjecture 1.4 for sums of exponentials (i.e., $b_k \in \overline{\mathbb{Q}}$ for any k), involving only a small number of exponential monomials.

This is fundamentally different from the methods that arose later (see, e.g., [16]), which depend on the rank of the subgroup $\Gamma(f)$ of the real line generated by the frequencies α_k.

2 What is Known in Connection with Results in Transcendental Number Theory

As mentioned in Sect. 1, besides the approach by C. Moreno in his thesis, most of the attempts toward Conjecture 1.4 rely on an additional hypothesis on the rank of the additive subgroup $\Gamma(f)$ of $\overline{\mathbb{Q}} \cap \mathbb{R}$ generated by the frequencies $\alpha_0, \dots, \alpha_M$, not on the number of monomials $e^{i\alpha_k z}$ involved. An easy case when Conjecture 1.4 holds is the case where the rank of $\Gamma(f)$ equals 2, and the b_k are constant [39]. The result means in that case that the analytic transcendental curve

$$t \in \mathbb{C} \mapsto (e^{it}, e^{i\gamma_1 t}), \ \gamma_1 \in (\overline{\mathbb{Q}} \cap \mathbb{R}) \setminus \mathbb{Q},$$

cannot approach a finite subset in $\overline{\mathbb{Q}}^2$. Explicitly, any finite linear combination of logarithms of r algebraic numbers ($r = 3$ here) α_ι with degrees at most D, logarithmic heights at most h, and with integer coefficients ν_ι having absolute values less than B is either 0 or bounded from below in absolute value,

$$\Big|\sum_{\iota=1}^{r} \nu_\iota \log \alpha_\iota\Big| \geq \frac{1}{B^{c(r)\times D^{r+2}\log D \times h^r}}. \tag{2.6}$$

This is a well-known fact originally due to A. Baker; see, e.g., [9, 10] or ([58], Sect. 4), for up-to-date results, references or conjectures. When the coefficients ν_ι are algebraic, with heights less than B, the following less explicit estimate continues to hold.

$$\Big|\sum_{\iota=1}^{r} \nu_\iota \log \alpha_\iota\Big| \geq \frac{1}{B^{c(r,D)\times h^{\kappa(r)}}} \tag{2.7}$$

for some constants $c(r, D)$ and $\kappa(r)$, D being the maximum of the degrees of the α_ι and ν_ι. The next natural step would be to show that, if γ_1, γ_2 are two real algebraic numbers such that $(1, \gamma_1, \gamma_2)$ are $\mathbb{Q}$-linearly independent, the transcendental curve

$$t \in \mathbb{C} \mapsto (e^{it}, e^{i\gamma_1 t}, e^{i\gamma_2 t})$$

cannot approach an algebraic curve in $\mathbb{C}^3$ which is defined over $\overline{\mathbb{Q}}$; That is, the set of common zeroes of polynomials belonging to $\overline{\mathbb{Q}}[X_1, X_2, X_3]$. Here we are close to a quantified version of the so-called Schanuel's conjecture (see [58], Sect. 4, for conjectures respect to its quantitative versions).

Conjecture 2.1 (Schanuel's conjecture, "numerical" version). *Given s complex numbers $y_1, \dots, y_s$ which are $\overline{\mathbb{Q}}$-linearly independent, the transcendence degree of the algebraic extension $\overline{\mathbb{Q}}[y_1, \dots, y_s, e^{y_1}, \dots, e^{y_s}]$ over $\overline{\mathbb{Q}}$ is at least equal to s.*

For $s = 1$, this is Gel'fond-Schneider's theorem. The $s = 2$ case would imply, for example, the algebraic independence over $\mathbb{Q}$ of the pair of numbers (e, π) or $(\log 2, \log 3)$, and is of course still open. When γ is an algebraic number with degree $D \geq 2$ and ζ a complex number such that $e^{i\zeta} \neq 1$, a result by G. Diaz [34] asserts that, among the exponentials $e^{i\gamma\zeta}, \dots, e^{i\gamma^{D-1}\zeta}$, at least $[(d+1)/2]$ are algebraically independent over $\mathbb{Q}$. This result covers Gel'fond's well-known result ($D = 3$) and even leads to a quantitative version of it. In fact, the quantitative formulation obtained by D. Brownawell in [14] for $D = 3$ (using Gel'fond-Schneider's method) implies the following (rather weak) result respect to Conjecture 1.4, when the rank of $\Gamma(f)$ equals 3.

Proposition 2.1 ([15]). *If f is an exponential sum in one variable with $b_k \in \overline{\mathbb{Q}}$ and $\Gamma(f) = \mathbb{Z} \oplus \gamma\mathbb{Z} \oplus \gamma^2\mathbb{Z}$, γ being an irrational cubic, then, for any $\epsilon > 0$, there is $c_\epsilon > 0$ depending on f such that*

$$\Big(f(z) = f(z') = 0 \; and \; z \neq z'\Big) \Longrightarrow |z - z'| \geq c_\epsilon e^{-|z|^{4+\epsilon}} \tag{2.8}$$

The methods introduced by Guy Diaz in [34] in fact allow one to replace $4 + \epsilon$ by $1 + \epsilon$ in (2.8). In any case, we are indeed very far from what would be the formulation of Conjecture 1.4 in the particular case where b_k are constant and the algebraic frequencies belong to the group $\mathbb{Z} \oplus \gamma\mathbb{Z} \oplus \gamma^2\mathbb{Z}$, γ being an irrational cubic. This is inherent to the approach of the problem *via* classical methods in diophantine approximation.

Besides these cases and the results of C. Moreno in his unpublished 1971 thesis when the number of monomial terms is small, to my knowledge nothing is really known about Conjecture 1.4, at least in connection with an approach based on transcendental number theory methods. For an up-to-date survey of Schanuel's conjecture and its quantitative versions, we refer to ([58], Sects. 3.1 and 4.3).

3 Using the Formal Counterpart of Schanuel's Numerical Conjecture

The point of view I developed with C. A. Berenstein in [16] and Sect. 2 of [15] relies on the fact that the formal analog of Schanuel's conjecture holds, despite the fact that very little is known about the numerical Schanuel conjecture. This is a result by J. Ax and B. Coleman [6, 31], following the ideas developed by C. Chabauty [28] and E. Kolchin [40], see also [22] for a modern up-to-date presentation. Here is a formulation.

Theorem 3.1 (Schanuel's conjecture, formal version). *Let $y_1, \dots, y_s$ be s formal power series in $\mathbb{C}[[t_1, \dots, t_k]]$ ($k \geq 1$), and I an ideal in $\mathbb{C}[X_1, \dots, X_s, Y_1, \dots, Y_s]$, defining in $\mathbb{C}^{2s}$ an algebraic subvariety $\mathcal{V}(I)$ with dimension less or equal to s, such that*

$$\forall\, P \in I,\ P(y_1(t), \dots, y_s(t), e^{y_1(t)}, \dots, e^{y_s(t)}) \equiv 0.$$

Then, there are rational numbers $r_1, \dots, r_s$ and a complex number[2] $\gamma \in \mathbb{C}$ *such that*

$$\sum_{j=1}^{s} r_j y_j(t) \equiv \gamma. \tag{3.9}$$

Here is a corollary of the last Theorem that shows the crucial role it plays when studying the *slowly decreasing conditions* introduced by Ehrenpreis (e.g., [36]) for ideals generated by exponential polynomials with frequencies in $(i\mathbb{Z})^n$. We ignore for the moment any condition of arithmetic type on the coefficients.

Corollary 3.1 ([16], Proposition 6.4 and Corollary 6.7). *Let $P_1, \dots, P_N$ be N polynomials in the 2n variables $(X_1, \dots, X_n, Y_1, \dots, Y_n)$, defining an algebraic variety $\mathcal{V}(P)$ in $\mathbb{C}^{2n}_{z,w}$. Let $\pi_z : (z, w) \in \mathbb{C}^{2n} \mapsto z$ be the projection on the factor $\mathbb{C}^n_z$. Let $W \subset \mathbb{C}^n_z$ be the subset defined by*

$$(z_1, \dots, z_n) \notin W \Longrightarrow \dim(\mathcal{V}(P) \cap \pi^{-1}(z)) = 0 \text{ or } -\infty.$$

Then, any irreducible component with strictly positive dimension of the analytic (transcendental) subset

$$V(F) = \{z \in \mathbb{C}^n\,;\ F_j(z) = P_j(z_1, \dots, z_n, e^{iz_1}, \dots, e^{iz_n}) = 0,\ j = 1, \dots, N\}$$

lies in $\overline{W}$. In particular, when $N \geq n$, any irreducible component with strictly positive dimension of $V(F)$ lies in the closure in $\mathbb{C}^n$ of the set $W' \subset \mathbb{C}^n_z$ defined as

$$z \notin W' \Longrightarrow \operatorname{rank}\left[\left(\frac{\partial P_j(z, w)}{\partial w_k}\right)_{\substack{1 \leq j \leq N \\ k \leq 1 \leq n}}\right] = n \quad \forall\, w \in \mathbb{C}^n.$$

The formal analog of Schanuel's conjecture also allows one to give refined versions of Ritt's theorem in several variables such as those formulated in [8]. Here is an example.

---

[2]Unfortunately, even when one specifies arithmetic conditions on the ideal I, such as the generating polynomials have algebraic coefficients, nothing more precise can be asserted about the constant γ. Indeed, this is the main stumbling block to such a result being an efficient tool in proving Conjecture 1.2 or even Conjecture 1.4.

Corollary 3.2 ([16], see also [52]). *Let*

$$F(z_1,\dots,z_n) = \sum_{k=0}^{M} b_k(z) \mathrm{e}^{\mathrm{i}\langle \alpha_k, z\rangle}$$

be an exponential polynomial in n complex variables which is identically zero on an algebraic irreducible curve $\mathcal{C}$. Then either all polynomial factors b_k vanish identically on $\mathcal{C}$ or else $\mathcal{C}$ is contained in some affine subspace $\langle \alpha_{k_1} - \alpha_{k_2}, z\rangle = \gamma$, where γ is a complex constant[3] *and $\alpha_{k_1} \neq \alpha_{k_2}$. If an irreducible polynomial $P \in \mathbb{C}[X_1,\dots,X_n]$ divides F (as an entire function) without dividing all the b_k, then P is necessarily of the form*

$$P(X) = \langle \alpha_{k_1} - \alpha_{k_2}, X\rangle - \gamma.$$

The main reason such analytic techniques arising from the formal analog of Schanuel's conjecture fail to imply Conjecture 1.2 (or more specifically Conjecture 1.4), is because they do not allow one to keep track of the arithmetic constraints. Though such a goal can be (partially) achieved when adapting Nœther Normalization's lemma to the frame of exponential polynomials $P(X_1,\dots,X_n,\mathrm{e}^{Y_1},\dots,\mathrm{e}^{Y_n})$ (as in Proposition 6.3 in [16]), it still seems far from providing enough information to make significant advances toward Conjectures 1.2 or 1.4.

4 Arithmetic Rigidity and the $\mathcal{D}$-Module Approach

4.1 Lindemann–Weierstrass Theorem Versus Ritt's Factorization

The ubiquity that was pointed out in [4, 5] with respect to the well-known Lindemann–Weierstrass theorem suggests how arithmetic rigidity is reflected in Ritt's factorization of exponential sums in the one variable setting. Let us recall the classical "numerical" formulation of Lindemann–Weierstrass theorem.

Theorem 4.1 (Lindemann–Weierstrass, "numerical" formulation). *Let $\alpha_1,\dots,\alpha_s$ be s algebraic numbers which are $\mathbb{Q}$-linearly independent. Then their exponentials $\mathrm{e}^{\alpha_1},\dots,\mathrm{e}^{\alpha_s}$ are algebraically independent over $\overline{\mathbb{Q}}$.*

Here is its equivalent "functional" formulation, which appears to be an arithmetic version of Ritt's factorization theorem. In this situation, arithmetic conditions indeed impose drastic rigidity constraints.

[3]Here again, additional arithmetic information on F does not impose any arithmetic constraint on γ.

Theorem 4.2 (Lindemann–Weierstrass, "functional formulation"). *Let $\mathfrak{f}$ be a formal power series in $\mathbb{Q}[[X]]$, which corresponds to the Taylor development about the origin of an exponential polynomial f with constant coefficients,*[4] *such that $f(1) = 0$, that is, f can be divided by $z-1$ as an entire function. Then the quotient*

$$z \mapsto \frac{\mathfrak{f}(X)}{X-1}$$

is also the formal power series at the origin of an exponential polynomial with constant coefficients.[5]

4.2 A First Ingredient for the Proof of Theorem 4.2: The Notion of "Size" for a Xd/dX Module over $\mathbb{K}(X)$

One of the major ingredients in the "modern" proof ([4, 5]) of Theorem 4.2 is the notion of "*being of finite size*" for a Xd/dX module over $\mathbb{K}(X)$, where $\mathbb{K}$ is a number field. We keep for the moment to the one variable setting.

Let $\mathbb{K}$ be such a number field, and $\mathcal{M}$ be a Xd/dX module over $\mathbb{K}(X)$. Assume $\mathcal{M}$ is such that the $\mathbb{K}(X)$ induced module is free with finite rank[6]. Thus, $\mathcal{M}$ can be represented in terms of a basis $\Upsilon = (\upsilon_0, \ldots, \upsilon_{\mu-1})$ with the action of the differential operator Xd/dX being represented as

$$(Xd/dX)[\upsilon_j] = \sum_{k=0}^{\mu-1} G_{jk}(X)[\upsilon_k].$$

Taking into account the fact that $\mathbb{K}$ is a number field (and thus the arithmetic rigidity), one can introduce a notion of *size* $\sigma(\mathcal{M})$ as

$$\sigma(\mathcal{M}) = \limsup_{N \to \infty} \frac{1}{N} \sum_{v \in \Sigma_{\text{finies}}(\mathbb{K})} \log^+ \max_{0 \le p \le N} \left\| \frac{G^{(p)}(X)}{p!} \right\|_v, \tag{4.10}$$

where Σ_{finies} denotes the set of non archimedian (conveniently normalized) absolute values on the number field $\mathbb{K}$, and G^p is the (μ, μ) matrix with entries in $\mathbb{K}(X)$, corresponding to the action of $X^p(d/dX)^p$, expressed within the basis Υ

[4]Certainly, the coefficients and frequencies of such an exponential polynomial f are in $\overline{\mathbb{Q}}$.

[5]That is, of course, is identically zero. Nevertheless, it seems better to keep this formulation to view the statement as the effect of arithmetic rigidity constraints in Ritt's factorization theorem.

[6]More generally, one may replace $\mathbb{K}(X)$ by some unitary $\mathbb{K}$-algebra containing $\mathbb{K}(X)$, such as $\mathbb{K}[[X]]$, and introduce then the notion of Xd/dX-module of finite type over $\mathbb{K}[[X]]$.

(see, e.g., [3, 33]). The size is in fact independent of the choice of the basis Υ. The module $\mathcal{M}$ is said to satisfy the *Galochkin condition* when its size $\sigma(\mathcal{M})$ is finite.

An important result by G. Chudnovsky [29, 30], one that relies on Siegel's lemma,[7] asserts that, if A is a (μ, μ) matrix with coefficients in $\mathbb{K}[X]$ such that the differential system

$$(d/dX - A)[Y] = 0 \tag{4.11}$$

admits a solution Y_0 in $(\mathbb{K}[[X]])^{\mu}$ with $\mathbb{K}(X)$-linearly independent components, then the *size* of the corresponding Xd/dX module $\mathcal{M}_A$ is bounded from above by $C(\Gamma)\, h(Y_0)$, where $h(Y_0)$ denotes the maximum of the heights of the coefficients of Y_0, the height being understood here as the height of a formal power series with coefficients in $\mathbb{K}$ (see [3]). In particular, $\mathcal{M}_A$ satisfies the Galochkin condition when the differential system admits a solution with $\mathbb{K}(X)$-linearly independent components, which are all G-functions (see [3] for various definitions[8] of such an arithmetic notion). Note that G. Chudnovsky's theorem has been extended to the several variable context by L. di Vizio in [32].

4.3 A Second Ingredient for the Proof of Theorem 4.2: A Theorem by N. Katz

Here again, one keeps to the one variable context. A differential operator with coefficients in $\mathcal{M}_{\mu,\mu}(\mathbb{C}[X])$

$$\mathcal{L} = \sum_{1}^{L} A_l(X)(d/dX)^q,$$

it is called *fuschian* if all its singularities $a \in \mathbb{C} \cup \{\infty\}$ are *regular* ones. That is, are such that

$$\min_{l \geq 1}(\mathrm{val}_a(A_l) - l) \geq \mathrm{val}_a(A_L).$$

A theorem by Katz [44] asserts that any Xd/dX module over $\mathbb{K}(X)$ ($\mathbb{K}$ being a number field) which satisfies Galochkin condition is necessarily fuschian.

This result has also an extension to the context of several variables ([32]). Such an extension can be combined with Chudnovsky's theorem in higher dimension, as formulated in geometric terms also in ([32]).

---

[7] See, for example, [33], Chap. VIII, for a pedestrian presentation and a proof.

[8] To say it briefly, a G-function is a formal power series in $\overline{\mathbb{Q}}[[X]]$ which is in the kernel of some element in $\overline{\mathbb{Q}}[X, d/dX]$ and, at the same time, has a finite logarithmic height, when considered as a power series in $\overline{\mathbb{Q}}[[X]]$ (see [3] for the notion of logarithmic height for a power series).

The proof of Theorem 4.2 ([5]) follows from such a combination between Chudnovsky's and Katz's theorems. It relies on the elementary proof proposed in [23], which bypasses the p-adic methods based on the Bézivin–Robba criterion that were previously introduced in [24].

4.4 The $\mathcal{D}$-Modules Approach

Let us start here with a few observations about division questions in multivariate complex analysis (see [1, 19]). This approach is reminiscent of pseudo-Wiener deconvolution methods that involve as deconvolutors filters with transfer functions

$$\omega \in \mathbb{R}^n \longmapsto \frac{\overline{F_j(\omega)}}{\|F(\omega)\|^2 + \epsilon^2},$$

where the F_j, $j = 1, \dots, N$, are the transfer functions of the convolutor filters, and $\epsilon^2 << 1$ stands here for a signal to noise ratio.

Let $F_1, \dots, F_N$ be N elements in the Paley–Wiener algebra $\widehat{\mathcal{E}'(\mathbb{R}^n)}$. Consider the holomorphic map $z \mapsto F(z) := (F_1(z), \dots, F_N(z))$ as an holomorphic section of the trivial bundle $\mathbb{C}^n \times \mathbb{C}^N \to \mathbb{C}^n$, equipped with its canonical basis. Let

$$\sigma(z) = \frac{\sum_{j=1}^{N} \overline{F_j(z)} \otimes e_j}{\|F(z)\|^2}, \quad z \in \mathbb{C}^n \setminus F^{-1}(0).$$

It can be shown that there are bundle-valued currents P_F and R_F in $\mathbb{C}^n$ defined by the formulas

$$\begin{aligned} P_F &:= \left[\|F(z)\|^{2\lambda} \sum_{r=1}^{n} \frac{\sigma(z) \wedge (\overline{\partial}[\sigma(z)])^{r-1}}{(2i\pi)^r} \right]_{\lambda=0} \\ R_F &:= \left[\overline{\partial}\,[\|F(z)\|^{2\lambda}] \wedge \sum_{r=1}^{n} \frac{\sigma(z) \wedge (\overline{\partial}\,[\sigma(z)])^{r-1}}{(2i\pi)^r} \right]_{\lambda=0}. \end{aligned} \tag{4.12}$$

That is, one analytically continues the complex parameter λ from $\{\mathrm{Re}\,\lambda \gg 1\}$ to some half-plane $\{\mathrm{Re}\,\lambda > -\eta\}$ for some $\eta > 0$. Note that $\mathrm{Supp}\, R_F \subset F^{-1}(0)$ and that P_F and R_F are related by $((2i\pi)\rfloor_F - \overline{\partial}) \circ P_F = 1 - R_F$, where $\rfloor_F$ denotes the interior product with F.

In order to justify such a construction, one takes a log resolution $\pi : \widetilde{\mathbb{C}^n} \to \mathbb{C}^n$ for the subvariety $\{F_1 = \cdots = F_N = 0\}$. Such a log resolution factorizes through the normalized blow-up of $\mathbb{C}^n$ along the coherent ideal sheaf $(F_1, \dots, F_N)\,\mathcal{O}_{\mathbb{C}^n}$. When $N \leq n$ and $F_1, \dots, F_N$ define a complete intersection in $\mathbb{C}^n$, the current R_F reduces to its $(0, N)$ component, which coincides in this case with the current realized in a neighborhhood of $\bigcup_1^N F_j^{-1}(0)$ as the value at $\lambda_1 = \cdots = \lambda_N = 0$ of the analytically continued current-valued holomorphic map

$$(\lambda_1, \dots, \lambda_N) \in \{\mathrm{Re}\,\lambda_1 >> 1, \dots, \mathrm{Re}\,\lambda_N >> 1\} \longmapsto \frac{1}{(2i\pi)^N} \bigwedge_{j=N}^{1} \overline{\partial}\left(\frac{|F_j|^{2\lambda_j}}{F_j}\right). \tag{4.13}$$

When $F_1, \dots, F_N$ are polynomials (i.e., Fourier transforms of distributions with support $\{0\}$), all distribution coefficients of the current P_F belong to $\mathcal{S}'(\mathbb{C}^n \simeq \mathbb{R}^{2n})$, in which case the ideal $(F_1, \dots, F_N)$ is of course closed in the Paley–Wiener algebra. The current P_F is said *to have Paley–Wiener growth* in $\mathbb{C}^n$ if and only if all its distribution coefficients T satisfy the weaker condition

$$\exists\, p \in \mathbb{N}, , \exists\, A > 0, \exists\, C > 0, \quad \text{such that}$$
$$|\langle T, \varphi\rangle| \le C \sup_{|\underline{l}|+|\underline{m}| \le p} \sup_{\mathbb{C}^n} \left[(1 + \|z\|)^p \mathrm{e}^{A\|\mathrm{Im}\, z\|} \left| \frac{\partial^{\underline{l}+\underline{m}} [\varphi]}{\partial \zeta^{\underline{l}}\, \partial \overline{\zeta}^{\underline{m}}} (z) \right| \right]. \tag{4.14}$$

If P_F has Paley–Wiener growth, so has R_F, since $((2i\pi)\rfloor_F - \overline{\partial}) \circ P_F = 1 - R_F$. Division methods such as developed in [1, 2, 19, 20], show that, if P_F (hence R_F) has Paley–Wiener growth in $\mathbb{C}^n$,

$$\left([I(F_1, \dots, F_N)]_{\mathrm{loc}}\right)^{\min(n,N)} \subset I(F_1, \dots, F_N). \tag{4.15}$$

In the particular case where $N \le n$ and $(F_1, \dots, F_N)$ define a complete intersection in $\mathbb{C}^n$, the fact that P_F (hence R_F) has Paley–Wiener growth in $\mathbb{C}^n$ implies that $I(F_1, \dots, F_N)$ is closed in the Paley–Wiener algebra (one can replace the exponent $\min(n, N)$ by 1 in (4.15)). When $(F_1, \dots, F_N)$ have no common zeroes in $\mathbb{C}^n$, it is therefore equivalent to say that $I(F_1, \dots, F_N)$ is closed in the Paley–Wiener algebra or to say that P_F has Paley–Wiener growth (here $R_F \equiv 0$ since $F^{-1}(0) = \emptyset$). Conjecture 1.2 suggests then the following conjecture.

Conjecture 4.1. *Let $F_1, \dots, F_N$ be N exponential polynomials such as in Conjecture 1.2. The current P_F (hence also R_F) has Paley–Wiener growth.*

Remark 4.1. Conjecture 4.1 implies Conjecture 1.4 : when $n = 1$, take N large enough and $F_1, \dots, F_N$ the list of successive derivatives of the exponential polynomial $f : z \mapsto \sum_{k=0}^M b_k(z)\, \mathrm{e}^{i\alpha_k z}$ (see, e.g., [15]).

In order to rephrase Conjecture 4.1 in more algebraic terms, let us recall the following trick. If $\mathrm{Re}\,\beta > 0$, and $t_1, \dots, t_N$ are N strictly positive numbers, then one has, for any $(\gamma_1, \dots, \gamma_{N-1}) \in]0, \infty[^{N-1}$ such that $\gamma_1 + \cdots + \gamma_{N-1} < \mathrm{Re}\,\beta$,

$$(t_1 + \cdots + t_N)^{-\beta} = \frac{1}{(2i\pi)^{N-1}\Gamma(\beta)} \int_{\gamma_1 + i\mathbb{R}} \cdots \int_{\gamma_{N-1}+i\mathbb{R}} \Gamma_N^*(\zeta)\, t_1^{-\zeta_1} \cdots t_{N-1}^{-\zeta_{N-1}} t_N^{\zeta^*}\, \mathrm{d}\zeta_1 \cdots \mathrm{d}\zeta_{N-1}, \tag{4.16}$$

where

$$\Gamma_N^*(\zeta) = \Gamma(\zeta_1)\cdots\Gamma(\zeta_{N-1})\Gamma(\beta-\zeta_1-\cdots-\zeta_{N-1}),\ \zeta^* = \sum_{k=1}^{N-1}\zeta_k - \beta\,.$$

Formula (4.16) allows the transformation of the *additive* operation between the t_j (namely $(t_1+\cdots+t_N)^{-\beta}$) into a *multiplicative* one (namely $t_1^{-\zeta_1}\cdots t_{N-1}^{-\zeta_{N-1}}t_N^{\zeta^*}$, once in the integrand). One can view it as a continuous version of the binomial formula (with negative exponent). Taking, for example, $t_j = |F_j(z)|^2$, $j = 1,\ldots,N$, it follows that one way then to tackle Conjecture 4.1 could be to study (first formally, then numerically in $\mathbb{C}^n$, pairing antiholomorphic coordinates with holomorphic ones in order to recover positivity) the analytic continuation of

$$\lambda = (\lambda_1,\ldots,\lambda_N) \longmapsto \prod_{j=1}^{N}(F_j(z_1,\ldots,z_n))^{\lambda_j}. \tag{4.17}$$

When $F_1,\ldots,F_N$ are polynomials in $\mathbb{K}[X_1,\ldots,X_n] = \mathbb{K}[X]$, where $\mathbb{K}$ is a number field, one may consider the $\mathbb{K}(\lambda)\langle X, d/dX\rangle$-module $\mathfrak{M}(F)$ freely generated by a single generator (formally denoted as $\mathfrak{F}^\lambda = \mathfrak{F}_1^{\lambda_1}\otimes\cdots\otimes\mathfrak{F}_N^{\lambda_N}$), namely

$$\mathfrak{M}(F) = \mathbb{K}(\lambda)[X]\left[\frac{1}{F_1},\ldots,\frac{1}{F_N}\right]\cdot\mathfrak{F}^\lambda.$$

This $\mathbb{K}(\lambda)\langle X, d/dX\rangle$-module is holonomic (i.e., $\dim\mathfrak{M}(F) = n$). A noetheriannity argument (see, e.g., [35]) implies then that there exists a set of *global Bernstein–Sato algebraic relations*

$$\mathcal{Q}_j(\lambda, X, d/dX)\big[F_j\cdot\mathfrak{F}^\lambda\big] = \mathcal{B}(\lambda)\cdot\mathfrak{F}^\lambda,\ j = 1,\ldots,N, \tag{4.18}$$

where $\mathcal{B}\in\mathbb{K}[\lambda]$ and $\mathcal{Q}_j\in\mathbb{K}[\lambda]\langle X, d/dX\rangle$, $j = 1,\ldots,N$. Such a set of algebraic relations (4.18) can be used in order to express (via (4.16) with $t_j = |F_j(z)|^2$, $t{=}1,\ldots,N$) the current P_F as a current with coefficients in $\mathcal{S}'(\mathbb{C}^n)$.

Local analytic analogs of global Bernstein–Sato algebraic relations (4.18) indeed exist. When $f_1,\ldots,f_N$ are N elements in $\mathcal{O}_{\mathbb{C}^n,0}$ and $\mathfrak{t}$ is an holonomic distribution about the origin in $\mathbb{C}^n$ (e.g., a distribution coefficient of some integration current $[V]$, or of some Coleff-Herrera current, see [27]), then there exists a set of *local Bernstein–Sato analytic equations*

$$q_{\mathfrak{t},j}(\lambda,\zeta,\partial/\partial\zeta)\big[f_j\cdot\mathfrak{f}^\lambda\otimes\mathfrak{t}\big] = b_{\mathfrak{t}}(\lambda)\cdot\mathfrak{f}^\lambda\otimes\mathfrak{t},\ j = 1,\ldots,N, \tag{4.19}$$

where $q_{\mathfrak{t},j}$ denotes a germ at the origin of a holomorphic differential operator with coefficients analytic in ζ and polynomial in λ, and $b_{\mathfrak{t}}$ is a finite product of affine forms $\kappa_0+\kappa_1\lambda_1+\cdots+\kappa_n\lambda_n$, with $\kappa_0\in\mathbb{N}^*$, $(\kappa_1,\ldots,\kappa_M)\in\mathbb{N}^M\setminus\{0\}$ ([25,26,42,54]).

Unfortunately, such a local result does not provide any algebraic information about the $q_{\mathfrak{t},j}$, when, for example, the f_j's represent the germs at the origin of exponential polynomials of the form (1.3), as in Conjecture 1.2 or Conjecture 1.3.

One intermediate way to proceed in this case is to consider the case of formal power series. For example, let us suggest an approach to tackle Conjecture 1.4 for exponential sums. Consider an exponential sum

$$f\ :\zeta\in\mathbb{C}\longmapsto\sum_{k=0}^{M}b_k\mathrm{e}^{\mathrm{i}\alpha_k\zeta},$$

with algebraic coefficients b_k, and purely imaginary algebraic distinct frequencies $i\alpha_k$. Let $\mathbb{K}$ be the number field generated by the b_k's, the α_k's, and i. Let $n\geq 1$ be the rank of the subgroup $\Gamma(f)=\mathbb{Z}\alpha_0+\cdots+\mathbb{Z}\alpha_M$, and $(\gamma_1,\dots,\gamma_n)$ be a basis of $\Gamma(f)$. For each $j=1,\dots,M$, let $P_j\in\mathbb{K}[X_1,\dots,X_n]$ such that

$$\frac{\mathrm{d}^{j-1}f}{\mathrm{d}\zeta^{j-1}}(z)=P_j(\mathrm{e}^{\mathrm{i}\gamma_1 z},\dots,\mathrm{e}^{\mathrm{i}\gamma_n z}),\quad \forall\, z\in\mathbb{C},$$

and $P:=(P_1,\dots,P_M)\ :\mathbb{C}^n\to\mathbb{C}^M$. Let $N=M+n-1$, and the exponential polynomials $F_1,\dots,F_N$ be defined as follows:

- For $j=1,\dots,M$, F_j is the exponential sum in n variables, with coefficients in $\mathbb{K}$,
$$(z_1,\dots,z_n)\longmapsto F_j(z)=P_j(\mathrm{e}^{\mathrm{i}z_1},\dots,\mathrm{e}^{\mathrm{i}z_n}).$$
- For $j=1,\dots,n-1$, F_{M+j} is the linear form, also with coefficients in $\mathbb{K}$,
$$(z_1,\dots,z_n)\longmapsto\gamma_n z_j-\gamma_j z_n.$$

Let ξ be a point in $\mathbb{C}^n$, such that $\mathrm{e}^{\mathrm{i}\xi}\in\mathbb{K}^n\cap\{P=0\}$. The Taylor developments of $F_1,\dots,F_M$ at ξ correspond to power series $\mathfrak{f}_{1,\xi},\dots,\mathfrak{f}_{M,\xi}$ in $\mathbb{K}[[X_1,\dots,X_n]]$, while the Taylor developments at ξ of $F_{M+1},\dots,F_N$ correspond to the affine power series

$$\mathfrak{f}_{M+j,\xi}\ :X=(X_1,\dots,X_n)\longmapsto\mathfrak{u}_j+(\gamma_n X_j-\gamma_j X_n),\ j=1,\dots,n-1,$$

where $\mathfrak{u}_j=\gamma_n\xi_j-\gamma_j\xi_n$ is a linear combination of logarithms of algebraic numbers with algebraic coefficients. Here $\mathfrak{u}_1,\dots,\mathfrak{u}_{n-1}$ can be interpreted as parameters. Inspired by [11], one could conjecture[9] the existence of a set of global formal generic Bernstein–Sato relations:

$$\mathfrak{Q}_{\xi,j}(\lambda,X,\mathfrak{u}_1,\dots,\mathfrak{u}_{n-1},d/dX)\big[\mathfrak{f}_{j,\xi}\cdot\mathfrak{F}_\xi^\lambda\big]=\mathfrak{g}_\xi(\mathfrak{u}_1,\dots,\mathfrak{u}_{n-1})\,\mathfrak{b}_\xi(\lambda)\cdot\mathfrak{F}_\xi^\lambda,$$
$$j=1,\dots,N,\qquad(4.20)$$

[9]The lines which follow intend just to sketch what could be a conjectural approach to Conjecture 1.4 for exponential sums f such that $\Gamma(f)$ has small rank.

where $\mathfrak{F}_\xi^\lambda = \mathfrak{f}_{1,\xi}^{\lambda_1} \otimes \ldots \mathfrak{f}_{N,\xi}^{\lambda_N}$, $\mathfrak{Q}_{\xi,j}$ is a differential operator with coefficients in $\mathbb{K}[\lambda][[\mathfrak{u}, X]]$, $\mathfrak{g}_\xi \in \mathbb{K}[[\mathfrak{u}]]$, $\mathfrak{b}_\xi \in \mathbb{K}[\lambda]$. Moreover, an argument based on Siegel's method (and principle), such as that developed by Ehrenpreis[10] in [37], could be then used in order to ensure then that the formal power series coefficients (in $X, \mathfrak{u}$) of the $\mathfrak{Q}_j$ (considered as polynomials in λ and d/dX) have a radius of convergence that is bounded from below by $\rho > 0$, independently of ξ, provided $\mathrm{e}^{\mathrm{i}\xi}$ belongs to a compact subset of $(\mathbb{C}^*)^n$. Then (4.20) would provide a *semi-global Bernstein–Sato set of relations.* The results quoted in Sect. 4, which rely on Siegel's lemma (see, e.g., the proof of Chudnovsky's theorem in [33], or the approach to Gelfand-Shidlovsky theorem as in [21]) give some credit to the conjectural existence of such a collection (indexed by ξ, with $\mathrm{e}^{\mathrm{i}\xi} \in \mathbb{K}^n \cap P^{-1}(0)$) of Berntein-Sato sets of semi-global relations $\mathcal{B}_\xi$ as (4.20). One could then identify terms with lower degree in $\mathfrak{u}$ in (4.20) and thus assume, in each set of relations $\mathcal{B}_\xi$ such as (4.20), that $\mathfrak{g}_\xi$ is homogeneous in $\mathfrak{u}$. In the particular case $n = 3$ (where we recall almost nothing is known concerning Conjecture 1.4, see Sect. 2), one could thus assume that $\mathfrak{g}_\xi$ factorizes as a product of linear factors $\beta_{\xi,1}\mathfrak{u}_1 + \beta_{\xi,2}\mathfrak{u}_2$, where $\beta_{\xi,1}$ and $\beta_{\xi,2}$ belong to $\mathbb{K}$. Combining this with A. Baker's theorem (take $(\mathfrak{u}_1, \mathfrak{u}_2) = (\log \xi_1 + 2ik_1\pi, \log \xi_2 + 2ik_2\pi)$, $(k_1, k_2) \in \mathbb{Z}^2$), one would get (with (4.20)) some way to control the analytic continuation procedure (4.17), leading to the conjectural lower estimates

$$\sum_{j=1}^{M} |P_j(\mathrm{e}^{\gamma_1 z}, \ldots, \mathrm{e}^{\gamma_n z})| = \sum_{1}^{M} \left| \frac{\mathrm{d}^{j-1} f}{\mathrm{d}\zeta^{j-1}}(z) \right| \geq c\, \frac{\mathrm{e}^{-A|\mathrm{Im}\, z|}}{(1+|z|)^p},$$

that ensure (1.5) (see [15]).

The conjectural approach proposed above can be seen as an attempt to take into account the intrinsic arithmetic rigidity of such problems that the results quoted in Sect. 4 suggest.

Another approach, one that would seem more direct, would be to try to mimic the algebraic construction that leads to the construction of a global set of Bernstein–Sato relations such as (4.18) when $F_1, \ldots, F_N$ belong to $\mathbb{K}[X_1, \ldots, X_n]$. That is, let $F_1, \ldots, F_N$ be N exponential polynomials of the form

$$F_j(z) = P_j(z_1, \ldots, z_n, \mathrm{e}^{\mathrm{i}\gamma_{1,1} z_1}, \ldots, \mathrm{e}^{\mathrm{i}\gamma_{1,N_1} z_1}, \ldots, \mathrm{e}^{\mathrm{i}\gamma_{n,1} z_n}, \ldots, \mathrm{e}^{\mathrm{i}\gamma_{n,N_n} z_n}),$$
$$j = 1, \ldots, N,$$

where $P_j \in \mathbb{K}[X_1, \ldots, X_n, Y_{1,1}, \ldots, Y_{1,N_1}, \ldots, Y_{n,1}, \ldots, Y_{n,N}]$, the $\gamma_{j,k}$ being also elements in $\mathbb{K}$ such that $\gamma_{j,1}, \ldots, \gamma_{j,N_j}$ are $\mathbb{Q}$-linearly independent for any $j = 1, \ldots, n$. Instead of the Weyl algebra $\mathbb{K}(\lambda)\langle X, d/dX \rangle$, one could introduce a non commutative algebra such as

[10]Note that this work of L. Ehrenpreis appeared in the Lecture Notes volume where appeared also the important results by Chudnovsky [29,30].

$$\mathbb{K}(\lambda_1,\dots,\lambda_n)\Big\langle X_1,\dots,X_n,Y_{1,1},\dots,Y_{1,N_1},\dots,Y_{n,1},\dots,Y_{n,N_n},\partial_1,\dots,\partial_n\Big\rangle,$$

with the following commutation rules: for any $j,k \in \{1,\dots,n\}$, for any $l \in \{1,\dots,N_j\}$,

$$[\partial_k, X_j] = -\delta_{jk}\,, \quad [X_k, Y_{j,l}] = 0\,, \quad [\partial_k, Y_{j,l}] = -\,\gamma_{j,l}\,\delta_{kl}\,Y_{j,l}\,.$$

One may consider, as in the Weyl algebra case, the $\mathbb{K}(\lambda)\langle X,Y,\partial\rangle$-module

$$\mathfrak{M}(F) = \mathbb{K}(\lambda)[X,Y,\partial]\left[\frac{1}{F_1},\dots,\frac{1}{F_N}\right]\cdot \mathfrak{F}^{\lambda}.$$

Nœtheriannity arguments based on the concept of *dimension*[11] for such a module lead (inspired by the argument described by Ehlers in [35]) to the existence, in some very particular cases, of what would be a substitute for a set of global Bernstein–Sato relations such as (4.18) (see [17, 18]). Unfortunately, the results obtained here cover only situations basically quite close of that of Conjecture 1.4 when rank $\Gamma(f) \leq 2$. Here are the results obtained that way:

- The current P_F attached to any system $F = (F_1,\dots,F_N)$, $F_j(z_1,\dots,z_n) = P_j(z_1,\dots,z_n,\mathrm{e}^{\mathrm{i}\,z_n})$, $j = 1,\dots,N$, where $P_j \in \mathbb{C}[X_1,\dots,X_n,Y]$, has Paley–Wiener growth in $\mathbb{C}^n$.
- The current P_F attached to any system $F = (F_1,\dots,F_N)$, $F_j(z_1,\dots,z_n) = P_j(z_1,\dots,z_{n-1},\mathrm{e}^{\mathrm{i}\,z_n},\mathrm{e}^{\mathrm{i}\,\gamma\,z_n})$, $j = 1,\dots,N$, where $P_j \in \overline{\mathbb{Q}}[X_1,\dots,X_{n-1},Y_1,Y_2]$ and $\gamma \in (\overline{\mathbb{Q}} \cap \mathbb{R}) \setminus \mathbb{Q}$, has Paley–Wiener growth in $\mathbb{C}^n$.

Note that only the second situation carries an arithmetic structure. The methods developed in [17, 18] failed, at least for their intended purpose of making progress toward Conjectures 1.2 or even 1.4. For example, they do not seem to be of any help toward Conjecture 1.4, when rank $(\Gamma(f)) = 2$ and f is a true exponential polynomial (not an exponential sum). The main reason for the failure is that these methods take into account only the concept of *dimension*, and ignore that of *logarithmic size*. On the other hand, the conjectural approach toward Conjecture 1.4 when rank $\Gamma(f) = 3$ (such as sketched above) was taking into account such concepts, basically through Siegel's lemma. It is natural to ask the following question: can some argument based on a filtration with respect to the size lead to what would be a substitute for a set of global Bernstein–Sato relations such as (4.18) or (4.20)? That would indeed provide a decisive step toward all conjectures mentioned here.

[11] That is on concepts of algebraic, not really arithmetic, nature, though arithmetics is deeply involved.

5 Some Other Miscellaneous Approaches

This paper has intended to give a brief, up-to-date discussion of the fascinating conjectures arising from arithmetic considerations added to L. Ehrenpreis' contributions to the study of the "*slowly decreasing condition*" in the Paley–Wiener algebra. One should add that recent developments in amœba theory [43, 48, 49], in relation with tropical geometry, might also be of some interest for such conjectures. Unfortunately, they usually are more adapted to the case of complex frequencies[12] than to the most delicate so-called "neutral case" where all frequencies are purely imaginary as in the questions discussed here. The most serious stumbling block is that, from the combinatorics point of view, when dealing with "algebraic" cones in $\mathbb{R}^n$, one is missing Gordon's lemma. One needs then to bypass such a difficulty; see, for example, [12] for the construction of toric varieties associated to non rational fans. In this connection, we mention some references that might inspire ideas for deciding such conjectures about exponential sums [38, 41, 43, 45, 46, 48–50, 53, 55]. Unfortunately, most of them do not really take into account the arithmetic constraints, and are more in the spirit of C. Moreno's papers [47].

References

1. M. Andersson, *Residue currents and ideals of holomorphic functions*, Bull. Sci. Math. **128** (2004), no. 6, pp. 481–512.
2. M. Andersson, E. Wulcan, Residue currents with prescribed annihilator ideals, Ann. Sci. École Norm. Sup. 40 (2007), pp. 985–1007.
3. Y. André, *G-Functions and Geometry*, Aspects of Mathematics, Vieweg, 1989.
4. Y. André, Séries Gevrey de type arithmétique, I. Théorèmes de pureté et de dualité, Annals of Mathematics 151 (2000), pp. 705–740.
5. Y. André, Séries Gevrey de type arithmétique, II. Transcendance sans transcendance, Annals of Mathematics 151 (2000), pp. 741–756.
6. J. Ax, On Schanuel's conjectures, Ann. of Math. 93 (1971), pp. 252–268.
7. R. Bellman, K. Cooke, *Differential difference equations*, Academic Press, New York, 1963.
8. C. A. Berenstein, A. Dostal, The Ritt theorem in several variables, Ark. Math. 12 (1974), pp. 267–280.
9. A. Baker, *Transcendental Number Theory*, Cambridge University Press, 1979.
10. A. Baker, G. Wülstholz, Logarithmic forms and group varieties, J. Reine Angew. Math. 442 (1993), pp. 19–62.
11. R. Bahloul, Global generic Bernstein–Sato polynomial on an irreducible algebraic scheme, Proc. Japan Acad. 79, Ser. A (2003), pp. 146–149.
12. F. Battaglia, E. Prato, Generalized toric varieties for simple non rational convex polytopes, Intern. Math. Res. Notices 24 (2001), pp. 1315–1337.
13. C. A. Berenstein, R. Gay, *Complex analysis and special topics in harmonic analysis*, Springer-Verlag, New York, 1995.

[12]Polya's theory, see also [7].

14. D. Brownawell, Pairs of polynomials small at a number to certain algebraic numbers, Séminaire Delange-Pisot-Poitou 11, 1975-1976, pp. 1–12.
15. C. A. Berenstein, A. Yger, On Lojasiewicz type inequalities for exponential polynomials, J. Math. Anal. Appl. 129 (1988), pp. 166–195.
16. C. A. Berenstein, A. Yger, Ideals generated by exponential polynomials, Advances in Mathematics 60 (1986), pp. 1–80.
17. C. A. Berenstein, A. Yger, Exponential polynomials and $\mathcal{D}$-modules, Compositio Mathematica 95 (1995), pp. 131–181.
18. C. A. Berenstein, A. Yger, Division interpolation methods and Nullstellensätze, pp. 41–59 in *Analysis, Geometry, Number Theory: The Mathematics of Leon Ehrenpreis*, E. L. Grinberg [et al.] (eds.), Contemporary Mathematics 251, AMS 1999.
19. C. A. Berenstein, R. Gay, A. Vidras, A. Yger, *Residue currents and Bézout identities*, Progress in Mathematics 114, Birkhäuser, 1993.
20. C. A. Berenstein, A. Vidras, A. Yger, *Multidimensional residue theory and applications*, manuscript in preparation.
21. D. Bertrand, On André's proof of the Siegel-Shidlovsky theorem, pp. 51–63 in Colloque Franco-Japonais: Théorie des Nombres Transcendants (Tokyo, 1998), Sem. Math. Sci., 27, Keio Univ., Yokohama, 1999.
22. D. Bertrand, Théories de Galois différentielles et transcendance, Annales Inst. Fourier 59 (2009), no. 7, pp. 2773–2803.
23. F. Beukers, J. P. Bézivin, P. Robba, An alternative proof of the Lindemann-Weierstrass theorem, American Monthly 97 (1990), no. 3, pp. 193–197.
24. J. P. Bézivin, P. Robba, A new p-adic method for proving irrationality and transcendence results, Annals of Mathematics 129 (1989) pp. 151–160.
25. J. E. Björk, *Rings of of differential operators*, North-Holland, Amsterdam, 1979.
26. J. E. Björk, *Analytic $\mathcal{D}$-modules and applications*, Mathematics and its Applications 247, Kluwer Academic Publishers Group, Dordrecht, 1993.
27. J. E. Björk, Residues and $\mathcal{D}$-modules, pp. 605–651 in *The legacy of Niels Henrik Abel*, O.A. Laudal, R. Piene (eds.), Springer-Verlag, Berlin (2004).
28. C. Chabauty, Sur les équations diophantiennes liées aux unités d'un corps de nombres algébrique fini, Ann. Math. Pura Appl. 17 (1938), pp. 127–168.
29. D. Chudnovsky, G. Chudnovsky, Applications of Padé approximations to diophantine inequalities in values of G-functions, pp. 9–51 in *Number theory (New York 1983-1984)*, Lecture Notes in Math. 1135, Springer-Verlag, Berlin (1985).
30. D. Chudnovsky, G. Chudnovsky, Applications of Padé approximation to the Grothendieck conjecture on linear differential equations, pp. 52–100 in *Number theory (New York 1983-1984)*, Lecture Notes in Math. 1135, Springer-Verlag, Berlin (1985).
31. R. Coleman, A generalization of the Ax-Schanuel theorem, Amer. J. Math. 102 (1980) pp. 595–624.
32. L. di Vizio, Sur la théorie géométrique des G-fonctions, le théorème de Chudnovsky à plusieurs variables, Math. Ann. 319 (2001), pp. 181–213.
33. B. Dwork, G. Gerotto, F. J. Sullivan, *An introduction to G-functions*, Annals of Mathematics Studies 133, Princeton University Press, 1994.
34. G. Diaz, Grands degrés de transcendance pour des familles d'exponentielles, C.R. Acad. SCi. Paris Sér. I Math. 305 (1987), no. 5, pp. 159–162.
35. F. Ehlers, The Weyl algebra, pp. 173–206 in *Algebraic $\mathcal{D}$-modules (Perspectives in Mathematics)*, A. Borel, J. Coates, S. Helgason (eds.), Academic Press, Boston, 1987.
36. L. Ehrenpreis, *Fourier Analysis in several complex variables*, Wiley, 1970.
37. L. Ehrenpreis, Transcendental numbers and partial differential equations, pp. 112–125 in *Number theory (New York, 1983-1984)*, Lecture Notes in Math. 1135, Springer-Verlag, Berlin (1985).
38. M. Forsberg, M. Passare, A. Tsikh, Laurent determinants and arrangements of hyperplane amoebas, Adv. in Maths. 151 (2000), pp. 45–70.

39. F. Gramain, Solutions indéfiniment dérivables et solutions presque périodiques d'une équation de convolution, Bull. Soc. Math. France 104 (1976), pp. 401–408.
40. E. Kolchin, Algebraic groups and algebraic dependence, Amer. J. Math. 90, 1968, pp. 1151–1164.
41. D. Gurevich, Closed ideals with zero dimensional root set in certain rings of holomorphic functions, J. Soviet Math. 9 (1978), pp. 172–182.
42. A. Gyoja, Bernstein–Sato's polynomial for several analytic functions, J. Math. Kyoto Univ. 33, (1993), no. 2, pp. 399–411.
43. A. Henriques, An analog of convexity for complements of amœbas of higher codimension, an answer to a question asked by B. Sturmfels, Adv. Geom. 4 (2004), no. 1, pp. 61–73.
44. N. Katz, Algebraic solutions of Differential Equations (p-curvature and the Hodge filtration), Invent. Math. 18 (1972), pp. 1–118.
45. B. Ya. Kazarnovskii, On zeros of exponential sums, Dokl. Akad. Nauk SSSR, 257 (1981), no. 4, pp. 804–808.
46. B. Ya. Kazarnovskii, Exponential analytic sets, Functional Analysis and its applications, 31 (1997), no. 2, pp. 86–94.
47. C. Moreno, The zeroes of exponential polynomials (I), Compositio Math. 26 (1973), no. 1, 69–78.
48. M. Passare, H. Rullgard, Amoebas,Monge-Ampère measures, and triangulations of the Newton polytope, Duke Math. J. 121 (2004), no. 3, pp. 481–507.
49. K. Purbhoo, A Nullstellensatz for amoebas, Duke Mathematical Journal 141 (2008), no. 3, 407–445.
50. A. Yu. Rashkovskii, Zeros of holomorphic almost periodic mappings with independent components. Complex Variables 44 (2001) 299–316.
51. J. F. Ritt, A factorization theory for functions $\sum_{i=1}^{n} a_i e^{\alpha_i x}$, Trans. Amer. Math. Soc. 29 (1927), no. 3, pp. 584–596.
52. L. I. Ronkin, *Functions of completely regular growth*, Mathematics and its Applications (Soviet Series), 81. Kluwer Academic Publishers, Dordrecht, 1992.
53. L. I. Ronkin, On zeros of almost periodic functions generated by holomorphic functions in multicircular domain, pp. 243–256 in *Complex Analysis in Modern Mathematics*, Fazis, Moscow, 2001.
54. C. Sabbah, Proximité évanescente II. Équations fonctionnelles pour plusieurs fonctions analytiques, Compositio Math 64 (1987), pp. 213–241.
55. J. Silipo, Amibes de sommes d'exponentielles, Canadian J. Math. 60, 1 (2008), pp. 222–240.
56. R. Tijdeman, A. Van der Poorten, On common zeroes of exponential polynomials, Enseignement Mathématique 21 (1975), no. 2, pp. 57–67.
57. A. Van der Poorten, A note on the zeroes of exponential polynomials, Compositio Math. 31 (1975), no. 2, pp. 109–113.
58. M. Waldschmidt, Open diophantine problems, Mosc. Math. J. 4, no. 1 (2004), pp. 245–305.

The Discrete Analog of the Malgrange–Ehrenpreis Theorem

Doron Zeilberger

In fond memory of Leon Ehrenpreis

Abstract One of the landmarks of the modern theory of partial differential equations is the Malgrange–Ehrenpreis theorem that states that every nonzero linear partial differential operator with constant coefficients has a Green function (alias fundamental solution). In this short note, I state the discrete analog and give two proofs. The first one is Ehrenpreis style, using duality, and the second one is constructive, using formal Laurent series.

Key words Formal Laurent series • Fundamental solution • Systems of constant-coefficient partial differential equations

Mathematics Subject Classification(2010): 35E05 (Primary), 39A06 (Secondary)

One of the *landmarks* of the modern theory of partial *differential* equations is the *Malgrange–Ehrenpreis*[E1, E2, M] theorem (see [Wi]) that states that every nonzero linear partial differential operator with constant coefficients has a Green's function (alias *fundamental solution*). Recently, Wagner[W] gave an elegant *constructive* proof.

In this short note, I will state the *discrete* analog and give two proofs. The first one is Ehrenpreis style, using *duality*, and the second one is *constructive*, using *formal Laurent series*.

D. Zeilberger (✉)
Department of Mathematics, Rutgers University (New Brunswick), Hill Center-Busch Campus, 110 Frelinghuysen Rd., Piscataway, NJ 08854-8019, USA
e-mail: http://www.math.rutgers.edu/~zeilberg/.

First version: July 21, 2011. This version: Sept. 7, 2011.

Let Z be the set of integers and n a positive integer. Consider functions $f(m_1, \dots, m_n)$ from Z^n to the complex numbers (or any field). A *linear partial difference operator with constant coefficients* $\mathcal{P}$ is anything of the form

$$\mathcal{P} f(m_1, \dots, m_n) := \sum_{\alpha \in A} c_\alpha f(m_1 + \alpha_1, \dots, m_n + \alpha_n),$$

where A is a *finite* subset of Z^n and $\alpha = (\alpha_1, \dots, \alpha_n)$ and the c_α are *constants*.

For example, the *discrete Laplace operator* in two dimensions:

$$f(m_1, m_2) \to f(m_1, m_2) - \frac{1}{4}(f(m_1+1, m_2) + f(m_1-1, m_2) + f(m_1, m_2+1) + f(m_1, m_2-1)).$$

The *symbol* of the operator $\mathcal{P}$ is the *Laurent polynomial*

$$P(z_1, \dots, z_n) = \sum_{\alpha \in A} c_\alpha z_1^{\alpha_1} \cdots z_n^{\alpha_n}.$$

The *discrete delta function* is defined in the obvious way

$$\delta(m_1, \dots, m_n) = \begin{cases} 1, \text{ if } (m_1, \dots, m_n) = (0, 0, \dots, 0); \\ 0, \text{ otherwise.} \end{cases}$$

Note that the beauty of the discrete world is that the delta function is a *genuine* function, not a "generalized" one, and one does not need the intimidating edifice of Schwartzian *distributions*.

We are now ready to state the

Discrete Malgrange–Ehrenpreis Theorem: Let $\mathcal{P}$ be any nonzero *linear partial difference operator with constant coefficients*. There exists a function $f(m_1, \dots, m_n)$ defined on Z^n such that

$$\mathcal{P} f(m_1, \dots, m_n) = \delta(m_1, \dots, m_n).$$

First Proof (Ehrenpreis style) Consider the infinite-dimensional vector space, $C[z_1, \dots, z_n, z_1^{-1}, \dots, z_n^{-1}]$, of *all* Laurent polynomials in $z_1, \dots, z_n$. Every function f on Z^n uniquely defines a *linear functional* T_f defined on monomials by

$$T_f [z_1^{m_1} \cdots z_n^{m_n}] := f(m_1, \dots, m_n),$$

and extended by linearity. Conversely, any linear functional gives rise to a function on Z^n. Let $P(z_1, \dots, z_n)$ be the symbol of the operator $\mathcal{P}$. We are looking for a linear functional T such that for every monomial $z_1^{m_1} \cdots z_n^{m_n}$

$$T [P(z_1, \dots, z_n) z_1^{m_1} \cdots z_n^{m_n}] = T_\delta (z_1^{m_1} \cdots z_n^{m_n}).$$

By linearity, for *any* Laurent polynomial $a(z_1, \ldots, z_n)$

$$T\,[\,P(z_1, \ldots, z_n)a(z_1, \ldots, z_n)\,] = T_\delta(a(z_1, \ldots, z_n)).$$

So T is defined on the (vector) *subspace* $P(z_1, \ldots, z_n)C\,[z_1, \ldots, z_n, z_1^{-1}, \ldots, z_n^{-1}]$ of $C\,[z_1, \ldots, z_n, z_1^{-1}, \ldots, z_n^{-1}]$. By elementary linear algebra, every linear functional on the former can be *extended* (in many ways!) to the latter. *QED.*

Before embarking on the second proof, we have to recall the notion of *formal power series* and, more generally, *formal Laurent series.*

A *formal power series* in one variable z is any creature of the form

$$\sum_{i=0}^{\infty} a_i z^i.$$

More generally, a *positive formal Laurent series* is any creature of the form

$$\sum_{i=m}^{\infty} a_i z^i,$$

where m is a (possibly negative) integer. On the other hand, a *negative formal Laurent series* is any creature of the form

$$\sum_{i=-\infty}^{m} a_i z^i,$$

where m is a (possibly positive) integer.

A *bilateral* formal Laurent series goes *both ways*

$$\sum_{i=-\infty}^{\infty} a_i z^i.$$

Note that the class of bilateral formal Laurent series is an abelian additive group, but one *cannot* multiply there. On the other hand, one can legally multiply two positive formal Laurent series by each other and two negative formal Laurent series by each other, but don't mix them! Of course, it is always legal to multiply any Laurent polynomial by any bilateral formal power series. *But watch out for zero divisors*, e.g.,

$$(1 - z) \sum_{i=-\infty}^{\infty} z^i = 0.$$

Any Laurent polynomial $p(z) = a_i z^i + \cdots + a_j z^j$ of low-degree i and high-degree j in z (so $a_i \neq 0$, $a_j \neq 0$) has *two* natural multiplicative inverses. One is in the ring of positive Laurent polynomials and the other in the ring of negative Laurent polynomials. Simply write $p(z) = z^i a_i p_0(z)$ and get $1/p(z) = z^{-i}(1/a_i)p_0(z)^{-1}$, and writing $p_0(z) = 1 + q_0(z)$, we form

$$p_0(z)^{-1} = (1 + q_0(z))^{-1} = \sum_{i=0}^{\infty}(-1)^i q_0(z)^i,$$

and this makes perfect sense and *converges* in the ring of formal power series. Analogously, one can form a multiplicative inverse in powers in z^{-1}.

It follows that every *rational function* $P(z)/Q(z)$ in one variable, z, has two natural inverses, one pointing positively and one negatively.

What about a rational function of several variables, $P(z_1, \ldots, z_n)/Q(z_1, \ldots, z_n)$? Here, we can form $2^n n!$ natural inverses. There are $n!$ ways to order the variables, and for each of these one can decide whether to do the positive-pointing inverse or the negative-pointing one. At each stage, we get a one-sided formal Laurent series whose coefficients are rational functions of the remaining variables, and one just keeps going.

Second Proof (Constructive): To every discrete function $f(m_1, \ldots, m_n)$ associate, the bilateral formal Laurent series

$$\sum_{(m_1,\ldots,m_n)\in Z^n} f(m_1, \ldots, m_n) z_1^{m_1} \cdots z_n^{m_n}.$$

We need to "solve" the equation

$$P(z_1^{-1}, \ldots, z_n^{-1}) \left(\sum_{(m_1,\ldots,m_n)\in Z^n} f(m_1, \ldots, m_n) z_1^{m_1} \cdots z_n^{m_n} \right) = 1.$$

So "explicitly"

$$\sum_{(m_1,\ldots,m_n)\in Z^n} f(m_1, \ldots, m_n) z_1^{m_1} \cdots z_n^{m_n} = 1/P(z_1^{-1}, \ldots, z_n^{-1}),$$

and we just described how to do it in $2^n n!$ ways.

The Maple Package LEON

This article is accompanied by a Maple package LEON. One of its numerous procedures is FS, that implements the above constructive proof. LEON can also compute *polynomial bases* to solutions of linear partial difference equations with constant coefficients, compute *Hilbert Series* for spaces of solutions of systems of linear differential equations, as well as find "multiplicity varieties" (in the style of Ehrenpreis) when they are zero-dimensional.

Leon Ehrenpreis (1930–2010): A Truly Fundamental Mathematician (A Videotaped Lecture)

I strongly urge readers to watch my lecture, available in six parts from `YouTube` and in two parts from `Vimeo`; see the following:

http://www.math.rutgers.edu/\char126\relaxzeilberg/mamarim/mamarimhtml/leon.html.

That page contains links to both versions, as well as numerous input and output files for the Maple package `LEON`.

Acknowledgements I'd like to thank an anonymous referee, and Hershel Farkas, for insightful comments. Accompanied by Maple package `LEON` available from http://www.math.rutgers.edu/~zeilberg/tokhniot/LEON. Supported in part by the USA National Science Foundation.

References

[E1] Leon Ehrenpreis, *Solution of some problems of division. I. Division by a polynomial of derivation*, Amer. J. Math. **76**(1954), 883–903.

[E2] Leon Ehrenpreis, *Solution of some problems of division. II. Division by a punctual distribution*, Amer. J. Math. **77**(1955), 286–292.

[M] Bernard Malgrange, *Existence et approximation des solutions des équations aux dérivées partielles et des équations de convolution*, Ann. Inst. Fourier, Grenoble **6**(1955-1956): 271–355.

[W] Peter Wagner, *A new constructive proof of the Malgrange–Ehrenpreis theorem*, Amer. Math. Monthly **116**(2009), 457–462.

[Wi] Wikipedia,the free Encyclopedia, *Malgrange–Ehrenpreis Theorem*, Retrieved 16:10, July 21, 2011.

The Legacy of Leon Ehrenpreis

Hershel M. Farkas, Robert C. Gunning, and B.A. Taylor

All those who knew Leon Ehrenpreis are well aware that he was a very multidimensional person. His interests went far beyond mathematics. Leon's interests included Bible and Talmud studies, music, sports (handball and marathon running), philosophy, and more. In this volume, we have only concentrated on his mathematical interests.

All the contributors to this volume have written on subjects that Leon either worked on actively or at least had a serious interest in. This is, on the one hand, to honor his memory and, on the other, to show his breadth.

When all is done, however, a person leaves memories, what he has built or written, and progeny (mathematical and physical). Memories are subject to interpretation and not all people remember things the same way. The mathematical works of Leon Ehrenpreis and the students he mentored are not subject to these vagaries.

In this final section, we include a list of Leon's Ph.D. students and we hope a complete list of his mathematical publications. This is his legacy.

H.M. Farkas et al. (eds.), *From Fourier Analysis and Number Theory to Radon Transforms and Geometry*, Developments in Mathematics 28, DOI 10.1007/978-1-4614-4075-8_28,

Phd Students of Leon Ehrenpreis

Mary Anderson	*New York University*	1970
Tong Banh	*Temple University*	1990
Carlos Berenstein	*New York University*	1970
Jane Friedman	*Temple University*	1989
Jongsook Han	*Temple University*	1993
Carlos Moreno	*New York University*	1972
Hannah Rosenbaum	*New York University*	1965
Carole Sirovich	*New York University*	1964
John Stevens	*New York University*	1972
Marvin Tretkoff	*New York University*	1971
Pierre Van Goethem	*New York University*	1964
Jinsong Wang	*Temple University*	1993

Publications of Leon Ehrenpreis

Special Functions, Inverse Probl. Imaging 4 (2010), no. 4 pp. 639–647.

Microglobal analysis, Adv. Nonlinear Stud. 10 (2010), no. 3 pp. 729–739.

Eisenstein and Poincare Series on $SL(3, R)$, Int. J. Number Theory 5 (2009) no. 8 pp. 1447–1475.

The Borel transform. Algebraic Analysis of Differential Equations from Microlocal Analysis to Exponential Asymtotics, Springer, Tokyo (2008) pp. 119–131.

The Radon Transform for Functions Defined on Planes, Integral Geometry and Tomography, Contemp. Math 405, AMS (2006) pp. 41–46.

Some novel aspects of the Cauchy Problem. Harmonic Analysis, Signal Processing, and complexity Progr. Math 238, Birkhauser Boston (2005), pp. 1–14.

The Universality of the Radon Transform (Appendix by Peter Kuchment, Tod Quinto). Oxford Mathematical Monographs, Oxford University Press (2003)

Three Problems at Mount Holyoke. Radon Transforms and Tomgraphy, Contemp. Math. 278, AMS (2001) pp. 123–130.

The Exponential X-ray Transform and Fritz John's Equation (with Kuchment, Peter and Panchenko, Alex) . Analysis Geometry Number Theory: the mathematics of Leon Ehrenpreis, Contemp. Math. 251, AMS (2000) pp. 173–188.

The Role of Payley Wiener Theory in Partial Differential Equations. The Legacy of Norbert Wiener. Proc. Sympos. Pure Math. 60 (1994) pp. 71–83.

Range Conditions for the Exponential Radon Transform (with Aguilar, Valentina and Kuchment, Peter), J. Analyse. Math. 68 (1996) pp. 1–13.

Parametric and Nonparametric Radon Transform. Conf. Proc. Lecture Notes Math. Physics , IV, Int. press, Cambridge (1994) pp. 110–122.

Some Nonlinear Aspects of the Radon Transform. Tomography, Impedance Imaging, and Integral Geometry. Lectures in applied Math. 30 AMS (1993) pp. 69–81.

Singularities, Functional Equations and the Circle Method. The Rademacher Legacy to Mathematics, Contemp. Math. 166 AMS (1992) pp. 35–80.

Exotic Parametrization Problems, Ann. Inst. Fourier (Grenoble) 43 no. 5 (1993) pp. 1253–1266.

Function Theory for Rogers-Ramanujan-like Partition Identities. A Tribute to Emil Grosswald, Contemp. Math. 143 (1993) pp. 259–320.

Nonlinear Fourier Transform. Geometric Analysi Contemp. Math. 140 AMS (1991) pp. 39–48.

The Schottky Relation in Genus 4. Curves, Jacobians and Abelian Varieties, Contem. Math. 136 AMS (1992) pp. 139–160.

Extensions of Solutions of Partial Differential Equations. Geometrical and Algebraical Aspects in Several Complex Variables. Sem. Conf., 8 EditEl Rende (1991) pp. 361–375.

Hypergeometric Functions. Special Functions, ICM -90 Satell. Conf. Proc. Springer, Tokyo (1991) pp. 78–89.

Lewy Unsolvability and Several Complex Variables. Michigan Math J. 38 (1991) pp. 417–439.

The Radon Transform and Tensor Products. Integral Geopmetry and Tomography. Contemp. Math 113 AMS pp. 57–63.

Hypergeometric Functions. Algebraic Analysis vol. 1 Academic Press (1988) pp. 85–128.

Reflection, Removable Singularities and Approximation for Partial differential Equations II. Trans. AMS 302 (1987) pp. 1–45.

Lewy's Operator and its Ramifications. J. Funct. Anal. 68 (1986) pp. 329–365.

Transcendental Numbers and Partial Differential Equations. Number Theory. Lecture Notes in Math. 1135 Springer (1985) pp. 112–125.

On Noether's Theorem Bull. Sci. Math. (2) 108 (1984) pp. 3–21.

Conformal Geometry. Differential Geometry Progr. Math. 32 Birkhauser (1983) pp. 73–88.

Poisson's Summation Formula and Hamburger's Theorem (with Kawai Takahiro) Publ. Res. Inst. Math. Sci. 18 no. 2 (1982) pp. 413–426.

Weak Solutions of the Massless Field Equations (with Wells, R. O. Jr.) Proc. Roy. Soc. London Ser A 384 (1982) pp. 403–425.

The Edge of the Wedge Theorem for Partial Differential Equations. Recent Devcelopments in Several Complex Variables . Annals of Math. Stud. 100 Princeton Univ. Press (1981) pp. 155–169.

Spectral Gaps and Lacunas. Bull. Sci. Math. (2) 105 (1981) pp. 17–28.

Reflection, Removable Singularities, and approximation for Partial Differential Equations I. Ann. of Math (2) 112 no. 1 (1980) pp. 1–20.

Edge of the Wedge Theorem for Partial Differential Equations . Harmonic Analysis in Euclidean Spaces part 2 (1979) Proc. Sympos. Pure Math. AMS (1979) pp. 203–212.

Hyperbolic Equations and Group Representations. Bull. AMS 81 (1975) pp. 1109–1111.

Invertible Operators and Interpolation in AU Spaces (with Malliavin, Paul). J. Math. Pures Appl. (9) (1974) pp. 165–182.

Some Refinements of the Poincare Period Relation (with Farkas, Hershel). Discontinuous Groups and Riemann surfaces. Ann. of Math. Studies no. 79 (1974) pp. 105–120.

On the Poincare Relation (with Farkas, H. Martens, H. Rauch, H.E.). Contributions to Analysis, Academic Press, (1974) pp.125–132.

The Use of Partial Differential Equations for the Study of Group Representations. Harmonic Analysis on Homogeneous Spaces. Proc. Sympos. Pure Math. vol 26 (1973) pp. 317–320.

An Eigenvalue Problem for Riemann surfaces. Advances in the Theory of Riemann Surfaces. Ann. of Math. Studies no. 66 Princeton (1971) pp. 131–140.

Analyse de Fourier sur des Ensembles Non-Covexes. Equations aux Derivees Partielles et Analyse fonctionnelle, Exp no. 17 Centre de Math. Ecole Polytech Paris (1971).

Fourier Analysis on Non-Convex Sets. Symposia Mathematica vol. 7 Rome Academic Press (1970) pp.413–419.

Fourier Analysis in Several Complex Variables. Wiley Interscience (1970).

Cohomology with Bounds. Symposia Mathematica vol. 4 Rome Academic Press (1970) pp. 389–375.

Complex Fourier Transform Technique in Variable Coefficient Partial Differential Equations. J. Analyse Math. 19 (1967) pp. 75–95.

On Spencer's Estimate for δ - Poincare (with Guillemin, Victor and Sternberg, Shlomo). Ann.of Math (2) 82 (1965) pp. 128–138.

Holomorphic Convexity of Teichmueller Spaces (with Bers, Lipman) Bull. AMS 70 (1964) pp. 761–764.

The Fundamental Principle and some of its Applications. Studia Math (Ser. Specjalna) Zeszyt 1 (1963) pp. 35–36.

Solutions of some Problems of Division V. Hyperbolic Operators. Amer. J. Math. 84 (1962) pp. 324–348.

A Fundamental Principle for Systems of linear Differential Equations with Constant Coefficients and some of its Applications. Jerusalem Academic Press (1960) pp. 161–174.

Analytically Uniform Spaces and some Applications. Trans. AMS 101 (1961) pp. 52–74.

A New Proof and an Extension of Hartog's Theorem. Bull AMS 67 (1961) pp. 507–509.

Solution of some Problems of Division IV. Invertible and Elliptic Operators. Amer. J. Math. 82 (1960) pp. 522–588.

Theory of Infinite Derivatives. Amer. J. Math. 81 (1959) pp. 799–845.

Some Properties of the Fourier Transform on Semisimple Lie Groups III (with Mautner, F.I). Trans. AMS 90 (1959) pp. 431–484.

Analytic Functions and the Fourier Transform of Distributions II. Trans. AMS 89 (1958) pp. 450–483.

Some Properties of the Fourier Transform on Semisimple Lie Groups II (with Mautner, F. I.). Trans. AMS 84 (1957) pp. 1–55.

Theory of Distributions for Locally Compact Spaces. Mem. AMS no. 21 (1956).

Some Properties of Distributions on Lie Groups. Pacific J. Math. 6 (1956) pp. 591–605.

Sheaves and Differential Equations. Proc. AMS 7 (1956) pp. 1131–1138.

Solutions of some Problems of Division III. Amer. J. Math. 78 (1956) pp. 685–715.

Cauchy's Problem for Linear Differential Equations with Constant Coefficients. Proc. Nat. Acad. Sci. USA 42 (1956) pp. 642–646.

On the Theory of Kernels of Schwartz. Proc. AMS 7 (1956) pp. 713–718.

Analytic Functions and the Fourier Transform of Distributions I. Ann. of Math. 2 63 (1956) pp. 129–159.

General Theory of Elliptic Equations. Proc. Nat. Acad. Sci. USA 42 (1956) pp. 39–41.

Appendix to the paper Mean Periodic Functions I. Amer. J. Math. 77 (1955) pp. 731–733.

Completely Inversible Operators. Proc. Nat. Acad. SCi USA 41 (1955) pp. 945–946.

The Division Problem for Distributions. Proc. Nat. Acad. Sci. USA 41 (1955) pp. 756–758.

Uniformally Bounded Representations of Groups (with Mautner, F. I.) Proc. Nat. Acad. Sci.USA 41 (1955) pp. 231–233.

Solution of some Problems of Division II. Amer. J. Math. 77 (1955) pp. 293–328.

Mean Periodic Functions I. Amer. J. Math. 77 (1955) pp. 293–328.

Some Properties of the Fourier Transform on Semisimple Lie Groups I (with Mautner, F.I.) Ann. of Math. 2 61 (1955) pp. 406–439.

Solution of some Problems of Division I. Amer. J. Math. 76 (1954) pp. 883–903.

Volumes edited by L. Ehrenpreis

Radon Transforms and Tomography. Proceedings of the AMS-IMS-SIAM joint Summer research Conference at Mt. Holyoke College, South Hadley , Ma. June 18–22, 2000. Contemporary Mathematics, 278 AMS (with Eric Quinto, Adel Faridani, Fulton Gonzalez, Eric Grinberg).

Theta Functions-Bowdoin 1987 parts 1,2. Proceedings of the thirty-fifth Summer Research Institute held at Bowdoin College, Brunswick, Maine, July 6–24 1987. Proceedings of Symposia in Pure Mathematics parts 1,2 49 AMS (with Robert C. Gunning).

Entire Functions and Related parts of Analysis. Proceedings of Symposia in Pure Mathematics vol. XI AMS 1968 (with J. Korevaar, S. Chern, W H. J. Fuchs, L. A. Rubel).

Vladimirov, Vasily Sergeyevich, Methods of thc Theory of Functions of many Complex Variables. M.I.T Press (1966).

Markusevich, A. I. Entire Functions. American Elsevier Publishing NY (1966).

MIX
Papier aus verantwortungsvollen Quellen
Paper from responsible sources
FSC® C105338

If you have any concerns about our products,
you can contact us on
ProductSafety@springernature.com

In case Publisher is established outside the EU,
the EU authorized representative is:
Springer Nature Customer Service Center GmbH
Europaplatz 3, 69115 Heidelberg, Germany

Printed by Libri Plureos GmbH
in Hamburg, Germany